Modern Electrical Communications

Modern Electrical Communications

Analog, Digital, and Optical Systems

Second Edition

HENRY STARK, D. Eng. Sc.
Professor of Electrical, Computer, and Systems Engineering
Rensselaer Polytechnic Institute

FRANZ B. TUTEUR, Ph.D.
Professor of Engineering and Applied Science
Yale University

JOHN B. ANDERSON, Ph.D.
Professor of Electrical, Computer, and Systems Engineering
Rensselaer Polytechnic Institute

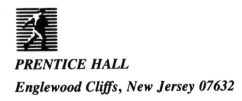

PRENTICE HALL
Englewood Cliffs, New Jersey 07632

Library of Congress Cataloging-in-Publication Data

Stark, Henry, 1938–
Modern electrical communications.

Includes bibliographies and index.
1. Telecommunication. I. Tuteur, Franz B.
II. Anderson, John B., 1945– III. Title.
TK5101.S67 19888 621.38 87–11485
ISBN 0–13–593112–6

Editorial/production supervision: *Raeia Maes*
Cover design: *Diane Saxe*
Manufacturing buyer: *Gordon Osbourne*

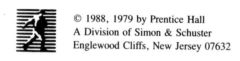

Printed in the United States of America

10 9 8 7 6 5 4 3 2

ISBN 0-13-593112-6 025

PRENTICE-HALL International (UK) Limited, *London*
PRENTICE-HALL of Australia Pty. Limited, *Sydney*
PRENTICE-HALL Canada Inc., *Toronto*
PRENTICE-HALL Hispanoamericana, S.A., *Mexico*
PRENTICE-HALL of India Private Limited, *New Delhi*
PRENTICE-HALL of Japan, Inc., *Tokyo*
SIMON & SCHUSTER Asia Pte. Ltd., *Singapore*
EDITORA PRENTICE-HALL DO BRASIL, LTDA., *Rio de Janeiro*

IN MEMORY OF
Anna Stark
Ruth Tuteur

TO
Janet Anderson

Contents

Preface to the First Edition

This book grew out of a set of notes prepared for a two-term senior-level course in electrical communication systems and a one-term graduate course in statistical communications and signal processing. In the five years or so that we have taught the undergraduate communication sequence at Yale, it has been our common experience that, because of the simultaneous proliferation of newer disciplines and student interests, we could not count on the students' having had the traditional background courses required to tackle a fairly mature course devoted exclusively to communication engineering. For this reason we had to develop a course that was relatively complete and self-contained.

We had to develop fundamental topics such as Fourier methods and linear-systems theory, but we also felt that the students deserved to be informed about advanced topics, such as statistical communication theory, signal processing, television, radar, and sonar, in as much depth as possible. We felt that, although theory had to be properly covered, it was necessary to show application of the theory in at least a few concrete modern devices. In short we attempted to make our course a broad and integrated study of the entire field of electrical communication engineering.

We believe that our experience with our own students is in no sense unique. The field of electrical engineering has changed so vastly in the last few decades and there are so many topics that our students must know today, that what used to be the ''traditional'' electrical engineering curriculum has ceased to exist in many places. Even schools with more traditional electrical engineering curricula have developed a greater range of options and have given their students additional freedom in crossing the boundaries between various disciplines. Faced with this greater loosening of the traditional course-of-study hierarchy in electrical engineering, we felt that a text that presented the essential background to communications theory in a few brief introductory chapters and then pushed on to more advanced topics would fill a general need. This is why we have written this book.

In teaching first-year graduate students we found that, if anything, the variations in students' preparedness, sophistication, and interests are even greater than they are for undergraduates. Some graduate students had not even been electrical engineering majors but came from such allied fields as computer science or physics. Others came from undergraduate schools that offered relatively unstructured programs that often left gaps in the students' appreciation of fundamental notions in electrical engineering. In many cases, the students were inadequately prepared in linear systems, random processes, and basic communication systems. In teaching students with such varied backgrounds, we found it convenient to have available a book that not only discussed signal processing and the central ideas in statistical communication theory, but also contained the necessary background material to enable the student to "catch up" on his or her own time.

In line with the experience that we have had in our course, we have tried to keep the prerequisite knowledge required from our readers to a mimimum; we expect basically college-level calculus and a first course in linear-system theory that includes some exposure to the Laplace transform. Beyond this base level, we have tried to make our explanations and derivations as complete and as self-contained as possible. Much use is made of mathematical arguments, and we have included very few results, curves, or tables that the reader could not generate by himself on the basis of the information furnished in the text. In some sections we have used elementary concepts of complex-variable theory, but there is little loss in either continuity or substance for the reader not familiar with this theory.

As might have been expected, the book contains more material than we normally covered in our undergraduate course. For instance in the introductory chapter that deals with Fourier methods, we included material on convergence, vector representation of signals, and concepts such as inner product and projection. We did this partly for completeness and partly because these concepts were needed in later chapters. The book contains a survey of linear systems and active filters going considerably beyond the treatment that we were able to give in our course. We also felt that with so much signal processing being done digitally today we would be amiss if we did not discuss digital systems and filters in some depth. We therefore included a chapter on digital filters that contains a discussion of the discrete Fourier transform (DFT) and the fast Fourier transform (FFT). Digital methods were also integrated into several of the other chapters, notably the one on signal processing.

The authors are extremely grateful to Dr. Reed Even who critically read the entire manuscript and made numerous excellent suggestions. Thanks are due to Helen Brady for doing the typing of most of the first draft, to Dan Tuteur, who helped with two of the chapters, and to Joy Breslauer and Michelle Gall, who exhibited patience and skill in typing the manuscript to completion. Thanks are due also to the administrations of the Department of Engineering and Applied Science at Yale and the School of Engineering at Rensselaer Polytechnic Institute, who permitted us to take some of the large amount of time needed to complete this project from our normal duties. Finally, thanks are due to our wives Alice and Ruth without whose unfailing patience and encouragement the work could not have been completed.

Henry Stark
Franz B. Tuteur

Preface to the Second Edition

The reason for this second edition is the same as stated in the Preface to the first edition. Indeed, if anything, our teaching experience since writing the original version has more than substantiated the need for a relatively broadly based and integrated text on the entire field of electrical communication engineering. On the other hand, the field of electrical communications had definitely moved toward the digital and optical. A modern textbook must reflect this trend and, consequently, we have significantly increased the scope and depth of the discussions on digital communications and added a new chapter on optical communications.

Our treatment of digital communications involves two levels: we offer basic discussions of sampling, quantizing, and elementary digital techniques in Chapter 4, suitable for undergraduates. We continue with a more sophisticated treatment involving optimum digital communication systems in Chapter 10. And since digital communications involves coding whose theoretical foundation is information theory, we devote an entire chapter (Chapter 11) to this subject.

Optical communications, especially fiber optical communications, has made significant inroads in many industralized countries. It cannot be ignored in the students' curriculum. In Chapter 12 we try to give a balanced discussion of this subject, dividing the discussion about equally between component operations and system design considerations.

As in the case of the first edition, the second edition is intended as a two-course sequence in electrical communications, the first offered, typically, as an upper-level undergraduate course and the second as a first-year graduate course. For the first course, Chapters 1 through 7 would be a suitable goal, the material to be covered in a relatively demanding semester in which classes meet for three class hours per week and homework assignments include from 5 to 10 problems per week. If this proves too taxing or the quarter system is used, we suggest omitting the material on active filters in Chapter 3, all of Chapter 5, and the material on spectrum analysis in Chapter 6. If the student is taking or has taken a parallel course in Fourier theory and/or linear systems, much of Chapter 2 can be omitted, as well as the first few sections of Chapter 3.

Chapters 8 through 12 comprise the second half of the two-course sequence. If probability is a prerequisite for the course, Chapter 8 can be omitted. Then a logical approach might be to start the course with a brief review of the types of modulation systems in use and compare these on the basis of bandwidth and signal-to-noise ratio, as is done in Chapter 9. Chapters 9 through 12 form the heart of the second course, and these can be straightforwardly understood by students with a background in probability and some knowledge of elementary systems theory and Fourier transforms.

Once again, the authors wish to acknowledge the administrations of the Schools of Engineering at Rensselaer Polytechnic Institute and Yale University for their recognition of the pedagogical and scholarly aspects of this project. Thanks are due to MeeLi Chew Leith and Ruth Houston for their expert typing skills and technical help. Finally, gratitude is due to our families and friends, who make it all worthwhile.

Henry Stark
Frank B. Tuteur
John B. Anderson

CHAPTER 1

Introduction

1.1 DEFINITIONS

There is a humorous expression that "pressing a suit" does not mean the same thing to a lawyer that it does to a tailor [1-1]. It's somewhat like that with the word *communication*. *Webster's New Collegiate Dictionary* offers, among others, these definitions: "a process by which information is exchanged between individuals through a common system of symbols . . . ," and also "a technique for expressing ideas effectively . . ." (one thinks of a public-speaking improvement course), and finally "a system of routes for moving troops, supplies, and vehicles."

From our point of view, the first definition is obviously closest to the subject matter of this book. The word *electrical* in electrical communication means that our concern is less with original symbols such as written or spoken speech than with the transmission and reception of the electrical symbols into which the original symbols have been translated. Not all electrical communication is concerned with communication between persons; in the modern world, computers also talk to each other.

1.2 PROBLEMS IN COMMUNICATION THEORY

In a classic work by Claude Shannon and Warren Weaver [1-2] it is argued that problems in communication theory fall into one of three levels. Calling these levels A, B, and C, the authors describe them as follows:

Level A. With what precision can the symbols in communication be transmitted?

Level B. How precisely do the transmitted symbols convey the desired meaning?

Level C. How effectively does the received meaning affect conduct in the desired way?

Problems in level A are basically technical and of primary concern to electrical and communication engineers. Problems in level B are semantic in nature and are related to how well the meaning of the symbols is interpreted at the receiver. Linguists and others concerned with the structure and theory of language are on comfortable ground here. Effectiveness problems (level C) are perhaps harder to define as they involve esthetic and psychological factors. The following question, however, arises: Does level A encompass the most superficial problem in communication? Shannon and Weaver write that this is decidedly not so. Initially, one might be led to feel that levels B and C, dealing with the philosophical foundations of communication and behavior, are somehow "deeper" than level A, whose primary concern is, ostensibly, good design, Shannon and Weaver, however, argue that any limitations discovered in level A must, assuredly, affect levels B and C. More basically still, the mathematical theory of communication discloses that level A influences the other levels far more fundamentally than one would initially suspect.

We shall not pursue these points further. They were made to suggest that the technological aspects of electrical communication transcend engineering considerations of economy, efficiency, and accuracy (all-important as these are in the real world). Indeed, fundamental results in communication technology have a profound influence on the broader aspects of communication. These thought-provoking ideas are discussed in Reference [1-2], and the reader is urged to consult this milestone work.

1.3 A GENERAL COMMUNICATION SYSTEM

Figure 1.3-1 is a symbolic representation of an electrical communication system. As it stands, it is a more general representation than is required for electrical communications. For example, if used to model a speaker–listener "system," the source might be the speaker's brain, the message is the thought to be expressed, the transmitter is the vocal system that encodes the message in a form suitable for transmission through air, the carrier signal is the variations in sound pressure, the channel is the air, the receiver is the listener's ear (and associated nerves), and the destination is the listener's brain.

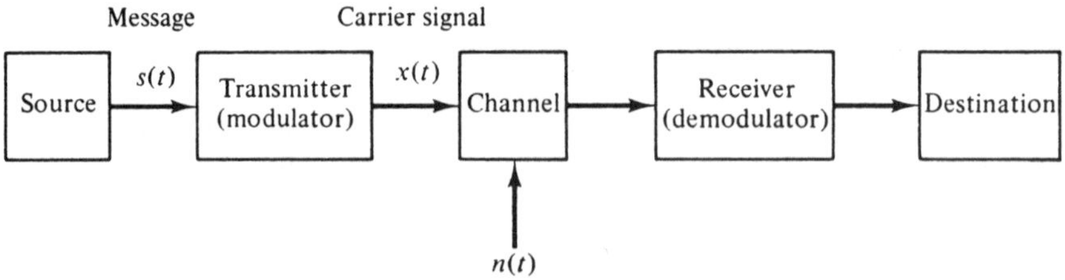

Figure 1.3-1. Block diagram of a communication system.

However, if Fig. 1.3-1 is used to model an electrical communication system, the message $s(t)$ will be taken to be an electrical signal, it being assumed that the original message (i.e., speech, written text, pictures, etc.) has already been converted into $s(t)$. We shall take it for granted that in electrical communications means exist to convert the original message into an electrical signal. In addition to the term *message*, $s(t)$ is also sometimes called intelligence, baseband signal, audio signal, envelope, information,[†] and so on. Examples of message signals are the signals out of a microphone, the induced voltage in a magnetic phonograph cartridge, the induced current in the winding of a tape recording head, the binary voltage levels in a computer, the video current out of a TV camera, and the current in a photoconductor exposed to light. Obviously, there are many more.

The unprocessed messages are generally not suited for transmission through the channel. This stems from the fact that $s(t)$ is generally a low-frequency waveform, which implies long wavelengths. The basic laws of electromagnetic radiation require that for efficient radiation the radiating element (i.e., the antenna) be a significant fraction of the wavelength of the signal. A 1000-hertz[‡] signal has a wavelength of 3×10^5 meters in air; a quarter-wave antenna therefore would be 75 kilometers long! This is much too impractical for ordinary broadcasting.[§] On the other hand, a 1.0-MHz[‖] signal has a wavelength of 300 meters for which a quarter-wave antenna would be only 75 meters long.

In addition to this mismatch between the medium and the message, there are other problems with the transmission of low-frequency signal components as well. In radar and microwave communications, it is frequently necessary to generate a narrow beam in order for the power intensity at the receiver to be sufficiently great to overcome electrical noise. However, waves undergo the phenomenon of *diffraction*, which results in a spreading of the beam. The diffraction angle is proportional to λ/D, where λ is the wavelength and D is the diameter of the radiator. The spreading therefore is seen to be *inversely* proportional to frequency, and at low frequencies the beam intensity may drop off too sharply for satisfactory communication.

There are several other important reasons why $s(t)$ must be processed before transmission. We shall mention only one more here. Consider ordinary commercial radio broadcasting. This is perhaps the electrical communication system that we know best. Most programs consist of speech or music in which the signals cover a band from tens of hertz to perhaps 15 kHz. How could we ever have more than one listening choice in one locality if we could not separate the signals? By modulating the various messages $\{s(t)\}$ onto different carriers, overlap and interference between them are avoided and

[†] As we shall see in later chapters, some of these terms have rather special technical meaning and should be used only in the proper context.

[‡] 1 cycle per second = 1 hertz, abbreviated Hz. The unit is named in honor of the great experimentalist Heinrich Hertz (1857–1894), who verified Maxwell's theory.

[§] Although very large antennas may be in use for low-frequency communications in the military.

[‖] 1 MHz = 1 megahertz = 10^6 Hz. A gigahertz is 10^9 Hz and is written 1 GHz. A kilohertz is 10^3 Hz and is written 1 kHz.

the listener is given a large selection of programs to choose from. The same holds true for TV and telephony. In the latter, it is desired to send many messages between two specific points on the same communication link. To prevent interference, the signals are *multiplexed*. *Frequency multiplexing* separates the various messages by modulating them onto different carriers and assigning them nonoverlapping frequency bands. In point of fact, the telephone company sends hundreds and even thousands of signals simultaneously over some of its links.

Henceforth we shall refer to the preprocessing of $s(t)$ for transmission simply as *modulation*. A precise meaning of this term will be given in Chapter 6. A significant portion of this book is devoted to modulation, its effect on the signal, and its ability to overcome noise. In Section 1.4 we shall furnish a brief survey of some common modulation schemes.

Returning now to Fig. 1.3-1, we designate the medium between transmitter and receiver as the channel. In most electrical communication systems the channel is characterized by a long transmission path. This means that signals become attenuated and, therefore, that contamination by random noise and interference is more likely. The longest channels are found in deep-space probes. When possible, relay stations are added to break up the overall channel length into more manageable sections. In microwave systems, relay stations are placed every 25 to 30 miles and reamplify the microwave beams. The use of relays in long-range communication channels is widespread, and they are extensively used in optical fiber communication systems, where they are sometimes called *repeaters*.

The carrier frequency must be matched to the channel medium. Equivalently, for a fixed carrier, the channel medium must be matched to the carrier. Ordinary copper wire is adequate for transmitting low-frequency signals, but it introduces excessive attenuation and radiation losses for signals in the microwave region. Parallel-wire transmission lines are better when the wavelength is less than 1 kilometer, but even here shielding and radiation losses become serious problems at high frequencies. Coaxial cables are still better because the fields are practically perfectly shielded inside the line and confined between the inner and outer conductors. For wavelengths shorter than 1 meter, metallic waveguides are the preferred medium for guiding electromagnetic waves. At optical frequencies, for which wavelengths are measured in microns (1 millionth of a meter), dielectric waveguides called *fibers* are useful for guiding energy. A comparison of the losses associated with various wires, cables, and waveguides is furnished in [1-3], p. 615. A useful table of preferred transmission media versus frequency is furnished in [1-4], p. 8. The underlying physics that governs the guiding and transmission of electrical energy is covered in many places, e.g., [1-5].

Returning once more to Fig. 1.3-1, we have somewhat unfairly attributed all forms of noise to the channel. Actually noise is introduced at the receiver and transmitter also. The term noise is used very broadly here and includes interference from other stations, nonlinear effects in the system itself, effects due to signal fading, multipath propagation, other problems, and finally random ''natural'' noise. The last reflects the two facts that we live in a warm universe (thermal noise) and that electromagnetic radiation as well as electrons are quantized and exhibit particle-type behavior (shot noise). Shot noise is less a problem for the more common communication systems than thermal noise. Although interference and nonlinear distortions can, conceivably at impractical costs, be reduced below any predetermined threshold, such is not the case with natural noise. The latter represents a "hard" constraint put in our way by an inflexible nature. Because thermal noise is pervasive in electrical communication systems, we shall refer to it simply as *noise* and use more specific terms to describe other distorting effects.

The ultimate quality of a communication channel is measured by its capacity C in bits per second. The term *bit* here has a rather special meaning and is defined as the information gained when the outcome of a binary experiment (e.g., an experiment involving a yes–no answer) with equiprobable outcomes is disclosed. A very basic theorem,† first proved by Shannon [1-2], says that, given a channel with capacity C and a source with information rate $R < C$ (C, R both in bits per second), then, by appropriate coding, the output of the source can be transmitted through the channel with *arbitrarily small probability of error even though random noise is present.* Another important result, also attributed to Shannon, says that the capacity of the band-limited *Gaussian* channel is given by

$$C = W \log_2 \left(1 + \frac{P_x}{P_n} \right), \tag{1.3-1}$$

where W is the channel bandwidth, P_x is the average power in the transmitted signal, and P_n is the average power in the noise. The latter is "white" thermal noise with uniform power density over the band W and fluctuations subject to the Gaussian probability law. In Chapter 11, Eq. (1.3-1) is restated in a form more useful for computing the capacity of a digital channel.

The study of Shannon's fundamental theorem, associated source coding schemes, and the determination of C for different channels is a branch of communication theory called *information theory*. Although, with rare exceptions, information theory has had little influence on the design of electrical communication schemes, it has aroused great interest among such diverse groups as electrical engineers, social scientists, and linguists. It is not uncommon in some professional circles to lump all electrical communication theory dealing with noise and other random processes under the heading of information theory.

Returning one final time to Fig. 1.3-1, we see that the attempted recovery of the message $s(t)$ is done with the receiver. We add the term *attempted* because the received carrier is invariably contaminated to some degree by noise. For high ratios of signal power to noise power, a very good replica of $s(t)$ can be recovered. The receiver strips away $s(t)$ from the carrier signal and basically does the inverse operation of the transmitter. If the transmitter is fundamentally a modulator, the receiver is fundamentally a demodulator. *Optimum* receivers try to recover the signal while minimizing noise effects. "Optimum" implies a criterion of sorts, and there is no single optimum receiver for all criteria.

† This theorem is frequently called the *fundamental theorem of information*.

Last, the demodulated message is absorbed at the destination. The latter is frequently a person, but could be a machine such as a robot or a digital computer.

1.4 MODULATION

We saw in Sec. 1.3 that modulation holds a key position in electrical communication systems. There are many modulation schemes, and they can be grouped, somewhat broadly, into the following categories: continuous-wave (CW), pulse, analog, and digital.

Continuous-Wave Modulation

In CW modulation, the carrier is most frequently a fixed-frequency sinusoid upon which the signal $s(t)$ is impressed. Common examples are double-sideband modulation (DSB), amplitude modulation (AM), single-sideband modulation (SSB), vestigial sideband modulation (VSB), frequency modulation (FM), and phase modulation (PM). In DSB,[†] the signal is directly impressed on the carrier amplitude. Conceptually, the simplest type of modulation, DSB, is not so easy to demodulate. AM is very similar to DSB except that the signal also contains a strong unmodulated carrier. The result is a wave that is extremely easy to demodulate, a fact that accounts in part for the widespread use of AM. SSB and VSB are bandwidth-conserving schemes that are generally not so easy to demodulate as AM. Basically, in SSB and VSB, only a portion of the modulated signal band is transmitted. The redundant part is omitted. The result is that more channels can be allocated in the same spectral band. VSB is widely used in TV systems. In all these schemes, the information is basically stored in the low-frequency amplitude fluctuations of the carrier envelope. FM and PM are examples of *angle* modulation. In FM the instantaneous frequency of the carrier is proportional to the signal. Unlike AM, DSB, VSB, and SSB, there is no information in the carrier envelope; the information is stored in the zero crossings of the carrier wave. PM is very much like FM except that it is the *instantaneous phase* that is proportional to the signal. FM is widely used in high-quality broadcasting and also is the method by which the TV sound signal is modulated.

Pulse Modulation

Pulse modulation differs from CW modulation in that the carrier signal exists only at certain intervals of time. The carrier in this case consists most often of a periodic sequence of pulses that is on only during a portion of the cycle. The signal $s(t)$ is used to vary some parameter of the pulse such as its amplitude (which leads to pulse amplitude modulation or PAM for short), its duration (pulse duration modulation, PDM), or its

[†] Double-sideband modulation is also abbreviated DSBSC, meaning "double-sideband suppressed carrier." We shall use both DSB and DSBSC. Still another term in use is AM-SC, meaning "AM with suppressed carrier."

displacement from a reference point on the time scale (pulse position modulation, PPM). In PAM, which is the pulse analog of AM, the signal is impressed on the *amplitude* of the pulse. In PDM, the duration of the pulse is made proportional to the signal amplitude during the instant when the pulse is on. In PPM, the *delay* of the pulse is proportional to the signal amplitude. PPM can be derived from PDM by differentiating and clipping. Unlike PAM, there is no information stored in the amplitude variations of the pulses in PDM and PPM.

The theoretical foundation for pulse modulation is the famous *sampling theorem*, to be discussed in Chapter 4. It might have occurred to the reader that a carrier consisting only of pulses that are not on at all times cannot transmit the message $s(t)$ continuously. Are we then throwing away valuable information? The answer, surprisingly, is *no* if the message is band limited and the pulse-repetition rate is high enough. The sampling theorem is not only important in pulse modulation; it also is invaluable for explaining numerous phenomena in communication theory and forms the basis for the discrete Fourier transform (DFT), discussed in Chapter 5.

Analog Modulation

The CW pulse modulation schemes just described all share the property that some parameter of the carrier wave is *proportional* to the instantaneous values of the message and that this parameter varies *continuously* (at least within a predetermined range). Such is not the case in digital modulation. We shall consider an example next.

Conversion to Digital Form: PCM

Pulse code modulation (PCM) starts out essentially as PAM with an important difference: The amplitudes of the pulses are altered so that, in the end, each of them acquires only one of N preselected levels. Here information is really thrown away. However, because even the finest receivers, including the human ear, are not sensitive to fine variations below some threshold, the throwing away of information need not be bothersome in a practical sense. The preselected levels are assigned numerical values, and these, properly coded, are transmitted over the channel. Thus, to generate PCM, three key steps are required: sampling, quantizing, and coding. Sampling is done as in PAM by using a periodic sequence of pulses for a carrier. Quantizing, however, generates a finite set of values from a possible infinite set of values and corresponds mathematically to mapping from an infinite-dimensional space [the set of all values $s(t)$ can take on] to a finite-dimensional space (the set of all quantized levels). The final step in PCM is to assign to each level a code word, preferably made up of simple binary symbols. For example, a quantized level of 7 volts might be coded into the code word 0111, which uses only the two symbols 0 and 1. Because *two* symbols are in use, this code is called *binary*. The conversion of these symbols to signals for transmission can be done in several ways. In amplitude-shift keying (ASK), a constant-frequency carrier is switched between two levels, say 0 volts and V volts. The 0-volt level might signify a 0, in which case the V-volt level would signify a 1. Frequency-shift keying (FSK) uses two frequencies to denote the 0, 1 symbols; phase-shift keying (PSK) switches a constant-frequency carrier between two phases, say 0 radians to signify a 0 and π radians to signify a 1. In Chapters 4, 10, and 11, the techniques for converting to a digital form and methods of modulating a carrier with digital signals are discussed in greater detail.

1.5 HISTORICAL REVIEW

One might begin a historical review of the development of communication with Volta's discovery of the voltaic battery. Volta's discovery was more than a useful device: It laid to rest Galvani's theory of "animal electricity," which proposed that the convulsions observed in severed frogs' legs were due to a mysterious vital force hidden in the tissues of the leg. In 1801, when Volta demonstrated his battery to an important audience that included Napoleon, he took great care to make his battery appear like an electric eel, probably to mock Galvani's theory of animal electricity.

At about the same time, Michael Faraday was an errand boy in a bookshop near London. Faraday's apprenticeship in the bookshop led to his becoming a bookbinder, a job he apparently despised. Faraday's interest in science ultimately led to a job with Sir Humphrey Davy. On August 29, 1831, Faraday produced an induced current by moving a magnet near a conductor. Conceivably one of the greatest discoveries of all time, electromagnetic induction did not strike Faraday the same way (at least immediately). He wrote to a friend, "It may be a weed instead of a fish that, after all my labor, I may have at last pulled up."

In 1834, Gauss and Weber designed one of the first telegraphs to be operated over any significant distance. Gauss, of course, is better known as one of the finest mathematicians who ever lived. In addition to mathematics and telegraphy, Gauss made fundamental contributions in astronomy, electromagnetic theory, and actuarial science. The Gauss–Weber telegraph receiver used a free-swinging magnetic needle inside the coil carrying the signal current. The direction in which the magnet swung depended on the direction of the current in the coil. The code that Gauss–Weber used is shown in Table 1.5-1.

The telegraph system of Cooke and Wheatstone, first demonstrated in 1837 and enthusiastically acclaimed in 1845 when it was used to capture a murderer who was subsequently hanged, led to the formation of the English Electric Telegraph Company in 1846. By 1852, the company had laid 4000 miles of telegraph lines in England.

In the United States, it was left to Samuel Morse, assisted by some very able people, to devise a revolutionary new telegraph system using the now-famous "dot–dash" Morse code. With Congress assisting via a $30,000 appropriation, the first telegraph line went into operation on May 24, 1844. It linked Washington to Baltimore, a distance of about 40 miles.

The first transatlantic telegraph cable was the result of a partnership between Cyrus Field, John Brett, and Charles Bright. Field, an American, and his two English partners faced incredible hardships in their attempts to lay a transatlantic cable. Finally,

TABLE 1.5-1 TELEGRAPH CODE OF GAUSS AND WEBER

A	*r*	M	*lrl*	0	*rlrl*
B	*ll*	N	*rll*	1	*rllr*
C, K	*rrr*	O	*rl*	2	*lrrl*
D	*rrl*	P	*rrrr*	3	*lrlr*
E	*l*	R	*rrrl*	4	*llrr*
F, V	*rlr*	S	*rrlr*	5	*lllr*
G	*lrr*	T	*rlrr*	6	*llrl*
H	*lll*	U	*lr*	7	*lrll*
I, J	*rr*	W	*lrrr*	8	*rlll*
L	*llr*	Z	*rrll*	9	*llll*

l = left, *r* = right.

SOURCE: W. R. Bennett and J. R. Davey, *Data Transmission*, McGraw-Hill Book Company, New York, 1954. Used with permission of McGraw-Hill Book Company.

on July 27, 1866, a successful and permanent telegraph link between Europe and the United States was established.

Although the first practical telephone system is attributed to Alexander Graham Bell, a German schoolteacher by the name of Philipp Reis is actually credited with developing the first telephone (1860). In 1877, Bell established the Bell Telephone Company, and the first telephone exchange was opened in New Haven, Connecticut, in 1878. It was not, however, until 1915 that, using the then new concept of electronic amplification, Bell held the first transcontinental telephone conversation with Thomas Watson (New York to San Francisco). And it was not until an almost unbelievably recent year (1953) that the first transatlantic underwater telephone line was completed.

Many people played important roles in the early development of wireless communications, but none played greater parts than James Maxwell, Henrich Hertz, Oliver Lodge, Marchesi Marconi, and A. S. Popov.

From all the electromagnetic phenomena discovered by Oersted, Faraday, and others, Maxwell synthesized a general theory that to this day forms the basis of radio communication. Although Maxwell published his results in 1864, it was not until 1887, when Hertz experimentally verified key predictions[†] in Maxwell's theory, that universal acceptance of the theory was achieved. Hertz died in 1894, at the age of 37. Impetus to the development of radio was furnished by the invention of the *coherer* by Oliver Lodge in 1877. This sensitive device could detect radio signals that were far fainter than any signals that could be picked up with other devices. Lodge demonstrated wireless signaling over a distance of 150 yards in 1894 at Oxford, England.

Marconi and Popov, working independently and at around the same time, put the finishing touches on the first practical wireless systems. In 1895, Marconi transmitted

[†] Which included a demonstration that radio waves and light waves were fundamentally the same entity.

radio signals through a distance of over 2 kilometers. By 1898, after having obtained a patent and having founded the Wireless Telegraph and Signal Company, Marconi could communicate using radio signals over 60-mile links. The future of the "wireless" was assured. On December 12, 1901, at 12:30 in the afternoon, a prearranged signal of three faint clicks was heard on Signal Hill in Newfoundland. The signals originated 1700 miles away in Cornwall, England.

The discovery of the vacuum diode by Fleming in 1904 and the triode by Lee De Forest in 1906 signaled the dawn of wireless voice communication. The triode, a device of tremendous importance, was, at least until the development of the transistor, the main device for electronic amplification.

At around this time developments came very rapidly. By 1907, speech was being transmitted over 200-mile channels in the eastern United States. By 1920, station KBKA in Pittsburgh began broadcasting on a scheduled basis. In 1923, over 500 transmitters were operating in the United States. They all used nearly the same wavelength. Shortly thereafter (1927), the Federal Radio Commission was formed to put things in order.

Lee De Forest, who was not exactly a modest man (his autobiography was titled *Father of Radio*), decried the intellectual mismatch between the high-level intelligence that produced radio and the low-level thinking that characterized the broadcasting industry. He called their programs a "stench in the nostrils of the gods of the ionosphere."

During World War I, a young electrical engineer named Edwin Armstrong designed a greatly improved version of the broadcast receiver, which became known as the superheterodyne. Almost all modern receivers are of this type. If the invention of the superheterodyne was all that Armstrong could claim, his name would still be honored in the history of communications. However, in 1933, Armstrong demonstrated a revolutionary new system of communication which he called *frequency modulation* (FM). Bitter squabbles with De Forest,[†] industry disinterest, and other factors delayed the wide acceptance of FM. By 1949, however, there were 600 FM stations operating in the United States.

In 1929, a Russian emigre by the name of V. K. Zworykin demonstrated the first television system using the now-obsolescent iconoscope. By 1939, the British Broadcasting Corp. (BBC) was broadcasting on a commercial basis, and over 20,000 TV sets had been sold in London. The iconoscope has been replaced by other cameras such as the image orthicon and vidicon. However, the underlying principle (i.e., scanning) remains the same. Color TV, a very complicated extension of black and white TV, began in the United States in 1954. Through some very ingenious engineering, color TV has been made *compatible* with black/white TV. This means that no increase in channel bandwidth is necessary and that black/white receivers can receive signals broadcast in color as black/white pictures, and vice versa.

The developments of solid-state devices and integrated circuits have their roots in the invention of the transistor by Brattain, Bardeen, and Shockley in 1948. The ultimate impact of large-scale integration (LSI) of circuits is still to be felt. Desk-sized

[†] These apparently went on for a good portion of his life and filled him with a great sense of failure and depression.

computers now give more computing capacity than house-sized computers of 20 years ago.

The possibility of almost unlimited bandwidth capacity was suggested by the observation of laser action by Maiman in the United States in 1960. The initial proposals for a practical laser came from Schawlow and Townes of the United States and Basov and Prokhorov in the U.S.S.R. in 1958. They were awarded the 1964 Nobel Prize for physics for their contributions to the development of the laser.

A new way to produce images, made practically possible by the invention of the laser, is *holography*. Originally invented by D. Gabor in 1947, holography is a way of generating three-dimensional imagery by reconstructing the original wavefronts of light scattered from the object. Very significant improvements in holography were obtained in the early 1960s by E. Leith and J. Upatnicks, who invented *off-axis holography*. Conceivably the ultimate in image reproduction, holography has encountered many difficulties in its practical incorporation in TV and movies. Gabor received the 1971 Nobel Prize in physics for his discovery of holography.

The last quarter of the twentieth century belongs to satellite and optical-fiber communications, massive computer communication networks, and hardware integration techniques allowing for the most sophisticated error-correcting coding techniques to be implemented in practice. Closely related to advances in communications and using similar types of mathematics are the many new imaging techniques of great value in astronomy, industry, and medicine. Included in this group are computer-aided tomography, magnetic resonance imaging, and extremely sophisticated imaging in astronomy. The future will see us interconnected through a network of optical fibers with Europe and beyond. Computers in distant corners of our continent will exchange data in even greater volumes, and this will not always be to our benefit. What about the *distant* future? Will the elusive tachyon enable nearly instant communication across the heavens? Will we become so interconnected with the infrastructure of our civilization that we need never leave our homes to get anything done? Or, perhaps, will we run out of things to say to each other and begin shutting down? Only time will tell.

A list of historical events in communications is given in Table 1.5-2. A great deal of the material in this section was obtained from a delightful book by P. Davidovits called *Communication* [1-6]. Written in an informal and humorous manner, the book gives the history of communication and surveys how communication systems work in a qualitative manner.

1.6 THE BOOK

Our quantitative study of electrical communication systems begins with Chapter 2 in which we discuss the basic mathematical tools required for the rest of the book. The most important mathematical theory required for the understanding of communication systems is *Fourier theory*. This is the heart of Chapter 2. After completing the discussion of Fourier theory, we continue with a quantitative description of *linear systems*. This

TABLE 1.5-2 DATES OF SIGNIFICANT EVENTS IN COMMUNICATION

Year	Event
1600–1750:	Initial studies of electric and magnetic phenomena started.
1780–1790:	Coulomb experiments on the force between electric charges.
1790–1800:	Voltaic cell discovered.
1820:	Oersted shows that electric currents produce magnetic fields.
1830–1840:	Faraday and others show that electric fields are produced by changing magnetic fields.
1834:	Gauss and Weber build electromagnetic telegraph.
1837:	Cooke and Wheatstone telegraph built.
1842:	Morse's telegraph installed between Baltimore and Washington.
1860:	First telephone built by Reis.
1864:	Maxwell's theory of electromagnetism published.
1866:	Permanent transatlantic telegraph cable installed.
1876:	Telephone developed by Bell.
1887:	Hertz experiments on electromagnetic radiation.
1894:	Lodge demonstrates electromagnetic wireless communication over a distance of 150 yards.
1898:	Marconi and Jackson transmit a signal over a distance of 60 miles.
1901:	Transatlantic wireless communication developed by Marconi.
1904:	Fleming develops vacuum-tube diode.
1906:	Lee De Forest invents vacuum-tube triode.
1907:	Fessenden broadcasts speech over a distance of 200 miles.
1908:	Campbell-Swinton publishes basic ideas on television broadcasting.
1920:	First scheduled broadcast by station KBKA, Pittsburgh.
1929:	Zworykin demonstrates his television system.
1933:	Armstrong develops frequency modulation.
1936:	Commercial television broadcasting started by BBC.
1948:	Transistor amplifier built by Brattain, Bardeen, and Shockley.
1950–1960:	Microwave communication links developed.
1953:	First transatlantic telephone cable.
1954:	Color television broadcasting started in the United States.
1954:	First maser built.
1960:	First laser built.
1960:	First communication satellite launched (Echo I).
1960:	First TV pictures from space (TIROSI weather satellite).
1962:	Era of practical satellite communication begins with Telstar I.
1970s to present:	Extensive use of satellite communications, fiber optics, computer communication networks; extensive use of very large scale integration (VLSI) in communications; integrated electrooptical and photonic devices; era of supercomputers and parallel computing devices; computer imaging in medicine such as magnetic resonance and computer tomography; advances in x-ray lasers; synthetic aperture imaging in radar and astronomy.

is the main subject of Chapter 3. Here we shall also discuss the theory of filters, including *active filters*. In Chapter 4, we discuss the sampling theorem and its implications and continue with an analysis of analog pulse modulation. Soon after we switch gears and consider the fundamentals of quantizing and conversion of signals to digital form. Our discussion here considers only basic pulse code modulation (PCM) and its variations. More complex schemes are discussed in Chapters 10 and 11. With the notions of sampling and quantizing under our belt, we continue our study with *digital systems* and fast Fourier transforms, which are the subject matter of Chapter 5. Digital filters, a very rapidly growing technology, are discussed here. In Chapters 6 and 7 we deal with amplitude (AM) and angle modulation, respectively. In Chapter 6, which is conceptually straightforward, we discuss AM and the systems derived from it. Also, the theory and practice of *television*, both black/white and color, are treated here. Because TV involves so much of the high technology of communication and because of its great impact on our civilization, it represents an almost perfect case study to illustrate the concepts dealt with in detail in this chapter. In Chapter 7, also conceptually simple, we use somewhat more mathematics because we discuss frequency (FM) and phase (PM) modulation. Chapter 7 concludes the first phase of the book.

The second phase of the book, Chapters 8 through 12, is more concerned with statistical and optical communications. The mathematical tools of statistical communications are *probability theory* and *random processes*. Chapter 8 is a brief discussion of the fundamental calculus of probability and random variables. In Chapter 9, we apply the material from Chapter 8 to random processes. The first part of Chapter 9 introduces the notion of power spectra and correlation functions and illustrates these for some simple random processes. There we also derive the elementary input–output relations for linear time-invariant systems. The second part applies the theory to computing signal-to-noise calculations for standard communication systems.

Chapter 10 is entirely devoted to digital communications. It begins by developing the maximum likelihood receiver when the source can transmit one of M symbolic messages. Various sophisticated digital communication circuits are described, including the complex and difficult task of synchronization. Chapter 11 discusses the ultimate limits of analog and especially digital communication systems from the point of view of *information theory*. How good is a particular system compared to the best system that can theoretically exist? Allowing that we cannot achieve error-free communications, how can we streamline a communication system without exceeding a given bound on the error rate? Here we discuss *rate-distortion theory*, the basic coding theorems, and source and channel codes.

Finally, we discuss optical communications in a somewhat lengthy Chapter 12. Here we furnish an overview of hardware and systems. About half the chapter is devoted to devices: fibers, sources, and detectors and how they work. The other half deals with system aspects of optical communications: the effect of source coherence, the capacity of the channel, signal-to-noise considerations, and optimum receiver design when Poisson noise is the ultimate limitation.

Readers familiar with the first edition will find here much that is new. Unfortunately, because of strong constraints on the size of the book, some material from the first edition had to be removed. All the materials on radar and sonar processing, as well as material on signal and image processing falls in this category. We hope, however, that the reader will see the changes as we do: that on the whole the gains exceed the losses.

Items marked with an asterisk are either more advanced, more specific, or more peripheral than is necessary for a first reading. They can be omitted without serious loss the first time around.

REFERENCES

[1-1] G. Raisbeck, *Information Theory*, M.I.T. Press, Cambridge, Mass., 1963.

[1-2] C. E. Shannon and W. Weaver, *The Mathematical Theory of Communication*, University of Illinois Press, Urbana, Ill., 1964.

[1-3] *Reference Data for Radio Engineers*, 4th ed., International Telephone and Telegraph Corporation, New York, 1956.

[1-4] A. B. Carlson, *Communication Systems*, 3rd ed., McGraw-Hill, New York, 1986.

[1-5] S. A. Schelkunoff, *Electromagnetic Fields*, Ginn/Blaisdell, Waltham, Mass., 1963.

[1-6] P. Davidovits, *Communication*, Holt, Rinehart and Winston, New York, 1972.

CHAPTER 2

Fourier Methods

2.1 INTRODUCTION

The interrelation between time and frequency plays a central role in communication theory. Signals can be characterized by such parameters as duration, period, voltage level, and time-history. The use of such parameters leads to the *time-domain* characterization of signals. But signals can also be characterized by bandwidth, spectral content, phase, and frequency. The use of these parameters leads to a *frequency-domain* characterization of signals. How are time-domain parameters related to frequency-domain parameters? This is the main concern of time–frequency analysis.

The analytical relations between time and frequency are established by the Fourier series and the Fourier integral, studied in this chapter. We start with a brief discussion of linear time-invariant systems, since Fourier analysis is particularly useful in input–output calculations for these systems. We consider periodic signals and their expansion into Fourier series and eventually extend the analysis to nonperiodic signals and the Fourier integral, whose properties we deal with in some detail. The discussion of the Fourier series is itself extended to incorporate general orthogonal-function expansions of which the Fourier series is only one example. This leads us to a brief examination of the minimum mean-square-error property of such expansions, as well as concepts such as completeness, L_2 spaces, vector-space representation of signals, orthogonal projections, and the Gram–Schmidt process for generating a set of orthonormal functions. The chapter concludes with a discussion of functions of two variables and their Fourier transformation.

2.2 THE SUPERPOSITION PRINCIPLE

Finding the response of a system to an arbitrary input signal is a central problem encountered in communication theory. It is a problem that arises in the description of small systems, such as amplifiers and filter circuits, as well as in large systems, such as telephone networks, television, or radio. For nonlinear or time-varying systems, this problem is generally quite difficult, and in most cases there are no analytic solutions. Fortunately, we are often able to deal with linear, time-invariant systems for which the input–output problem has well-known solutions.

Practical systems are, of course, never exactly linear or time invariant. Aside from the fact that certain circuits such as the rectifiers, modulators, and demodulators considered in later chapters are intrinsically nonlinear, time varying, or both, even a circuit that is nominally linear will generally exhibit some nonlinear behavior. For instance, a "linear" amplifier is linear only for signal amplitudes up to a certain maximum level and will saturate if one tries to go beyond this. Saturation is a nonlinear effect. Similarly, heating of the components of an electric circuit will cause parameter values to change and thereby destroy strict time invariance. However, in practice it is often possible to disregard these effects, to deal with nominally linear systems as though they were exactly linear, and to treat departures from linearity and time invariance as distortions that can be examined separately.

Consider the linear system represented in Fig. 2.2-1. The input is an arbitrary signal $x(t)$; the output is $y(t)$. This system is sometimes referred to as a *two-port* because it has an input port and an output port. Systems may have several input ports or several output ports or both, in which case they are called multiport systems; we shall not deal with these here.

Without inquiring into exactly what is inside the box shown in Fig. 2.2-1, we can represent the operation of the system by

$$y(t) = \mathcal{H}_t[x(\tau)], \qquad \tau \in I \tag{2.2-1}$$

In words, the system $\mathcal{H}_t$ operates on the input $x(\tau)$ over some time segment I to produce the output at time t. The symbol $\mathcal{H}_t$ stands simply for an arbitrary transformation or operator producing an output at time t. The operator is *real* if a real input x results in a real output y. For a system with a *finite memory*, the time interval I is finite; for a *causal* or *nonanticipative* system, $\tau \leq t$.

A linear system is defined as a system for which superposition holds. Thus, let $x_1(t)$ and $x_2(t)$ be two arbitrary inputs and $y_1(t)$ and $y_2(t)$ be the corresponding outputs. Let a_1, a_2 be two arbitrary constants. Then $\mathcal{H}_t$ is a *linear operator* if and only if

$$\begin{aligned}\mathcal{H}_t[a_1x_1(\tau) + a_2x_2(\tau)] &= a_1\mathcal{H}_t[x_1(\tau)] + a_2\mathcal{H}_t[x_2(\tau)] \\ &= a_1y_1(t) + a_2y_2(t).\end{aligned} \tag{2.2-2}$$

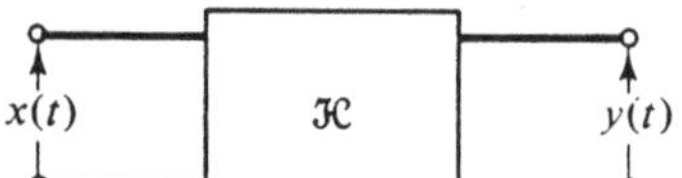

Figure 2.2-1. Linear system.

This is the *superposition principle*. It implies, for instance, that doubling the input doubles the output. By letting $x_1(\tau) = x_2(\tau)$ and $a_1 = -a_2$, we also observe that linearity (or superposition) implies that zero input gives zero output. Thus linearity means more than just straight line: the straight line must pass through the original as well.

If there are three inputs and three constants, we can group any two of them and add the third:

$$\begin{aligned}\mathcal{H}_t[a_1x_1(\tau) + a_2x_2(\tau) + a_3x_3(\tau)] &= \mathcal{H}_t\{[a_1x_1(\tau) + a_2x_2(\tau)] + a_3x_3(\tau)\}\\ &= a_1\mathcal{H}_t[x_1(\tau)] + a_2\mathcal{H}_t[x_2(\tau)]\\ &\quad + a_3\mathcal{H}_t[x_3(\tau)].\end{aligned} \tag{2.2-3}$$

This process can obviously be repeated, and therefore the basic definition of superposition easily extends to any finite number of inputs. The definition is, in fact, usually extended to encompass an infinite number of inputs, even though this does not follow quite so obviously from Eq. (2.2-3). Thus, if

$$y_i(t) = \mathcal{H}_t[x_i(\tau)],$$

linearity implies that

$$\mathcal{H}_t\left[\sum_{i=1}^{\infty} a_ix_i[\tau]\right] = \sum_{i=1}^{\infty} a_iy_i(t). \tag{2.2-4}$$

In addition to being linear, a system or operator may or may not be time invariant. For a time-invariant operator,

$$\mathcal{H}_t[x(\tau - \sigma)] = y(t - \sigma) \tag{2.2-5}$$

for all τ; in other words the response at any particular time t depends on the difference between this time and some arbitrary reference; it does not depend on absolute time. This is illustrated in Fig. 2.2-2.

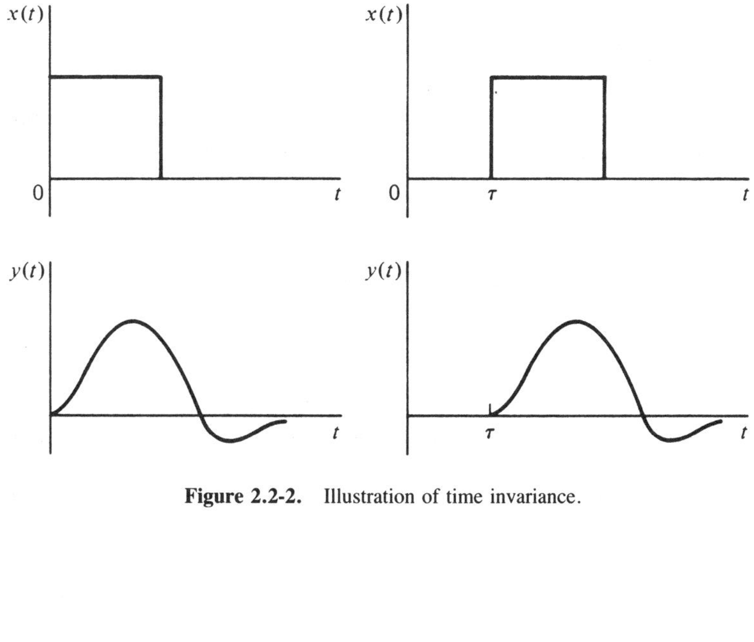

Figure 2.2-2. Illustration of time invariance.

2.3 LINEAR SYSTEM RESPONSE

To find out something about a linear system, one may apply certain standard test signals $z_n(t)$ to the input and observe the output. Typical of these test signals are impulses, steps, or sine waves. The response of the system to such a test signal is a characteristic description of the system's more general input–output behavior. In fact, vis-à-vis input–output behavior, the test-signal response may be a better characteristic than a complete circuit diagram of the insides of the box because it tells us more readily what the system actually does.[†]

If we want to find the response of the system to an arbitrary input $x(t)$, we can try to represent $x(t)$ in terms of a sum of elementary test functions and then use superposition to find the output. Typically, such a representation has the form

$$x(t) = \sum_{n=1}^{\infty} a_n z_n(t). \tag{2.3-1}$$

Then, if the response to the elementary signal $z_n(t)$ is $h_n(t)$, the superposition principle gives

$$y(t) = \sum_{n=1}^{\infty} a_n h_n(t). \tag{2.3-2}$$

We should point out that the right-hand side of Eq. (2.3-1) is not always an exact representation of $x(t)$. If so, Eq. (2.3-2) is also an approximation, but in practice the approximation is usually excellent.

In the following sections we consider a particular signal representation, the Fourier series. Some of the details of the superposition procedure are dealt with in Chapter 3. Signal representation is one of the central issues in communications theory, and it will come up again in various guises throughout this book.

2.4 PERIODIC SIGNALS

The Fourier series is a representation for periodic signals. A real function of time $x(t)$ is periodic if there is some time interval τ for which

$$x(t + \tau) = x(t) \tag{2.4-1}$$

for all t in $(-\infty, \infty)$.[‡] The smallest value of τ for which Eq. (2.4-1) is true is called the period T. The fundamental frequency is defined by

$$f_0 = \frac{1}{T} \tag{2.4-2a}$$

[†] The detailed internal structure and inner working of the system, furnished by the circuit diagram, are not generally conveyed by the test-signal response.

[‡] Note that the meanings given to t and τ are no longer those of the previous two sections.

in cycles per second or hertz (Hz), or

$$\omega_0 = \frac{2\pi}{T} \tag{2.4-2b}$$

in radians per second.

The Fourier series representation is given by

$$x(t) = \sum_{n=0}^{\infty} (a_n \cos n\omega_0 t + b_n \sin n\omega_0 t). \tag{2.4-3}$$

Thus the test functions $z_n(t)$ referred to earlier are sines and cosines. If the input to a linear time-invariant system is a sinusoid with frequency ω, the output is a sinusoid of the same frequency, but the amplitude is multiplied by an amplitude factor $A(\omega)$ and the phase is shifted by a phase angle $\theta(\omega)$. Therefore, if the input is represented by a Fourier series as in Eq. (2.4-3), the output will be

$$y(t) = \sum_{n=0}^{\infty} A(n\omega_0)\{a_n \cos [n\omega_0 t + \theta(n\omega_0)] + b_n \sin [n\omega_0 t + \theta(n\omega_0)]\}. \tag{2.4-4}$$

This relation is the main reason for the use of the Fourier series in the analysis of linear systems. The complex number

$$H(\omega) = A(\omega) \exp [j\theta(\omega)]$$

is called the *transfer function* or *frequency response* and is probably the most commonly used characteristic of linear time-invariant systems. The Fourier series permits the transfer function to be used in studying the behavior of the system with arbitrary (i.e., nonsinusoidal) periodic input signals. The restriction to periodic signals can be removed, as shown in Sec. 2.8. The transfer function is considered in more detail in Chapter 3.

An equivalent and somewhat more convenient form of the Fourier series is

$$x(t) = \sum_{n=-\infty}^{\infty} c_n e^{jn\omega_0 t}. \tag{2.4-5}$$

The equivalence between Eqs. (2.4-3) and (2.4-5) is easily demonstrated by using the exponential representation for the sine and cosine functions:

$$\sin \theta = \frac{-j}{2}(e^{j\theta} - e^{-j\theta}), \tag{2.4-6}$$

$$\cos \theta = \frac{1}{2}(e^{j\theta} + e^{-j\theta}). \tag{2.4-7}$$

Substituting this into Eq. (2.4-3) results in

$$\begin{aligned} x(t) &= \sum_{n=0}^{\infty} \frac{a_n}{2}(e^{jn\omega_0 t} + e^{-jn\omega_0 t}) - \frac{jb_n}{2}(e^{jn\omega_0 t} - e^{-jn\omega_0 t}) \\ &= \sum_{n=0}^{\infty} \left(\frac{a_n - jb_n}{2}\right) e^{jn\omega_0 t} + \left(\frac{a_n + jb_n}{2}\right) e^{-jn\omega_0 t}. \end{aligned} \tag{2.4-8}$$

We now define

$$c_n = \frac{a_n - jb_n}{2}$$
$$c_n^* = \frac{a_n + jb_n}{2} \qquad (n \neq 0) \tag{2.4-9}$$
$$c_0 = a_0,$$

where the asterisk means complex conjugation. Because $x(t)$ is assumed real, $x(t) = x^*(t)$. It then follows that $c_n^* = c_{-n}$, because

$$\begin{aligned} x^*(t) &= \sum_{n=-\infty}^{\infty} c_n^* e^{-jn\omega_0 t} \\ &= \sum_{n=-\infty}^{\infty} c_{-n}^* e^{jn\omega_0 t} \qquad \text{(if } n \text{ is replaced by } -n) \\ &= \sum_{n=-\infty}^{\infty} c_n e^{jn\omega_0 t} = x(t). \end{aligned}$$

Equation (2.4-8) can be written in the form

$$x(t) = \sum_{n=1}^{\infty} c_n e^{jn\omega_0 t} + \sum_{n=1}^{\infty} c_n^* e^{-jn\omega_0 t} + c_0,$$

which, when use is made of $c_n^* = c_{-n}$, can be written as

$$\begin{aligned} x(t) &= \sum_{n=1}^{\infty} c_n e^{jn\omega_0 t} + \sum_{n=-1}^{-\infty} c_n e^{jn\omega_0 t} + c_0 \\ &= \sum_{n=-\infty}^{\infty} c_n e^{jn\omega_0 t}. \end{aligned} \tag{2.4-10}$$

Some important results that follow from the fact that $c_n^* = c_{-n}$ are that

$$|c_n| = |c_{-n}|$$

and

$$\angle c_n = -\angle c_{-n};$$

i.e., the magnitude has even symmetry and the phase has odd symmetry.

Simply writing down the series as we have done in Eq. (2.4-3) or (2.4-5) is, of course, no guarantee that it converges to the desired periodic function $x(t)$, or that it converges at all. We shall investigate the convergence problem in the next section. However, these expressions suggest a simple way to calculate the coefficients. One reason for preferring the exponential form of the Fourier series is that the procedure for doing this is slightly simpler. We multiply Eq. (2.4-10) by $e^{-jm\omega_0 t}$, giving

$$x(t)t^{-jm\omega_0 t} = \sum_{n=-\infty}^{\infty} c_n e^{j(n-m)\omega_0 t}. \tag{2.4-11}$$

Assuming convergence of the series, we can integrate term by term over one period with the result

$$\begin{aligned}\int_{-T/2}^{T/2} x(t)e^{-jm\omega_0 t}\,dt &= \sum_{n=-\infty}^{\infty} c_n \int_{-T/2}^{T/2} e^{j(n-m)\omega_0 t}\,dt \\ &= \sum_{n=-\infty}^{\infty} c_n \left[\frac{e^{j(n-m)\omega_0 T/2} - e^{-j(n-m)\omega_0 T/2}}{j(n-m)\omega_0}\right] \\ &= T \sum_{n=-\infty}^{\infty} c_n \frac{\sin (n-m)\omega_0 T/2}{(n-m)\omega_0 T/2}.\end{aligned} \tag{2.4-12}$$

By Eq. (2.4-2b) $\omega_0 T/2 = \pi$, and therefore Eq. (2.4-12) can be written in the form

$$\int_{-T/2}^{T/2} x(t)e^{-jm\omega_0 t}\,dt = T \sum_{n=-\infty}^{\infty} c_n \,\text{sinc}\,(n-m), \tag{2.4-13}$$

where the function sinc (•) is defined by

$$\text{sinc}\,(z) \equiv \frac{\sin \pi z}{\pi z}. \tag{2.4-14}$$

The sinc function is an important and useful function which will reappear many more times in this book. It is plotted in Fig. 2.4-1. For the purposes of this section the important property to note is that it is zero for all nonzero integer values of z and that its value for $z = 0$ is 1. Hence in Eq. (2.4-13) all the elements of the sum vanish except the one for which $n = m$. This then results in the formula for the Fourier coefficients:

$$c_m = \frac{1}{T}\int_{-T/2}^{T/2} x(t)e^{-jm\omega_0 t}\,dt. \tag{2.4-15}$$

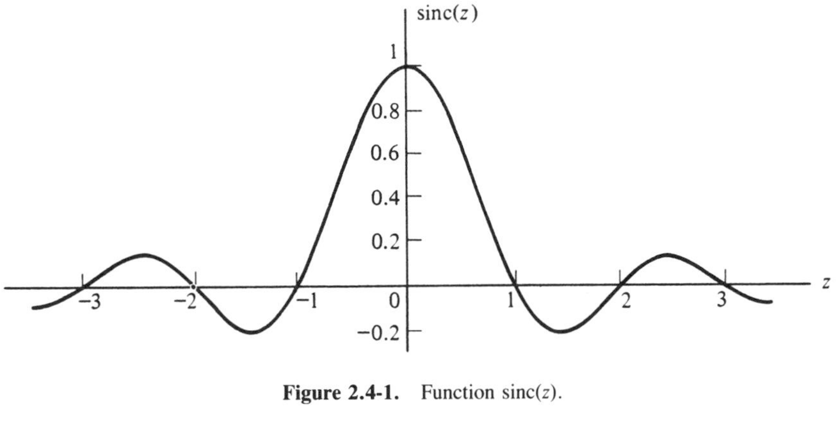

Figure 2.4-1. Function sinc(z).

The formulas for the a_n and b_n used in Eq. (2.4-4) follow directly from Eqs. (2.4-15) and (2.4-9). Thus, for $n \neq 0$,

$$a_n = c_n + c_{-n} = \frac{1}{T}\int_{-T/2}^{T/2} x(t)(e^{-jn\omega_0 t} + e^{jn\omega_0 t})\, dt = \frac{2}{T}\int_{-T/2}^{T/2} x(t)\cos n\omega_0 t\, dt, \tag{2.4-16}$$

$$a_0 = c_0 = \frac{1}{T}\int_{-T/2}^{T/2} x(t)\, dt, \tag{2.4-17}$$

$$b_n = \frac{1}{j}(c_{-n} - c_n) = \frac{1}{T}\int_{-T/2}^{T/2} x(t)\left(\frac{e^{jn\omega_0 t} - e^{-jn\omega_0 t}}{j}\right) dt = \frac{2}{T}\int_{-T/2}^{T/2} x(t)\sin n\omega_0 t\, dt. \tag{2.4-18}$$

Another form of the Fourier series, which is often convenient, is the cosine series:

$$x(t) = c_0 + \sum_{n=1}^{\infty} x_n \cos(\omega_n t + \phi_n), \tag{2.4-19}$$

where $x_n = 2|c_n| = \sqrt{a_n^2 + b_n^2}$,
$\phi_n = \angle c_n = -\tan^{-1}(b_n/a_n)$,
$\omega_n = n\omega_0$,

The phase angle† ϕ_n is zero for even functions, i.e., for functions $x(t)$ such that $x(-t) = x(t)$. Odd functions, for which $x(t) = -x(-t)$, can be represented by a sine series $\sum_{n=1}^{\infty} x_n \sin \omega_n t$.

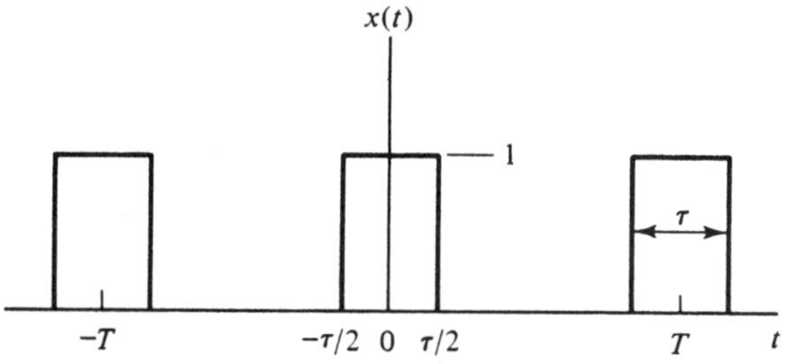

Figure 2.4-2. Square pulse train.

Examples

1. *Square pulse train* (Fig. 2.4-2). This function has the value 1 in the interval $-\tau/2, \tau/2$; hence

$$c_n = \frac{1}{T}\int_{-\tau/2}^{\tau/2} e^{-j2\pi nt/T}\, dt = \frac{1}{-j2\pi n}(e^{-j2\pi nt/T})\Big|_{-\tau/2}^{\tau/2} = \frac{\tau}{T}\operatorname{sinc} n\frac{\tau}{T} = d \operatorname{sinc} nd,$$

† The phase or angle of c_n is sometimes called its argument, written arg c_n.

where $d \equiv \tau/T$ is the duty ratio. The magnitude and phase of the first few c_n's are shown in Fig. 2.4-3(a) for $d = \frac{1}{4}$. This is called a *line spectrum* since only discrete frequencies appear in it. Line spectra are characteristic of periodic functions. In this particular example the c_n's are all real. This means that the trigonometric form of the series [Eq. (2.4-3)] has only cosine terms; i.e., the b_n's of Eq. (2.4-18) are all zero. This is so because we chose the time origin to make $x(t)$ an even function. The effect of a shift in the time origin is considered in another example which we shall consider shortly.

We have already seen that for real $x(t)$ the magnitude of c_n has even symmetry and

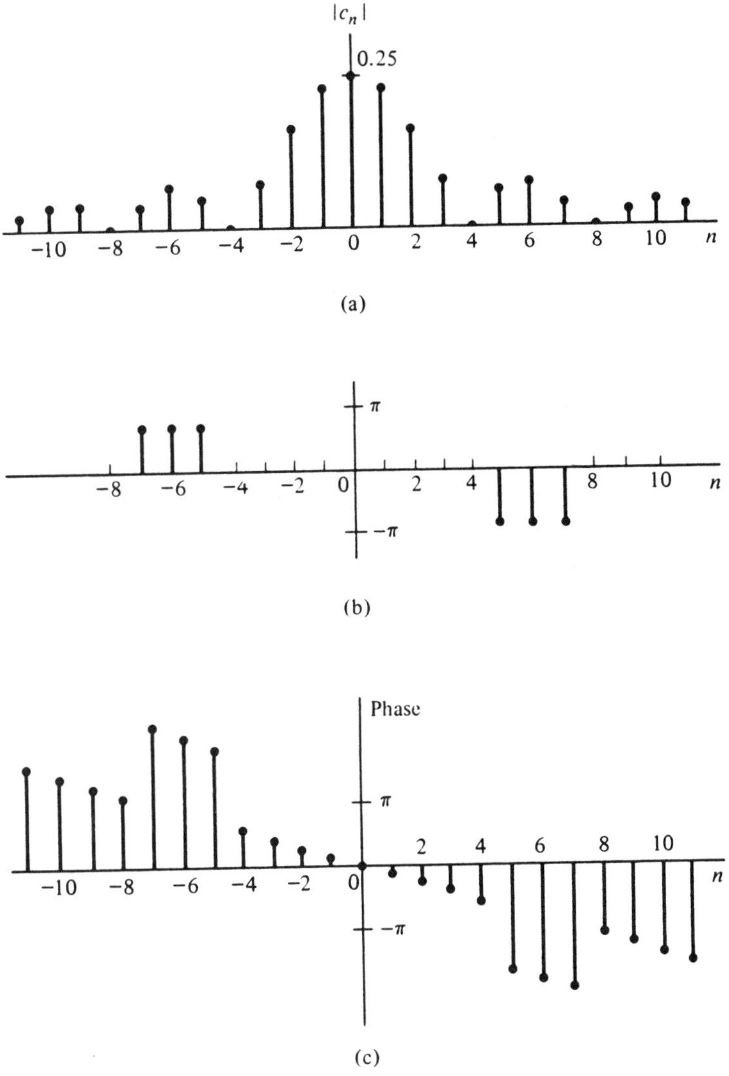

Figure 2.4-3. Line spectrum for the square pulse train with duty ratio $d = \frac{1}{4}$: (a) magnitude; (b) phase for even function; (c) phase for shifted pulse train.

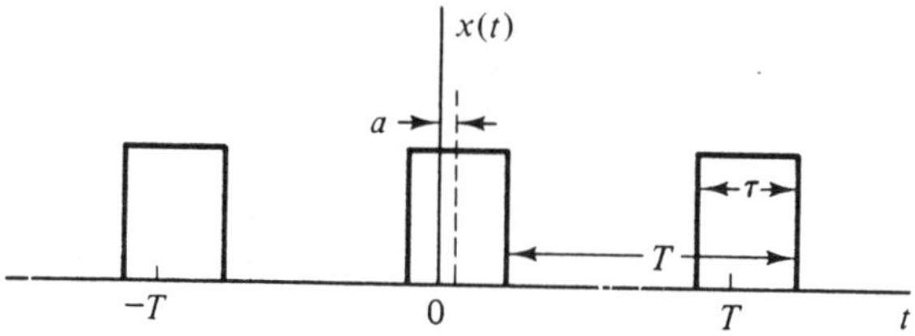

Figure 2.4-4. Shifted square pulse train.

the phase of c_n has odd symmetry. The odd phase symmetry is shown in Figs. 2.4-3(b) and (c).

Observe that $c_0 = d$ and that the smaller the value of d, the larger the numbers n for which sinc nd has significantly large values. In fact we can define the significant width of the spectrum in terms of those values of n around $n = 0$ for which sinc nd exceeds zero.† These are given by $-1 \leq nd \leq 1$ or $-1/d \leq n \leq 1/d$.

The corresponding bandwidth is f_0/d, where f_0 is the fundamental frequency in hertz. In Fig. 2.4-3 the significant values of n go from -4 to 4, and therefore the bandwidth is $4f_0$. If d were smaller than $\frac{1}{4}$, the bandwidth would be larger. The reciprocal relation between duration and bandwidth is one of the most important properties of the Fourier expansion.

2. *Shifted square pulse train* (Fig. 2.4-4). Here

$$c_n = \frac{1}{T}\int_{a-\tau/2}^{a+\tau/2} e^{-j2\pi nt/T}\,dt = \frac{1}{-j2\pi n}\left(e^{-j2\pi n(a+\tau/2)/T} - e^{-j2\pi n(a-\tau/2)/T}\right)$$

$$= e^{-j2\pi na/T}\frac{\tau}{T}\operatorname{sinc} n\frac{\tau}{T}. \tag{2.4-20}$$

The magnitude spectrum (and therefore the bandwidth) is not changed by the time shift, but there is now an additional phase shift of $-2\pi na/T$ radians in the c_n's. This is shown in Fig. 2.4-3(c).

3. *Cosine pulse train* (Fig. 2.4-5). The pulse centered about the origin is described by

$$x(t) = \begin{cases} \cos 2\pi f_1 t, & -\dfrac{\tau}{2} < t < \dfrac{\tau}{2}, \\ 0, & \text{for the remainder of the period.} \end{cases}$$

The cosine pulse train is given by $\Sigma_{n=-\infty}^{\infty} x(t - nT)$. Here c_n is computed from

$$c_n = \frac{1}{T}\int_{-\tau/2}^{\tau/2} e^{-j2\pi nt/T}\cos 2\pi f_1 t\,dt. \tag{2.4-21}$$

This is most conveniently evaluated by expressing the cosine itself in terms of exponentials:

$$c_n = \frac{1}{T}\int_{-\tau/2}^{\tau/2} e^{-j2\pi nt/T}\frac{e^{j2\pi f_1 t} + e^{-j2\pi f_1 t}}{2}\,dt$$

$$= \frac{1}{2T}\int_{-\tau/2}^{\tau/2}\left(e^{-j2\pi(n/T+f_1)t} + e^{-j2\pi(n/T-f_1)t}\right)dt \tag{2.4-22}$$

$$= \frac{\tau}{2T}\left[\operatorname{sinc}\left(\frac{n}{T}+f_1\right)\tau + \operatorname{sinc}\left(\frac{n}{T}-f_1\right)\tau\right].$$

† This is, obviously, an arbitrary definition. The lines around $n = 0$ for which $|n| \leq d^{-1}$ are sometimes called the *main lobe* of the sinc function.

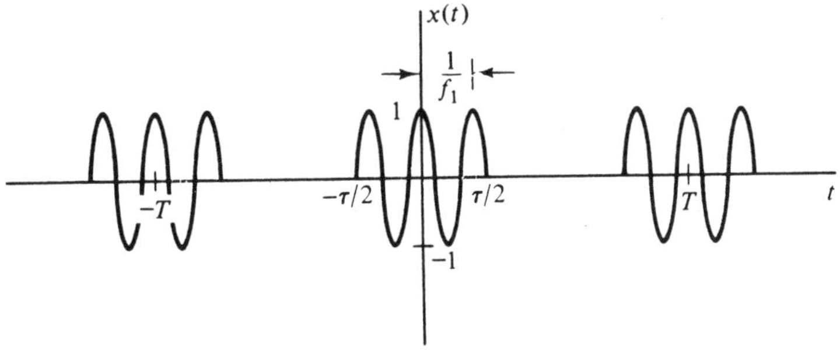

Figure 2.4-5. Cosine pulse train.

The line spectrum (i.e., values of $|c_n|$) for $d \equiv \tau/T = \frac{1}{4}$, $f_1 = 4/T$ is shown in Fig. 2.4-6.

We now have two spectral lobes centered at $\pm f_1$. The significant width of these spectral lobes is still given by $n = \pm T/\tau$; i.e., it depends on the duty cycle and not on the details of the pulse shape. For the particular values of d and f_1 used in Fig. 2.4-6, c_0 is zero. This reflects the fact that the cosine pulse, with these values of d and f_1, has no dc value. The pulse shown in Fig. 2.4-5 *does* have a dc value (since the area above the t axis is greater than that below); hence for that cosine pulse c_0 would be positive.

4. *Square Wave* (Fig. 2.4-7). This function has the value 1 for one half-period and the value -1 for the other half-period. It is convenient to locate the time origin at the center of one of the positive half-periods, as shown in the figure. The square wave can then be regarded as the difference between the square pulse train of the first example,

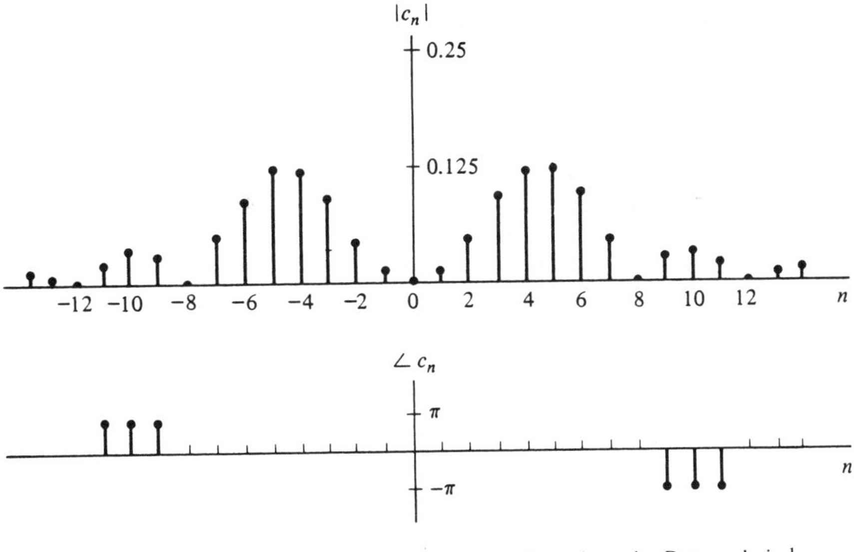

Figure 2.4-6. Line spectrum corresponding to a cosine pulse train. Duty cycle is $\frac{1}{4}$, $f_1 = 4/T$.

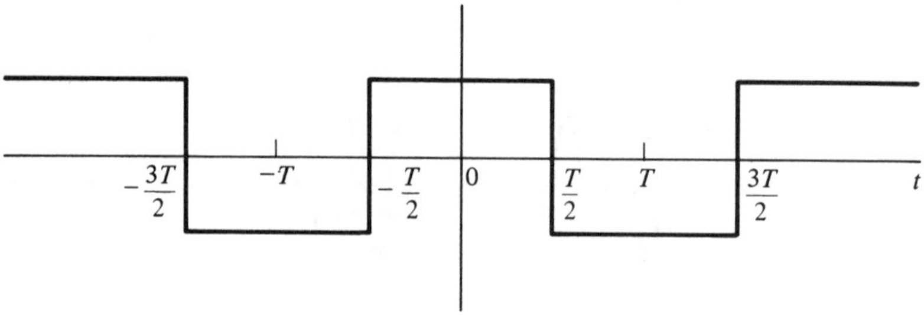

Figure 2.4-7. Square wave.

with a duty ratio $d = \frac{1}{2}$, and the same square pulse train shifted by half a period. Hence, using the results of Examples 1 and 2, we get

$$c_n = \tfrac{1}{2}\,\text{sinc}\left(\frac{n}{2}\right) - \tfrac{1}{2}\,\text{sinc}\left(\frac{n}{2}\right)\exp(-j2\pi n/2)$$

$$= \tfrac{1}{2}\,\text{sinc}\left(\frac{n}{2}\right)[1 - \exp(-j\pi n)]$$

$$= \begin{cases} 0, & \text{for even } n, \text{ including } n = 0 \\ \dfrac{2}{\pi}, -\dfrac{2}{3\pi}, \dfrac{2}{5\pi}, -\dfrac{2}{7\pi}, \ldots, & \text{for } n = 1, 3, 5, 7, \ldots. \end{cases}$$

Also, $c_{-n} = c_n$.

Another approach is to regard the square wave as the square pulse train of Example 1, with a duty ratio $d = \frac{1}{2}$ and an amplitude of 2, and subtract a constant value of 1. The constant 1 can in turn be regarded as the square pulse train of Example 1 with a duty ratio $d = 1$. Using this approach, we get

$$c_n = \text{sinc}\left(\frac{n}{2}\right) - \text{sinc}\, n, \tag{2.4-23}$$

which is the same as the result obtained with the first approach.

Because of our choice of time origin, the square wave is an even function, and the trigonometric form of the Fourier series (Eq. 2.4-3) therefore contains only cosine terms. By combining c_n's for positive and negative n, we can write the Fourier cosine series for the square wave as

$$x(t) = \frac{4}{\pi}\left[\cos\frac{2\pi t}{T} - \tfrac{1}{3}\cos\frac{6\pi t}{T} + \tfrac{1}{5}\cos\frac{10\pi t}{T}\cdots\right. \tag{2.4-24}$$

We see that the square wave has only odd harmonics whose amplitude is inversely proportional to the harmonic order. Also, the amplitude of the fundamental is about 30% larger than the amplitude of the square wave.

In Fig. 2.4-8 we show how the first few harmonics combine to approximate a square wave.

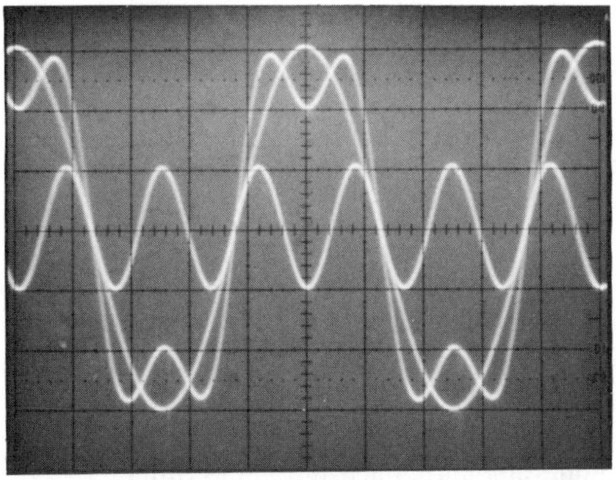

(a)

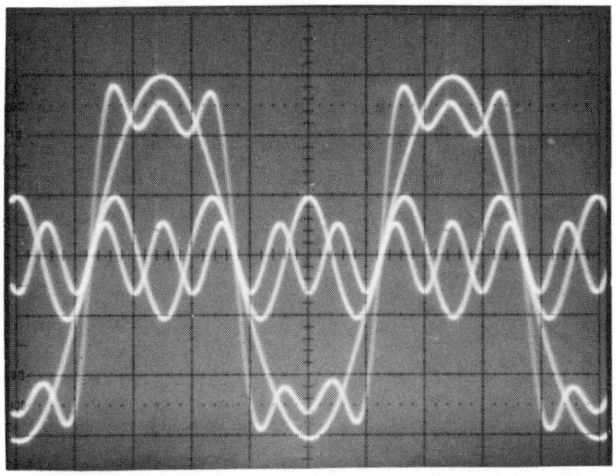

(b)

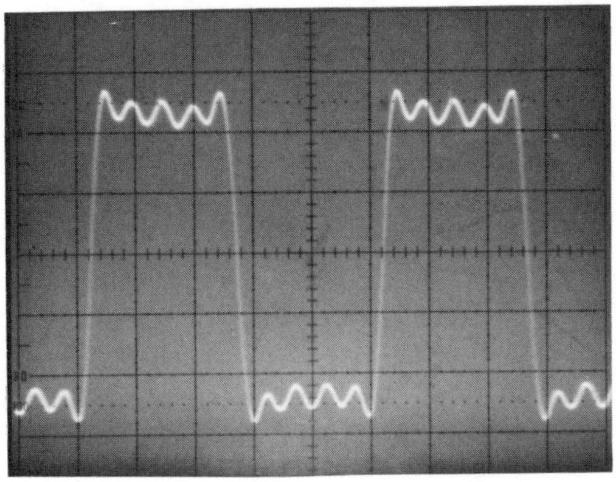

(c)

Figure 2.4-8. Square wave approximated by the first few harmonics. (a) Fundamental, third harmonic, and sum. (b) Fundamental, third and fifth harmonics, and sum. (c) Square wave approximation by the sum of the first to the ninth harmonic.

2.5 CONVERGENCE

The formal procedure of the previous section for calculating the coefficients was based on the assumption that Eqs. (2.4-3) and (2.4-5) are truly equalities, i.e., that an arbitrary periodic function can be represented in terms of a sum of sines and cosines. There is a priori no reason to think that this is always possible. However, if $x(t)$ is integrable, the coefficients as defined in Eqs. (2.4-15)–(2.4-18) will exist. We can therefore start with the function $x(t)$, find the coefficients, and then examine the properties of the series formed with these coefficients. We shall see that the series does converge to $x(t)$ after a fashion, provided that $x(t)$ is not too complicated. We should not expect, however, that the series will converge to $x(t)$ at every t, for arbitrary $x(t)$.

To see this, consider a function $y(t)$ which is equal to $x(t)$ except at a single point, where it differs by a finite amount. Since this finite difference does not affect the calculation of c_n, the Fourier coefficients for $x(t)$ and $y(t)$ will be identical; hence the series cannot represent both functions at the point in question. More generally, functions that are equal "almost everywhere"—that is, everywhere except at a finite number of isolated points—will have the same Fourier series representation, and the series cannot represent all of them at every point. Convergence of the Fourier series is therefore, in general, *not pointwise*.

To see how the Fourier series converges to the desired time function, consider the truncated series

$$x_N(t) = \sum_{n=-N}^{N} c_n e^{jn\omega_0 t}, \tag{2.5-1}$$

and observe how this series behaves as N becomes very large. Substituting from Eq. (2.4-15) and exchanging the order of summation and integration, we can write

$$\begin{aligned} x_N(t) &= \sum_{n=-N}^{N} \left[\frac{1}{T}\int_{-T/2}^{T/2} x(\tau)e^{-jn\omega_0\tau}\,d\tau\right]e^{jn\omega_0 t} \\ &= \frac{1}{T}\int_{-T/2}^{T/2} x(\tau) \sum_{n=-N}^{N} e^{jn\omega_0(t-\tau)}\,d\tau. \end{aligned} \tag{2.5-2}$$

The summation is a geometric series which is easily put into a closed form (Prob. 2-10):

$$S_N(t-\tau) \equiv \sum_{n=-N}^{N} e^{jn\omega_0(t-\tau)} = \frac{\sin\,[(N+\frac{1}{2})\omega_0(t-\tau)]}{\sin\,[(\omega_0/2)(t-\tau)]}. \tag{2.5-3}$$

Then Eq. (2.5-2) becomes

$$x_N(t) = \frac{1}{T}\int_{-T/2}^{T/2} x(\tau)S_N(t-\tau)\,d\tau. \tag{2.5-4}$$

A plot of the function $S_N(t-\tau)$ for fixed t and $N = 5$ is shown in Fig. 2.5-1. In performing the integration, t is treated as a constant. Since the integrand is periodic,

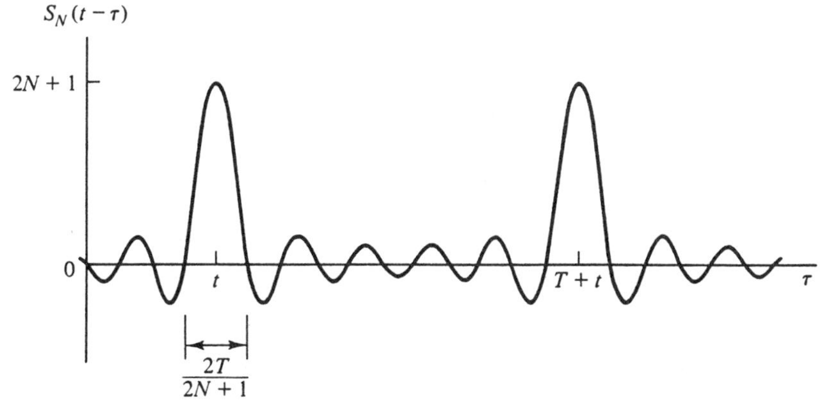

Figure 2.5-1. Plot of the function $S_N(t - \tau)$ for $N = 5$.

we can always arrange the interval $[-T/2, T/2]$ so that the peak of $S_N(t - \tau)$ at $\tau = t$ is well within the range of integration. We see that $x_N(t)$ is the weighted average of $x(\tau)$, with the function $S_N(t - \tau)$ placing most of the weight at $\tau = t$. As N becomes larger the peak and the number of oscillations in one period of $S_N(t - \tau)$ increase so that the weighting of $x(\tau)$ becomes more and more concentrated in the neighborhood of $\tau = t$. Formally it is easily shown (Prob. 2-11) that for τ near t and for large N

$$S_N(t - \tau) \approx 2N \operatorname{sinc} 2Nf_0(t - \tau), \tag{2.5-5}$$

where $f_0 = \omega_0/2\pi$. This function has a large and narrow peak at $\tau = t$ and negligible magnitude for $|\tau - t| \gg 1/Nf_0$. Hence if $x(\tau)$ is "smooth" near $\tau = t$ (this means that it is continuous and has a derivative), we can approximate Eq. (2.5-4) by

$$\begin{aligned} x_N(t) &\approx \frac{1}{T} x(t) \int_{-T/2}^{T/2} S_N(t - \tau)\, d\tau \\ &\approx x(t) \int_{-\infty}^{\infty} 2Nf_0 \operatorname{sinc} 2Nf_0(t - \tau)\, d\tau, \end{aligned} \tag{2.5-6}$$

where we have replaced $1/T$ by f_0 and extended the limits of integration to $\pm\infty$ because $S_N(t - \tau)$ is a narrowly peaked function. An elementary property of the sinc function is

$$\int_{-\infty}^{\infty} \operatorname{sinc} z\, dz = 1 \tag{2.5-7}$$

(see, for instance, formula 499 on p. 68 of Peirce and Foster's table of integrals [2-1]). Hence, by making the change of variable $2Nf_0(t - \tau) = z$ we find that if $x(t)$ is a smooth function,

$$\lim_{N\to\infty} x_N(t) = x(t).$$

Thus we see that the finite Fourier series does indeed converge to the function $x(t)$, provided that $x(t)$ is sufficiently well behaved.

Convergence at a Discontinuity

Functions with isolated discontinuities, such as the square pulses considered in the examples, occur frequently. The Fourier series cannot represent such functions at the points of discontinuity, and the question therefore arises as to what the series does there.

Accordingly, consider a periodic function $x(t)$, discontinuous at $t = t_0$ but possessing a derivative on each side of the point t_0. A piece of the function might look like Fig. 2.5-2. In the vicinity of $t = t_0$, we can regard $x(t)$ as the sum of a continuous function and a step; i.e.,

$$x(t) = x_c(t) + [x(t_+) - x(t_-)]u(t - t_0), \tag{2.5-8}$$

where the function $u(\bullet)$ is defined by

$$u(t - t_0) = \begin{cases} 1, & t > t_0, \\ 0, & t \le t_0. \end{cases} \tag{2.5-9}$$

We do not have to be concerned about points at some distance from t_0 because the weighting function $S_N(t - \tau)$ of Eq. (2.5-3) will not give them any appreciable weight. Substituting Eq. (2.5-8) into Eq. (2.5-4) results in

$$x_N(t) = \frac{1}{T}\int_{-T/2}^{T/2} \{x_c(\tau) + [x(t_+) - x(t_-)]u(\tau - t_0)\}S_N(t - \tau)\, d\tau.$$

We assume that the N-term Fourier series is a good approximation of the continuous part of $x(t)$, and therefore the integration of the component $x_c(\tau)S_N(t - \tau)$ results approximately in $x_c(t)$. This leaves

$$\begin{aligned} x_N(t) &\approx x_c(t) + [x(t_+) - x(t_-)]\frac{1}{T}\int_{-T/2}^{T/2} u(\tau - t_0)S_N(t - \tau)\, d\tau \\ &= x_c(t) + [x(t_+) - x(t_-)]\frac{1}{T}\int_{t_0}^{T/2} S_N(t - \tau)\, d\tau. \end{aligned}$$

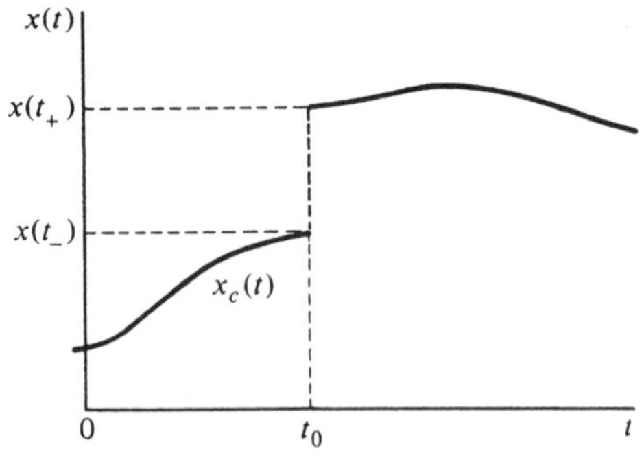

Figure 2.5-2. Discontinuous function.

We define

$$u_N(t - t_0) \equiv \frac{1}{T}\int_{t_0}^{T/2} S_N(t - \tau)\, d\tau \tag{2.5-10}$$

and obtain, finally,

$$x_N(t) = x_c(t) + [x(t_+) - x(t_-)]u_N(t - t_0). \tag{2.5-11}$$

We now investigate the function $u_N(t - t_0)$.

Without materially affecting the discussion we can assume that t_0 is not near the end points $\pm T/2$. (If it is, one can simply redefine the starting point of the period.) For large N, $S_N(t - \tau)$ is sharply peaked near t and essentially negligible elsewhere. Hence it is permissible to replace the upper limit of integration in the definition of $u_N(t - t_0)$ by ∞. Also, instead of $S_N(t - \tau)$ we use $2N$ sinc $2Nf_0(t - \tau)$ as in Eq. (2.5-5). These changes result in

$$u_N(t - t_0) = \frac{2N}{T}\int_{t_0}^{\infty} \operatorname{sinc} 2Nf_0(t - \tau)\, d\tau. \tag{2.5-12}$$

We now make the change of variable $\lambda = 2\pi Nf_0(t - \tau) = N\omega_0(t - \tau)$, and this results in

$$u_N(t - t_0) = \frac{1}{\pi}\int_{-\infty}^{N\omega_0(t-t_0)} \frac{\sin\lambda}{\lambda}\, d\lambda = \frac{1}{2} + \frac{1}{\pi}\operatorname{Si}[N\omega_0(t - t_0)], \tag{2.5-13}$$

where

$$\operatorname{Si}(x) \equiv \int_0^x \frac{\sin\lambda}{\lambda}\, d\lambda \tag{2.5-14}$$

is the sine-integral function. This is a tabulated function (see, for instance, Jahnke and Emde [2-2], pp. 1–9). A plot of this function and of $u_N(t)$ is shown in Fig. 2.5-3. We see that the step at $t = t_0$ is replaced by the oscillating function $u_N(t - t_0)$, scaled by the magnitude of the step. The N-term Fourier series representation of the discontinuous function $x(t)$ shown in Fig. 2.5-2 will therefore appear as in Fig. 2.5-4. At the point $t = t_0$, $u_N(t - t_0) = \frac{1}{2}$; therefore, since $x(t_-) = x_c(t_0)$,

$$\begin{aligned} x_N(t_0) &= x(t_-) + \tfrac{1}{2}[x(t_+) - x(t_-)] \\ &= \frac{x(t_+) + x(t_-)}{2}. \end{aligned} \tag{2.5-15}$$

This is independent of N and will hold for $N \to \infty$. Thus we find that at points of discontinuity the Fourier series gives the *average* between the two extreme points.

At points near t_0 we see that the series representation oscillates around the true value. The period of the oscillation is equal to $\pi/N\omega_0$, and it decreases as N becomes larger. However, the amplitude of the oscillation is independent of N. In particular, the size of the first overshoot is 18% of the step size, and this does not decrease. This effect is called the *Gibbs phenomenon*. Although usually of little consequence, it shows

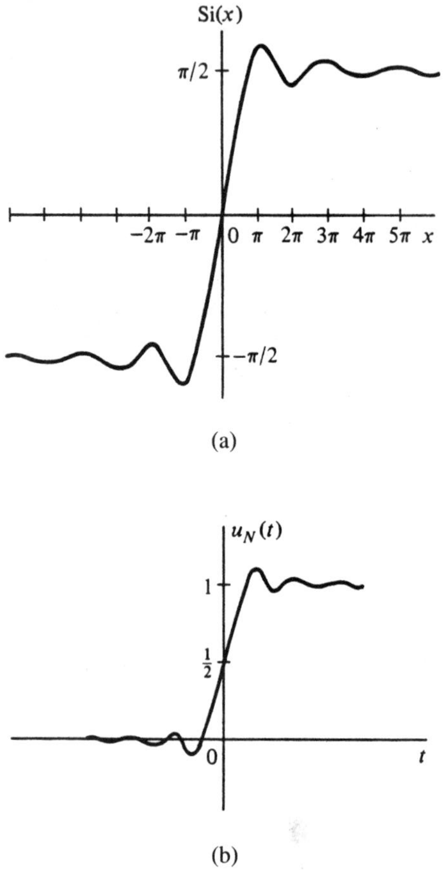

Figure 2.5-3. (a) Function Si(x). (b) Function $u_N(t)$.

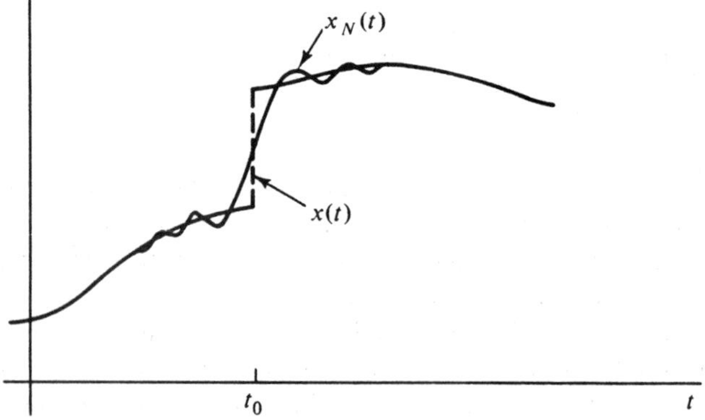

Figure 2.5-4. N-Term representation of a discontinuous function near the discontinuity.

that the Fourier series does not converge uniformly in the vicinity of a discontinuity. In other words, the finite series $x_N(t)$ does not converge to a definite limit with increasing N near the discontinuity point. There are modifications of the Fourier series, achieved by the inclusion of weighting functions called the *Fejér polynomials*, that will eliminate this phenomenon ([2-3] and [2-4], p. 46). However, besides resulting in a more complicated expression for the coefficients, the Fejér modification also destroys the minimum mean-square-error property of the Fourier transform, about which more will be said in Sec. 2.6.

2.6 ORTHOGONAL FUNCTION EXPANSION

The Fourier expansion is only a particular example of expanding a given function in terms of a set of orthogonal functions.

A set of complex-valued functions $\{\phi_n(t)\}$, $n = 1, 2, \ldots$, is said to be *orthogonal* over the interval $(0, T)$ if for all integers m, n

$$\int_0^T \phi_n(t)\phi_m^*(t)\,dt = \begin{cases} C & \text{if } m = n \quad (C \text{ a constant}), \\ 0 & \text{if } m \neq n. \end{cases} \tag{2.6-1}$$

The set is *orthonormal* if

$$\int_0^T \phi_n(t)\phi_m^*(t)\,dt = \delta_{mn} = \begin{cases} 1 & \text{if } m = n, \\ 0 & \text{if } m \neq n. \end{cases} \tag{2.6-2}$$

The number δ_{mn} is called the *Kronecker delta* and appears frequently in analyses. An orthogonal set of functions can obviously be made orthonormal by dividing each function by $\sqrt{C}$. Thus, the functions $e^{j2\pi nt/T}$, $n = 0, \pm 1, \pm 2, \ldots$, are orthogonal over $[0, T]$, and the corresponding orthonormal functions are $\{(1/\sqrt{T})e^{j2\pi nt/T}\}$.

An orthogonal function expansion has the form

$$x(t) = \sum_{n=1}^{\infty} c_n \phi_n(t). \tag{2.6-3}$$

This is sometimes referred to as a *generalized Fourier series*. It is identical with the Fourier expansion given in Eq. (2.4-5) if we make the correspondence $\phi_1(t) = 1$, $\phi_2(t) = e^{j\omega_0 t}$, $\phi_3(t) = e^{-j\omega_0 t}$, $\phi_4(t) = e^{j2\omega_0 t}$, $\phi_5(t) = e^{-j2\omega_0 t}, \ldots$. The fact that the index in one sum runs from $-\infty$ to $+\infty$ and the other from 1 to ∞ is seen to make no real difference. The set of c_n's is called the *spectrum* of the function $x(t)$.

As pointed out earlier, writing down the series is no guarantee that it converges. However, Eq. (2.6-3) suggests that we can find the coefficients c_n by multiplying by $\phi_m^*(t)$ and integrating term by term. If we take the $\phi_n(t)$ to be orthonormal, this gives the formula

$$\begin{aligned} \int_0^T x(t)\phi_m^*(t)\,dt &= \sum_{n=1}^{\infty} c_n \int_0^T \phi_n(t)\phi_m^*(t)\,dt \\ &= c_n. \end{aligned} \tag{2.6-4}$$

Note the similarity to the procedure for finding the Fourier coefficients. We can define the partial sum

$$x_N(t) = \sum_{n=1}^{N} c_n \phi_n(t), \tag{2.6-5}$$

where the c_n are given by Eq. (2.6-4). This will exist if the c_n exist, i.e., if $x(t)$ is integrable. The mean-square error between $x(t)$ and $x_N(t)$ is defined by

$$\begin{aligned} \langle \epsilon^2 \rangle &= \int_0^T |x(t) - x_N(t)|^2 \, dt \\ &= \int_0^T |x(t) - \sum_{n=1}^{N} c_n \phi_n(t)|^2 \, dt. \end{aligned} \tag{2.6-6}$$

(We use the absolute-value bars so that complex signals can be included in the definition.)

The set of functions $\phi_n(t)$, $n = 1 \ldots$, is often referred to as the *basis set* and the functions themselves as the *basis*. The choice of a suitable basis is generally dictated by the type of problem being solved. For instance, if one deals with linear, time-invariant systems, the exponential or trigonometric functions are convenient because they form the general solution of the differential equations describing such systems. (A somewhat more technical way of saying the same thing is that the exponential functions are eigenfunctions of time-invariant linear operators.) In other applications, other basis sets such as the Bessel or Laguerre functions may be more appropriate. In digital applications, functions such as the Walsh and Haar functions [2-5] which can take on only two values (e.g., zero and one) are sometimes convenient.

An important requirement of a basis is that it should be *complete*. A set of orthonormal functions $\{\phi_n(t)\}$ is said to be *complete* if, by choosing N large enough, the mean-square error can be made arbitrarily small, i.e., if

$$\lim_{N\to\infty} \int_0^T |x(t) - \sum_{n=1}^{N} c_n \phi_n(t)|^2 \, dt = 0. \tag{2.6-7}$$

If the $\phi_n(t)$ form a complete set, the generalized Fourier series is said to converge *in the mean* to the function $x(t)$. Note that this does not imply pointwise convergence unless the series converges uniformly. It can be shown [2-6] that the exponential functions

$$\frac{1}{\sqrt{T}}, \frac{e^{j2\pi t/T}}{\sqrt{T}}, \frac{e^{-j2\pi t/T}}{\sqrt{T}}, \ldots \quad \text{as well as} \quad \frac{1}{\sqrt{T}}, \sqrt{\frac{2}{T}}\cos\frac{2\pi t}{T},$$

$$\sqrt{\frac{2}{T}}\sin\frac{2\pi t}{T}, \sqrt{\frac{2}{T}}\cos\frac{4\pi t}{T}, \sqrt{\frac{2}{T}}\sin\frac{4\pi t}{T} \ldots$$

form complete orthonormal sets. Other complete sets include the Bessel functions, Legendre polynomials, etc. The taks of demonstrating completeness of a given set of orthonormal functions is not trivial, and we shall not pursue it here. We assume completeness of the set of orthogonal functions $\{\phi_n(t)\}$ in the following discussions.

By expanding the square and integrating term by term, we can write Eq. (2.6-7) as

$$\lim_{N\to\infty} \int_0^T |x(t)|^2\, dt - 2\,\mathrm{Re}\left[\sum_{n=1}^{N} c_n^* \int_0^T x(t)\phi_n(t)\, dt\right] + \sum_{n=1}^{N}\sum_{m=1}^{N} c_n c_m^* \int_0^T \phi_n(t)\phi_m^*(t)\, dt$$

$$= \lim_{N\to\infty} \int_0^T |x(t)|^2\, dt - 2\sum_{n=1}^{N} |c_n|^2 + \sum_{n=1}^{N} |c_n|^2 = \int_0^T |x(t)|^2\, dt - \lim_{N\to\infty} \sum_{n=1}^{N} |c_n|^2 = 0$$

or

$$\int_0^T |x(t)|^2\, dt = \sum_{n=1}^{\infty} |c_n|^2. \tag{2.6-8}$$

Equation (2.6-8) is a form of *Parseval's theorem*. If $x(t)$ is regarded as the current in a 1-ohm resistor, then $|x(t)|^2$ is the instantaneous power in watts. Therefore the expression on the left is the *energy* per period. One can regard $|c_n|^2$ as the energy associated with the orthonormal function $\phi_n(t)$. For instance, consider the trigonometric Fourier series for an even function $x(t)$. This series has only cosine terms, and if we want it to be in terms of *orthonormal* functions, it will have the form

$$x(t) = \frac{c_0}{\sqrt{T}} + \sum_{n=1}^{\infty} \sqrt{\frac{2}{T}}\, c_n \cos n\omega_0 t.$$

The energy, E_n, in the nth term is given by

$$E_n = \frac{2}{T}|c_n|^2 \int_0^T \cos^2 n\omega_0 t\, dt = |c_n|^2.$$

Thus $|c_n|^2$ is the energy per period of the nth frequency component. Since the c_n's constitute the spectrum of the function $x(t)$, Parseval's theorem says that the energy in the spectrum is identical to the energy in the time function.

Signals having finite energy in the time interval T are referred to as being in the space $L_2(T)$; that is, for $x \in L_2(T)$

$$\int_0^T |x(t)|^2\, dt < \infty.$$

Every signal in $L_2(T)$ can be represented by a Fourier series of the form of Eq. (2.6-3).

Optimum Property of the Fourier Series

As with any infinite series, the usefulness of the (generalized) Fourier expansion depends on a small number of terms furnishing an adequate approximation. This raises the question of whether the Fourier series using a given basis is necessarily the best possible way to represent a periodic function by a few terms.†

† It is clear that some bases will fit a specified function better than others. However, since the choice of basis is generally determined by factors external to the problem of approximating a specific function, we assume the basis to be given.

This question is easily answered by supposing that the finite sum $\Sigma_{n=1}^{N} \gamma_n \phi_n(t)$ is a better approximation to $x(t)$ than the sum using the c_n's defined in Eq. (2.6-4). If we use the mean-square error as our criterion, we get

$$\langle \epsilon^2 \rangle = \int_0^T |x(t) - \sum_{n=1}^{N} \gamma_n \phi_n(t)|^2 \, dt,$$

which by adding and subtracting $c_n \phi_n(t)$ can be written as

$$\langle \epsilon^2 \rangle = \int_0^T |x(t) - \sum_{n=1}^{N} (\gamma_n - c_n)\phi_n(t) - \sum_{n=1}^{N} c_n \phi_n(t)|^2 \, dt. \tag{2.6-9}$$

If we expand the square, there will be cross-product terms of the form

$$\int_0^T x^*(t) \sum_{n=1}^{N} c_n \phi_n(t) \, dt = \sum_{n=1}^{N} c_n \int_0^T x^*(t)\phi_n(t) \, dt = \sum_{n=1}^{N} |c_n|^2 \tag{2.6-10}$$

as well as terms like

$$\int_0^T \sum_{n=1}^{N} \sum_{m=1}^{N} c_n c_m^* \phi_n(t) \phi_m^*(t) \, dt = \sum_{n=1}^{N} |c_n|^2. \tag{2.6-11}$$

The first of these follows from the definition of the c_n's and the second from the orthonormality of the ϕ_n's. We leave the details to the reader, but after all reductions of this sort are made one is left with

$$\langle \epsilon^2 \rangle = \int_0^T |x(t)|^2 \, dt + \sum_{n=1}^{N} |\gamma_n - c_n|^2 - \sum_{n=1}^{N} |c_n|^2, \tag{2.6-12}$$

and this is clearly minimized if $\gamma_n = c_n$. Thus the partial Fourier series is indeed an optimum approximation, at least in the mean-square-error sense.

The fact that the summation index goes from 1 to N doesn't necessarily mean that one gets the best approximation by using the first N harmonic terms. We see from Eq. (2.6-8) that if the $\phi_n(t)$ form a complete set, then

$$\int_0^T |x(t)|^2 \, dt = \sum_{n=1}^{\infty} |c_n|^2. \tag{2.6-13}$$

Therefore the minimum mean-square error becomes

$$\langle \epsilon^2 \rangle_{\text{min}} = \sum_{n=1}^{\infty} |c_n|^2 - \sum_{n=1}^{N} |c_n|^2 = \sum_{n=N+1}^{\infty} |c_n|^2. \tag{2.6-14}$$

Thus the mean-square error is equal to the sum of the squared magnitude of all the coefficients "left out" of the expansion. This suggests that the best approximation will be obtained by including in the finite sum the terms with the largest coefficients. It also shows that the mean-square error of the approximation becomes negligibly small if the magnitude of the left-out coefficients is small relative to the ones retained in the sum.

*2.7 Representation of Signals by Vectors†

It is sometimes convenient to think of the orthonormal functions $\phi_n(t)$ as unit vectors in an n-dimensional vector space. As an example consider the function

$$x_1(t) = 0.5\sqrt{\frac{2}{T}}\sin\frac{2\pi t}{T} + 0.3\sqrt{\frac{2}{T}}\sin\frac{6\pi t}{T}, \tag{2.7-1}$$

which we may regard as a Fourier series consisting of two terms. The functions $\phi_1(t) = \sqrt{2/T}\sin(2\pi t/T)$ and $\phi_2(t) = \sqrt{2/T}\sin(6\pi t/T)$ are orthonormal over the period T. We can represent this function in a two-dimensional vector space as shown in Fig. 2.7-1(a). Another signal, $x_2(t) = \sqrt{2/T}\sin(2\pi t/T) - 0.5\sqrt{2/T}\sin(6\pi t/T)$ is also shown. The corresponding waveforms are shown in Figs. 2.7-1(b) and (c).

The characterization of signals by finite-dimensional vectors permits us to use such geometrical and vectorial concepts as distance, length, and dot product. Of particular importance in some of the later chapters is the concept of closeness: "How close is $x_1(t)$ to $x_2(t)$?" We shall find that the vector characterization of signals facilitates our visualization of this concept.

Consider all the signals $x(t)$ in $L_2(T)$ that can be *exactly* represented by the finite sum

$$x(t) = \sum_{n=1}^{N} x_n\phi_n(t), \tag{2.7-2}$$

where the $\phi_n(t)$, $n = 1, \ldots, N$, constitute a particular set of orthonormal functions. Each of these signals can be represented by a single point or vector in an *N-dimensional vector subspace* M_N. This vector space is defined by the orthonormal functions $\phi_n(t)$ which constitute its unit vectors. The functions $\phi_n(t)$ are a *basis* for the vector space M_N, and one says that the space M_N is *spanned* by the basis set $\{\phi_n\}$. Because Eq. (2.7-2) is an exact relation, we know that the components are

$$x_n = \int_0^T x(t)\phi_n^*(t)\,dt. \tag{2.7-3}$$

Thus every signal $x(t)$ in M_N is completely determined by the vector $\mathbf{x} = (x_1, x_2, \ldots, x_N)$; in fact, is equivalent to it.

The energy of $x(t) \in M_N$ is given by

$$E = \int_0^T |x(t)|^2\,dt = \sum_{n=1}^{N} |x_n|^2 \equiv \|\mathbf{x}\|^2, \tag{2.7-4}$$

where $\|\mathbf{x}\|$ is called the *norm* (more precisely, *Euclidean norm*) of the vector $\mathbf{x}$. The energy E is essentially the Euclidean length squared.

† This material is used in Chapters 10 and 11.

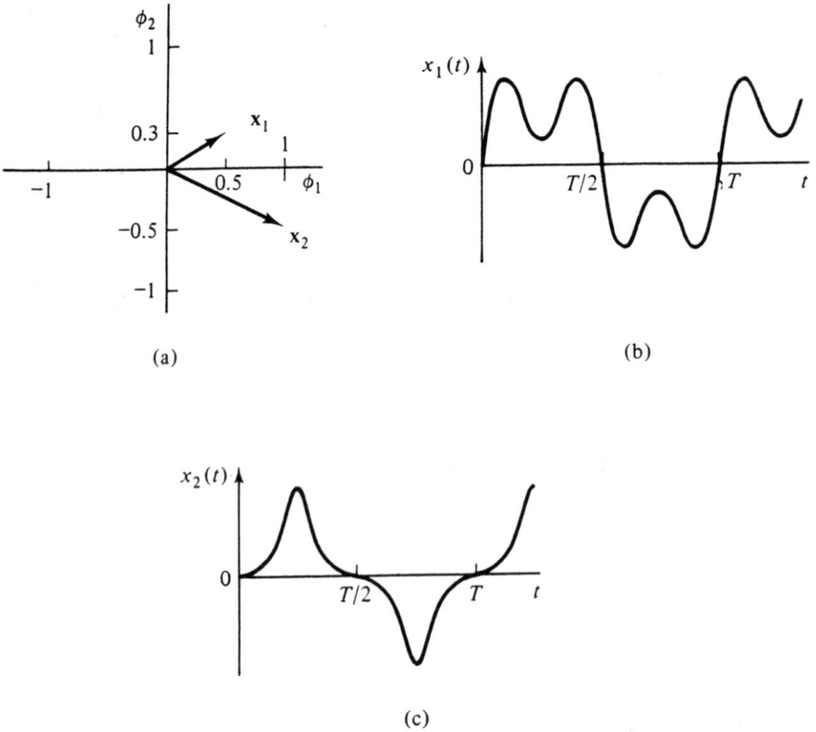

Figure 2.7-1. (a) Vector-space representation of two signals. (b) $x_1(t)$ as a time function. (c) $x_2(t)$ as a time function.

The integral or inner product of two functions in M_N is denoted (x, y) and is defined by

$$(x, y) \equiv \int_0^T x(t)y^*(t)\,dt \tag{2.7-5}$$

$$= \sum_{n=1}^{N}\sum_{n=1}^{N} x_n y_m^* \int_0^T \phi_n(t)\phi_m^*(t)\,dt$$

$$= \sum_{n=1}^{N} x_n y_n^*. \tag{2.7-6}$$

The quantity $\Sigma_{n=1}^{N} x_n y_n^*$ is recognized as the dot or inner product of the two N-vectors $\mathbf{x} = (x_1, \ldots, x_N)$ and $\mathbf{y} = (y_1, \ldots, y_N)$. Hence the inner product of two functions is the dot product of their vector representations in an orthogonal space. The dot product is frequently symbolized by $\mathbf{x} \cdot \mathbf{y}$. Hence Eq. (2.7-6) is conveniently written $(x, y) = \mathbf{x} \cdot \mathbf{y}$. If we define $\boldsymbol{\phi}_i$ $(i = 1, \ldots, N)$ as a unit vector with a one in the ith position and zeros in all other positions, then the vector $\mathbf{x}$ can be represented by

$$\mathbf{x} = \sum_{i=1}^{N} x_i \boldsymbol{\phi}_i. \tag{2.7-7}$$

In this form **x** is expressed as the vectorial sum of its components along the orthogonal axes.

It is useful to talk about distances between vectors. It seems reasonable to expect a distance function d to satisfy

1. $d(\mathbf{x}, \mathbf{y}) = 0 \Rightarrow \mathbf{x} = \mathbf{y}$.
2. $d(\mathbf{x}, \mathbf{y}) \geq 0$ (negative distance is a meaningless concept).
3. $d(\mathbf{x}, \mathbf{y}) = d(\mathbf{y}, \mathbf{x})$ (the distance from **x** to **y** should be the same as the distance from **y** to **x**).
4. $d(\mathbf{x}, \mathbf{y}) \leq d(\mathbf{x}, \mathbf{z}) + d(\mathbf{y}, \mathbf{x})$ (triangle inequality).

A useful distance is the generalized Euclidean distance, given by

$$d(\mathbf{x}, \mathbf{y}) = \left[\sum_{n=1}^{N} |x_n - y_n|^2\right]^{1/2}. \tag{2.7-8}$$

With $\mathbf{z} = \mathbf{x} - \mathbf{y}$ and $z_n = x_n - y_n$, we obtain

$$\sum_{n=1}^{N} |x_n - y_n|^2 = \sum_{n=1}^{N} |z_n|^2 = \|\mathbf{z}\|^2. \tag{2.7-9}$$

It follows therefore from Eq. (2.7-4) that

$$\int_0^T |x(t) - y(t)|^2 \, dt = \|\mathbf{x} - \mathbf{y}\|^2. \tag{2.7-10}$$

The relation between norm, distance, and dot products is summarized by

$$\|\mathbf{x} - \mathbf{y}\| = d(\mathbf{x}, \mathbf{y}),$$
$$(\mathbf{x} - \mathbf{y}) \cdot (\mathbf{x} - \mathbf{y}) = d^2(\mathbf{x}, \mathbf{y}).$$

Projection

In Sec. 2.6 we considered the problem of how to best approximate a given finite energy signal $x(t)$ by a finite sum $x_N(t)$. In terms of the vector-space point of view we regard $x(t)$ as being in the infinite-dimensional vector space $L_2(T)$, and what we are doing when we approximate it is to map it into the finite subspace M_N which contains $x_N(t)$. We have already seen that the best approximation in the minimum mean-square-error sense is given by

$$x_N(t) = \sum_{n=1}^{N} x_n \phi_n(t), \tag{2.7-11}$$

where

$$x_n = (x, \phi_n) = \int_0^T x(t)\phi_n^*(t)\, dt. \tag{2.7-12}$$

We regard $x(t)$ as being represented by the infinite-dimensional vector $\mathbf{x} = \Sigma_{n=1}^{\infty} x_n \boldsymbol{\phi}_n$ and $x_N(t)$ by the N-dimensional vector $\mathbf{x}_N$. The error in the approximation corresponds to the vector $\boldsymbol{\epsilon} = \mathbf{x} - \mathbf{x}_N$. The dot product of $\mathbf{x}_N$ and $\boldsymbol{\epsilon}$ is

$$\mathbf{x}_N \bullet \boldsymbol{\epsilon} = \mathbf{x}_N \bullet \mathbf{x} - \mathbf{x}_N \bullet \mathbf{x}_N = 0 \tag{2.7-13}$$

because

$$\begin{aligned} \mathbf{x}_N \bullet \mathbf{x} &= \int_0^T x(t) x_N^*(t)\, dt \\ &= \int_0^T x(t) \sum_{n=1}^{N} x_n^* \phi_n^*(t)\, dt \\ &= \sum_{n=1}^{N} |x_n|^2 = \mathbf{x}_N \bullet \mathbf{x}_N. \end{aligned}$$

Thus we see that $\mathbf{x}_N$ and $\boldsymbol{\epsilon}$ are orthogonal, i.e., perpendicular.

A useful geometric interpretation is obtained by considering a three-dimensional vector $\mathbf{x}_3$ which is optimally approximated by a two-dimensional vector $\mathbf{x}_2$. This is shown in Fig. 2.7-2. Observe that the best approximation is given by $\mathbf{x}_2$ such that $\mathbf{x}_2$ and $\boldsymbol{\epsilon}$ are perpendicular. We therefore refer to $\mathbf{x}_2$ as the *orthogonal projection*, or simply the *projection*, of $\mathbf{x}_3$ into the two-dimensional vector subspace spanned by ϕ_1 and ϕ_2. Similarly, we can think of

$$x_N(t) = \sum_{n=1}^{N} x_n \phi_n(t)$$

as the *projection* of $x(t)$ into the N-dimensional subspace M_N if $x_n = (x, \phi_n)$. These ideas are formalized in the *projection theorem*, one form of which is the following: Let x be an element of the vector space M spanned by the orthogonal basis set $\{\phi_n\}$.

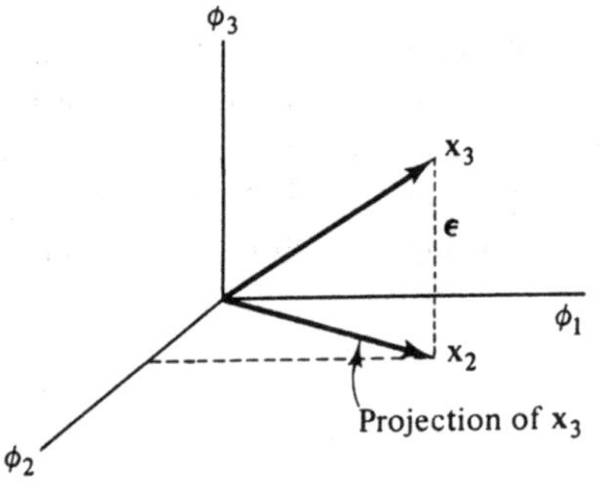

Figure 2.7-2. Projection of a three-dimensional vector onto two dimensions.

Let x_N be an element in a lower-dimensional subspace $M_N \subset M$.[†] Then the norm of the error $\|x - x_N\|$ is a minimum if and only if x_N is the projection of x into M_N. Furthermore, the minimum error and the projection are orthogonal.

In the statement of the projection theorem we have omitted the boldface notation for vectors because the elements $\{x\}$ can be signals $\{x(t)\}$ in $L_2(T)$ function space or N-tuples $\{\mathbf{x}\}$ in the space R^N. It is not difficult to show that the set of real or complex functions in $L_2(T)$ is a linear vector space and, therefore, that treating $x(t)$ as a vector is justified. The norm of $x(t)$ in function space is defined by

$$\|x\| = \left[\int_0^T |x(t)|^2\, dt\right]^{1/2}.$$

Two vectors $x(t)$, $y(t) \in L_2(T)$ are orthogonal if their inner product is zero, i.e.,

$$\int_0^T x(t)y^*(t)\, dt = 0.$$

The close relationship between $x_N(t) \in M_N$ and $\mathbf{x}_N \in R^N$ is called an *isomorphism*. Specifically, the isomorphism is a one-to-one and onto map, T, from M_N to R^N described by

$$T[x_N(t)] = [(x_N, \phi_1), (x_N, \phi_2), \ldots, (x_N, \phi_N)].$$

The construction of an orthonormal basis for an arbitrary N-dimensional subspace is possible through use of the Gram-Schmidt procedure. It is discussed below.

Gram-Schmidt Procedure for Generating Orthonormal Functions

We have considered orthonormal function expansions such as the Fourier series that use "naturally" orthogonal functions such as the exponential or the trigonometric functions. However, it is possible to take any *arbitrary* set of functions $s_1(t), s_2(t), \ldots, s_M(t)$ and convert it into a set of orthonormal functions. We shall find use for this concept later in this book, especially in the material on digital communications.

A simple method for generating a set of orthonormal functions is the Gram-Schmidt procedure, explained below. For simplicity we assume that the functions $s_i(t)$ are real; then the procedure is as follows:

1. *Construction of* $\phi_1(t)$. Any one of the functions, say $s_1(t)$, can be used to start the process. Thus

$$\phi_1(t) = \frac{s_1(t)}{\sqrt{E_1}}, \tag{2.7-14}$$

[†] The set inclusion symbol $\subset$ implies that M_N is spanned by a subset of the same basis as M.

where

$$E_1 = \int_0^T s_1^2(t)\,dt. \tag{2.7-15}$$

It is clear that ϕ_1 is properly normalized since

$$\int_0^T \phi_1^2(t)\,dt = \frac{1}{E_1}\int_0^T s_1^2(t)\,dt = 1. \tag{2.7-16}$$

2. *Construction of* $\phi_2(t)$. First define an auxiliary function $\theta_2(t)$ by

$$\theta_2(t) = s_2(t) - (s_2, \phi_1)\phi_1(t), \tag{2.7-17}$$

where the notation

$$(s_2, \phi_1) = \int_0^T s_2(t)\phi_1(t)\,dt$$

signifies the inner product as in Eq. (2.7-5). This subtracts out any component of $s_2(t)$ in the direction of $\phi_1(t)$ and leaves a residual θ_2 which is orthogonal to ϕ_1. We again normalize by defining

$$E_2 = \int_0^T \theta_2^2(t)\,dt. \tag{2.7-18}$$

Then

$$\phi_2(t) = \frac{\theta_2(t)}{\sqrt{E_2}}. \tag{2.7-19}$$

From Eqs. (2.7-17) and (2.7-19) we obtain

$$\sqrt{E_2}\,\phi_2(t) = s_2(t) - (s_2, \phi_1)\phi_1(t). \tag{2.7-20}$$

Taking the inner product with ϕ_2 and noting that $(\phi_2, \phi_2) = 1$ and $(\phi_2, \phi_1) = 0$, we get

$$\sqrt{E_2} = (s_2, \phi_2). \tag{2.7-21}$$

Hence

$$s_2(t) = (s_2, \phi_1)\phi_1(t) + (s_2, \phi_2)\phi_2(t). \tag{2.7-22}$$

3. *General steps*. Assume that $\phi_1(t), \ldots, \phi_{l-1}(t)$ have been computed. To compute $\phi_l(t)$, first form the residual

$$\theta_l(t) = s_l(t) - \sum_{j=1}^{l-1} (s_l, \phi_j)\phi_j(t) \tag{2.7-23}$$

and then normalize by

$$\phi_l(t) = \frac{\theta_l(t)}{\sqrt{E_l}}. \tag{2.7-24}$$

It is easily shown that the set $\{\phi_l\}$ is an orthogonal set. For example, for $l \neq i$,

$$\begin{aligned}\int_0^T \phi_l(t)\phi_i(t)\,dt &= \frac{1}{\sqrt{E_l}}\int_0^T \theta_l(t)\phi_i(t)\,dt \\ &= \frac{1}{\sqrt{E_l}}\left[\int_0^T s_l(t)\phi_i(t)\,dt - \sum_{j=1}^{l-1}(s_l, \phi_j)\int_0^T \phi_j(t)\phi_i(t)\,dt\right] \\ &= \frac{1}{\sqrt{E_l}}[(s_l, \phi_i) - (s_l, \phi_i)] = 0,\end{aligned} \tag{2.7-25}$$

where the last step follows from the orthonormality of the ϕ_i which causes all terms in the sum except the ith to vanish.

We see that each new function introduces at most one new orthonormal function. If $s_l(t)$ is not independent, it can be expressed as a linear combination of the earlier orthonormal functions. Then $\theta_l(t) = 0$, and no new function is created. Hence the number N of orthonormal functions satisfies $N \leq M$, where M is the number of distinct functions $s_k(t)$.

To illustrate the procedure, consider the three waveforms $s_1(t)$, $s_2(t)$, and $s_3(t)$ shown in Fig. 2.7-3. We leave the details of the computation to the reader (Prob. 2-18), but it is easily demonstrated by following steps 1, 2, and 3 above that there are only two orthonormal functions:

$$\phi_1(t) = \frac{1}{\sqrt{2}} s_1(t), \tag{2.7-26}$$

$$\phi_2(t) = \frac{1}{\sqrt{6}} [s_1(t) + 2s_2(t)]. \tag{2.7-27}$$

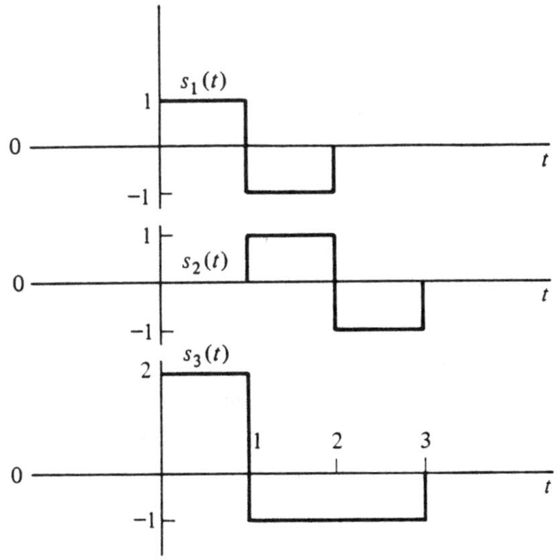

Figure 2.7-3. Set of three waveforms from which to construct an orthonormal set.

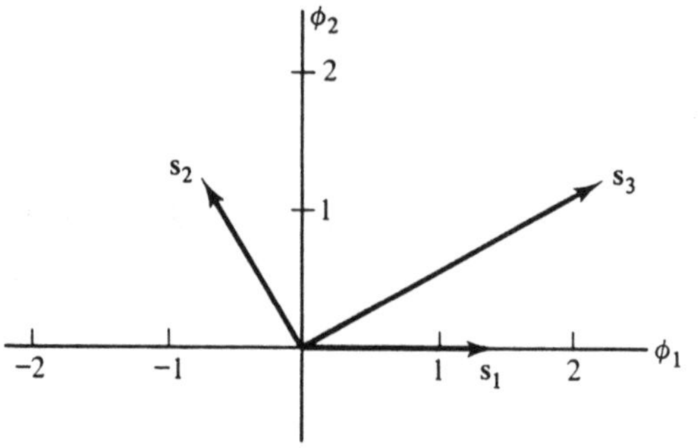

Figure 2.7-4. Vector representation of $s_i(t)$, $i = 1, 2, 3$.

The three signals are seen to be linearly dependent; in fact,

$$s_1(t) = \sqrt{2}\,\phi_1(t), \tag{2.7-28a}$$

$$s_2(t) = -\frac{1}{\sqrt{2}}\,\phi_1(t) + \frac{3}{\sqrt{6}}\,\phi_2(t), \tag{2.7-28b}$$

$$s_3(t) = \frac{3}{\sqrt{2}}\,\phi_1(t) + \frac{3}{\sqrt{6}}\,\phi_2(t). \tag{2.7-28c}$$

A two-dimensional space is sufficient to represent the three signals, and the three vectors representing $s_1(t)$, $s_2(t)$, and $s_3(t)$ in the ϕ_1, ϕ_2, coordinate system are

$$\begin{aligned} s_1 &= (\sqrt{2}, 0) \\ s_2 &= \left(-\frac{1}{\sqrt{2}}, \frac{3}{\sqrt{6}}\right) \\ s_3 &= \left(\frac{3}{\sqrt{2}}, \frac{3}{\sqrt{6}}\right). \end{aligned} \tag{2.7-29}$$

These are shown in Fig. 2.7-4.

The reader is invited to sketch the functions ϕ_1 and ϕ_2 to convince himself that Eqs. (2.7-28) give the original signal set.

2.8 THE FOURIER INTEGRAL

Fourier analysis can be extended to nonperiodic functions by regarding a nonperiodic function as a limit of a periodic signal with an infinitely long period. We first rewrite the Fourier series in the form

$$x(t) = \frac{1}{2\pi} \sum_{n=-\infty}^{\infty} X(\omega_n) e^{j\omega_n t}\,\Delta\omega, \tag{2.8-1}$$

where

$$X(\omega_n) = \int_{-T/2}^{T/2} x(t) e^{-j\omega_n t}\,dt. \tag{2.8-2}$$

Equation (2.8-2) is basically the same as Eq. (2.4-15) except that we have chosen to replace c_n by $X(\omega_n)/T$. Also, Eq. (2.8-1) is the same as Eq. (2.4-5), but we use the fact that $1/T = f_0 = \Delta\omega/2\pi$; i.e., the fundamental frequency is regarded as the frequency increment between successive harmonics.

As the period T increases, the frequency increment $\Delta\omega$ shrinks, and the discrete frequencies ω_n come closer together. Equation (2.8-1) can be approximated by an integral; in fact, this process is similar to the standard Riemann definition of the integral. Thus,

$$x(t) = \frac{1}{2\pi}\sum_{n=-\infty}^{\infty} X(\omega_n)e^{j\omega_n t}\,\Delta\omega \underset{T\to\infty}{\longrightarrow} \frac{1}{2\pi}\int_{-\infty}^{\infty} X(\omega)e^{j\omega t}\,d\omega, \tag{2.8-3}$$

or†

$$x(t) = \int_{-\infty}^{\infty} X(f)e^{j2\pi ft}\,df, \qquad \text{where } f = \frac{\omega}{2\pi}. \tag{2.8-4}$$

The expression given in Eq. (2.8-4) is sometimes preferred since it eliminates the factor 2π and results in a more symmetrical formulation.

The same limiting operation can formally be applied to Eq. (2.8-2) by simply extending the limits of integration to $\pm\infty$ and by replacing ω_n by ω. When this is done, however, we must make sure that the resulting infinite integral exists. Sufficient conditions under which this is true are known as the *Dirichlet conditions*. Essentially, they require the function $x(t)$ to have only a finite number of discontinuities for $-\infty < t < \infty$ and to satisfy

$$\int_{-\infty}^{\infty} |x(t)|dt < \infty.^{\ddagger} \tag{2.8-5}$$

Then we have the Fourier integral pair:

$$x(t) = \int_{-\infty}^{\infty} X(f)e^{j2\pi ft}\,df, \tag{2.8-6}$$

$$X(f) = \int_{-\infty}^{\infty} x(t)e^{-j2\pi ft}\,dt. \tag{2.8-7}$$

We refer to the second of these two equations as the Fourier transform and the first as the inverse Fourier transform. Instead of writing the integral expression, we frequently use the symbolism

$$\begin{aligned} X(f) &= \mathcal{F}[x(t)], \\ x(t) &= \mathcal{F}^{-1}[X(f)]. \end{aligned} \tag{2.8-8}$$

† Observe that $X(\omega) = X(2\pi f) = \tilde{X}(f)$, where the $\tilde{X}$ indicates a different function. However, because $\tilde{X}(\cdot)$ differs from $X(\cdot)$ only by an argument factor of 2π, it is standard to use the same notation for both.

‡ The precise form of the conditions is slightly more complicated. See, for instance, [2-7].

Also, we generally use lowercase symbols to denote time functions and uppercase for Fourier transforms.

Observe that Eq. (2.8-5) is not satisfied by periodic time functions or functions having finite power at all times. However, by introducing the concept of the Dirac delta function (see Sec. 2.11), the Fourier integral definition can be extended to such functions as well.

Although a completely rigorous proof of the Fourier integral relations is beyond the scope of this text, we can present a simple demonstration of their validity. This is quite similar to the one given for the Fourier series in Sec. 2.5. We assume that $x(t)$ is absolutely integrable so that $X(f)$ exists. Then we write the inverse Fourier transform in the form

$$x(t) = \lim_{A\to\infty} \int_{-A}^{A} X(f)e^{j2\pi ft}\, df \tag{2.8-9}$$

and investigate the behavior of this limit. Substituting Eq. (2.8-7) for $X(f)$ results in

$$x(t) = \lim_{A\to\infty} \int_{-A}^{A} df e^{j2\pi ft} \int_{-\infty}^{\infty} x(\tau)e^{-j2\pi f\tau}\, d\tau. \tag{2.8-10}$$

Exchanging the orders of integration is generally permissible† and results in

$$\begin{aligned} x(t) &= \lim_{A\to\infty} \int_{-\infty}^{\infty} x(\tau)\, d\tau \int_{-A}^{A} e^{j2\pi f(t-\tau)}\, df \\ &= \lim_{A\to\infty} \int_{-\infty}^{\infty} x(\tau)\cdot 2A \operatorname{sinc}\,[2A(t-\tau)]\, d\tau. \end{aligned} \tag{2.8-11}$$

This expression is very similar to Eq. (2.5-4). As before, we can regard the integral as a weighted average of $x(\tau)$ with the weighting function $2A \operatorname{sinc} 2A(t-\tau)$. This function is shown in Fig. 2.8-1 for relatively large A. It has a large narrow peak near $\tau = t$ and is otherwise oscillatory. As $A \to \infty$ the peak becomes infinitely large and infinitely narrow; hence the weighting is more and more concentrated near $\tau = t$. Then if $x(\tau)$ is sufficiently smooth (continuous and possessing a derivative), it can be replaced by $x(t)$ and taken outside of the integral as a constant. The remaining integration yields unity by Eq. (2.5-7), and Eq. (2.8-11) is shown to be an identity.

If $x(\tau)$ has an isolated discontinuity at $\tau = t$, the effect is exactly the same as in the Fourier series. At the discontinuity point the inverse Fourier transform has the value $\frac{1}{2}[x(t_+) + x(t_-)]$. For finite A in Eq. (2.8-11) the step in $x(\tau)$ becomes an oscillatory function whose frequency increases with A but whose amplitude is independent of A. This is the Gibbs phenomenon already referred to.

† A sufficient condition is that $x(\tau)$ be piecewise continuous and that the integral converge uniformly with respect to its upper limit [2-7].

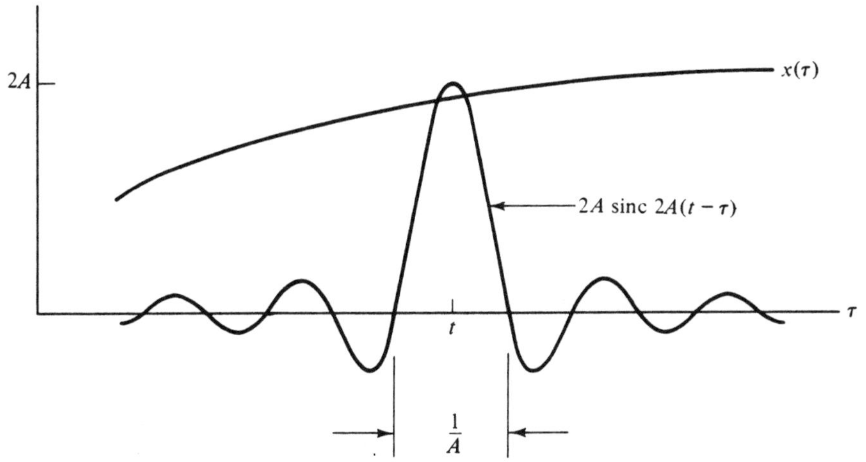

Figure 2.8-1. Weighting function $2A$ sinc $2A(t - \tau)$ together with the smooth function $x(\tau)$.

2.9 ELEMENTARY PROPERTIES OF THE FOURIER TRANSFORM

Parseval's Theorem

This theorem has already been discussed in connection with the Fourier series. If $X(f) = \mathcal{F}[x(t)]$, Parseval's theorem takes the form

$$\int_{-\infty}^{\infty} |x(t)|^2 \, dt = \int_{-\infty}^{\infty} |X(f)|^2 \, df. \tag{2.9-1}$$

The absolute-value signs are used since both $x(t)$ and $X(f)$ may be complex. As noted earlier, $|x(t)|^2$ can be identified with the instantaneous power in the signal $x(t)$, and therefore the expression on the left is the signal energy. This is assumed to be finite; i.e., the signal is assumed to be in class L_2. The expression on the right must then also be the signal energy, and the integrand $|X(f)|^2$ is called the *energy density*, expressed in joules per hertz. If $|X(f)|$ is larger in a certain range of frequencies $[f_1, f_2]$ than in the range $[f_3, f_4]$, we say that $x(t)$ has more energy in the band $[f_1, f_2]$ than in the band $[f_3, f_4]$. A useful generalization of Parseval's theorem is

$$\int_{-\infty}^{\infty} x(t)y^*(t) \, dt = \int_{-\infty}^{\infty} X(f)Y^*(f) \, df. \tag{2.9-2}$$

The derivation of both Eqs. (2.9-1) and (2.9-2) is left as an exercise.

Linearity

If $X(f) = \mathcal{F}[x(t)]$, $Y(f) = \mathcal{F}[y(t)]$, then

$$\mathcal{F}[ax(t) + by(t)] = aX(f) + bY(f). \tag{2.9-3}$$

This follows directly from the definition and the linearity of the integral operation.

Time Shift

If $X(f) = \mathcal{F}[x(t)]$, then

$$\mathcal{F}[x(t - a)] = X(f)e^{-j2\pi fa}. \tag{2.9-4}$$

Proof:

$$\begin{aligned}\mathcal{F}[x(t - a)] &= \int_{-\infty}^{\infty} x(t - a)e^{-j2\pi ft}\,dt \\ &= \int_{-\infty}^{\infty} x(\tau)e^{-j2\pi f(\tau + a)}\,d\tau \\ &= e^{-j2\pi fa}\int_{-\infty}^{\infty} x(\tau)e^{-j2\pi f\tau}\,d\tau = X(f)e^{-j2\pi fa}\end{aligned} \tag{2.9-5}$$

by a simple change of variable in line 2. Similarly, if $x(t)$ and $X(f)$ form a Fourier pair, then the inverse transform of $X(f - f_0)$ is $x(t)e^{j2\pi f_0 t}$. This result is called the *frequency-shift* property of the Fourier transform.

Scale Change

If $X(f) = \mathcal{F}[x(t)]$,

$$\mathcal{F}[x(at)] = \frac{1}{|a|}X\left(\frac{f}{a}\right). \tag{2.9-6}$$

Proof: Let $a > 0$; then

$$\int_{-\infty}^{\infty} x(at)e^{j2\pi ft}\,dt = \int_{-\infty}^{\infty} x(\tau)e^{j2\pi f(\tau/a)}\frac{d\tau}{a} = \frac{1}{a}X\left(\frac{f}{a}\right).$$

If $a < 0$, the change of variable also results in a reversal of the limits of integration, which causes a further sign change; Eq. (2.9-6) results.

The scale-change property is another manifestation of the reciprocal relation between time and frequency, already mentioned. It has the important consequence that short-duration signals have wide spectral widths and that, conversely, narrow spectra correspond to long-duration signals. This is demonstrated by the Fourier transform of the square pulse:

$$x(t) = \text{rect}\,(t) = \begin{cases} 1, & |t| \le \frac{1}{2}, \\ 0, & |t| > \frac{1}{2}. \end{cases} \tag{2.9-7}$$

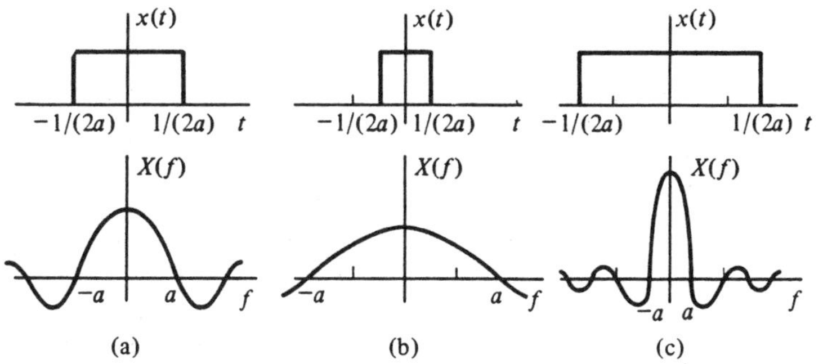

Figure 2.9-1. Reciprocal relation between time and frequency: (a) medium-length pulse, medium-width spectrum; (b) short pulse, wide spectrum; (c) long pulse, narrow spectrum.

The Fourier transform of this function is

$$X(f) = \int_{-1/2}^{1/2} e^{-j2\pi ft}\,dt = \text{sinc}\,(f). \tag{2.9-8}$$

By the scale-change property the spectrum of rect (at) is $(1/|a|)$ sinc $(f/|a|)$. The time function and its spectrum are shown in Fig. 2.9-1 for three different values of a.

Duality

If $X(f) = \mathcal{F}[x(t)]$, then

$$\mathcal{F}[X(t)] = x(-f).$$

This property, which follows directly from the defining integrals of the Fourier transformation, generates additional simple relationships. For example, we showed that

$$\mathcal{F}[\text{rect}\,(t)] = \text{sinc}\,(f). \tag{2.9-9}$$

Hence, by duality

$$\mathcal{F}[\text{sinc}\,(t)] = \text{rect}\,(f). \tag{2.9-10}$$

Modulation Property

If $X(f) = \mathcal{F}[x(t)]$, then

$$\mathcal{F}[x(t)\cos 2\pi f_0 t] = \tfrac{1}{2}[X(f - f_0) + X(f + f_0)]. \tag{2.9-11}$$

This follows by writing $\cos 2\pi f_0 t$ as $\frac{1}{2}(e^{j2\pi f_0 t} + e^{-j2\pi f_0 t})$ and then using the frequency-shift property. The effect is illustrated in Fig. 2.9-2.

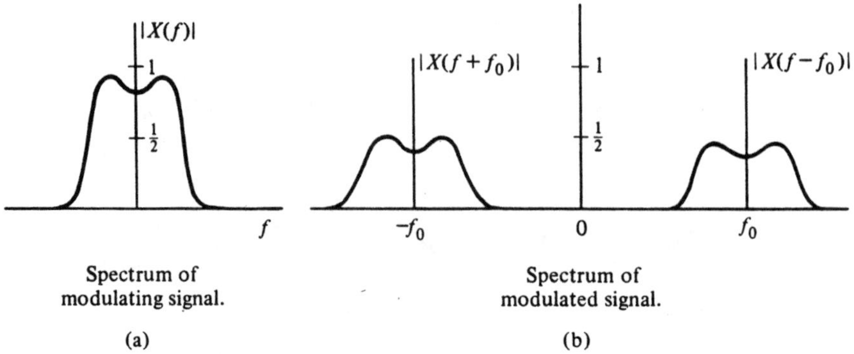

Figure 2.9-2. Illustration of the modulation property: (a) spectrum of modulating signal; (b) spectrum of modulated signal.

Differentiation and Integration

If $X(f) = \mathcal{F}[x(t)]$, then

$$\mathcal{F}\left[\frac{dx}{dt}\right] = j2\pi f X(f). \tag{2.9-12}$$

This follows directly by differentiating the Fourier integral under the integral sign. The process can obviously be repeated, so that if the nth derivative exists, then

$$\mathcal{F}\left[\frac{d^n x}{dt^n}\right] = (j2\pi f)^n X(f). \tag{2.9-13}$$

If the nth derivative does not exist, then $(j2\pi f)^n X(f)$ will generally not satisfy the Dirichlet conditions [Eq. (2.8-5)].

Equation (2.9-12) shows that if we set

$$x(t) = \frac{dy(t)}{dt}, \tag{2.9-14}$$

so that $X(f) = j2\pi f Y(f)$, then

$$Y(f) = \frac{X(f)}{j2\pi f}. \tag{2.9-15}$$

It is tempting to conclude from Eq. (2.9-15) that

$$\mathcal{F}\left[\int_{-\infty}^{t} x(\tau)\, d\tau\right] = \frac{X(f)}{j2\pi f}. \tag{2.9-16}$$

This approach, although it sometimes results in the right answer, does not consider arbitrary constants of integration.

A better method to obtain the formula for the transform of the integral is to evaluate it formally using integration by parts:

$$\mathcal{F}\left[\int_{-\infty}^{t} x(\tau)\, d\tau\right] = \int_{-\infty}^{\infty} e^{-j2\pi ft} \int_{-\infty}^{t} x(\tau)\, d\tau\, dt \tag{2.9-17}$$

$$= -\frac{e^{-j2\pi ft}}{j2\pi f} \int_{-\infty}^{t} x(\tau)\, d\tau \Bigg|_{-\infty}^{\infty} + \frac{X(f)}{j2\pi f}. \tag{2.9-18}$$

The first term on the right will be zero if

$$X(0) = \int_{-\infty}^{\infty} x(\tau)\, d\tau = 0, \tag{2.9-19}$$

and if this is true, Eq. (2.9-16) holds. If it is not true, then $y(t)$ does not satisfy the Dirichlet conditions, and formally, therefore, the Fourier transform of the integral does not exist. We shall find, however, that this difficulty, as well as several others in the formal Fourier theory, can be circumvented by the use of delta functions. We shall see below (Sec. 2.11) that with this addition to the theory, the Fourier transform of the integral does exist as long as $X(0)$ is finite, and its form is

$$\mathcal{F}\left[\int_{-\infty}^{t} x(\tau)\, d\tau\right] = X(f)\left[\frac{\delta(f)}{2} + \frac{1}{j2\pi f}\right]. \tag{2.9-20}$$

The duality theorem can be applied to these results to generate two additional theorems: If $X(f) = \mathcal{F}[x(t)]$, and if the indicated Fourier transforms exist, then

$$\mathcal{F}[t^n x(t)] = (-j2\pi)^{-n} \frac{d^n X(f)}{df^n}, \tag{2.9-21}$$

and

$$\mathcal{F}\left[\frac{x(t)}{t}\right] = -j2\pi \int_{-\infty}^{f} X(u)\, du. \tag{2.9-22}$$

Symmetry Properties

Although both $x(t)$ and its Fourier transform $X(f)$ can be complex functions, important simplications are possible when these functions are either purely real or imaginary.

Thus, suppose that $x(t)$ is real. Then by inspection of the Fourier integral it follows that

$$X(-f) = X^*(f). \tag{2.9-23}$$

The function $X(f)$ is said to have *Hermitian* symmetry. It implies that

$$\begin{aligned} |X(-f)| &= |X(f)|, \\ \arg X(-f) &= -\arg X(f). \end{aligned} \tag{2.9-24}$$

Thus for a real function of time we find that the amplitude of the Fourier transform is an even function, while the phase is odd.

If we set

$$X(f) = A(f) + jB(f), \tag{2.9-25}$$

it follows from Eq. (2.9-23) that

$$\begin{aligned} A(f) &= A(-f) \\ B(f) &= -B(-f); \end{aligned} \tag{2.9-26}$$

i.e., the real part of $X(f)$ is even and the imaginary part odd. Analogous relations hold if $x(t)$ is pure imaginary. In this case it is easily seen that $X(f)$ is *skew Hermitian*; i.e.,

$$X(-f) = -X^*(f).$$

Multiplication and Convolution

Let $X(f)$, $Y(f)$, and $Z(f)$ be the Fourier transforms of $x(t)$, $y(t)$, and $z(t)$, respectively. Then if

$$Z(f) = X(f)Y(f),$$

it follows that

$$z(t) = \int_{-\infty}^{\infty} x(\tau)y(t-\tau)\,d\tau. \tag{2.9-27}$$

Proof: Fourier-transforming Eq. (2.9-27) gives

$$\begin{aligned} \mathcal{F}[z(t)] &= \int_{-\infty}^{\infty}\int_{-\infty}^{\infty} x(\tau)y(t-\tau)\,d\tau\, e^{-j2\pi ft}\,dt \\ &= \int_{-\infty}^{\infty}\int_{-\infty}^{\infty} x(\tau)e^{-j2\pi f\tau}\, y(t-\tau)e^{-j2\pi f(t-\tau)}\,d\tau\,dt, \end{aligned} \tag{2.9-28}$$

where we have simply multiplied the integrand by $e^{j\pi f(\tau-\tau)} = 1$. Exchanging orders of integration and making the change of variable $s = t - \tau$ in the second integral, we obtain

$$\begin{aligned} \mathcal{F}[z(t)] &= \int_{-\infty}^{\infty} x(\tau)e^{-j2\pi f\tau}\,d\tau \int_{-\infty}^{\infty} y(s)e^{-j2\pi fs}\,ds \\ &= X(f)Y(f), \end{aligned} \tag{2.9-29}$$

which is seen to be an identity.

The integral on the right in Eq. (2.9-27) is referred to as the *convolution* integral, and we say that $x(t)$ is convolved with $y(t)$. A simple notation for Eq. (2.9-27) is

$$z(t) = x(t) * y(t). \tag{2.9-30}$$

This notation symbolizes the fact that the operation of convolution is similar to that of multiplication, i.e., that it is commutative, associative, and distributive: $x * y = y * x$; $x * (y * z) = (x * y) * z$; $(x + y) * z = x * z + y * z$. These relations are easily derived; see Prob. 2-23.

Generally speaking, convolution is a kind of smoothing or averaging operation in which one of the functions is weighted by the other. Of course the nature of the smoothing depends on the functions involved, and if one of the functions is "badly behaved," e.g., the derivative of the Dirac delta of Sec. 2.11, the signal $z(t)$ may be anything but smooth. As a simple example, if in Eq. (2.9-27) we set $y(t) = (1/T)\ \mathrm{rect}\ (t/T)$,

$$z(t) = \frac{1}{T}\int_{t-T/2}^{t+T/2} x(\tau)\, d\tau. \tag{2.9-31}$$

This shows that $z(t)$ is simply a running average of $x(t)$ with averaging time T.

By use of the duality theorem it is easily shown that if

$$z(t) = x(t)y(t),$$

then

$$Z(f) = \int_{-\infty}^{\infty} X(\lambda)Y(f - \lambda)\, d\lambda$$

or

$$Z(f) = X(f) * Y(f). \tag{2.9-32}$$

The fact that convolution of two time functions is equivalent to multiplication of their Fourier transforms [Eq. (2.9-27)] is doubtlessly the most important property of the Fourier integral. It permits the complicated process of convolution to be replaced by the much simpler process of multiplication. We shall see in the next chapter that the output of a linear time-invariant system is the convolution of the input with the impulse response of the system.† The Fourier transform of the impulse response is the *transfer function* already briefly mentioned in Sec. 2.4. Because of the convolution-multiplication property of the Fourier integral, the transform of the output is the product of the transform of the input and the transfer function. This is the extension of the important result given in Eq. (2.4-4) for periodic signals to nonperiodic signals.

The properties of the Fourier transform discussed in this chapter are summarized in Table 2.9-1.

† The impulse response is precisely that: the response of a system to an input that consists of an impulse.

TABLE 2.9-1 PROPERTIES OF THE FOURIER TRANSFORMATION

Name	Time function	Fourier spectrum
1. Parseval's theorem	$\int_{-\infty}^{\infty} x(t)y^*(t)\,dt$	$\int_{-\infty}^{\infty} X(f)Y^*(f)\,df$
2. Linearity	$ax(t) + by(t)$	$aX(f) + bY(f)$
3. Time shift	$x(t-a)$	$X(f)e^{-j2\pi fa}$
4. Scale change	$x(at)$	$\frac{1}{\lvert a\rvert}X(f/a)$
5. Duality	$X(t)$	$x(-f)$
6. Frequency shift	$x(t)e^{j2\pi f_0 t}$	$X(f-f_0)$
7. Modulation property	$x(t)\cos 2\pi f_0 t$	$\frac{1}{2}[X(f-f_0) + X(f+f_0)]$
8. Differentiation	dx/dt	$j2\pi f X(f)$
9. Integration	$\int_{-\infty}^{t} x(\tau)\,d\tau$	$X(f)\left[\frac{\delta(f)}{2} + \frac{1}{j2\pi f}\right]$
10. Multiplication by t^n (also called frequency differentiation)	$t^n x(t)$	$(-j2\pi)^{-n}\, d^nX(f)/df^n$
11. Symmetry	Real $x(t)$ Imaginary $x(t)$	$X(-f) = X^*(f)$ $X(-f) = -X^*(f)$
12. Convolution	$\int_{-\infty}^{\infty} x(\tau)y(t-\tau)\,d\tau$	$X(f)\cdot Y(f)$
13. Multiplication	$x(t)y(t)$	$\int_{-\infty}^{\infty} X(\lambda)Y(f-\lambda)\,d\lambda$

2.10 SOME USEFUL FOURIER PAIRS

Square Pulse

This function and its Fourier transform have already been discussed; cf. Eqs. (2.9-7) and (2.9-8). Thus we have the pair

$$\text{rect}\,(t) \longleftrightarrow \text{sinc}\,(f) \tag{2.10-1}$$

The two functions are shown in Fig. 2.9-1. They are both real and even. The notation rect (·) and sinc (·) is due to P. M. Woodward [2-9].

Triangular Pulse

The triangular pulse (see Fig. 2.10-1) is given by

$$x(t) = \begin{cases} 1 - |t|, & |t| \le 1, \\ 0, & |t| > 1. \end{cases} \tag{2.10-2}$$

One way to compute the Fourier transform of $x(t)$ is to think of it as the convolution of two rectangular pulses:

$$x(t) = \int_{-\infty}^{\infty} \text{rect}\,(\tau)\,\text{rect}\,(t-\tau)\,d\tau. \tag{2.10-3}$$

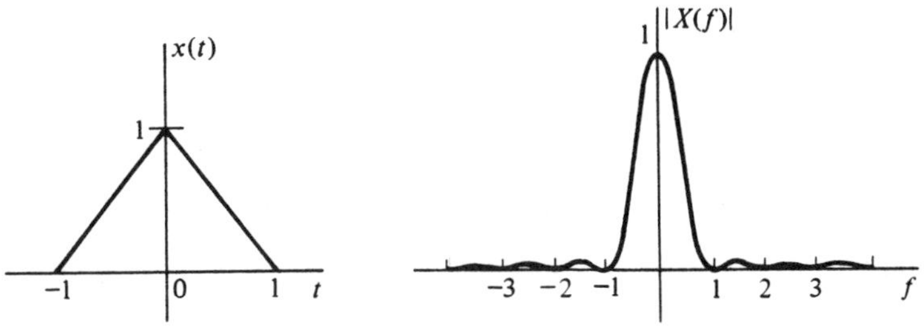

Figure 2.10-1. Triangular pulse and its Fourier spectrum.

The integrand of this expression is shown for several values of t in Fig. 2.10-2; the area of the shaded overlap region is the value of the integral. Observe that there is no overlap for $|t| > 1$ so that the area is zero; for $|t| < 1$ the overlap is a linear function of t which is maximum when $t = 0$. Thus the triangular shape results. Because of the convolution property and Eq. (2.10-1), the corresponding spectrum is immediately written as

$$X(f) = \text{sinc}^2\,(f). \tag{2.10-4}$$

The triangular pulse and its spectrum are shown in Fig. 2.10-1. The triangular pulse is "smoother" than the rectangular pulse since it is continuous, whereas the rectangular pulse is not. This greater smoothness is reflected in a reduced high-frequency content—the amplitude of $\text{sinc}^2\,(f)$ is much smaller than that of $|\text{sinc}\,(f)|$ for $f > 1$. This is a general and very important observation—the smoother the time function, the smaller its high-frequency content.

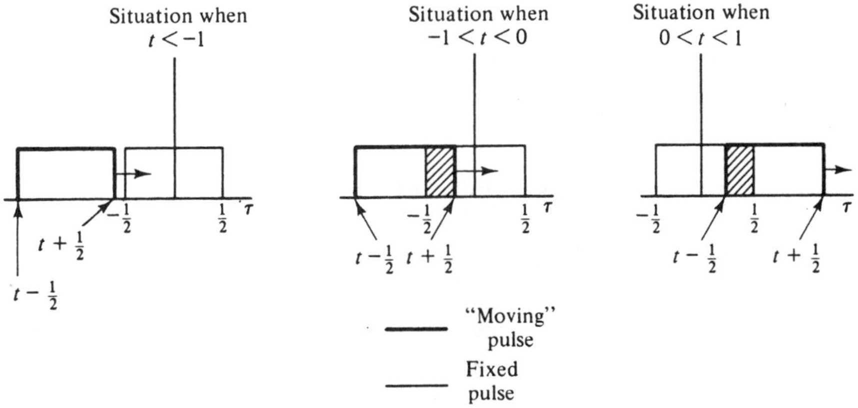

Figure 2.10-2. Convolution of two rectangular pulses.

Gaussian Pulse

By using the identity

$$\int_{-\infty}^{\infty} e^{-(ax^2+bx)}\,dx = \sqrt{\frac{\pi}{a}}\,e^{b^2/4a}, \tag{2.10-5}$$

which holds for complex a and b as long as Re $a > 0$, the following Fourier pair is easily derived:

$$e^{-\pi t^2} \longleftrightarrow e^{-\pi f^2}. \tag{2.10-6}$$

The pulse $e^{-\pi t^2}$ has a Gaussian, i.e., bell, shape and is known as the Gaussian pulse. Thus the Gaussian pulse and its Fourier transform are identical in form. By using the scale-change theorem (property 4 in Table 2.9-1) it is easily shown that for

$$x(t) = e^{-at^2},$$

we have

$$X(f) = \sqrt{\left|\frac{\pi}{a}\right|}\; e^{-\pi^2 f^2/a},$$

or

$$X(\omega) = \sqrt{\left|\frac{\pi}{a}\right|}\; e^{-\omega^2/4a}. \tag{2.10-7}$$

Exponential Pulse

In linear systems theory the asymmetrical exponential pulse occurs frequently. Its form is

$$x(t) = e^{-at}u(t), \quad \text{Re}\, a > 0, \tag{2.10-8}$$

where $u(t)$ is the unit step function:

$$u(t) = \begin{cases} 1, & t > 0, \\ 0, & t < 0. \end{cases} \tag{2.10-9}$$

The Fourier transform of this pulse is easily shown to be

$$X(f) = \frac{1}{j\omega + a} = \frac{1}{\sqrt{a^2+\omega^2}}\, e^{-j\tan^{-1}(\omega/a)}. \tag{2.10-10}$$

See Fig. 2.10-3. Since

$$\frac{d^n}{d\omega^n}\left(\frac{1}{j\omega + a}\right) = \frac{n!(-j)^n}{(j\omega + a)^{n+1}},$$

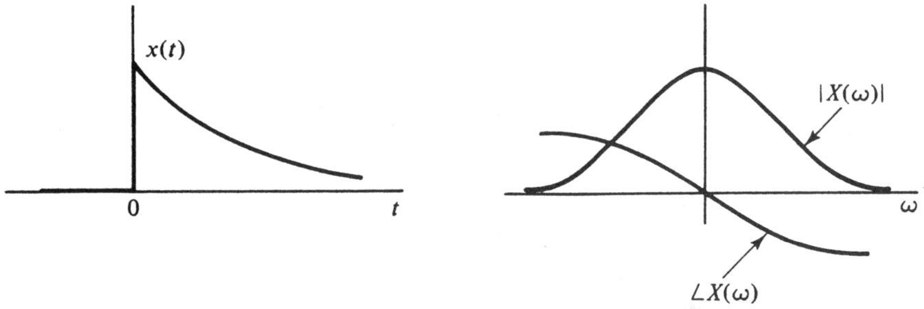

Figure 2.10-3. Asymmetrical exponential pulse exp(-at)·$u(t)$ and its Fourier transform.

it easily follows from the frequency-differentiation rule (property 10 in Table 2.9-1) that

$$\mathcal{F}\left[\frac{t^n}{n!}e^{-at}u(t)\right] = \frac{1}{(j\omega + a)^{n+1}}. \tag{2.10-11}$$

Also, since the symmetrical exponential pulse $e^{-a|t|}$ can be written as $x(t) + x(-t)$, where $x(t)$ is given by Eq. (2.10-8), we have the additional pair

$$\mathcal{F}[e^{-a|t|}] = \frac{1}{j\omega + a} + \frac{1}{-j\omega + a} = \frac{2a}{\omega^2 + a^2}. \tag{2.10-12}$$

Many additional Fourier pairs can be derived from the exponential pulse pair. For instance, since a in Eq. (2.10-8) can be complex, Fourier transforms for functions of the form e^{-at} sin $(bt)\cdot u(t)$ or e^{-at} cos $(bt)\cdot u(t)$ are straightforward extensions. (See Prob. 2-27.)

Signum Function

The signum function, written sgn (t), is defined by

$$\text{sgn}\,(t) = \begin{cases} 1, & t > 0, \\ 0, & t = 0, \\ -1, & t < 0. \end{cases} \tag{2.10-13}$$

This function does not satisfy the Dirichlet conditions and therefore, strictly speaking, has no Fourier transform. The usual approach is to deal with the function $e^{-a|t|}$ sgn (t), which for $a > 0$ does satisfy the Dirichlet condition [see Fig. 2.10-4(a)]. This function is very similar to the symmetrical exponential pulse of Eq. (2.10-12) except that the portion for $t < 0$ is negative. Therefore

$$\mathcal{F}[e^{-a|t|}\,\text{sgn}\,(t)] = \frac{1}{j\omega + a} - \frac{1}{-j\omega + a} = \frac{-2j\omega}{\omega^2 + a^2}. \tag{2.10-14}$$

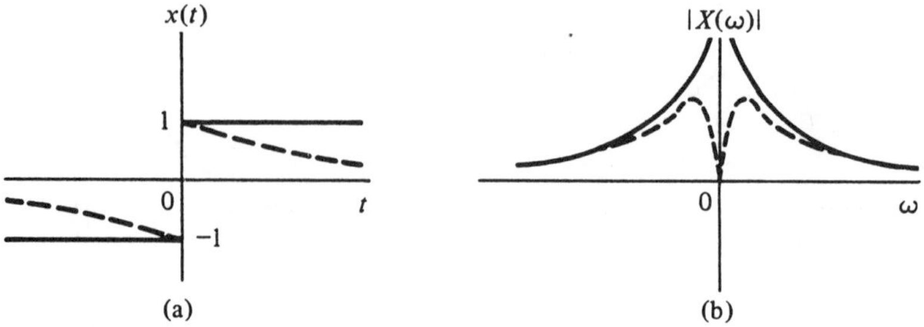

Figure 2.10-4. Signum function and its Fourier spectrum. (a) The solid line shows the signum function, and the dashed line shows the approximation $\exp(-a|t|)\cdot\text{sgn}(t)$, which satisfies the Dirichlet conditions. (b) The Fourier spectrum of the signum function (solid lines) and of the approximation (dashed lines). Observe that the approximation is excellent except at the frequency origin.

It is now possible to go to the limit of $a \to 0$, so that

$$\mathcal{F}[\text{sgn}\,(t)] = \lim_{a\to 0} \frac{-2j\omega}{\omega^2 + a^2} = \frac{2}{j\omega}. \tag{2.10-15}$$

Spectra for finite a and for the limit as $a \to 0$ are shown in Fig. 2.10-4(b). Observe that the approximation for small a is very good except near the frequency origin.

Equation (2.10-15) implies that the inverse Fourier transform of $1/j\omega$ is 1/2 sgn (t). The formal demonstration of this is illuminating. Thus consider

$$\begin{aligned} x(t) &= \frac{1}{2\pi}\int_{-\infty}^{\infty} \frac{e^{j\omega t}}{j\omega}\, d\omega \\ &= \frac{1}{2\pi j}\int_{-\infty}^{\infty} \frac{\cos \omega t}{\omega}\, d\omega + \frac{1}{2\pi}\int_{-\infty}^{\infty} \frac{\sin \omega t}{\omega}\, d\omega. \end{aligned} \tag{2.10-16}$$

The second term gives the desired $\frac{1}{2}$ sgn (t); this follows from the fact that

$$\int_{-\infty}^{\infty} \frac{\sin mx}{x}\, dx = \begin{cases} \pi, & m > 0, \\ 0, & m = 0, \\ -\pi, & m < 0. \end{cases} \tag{2.10-17}$$

(See any standard table of integrals, e.g., Peirce and Foster [2-1], formula 499.) Thus we get the correct result if we interpret the first term as being zero. This is true if the improper integral is defined as the *Cauchy principal value* (CPV):

$$\int_{-\infty}^{\infty} \frac{\cos \omega t}{\omega}\, d\omega \equiv \lim_{\substack{T\to\infty \\ \epsilon\to 0}} \left(\int_{-T}^{-\epsilon} \frac{\cos \omega t}{\omega} + \int_{\epsilon}^{T} \frac{\cos \omega t}{\omega}\, d\omega \right). \tag{2.10-18}$$

Since the integrand is an odd function of ω and the limits are symmetrical (by definition of CPV), the integral vanishes.

More generally, use of the CPV integration in evaluating inverse Fourier transforms makes it possible to obtain transforms of functions that do not satisfy the Dirichlet conditions. Therefore when the defining integrals of the Fourier transform or its inverse are improper, the CPV definition is always implied.

2.11 THE DIRAC DELTA FUNCTION

The restriction of formal Fourier integral theory to functions satisfying the Dirichlet conditions is inconvenient in applications, since one often wants to deal with signals that have finite average power at all times and hence infinite energy. Examples of such signals are dc and periodic waveforms and certain random signals such as continuous noise, speech, music, or television signals. Also, it is desirable to combine the Fourier series and the Fourier integral into a unified theory so that the Fourier series can be regarded as just a special case of the Fourier integral. It turns out that the proper use of the Dirac delta function takes care of practically all of these objectives. We emphasize the words "proper use," because the delta function is not really a function in the normal sense, and some of its properties must be obtained by means of complicated, and not always clearly rigorous, limiting arguments. Many of the theoretical difficulties surrounding the delta function can be eliminated by using the theory of distributions which puts many intuitive arguments on a rigorous foundation ([2-4], p. 269, and [2-8]). For our purposes a fairly elementary treatment will suffice, but the interested reader is referred to the book by Friedman [2-8].

The delta function $\delta(t)$ is often "defined" as a function that is zero everywhere except for $t = 0$ where it is infinite and such that

$$\int_{-\infty}^{\infty} \delta(t)\, dt = 1. \tag{2.11-1}$$

A more satisfactory definition is to regard $\delta(t)$ as the limit of one of several pulses; for instance,

$$\delta(t) = \lim_{a\to\infty} a \operatorname{rect}(at), \tag{2.11-2}$$

or

$$\delta(t) = \lim_{a\to\infty} a \operatorname{sinc}(at), \tag{2.11-3}$$

or

$$\delta(t) = \lim_{a\to\infty} a e^{-\pi a^2 t^2}. \tag{2.11-4}$$

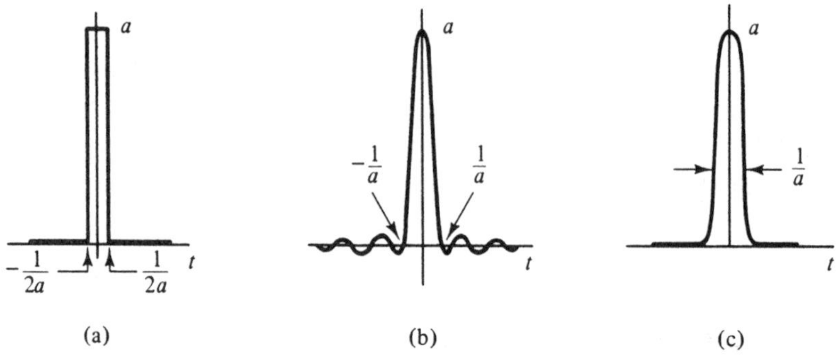

Figure 2.11-1. Pulses that approach the delta function as a limit: (a) rectangular pulse; (b) sinc pulse; (c) Gaussian pulse.

These are shown, for large a, in Fig. 2.11-1. The exact shape of these functions is immaterial. Their important features are (1) unit area and (2) rapid decrease to zero for $t \neq 0$.

Still another definition is to call any function a delta function if for arbitrary continuous $f(t)$ it satisfies the integral equation

$$\int_{-\infty}^{\infty} f(\tau)\delta(\tau - t)\, d\tau = f(t). \qquad (2.11\text{-}5)$$

This definition can, of course, be related to the previous one, since any of the pulses in Eqs. (2.11-2)–(2.11-4), when substituted for $\delta(t)$ in Eq. (2.11-5), will essentially furnish the same result when a is very large. This follows because the integrand is significantly nonzero only for $\tau \simeq t$. It can therefore be approximately evaluated by replacing $f(\tau)$ by the constant $f(t)$. Then since the delta function has unit area, the result follows. The way in which the delta function picks out a particular value of $f(t)$ is sometimes referred to as *sifting*, and Eq. (2.11-5) is therefore called a *sifting integral*. Because delta functions ordinarily appear in such sifting integrals, Eq. (2.11-5) is often regarded as a more "fundamental" definition than the ones given earlier in this paragraph.[†]

Parenthetically, it might be noted that integral equations similar to Eq. (2.11-5), i.e., of the form

$$\int_{a}^{b} f(\tau)g(t, \tau)\, d\tau = \lambda f(t), \qquad (2.11\text{-}6)$$

play an important role in analysis [2-6]. The function $g(t, \tau)$ is referred to as the *kernel* of the integral equation, and for a given kernel such an equation generally has solutions only for particular values of λ called *eigenvalues*. The corresponding solutions are the

[†] It also can be shown that the earlier definitions do not uniquely specify $\delta(t)$. See, for example, [2-4], p. 270.

eigenfunctions. According to this point of view, we find that when the kernel is a delta function any continuous function is an eigenfunction, and the eigenvalues are all equal to 1.

Observe that $\delta(t) = \delta(-t)$; therefore Eq. (2.11-5) is a convolution of $f(t)$ and $\delta(t)$. Hence if the Fourier transform of $f(t)$ is $F(\omega)$, we should have, from property 12 in Table 2.9-1, that

$$F(\omega)\mathcal{F}[\delta(t)] = F(\omega),$$

or

$$\mathcal{F}[\delta(t)] = \int_{-\infty}^{\infty} \delta(t)e^{-j\omega t}\,dt = 1. \tag{2.11-7}$$

This can be regarded as still another definition of the delta function: A delta function is the (inverse) Fourier transform of 1. This gives rise to the further useful relation

$$\int_{-\infty}^{\infty} e^{\pm j2\pi ft}\,df = \int_{-\infty}^{\infty} \cos 2\pi ft\,df = \delta(t), \tag{2.11-8}$$

and similarly,

$$\int_{-\infty}^{\infty} e^{\pm j2\pi ft}\,dt = \int_{-\infty}^{\infty} \cos 2\pi ft\,dt = \delta(f). \tag{2.11-9}$$

Note that if the infinite integral is regarded as a limit, one has

$$\lim_{a\to\infty} \int_{-a/2}^{a/2} e^{j2\pi ft}\,df = \lim_{a\to\infty} a \operatorname{sinc}(at) = \delta(t). \tag{2.11-10}$$

Thus Eq. (2.11-8) is consistent with Eq. (2.11-3).

There is a possible source of error in using these definitions with the variable ω rather than f. If in Eq. (2.11-8) we make the change of variable $\omega = 2\pi f$, we get

$$\int_{-\infty}^{\infty} e^{\pm j\omega t}\,d\omega = 2\pi\delta(t), \tag{2.11-11}$$

$$\int_{-\infty}^{\infty} e^{\pm j\omega t}\,dt = 2\pi\delta(\omega). \tag{2.11-12}$$

Equation (2.11-12) is exactly the same as Eq. (2.11-11) except that the roles of t and ω are interchanged. However, we see from this that the Fourier transform of 1 is $\delta(f)$ or $2\pi\delta(\omega)$; it is *not* $\delta(\omega)$.

One occasionally encounters derivatives of delta functions. By using Eq. (2.11-7) with the differentiation rule (property 8 in Table 2.9-1), we find that the Fourier transform of $\delta'(t)$ is $j\omega$. Therefore

$$\int_{-\infty}^{\infty} f(\tau)\delta'(\tau - t)\,d\tau \quad \longleftrightarrow \quad j\omega F(\omega) \quad \longleftrightarrow \quad \int_{-\infty}^{\infty} f'(\tau)\delta(\tau - t)\,d\tau, \tag{2.11-13}$$

or, in general,

$$\int_{-\infty}^{\infty} f(\tau)\frac{d^n}{d\tau^n}[\delta(\tau - t)]\,d\tau = \int_{-\infty}^{\infty} \frac{d^n f(\tau)}{d\tau^n}\delta(\tau - t)\,d\tau. \qquad (2.11\text{-}14)$$

This result can also be obtained by using integration by parts to evaluate the left-hand side of Eq. (2.11-13).

2.12 APPLICATIONS OF THE DELTA FUNCTION

Unit Step Function

The unit step function is commonly used in circuit analysis. It has the value 1 for $t > 0$ and 0 for $t < 0$ and can be written in the form

$$u(t) = \tfrac{1}{2} + \tfrac{1}{2}\operatorname{sgn}(t), \qquad (2.12\text{-}1)$$

where sgn (t) is the signum function introduced in Sec. 2.10 [Eq. (2.10-13)]. The transform of the signum function was shown in Eq. (2.10-15) to be given by $2/j\omega$; hence

$$\mathcal{F}[u(t)] = \pi\delta(\omega) + \frac{1}{j\omega} = \frac{\delta(f)}{2} + \frac{1}{j2\pi f}. \qquad (2.12\text{-}2)$$

This result can be used to show that the formula for the Fourier transform of the integral is as given in Eq. (2.9-20). Thus let

$$y(t) = \int_{-\infty}^{t} x(\tau)\,d\tau. \qquad (2.12\text{-}3)$$

This can be regarded as the convolution of $x(t)$ with a unit step function; i.e.,

$$y(t) = \int_{-\infty}^{\infty} x(\tau)u(t - \tau)\,d\tau = x(t) * u(t). \qquad (2.12\text{-}4)$$

Then the multiplication-convolution rule (property 12 in Table 2.9-1) can be applied to give

$$Y(\omega) = X(\omega)\left[\pi\delta(\omega) + \frac{1}{j\omega}\right] = X(f)\left[\frac{\delta(f)}{2} + \frac{1}{j2\pi f}\right], \qquad (2.12\text{-}5)$$

which is the same result given in Eq. (2.9-20).

Observe the fallacy of arguing that since the unit step function is the integral of the delta function its Fourier transform should just be $1/j\omega$. This fallacy results from neglecting a constant of integration.

Ramp Function

This function is obtained directly as the integral of a square pulse or equivalently by convolution of the square pulse with the unit step function:

$$r(t) = \int_{-\infty}^{\infty} \text{rect}\,(\tau)u(t - \tau)\,d\tau. \tag{2.12-6}$$

The Fourier transform is therefore

$$\begin{aligned} R(f) &= \text{sinc}\,(f)\left[\frac{\delta(f)}{2} + \frac{1}{j2\pi f}\right] \\ &= \frac{\delta(f)}{2} + \frac{\text{sinc}\,(f)}{j2\pi f} \\ &= \pi\delta(\omega) + \frac{\sin\,(\omega/2)}{j\omega^2/2}. \end{aligned} \tag{2.12-7}$$

In going from the first line to the second we use the fact that for any continuous function $\phi(t)$, $\delta(t)\phi(t) = \delta(t)\phi(0)$, which is easily checked by use of the limiting relations for the delta function, i.e., Eqs. (2.11-2)–(2.11-4). In the third line we have used the fact that $\omega = 2\pi f$ and $\pi\delta(\omega) = \frac{1}{2}\delta(\omega/2\pi)$. The ramp and its Fourier transform are shown in Fig. 2.12-1.

Periodic Signals

The use of delta functions permits us to easily extend the definition of the Fourier integral to periodic functions even though periodic functions do not satisfy the Dirichlet conditions. In particular, for

$$x(t) = \cos 2\pi f_0 t, \tag{2.12-8}$$

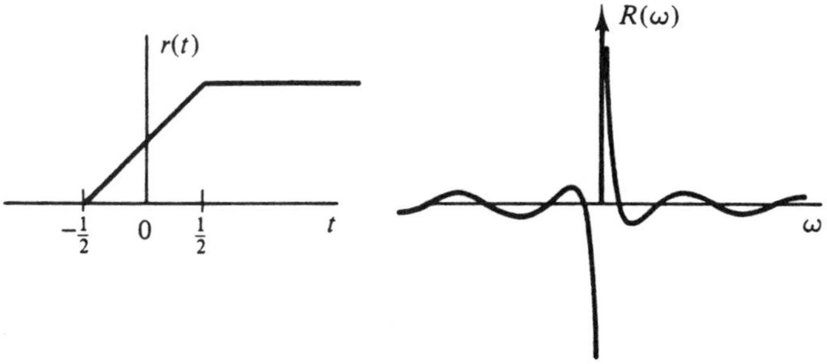

Figure 2.12-1. Ramp function and its Fourier transform.

we have

$$\begin{aligned} X(f) &= \int_{-\infty}^{\infty} \cos 2\pi f_0 t e^{-j2\pi ft}\, dt \\ &= \tfrac{1}{2}\int_{-\infty}^{\infty} (e^{j2\pi(f_0-f)t} + e^{-j2\pi(f_0+f)t})\, dt \qquad (2.12\text{-}9) \\ &= \tfrac{1}{2}[\delta(f-f_0) + \delta(f+f_0)], \end{aligned}$$

by Eq. (2.11-9). This is pictured in Fig. 2.12-2. The spectrum of the cosine consists of two *lines* or *spikes*. Similarly, for $x(t) = \sin \omega_0 t$, $X(f) = \frac{1}{2}[-j\delta(f - f_0) + j\delta(f + f_0)]$; thus the spectrum consists of the same two lines, but they are associated with 90° phase shifts. For a zero frequency cosine wave, i.e., a dc signal, the two spikes coalesce, and therefore the spectrum of the dc signal is a single line at the origin of the frequency scale.

Observe that for a cosine wave lasting for a finite time T we have

$$x(t) = \cos 2\pi f_0 t \cdot \text{rect}\left(\frac{t}{T}\right), \qquad (2.12\text{-}10)$$

for which the spectrum, by the convolution theorem, is given by

$$\begin{aligned} X(f) &= \frac{1}{2}\int_{-\infty}^{\infty} [\delta(f' - f_0) + \delta(f' + f_0)]T \,\text{sinc}\, T(f - f')\, df' \\ &= \frac{T}{2}[\text{sinc}\, T(f - f_0) + \text{sinc}\, T(f + f_0)]. \end{aligned} \qquad (2.12\text{-}11)$$

This is shown in Fig. 2.12-3 for $Tf_0 >> 1$. The two infinitely narrow lines in Fig. 2.12-2 have become spectral lobes of finite width. Equation (2.12-11) is a particular instance of a sifting integral. It illustrates the ease with which convolutions involving delta functions can be performed.

The Fourier transform of an arbitrary periodic function $x_p(t)$ can be obtained by first expanding $x_p(t)$ in a Fourier series and then taking the Fourier transform. The transform of each term of the series produces a delta function so that the final result is an infinite series of delta functions at each of the harmonic frequencies.

This procedure can be generalized somewhat by first defining the periodic function $x_p(t)$ in terms of a nonperiodic function $x(t)$ by the series expansion

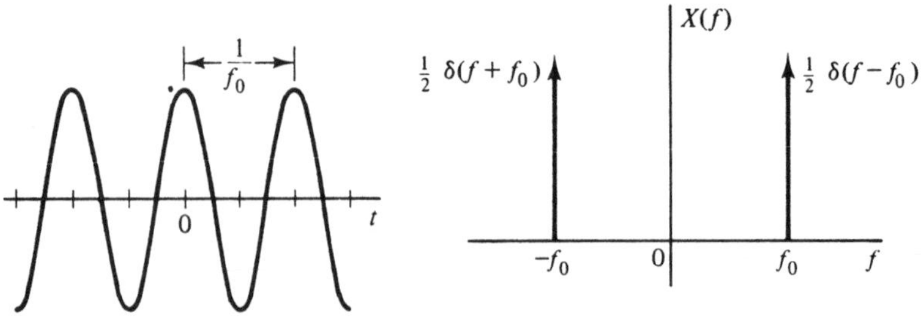

Figure 2.12-2. Cosine wave and its Fourier spectrum.

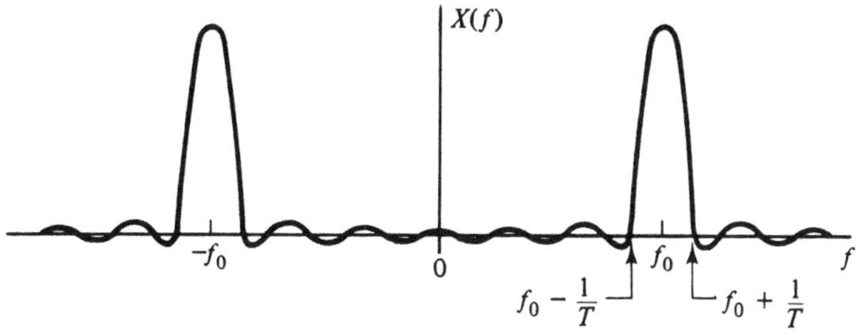

Figure 2.12-3. Spectrum of finite-length cosine wave. The frequency is f_0, and the length of the pulse is T.

$$x_p(t) = \sum_{n=-\infty}^{\infty} x(t - nT). \tag{2.12-12}$$

It is assumed that $x(t)$ is chosen so that this series converges to $x_p(t)$. It is easily seen that $x_p(t)$ as defined in Eq. (2.12-12) satisfies the basic definition of a periodic function given in Eq. (2.4-1):

$$x_p(t + T) = \sum_{n=-\infty}^{\infty} x[t - (n - 1)T] = x_p(t). \tag{2.12-13}$$

A possible and obvious choice for $x(t)$ is a function which equals $x_p(t)$ over one period and is zero elsewhere. However, this is clearly not the only possibility; in fact, there is generally an infinite number of ways of choosing $x(t)$ in Eq. (2.12-12) to yield a given $x_p(t)$. According to a convenient notation introduced by Woodward [2-9], Eq. (2.12-12) can be written as

$$x_p(t) = \text{rep}_T x(t). \tag{2.12-14}$$

We can think of $x(t)$ as being the *generating function* which generates the periodic function $x_p(t)$.

We now expand $x_p(t)$ in a Fourier series:

$$x_p(t) = \sum_{m=-\infty}^{\infty} X_m e^{j2\pi mt/T}, \tag{2.12-15}$$

where the coefficients are given by

$$\begin{aligned} X_m &= \frac{1}{T}\int_{-T/2}^{T/2} x_p(t)e^{-j2\pi mt/T}\,dt \\ &= \frac{1}{T}\int_{-T/2}^{T/2} \sum_{n=-\infty}^{\infty} x(t - nT)e^{-j2\pi mt/T}\,dt \\ &= \sum_{n=-\infty}^{\infty} \frac{1}{T}\int_{-T/2}^{T/2} x(t - nT)e^{-j2\pi m(t-nT)/T}\,dt \\ &= \sum_{n=-\infty}^{\infty} \frac{1}{T}\int_{-T/2-nT}^{T/2-nT} x(\tau)e^{-j2\pi m\tau/T}\,d\tau. \end{aligned} \tag{2.12-16}$$

In going from the second line to the third line we have exchanged the orders of summation and integration and multiplied by $e^{j2\pi mn} = 1$. In going from the third line to the fourth line we have made the change of variable $\tau = t - nT$. Observe now that the summands in Eq. (2.12-16) consist of integrals over adjacent segments of the τ axis; hence performing the summation is equivalent to integrating over infinite limits. Hence

$$X_m = \frac{1}{T}\int_{-\infty}^{\infty} x(\tau)e^{-j2\pi m\tau/T}\,d\tau = \frac{1}{T}X\left(\frac{m}{T}\right), \tag{2.12-17}$$

where $X(f)$ is the Fourier transform of $x(t)$. Thus we obtain the Fourier series representation of $x_p(t)$ in terms of the Fourier transform of the generating function $x(t)$:

$$x_p(t) = \sum_{n=-\infty}^{\infty} x(t - nT) = \frac{1}{T}\sum_{m=-\infty}^{\infty} X\left(\frac{m}{T}\right)e^{j2\pi mt/T}. \tag{2.12-18}$$

This equation is one form of *Poisson's sum formula*. The coefficients $(1/T)X(m/T)$ are identical with the coefficients c_m of the exponential form of the Fourier series [cf. Eq. (2.4-5)]; in fact, Eq. (2.12-18) is nothing more than a Fourier series. However, we now have a relation between the standard Fourier coefficients c_n and the Fourier transform of the nonperiodic generating function $x(t)$. We have found, in fact, that the Fourier coefficients are proportional to the sampled Fourier transform at points m/T, i.e., at integral multiples of the fundamental frequency. This is the principle on which much of the theory of signal sampling is based, and we shall see it again in Chapter 4.

We now proceed formally to Fourier transform equation (2.12-18) with the result

$$X_p(f) = \int_{-\infty}^{\infty} x_p(t)e^{-j2\pi ft}\,dt = \frac{1}{T}\int_{-\infty}^{\infty}\sum_{m=-\infty}^{\infty} X\left(\frac{m}{T}\right)e^{-j2\pi[f-(m/T)]t}\,dt. \tag{2.12-19}$$

Exchanging orders of integration and summation and using the definition for the delta function given in Eq. (2.11-8), we obtain the Fourier pair

$$\sum_{n=-\infty}^{\infty} x(t - nT) \quad \longleftrightarrow \quad \frac{1}{T}\sum_{m=-\infty}^{\infty} X\left(\frac{m}{T}\right)\delta\left(f - \frac{m}{T}\right). \tag{2.12-20}$$

This can be written in simplified notation using the rep operation defined in Eq. (2.12-14) and a *comb* operation also defined by Woodward [2-9] as follows:

$$\text{comb}_F X(f) = \sum_{m=-\infty}^{\infty} X(mF)\delta(f - mF), \tag{2.12-21}$$

where F is the repetition frequency of the delta functions. In simplified notation Eq. (2.12-20) is

$$\text{rep}_T x(t) \quad \longleftrightarrow \quad \frac{1}{T}\text{comb}_{1/T} X(f). \tag{2.12-22}$$

The comb of a sinc function is shown in Fig. 2.12-4. Note the essential similarity to Fig. 2.4-3(a). The figure does, in fact, represent the spectrum of a train of rectangular

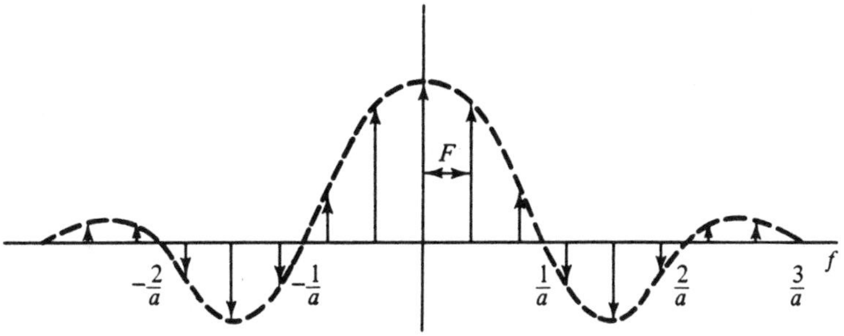

Figure 2.12-4. The function $\text{comb}_F \operatorname{sinc}(af)$.

pulses with pulse length a and repetition frequency F. It can therefore also be regarded as a representation of the Fourier series coefficients of such a pulse train. Equations (2.12-20) and (2.12-21) are the desired Fourier transforms of a periodic function.

A periodic function of considerable interest in later work is the ideal sampling function given by

$$\text{rep}_T\, \delta(t) = \sum_{n=-\infty}^{\infty} \delta(t - nT) = \text{comb}_T(1). \tag{2.12-23}$$

This is simply a comb of delta functions spaced T seconds apart. The Fourier transform of the delta function is 1 by Eq. (2.11-7); hence from Eq. (2.12-20) we get

$$\sum_{n=-\infty}^{\infty} \delta(t - nT) \quad \longleftrightarrow \quad \frac{1}{T} \sum_{m=-\infty}^{\infty} \delta\left(f - \frac{m}{T}\right), \tag{2.12-24}$$

and we find that except for scale the Fourier transform of a comb of delta functions is another comb of delta functions†:

$$\text{comb}_T(1) \quad \longleftrightarrow \quad \frac{1}{T} \text{comb}_{1/T}(1). \tag{2.12-25}$$

We observe that the derivation of Eq. (2.12-20) could have been simplified by introducing the comb operation initially and deriving Eq. (2.12-25). Then use of the multiplication-convolution rule would have given the equivalent of Eq. (2.12-20) directly:

$$x(t) * \text{comb}_T(1) \quad \longleftrightarrow \quad \frac{1}{T} X(f)\, \text{comb}_{1/T}(1) = \frac{1}{T} \text{comb}_{1/T}\, X(f). \tag{2.12-26}$$

We summarize Secs. 2.10-2.12 by the Fourier pairs in Table 2.12-1. The entries in this table together with the ones in Table 2.9-1 make it possible to find the Fourier transforms of most of the functions of practical interest almost by inspection. This process has already been used in several places. For instance, we obtained the spectrum

† This formula is so useful in deriving various results in Fourier theory that it probably should be committed to memory.

TABLE 2.12-1 FOURIER PAIRS

Name	$x(t)$	$X(f)$
1. Square pulse	rect (t)	sinc f
2. Triangular pulse	$1 - \lvert t \rvert$ for $\lvert t \rvert \le 1$, 0 for $\lvert t \rvert > 1$	$\text{sinc}^2 f$
3. Gaussian pulse	$e^{-\pi t^2}$	$e^{-\pi f^2}$
4. Asymmetric exponential pulse	$e^{-at}u(t)$	$1/(j\omega + a)$
5. Symmetric exponential pulse	$e^{-a\lvert t \rvert}$	$2a/(\omega^2 + a^2)$
6. Signum function	sgn (t)	$2/j\omega$
7. Delta function	$\delta(t)$	1
8. Unit step	$u(t)$	$\dfrac{\delta(f)}{2} + \dfrac{1}{j2\pi f}$
9. Cosine	$\cos 2\pi f_0 t$	$\frac{1}{2}[\delta(f - f_0) + \delta(f + f_0)]$
10. Periodic functions	$\text{rep}_T x(t) = \sum_{n=-\infty}^{\infty} x(t - nT)$	$\dfrac{1}{T}\text{comb}_{1/T} X(f) = \dfrac{1}{T}\sum_{n=-\infty}^{\infty} X\left(\dfrac{n}{T}\right)\delta\left(f - \dfrac{n}{T}\right)$

for the triangular pulse by observing that this pulse could be obtained by convolving two square pulses, and the spectrum then followed from property 12 in Table 2.9-1. We obtained the transform of the unit step function by noting that it was the sum of a constant ($\frac{1}{2}$) and $\frac{1}{2}$ sgn (t), etc. For more complicated examples, see Probs. 2-31 and 2-32.

2.13 TIME FUNCTIONS WITH ONE-SIDED SPECTRA

In applications dealing with modulation, narrowband filters, or narrowband signals it is often very convenient to use time functions having one-sided spectra. We observed earlier (Sec. 2.9) that the Fourier transform $X(f)$ of a real signal $x(t)$ has Hermitian symmetry; i.e., $X(-f) = X^*(f)$. Thus, a signal having a nonzero spectrum only for positive frequencies must be a complex signal. However, since for real signals the spectrum for negative frequencies is completely determined by the spectrum for positive frequencies, there cannot be any loss of information if we simply remove the negative-frequency portion; i.e., the complex signal that results from this operation must somehow be equivalent to the original real signal.

To investigate the implications of one-sided spectra, consider the real function of time $x(t)$ having the spectrum $X(f)$ with the usual symmetry properties. To convert this to a one-sided spectrum we define the new spectrum

$$\Xi(f) = X(f)[1 + \text{sgn}(f)]; \tag{2.13-1}$$

that is,

$$\Xi(f) = \begin{cases} 0, & f < 0, \\ 2X(f), & f > 0. \end{cases} \tag{2.13-2}$$

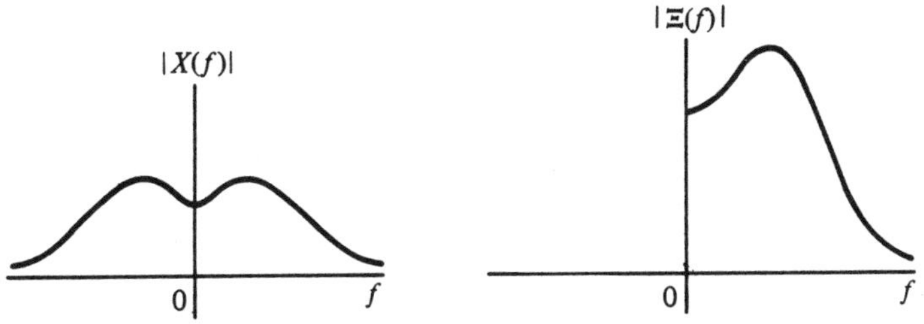

Figure 2.13-1. Symmetric spectrum and the one-sided form obtained by removing the negative-frequency portion and adding it to the positive-frequency side.

This is shown in Fig. 2.13-1; in effect we remove the spectrum from the negative-frequency side and pile it on top of the spectrum on the positive-frequency side; the area under the magnitude of the spectrum remains the same.

The resulting time function is obtained by inverse-Fourier-transforming Eq. (2.13-1); the result is the complex signal

$$\xi(t) = x(t) * \left[\delta(t) - \frac{1}{j\pi t} \right], \tag{2.13-3}$$

where we have used property 12 of Table 2.9-1 (convolution-multiplication) and pairs 6 and 7 in Table 2.12-1. The negative sign is one of the consequences of the duality property. Performing the indicated convolution results in

$$\xi(t) = x(t) + j\hat{x}(t), \tag{2.13-4}$$

where

$$\hat{x}(t) = \frac{1}{\pi} \int_{-\infty}^{\infty} \frac{x(\tau)}{t - \tau} \, d\tau \tag{2.13-5}$$

is the *Hilbert transform* of $x(t)$. A complex waveform whose imaginary part is the Hilbert transform of the real part is referred to as a *complex analytic waveform*. The term ''analytic'' is used in the sense of complex-variable theory. A complex function $f(p)$ of the complex variable $p = t + ju$ is analytic in a region R if it has no poles or other singularities in that region (more precisely if it satisfies the Cauchy-Riemann equations at every point in R). For a complex function $\xi(p)$ that is analytic over the upper half-plane (i.e., for all $p = t + ju$ such that $u > 0$) it can be shown that the real and imaginary parts of $\xi(p)$ on the t axis are related by the Hilbert transformation. A demonstration is given in LePage [2–7], Sec. 7–10.

The defining equation [Eq. (2.13-5)] for the Hilbert transform is an improper integral since the integrand generally goes to infinity for $\tau = t$. The transform is therefore defined as the Cauchy principal value:

$$\hat{x}(t) = \lim_{\substack{\epsilon \to 0 \\ A \to \infty}} \frac{1}{\pi} \left[\int_{-A}^{-\epsilon} + \int_{\epsilon}^{A} \right] \frac{x(\tau)}{t - \tau} \, d\tau. \tag{2.13-6}$$

The use of the Cauchy principal value here is consistent with its use in the inverse transformation of $1/j\omega$, i.e., Eq. (2.10-18).

We see from Eq. (2.13-5) that the Hilbert transform of $x(t)$ is the convolution of $x(t)$ with $1/\pi t$:

$$\hat{x}(t) = x(t) * \frac{1}{\pi t}. \tag{2.13-7}$$

Therefore the Fourier transform of the Hilbert transform is given by

$$\mathscr{F}[\hat{x}(t)] \equiv \hat{X}(f) = -jX(f) \operatorname{sgn}(f). \tag{2.13-8}$$

Thus it is theoretically possible to generate the Hilbert transform of a signal by passing it through a linear circuit that provides a 90° phase *lag* at all positive frequencies and a 90° phase *lead* at all negative frequencies. As will be seen in the next chapter, such a circuit cannot be exactly realized because of constraints imposed by causality. However, for narrowband signals (i.e., where the Fourier spectrum has significant magnitude only in a narrow range of frequencies around some frequency f_0), a circuit that provides 90° phase lag in the band around f_0 and 90° phase lead in the band around $-f_0$ can be reasonably well approximated. Such circuits are used in connection with single-sideband modulation systems and will be discussed in Chapter 6.

Some of the properties of the Hilbert transform and of complex analytic signals are best illustrated by means of examples.

Examples

1. Consider the signal

$$x(t) = \cos \omega_0 t. \tag{2.13-9}$$

The Hilbert transform is

$$\begin{aligned}\hat{x}(t) &= \frac{1}{\pi}\int_{-\infty}^{\infty} \frac{\cos \omega_0 \tau}{t - \tau}\, d\tau \\ &= \frac{1}{\pi}\int_{-\infty}^{\infty} \frac{\cos \omega_0(\tau - t + t)}{t - \tau}\, d\tau \\ &= \frac{\cos \omega_0 t}{\pi}\int_{-\infty}^{\infty} \frac{\cos \omega_0(t-\tau)}{t-\tau}\, d\tau + \frac{\sin \omega_0 t}{\pi}\int_{-\infty}^{\infty} \frac{\sin \omega_0(t-\tau)}{t-\tau}\, d\tau \\ &= \sin \omega_0 t.\end{aligned} \tag{2.13-10}$$

The result in Eq. (2.13-10) follows from the fact that the first CPV integral in the third line is zero, i.e., the integrand is odd, and the second CPV integral has value π for $\omega_0 > 0$. The complex analytic signal corresponding to $\cos \omega_0 t$ is

$$\xi(t) = \cos \omega_0 t + j \sin \omega_0 t = e^{j\omega_0 t}. \tag{2.13-11}$$

Thus we find that the complex analytic signal corresponding to a cosine (or sine) wave is just the complex phasor used in circuit theory. Note that the spectrum of the cosine wave consists of two delta functions at $\pm\omega_0$ having an area of $\frac{1}{2}$; the spectrum of the complex analytic signal is a single delta function at ω_0 having an area of 1. This

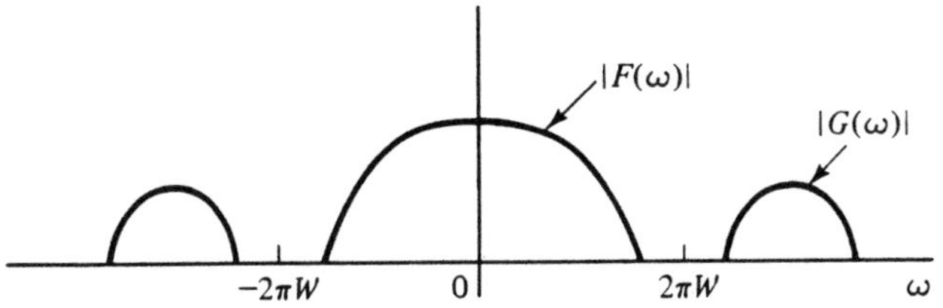

Figure 2.13-2. Nonoverlapping spectra of two functions $f(t)$ and $g(t)$.

illustrates how the two-sided spectrum of the real time function is converted to a one-sided spectrum as discussed earlier; cf. Fig. 2.13-1.

2. Communications systems often utilize signals of the form $f(t)\cos(\omega_0 t + \phi)$ where the spectrum of $f(t)$ contains only low frequencies and ω_0 is a relatively high frequency. We shall be interested in the complex analytic form for this kind of signal.

A more general version of this problem is obtained by considering the signal

$$x(t) = f(t)g(t), \tag{2.13-12}$$

where $f(t)$ is a low-frequency signal and $g(t)$ a high-frequency signal. Stated more precisely, let $F(\omega)$ and $G(\omega)$ be the Fourier transforms of $f(t)$ and $g(t)$, respectively; then these transforms have the property

$$\begin{aligned} F(\omega) &= 0, \quad |\omega| > 2\pi W \\ G(\omega) &= 0, \quad |\omega| < 2\pi W. \end{aligned} \tag{2.13-13}$$

See Fig. 2.13-2. The desired Hilbert transform of the product is obtained by an indirect method. By use of the multiplication-convolution property of the Fourier transform and Eq. (2.13-8) expressed in terms of ω rather than f, we find that the Fourier transform of $\hat{x}(t)$ is

$$\mathcal{F}[\hat{x}(t)] = \mathcal{F}[\widehat{f(t)g(t)}] = -\frac{j}{2\pi}\operatorname{sgn}(\omega)\int_{-\infty}^{\infty} F(\omega - u)G(u)\,du. \tag{2.13-14}$$

Also,

$$\mathcal{F}[f(t)\hat{g}(t)] = -\frac{j}{2\pi}\int_{-\infty}^{\infty} \operatorname{sgn}(u)F(\omega - u)G(u)\,du. \tag{2.13-15}$$

These two results are identical. To show this, consider their difference:

$$\frac{j}{2\pi}\int_{-\infty}^{\infty} [\operatorname{sgn}(\omega) - \operatorname{sgn}(u)]F(\omega - u)G(u)\,du.$$

The product $F(\omega - u)G(u)$ is shown in Fig. 2.13-3 for $\omega > 0$. Note that if the two spectra overlap they do so for $u > 0$, but then $[\operatorname{sgn}(\omega) - \operatorname{sgn}(u)] = 0$. A similar argument can be made for $\omega < 0$. Thus as long as the spectra $F(\omega)$ and $G(\omega)$ have no region of overlap we find that

$$\hat{x}(t) = f(t)\hat{g}(t). \qquad 2.13\text{-}16)$$

Thus we have the important result that *only the high-frequency factor in the signal $f(t)g(t)$ is Hilbert transformed* when the Hilbert transform of the product is computed.[†]

[†] Generalizations of this theorem to two- and higher-dimensional functions exist. See [2-10] and [2-11].

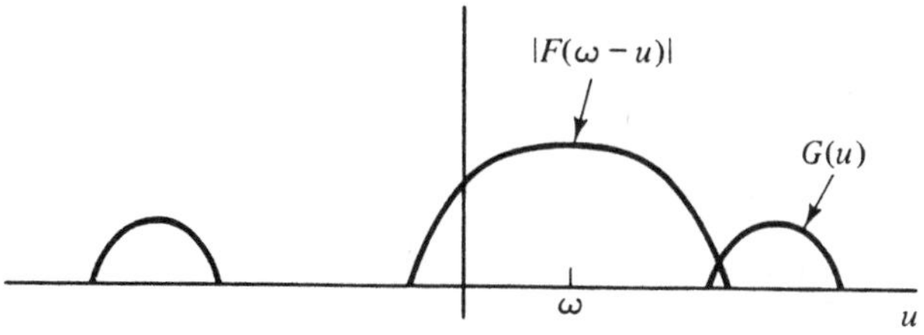

Figure 2.13-3. Overlap of the spectra $F(\omega - u)$ and $G(u)$ for positive ω.

For the signal of the form

$$x(t) = f(t) \cos(\omega_0 t + \phi), \tag{2.13-17}$$

where $f(t)$ is the low-frequency envelope of a high-frequency carrier, we therefore find that the complex analytic form is

$$\xi(t) = f(t)e^{j(\omega_0 t + \phi)}. \tag{2.13-18}$$

Thus the complex analytic signal is again identical with the familiar phasor form. In some cases $e^{j\phi}$ is also a low-frequency time function. Then $\mu(t) = f(t)e^{j\phi(t)}$ is called the *complex envelope* of the signal. Its modulus $|\mu(t)| = |\xi(t)|$ is the *amplitude modulation*, and the phase $\phi(t)$ is the *phase modulation* of the signal $x(t)$. This simple representation of amplitude and phase modulation constitutes one of the main advantages of the complex signal notation. If the Fourier spectrum of the complex envelope is

$$M(f) = \int_{-\infty}^{\infty} \mu(t)e^{-j2\pi ft}\, dt, \tag{2.13-19}$$

then

$$\Xi(f) = M(f - f_0),$$

where $f_0 = \omega_0/2\pi$ and $\Xi(f) = \mathcal{F}[\xi(t)]$. The complex envelope function $\mu(t)$ is generally not complex analytic, and therefore $M(f)$ will not generally be zero for negative argument. However, if $\xi(t)$ is to be complex analytic, $\Xi(f)$ must be zero for negative f; this requires that $M(f)$ be zero for $f < -f_0$. (See Fig. 2.13-4.) This is approximately satisfied if $x(t)$ is *narrowband*, i.e., if the extent of $M(f)$ along the frequency axis is much smaller than the center frequency f_0. For this reason the concepts of the complex analytic signal and the complex envelope are generally useful mainly for narrowband signals.

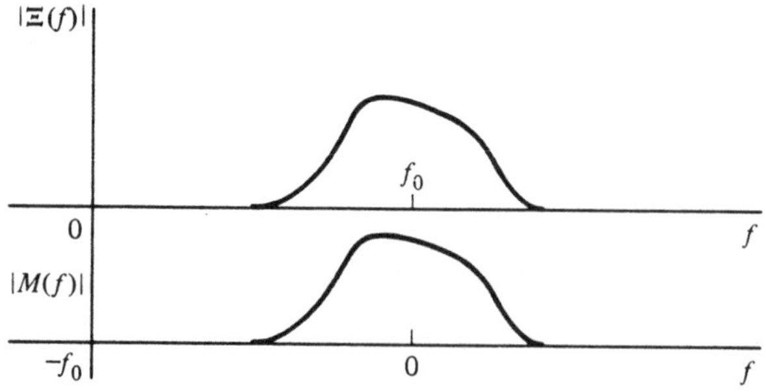

Figure 2.13-4. Spectrum of a narrowband complex analytic signal. For center frequency $f = f_0$ it is necessary that $M(f) = 0$ for $f < -f_0$.

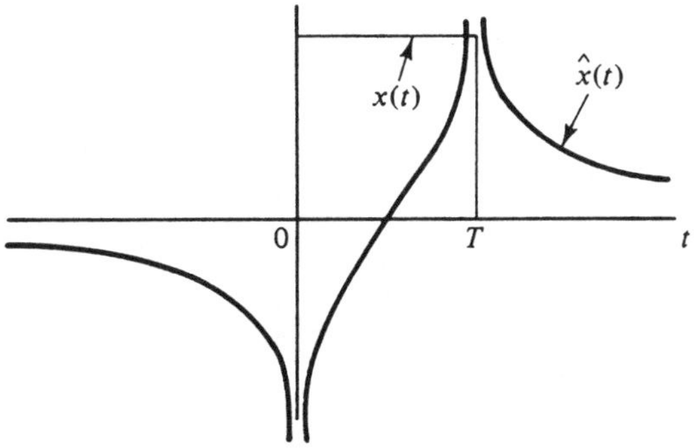

Figure 2.13-5. Square pulse and its Hilbert transform. The places where $x(t)$ goes to infinity are sometimes called "horns," and they may cause problems in communication systems in which the Hilbert transforms of signals are involved, such as in single sideband.

3. As a final example, we consider the Hilbert transform of the square pulse:

$$x(t) = \text{rect}\left(\frac{t - T/2}{T}\right). \tag{2.13-20}$$

The computation of the transform is straightforward, although as with all convolutions involving discontinuous functions it is necessary to consider several cases in performing the integration. The result can be put in the form

$$\hat{x}(t) = \frac{1}{\pi} \ln\left|\frac{t}{T - t}\right|. \tag{2.13-21}$$

The signal and its Hilbert transform are shown in Fig. 2.13-5.

2.14 TWO-DIMENSIONAL FOURIER TRANSFORMS

The one-dimensional Fourier transform can easily be extended to higher-dimensional space, and most of its properties either remain unchanged or have higher-dimensional analogs. We consider here in detail only the two-dimensional (2-D) Fourier transform, which is used in image description and processing, television, and other 2-D applications.

The 2-D Fourier transform of a 2-D function $f(x, y)$ is defined by

$$F(u, v) = \int_{-\infty}^{\infty} \int_{-\infty}^{\infty} f(x, y) e^{-j2\pi(ux+vy)}\, dx\, dy, \tag{2.14-1}$$

where $f(x, y)$ may be real or complex. The inverse Fourier transform recovers $f(x, y)$ from $F(u, v)$ and is given by

$$f(x, y) = \int_{-\infty}^{\infty} \int_{-\infty}^{\infty} F(u, v) e^{j2\pi(ux+vy)}\, du\, dv. \tag{2.14-2}$$

Equations (2.14-1) and (2.14-2) can be written symbolically as

$$F(u, v) = \mathcal{F}_2[f(x, y)],$$

and

$$f(x, y) = \mathcal{F}_2^{-1}[F(u, v)],$$

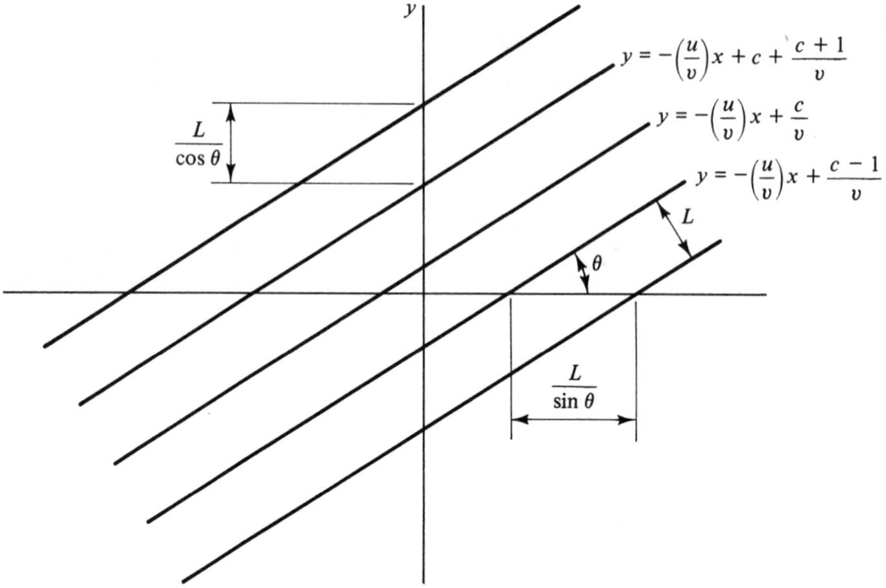

Figure 2.14-1. Elementary spatial sinusoid at an angle θ with the x axis.

where $\mathcal{F}_2(\cdot)$ and $\mathcal{F}_2^{-1}(\cdot)$ are the two-dimensional Fourier transform and inverse Fourier transform operators, respectively.

We see from Eq. (2.14-2) that the 2-D Fourier transform is a decomposition of a 2-D signal, sometimes referred to as a *scene*, into a sum of elementary complex sinusoids of the form $F(u, v) \exp[j2\pi (ux + vy)]$. If the dimensions of x and y are some units of length, such as centimeters, then u and v have units of cycles per centimeter (simply to keep the argument of the exponential function dimensionless). For this reason, u and v are referred to as spatial frequencies.

To gain further insight into the meaning of Eq. (2.14-2), we investigate some of the properties of the function $\exp[j2\pi(ux + vy)]$. Note that for fixed u and v the argument $2\pi(ux + vy)$ is constant along lines described by

$$y = -\left(\frac{u}{v}\right)x + \frac{c}{v}, \tag{2.14-3}$$

where c is an arbitrary constant. These are parallel lines making an angle θ with the x axis, with $\theta = \arctan(u/v)$ (see Fig. 2.14-1).

The value of the function is the same for all c's differing by an integer. The function is therefore periodic, and the period L is the distance between any two lines for which c differs by 1. It is a simple exercise in trigonometry to show that L is given by

$$L = \frac{1}{\sqrt{u^2 + v^2}}. \tag{2.14-4}$$

Along the x axis ($y = 0$) the period is seen to be $1/u$ from Eq. (2.14-2), and from Fig. 2.14-1 it is seen to be $L/\sin\theta$. Thus $u = (\sin\theta)/L$ is the component frequency in the x direction. Note that this frequency is low (wave crests far apart) for small θ and large L. Similarly, $v = (\cos\theta)/L$ is the component frequency in the y direction.

Since the exponential function can be written as the sum of two sinusoidal functions in phase quadrature, the complex sinusoid can be regarded as composed of two sinusoidal waves whose crests and troughs make an angle θ with the x axis and whose period is given by Eq. (2.14-4). The Fourier transform is a decomposition of the 2-D function $f(x, y)$ into a spectrum of such elementary sinusoids having different periods and directions.

Alternatively, one can view the 2-D Fourier transform as a succession of 1-D transforms (spatial spectra) along the x and y axes. This follows from Eq. (2.14-1) by writing it in the form

$$\begin{aligned} F(u, v) &= \int_{-\infty}^{\infty} dy\, e^{-j2\pi vy} \int_{-\infty}^{\infty} dx\, e^{-j2\pi ux} f(x, y) \\ &= \int_{-\infty}^{\infty} dy\, e^{-j2\pi vy}\, Q(u, y), \end{aligned}$$

where $Q(u, y)$ is the 1-D Fourier transform of $f(x, y)$ along the x axis only. This point of view easily generalizes to any number of dimensions. When, as in television, an image is scanned by a beam moving at a constant velocity, then $x = V_x t$ and $y = V_y t$, where V_x and V_y are the velocities of the scanning spot in the x (horizontal) and y (vertical) directions. Then the inverse 2-D Fourier transform becomes

$$\begin{aligned} f(x, y) &= \int_{-\infty}^{\infty} \int_{-\infty}^{\infty} F(u, v)\, e^{j2\pi(uV_x + vV_y)t}\, du\, dv \\ &= \frac{1}{V_x V_y} \int_{-\infty}^{\infty} \int_{-\infty}^{\infty} F(u, v)\, e^{j2\pi(f_x + f_y)t}\, df_x\, df_y \\ &= g(t), \end{aligned} \tag{2.14-5}$$

where $f_x = uV_x$ and $f_y = vV_y$ are, respectively, the horizontal and vertical scanning frequencies in hertz, and where $g(t)$ symbolizes the time function that is generated by uniformly scanning the picture. This is the video signal that is transmitted to the TV receiver.

As in the 1-D case, the integrals defining the 2-D Fourier transformation (Eqs. 2.14-1 and 2.14-2) do not always converge. The Dirichlet conditions mentioned in Sec. 2.7 can be directly extended to two or more dimensions. For two dimensions these conditions are that the functions f or F should have at most a finite number of discontinuities in any finite rectangle, should have no infinite discontinuities, and should be absolutely integrable on the infinite square; that is,

$$\int_{-\infty}^{\infty} \int_{-\infty}^{\infty} |f(x, y)|\, dx\, dy = M < \infty. \tag{2.14-6}$$

These are sufficient conditions; 2-D delta functions (which are infinite at one point) or sinusoids (which don't satisfy Eq. 2.14-6) do have well-defined Fourier transforms. Specifically,

$$\int_{-\infty}^{\infty}\int_{-\infty}^{\infty} \delta(x, y)e^{-j2\pi(ux+vy)}\, dx\, dy = 1, \tag{2.14-7}$$

and, for $f(x, y) = \cos 2\pi(u_0x + v_0y)$,

$$\int_{-\infty}^{\infty}\int_{-\infty}^{\infty} \cos 2\pi(u_0x + v_0y)\, e^{-j2\pi(ux+vy)}\, dx\, dy$$

$$= \frac{1}{2}\int_{-\infty}^{\infty}\int_{-\infty}^{\infty} e^{j2\pi[(u_0-u)x+(v_0-v)y]} + e^{-j2\pi[(u_0+u)x+(v_0+v)y]}\, dx\, dy \tag{2.14-8}$$

$$= \frac{1}{2}[\delta(u - u_0)\delta(v - v_0) + \delta(u + u_0)\delta(v + v_0)].$$

2.15 TWO-DIMENSIONAL FOURIER TRANSFORM PROPERTIES

The following properties are either direct extensions of the corresponding 1-D results or as easily derived from first principles.

Linearity

The 2-DFT is a linear operator. This means that if α and β are any complex constants and $f_1(\bullet)$ and $f_2(\bullet)$ are any two functions satisfying the sufficiency conditions given in Sect 2.14, then

$$\mathcal{F}_2(\alpha f_1 + \beta f_2) = \alpha F_1(u, v) + \beta F_2(u, v), \tag{2.15-1}$$

where $\mathcal{F}_2$ denotes the 2-D Fourier operator.

Scaling

If $f(x, y) \leftrightarrow F(u, v)$, then

$$f(ax, by) = \frac{1}{|ab|} F\left(\frac{u}{a}, \frac{v}{b}\right). \tag{2.15-2}$$

This is a direct extension of Eq. (2.9-6).

Parseval's Theorem

Parseval's theorem is a special case of the two-dimensional convolution theorem. It states that

$$\int_{-\infty}^{\infty}\int_{-\infty}^{\infty} |f(x, y)|^2\, dx\, dy = \int_{-\infty}^{\infty}\int_{-\infty}^{\infty} |F(u, v)|^2\, du\, dv, \tag{2.15-3}$$

which, in the notation of Eq. (2.7-4), can be succinctly written as

$$\|f\|^2 = \|F\|^2 \quad \text{or} \quad \|f\| = \|F\|. \tag{2.15-4}$$

In texts on Fourier optics [e.g., (2-12) or (2-13)], it is shown that a convex lens (as used in a magnifying glass) acts as a 2-D Fourier transformer in a monochromatic illumination; that is, an image in one focal plane of the lens is converted to its Fourier transform in the other focal plane. Hence Parseval's theorem applied to Fourier optics states that the total illumination power on one side of a lossless lens is the same as the total illumination power in focal plane on the other side.

Shift Theorem

If $f(x, y) \leftrightarrow F(u, v)$, then

$$f(x - a, y - b) \quad \longleftrightarrow \quad F(u, v)e^{-j2\pi(ua+vb)}.$$

This result is a direct extension of Eq. (2.9-5). Note that

$$|\mathcal{F}_2[f(x - a, y - b)]| = |\mathcal{F}_2[f(x, y)]|, \tag{2.15-5}$$

which is a result of considerable utility in optics, where it is somewhat inaccurately called the ''shift-invariance property of the Fourier transform.''

Symmetry Properties

Let $f(x, y)$ be a real function of (x, y). Then

$$\begin{aligned} F(u, v) &= F^*(-u, -v), \\ F(u, -v) &= F^*(-u, v), \\ F(-u, v) &= F^*(u, -v). \end{aligned} \tag{2.15-6}$$

This is the 2-D extension of the Hermitian property of Eq. (2.9-23). For other relationships implied by this property, see Problem 2-35. If $f(x, y)$ is real and equal to $f(-x, -y)$, then it is easy to show, either from duality or from the definition of the inverse transform, that $F(u, v)$ is real and $F(u, v) = F(-u, -v)$. If $f(x, y)$ is pure imaginary, then certain skew Hermitian properties hold (see Problem 2-36).

Convolution

Let $f(x, y)$ and $h(x, y)$ be two functions with Fourier transforms $F(u, v)$ and $H(u, v)$, respectively. If $g(x, y)$ denotes their convolution† product, that is,

$$\begin{aligned} g(x, y) &= \int_{-\infty}^{\infty}\int_{-\infty}^{\infty} f(\xi, \eta)h(x - \xi, y - \eta)\, d\xi\, d\eta \\ &\equiv f(x, y) * h(x, y), \end{aligned} \tag{2.15-7}$$

† The centered asterisk denotes convolution, whereas the asterisk on a function denotes complex conjugation.

then $G(u, v) = F(u, v)H(u, v)$, where $G(u, v) = \mathcal{F}_2[g]$. This is a direct extension of the very important convolution theorem introduced in Sec. 2.9.

Fourier Identity

At points where the function $f(x, y)$ is continuous, it can be shown that

$$\mathcal{F}_2^{-1}\mathcal{F}_2[f(x, y)] = \mathcal{F}_2\mathcal{F}_2^{-1}[f(x, y)] = f(x, y)$$

and

$$\mathcal{F}_2\mathcal{F}_2[f(x, y)] = f(-x, -y). \tag{2.15-8}$$

At points where $f(x, y)$ is discontinuous, the two successive transforms give an average of $f(x, y)$ in a small neighborhood of the point of discontinuity.

2.16 SEPARABLE AND CIRCULARLY SYMMETRIC FUNCTIONS

A function f is said to be separable in rectangular coordinates if

$$f(x, y) = f_1(x)f_2(y). \tag{2.16-1}$$

It is separable in polar coordinates (r,θ) if

$$f(r,\theta) = f_1(r)f_2(\theta). \tag{2.16-2}$$

By applying the definition of the 2-DFT to Eq. (2.16-1), we obtain

$$F(u, v) = F_1(u)F_2(v), \tag{2.16-3}$$

where

$$F_1(u) = \int_{-\infty}^{\infty} f_1(x)e^{-j2\pi ux}\,dx = \mathcal{F}[f_1]$$

and

$$F_2(v) = \int_{-\infty}^{\infty} f_2(y)e^{-j2\pi vy}\,dy = \mathcal{F}[f_2].$$

Thus the 2-DFT of a function separable in x and y is itself separable into a product of two transforms, one depending only on u and the other depending only on v.

Functions $f(x, y)$ that depend only on $r = \sqrt{x^2 + y^2}$ are said to possess *circular symmetry*. This is a special case of separability in polar coordinates (Eq. 2.16-2), but with constant $f_2(\theta)$. The 2-D Fourier transform of such functions is also circularly symmetric. This is easily shown by converting both the x, y and u, v coordinate system into polar form:

$$\begin{aligned} x &= r\cos\theta, & u &= \rho\cos\phi, \\ y &= r\sin\theta, & v &= \rho\sin\phi, \\ r^2 &= x^2 + y^2, & \rho^2 &= u^2 + v^2, \\ \theta &= \tan^{-1}\left(\frac{y}{x}\right), & \phi &= \tan^{-1}\left(\frac{v}{u}\right). \end{aligned} \tag{2.16-4}$$

Making these transformations in Eq. (2.14-1), we obtain

$$\begin{aligned} F(\rho\cos\phi, \rho\sin\phi) &= \int_0^{2\pi} d\theta \int_0^{\infty} f(r) e^{-j2\pi r\rho(\cos\theta\cos\phi + \sin\theta\sin\phi)} r\, dr \\ &= \int_0^{\infty} dr\, rf(r) \int_0^{2\pi} e^{-j2\pi r\rho\cos(\theta-\phi)}\, d\theta. \end{aligned} \tag{2.16-5}$$

The inner integral in this expression looks like the defining integral for the zero-order Bessel function of the first kind. One of the definitions for this function is

$$J_0(x) = \frac{1}{2\pi}\int_0^{2\pi} e^{-jx\cos(\theta-\phi)}\, d\theta. \tag{2.16-6}$$

Note that the integral does not depend on ϕ; in fact, ϕ is usually omitted from the definition. Substituting in Eq. (2.16-5) results in

$$\begin{aligned} F(\rho\cos\phi, \rho\sin\phi) &= 2\pi \int_0^{\infty} rf(r) J_0(2\pi r\rho)\, dr \\ &\equiv \tilde{F}(\rho), \end{aligned} \tag{2.16-7}$$

which is, of course, again independent of ϕ and therefore circularly symmetric. Equation (2.16-7) is known as a *Hankel transform*, and we use the shorthand notation

$$\tilde{F}(\rho) = Hf(r). \tag{2.16-8}$$

It is easily shown that the inverse transform is also a Hankel transform:

$$f(r) = H\tilde{F}(\rho). \tag{2.16-9}$$

Although the Bessel functions are not as familiar as, say, the exponential or trigonometric functions, they have been extensively studied and tabulated. Some Bessel function properties are given in Chapter 7; a much more extensive list, especially of integrals involving Bessel functions, is contained in the mathematical tables of Ref. [2-15] and elsewhere. Thus, although the Hankel transform looks somewhat formidable, it can, in fact, be easily evaluated for simple functions $f(r)$.

Example 1

$$f(r) = \frac{1}{r}, \qquad r > 0.$$

Then

$$\tilde{F}(\rho) = 2\pi \int_0^\infty \left(\frac{1}{r}\right) rJ_0(2\pi r\rho)\, dr$$
$$= \frac{1}{\rho}\int_0^\infty J_0(\alpha)\, d\alpha = \frac{1}{\rho}. \tag{2.16-10}$$

The last result follows from [2-15, Eq. (11.4-17)], by which

$$\int_0^\infty J_v(\alpha)\, d\alpha = 1, \qquad \text{for Re } v > -1.$$

Example 2

$$f(r) = \text{circ}\left(\frac{r}{a}\right) = \begin{cases} 1, r < a, \\ 0, \text{otherwise}. \end{cases} \tag{2.16-11}$$

This is the circularly symmetric analogue of the rect function of Sec. 2.9-7. The Hankel transform is

$$\tilde{F}(\rho) = 2\pi \int_0^a rJ_0(2\pi r\rho)\, dr. \tag{2.16-12}$$

We find in [2-15, Eq. (11.3-20)] the relation

$$\int_0^x rJ_0(x)\, dx = xJ_1(x), \tag{2.16-13}$$

where $J_1(x)$ is the first-order Bessel function of the first kind. Hence, the 2-D Fourier transform of the circ function is

$$\tilde{F}(\rho) = 2\pi a^2 \frac{J_1(2\pi\rho a)}{2\pi\rho a}. \tag{2.16-14}$$

The function $J_1(x)$ is an oscillatory function that looks a little like sin x near $x = 0$; it is an odd function of x and it is zero for $x = 0$ (see Fig. 7.3-2). The function $J_1(x)/x$ is therefore somewhat similar to sinc x; in particular, it has a peak of 1 at $x = 0$ and then decays in an oscillatory fashion. A sketch of $f(r)$ and $\tilde{F}(\rho)$ is shown in Fig. 2.16-1. Note the reciprocal relation between the variation in r and ρ that follows from the Fourier scaling rule: for large a, the Hankel transform has a narrow tall peak; for small a, it has a wide, low peak. This is analogous to the 1-D results.

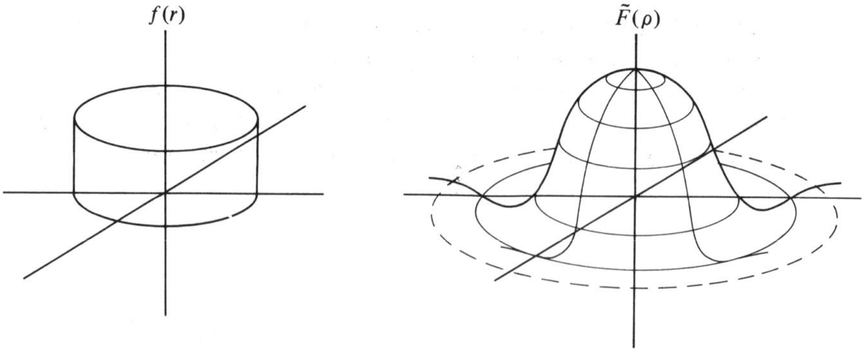

Figure 2.16-1. The circ function and its Hankel transform.

Another simple example is

$$f(r) = \delta(r - r_0) \tag{2.16-15}$$

This is a cylindrical surface of infinite height at the radius $r = r_0$. The Hankel transform is easily seen to be $2\pi r_0 J_0(2\pi r_0 \rho)$. Reference to Fig. 7.3-2 shows that $J_0(x)$ behaves a little like a damped cosine; hence the Hankel transform is like a damped cosine wave whose "frequency" is proportional to r_0. Other simple examples are left as exercises.

2.17 SUMMARY

In this chapter we explored the relationship between time and frequency as embodied in the Fourier series and the Fourier integral. Conversion to the frequency domain by one of these two methods is convenient because, if one knows the Fourier spectrum of the input to a linear system, one can find the output spectrum by simply multiplying the input spectrum by the transfer function or frequency response of the system. The use of both the Fourier series and the Fourier integral is simplified by the use of a number of standard properties and theorems, which generally make it relatively easy to find the Fourier transforms of arbitrary functions. These theorems and some of the more commonly used Fourier transform pairs were summarized in Tables 2.9-1 and 2.12-1. Although the Fourier series is generally an expansion of a function in terms of trigonometric functions (sines and cosines) or exponential functions, the idea can be generalized to arbitrary orthogonal or orthonormal functions. The implications of this generalization and some of the more useful properties of orthogonal-function expansions were discussed in Secs. 2.6 and 2.7. Impulse functions are important in Fourier analysis because, among other things, they provide a bridge between the Fourier series and the Fourier integral. These functions and their applications were discussed in Secs. 2.11 and 2.12. We discussed the complex analytic signal representation and the Hilbert transform, which permit the two-sided spectra of ordinary (real) signals to be replaced by single-sided spectra. Finally, we showed how the one-dimensional Fourier transform can be generalized to two or more dimensions.

PROBLEMS

2-1. Assume that a system is characterized by an operator $\mathcal{H}_t$ that relates output $y(t)$ to input $x(t)$ according to

$$y(t) = \mathcal{H}_t[x(\tau)].$$

If $\mathcal{H}_t$ is causal, meaning that $y(t)$ doesn't depend on future values of $x(t)$, what constraints must be applied to the time interval during which values of $x(t)$ contribute to $y(t)$?

2-2. A given system, described by an operator $\mathcal{H}_t$ gives the response

$$\begin{aligned} y(t) &= \mathcal{H}_t[x(\tau)] \\ &= x(\tau^3). \end{aligned}$$

Is the system time invariant?

2-3. A particular zero-memory system is described by

$$y(t) = \mathcal{H}_t[x(\tau)]$$
$$= \alpha x(\tau) + \beta.$$

Is the system linear? Is it time invariant?

2-4. Compute the Fourier coefficients c_n for all n in the representation

$$x(t) = \sum_{n=-\infty}^{\infty} c_n e^{jn\omega_0 t}$$

for the waveform shown in Fig. P2-4. Plot the magnitude and phase of c_n.

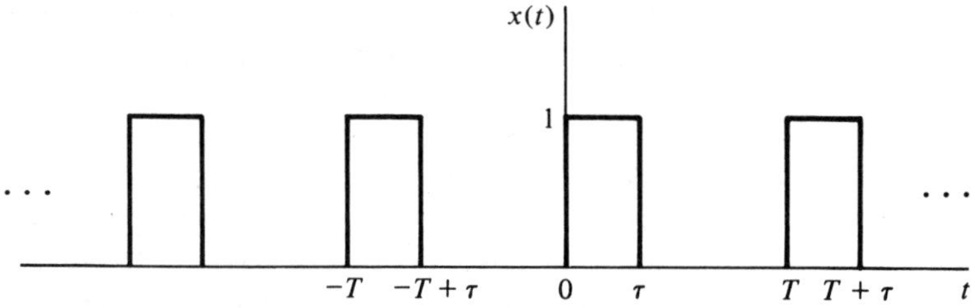

Figure P2-4

2-5. Repeat Prob. 2-4 for the waveform shown in Fig. P2-5.

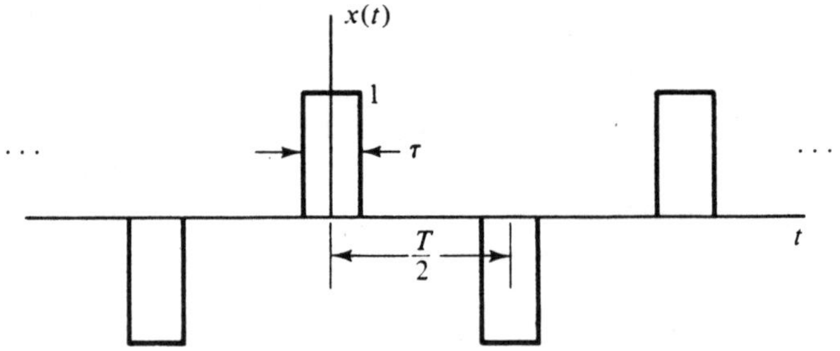

Figure P2-5

2-6. Repeat Prob. 2-4 for the waveform shown in Fig. P2-6. Use any properties of the Fourier series that can simplify the computations. *Hint*: Consider differentiating $x(t)$.

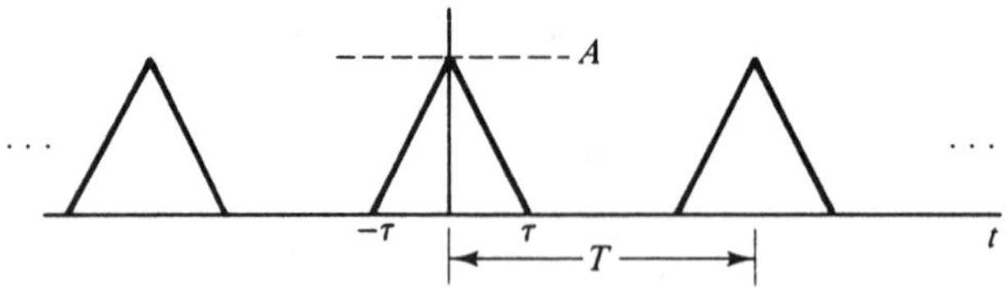

Figure P2-6

2-7. Use Eq. (2.4-22) and the appropriate Fourier series property to rapidly determine the Fourier coefficients (magnitude and phase) of the waveform shown in Fig. P2-7.

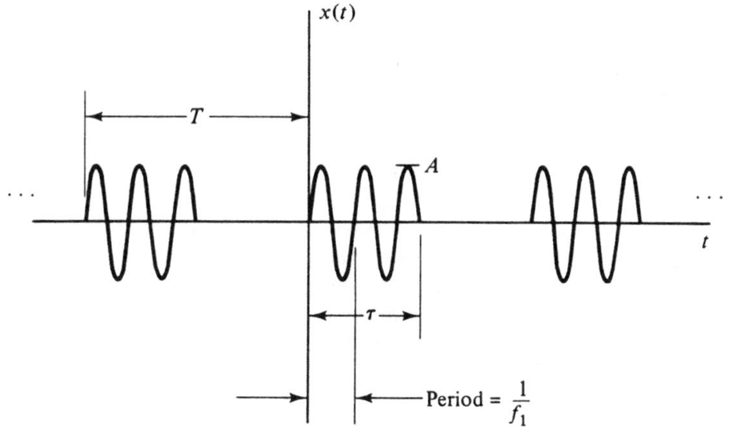

Figure P2-7

2-8. Consider the following three forms of the Fourier series,

$$x(t) = \sum_{n=-\infty}^{\infty} c_n e^{jn\omega_0 t} \qquad \text{(form 1)}$$

$$x(t) = c_0 + \sum_{n=1}^{\infty} x_n \cos(n\omega_0 t + \phi_n) \qquad \text{(form 2)}$$

$$x(t) = a_0 + \sum_{n=1}^{\infty} a_n \cos n\omega_0 t + b_n \sin n\omega_0 t \qquad \text{(form 3),}$$

and establish the relations between the coefficients. Suggest which form might be most convenient under what circumstance.

2-9. In form 3 of Prob. 2-8, what simplifications occur if

(a) $x(t)$ is odd: $x(t) = -x(-t)$?

(b) $x(t)$ is even: $x(t) = x(-t)$?

(c) $x(t)$ has half-wave odd symmetry:
$x(t) = -x[t + (T/2)]$?

2-10. Prove that the function

$$S_N(t) \equiv \sum_{n=-N}^{N} e^{jn\omega_0 t}$$

can be written in closed form as

$$S_N(t) = \frac{\sin(N + \frac{1}{2})\omega_0 t}{\sin(\omega_0/2)t}.$$

Carefully sketch the result and show the period. What happens when $N \to \infty$?

2.11. Show that $S_N(t)$ in Prob. 2-10 is approximately given by

$$S_N(t) \simeq 2N \operatorname{sinc} 2Nf_0 t$$

when t is near the origin, i.e., $t \simeq 0$ and N is large

2.12. In the circuit shown in Fig. P2-12 the diode is assumed ideal.

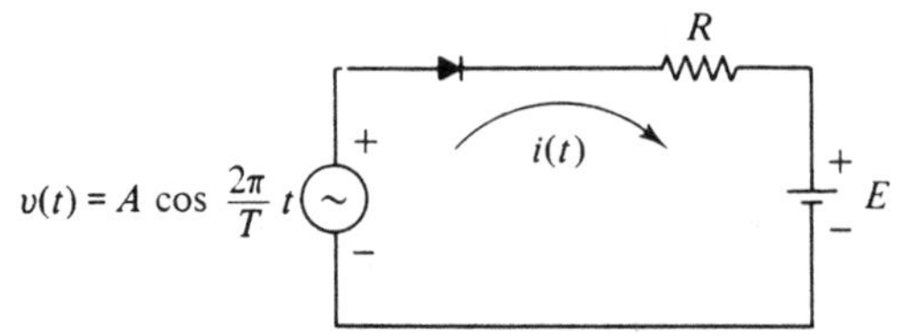

Figure P2-12

(a) With $A > E$, write an expression for the current $i(t)$.

(b) Compute the Fourier coefficients $\{c_n\}$ in the Fourier series representation of $i(t)$.

(c) Compute the average power dissipated in the resistor at frequencies n/T, $n = 0, 1, 2, \ldots$. How many of the first n terms account for 90% of the total power dissipated if $A = 100$ volts and $E = 50$ volts?

2-13. Obtain by inspection, a Fourier series representation for

$$x(t) = 2 \cos \omega_0 t + 5 \cos 2\omega_0 t + 4.$$

2-14. A complete orthonormal set $\{\phi_i(t)\}$ over the interval $T = [-1, 1]$ can be obtained by applying the Gram-Schmidt procedure to the sequence $1, t, t^2, t^3, \ldots$. Show that

$$\phi_0(t) = \frac{1}{\sqrt{2}}, \quad \phi_1(t) = \left(\frac{3}{2}\right)^{1/2} t, \quad \phi_2(t) = \left(\frac{5}{2}\right)^{1/2}\left(\frac{3}{2}t^2 - \frac{1}{2}\right), \quad \text{etc.}$$

The resultant functions

$$P_n(t) \equiv \left(\frac{2}{2n+1}\right)^{1/2} \phi_n(t)$$

are known as the Legendre polynomials.

2-15. Work out the details leading up to Eq. (2.6-12), and show that the generalized Fourier coefficients are indeed optimum for minimizing the mean-square error in representing a function $x(t)$.

2-16. **(a)** Show that for the Fourier coefficients

$$c_n = \frac{1}{T}\int_{-T/2}^{T/2} x(t)e^{-jn\omega_0 t}\, dt, \qquad n = 0, \pm 1, \ldots,$$

Parseval's theorem takes the form

$$\frac{1}{T}\int_{-T/2}^{T/2} |x(t)|^2\, dt = \sum_{n=-\infty}^{\infty} |c_n|^2,$$

where $x(t)$ has period T.

(b) What is the average power per cycle for the waveform

$$x(t) = \sum_{n=-\infty}^{\infty} \text{rect}\left(\frac{t - nT}{\tau}\right), \qquad \tau < T?$$

(c) Can you suggest a way in which Parseval's theorem can be used to evaluate the series $\sum_{n=-\infty}^{\infty} |c_n|^2$?

2-17. Instead of the Euclidean distance given in Eq. (2.7-8), consider the quantity

$$q(\mathbf{x}, \mathbf{y}) = \sum_{n=1}^{N} |x_n - y_n|.$$

Show that $q(\mathbf{x}, \mathbf{y})$ satisfies the properties of a distance and hence can be considered as such.

2.18. Starting with the three waveforms shown in Fig. 2.7-3, show that

$$\phi_1(t) = \frac{1}{\sqrt{2}} s_1(t)$$

$$\phi_2(t) = \frac{1}{\sqrt{6}} [s_1(t) + 2s_2(t)]$$

and that the three signals are given by

$$s_1(t) = \sqrt{2}\phi_1(t)$$

$$s_2(t) = -\frac{1}{\sqrt{2}} \phi_1(t) + \frac{3}{\sqrt{6}} \phi_2(t)$$

$$s_3(t) = \frac{3}{\sqrt{2}} \phi_1(t) + \frac{3}{\sqrt{6}} \phi_2(t).$$

2-19. Let $x(t)$ and $X(f)$ form a Fourier transform pair. Prove Parseval's theorem,

$$\int_{-\infty}^{\infty} |x(t)|^2 \, dt = \int_{-\infty}^{\infty} |X(f)|^2 \, df,$$

and its generalization

$$\int_{-\infty}^{\infty} x(t) y^*(t) \, dt = \int_{-\infty}^{\infty} X(f) Y^*(f) \, df.$$

2-20. Consider the two signals

$$x_1(t) = \sqrt{2a_1}\, e^{-a_1 t} u(t)$$

$$x_2(t) = \sqrt{2a_2}\, e^{-a_2 t} u(t), \qquad a_1, a_2 > 0.$$

Assume that $a_1 \gg a_2$. How do the signals compare on the basis of energy? How do they compare on the basis of *energy density* $|X(f)|^2$? Which of the two signals would require greater bandwidth for transmission without distortion? Assume that distortion is insignificant when 90% or more of the energy is transmitted.

2-21. Consider the signal $x_T(t)$ defined by

$$x_T(t) = \begin{Bmatrix} x(t), & |t| < \dfrac{T}{2} \\ 0, & \text{otherwise} \end{Bmatrix} = x(t) \operatorname{rect}\left(\frac{t}{T}\right).$$

Let $X_T(f)$ denote the Fourier transform of $x_T(t)$. Write an expression for $X_T(f)$ in terms of $X(f)$. The above is known as *truncation in time*. Truncation in frequency is described by

$$X_W(f) = \begin{cases} X(f), & |f| < W \\ 0, & \text{otherwise} \end{cases} = X(f)\,\text{rect}\left(\frac{f}{2W}\right).$$

If $x_W(t)$ denotes the inverse transform of $X_W(f)$, describe $x_W(t)$ in terms of $x(t)$. Justify the term "smoothing" for these operations, i.e., that $X_T(f)$ is a smoothed form of $X(f)$ and $x_W(t)$ is a smoothed form of $x(t)$.

2-22. A generalization of the notion introduced in Prob. 2-21 is the following. Describe $x_s(t)$ by

$$x_s(t) = \begin{cases} x(t)s(t), & |t| < T, \\ 0, & \text{otherwise.} \end{cases}$$

Let $s(t) = [1 - (|t|/T)]\,\text{rect}\,(t/2T)$. Describe $X_s(f)$ in terms of $X(f)$ if $X_s(f)$ is the Fourier transform of $x_s(t)$.

2-23. Prove that the convolution operation is commutative, associative, and distributive; that is,

$$x * y = y * x,$$

$$x * (y * z) = (x * y) * z,$$

$$(x + y) * z = x * z + y * z.$$

2-24. Let the nth moment M_n, of $x(t)$ be defined by

$$M_n = \int_{-\infty}^{\infty} t^n x(t)\,dt, \qquad n = 0, 1, 2, \ldots .$$

Assume that all the moments exist. Show that

$$\left.\frac{d^n X(\omega)}{d\omega^n}\right|_{\omega=0} = (-j)^n M_n, \qquad n = 0, 1, 2, \ldots .$$

2-25. Consider a signal $x(t)$ that is zero for $t < 0$. Show that, if its transform is written as

$$X(f) = A(f) + jB(f),$$

then $x(t)$ can be written as

$$x(t) = 2\int_{-\infty}^{\infty} A(f) \cos 2\pi ft\,df$$

$$= -2\int_{-\infty}^{\infty} B(f) \sin 2\pi ft\,df.$$

What does this say about the dependency relationship between $A(f)$ and $B(f)$?

2-26. An interpolating function $p(t)$ has the property that $p(0) = 1$; $p(t) = p(t - k\tau)$, $k = 0, \pm 1, \ldots$; τ is the period; and $p(k\tau) = 0$ for $k \neq 0$. Consider the representation

$$x(t) = \sum_{k=-\infty}^{\infty} x(k\tau)p(t - k\tau),$$

where $\tau = [2W]^{-1}$, W is a parameter, and $p(t) = \text{sinc}\ 2Wt$. Show that this representation is adequate only if the Fourier transform of $x(t)$ is strictly zero outside $|f| > W$. *Hint*: Consider writing $x(t)$ as

$$x(t) = [x(t) \cdot \sum_{k=-\infty}^{\infty} \delta(t - k\tau)] * \text{sinc } 2Wt,$$

and consider its transform.

2-27. Compute the Fourier transforms of the following:

(a) $x(t) = e^{-at} \sin (bt) \cdot u(t),\ a > 0.$

(b) $x(t) = e^{-at} \cos (bt) \cdot u(t),\ a > 0.$

2-28. Let $x(t)$ be an arbitrary function, and let $x_c(t)$ be its "causal" part; that is,

$$x_c(t) = \begin{cases} x(t), & t \geq 0, \\ 0, & t < 0. \end{cases}$$

(a) Show that $x_c(t)$ can be written as

$$x_c(t) = \tfrac{1}{2}x(t)[1 + \text{sgn } t].$$

(b) Compute $X_c(f) \equiv \mathcal{F}[x_c(t)]$ in terms of the Fourier transform of $x(t)$.

(c) Let $x_c(t) = x(t)$; that is, $x(t) = 0$ for $t < 0$. Derive a relation between the real and imaginary part of $X(f)$.

2-29. Let $X_1(f) = \text{sinc } fT_1$ and $X_2(f) = \text{sinc } fT_2$, where $T_1 > T_2$. Compute $X_1(f) * X_2(f)$. What can you conclude with respect to smoothing one sinc function with another. *Hint*: Use the property that

$$\int_{-\infty}^{\infty} X_1(\lambda)X_2(f - \lambda)\, d\lambda = \int_{-\infty}^{\infty} x_1(t)x_2(t)e^{-j2\pi ft}\, dt.$$

2-30. Prove Eq. (2.12-24) by writing

$$g(t) \equiv \sum_{n=-\infty}^{\infty} \delta(t - nT)$$

and expanding $g(t)$ into an ordinary Fourier series; that is,

$$g(t) = \sum_{n=-\infty}^{\infty} c_n e^{j(2\pi/T)nt}.$$

Finally, take the Fourier transform of the Fourier series of $g(t)$ term by term.

2-31. Compute the Fourier transform of the periodic waveform shown in Fig. P2-31. There are $N = 50$ pulses altogether. Sketch the results carefully. Put the time origin where it might be most convenient, and assume that exactly the same cosine pattern appears in each pulse.

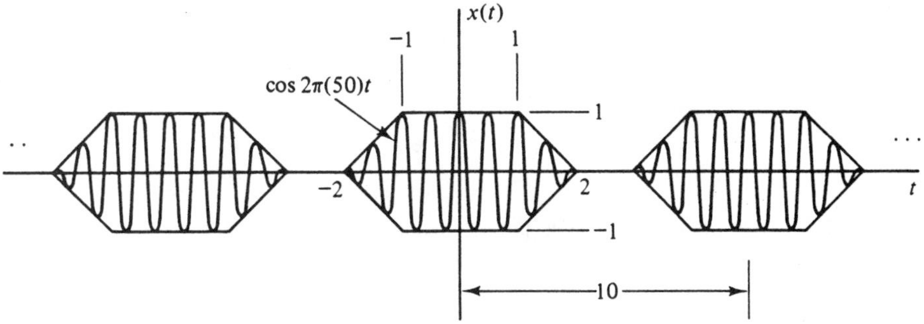

Figure P2-31

2-32. A sequence of N cosine pulses is shown in Fig. P2-32. The frequency of the cosine is f_0, the length of the pulse is a, and the pulse period is T. The string of pulses can be written in the form

$$x(t) = \text{rect}\left(\frac{t}{NT}\right)\left\{\text{rep}_T\left[\text{rect}\left(\frac{t}{a}\right)\cos 2\pi f_0 t\right]\right\}.$$

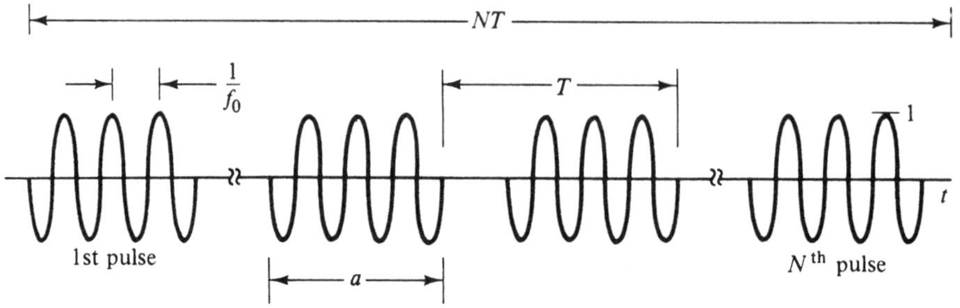

Figure P2-32

Find the Fourier transform, using the appropriate rules and Fourier pairs. Sketch the result if N is assumed to be fairly large and if $T \gg a \gg 1/f_0$.

2-33. Prove Eqs. (2.15-1) to (2.15-4).

2-34. Prove Eq. (2.15-6).

2-35. The two-dimensional Fourier transform $F(u, v)$ of a function $f(x, y)$ can be written in the form $F(u, v) = A(u, v) + jB(u, v)$, where $A(u, v)$ and $B(u, v)$ are both real. Show that if $f(x, y)$ is real, then

$$A(u, v) = A(-u, -v),$$

$$B(u, v) = -B(-u, -v),$$

$$A(-u, v) = A(u, -v),$$

$$B(-u, v) = -B(u, -v).$$

2-36. Instead of writing $F(u, v)$ in terms of real and imaginary components, as in Problem 2-35, one can also write it in terms of magnitude and phase; that is $F(u, v) = R(u, v) \exp[j\theta(u, v)]$. Show that for real $f(x, y)$

$$R(u, v) = R(-u, v),$$

$$R(-u, v) = R(u, -v).$$

Also show that

$$\theta(u, v) = -\theta(-u, v),$$

$$\theta(u, -v) = -\theta(-u, v).$$

2-37. Show that if $f(x, y)$ is purely imaginary then the real part, $A(u, v)$, and the imaginary part, $B(u, v)$, of its Fourier transform obey

$$A(u, v) = -A(-u, -v),$$

$$B(u, v) = B(-u, -v).$$

2-38. Prove Eq. (2.15-7).

2-39. Prove the Fourier identity, Eq. (2.15-8), and investigate the nature of the Gibbs phenomenon in two-dimensional space.

2-40. Show that the Hankel transfom of the function $\exp(-ar^2)$, Re $a > 0$, is given by $(\pi/a) \exp(-\pi^2\rho^2/a)$.

2-41. Show that the Hankel transform of the function $\exp(jar^2)$, Im $a > 0$, is given by $(j\pi/a) \exp(-j\pi^2\rho^2/a^2)$.

2-42. **(a)** Show that the convolution of the function circ (r/a) with itself is given by

$$\text{circ}\left(\frac{r}{a}\right) * \text{circ}\left(\frac{r}{a}\right) = \begin{cases} 2a^2\left[\cos^{-1}\left(\frac{r}{2a}\right) - \frac{r}{2a}\sqrt{1-\left(\frac{r}{2a}\right)^2}\right], & r \leq 2a. \\ 0, \quad r > 2a \end{cases}$$

(b) Sketch this function.

(c) What is the Hankel transform of this function?

REFERENCES

[2-1] B. O. Peirce and R. M. Foster, *A Short Table of Integrals*, 4th ed., Ginn, Boston, 1956.

[2-2] E. Jahnke and F. Emde, *Table of Functions*, Dover, New York, 1945.

[2-3] E. A. Guillemin, *The Mathematics of Circuit Analysis*, Wiley, New York, 1959, p. 496.

[2-4] A. Papoulis, *The Fourier Integral and Its Applications*, McGraw-Hill, New York, 1962.

[2-5] J. L. Hammond and R. S. Johnson, "Review of Orthogonal Square-Wave Functions and Their Application to Linear Networks," *J. Franklin Inst.*, **273,** pp. 211–225, March 1962.

[2-6] R. Courant and D. Hilbert, *Methods of Mathematical Physics*, Vol. 1, Wiley-Interscience, New York, 1953.

[2-7] W. R. LePage, *Complex Variables and the Laplace Transform for Engineers*, McGraw-Hill, New York, 1952, Chap. 9.

[2-8] B. Friedman, *Principles of Applied Mathematics*, Wiley, New York, 1956.

[2-9] P. M. Woodward, *Probability and Information Theory with Application to Radar*, McGraw-Hill, New York, 1952, Chap. 2.

[2-10] H. Stark, "An Extension of the Hilbert Transform Product Theorem," *Proc. IEEE*, **59,** No. 9, pp. 1359–1360, Sept. 1971.

[2-11] P. Bedrosian and H. Stark, "An Extension of the Hilbert Transform Product Theorem," *Proc. IEEE*, **60,** No. 2, pp. 228–229, Feb. 1972.

[2-12] J. W. Goodman, *Introduction to Fourier Optics*, McGraw-Hill, New York, 1968.

[2-13] H. Stark, ed. *Applications of Optical Fourier Transforms*, Academic, Orlando, Fla., 1982, Chapter 1.

[2-14] H. Bateman, *Tables of Integral Transforms*, Vol. II, McGraw-Hill, New York, 1954.

[2-15] M. Abramowitz and I. A. Stegun, eds. *Handbook of Mathematical Functions*, Dover, New York, 1965.

CHAPTER 3

Linear Circuits and Filters

3.1 INTRODUCTION

A filter is a device that passes certain frequencies of the input signal and rejects or attenuates others. In this chapter we deal with linear, lumped-parameter, time-invariant filters constructed from electric circuit elements such as resistors, capacitors, inductors, or operational amplifiers (op-amps). These devices are all assumed to be substantially linear and time invariant, and we shall ignore effects of distributed parameters (e.g., transmission-line effects in the lines connecting the components).

We start by reviewing some simple concepts from elementary linear circuit theory. We then consider the various filter types, such as low-pass, high-pass, and band-pass, and the theoretical limits placed on their performance by physical constraints such as realizability and causality. We then briefly describe a few of the standard ways in which ideal filters are approximated by practical circuits.

3.2 INPUT–OUTPUT USING SUPERPOSITION

In Chapter 2 we defined a linear system as one obeying the superposition principle. We can use this principle to obtain the output for any arbitrary input if we know the response of the system to certain test functions. To do this we express the arbitrary input in terms of these test functions and then use the superposition principle to express the output in terms of the elementary responses. In Chapter 2 we dealt with the expansion of the input signal in terms of sinusoids, in which case the output is determined from the frequency response of the system. This is the *frequency-domain* point of view.

Another, and in some ways more general, expansion of signals is in terms of damped sinusoids, or exponentials of the form e^{-st}, where $s = \sigma + j\omega$ (σ, ω real) is called a *complex frequency*. Such an expansion results in the *Laplace transform*. For functions $x(t)$ that are zero for $t < 0$, this has the one-sided form given by

$$X(s) = \int_0^\infty x(t)e^{-st}\,dt \tag{3.2-1}$$

and is seen to be a direct generalization of the Fourier transform considered in Chapter 2.

The Laplace transform is useful in the solution of differential equations and transient problems, while the Fourier transform is more appropriate for dealing with the kind of signals encountered in communication systems. This is the reason for emphasizing Fourier transform methods in this book. We assume that the reader is generally familiar with the Laplace transform and with the notion of complex frequency. We shall find the latter useful in Sec. 3.4.

In this section we consider a representation of the signal in terms of impulses, so the output is obtained from the superposition of impulse responses of the system. This is referred to as the *time-domain* approach.

In general, the response of a linear system to an impulse depends on the time at which the impulse was applied and the time elapsed. It is therefore necessary to use three different time parameters: the present time t, the time τ at which the impulse was applied, and a running time variable σ. For a linear system characterized by the transformation $\mathcal{H}_t$, the impulse response is

$$h(t, \tau) = \mathcal{H}_t[\delta(\sigma - \tau)] \equiv \int_{-\infty}^{\infty} h(t, \sigma)\delta(\sigma - \tau)\,d\sigma.$$

This is the response at time t to an impulse applied at time τ. If the system is *time invariant*, the response depends only on the time elapsed since the *application of the pulse*; therefore, for such a system

$$\mathcal{H}_t[\delta(\sigma - \tau)] = h(t - \tau), \tag{3.2-2}$$

or since the time origin is immaterial,

$$h(t) = \mathcal{H}_t[\delta(\sigma)]. \tag{3.2-3}$$

A *causal system* is one that cannot respond to a signal prior to its application. Hence, for causal systems

$$h(t, \tau) = 0, \qquad \text{for } t < \tau. \tag{3.2-4}$$

If an input $x(t)$ is applied to a linear system with impulse response $h(t, \tau)$, the output is given by the *superposition integral*

$$y(t) = \int_{-\infty}^{\infty} x(\sigma)h(t, \sigma)\, d\sigma. \tag{3.2-5}$$

This important relation is easily derived by considering the sifting-integral expression

$$x(\sigma) = \int_{-\infty}^{\infty} x(\xi)\delta(\xi - \sigma)\, d\xi. \tag{3.2-6}$$

Now replace $x(\sigma)$ in Eq. (3.2-5) by Eq. (3.2-6) to obtain

$$\begin{aligned} y(t) &= \int_{-\infty}^{\infty}\int_{-\infty}^{\infty} x(\xi)\delta(\xi - \sigma)h(t, \sigma)\, d\xi\, d\sigma \\ &= \int_{-\infty}^{\infty} d\xi x(\xi) \int_{-\infty}^{\infty} h(t, \sigma)\delta(\xi - \sigma)\, d\sigma. \end{aligned}$$

But, by the sifting property of the δ function,

$$\int_{-\infty}^{\infty} h(t, \sigma)\delta(\xi - \sigma)\, d\sigma = h(t, \xi). \tag{3.2-7}$$

Hence

$$y(t) = \int_{-\infty}^{\infty} x(\xi)h(t, \xi)\, d\xi \tag{3.2-8}$$

which is identical with Eq. (3.2-5) since ξ and σ are merely dummy variables inside the integral. In Eq. (3.2-6) the integrand $x(\xi)\delta(\xi - \sigma)$ can be regarded as an impulse at time σ with area proportional to $x(\xi)$. Then Eq. (3.2-8) can be regarded as the summation or superposition of the responses to all of these impulses; hence the name *superposition integral*.

Strictly speaking, neither Eq. (3.2-6) nor (3.2-8) is valid at points where $x(\xi)$ is discontinuous. However, both equations hold on either side of any discontinuity. The failure of the representation at points of discontinuity is therefore, in general, unimportant. In particular, it often happens that $y(t)$ is continuous even though $x(\xi)$ is not. In this case, Eq. (3.2-8) is valid even at points of discontinuity of $x(\xi)$.

If the linear system $\mathcal{H}_t$ is time invariant, we can use Eq. (3.2-2) in Eq. (3.2-5) to write

$$y(t) = \int_{-\infty}^{\infty} x(\tau)h(t - \tau)\, d\tau \tag{3.2-9}$$

where we have changed the variable of integration from σ to the more traditional τ. This will be our practice in the remainder of this section since the three different time parameters are no longer needed.

Thus, in this case the output is the *convolution* of the input and the impulse response, and we can use the shorthand notation

$$y(t) = x(t) * h(t). \tag{3.2-10}$$

If the linear system is causal, the integrand in Eq. (3.2-8) is zero for all $\tau > t$. Hence the upper limit of integration is effectively t, giving the result

$$y(t) = \int_{-\infty}^{t} x(\tau)h(t, \tau)\, d\tau,$$

or, for time-invariant, causal systems,

$$y(t) = \int_{-\infty}^{t} x(\tau)h(t - \tau)\, d\tau. \tag{3.2-11}$$

Finally, it is often assumed that the input is zero prior to some time, say, $t = 0$ so that $x(\tau) = 0$ for $\tau < 0$. This leads to

$$y(t) = \int_{0}^{t} x(\tau)h(t, \tau)\, d\tau \tag{3.2-12}$$

or

$$y(t) = \int_{0}^{t} x(\tau)h(t - \tau)\, d\tau \tag{3.2-13}$$

for time-varying and time-invariant systems, respectively.

Example

See Fig. 3.2-1. Suppose that

$$x(t) = a \operatorname{rect}\left[\frac{t - T/2}{T}\right], \qquad T < 1$$

$$h(t) = \begin{cases} 0, & t \le 0, \\ 1 - t, & 0 < t \le 1, \\ 0, & t > 1. \end{cases}$$

The factors $x(\tau)$ and $h(t - \tau)$ in the integrand of Eq. (3.2-13) are shown in Fig. 3.2-1(c) for $0 < t < T$. There are five distinct regimes:

For $t < 0$: $y(t) = 0$;

For $0 \leq t \leq T$:
$$y(t) = a \int_0^t (1 - t + \tau)\, d\tau = a\left(t - \frac{t^2}{2}\right);$$

For $T \leq t \leq 1$:
$$y(t) = a \int_0^T (1 - t + \tau)\, d\tau = a\left(T - tT + \frac{T^2}{2}\right);$$

For $1 \leq t \leq T + 1$:
$$y(t) = a \int_{t-1}^T (1 - t + \tau)\, d\tau = \left(\frac{a}{2}\right)[t - (T + 1)]^2;$$

For $T + 1 \leq t$: $y(t) = 0$.

The result is shown in Fig. 3.2-1(d).

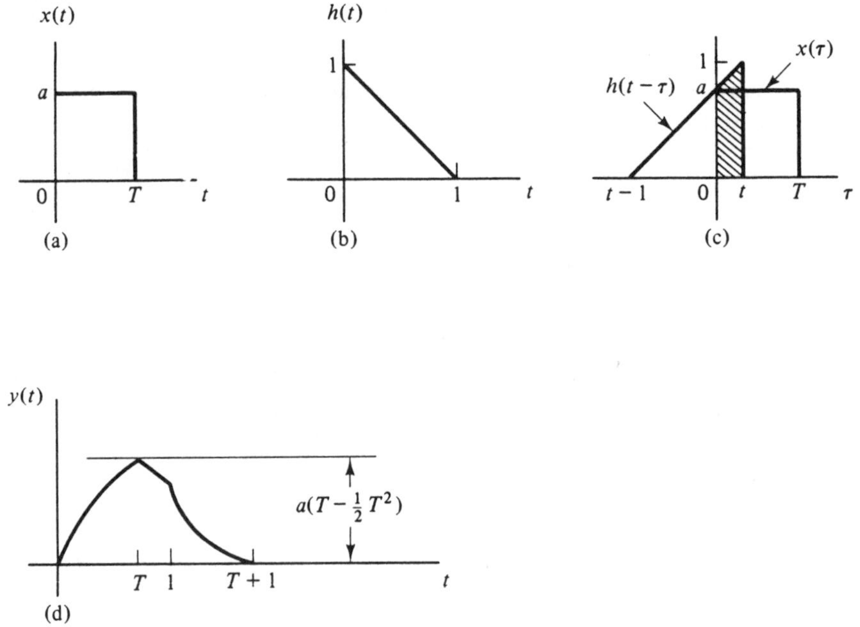

Figure 3.2-1. Example of an input $x(t)$ applied to a time-invariant linear system with impulse response $h(t)$. (a) The input. (b) The impulse response. (c) The two factors $x(\tau)$ and $h(t - \tau)$ involved in the convolution integral for $0 < t < T$ (the output at time t is proportional to the shaded area). (d) The output.

3.3 DIFFERENTIAL EQUATIONS FOR LINEAR TIME-INVARIANT SYSTEMS

The representation of linear systems in terms of the impulse response is not the only possibility, and for many purposes a representation using differential equations is more convenient.

A lumped-parameter, linear, time-invariant system is described by an ordinary linear differential equation with constant coefficients. By contrast, a distributed-parameter system gives rise to a system description in terms of partial differential equations, and a time-varying system is represented by a differential equation with nonconstant (i.e., time-dependent) coefficients.

In practice, a very wide class of electric circuits can be modeled as consisting of essentially fixed resistors, capacitors, inductors, and operational amplifiers that are linear, time-invariant, lumped-parameter elements. Although we expect the reader to be generally familiar with the analysis of such circuits, we present a brief review here. This will serve to introduce our notation and also permit us to discuss some of the properties of the resulting differential equations.

A simple circuit consisting of a resistor, inductor, and capacitor in series is shown in Fig. 3.3-1. We assume that the circuit is driven by a source of voltage $x(t)$. The source has zero impedance; that is, its voltage is not affected by any current drawn by the network. We are interested in the open-circuit voltage across the capacitor C.

The integrodifferential equation for the current in this circuit is obtained from the Kirchhoff voltage law, which gives

$$x = Ri + L\frac{di}{dt} + \frac{1}{C}\int i\,dt. \tag{3.3-1}$$

Also, the desired output voltage is

$$y = \frac{1}{C}\int i\,dt. \tag{3.3-2}$$

(Note that x, y, and i are all functions of time; the time dependence is suppressed to simplify the appearance of the equations.) We can eliminate the variable i from the two equations by differentiating Eq. (3.3-2). This gives

$$i = C\frac{dy}{dt}, \tag{3.3-3}$$

$$\frac{di}{dt} = C\frac{d^2y}{dt^2}. \tag{3.3-4}$$

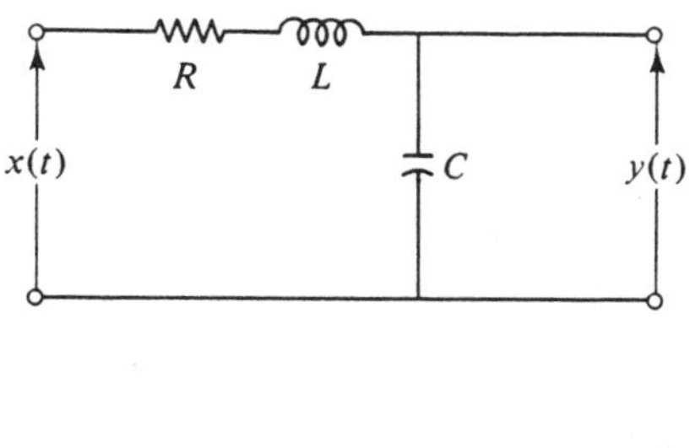

Figure 3.3-1. Series *RLC* circuit.

Substituting these back into Eq. (3.3-1) results in the second-order differential equation

$$LC\frac{d^2y}{dt^2} + RC\frac{dy}{dt} + y = x. \tag{3.3-5}$$

The differential equations for more complicated circuits are obtained in a similar way, and all such circuits can be represented by a differential equation of the general form

$$\sum_{i=0}^{n} a_i \frac{d^iy}{dt^i} = \sum_{i=0}^{k} b_i \frac{d^ix}{dt^i}, \tag{3.3-6}$$

where y is the output and x the input. The *degree* of the equation is n, and for most practical systems $k \leq n$. (A reason for this will appear later.)

Solution in Terms of Exponential Functions

The general solution of differential equations of the form of Eq. (3.3-6) consists of two parts: the complementary or transient solution and the particular or driven solution. The transient solution is obtained by setting the right side of Eq. (3.3-6) equal to zero and assuming a solution in the form

$$y(t) = Ce^{st}, \tag{3.3-7}$$

where C and s are arbitrary complex numbers. Substitution of Eq. (3.3-7) into the equation results in

$$C\sum_{i=0}^{n} a_i s^i e^{st} = 0. \tag{3.3-8}$$

If this is to be true for all t, then $e^{st} \neq 0$. Also, $C = 0$ yields $y(t) = 0$, which, although perfectly valid, is usually not of interest. Hence, for nontrivial solutions

$$\sum_{i=0}^{n} a_i s^i = a_n s^n + \cdots + a_0 = 0. \tag{3.3-9}$$

This algebraic equation is called the *characteristic equation*. It has n roots $s = \lambda_1, s = \lambda_2, \ldots, s = \lambda_n$, called *characteristic roots*. The function $Ce^{\lambda_i t}$ is a solution of the differential equation if λ_i is one of the characteristic roots. Since the differential equation is linear, any linear combination of functions $C_i e^{\lambda_i t}$, where the λ_i, $i = 1, \ldots, n$, are different characteristic roots, will also be a solution. Hence, the complete complementary solution is

$$y = C_1 e^{\lambda_1 t} + C_2 e^{\lambda_2 t} + \cdots + C_n e^{\lambda_n t}. \tag{3.3-10}$$

The coefficients C_i, $i = 1, \ldots, n$, are arbitrary and are not directly determined by the differential equation. They are instead determined by the initial or "boundary" conditions. If there is to be a unique solution for all the n coefficients $\{C_i\}$, there must be n independent boundary conditions.

The nature of the complementary solution of the differential equation is determined by the characteristic roots. If any one root λ_i has a positive real part, the corresponding $e^{\lambda_i t}$ will grow exponentially with time. Although practical systems cannot permit such growth indefinitely and eventually saturate, any system in which a small initial disturbance grows with time is termed *unstable*. Thus in a *stable* system none of the characteristic roots can have positive real values.

The coefficients a_i and b_i of real systems are real [e.g., combinations of R, L, or C as in Eq. (3.3-5)]. Hence any complex characteristic roots must appear in complex conjugate pairs: If λ_i is a complex root, λ_i^* is also a root. The corresponding coefficients $\{C_i\}$ must also occur as complex conjugate pairs if the boundary conditions and the subsequent system response are to be real functions of time. Suppose that a pair of complex conjugate characteristic roots is $\alpha \pm j\beta$ and that the respective coefficients are $A \pm jB$. The corresponding component of the response will be

$$\begin{aligned}(A + jB)e^{(\alpha+j\beta)t} + (A - jB)e^{(\alpha-j\beta)t} &= 2e^{\alpha t}(A \cos \beta t - B \sin \beta t)\\ &= 2\sqrt{A^2 + B^2}\, e^{\alpha t} \cos\left(\beta t + \tan^{-1}\frac{B}{A}\right). \qquad (3.3\text{-}11)\end{aligned}$$

If α is positive, this kind of response represents an oscillation whose amplitude grows exponentially with time, and it constitutes a form of instability, as already mentioned. If $\alpha = 0$, the output contains an oscillating component with constant amplitude, which is generally also regarded as a form of instability. Hence one requires all the roots to have negative real parts for stability, or equivalently the roots should lie in the *left half* of the complex plane.

3.4 THE TRANSFER FUNCTION

The response of the system to the input is given by the *particular solution* of the differential equation. For stable systems the transient response given by the complementary solution eventually dies out, and therefore the particular solution is also the *steady-state* solution. In most applications to communications, one is interested mainly in the steady-state solution.

If the Fourier transforms of the functions $x(t)$ and $y(t)$ exist, the particular solution can be obtained by Fourier-transforming both sides of Eq. (3.3-6). This results in[†]

$$Y(j\omega) \sum_{i=0}^{n} a_i(j\omega)^i = X(j\omega) \sum_{i=0}^{k} b_i(j\omega)^i$$

[†] In this chapter we frequently write Fourier transforms with arguments of $j\omega$ rather than ω, that is, $H(j\omega)$ instead of $H(\omega)$, because we often switch between variables s and $j\omega$. Use of ω would necessitate switching between ω and $-js$. Also, instead of $H(j2\pi f)$ we frequently simply use $H(f)$. The meaning of the notation can generally be inferred from the context, and there should be no confusion.

or

$$\frac{Y(j\omega)}{X(j\omega)} \equiv H(j\omega) = \frac{\sum_{i=0}^{k} b_i(j\omega)^i}{\sum_{i=0}^{n} a_i(j\omega)^i}. \tag{3.4-1}$$

In communication theory, $H(j\omega)$ is frequently taken as the transfer function of the system. More generally, however, the transfer function is defined in terms of the complex variable $s = \sigma + j\omega$; that is,

$$\frac{Y(s)}{X(s)} \equiv H(s) = \frac{\sum_{i=0}^{k} b_i s^i}{\sum_{i=0}^{n} a_i s^i}. \tag{3.4-2}$$

Note that this result can be obtained directly by taking Laplace [Eq. (3.2-1)] rather than Fourier transforms of $x(t)$ and $y(t)$ in Eq. (3.3-6). For linear, time-invariant, lumped-parameter systems, the transfer function is always in the form of a ratio of two polynomials in the variable $s = j\omega$. The denominator polynomial can be represented in the form

$$\sum_{i=0}^{n} a_i s^i = a_n(s - \lambda_1)(s - \lambda_2), \ldots, (s - \lambda_n), \tag{3.4-3a}$$

where the numbers $\lambda_1, \lambda_2, \ldots, \lambda_n$ are the n characteristic roots already referred to. Similarly, the numerator polynomial can be put into the form

$$\sum_{i=0}^{k} b_i s^i = b_k(s - \mu_1)(s - \mu_2) \cdots (s - \mu_k), \tag{3.4-3b}$$

where the numbers $\mu_1, \ldots, \mu_k$ are the roots of the equation $\sum_{i=0}^{k} b_i s^i = 0$. Since $H(s) = 0$ for $s = \mu_i$, the μ's are referred to as the *zeros* of the transfer function. Similarly, $H(s)$ will be infinite if s takes on one of the values $\lambda_1, \ldots, \lambda_n$; these are referred to as the *poles* of the transfer function. The poles are identical to the characteristic roots.

If we substitute Eqs. (3.4-3a) and (3.4-3b) with $s = j\omega$ into Eq. (3.4-1), we find that

$$H(j\omega) = \frac{b_k(j\omega - \mu_1) \cdots (j\omega - \mu_k)}{a_n(j\omega - \lambda_1) \cdots (j\omega - \lambda_n)}.$$

It follows that, except for the scale factor b_k/a_n, the transfer function, and therefore the steady-state behavior of the system, is completely determined by the zeros and poles.

Qualitatively, we see that if a pole λ_i is complex (i.e., $\lambda_i = \sigma_i + j\omega_i$) and if $\sigma_i \ll \omega_i$, then $|H(j\omega)|$ will tend to peak for $\omega = \omega_i$. Similarly, a complex zero in which the imaginary part is much larger than the real part tends to cause a notch in the frequency response. The effect of pole and zero location on frequency response is discussed in more detail in Sec. 3.5.

We have already seen that if the poles of a transfer function of a real system are complex they must occur as complex conjugate pairs. This is true also of the zeros and for the same reason. A stable system must have all its poles in the left half-plane. There is no such restriction on the location of the zeros; but if none of the zeros is in the right half-plane, the transfer function is of the *minimum-phase* type. The meaning of this notion will be explained shortly.

Properties of the Transfer Function

The transfer function of a real system (i.e., one where the coefficients of the differential equation are real) evaluated for real frequencies ($s = j\omega$) must have Hermitian symmetry; that is,

$$|H(j\omega)| = |H(-j\omega)| \tag{3.4-4}$$

and

$$\angle H(j\omega) = -\angle H(-j\omega). \tag{3.4-5}$$

This follows directly from the fundamental definition, Eq. (3.4-1). Hence, if we set

$$H(j\omega) = A(j\omega) + jB(j\omega), \qquad A, B \text{ real}, \tag{3.4-6}$$

then

$$A(j\omega) = A(-j\omega), \qquad B(j\omega) = -B(-j\omega), \tag{3.4-7}$$

and

$$|H(j\omega)| = [A^2(j\omega) + B^2(j\omega)]^{1/2}. \tag{3.4-8}$$

The transfer function $H(j\omega)$ is identical with the frequency response of the system. We show this formally by considering an input $x(t) = \cos 2\pi f_0 t$. For simplicity we write $H(f)$ instead of $H(j2\pi f)$. Then, as shown in Chapter 2,

$$X(f) = \tfrac{1}{2}[\delta(f - f_0) + \delta(f + f_0)], \tag{3.4-9}$$

and therefore

$$\begin{aligned} Y(f) &= \frac{H(f)}{2}[\delta(f - f_0) + \delta(f + f_0)] \\ &= \frac{H(f_0)}{2}\delta(f - f_0) + \frac{H(-f_0)}{2}\delta(f + f_0). \end{aligned} \tag{3.4-10}$$

Because of the Hermitian symmetry of the transfer function, this can be put into the form

$$Y(f) = \frac{A(f_0)}{2}[\delta(f - f_0) + \delta(f + f_0)] - \frac{B(f_0)}{2j}[\delta(f - f_0) - \delta(f + f_0)]. \tag{3.4-11}$$

Finally, to get the steady-state time response, we inverse-transform to get

$$\begin{aligned} y(t) = \mathcal{F}^{-1}[Y(f)] &= A(f_0)\cos 2\pi f_0 t - B(f_0)\sin 2\pi f_0 t \\ &= |H(f_0)|\cos[2\pi f_0 t + \angle H(f_0)]. \end{aligned} \tag{3.4-12}$$

Thus the steady-state output is a sinusoid of the same frequency as the input, but with amplitude and phase determined by the magnitude and phase, respectively, of the transfer function evaluated at the frequency of the applied signal.

The fact that the transfer function and the impulse response of a linear time-invariant system are Fourier transforms of each other, that is, that

$$H(j\omega) = \mathcal{F}[h(t)] \tag{3.4-13a}$$

and

$$h(t) = \mathcal{F}^{-1}[H(j\omega)] \tag{3.4-13b}$$

follows directly from Eq. (3.4-1) by assuming $X(j\omega)$ to be the Fourier transform of the impulse function. This gives $X(j\omega) = 1$, $Y(j\omega) = H(j\omega)$, and therefore

$$y(t) = h(t) = \mathcal{F}^{-1}[H(j\omega)].$$

Equation (3.4-13a) follows by the uniqueness of the Fourier transform.

Bode Plots

There are a number of simple methods for obtaining the frequency response from the poles and zeros of the transfer function. Perhaps the most convenient is the asymptotic log-amplitude or Bode plot.† To develop this, we write the transfer function in the form

$$\begin{aligned} H(j\omega) &= \frac{b_k(j\omega - \mu_1)(j\omega - \mu_2)\cdots(j\omega - \mu_k)}{a_n(j\omega - \lambda_1)(j\omega - \lambda_2)\cdots(j\omega - \lambda_n)} \\ &= K\frac{\left(\frac{j\omega}{-\mu_1}+1\right)\left(\frac{j\omega}{-\mu_2}+1\right)\cdots\left(\frac{j\omega}{-\mu_k}+1\right)}{\left(\frac{j\omega}{-\lambda_1}+1\right)\left(\frac{j\omega}{-\lambda_2}+1\right)\cdots\left(\frac{j\omega}{-\lambda_n}+1\right)}, \end{aligned} \tag{3.4-14}$$

where

$$K = \frac{b_k(-\mu_1)(-\mu_2)\cdots(-\mu_k)}{a_n(-\lambda_1)(-\lambda_2)\cdots(-\lambda_n)}$$

is a constant. Because Eq. (3.4-14) involves a ratio of products of simple factors, it is convenient to use a logarithmic representation of the form

$$H(j\omega) = e^{\alpha(\omega)+j\beta(\omega)}$$

so that

$$\alpha(\omega) = \ln|H(j\omega)| \tag{3.4-15}$$

† After H. W. Bode [3-2].

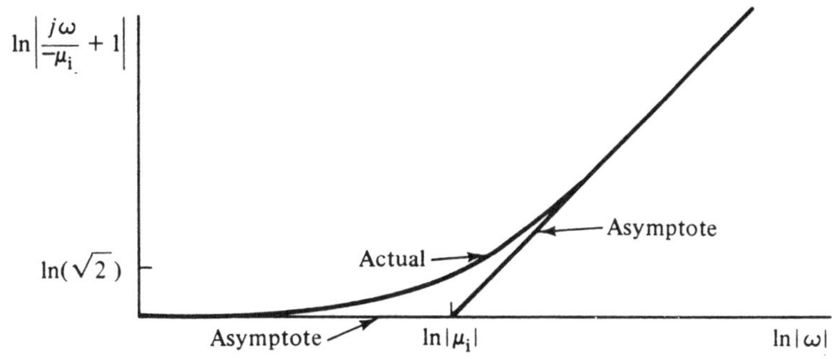

Figure 3.4-1. Asymptotic and actual behavior of the term $\ln |(j\omega/|\mu_i|) + 1|$.

and

$$\beta(\omega) = \angle H(j\omega).^{\dagger} \tag{3.4-16}$$

If we take the logarithm of both sides of Eq. (3.4-14), we find that

$$\alpha(\omega) = \ln |H(j\omega)| = \ln |K| + \sum_{i=1}^{k} \ln \left| \frac{j\omega}{-\mu_i} + 1 \right| - \sum_{i=1}^{n} \ln \left| \frac{j\omega}{-\lambda_i} + 1 \right| \tag{3.4-17}$$

and

$$\beta(\omega) = \angle H(j\omega) = \sum_{i=1}^{k} \angle \left(\frac{j\omega}{-\mu_i} + 1 \right) - \sum_{i=1}^{n} \angle \left(\frac{j\omega}{-\lambda_i} + 1 \right) \cdot \tag{3.4-18}$$

Suppose initially that the μ_i and λ_i are negative real numbers. A typical term of the form $\ln |(j\omega/-\mu_i) + 1|$ approaches zero as $|\omega/\mu_i| \to 0$, and it approaches $\ln |\omega/\mu_i| = \ln |\omega| - \ln |\mu_i|$ asymptotically for very large $|\omega/\mu_i|$. The two asymptotes and the actual magnitude are shown in Fig. 3.4-1. If logarithmic scales are used for both the magnitude and the frequency, the asymptotic plot consists of two straight lines that meet at the point $\omega = |\mu_i|$. For frequencies greater than $|\mu_i|$, the line has a slope of $+1$. For frequencies below $|\mu_i|$, the slope is zero. The frequency $\omega = |\mu_i|$ is called the *break frequency*. If $-\mu_i$ is a positive real number, the actual magnitude of the term $[(j\omega/-\mu_i) + 1]$ never departs very far from the asymptotes, and at the break point it is equal to $\sqrt{2}$. Hence, for many purposes a plot of only the asymptotes gives an adequate picture of the magnitude of the frequency response. Such a plot is easily sketched on log-log paper by adding lines of the type shown in Fig. 3.4-1, paying due attention to sign. As an example we show a Bode diagram for the function

$$H(j\omega) = 10 \frac{(j\omega/1) + 1}{[(j\omega/2) + 1][(j\omega/10 + 1]} \tag{3.4-19}$$

in Fig. 3.4-2.

† In transmission-line theory, the function $\gamma(\omega) \equiv \alpha(\omega) + j\beta(\omega)$ is called the *propagation function*; $\alpha(\omega)$ is the *attenuation function* and $\beta(\omega)$ the *phase function*.

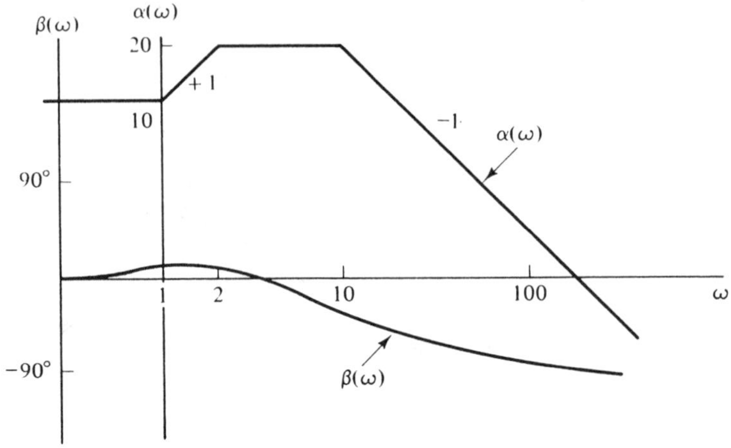

Figure 3.4-2. Asymptotic log-magnitude plot (Bode diagram) and the corresponding minimum phase lag curve for the transfer function $H(j\omega) = 10\ [(j\omega/1) + 1]/\{[(j\omega/2) + 1][(j\omega/10) + 1]\}$.

For negative real μ_i, the phase shift contributed by a typical term $[(j\omega/-\mu_i) + 1]$ is $\tan^{-1}(\omega/|\mu_i|)$. Thus the phase-shift curve can easily be sketched on semilog paper (logarithmic abscissa, linear ordinate scale) by adding typical terms. The phase curve for the function $H(j\omega)$ in Eq. (3.4-19) is also shown in Fig. 3.4-2. We see that real zeros in the transfer function cause the magnitude of the frequency response to go up, whereas real poles cause it to go down. The frequencies where the increases or decreases take place are substantially equal to the magnitude of the zeros or poles, respectively.

Complex Poles and Zeros

The basic argument leading to the Bode diagram is unaffected if the zeros and poles are complex. As noted earlier, complex poles or zeros must occur in complex conjugate pairs if the impulse response is real. A typical pair of factors with complex conjugate poles is

$$\left(\frac{j\omega}{-\lambda} + 1\right)\left(\frac{j\omega}{-\lambda^*} + 1\right) = -\left|\frac{\omega}{\lambda}\right|^2 + 2j\zeta\left|\frac{\omega}{\lambda}\right| + 1, \tag{3.4-20}$$

where

$$\zeta = \frac{\text{Re}\,(-\lambda)}{|\lambda|}. \tag{3.4-21}$$

The parameter ζ is called the damping constant, and for complex roots ζ is restricted to $0 < \zeta < 1$. The log-magnitude and phase diagram for a transfer function having only a single pair of such complex poles is shown in Fig. 3.4-3 for various values of ζ. The asymptotic log-magnitude diagram has a second-order break point at $\omega = |\lambda|$;

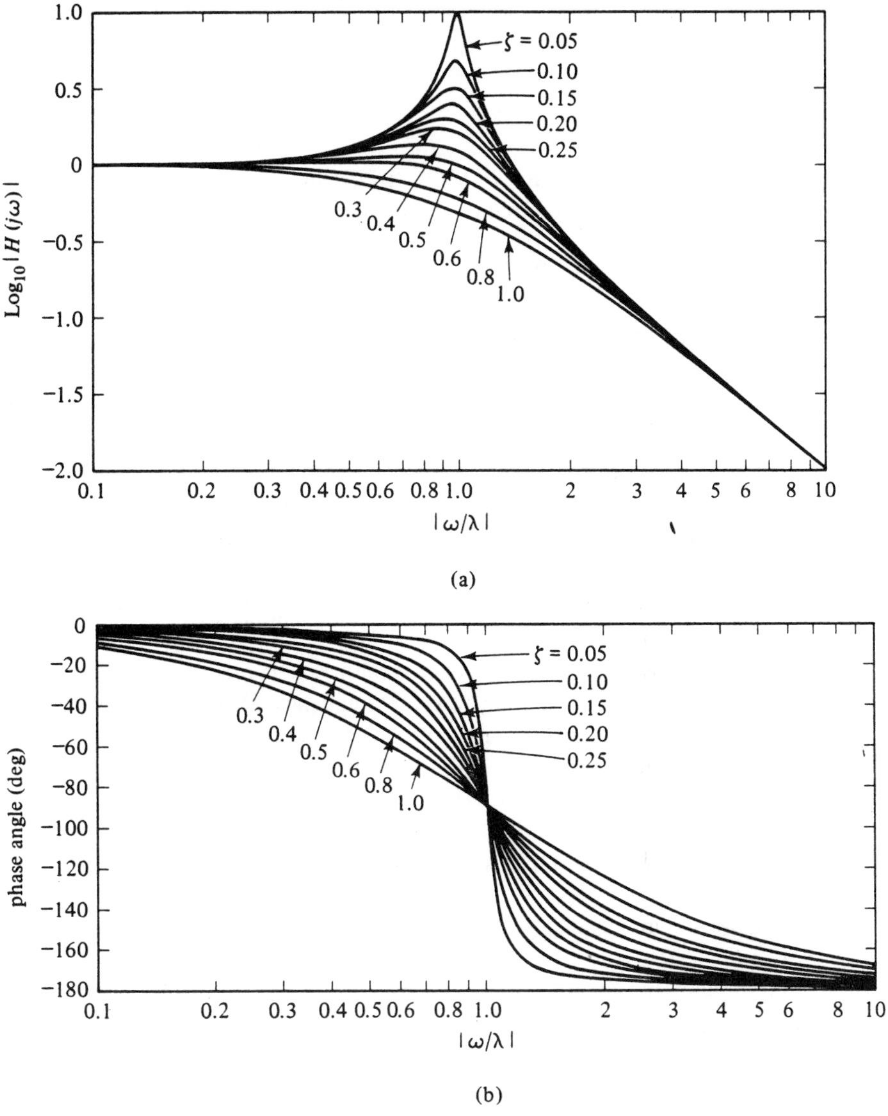

Figure 3.4-3. Frequency-response diagram for the transfer function $H(j\omega) = (-|\omega/\lambda|^2 + 2j\zeta|\omega/\lambda| + 1)^{-1}$. (a) Actual log-magnitude curves for various values of ζ. (b) Corresponding phase characteristic.

that is, the slope changes from 0 to -2. For small ζ the actual magnitude curve departs considerably from the asymptotic curve. This is due to the resonance phenomenon. Observe that complex poles in the transfer function generally result in resonant peaks in the frequency response for frequencies close to the pole magnitude. If the real part of the pole is small, the peak is large.

By way of example, the transfer function of the series RLC circuit considered in the last section (Fig. 3.3-1) has two poles:

$$\lambda = -\frac{R}{2L} \pm \sqrt{\left(\frac{R}{2L}\right)^2 - \frac{1}{LC}}. \tag{3.4-22}$$

These poles are complex if $(R/2L)^2 < 1/LC$, and in this case $|\lambda| = \omega_0 = 1/\sqrt{LC}$, the resonant frequency of the circuit. The parameter ζ is given by $\zeta = R/(2L\omega_0)$.

The Quality Factor Q

For values of $\zeta \ll 1$, the peak of the magnitude of

$$H(j\omega) = \frac{1}{1 + 2j\zeta(\omega/\omega_0) - (\omega/\omega_0)^2} \tag{3.4-23}$$

occurs near $\omega = \omega_0$ and has a value very nearly equal to $1/(2\zeta)$. The sharpness of the peak is often a measure of the selectivity of filters, tuned circuits, and oscillators. A measure of the sharpness of the peak is the quality factor Q, defined by [3-3]:

$$Q = 2\pi \frac{\text{energy stored in the circuit}}{\text{energy dissipated per cycle}}. \tag{3.4-24}$$

This is also proportional to the peak of the resonance curve divided by the half-power bandwidth (which is determined by the frequency at which the peak has decayed to $1/\sqrt{2}$ times its maximum). Clearly, the smaller the energy loss, the sharper the resonance and the larger is Q. It can be shown (Prob. 3-8) that the series RLC circuit has a Q given by $Q = \omega_0 L/R = (2\zeta)^{-1}$. Hence Eq. (3.4-23) can also be written as

$$H(j\omega) = \frac{1}{1 + (j/Q)(\omega/\omega_0) - (\omega/\omega_0)^2}. \tag{3.4-25}$$

The peak of the resonance curve is then seen to be approximately equal to Q (for $\zeta \ll 1$).

Decibels

As we have already seen, logarithmic scales are often used in plots of $|H(j\omega)|$. The commonly used unit for measuring power ratios is the *decibel*. The gain or loss of power expressed in decibels is 10 times the common logarithm of the power ratio. If P is the power level to be compared to the reference level P_0, the number of decibels is given by $10 \log_{10} (P/P_0)$. Thus, a power ratio of 10 is 10 dB, and a power ratio of 2 is 3 dB (since $\log_{10} 2 = 0.30103$). When the decibel measure is applied to transfer-function magnitude levels, one identifies $|H(j\omega)|^2$ with power, and therefore the number of decibels corresponding to a magnitude ratio $|H(j\omega)/H(j\omega_0)|$ is $20 \log_{10} |H(j\omega)/H(j\omega_0)|$. For instance, the magnitude of the transfer function $H(j\omega) = \alpha/(j\omega + \alpha)$ is $1/\sqrt{2}$ at the break frequency α; this is 3 dB down from the value at zero frequency. For this

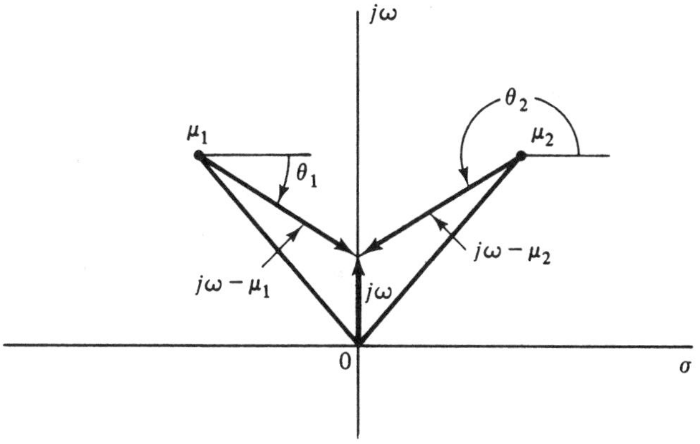

Figure 3.4-4. Effect of moving a zero from the left half-plane to its mirror location in the right half-plane. Only one set of the complex conjugate pair is shown.

reason one sometimes refers to the break frequency as the 3-dB point. A slope of ± 1 in the Bode diagram means that the magnitude doubles (or halves) for each doubling of the frequency; therefore, this is sometimes referred to as 6 dB per octave. Also, a 10% change in transfer-function magnitude is commonly referred to as a 1-dB change; this is an approximation since 1 dB is about 12.2%.

The decibel is one-tenth of a larger unit, the *bel*,† named after Alexander Graham Bell, who in addition to inventing the telephone was also an early pioneer in sound, speech, and hearing research. The unit is appropriate to measuring sound, because the ear responds approximately logarithmically, and equal decibel increments are perceived as equal increments in sound. One decibel is the smallest change in sound intensity that the ear can normally detect. In applications to sound measurement, 0 dB is a reference power of 10^{-16} watts, supposedly the smallest detectable sound for a normal ear, but other reference levels are also used.

Minimum Phase Property

The magnitude versus frequency characteristic of the transfer function $H(j\omega)$ is unaffected if a zero is moved from the left half-plane to its mirror location in the right half-plane (i.e., if only the sign of the real part is changed). This is made clear by Fig. 3.4-4, where μ_1 is a zero in the left half-plane and μ_2 the corresponding zero in the right half-plane. The magnitude of the "vector" $j\omega - \mu_1$ is simply its length, and this is seen to be the same as the length of $j\omega - \mu_2$. However, the phase change $d\theta_1/d\omega$, contributed by μ_1, is positive, while that contributed by μ_2 (i.e., $d\theta_2/d\omega$) is negative. Thus the zero in the right half-plane contributes more negative phase (i.e., phase lag) to the transfer function than the corresponding left half-plane zero. The minimum amount of phase lag for a given magnitude characteristic is obtained if all the zeros lie in the left half-plane. For this reason a transfer function having no zeros in the right half-plane is called a minimum phase-lag (minimum phase for short) function.

† The decibel, rather than the bel, is widely used because for most applications the bel is too large a unit.

For a minimum phase transfer function, the phase function is completely determined by the magnitude function. To see why this is so, let $H(s) = e^{\alpha(s)+j\beta(s)}$. Then $\gamma(s) \equiv \alpha(s) + j\beta(s) = \ln H(s)$ is a complex function of the complex variable s, and it will be analytic in the right half-plane if $H(s)$ is stable and minimum phase [i.e., if all the zeros and poles of $H(s)$ are in the left half-plane]. We saw in Sec. 2.13 that if a complex function is analytic in one of the half-planes formed by the axis of reals then its real and imaginary parts for real argument are related by a Hilbert transform. A very similar argument holds if the complex function is analytic in one of the half-planes formed by the imaginary axis, as is the case here. Then the real and imaginary parts evaluated for $s = j\omega$ are related by a Hilbert transform; in fact, we have

$$\beta(\omega) = \frac{1}{\pi}\int_{-\infty}^{\infty} \frac{\alpha(\lambda)}{\omega - \lambda}\, d\lambda, \tag{3.4-26}$$

$$\alpha(\omega) = -\frac{1}{\pi}\int_{-\infty}^{\infty} \frac{\beta(\lambda)}{\omega - \lambda}\, d\lambda. \tag{3.4-27}$$

Equations (3.4-26) and 3.4-27) can be used to establish further transformations and lead to the famous Bode phase-integral theorem ([3-2] and [3-4], Chap. 6):

$$\beta(\omega_0) = \frac{1}{\pi}\int_{-\infty}^{\infty} \left(\frac{d\alpha}{du}\right) \ln \coth \frac{|u|}{2}\, du, \tag{3.4-28}$$

where $u = \ln(\omega/\omega_0)$ is the frequency on a logarithmic scale so that $d\alpha/du$ is the slope of the α curve on a log-log scale. Equation (3.4-28) therefore is a statement of the fact that for a minimum phase transfer function the phase depends on the slope of the α curve. The factor $\ln \coth |u/2|$ is sharply peaked at $u = 0$ (or $\omega = \omega_0$) and shows that the phase at a given frequency ω_0 is most heavily influenced by the slope of the α curve in the immediate vicinity of ω_0.

Bode's phase integral is one of the few mathematical formulas known to the authors that is covered by a patent; it is U.S. Patent No. 2,123,178.

Behavior of $|H(\omega)|$ for Infinite ω

If a transfer function has more zeros than poles, the magnitude of the transfer function goes to infinity for infinite frequency. Practical circuits do not behave this way; in fact, $|H(\omega)|$ generally goes to zero for infinite ω. At most one encounters systems having finite response at very high frequency. Thus the transfer function of a simple resistive voltage-divider circuit is $R_2/(R_1 + R_2)$, where R_2 is the shunt resistor and R_1 the series resistor; this remains constant with increasing frequency as long as shunt capacitances generally present in such circuits have negligible effect. For this reason, one generally assumes that the number of zeros of the transfer function is no larger than the number of poles.†

† Since a zero or a pole at infinity has no effect on the transfer function at finite frequencies, one sometimes regards transfer functions with fewer zeros than poles as having enough infinite zeros to make the total number of zeros equal to the number of poles. With this convention, all transfer functions have the same number of zeros as poles.

3.5 INPUT–OUTPUT

From the definition of the transfer function, Eq. (3.4-1), we obtain immediately the important formula

$$Y(j\omega) = H(j\omega)X(j\omega), \tag{3.5-1}$$

where

$$X(j\omega) = \int_{-\infty}^{\infty} x(t)e^{-j\omega t}\, dt. \tag{3.5-2}$$

Thus we can compute the output $y(t)$ of a linear time-invariant system resulting from any arbitrary input $x(t)$ by first Fourier-transforming the input, multiplying the transform by the transfer function, and then inversely transforming to obtain $y(t)$.

In Sec. 3.2 we found that the output can also be obtained by the superposition integral:

$$y(t) = \int_{-\infty}^{\infty} h(t, \tau)x(\tau)\, d\tau, \tag{3.5-3}$$

where $h(t, \tau)$ is the response of the system at time t to an impulse applied at time τ. For time-invariant systems this becomes the convolution

$$y(t) = \int_{-\infty}^{\infty} h(t - \tau)x(\tau)\, d\tau. \tag{3.5-4}$$

From the convolution-multiplication property of the Fourier transform (property 12 in Table 2.9-1), we see that Eq. (3.5-4) can be transformed to

$$Y(j\omega) = \mathcal{F}[h(t)]X(j\omega), \tag{3.5-5}$$

which is, of course, identical with Eq. (3.5-1).

We see that we have two different methods for obtaining the output of a linear, time-invariant system to an arbitrary input. One is the direct method, using the superposition integral; the other is an indirect method where we first Fourier-transform the input, multiply by the transfer function, and then inverse-transform the product. It would seem that the direct method should be preferable, but this is not necessarily so. For many standard signals (e.g., sine waves) the Fourier transform of the input is known, and therefore the first step in the procedure is not necessary; it has, in effect, been done once and for all. Also, it frequently turns out that the output in the form $Y(j\omega)$ is as useful and informative as $y(t)$; thus the inverse transformation is also not needed. Then the advantage of the transfer-function method is the substitution of a simple multiplication for an integration. Another point in favor of the transfer-function method is that it is usually much easier to derive the transfer function from a circuit diagram, whereas derivation of the impulse response for other than the simplest circuits is not straightforward. The impulse response is in fact best obtained as the inverse Fourier transform of the transfer function.

The superposition integral is easier to use when the input signal and impulse response have simple forms for which the Fourier transforms may be complicated. Also, the superposition integral can accommodate time-varying systems, at least in principle. As should be clear from our derivation here, the transfer function is, strictly speaking, defined only for time-invariant systems. Time-varying transfer functions $H(\omega, t)$ can be defined, but they give only approximate results, and generally work only if the time variation is slow compared to the frequencies ω of interest [3-5].

Example

To illustrate the two methods for calculating the output, we consider the simple RC circuit shown in Fig. 3.5-1, driven by a square pulse. The transfer function of this circuit can be obtained by the application of Kirchhoff's law as explained in Sec. 3.3. However, with simple circuits such as the one in this example, one can obtain the transfer function by inspection, using the principle of the voltage divider:

$$H(j\omega) = \frac{Z_2(j\omega)}{Z_1(j\omega) + Z_2(j\omega)}, \tag{3.5-6}$$

where $Z_2(j\omega)$ is the impedance of the shunt branch and $Z_1(j\omega)$ the impedance of the series branch. In our example, $Z_2(j\omega) = 1/j\omega C$ and $Z_1(j\omega) = R$; hence

$$H(j\omega) = \frac{1/j\omega C}{1/j\omega C + R}, \tag{3.5-7}$$

or, if we use the frequency variable f and, for simplicity, write $H(f)$ instead of $H(j2\pi f)$,

$$H(f) = \frac{a}{a + j2\pi f}, \tag{3.5-8}$$

where

$$a \equiv \frac{1}{RC}. \tag{3.5-9}$$

The input signal is

$$x(t) = \text{rect}\left(\frac{t}{T}\right), \tag{3.5-10}$$

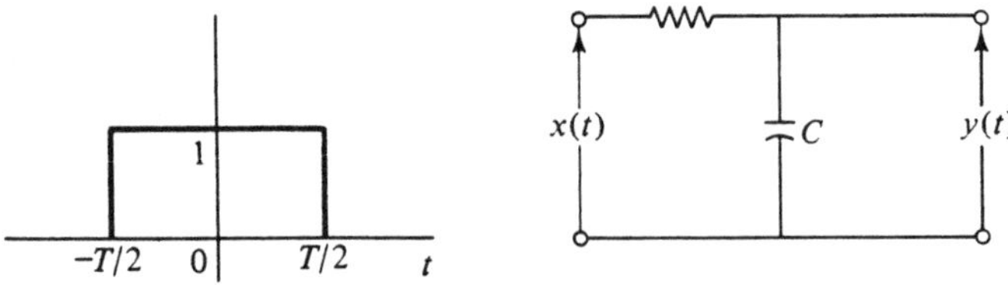

(a) Input (b) Circuit

Figure 3.5-1. RC circuit with square pulse input signal.

and its Fourier transform is

$$X(f) = T \operatorname{sinc}(Tf). \tag{3.5-11}$$

To get the output by means of the transfer-function method, we multiply $X(f)$ by $H(f)$:

$$Y(f) = H(f)X(f) = \frac{aT \operatorname{sinc}(Tf)}{a + j2\pi f}. \tag{3.5-12}$$

The input spectrum, transfer function, and output spectrum are shown in Fig. 3.5-2 for $aT/2\pi = 0.5$.

To obtain the output as a time function, we perform the inverse transformation of $Y(f)$:

$$\begin{aligned} y(t) &= \int_{-\infty}^{\infty} \frac{aT \operatorname{sinc}(Tf) e^{j2\pi ft}}{a + j2\pi f}\, df \\ &= a\int_{-\infty}^{\infty} \frac{(e^{j\pi fT} - e^{-j\pi fT}) e^{j2\pi ft}}{(a + j2\pi f) j2\pi f}\, df \\ &= a\int_{-\infty}^{\infty} \frac{e^{j2\pi(t+T/2)f} - e^{j2\pi(t-T/2)f}}{j2\pi f(a + j2\pi f)}\, df \\ &= \int_{-\infty}^{\infty} \left(\frac{1}{j2\pi f} - \frac{1}{a + j2\pi f}\right) e^{j2\pi(t+T/2)f}\, df \\ &\quad - \int_{-\infty}^{\infty} \left(\frac{1}{j2\pi f} - \frac{1}{a + j2\pi f}\right) e^{j2\pi(t-T/2)f}\, df. \end{aligned} \tag{3.5-13}$$

In the last step the partial fraction expansion

$$\frac{a}{j2\pi f(a + j2\pi f)} = \frac{1}{j2\pi f} - \frac{1}{a + j2\pi f} \tag{3.5-14}$$

is used.

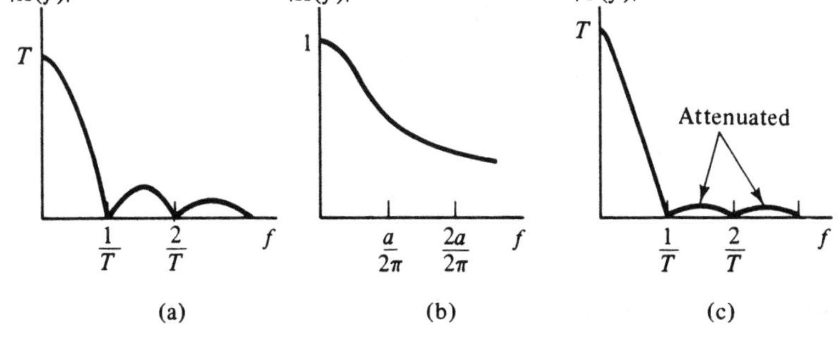

Figure 3.5-2. (a) Fourier spectrum of the input signal. (b) Magnitude of the transfer function. (c) Output spectrum.

The inverse transformation is now easily completed by using some of the Fourier pairs in Table 2.12-1; thus, from pairs 6 and 4 we have

$$\begin{aligned} \frac{1}{j2\pi f} &\longleftrightarrow \frac{1}{2}\operatorname{sgn}(t) \\ \frac{1}{a + j2\pi f} &\longleftrightarrow e^{-at}u(t). \end{aligned} \tag{3.5-15}$$

Therefore,

$$\begin{aligned} y(t) &= \frac{1}{2}\operatorname{sgn}\left(t + \frac{T}{2}\right) - e^{-a(t+T/2)}u\left(t + \frac{T}{2}\right) \\ &\quad - \frac{1}{2}\operatorname{sgn}\left(t - \frac{T}{2}\right) + e^{-a(t-T/2)}u\left(t - \frac{T}{2}\right) \\ &= \operatorname{rect}\left(\frac{t}{T}\right) - e^{-a(t+T/2)}u\left(t + \frac{T}{2}\right) + e^{-a(t-T/2)}u\left(t - \frac{T}{2}\right), \end{aligned} \tag{3.5-16}$$

where the second step follows by combining the two sgn functions. The three components and the result are shown in Fig. 3.5-3.

An example of the use of the superposition integral has already been presented in Sec. 3.2. In the present example the impulse response is given by

$$h(t) = ae^{-at}u(t),$$

and therefore

$$\begin{aligned} y(t) &= \int_{-\infty}^{\infty} x(\tau)h(t - \tau)\, d\tau \\ &= \int_{-\infty}^{\infty} \operatorname{rect}\left(\frac{\tau}{T}\right) ae^{-a(t-\tau)}u(t - \tau)\, d\tau. \end{aligned} \tag{3.5-17}$$

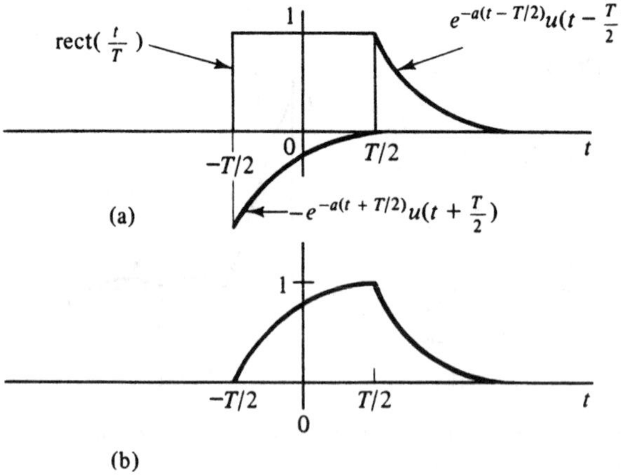

Figure 3.5-3. Output of the *RC* circuit by the transfer-function method. (a) The three components of Eq. (3.5-16). (b) The complete response.

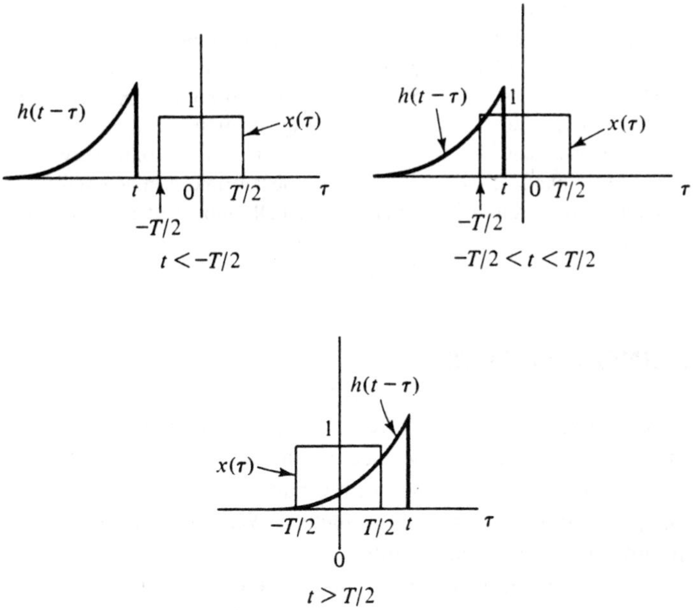

Figure 3.5-4. Input signal $x(\tau) = \text{rect}(\tau/T)$ and reversed and shifted impulse response $h(t-\tau) = ae^{-a(t-\tau)}u(t-\tau)$ for three different values of t.

The factors making up the integrand are shown in Fig. 3.5-4 for three different values of t. The output for the three values is as follows:

$$\text{For } t < -\frac{T}{2}: \qquad y(t) = 0,$$

$$\text{For } -\frac{T}{2} \leq t \leq \frac{T}{2}: \qquad y(t) = ae^{-at}\int_{-T/2}^{t} e^{a\tau}\,d\tau = 1 - e^{-a(t+T/2)}, \tag{3.5-18}$$

$$\text{For } \frac{T}{2} \leq t: \qquad y(t) = ae^{-at}\int_{-T/2}^{T/2} e^{a\tau}\,d\tau = e^{-a(t-T/2)}(1 - e^{-aT}).$$

The three solutions are put together in Fig. 3.5-5. The result is identical with the one shown in Fig. 3.5-3. The same result would be obtained by using the alternative form of the superposition integral:

$$y(t) = \int_{-\infty}^{\infty} x(t-\tau)h(\tau)\,d\tau. \tag{3.5-19}$$

The reader is invited to try this for himself.

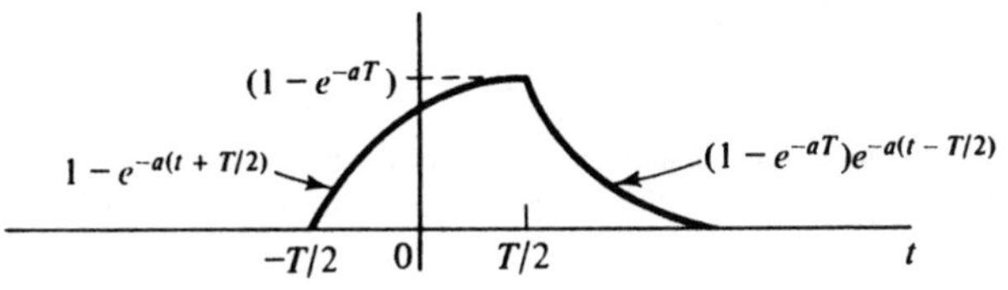

Figure 3.5-5. Output by superposition integral.

While this example has been presented mainly to illustrate two methods of input–output calculations, it illustrates several other things as well. The *RC* circuit in the example is an example of a low-pass filter since it selectively attenuates the higher frequencies. This is shown in Fig. 3.5-2. The effect is to convert the discontinuous steps of the input into a more gradually rising and falling function. Note that for $aT \ll 1$ this effect is very pronounced, and the square pulse is converted approximately into a triangular pulse. On the other hand, if $aT \gg 1$, the filter decay time is much shorter than the duration of the input, and therefore it only causes a small amount of rounding of the corners of the input function.

3.6 IDEAL LINEAR FILTERS

We have defined a filter as a circuit that passes certain desired frequencies of the input and rejects or attenuates unwanted frequencies. An ideal filter performs this function perfectly; that is, it passes the desired parts of the input spectrum with no change, and it rejects the unwanted parts completely. We shall see that ideal filters can only be approximated by practical circuits.

In communication systems, small time delays between input and output are generally of no consequence;† therefore, if $x(t)$ is the input, the output is ideally

$$y(t) = Kx(t - t_0), \tag{3.6-1}$$

where t_0 is the time delay and K is a gain constant. Fourier-transforming this expression results in

$$Y(f) = KX(f)e^{-j2\pi f t_0}. \tag{3.6-2}$$

Thus, if $x(t)$ has frequency components only in a certain band, then the ideal filter should have a constant amplitude and linear phase shift in that band. Any filter that does not satisfy these conditions introduces distortion into the signal. Nonuniform magnitude of the filter frequency response gives rise to *amplitude distortion*; departure from linearity of the phase causes *phase distortion*. The effects of these two kinds of distortion on the output signal are similar; however, since the human ear is relatively insensitive to phase, phase distortion is less serious in audio systems than amplitude distortion.

A *low-pass* filter is a filter that passes all frequencies whose magnitudes lie between zero and a certain cutoff frequency B. The frequency range $-B \leq f \leq B$ is called the *passband* of the filter. The transfer function of the *ideal* low-pass filter is given by

$$H(f) = K \operatorname{rect}\left(\frac{f}{2B}\right) e^{-j2\pi f t_0}. \tag{3.6-3}$$

† Large delays are sometimes of no consequence either. For instance, the unavoidable delays in the signals transmitted from deep-space probes are generally of little consequence. On the other hand, in telephony large delays are intolerable.

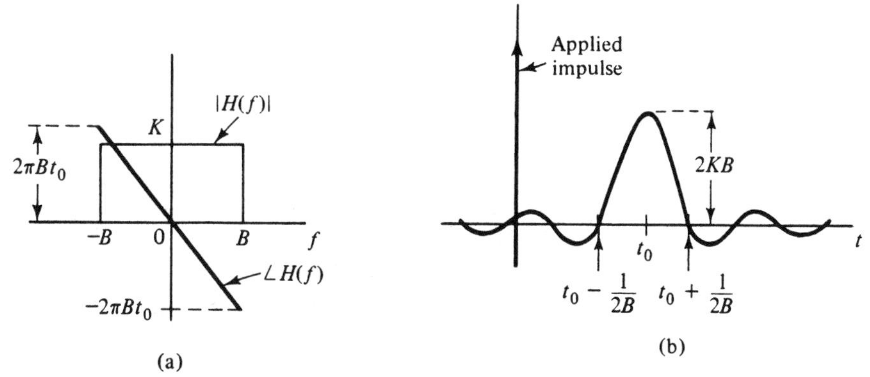

Figure 3.6-1. (a) Frequency response and (b) impulse response of an ideal low-pass filter.

This is illustrated in Fig. 3.6-1(a). This filter characteristic is clearly noncausal, as can be seen from the impulse response:

$$h(t) = \mathcal{F}^{-1}[H(f)] = 2KB \operatorname{sinc}[2B(t - t_0)], \tag{3.6-4}$$

shown in Fig. 3.6-1(b). The tails of the sinc function go to infinity in both directions and therefore extend beyond $t < 0$. However, if $Bt_0 \gg 1$, the tail extending to negative t is very small and can be neglected. Thus, although the ideal low-pass characteristic can never be exactly causal, it can be fairly well approximated by a causal function if one makes the delay t_0 large enough.

The ideal *band-pass filter* is given by

$$H(f) = K\left[\operatorname{rect}\left(\frac{f - f_0}{B}\right) + \operatorname{rect}\left(\frac{f + f_0}{B}\right)\right]e^{-j2\pi f t_0}. \tag{3.6-5}$$

This characteristic is shown in Fig. 3.6-2. The bandwidth of the filter is B Hz. If this is to be a band-pass transfer function, it is necessary that $f_0 > B/2$; generally, one assumes that $f_0 \gg B/2$.

The impulse response of the ideal band-pass filter is the inverse Fourier transform of Eq. (3.6-5):

$$h(t) = 2KB \operatorname{sinc}[B(t - t_0)] \cos 2\pi f_0(t - t_0). \tag{3.6-6}$$

This is shown in Fig. 3.6-3 for $f_0 \gg B/2$. The 3-dB width ($\simeq B^{-1}$) of the main lobe may be regarded as the *response time* τ_c of the filter. The envelope of the response (i.e., the locus of the peaks) is similar to the response of the low-pass filter, except that the envelope is superimposed on the high-frequency signal $\cos 2\pi f_0(t - t_0)$. As will be seen in Chapter 6, this is a form of amplitude modulation. The response is noncausal, just like the low-pass response; in fact, except for the high-frequency carrier it has the same characteristics as the low-pass response. In particular, for very large Bt_0 the response can be approximately realized with a causal system.

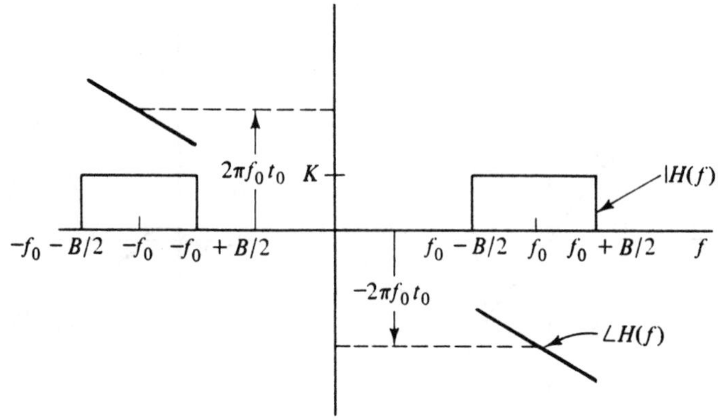

Figure 3.6-2. Ideal band-pass characteristic.

Two other generic filter types are the *high-pass* filter and the *band-stop* filter. These are complementary forms of the low-pass and band-pass filters, respectively. Thus the ideal high-pass transfer function is given by

$$H(f) = K\left[1 - \text{rect}\left(\frac{f}{2B}\right)\right]e^{-j2\pi f t_0}, \tag{3.6-7}$$

and the ideal band-stop transfer function is

$$H(f) = K\left[1 - \text{rect}\left(\frac{f - f_0}{B}\right) - \text{rect}\left(\frac{f + f_0}{B}\right)\right]e^{-j2\pi f t_0}. \tag{3.6-8}$$

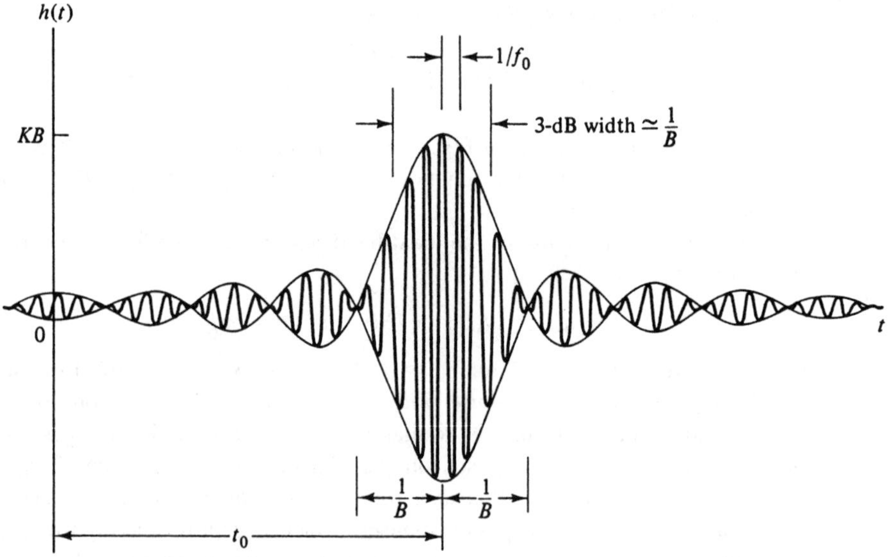

Figure 3.6-3. Impulse response of a band-pass filter with $f_0 \gg B/2$.

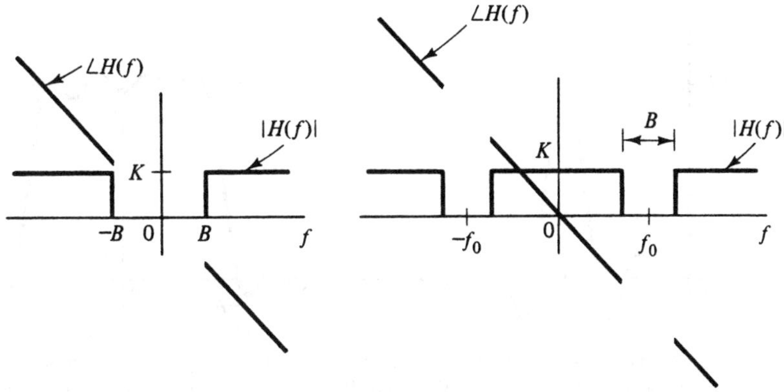

Figure 3.6-4. (a) Ideal high-pass characteristic and (b) ideal band-stop characteristic.

These are illustrated in Fig. 3.6-4. Both of these filters are modeled as having constant gain, K, even as f goes to infinity. Therefore, besides being noncausal they are also not realizable in practice because small capacitances and inductances that may be negligible at normal operating frequencies will generally reduce the gain at sufficiently high frequencies. Nevertheless, in this as well as the other cases, approximations to ideal behavior can be realized that are satisfactory for most applications.

One other filter type that should be mentioned is the *all-pass* filter. This has constant gain at all frequencies (and hence is also unrealizable), but an arbitrary phase shift. All-pass filters are used as *phase equalizers* (i.e., to correct the phase characteristic of some other filter to give it a desired form). Since the phase shift associated with a minimum phase transfer function having constant gain is zero, all-pass networks are necessarily nonminimum phase, and they therefore can only produce phase lag.

3.7 CAUSAL FILTERS

None of the ideal filters considered in Sec. 3.6 are causal because the sharp cutoff of the frequency response results in an impulse response containing sinc functions. The general question of whether a particular form of $|H(\omega)|$ can be realized by a causal filter is answered by the Paley–Wiener criterion [3-6]. This says that an even-magnitude function $|H(\omega)|$ can be realized by a causal filter if and only if it satisfies

$$\int_{-\infty}^{\infty} \frac{|\ln|H(\omega)||}{1+\omega^2}\,d\omega < \infty. \tag{3.7-1}$$

If the integral is bounded, a phase function $\theta(\omega)$ exists such that the impulse response associated with $|H(\omega)|e^{j\theta(\omega)}$ has zero response for $t < 0$. A proof of this relation is given in [3-7]; it is not trivial.

If one attempts to generate a causal response from noncausal responses such as those given in Eqs. (3.6-4) or (3.6-6) by simply setting $h(t) = 0$ for $t < 0$, the resulting

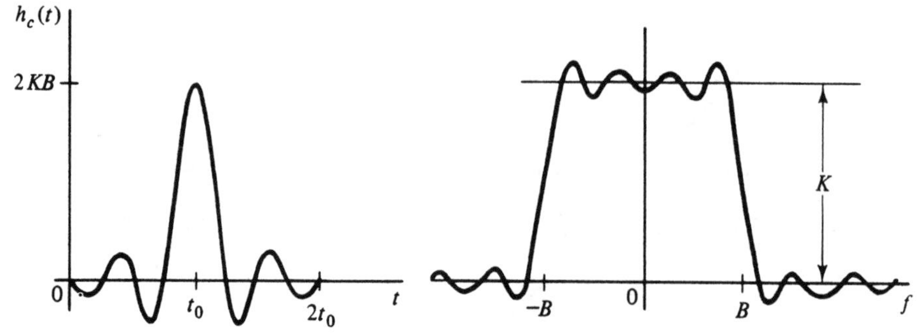

Figure 3.7-1. (a) Causal impulse response obtained by symmetric truncation of the response of Fig. 3.6-1. (b) Corresponding frequency response ($Bt_0 = 2$).

frequency response will extend beyond the passband and will generally have ripples in the passband. This is illustrated in Fig. 3.7-1; for computational details, see Prob. 3-27.

The ripples in the transfer function result in amplitude distortion. We can study this distortion by considering a transfer function of the form

$$H(f) = K(1 + \gamma \cos 2\pi\alpha f)e^{-j2\pi f t_0}. \tag{3.7-2}$$

This is a constant with a superimposed cosinusoidal ripple of frequency α and relative amplitude γ, together with a linear phase-lag term. Equation (3.7-2) can be written in the expanded form

$$H(f) = K\left(e^{-j2\pi f t_0} + \frac{\gamma}{2}e^{j2\pi f(\alpha - t_0)} + \frac{\gamma}{2}e^{-j2\pi f(\alpha + t_0)}\right). \tag{3.7-3}$$

Then if the input spectrum is $X(f)$, the output spectrum is

$$\begin{aligned} Y(f) &= X(f)H(f) \\ &= KX(f)\left(e^{-j2\pi f t_0} + \frac{\gamma}{2}e^{-j2\pi f(t_0 - \alpha)} + \frac{\gamma}{2}e^{-j2\pi f(t_0 + \alpha)}\right). \end{aligned} \tag{3.7-4}$$

The resulting time function is

$$y(t) = Kx(t - t_0) + \frac{K\gamma}{2}x(t - t_0 + \alpha) + \frac{K\gamma}{2}x(t - t_0 - \alpha). \tag{3.7-5}$$

Note that the amplitude ripple results in the generation of echoes. If $x(t)$ is a short pulse, these echoes may appear as two additional distinct pulses, but for an input of duration long compared to the time α, the effect of the echoes is simply to distort the output. This is illustrated in Fig. 3.7-2.

This method for analyzing amplitude distortion is referred to as the *paired-echo* method. It was invented by H. A. Wheeler in 1939 [3-8] and can be extended to more complicated transfer functions. It can be shown that phase distortion can be treated in a similar fashion and gives rise to similar kinds of distortion (Prob. 3-20).

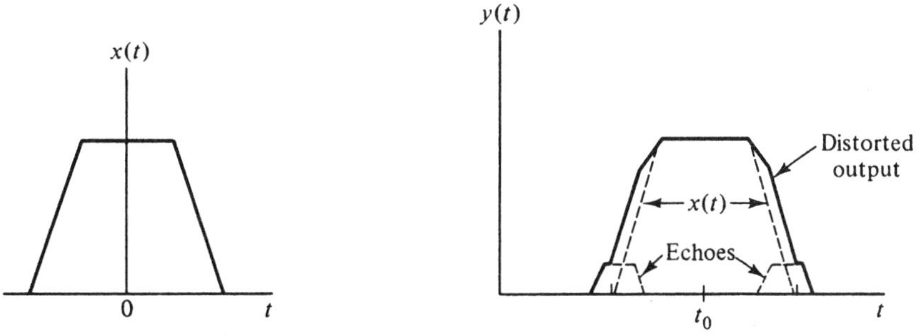

Figure 3.7-2. Distortion resulting from ripple in the magnitude of the frequency response.

3.8 REALIZABLE LOW-PASS FILTERS

The transfer function of a filter constructed of linear lumped-parameter circuit elements has the form

$$H(j\omega) = \frac{N(j\omega)}{D(j\omega)} = \frac{b_0 + b_1(j\omega) + \cdots + b_k(j\omega)^k}{a_0 + a_1(j\omega) + \cdots + a_n(j\omega)^n}. \tag{3.8-1}$$

However, it can be shown from complex-variable theory that the ideal transfer characteristic given in Eq. (3.6-3) *cannot* be represented by a ratio of polynomials. The main problem in practical filter design is therefore to determine, for all i, the coefficients a_i and b_i in Eq. (3.8-1) so that the transfer function best approximates the ideal characteristic in some sense. One usually imposes the constraint that the *order n* of the resulting filter, that is, the number of poles, should be fixed. The order of a filter is directly related to its complexity, to the number of components needed in its construction, and to the difficulty of making it work as designed. Recall that in general there are at least as many poles as zeros; hence the zeros do not affect the order of the filter.

Network synthesis was at one time one of the most challenging and interesting problems facing communications engineers, and many thick books have been written on the subject. A list of some of the foremost names associated with this effort, with photographs, is given in [3-9]. It is clear that in our very brief treatment we can present only a small selection of some of the many results that have been obtained in all this work. We hope to have chosen the most important, but our judgment is necessarily subjective.

Butterworth Filters

One of the first questions that arises in the design of practical filters is in what sense the approximation to the ideal is to be "best." One common criterion is that the transfer function be *maximally flat*. This means that as many derivatives of $|H(j\omega)|$ as possible should go to zero as ω goes to zero. This criterion leads to the *Butterworth* characteristic (Prob. 3-21):

$$|H(j\omega)| = \frac{1}{\sqrt{1 + (\omega/\omega_0)^{2n}}}. \tag{3.8-2}$$

This is shown in Fig. 3.8-1 for several values of n. The Butterworth magnitude characteristic is monotonically decreasing both inside and outside the passband (i.e., the stopband). It has relatively small attenuation in the passband, particularly for $\omega \ll \omega_0$, but the transition between passband and stopband is relatively slow, and the attenuation in the stopband is not so complete as in some other filter types to be described later.

The transfer function $H(j\omega)$ of Eq. (3.8-2) has n poles. To find them, consider the function $|H(j\omega)|^2$, and replace $j\omega$ by the complex frequency s. Since $|H(j\omega)|^2 = H(j\omega)H(-j\omega)$, replacing $j\omega$ by s results in $H(s)H(-s)$. If s_i is a pole of $H(s)$ lying in the left half-plane, then $-s_i$, a pole of $H(-s)$, is its mirror image in the right half-plane. The function $H(s)H(-s)$ has $2n$ poles of which n lie in the left half-plane and give rise to a stable filter. Hence the desired poles are the values of s lying in the left half-plane for which $H(s)H(-s)$ is infinite. The function

$$H(s)H(-s) = \frac{1}{1 + (-1)^n (s/\omega_0)^{2n}}$$

goes to infinity if s is one of the $2n$ roots of the equation

$$\left(\frac{s}{\omega_0}\right)^{2n} = \begin{cases} +1, & n \text{ odd}, \\ -1, & n \text{ even}. \end{cases}$$

The roots are

$$s_i = \omega_0 e^{j\pi(2i/n)/2}, \qquad \text{for odd } n,$$
$$s_i = \omega_0 e^{j\pi[(2i-1)/n]/2}, \qquad \text{for even } n,\ i = 1, \ldots, 2n.$$

These roots are equally spaced points on a circle of radius ω_0, as shown in Fig. 3.8-2. The poles of interest are the roots lying in the left half-plane. The first few Butterworth polynomials [i.e., denominators of $H(s)$] are given in Table 3.8-1.

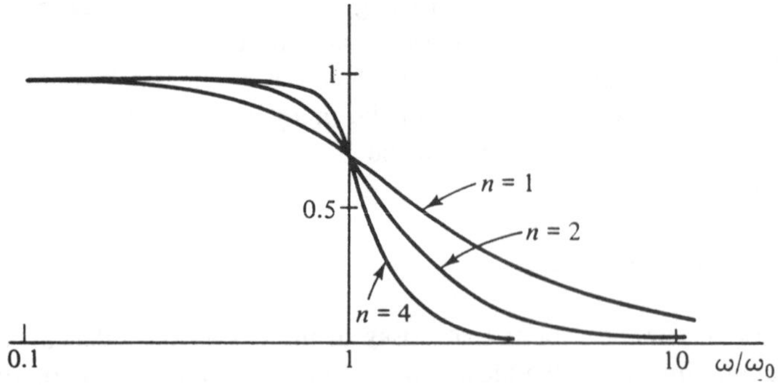

Figure 3.8-1. Butterworth amplitude characteristic for several values of n.

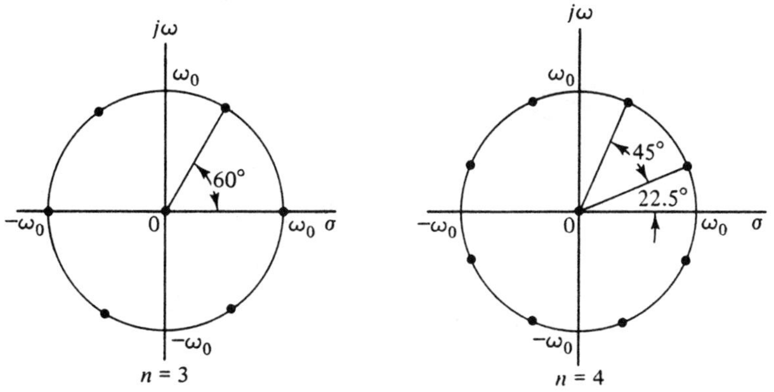

Figure 3.8-2. Roots of the equation $(s/\omega_0)^{2n} = (-1)^n = 0$ for $n = 3$ and $n = 4$. The poles of the Butterworth transfer function are the roots lying in the left half-plane.

TABLE 3.8-1

n	Butterworth polynomial ($x = s/\omega_0$)
1	$1 + x$
2	$\left(x + \frac{\sqrt{2} + j\sqrt{2}}{2}\right)\left(x + \frac{\sqrt{2} - j\sqrt{2}}{2}\right) = x^2 + x\sqrt{2} + 1$
3	$\left(x + \frac{1 + j\sqrt{3}}{2}\right)\left(x + \frac{1 - j\sqrt{3}}{2}\right)(x + 1) = (x + 1)(x^2 + x + 1)$
4	$(x^2 + 0.765x + 1)(x^2 + 1.848x + 1)$
5	$(x + 1)(x^2 + 0.618x + 1)(x^2 + 1.618x + 1)$

Chebyshev Filters

By permitting the magnitude of the transfer function to have small ripples in the passband, one can obtain a narrower transition between the pass- and stopbands than that achievable with a Butterworth characteristic of the same order. One can also increase the attenuation in the stopband. The usual constraint is that the maximum extent of the ripples be less than some predetermined amount, say 10%. This leads to the Chebyshev filter characteristic illustrated in Fig. 3.8-3.

The magnitude of the transfer function oscillates in the passband between two limits, shown as $|H_{\text{max}}|$ and $|H_{\text{min}}|$ in Fig. 3.8-3, and then drops off monotonically outside the passband. The Chebyshev characteristic has the property that for a *given maximum peak-to-peak ripple inside the passband and monotonic attenuation outside the passband*, the transition from passband to stopband is *fastest for any nth order filter*. The magnitude characteristic is given by

$$|H(j\omega)|^2 = \frac{1}{1 + \epsilon^2 T_n^2(\omega/\omega_0)}, \tag{3.8-3}$$

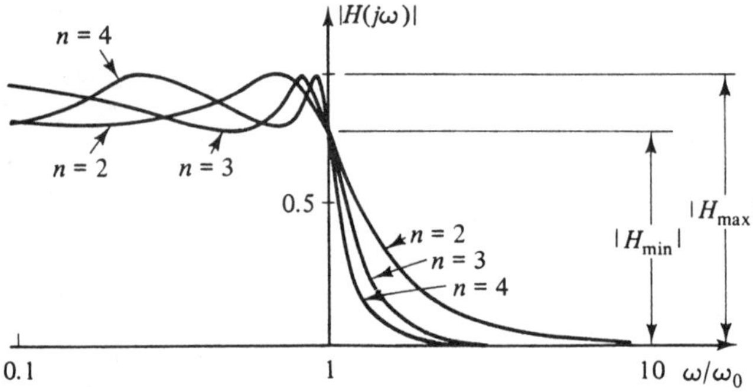

Figure 3.8.3. Chebyshev magnitude responses for $n = 2$, 3, and 4.

where ϵ is a constant and $T_n(\cdot)$ is the Chebyshev polynomial of the first kind of degree n, defined for $|x| < 1$ by

$$T_n(x) = \cos(n \cos^{-1} x).$$

Although it may not be obvious at first glance, this is indeed a polynomial because one can express $\cos n\theta$ as a polynomial in $\cos\theta$ [3-10]. For example, $\cos 2\theta = 2\cos^2\theta - 1$, $\cos 3\theta = 4\cos^3\theta - 3\cos\theta$, and so on. Thus, by identifying θ with $\cos^{-1} x$, one can readily find the entries for the Chebyshev polynomials in Table 3.8-2. The polynomials are defined for all x even though $\cos^{-1}x$ is defined only for $|x| \leq 1$. Observe that $-1 \leq T_n(x) \leq 1$ for $|x| \leq 1$ and that therefore $|H(j\omega)|^2$ oscillates between $(1 + \epsilon^2)^{-1}$ and 1 for $|\omega/\omega_0| \leq 1$. Hence ϵ is determined by the specified value of $|H_{max}| - |H_{min}|$. At the cutoff frequency, $x = \omega/\omega_0 = 1$ and $T_n(x) = 1$; hence $|H(j\omega_0)|$ is exactly equal to $|H_{min}|$ for all orders (i.e., the characteristics for all n go through the same point). It can be shown [3-11] that the poles of the Chebyshev transfer function lie on an ellipse whose major and minor diameters depend on n and ϵ.

Another standard class of filters is the class of *elliptic filters*, so called because elliptic functions are used in their design. The transfer function of elliptic filters can be put into the form of Eq. (3.8-3) except that the polynomial $T_n^2(\omega/\omega_0)$ is replaced by a ratio of polynomials. The transfer function of the elliptic filter therefore has zeros as well as poles. The zeros are located in the stopband and result in nulls in the frequency

TABLE 3.8-2

n	Chebyshev polynomials
1	x
2	$2x^2 - 1$
3	$4x^3 - 3x$
4	$8x^4 - 8x^2 + 1$
5	$16x^5 - 20x^3 + 5x$
6	$32x^6 - 48x^4 + 18x^2 - 1$

response. The transfer function therefore has ripples in the stopband as well as in the passband. The advantage is that this permits a still more rapid transition between the pass- and stopbands than is achieved by a Chebyshev filter. In fact elliptic filters have the most rapid transition between pass- and stopbands of any filter of order n. Elliptic filters and other equiripple filters are discussed in some detail in [3-11] and [3-12].

A number of other filter types are used for somewhat more special applications. For instance, one can specify a filter to have a maximally linear phase characteristic. This leads to a class of filters called Bessel filters since their design is based on Bessel polynomials. These filters have excellent transient response, but their amplitude response is generally poorer than any of the other filter types mentioned above. Additional filter types are given in [3-13].

3.9 ACTIVE FILTER CIRCUITS

Practical analog filters are constructed from standard circuit elements: resistors, capacitors, inductors, and operational amplifiers (op-amps). Thus there still remains the task of showing how to translate a given realizable arrangement of poles and zeros into a hardware configuration. A convenient procedure for doing this is to establish a small number of canonical circuits that provide certain standard combinations of poles and zeros. Then the desired arrangement of the poles and zeros in the filter is obtained by cascading or otherwise combining a number of these canonical circuits. The proper location of the poles and zeros is achieved by adjusting the parameters in the canonical circuits.

As a simple example, we can regard the *RC* and *RLC* filters shown in Fig. 3.9-1 as canonical circuits. Numerous other simple circuits could be added to this list. The transfer functions of the circuits given in (a) and (b) have been obtained earlier, and those in (c) and (d) are easily derived. In principle, any transfer function having real and complex poles and real or complex zeros can be synthesized by cascading such circuits. In making up such a cascade, one has to make sure that the assumption of zero source impedance and infinite load impedance is justified for each element of the cascade. This can be accomplished by raising the impedance level of each successive element of the cascade by about an order of magnitude over that of its predecessor, but this procedure can generally be used only in cascades of two or three elements. Alternatively, the elements of the cascade can be separated by isolating amplifiers.

Simple passive low-pass and band-pass filters employing mainly *LC* resonant circuits are commonly used in radio and television sets. However, for more demanding applications, passive circuits have a number of serious drawbacks. For one thing, the range of permissible pole and zero locations achievable with practical passive circuits is fairly limited. Also, the use of inductors is generally undesirable because they are bulky, they cannot easily be implemented in integrated circuits, and their characteristics are often far from ideal.

For these reasons, modern filter designs frequently use active circuits. In an active circuit a high-gain op-amp is used in a feedback configuration together with resistors

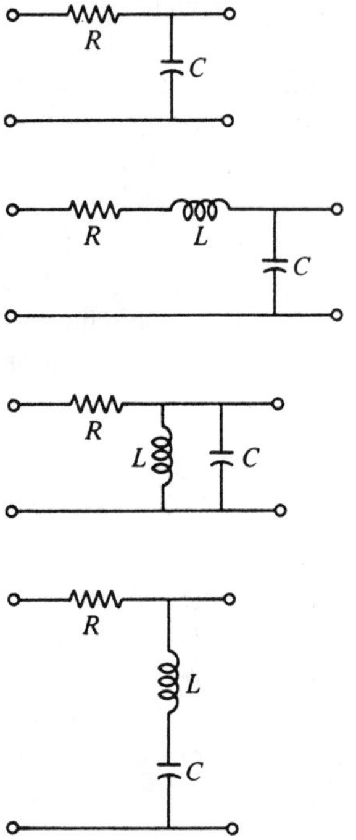

(a) $$H(s) = \frac{1}{1 + RCs}$$

(b) $$H(s) = \frac{1}{LCs^2 + RCs + 1}$$

(c) $$H(s) = \frac{\frac{L}{R}s}{LCs^2 + \frac{L}{R}s + 1}$$

(d) $$H(s) = \frac{LCs^2 + 1}{LCs^2 + RCs + 1}$$

Figure 3.9-1. Canonical circuits. (a) *RC* low-pass filter. (b) *RLC* low-pass filter. (c) *RCL* band-pass filter. (d) *RLC* notch filter.

and capacitors. Active circuits are easily designed with transfer functions having poles and zeros located anywhere in the complex plane. A large variety of op-amps is available as inexpensive integrated circuits, and therefore active circuits are actually simpler and cheaper to construct than passive circuits.

Operational Amplifiers

A symbol for an op-amp is shown in Fig. 3.9-2(a). It is basically a very high gain linear amplifier, normally used with output feedback, having two input terminals and one output terminal. Practical op-amps have additional terminals to provide power supply, frequency compensation, and dc compensation, but we need not be concerned with these. The terminal marked $-$ is the inverting input, and that marked $+$ the noninverting input terminal. Hence for low-frequency input signals† X_1 and X_2 connected as shown in Fig. 3.9-2, the output is ideally $Y = G(X_1 - X_2)$. In practice, zero output voltage may require a small nonzero value of $X_1 - X_2$, called the *offset voltage*. In many IC op-amps the offset is negligible, while in others it can be compensated. We shall therefore neglect it in our discussion.

A typical gain versus frequency curve is shown in Fig. 3.9-2(b). For stable operation in a feedback loop, op-amps require frequency compensation to control the high-frequency

† Capital letters indicate the frequency-domain representations of signals, that is, $X(j\omega)$ or $X(s)$.

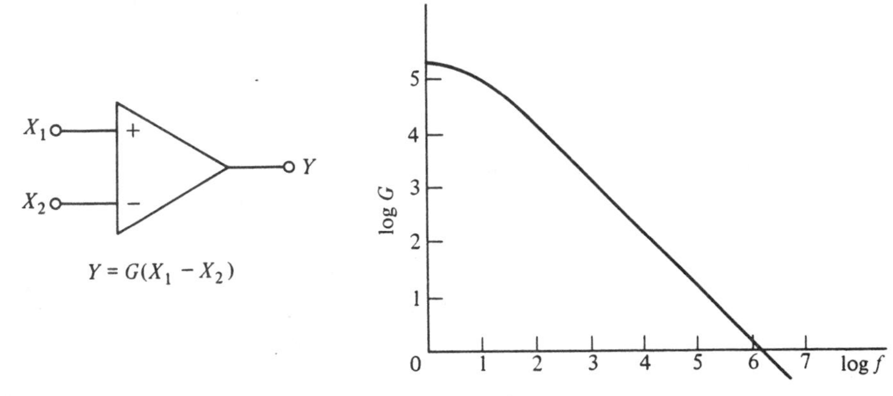

(a) (b)

Figure 3.9-2. Operational amplifier. (a) Elementary circuit operation. (b) Open-loop frequency response.

roll-off characteristic. Some IC op-amps, such as the 741, are internally compensated.† Others, such as the 101, require external compensation but can be used to higher frequencies. Typically, the low-frequency open-loop gain for these units is 2×10^5. The crossover frequency (i.e., the frequency where the gain is down to unity) for the 741 is about 1 MHz; for the 101 it may go as high as 3 MHz. There are special wideband units that generally have lower dc gain (about 3000), but crossover frequencies as high as 30 MHz.

A simple op-amp circuit with negative feedback is shown in Fig. 3.9-3. Assume that the input frequency is low enough so that the open-loop gain of the amplifier is $G \gg 1$. Also, we assume that the amplifier has infinite input impedance and zero output impedance.

The input–output relation for the circuit is most easily obtained by using the fact that, except for a constant offset (which we ignore), the input voltage $V_1 - X_2 \approx 0$. This follows since, for linear operation of the circuit, the absolute value of the output $|Y|$ cannot exceed some limit $|Y_{max}|$, which is fixed by supply voltages but is typically around 5 V. This means that $|V_1 - X_2| \leq |Y_{max}|/G$; hence, for typical values of G, $|V_1 - X_2|$ is in the microvolt range.

We assume, therefore, that

$$V_1 = X_2. \tag{3.9-1}$$

Also, if V_2 is assumed to be a zero-impedance source,

$$X_2 = \frac{Z_1 Y + Z_2 V_2}{Z_1 + Z_2}. \tag{3.9-2}$$

† The complete number may be something like SN72 741 or MCCI 741; the prefix identifies the manufacturer. Sometimes a suffix is added for further description.

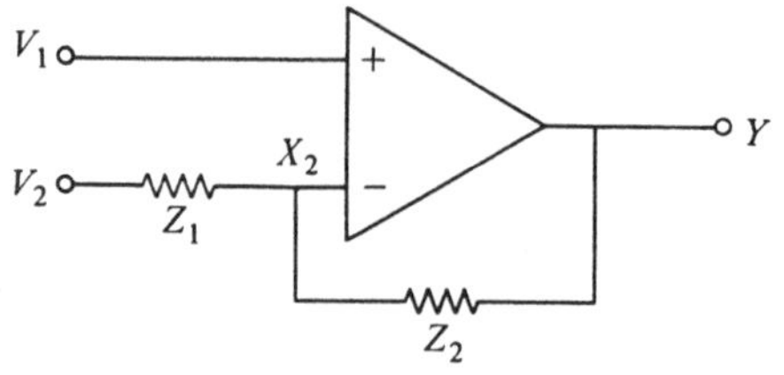

Figure 3.9-3. Operational amplifier with negative feedback.

Solving these equations simultaneously results in

$$Y = \frac{Z_1 + Z_2}{Z_1} V_1 - \frac{Z_2}{Z_1} V_2. \tag{3.9-3}$$

Observe particularly that the gain depends only on the ratios of passive impedances Z_1 and Z_2 and is essentially independent of the open-loop gain G. Therefore, large variations in open-loop gain have almost no effect on the circuit.

If the input signal V_1 is grounded, the signal X_2 must also be essentially at ground potential. For this reason the X_2 terminal is sometimes referred to as a *virtual ground.* The principle of the virtual ground can be used to show that the circuit shown in Fig. 3.9-4 acts as an adding circuit for the input signals X_1, X_2, X_3. Since the inverting terminal is a virtual ground and the sum of the currents flowing into the inverting terminal must be zero, we have

$$\frac{X_1}{R_1} + \frac{X_2}{R_2} + \frac{X_3}{R_3} = -\frac{Y}{R_f}$$

or

$$Y = -\frac{R_f}{R_1} X_1 - \frac{R_f}{R_2} X_2 - \frac{R_f}{R_3} X_3.$$

If all the R's are equal, Y is the sum of the X's; if the R's are not equal, Y is a weighted sum of the X's.

A general active filter prototype is shown in Fig. 3.9-5. We see from Eq. (3.9-3) that the gain from the noninverting input X_1 to the output Y is given by

$$\frac{Y}{X_1} = K = \frac{R_a + R_b}{R_b}. \tag{3.9-4}$$

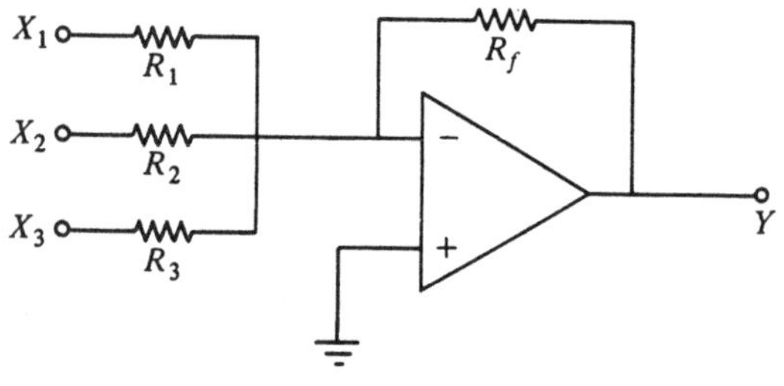

Figure 3.9-4. Op-amp used as an adding circuit.

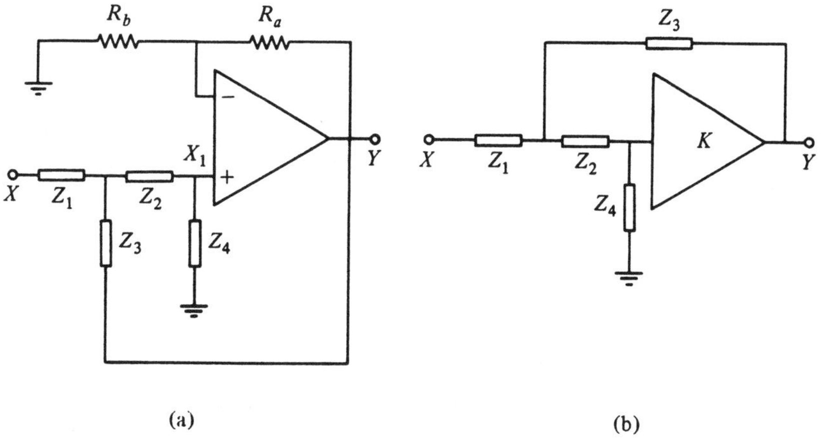

Figure 3.9-5. (a) Active filter prototype. (b) Equivalent circuit where $K = (R_a + R_b)/R_b$.

To obtain this result, set $V_2 = 0$, $Z_1 = R_b$, and $Z_2 = R_a$ in Eq. (3.9-3). Hence an equivalent circuit can be drawn in the form shown in Fig. 3.9-5(b). The transfer function of this circuit is easily obtained by standard methods and can be put into the form

$$H(s) = \frac{Y}{X} = \frac{KZ_3Z_4}{Z_1Z_2 + Z_3(Z_1 + Z_2 + Z_4) + Z_1Z_4(1 - K)}. \tag{3.9-5}$$

The general prototype is converted into a low-pass filter by making Z_1 and Z_2 resistances and Z_3 and Z_4 capacitances. A simple choice of values is $Z_1 = Z_2 = R$ and $Z_3 = Z_4 = 1/Cs$, as in Fig. 3.9-6. If we substitute these values into Eq. (3.9-5) and let $\omega_0 = 1/RC$, we obtain

$$H(s) = \frac{K\omega_0^2}{s^2 + s\omega_0(3 - K) + \omega_0^2}. \tag{3.9-6}$$

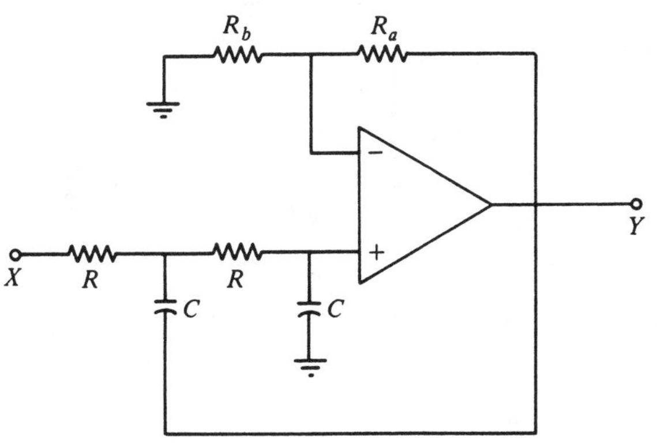

Figure 3.9-6. Second-order low-pass filter.

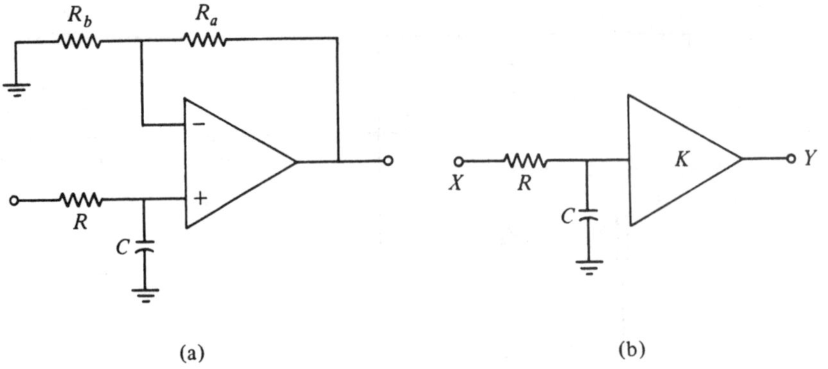

Figure 3.9-7. (a) First-order low-pass filter. (b) Equivalent circuit where $K = (R_b + R_a)/R_b$.

We see that the transfer function has two poles that will be complex if $1 < K < 5$ and that can be made purely imaginary if $K = 3$ (for $K = 3$, the circuit will oscillate at $\omega = \omega_0$). For a second-order (i.e., $n = 2$) Butterworth response, we see from Table 3.8-1 that $3 - K = \sqrt{2}$ or $K = 1.586$. Any even-order Butterworth filter can be constructed by cascading several sections having the same RC product. Thus, for a fourth-order filter, we see from Table 3.8-1 that we cascade a section in which $K = 2.235$ with another one in which $K = 1.152$. Odd-order filters require a first-order prototype of the type shown in Fig. 3.9-7. A fifth-order Butterworth filter is shown in Fig. 3.9-8. The cutoff frequency $\omega_0 = 1/RC$. If R is set to 1000 Ω, then $C = 1000/\omega_0\ \mu F$.

A feature of this design is that the filter does not have unity gain. If unity gain is desired, the gains of all the op-amps can be made equal to unity by connecting the output terminal directly to the inverting input terminal and omitting R_b. The filter parameters can then be adjusted by using either two different capacitances or two different resistors. We leave it to the reader to show that, if the resistors are both equal and if

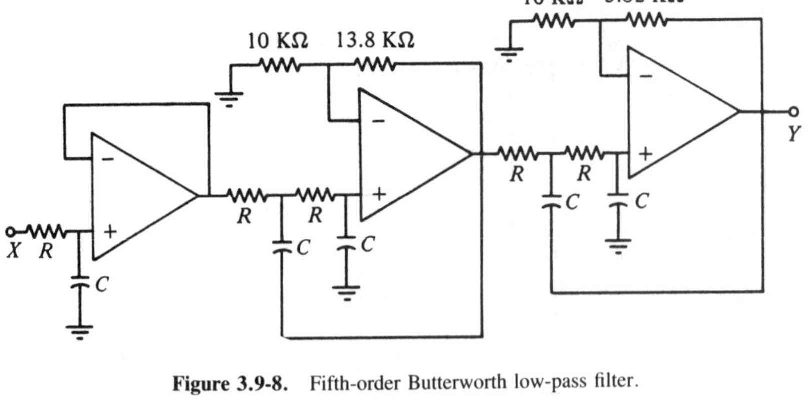

Figure 3.9-8. Fifth-order Butterworth low-pass filter.

the capacitors are C_1 and C_2 (from left to right in Fig. 3.9-6), then $\omega_0^2 = 1/R^2C_1C_2$, and the denominator polynomial of Eq. (3.9-6) is changed to $s^2 + 2\sqrt{C_2/C_1}\ \omega_0 s + \omega_0^2$. Thus any positive value of the coefficient of the middle term can be obtained by properly adjusting the ratio C_2/C_1.

Although active filter circuits will work well with a wide range of parameter values, it is important that the impedance levels of the R's are such that the input impedance of the op-amp is much larger and the output impedance much smaller than R. A value for R of about 1000 Ω is usually satisfactory. Also, the filter will not perform in accordance with the expressions derived here if the open-loop gain of the amplifier is not very large. Since this gain decreases with increasing frequency, there is always a maximum frequency for which these circuits should be used. A typical upper cutoff frequency is 20 kHz, but higher frequencies can be used.

Since the transfer functions of Chebyshev and Butterworth filters are very similar, the active prototypes presented in Figs. 3.9-6 and 3.9-7 can also be used to construct Chebyshev filters by appropriate choice of the feedback resistors R_a and R_b. The values of these resistors depend on the choice of the ripple-amplitude parameter ϵ, and although the calculations needed for their determination are basically straightforward, they are also quite tedious and are therefore omitted. In fact, the easiest way to construct any of these filters is by the use of standardized charts or nomographs, such as are given in several publications ([3-14] and [3-15]). These references also give design charts for some of the other filter types mentioned in Sec. 3.8, and they contain ranges of parameter values and frequencies for which the circuits perform properly.

3.10 FREQUENCY TRANSFORMATIONS

The discussion thus far has dealt exclusively with low-pass filters, but the results can be generalized to some of the other filter types discussed in Sec. 3.6 by simple transformations. For example, the transfer function of a high-pass filter can be obtained from the low-pass transfer function by using the transformation ([3-2], p. 209)

$$\frac{j\omega}{\omega_0} \longrightarrow \frac{\omega_0}{j\omega}. \tag{3.10-1}$$

Thus the second-order low-pass Butterworth transfer function

$$H(j\omega) = \frac{1}{1 + j\sqrt{2}(\omega/\omega_0) - (\omega/\omega_0)^2} \tag{3.10-2}$$

is transformed into the high-pass form:

$$H(j\omega) = \frac{-(\omega/\omega_0)^2}{1 + j\sqrt{2}(\omega/\omega_0) - (\omega/\omega_0)^2}. \tag{3.10-3}$$

Also, the first- and second-order low-pass prototype given in Figs. 3.9-7 and 3.9-6 can be converted to high-pass prototypes by simply exchanging R's and C's in the noninverting branch. This is shown in Fig. 3.10-1.

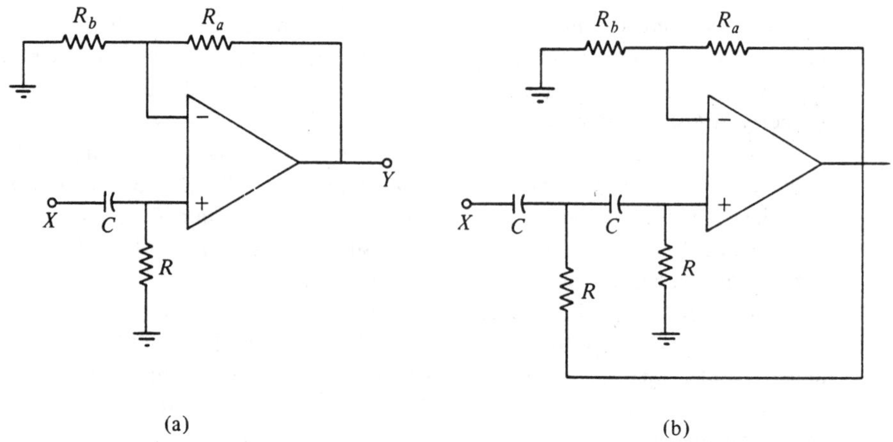

Figure 3.10-1. Active high-pass filters. (a) First order. (b) Second order.

A transformation that converts a low-pass transfer function to band-pass is ([3-2], p. 209)

$$\omega \longrightarrow \frac{\omega^2 - \omega_c^2}{\omega} \tag{3.10-4a}$$

or, when complex frequencies are used,

$$s \longrightarrow \frac{s^2 + \omega_c^2}{s}. \tag{3.10-4b}$$

If the original transfer function had a pole on the real axis at, say, $s = -\omega_0$, and $\omega_c > \omega_0/2$, then the effect of the transformation is to replace the pole at $s = -\omega_0$ with a set of poles at $s = -(\omega_0/2) \pm j\sqrt{\omega_c^2 - (\omega_0/2)^2}$. If $\omega_c \gg \omega_0$, then $s \simeq -(\omega_0/2) \pm j\omega_c$. The configuration is shown in Fig. 3.10-2.

The simplest example illustrating this transformation is the first-order low-pass transfer function

$$H(j\omega) = \frac{1}{j(\omega/\omega_0) + 1}. \tag{3.10-5}$$

After we apply the transformation of Eq. (3.10-4), this becomes

$$H(j\omega) = \frac{j\omega\omega_0}{\omega_c^2 + j\omega\omega_0 - \omega^2} = \frac{j(\omega/\omega_c)(\omega_0/\omega_c)}{1 + j(\omega/\omega_c)\cdot(\omega_0/\omega_c) - [\omega/\omega_c]^2}. \tag{3.10-6}$$

This is a typical second-order resonant filter response with $Q = \omega_c/\omega_0$ [cf. Eq. (3.4-25)].

For large Q, the resonant peak is very close to $\omega = \omega_c$ and the 3-dB bandwidth is the frequency interval for which $|\omega - \omega_c| \leq \omega_0/2$. The width of this interval along the positive-frequency axis is ω_0, and there is another interval of the same length along

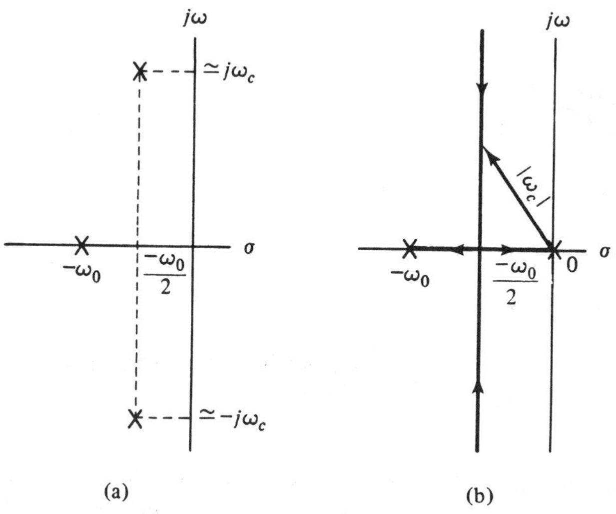

Figure 3.10-2. (a) Example of a low-pass to band-pass transformation. The original poles have been transformed to a set of poles at $s \simeq -(\omega_0/2) \pm j\omega_c$ if $\omega_0 \ll \omega_c$. (b) As ω_c decreases, the locus of the poles is a line intersecting the negative real axis at $\sigma = -(\omega_0/2)$. When $\omega_c \leq \omega_0/2$, the poles are purely real and move toward the points $\sigma = 0$ and $\sigma = -\omega_0$ as ω_c 0. In practice, the band-pass property refers to the situation in (a) when $\omega_c \ll \omega_0$.

the negative-frequency axis. Thus the total interval on the whole frequency axes for which $|H(j\omega)|$ exceeds $1/\sqrt{2}$ is $2\omega_0$, just as for the low-pass case.

The low-pass transfer function can be converted to the *band-stop* form by applying the high-pass and band-pass transformation in sequence ([3-2], p. 209). The transformation is therefore

$$\frac{j\omega}{\omega_0} \longrightarrow \frac{\omega\omega_0}{j(\omega^2 - \omega_c^2)}. \tag{3.10-7}$$

The effect of this transformation on the first-order low-pass transfer function of Eq. (3.10-5) is to produce

$$H(j\omega) = \frac{\omega_c^2 - \omega^2}{\omega_c^2 + j\omega_0\omega - \omega^2} = \frac{1 - (\omega^2/\omega_c^2)}{1 - (\omega^2/\omega_c^2) + j(\omega\omega_0/\omega_c^2)}. \tag{3.10-8}$$

This is seen to have zeros on the imaginary axis and therefore a null at $\omega = \pm\omega_c$. It has not so much a band-stop as a single-frequency-stop characteristic, and filters with this kind of transfer function are sometimes called *notch filters*. They are very commonly used as *frequency traps* in radio or television to eliminate interference from carrier-frequency components near the channel of interest.† The magnitude $|H(j\omega)|$ of a second-order notch filter is plotted in Fig. 3.10-3 for various values of ω_0/ω_c.

There are a number of active circuits having band-pass or band-stop characteristics. Perhaps the simplest arrangement is to place a high-pass filter in series with a low-pass filter to produce a band-pass filter and to use a parallel combination of high-pass and low-pass filters to obtain a band-stop characteristic. These two methods are illustrated in Fig. 3.10-4. An advantage of this method is its inherent flexibility. The high-pass and low-pass sections can be designed separately and need not be "matched" in any

† In commercial TV sets, simple *LC* traps rather than circuits involving op-amps are ordinarily used.

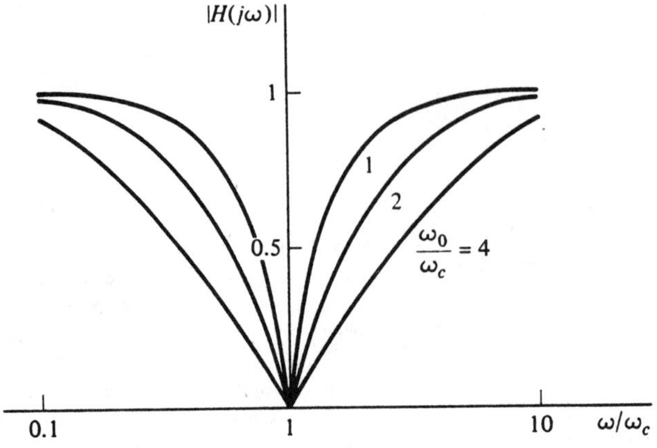

Figure 3.10-3. First-order band-stop characteristic.

sense. By separating the cutoff frequencies of the two sections, one can also easily obtain filters that pass or reject relatively wide frequency bands. On the other hand, the complexity (i.e., number of op-amps, resistors, and capacitors) of these combination filters is generally greater than for filters specifically designed for band-pass or band-stop.

An example of a cascadable second-order band-pass filter is shown in Fig. 3.10-5(a). This is one version of a standard class of related circuits called *single-amplifier biquad* (SAB) circuits. A more general circuit is shown in Fig. 3.10-5(b). We leave it to the reader to show that the transfer function for this circuit is in the form of a ratio of two quadratic polynomials in s:

$$H(s) = K\frac{b_0 + b_1 s + b_2 s^2}{a_0 + a_1 s + a_2 s^2}; \tag{3.10-9}$$

hence the designation biquad (for biquadratic). In the band-pass version of this circuit [Fig. 3.10-5(a)], resistors R_3 and R_4 are eliminated, and this causes b_0 and b_2 in Eq. (3.10-9) to vanish. If resistance R_3 is present but R_4 is left disconnected, it can be shown (Prob. 3-31) that $a_0 = b_0$ and $a_2 = b_2$. By proper choice of component values one can then obtain zero b_1 and finite positive a_1 with the result that the transfer function is that of the notch filter, Eq. (3.10-8). With R_4 connected, one can obtain a combination of notch and low-pass characteristics. SAB circuits are discussed in detail in [3-16] and [3-17]. The latter reference describes a general biquad circuit contained entirely on a hybrid integrated circuit (HIC) chip. Changes in the circuit topography are made by changes in the external pin connections, and resistance values are adjusted by burning away certain sections with a laser beam.

In this necessarily brief treatment of the subject of active filters we have been able to do little more than scratch the surface. We hope, however, that the examples presented will have given the reader a general overview of the subject. We have already given several references to more extensive treatments; additional references are given at the end of the chapter ([3-18]–[3-20]).

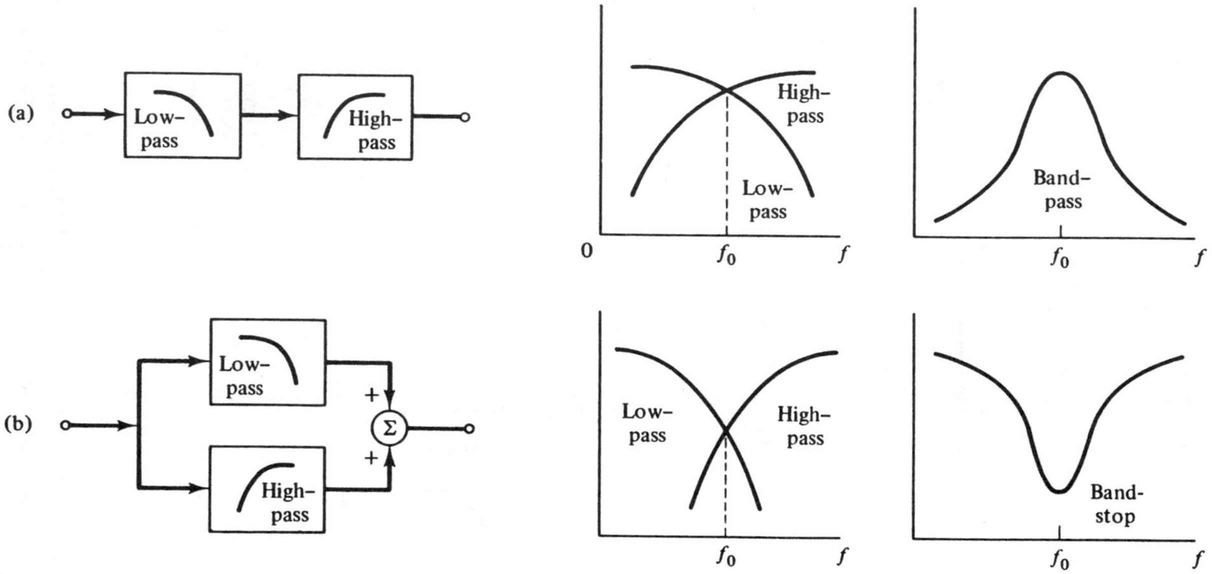

Figure 3.10-4. Combining high-pass and low-pass filters to achieve (a) band-pass and (b) band-stop filters.

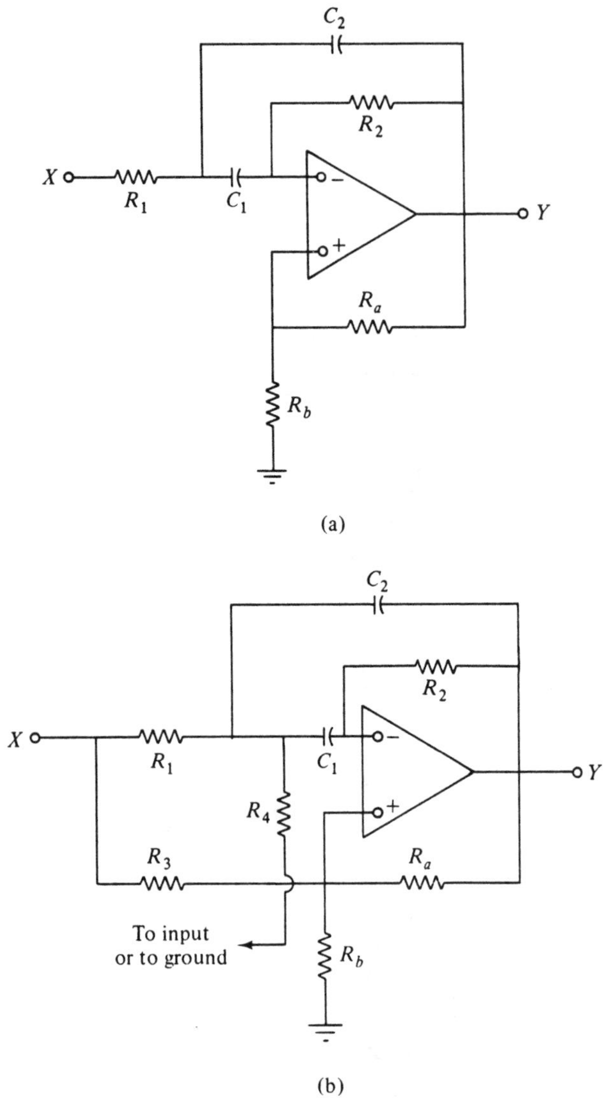

Figure 3.10-5. Single amplifier biquad (SAB) circuits. (a) Band-pass filter. (b) General form.

3.11 SUMMARY

In this chapter we have examined some of the properties of linear lumped-parameter electric circuits and filters. We have shown that the response of such circuits can be obtained both in the time domain (i.e., by the superposition integral) and in the frequency domain using the transfer function. We have investigated some of the properties of transfer functions, such as the poles and zeros, Bode plots, and phase-shift curves. We then defined a number of ideal filter characteristics: low-pass, high-pass, band-pass, and band-stop. We showed that the ideal forms are noncausal, and we considered several practical transfer functions, such as the Butterworth, Chebyshev, and elliptic forms, that are commonly used to approximate the ideal filter response functions. We showed finally how to design hardware realizations in the form of active filters.

PROBLEMS

3-1. ([3-21], p. 4) Consider a linear, time-invariant system $\mathcal{H}_t$ that furnishes an output $y(t)$ when the input is $x(t)$.

(a) Which properties of the system are involved to argue that the response to $(1/\Delta t)[x(t + \Delta t) - x(t)]$ is $(1/\Delta t)[y(t + \Delta t) - y(t)]$?

(b) Extend the result in part (a) to argue that the response to dx/dt is dy/dt.

(c) Take $x(t) = Ae^{j\omega t}$. Use the result in part (b) and the fact that the system is linear to show that $y(t)$ satisfies the differential equation

$$\frac{dy(t)}{dt} = j\omega y(t).$$

Show that the solution is $y(t) = Be^{j\omega t}$, where B is an arbitrary constant.

(d) Compute the transfer function $H(\omega)$ defined by

$$H(\omega) = \frac{\mathcal{H}_t(Ae^{j\omega\tau})}{Ae^{j\omega t}}.$$

That is, $H(\omega)$ is the response of $\mathcal{H}_t$ to $Ae^{j\omega\tau}$ divided by $Ae^{j\omega\tau}$.

3-2. Assume that $x(t)$ is a continuous function over $[t_0, t_0 + T]$ and continuous at $t = t_0$. If $\Delta\xi$ is a small increment of time, then, to a good approximation, $x(t)$, for $t_0 < t < T + t_0$, can be approximated by a finite number of unit steps according to

$$x(t) \simeq x(t_0)u(t - t_0) + \sum_{i=1}^{N} \{x(t_0 + i\Delta\xi) - x[t_0 + (i - 1)\Delta\xi]\}u(t - t_0 - i\Delta\xi), \qquad t_0 \le t \le t_0 + T,$$

where $\Delta\xi = T/N$. Derive this result.

3-3. Extend the result in Prob. 3-2, by letting $\Delta\xi \to 0$ and $N \to \infty$ with $N\,\Delta\xi = T$, to

$$x(t) = x(t_0)u(t - t_0) + \int_0^T \dot{x}(t_o + \xi)u(t - t_0 - \xi)\,d\xi.$$

3-4. In Prob. 3-3, assume that $x(t_0) = 0$, and let $\lambda = t_0 + \xi$. Show that $x(t)$ can be written, with $T \to \infty$, as

$$x(t) = \int_{t_0}^{\infty} \dot{x}(\lambda)u(t - \lambda)\,d\lambda, \qquad t_0 > 0.$$

3-5. Show that, if $\mathcal{H}_t$ is a causal system and $a(t)$ is defined to be the response of $\mathcal{H}_t$ to a unit step applied at $t = 0$, then the output $y(t)$ can be written as

$$y(t) = \int_0^{\infty} \dot{x}(t - \lambda)a(\lambda)\,d\lambda.$$

3-6. Use the result in Prob. 3-5 to compute the response $y(t)$ to an input $x(t) = \text{rect}\,[(t - T/2)/T]$ when $a(t) = (1 - e^{-t/T})u(t)$.

3-7. Use Eq. (3.2-9) to compute the response to the input $x(t) = \text{rect}\,[(t - T/2)/T]$ when $h(t)$ is $e^{-t}u(t)$.

3-8. Show that the parameter Q, as defined in Eq. (3.4-24), is given by $Q = \omega_0 L/R$ for the series RLC circuit in Fig. P3-8. *Hint*: Take $v(t) = V \cos \omega_0 t$, and use the definition

$$Q = \frac{2\pi(\text{energy stored in the circuit})}{\text{energy dissipated per cycle}}.$$

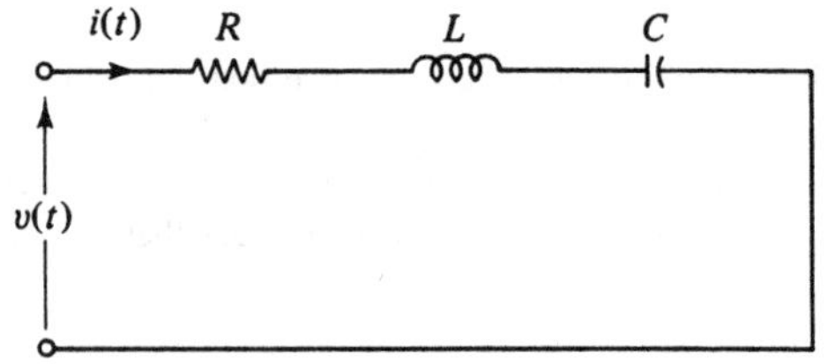

Figure P3-8

3-9. Consider the parallel RLC circuit shown in Fig. P3-9.

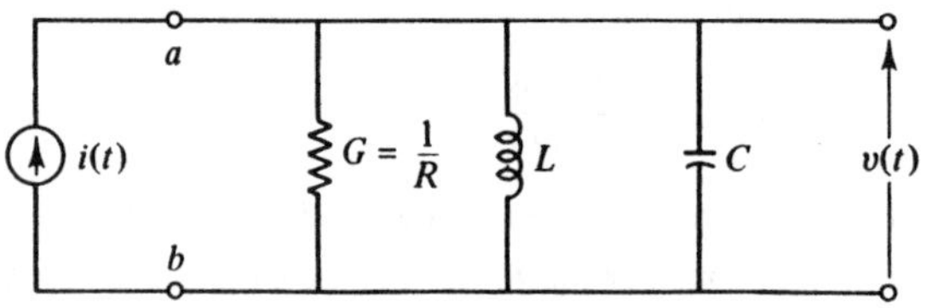

Figure P3-9

(a) Show that the differential equation relating the voltage $v(t)$ to the input current $i(t)$ is

$$C\frac{d^2v}{dt^2} + G\frac{dv}{dt} + \frac{1}{L}v(t) = \frac{di}{dt}.$$

(b) Show that when $v(t)$ and $i(t)$ are replaced by their Fourier transforms the impedance between a and b is

$$Z(j\omega) = \frac{V(j\omega)}{I(j\omega)} = \frac{j(\omega/\omega_0)}{QG[1 - (\omega/\omega_0)^2 + j(\omega/\omega_0)(1/Q)]},$$

where $Q = [\omega_0 LG]^{-1}$ and $\omega_0 = (LC)^{-1/2}$.

(c) Let $s = \sigma + j\omega$, and consider $Z(s)$. Discuss the locations of the poles in the complex s plane when $R \leq \omega_0 L/2$ and when $R > \omega_0 L/2$.

3-10. In this problem we consider some of the implications of Q. Assume where needed that $R > \omega_0 L/2$.

(a) Show that $Z(j\omega)$ in Prob. 3-9 can be written as

$$Z(j\omega) = \frac{R}{1 + jQ[(\omega^2 - \omega_0^2)/\omega\omega_0]}.$$

The half-power frequencies ω_1, ω_2 are determined from the points where $|Z(j\omega)|^2 = \frac{1}{2}|Z(j\omega)|^2_{max}$. Show that ω_1, ω_2 can be found from the equation

$$Q\frac{\omega^2 - \omega_0^2}{\omega\omega_0} = \begin{cases} 1, & \text{for } \omega = \omega_1, \\ -1, & \text{for } \omega = \omega_2, \end{cases}$$

and that $\omega_1 - \omega_2 = [RC]^{-1}$ and $\omega_1\omega_2 = \omega_0^2$.

(b) Let $\alpha \equiv [2RC]^{-1}$. Show that the 3-dB bandwidth B satisfies

$$\frac{B}{\omega_0} = \frac{2\alpha}{\omega_0} = \frac{1}{Q} \qquad (B \text{ in radians per second})$$

or, equivalently,

$$Q = \frac{f_0}{B} \qquad (B \text{ in hertz}).$$

For extracting the fundamental component of a square wave, would you choose a small or a large Q circuit?

(c) As Q increases, what happens to the poles of $Z(s)$? How would you expect the circuit to deal with transients?

(d) Take $i(t)$ in Prob. 3-9 as $i(t) = I\cos\omega_0 t$. Show that

$$\frac{2\pi(\text{peak energy stored})}{\text{energy dissipated per cycle}} = \frac{2\pi(\frac{1}{2}CV_1^2)}{(2\pi/\omega_0)(V_1^2/2R)} = \omega_0 CR = Q,$$

where $V_1 \equiv IR$.

3-11. Compute the Bode diagram (amplitude and phase curve) for the transfer function given by

$$H(j\omega) = \frac{j\omega + 100}{(j\omega + 10)(j\omega + 20)}.$$

3-12. Show for a causal system whose transfer function is $H(j\omega) = A(\omega) + jB(\omega)$ that

$$h(t) = \frac{2}{\pi}\int_0^\infty A(\omega)\cos\omega t\, d\omega$$
$$= -\frac{2}{\pi}\int_0^\infty B(\omega)\sin\omega t\, d\omega.$$

Hint: For a causal system, $h(t) = 0$ for $t < 0$. Therefore, by using the dual of the problem considered in Sec. 2.13 it can be shown that $A(\omega)$ and $B(\omega)$ are Hilbert transforms of each other.

3-13. Find the transfer function of the system shown in Fig. P3-13. Sketch the output for a rectangular pulse input of duration $\tau = T$, $\tau \ll T$, and $\tau \gg T$.

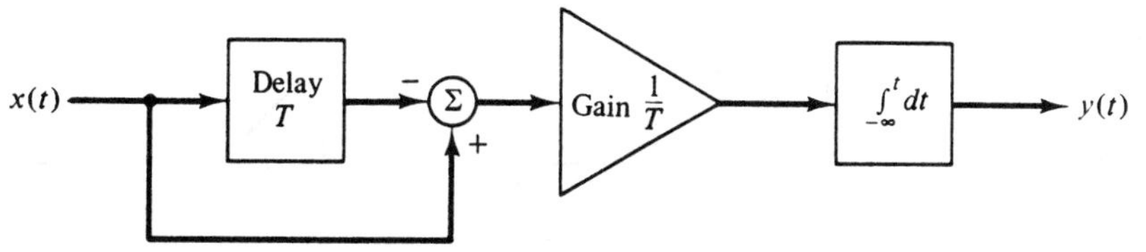

Figure P3-13

3-14. Compute the response of the circuit shown in Fig. P3-14 to an input $x(t) = \text{rect}(t/T)$. Obtain the solution with the help of the transfer function and the Fourier inversion theorem.

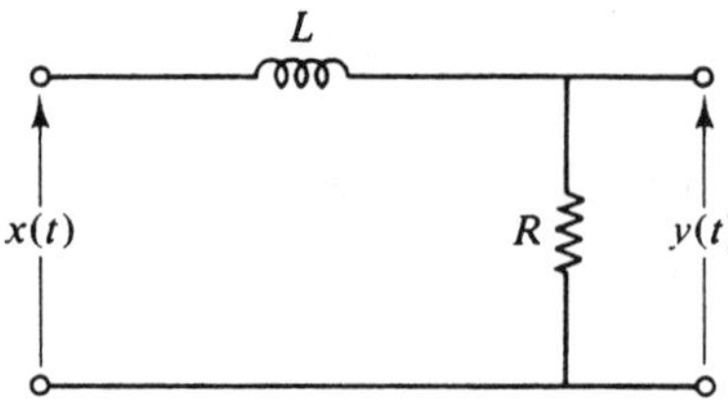

Figure P3-14

3-15. A series LC network with losses is sometimes modeled as shown in Fig. P3-15. Compute $Y(j\omega)/X(j\omega)$. Take $L/R_1 = R_2C$; show that the transfer function is then given by

$$H(s) = \frac{\alpha/(1+\alpha)}{(s^2/\omega_0^2) + (2/\sqrt{1+\alpha})(s/\omega_0) + 1},$$

where $\alpha = R_2/R_1$ and $\omega_0^{-2} = R_2LC/(R_1 + R_2)$. Plot the loci of the poles of $H(s)$ as α varies from ∞ to 0.

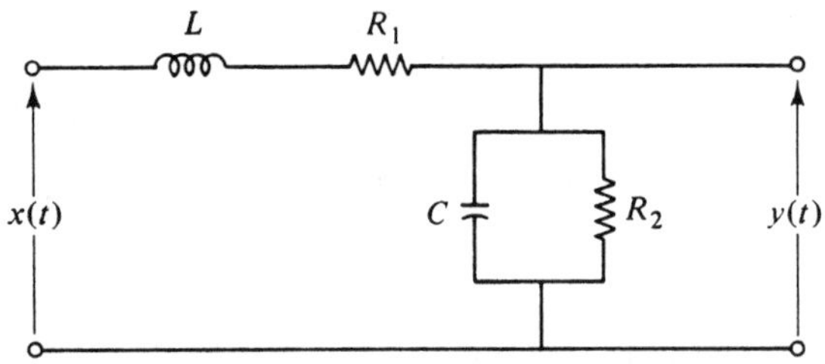

Figure P3-15

3-16. The input $x(t)$ (Fig. P3-16) is applied in turn to two systems with impulse responses as shown in Figs. P3-16(a) and (b), respectively. Sketch the output time function in each

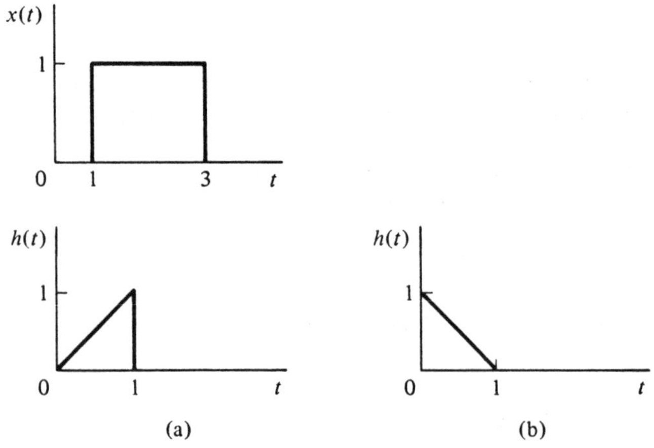

Figure P3-16

case, indicating the amplitude of the output pulse and the quantitative shapes of leading and trailing edges.

3-17. A linear system has the impulse response shown in Fig. P3-17(a) and is subjected to the indicated input. Find the output time function as follows:

(a) By formal evaluation of the convolution integral.

(b) By use of Fourier transforms.

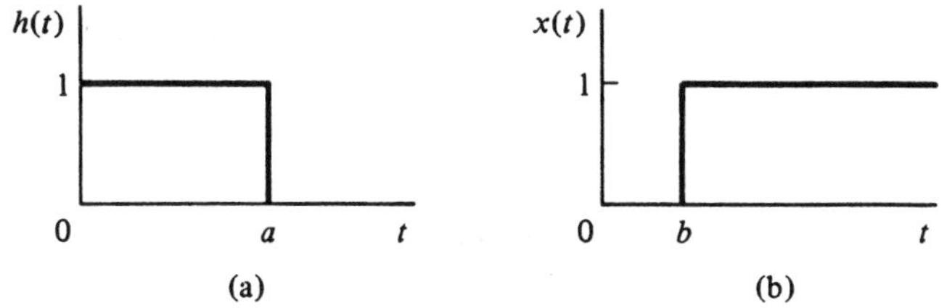

Figure P3-17

3-18. Use the alternative form of the convolution theorem given in Eq. (3.5-19) to compute the waveform shown in Fig. 3.5-5.

3-19. Assume that a time-invariant system is stable if it satisfies

$$\int_{-\infty}^{\infty} |h(t-\lambda)|d\lambda < \infty.$$

(a) Is the ideal integrator, that is, a system that furnishes an output $y(t)$ when the input is $x(t)$ according to

$$y(t) = \int_{-\infty}^{t} x(\lambda)\, d\lambda,$$

stable?

(b) Is the system in Fig. P3-19 stable? If the integration time is T, what inequality must the product RC satisfy for the circuit to behave as an integrator?

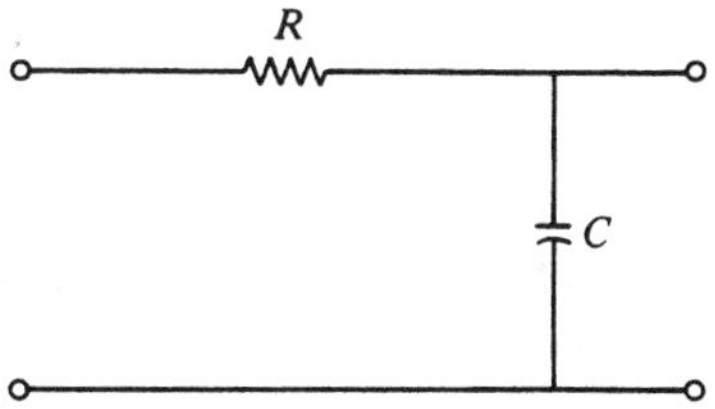

Figure P3-19

3-20. A network whose phase shift has ripples can produce echoes. Show this by considering

$$|H(f)| = 1, \qquad \theta(f) = -\omega t_0 + b \sin \omega t_d, \qquad \text{where } b \ll 1.$$

Hint: Use a series expansion for exp $(jb \sin \omega t_d)$.

3-21. A Butterworth low-pass filter has

$$|H(f)| = \left[1 + \left(\frac{f}{f_0}\right)^{2n}\right]^{-1/2},$$

where n is the number of reactive components (i.e., inductors or capacitors).

(a) Show that as $n \to \infty$ the amplitude response of the Butterworth filter approaches that of the ideal low-pass filter.

(b) Find n so that $|H(f)|^2$ is constant to within 1 dB over the range $|f| \leq 0.8 f_0$. Repeat for $|f| \leq 0.9 f_0$.

3-22. A Gaussian filter has

$$H(f) = \exp(-af^2).$$

Calculate the 3-dB bandwidth and

$$B_{eq} = \frac{1}{2H(0)} \int_{-\infty}^{\infty} |H(f)|\, df.$$

3-23. Compute and sketch the poles of a Butterworth filter of order 5. Repeat for a Butterworth filter of order 6.

3-24. Derive the transfer functions given in the text for the circuits shown in Fig. 3.9-1.

3-25. Derive Eq. (3.9-5), which gives the transfer function of the general prototype in Fig. 3.9-5(b).

3-26. Prove for the second-order low-pass filter whose transfer function is given in Eq. (3.9-6) that the roots will be complex when $1 < K < 5$.

3-27. It is possible to convert the noncausal low-pass impulse response given in Eq. (3.6-4) to a causal response by the simple expedient of multiplying $h(t)$ by the step function $u(t)$ and thereby removing the response for negative t. An almost equivalent procedure leading to a much simpler transfer function is to use a square pulse instead of a step; that is, consider the causal impulse response

$$h_c(t) = 2KB \operatorname{sinc}[2B(t - t_0)] \operatorname{rect}\left(\frac{t - t_0}{2t_0}\right).$$

(a) Sketch this function for $Bt_0 = 2$.

(b) Show that the corresponding transfer function is

$$H_c(f) = \frac{K}{\pi} e^{-j2\pi f t_0}\{\mathrm{Si}[2\pi t_0(f + B)] - \mathrm{Si}[2\pi t_0(f - B)]\},$$

where $\mathrm{Si}(x) = \int_0^x [(\sin t)/t]\, dt$ is the sine-integral function.

(c) Sketch $|H_c(f)|$ for $Bt_0 = 2$; observe that, instead of the square pulse corresponding to the noncausal impulse response, we now have an oscillatory approximation to a square pulse.

(d) Show for large Bt_0 that $H_c(t)$ approximates the noncausal frequency response except at the transition frequencies.

3-28. Consider the transformation in Eq. (3.10-4b), that is,

$$s \longrightarrow \frac{s^2 + \omega_c^2}{s},$$

in the transfer function of the low-pass filter shown in Fig. P3-28. Compare the position of the original poles with the position of the new poles.

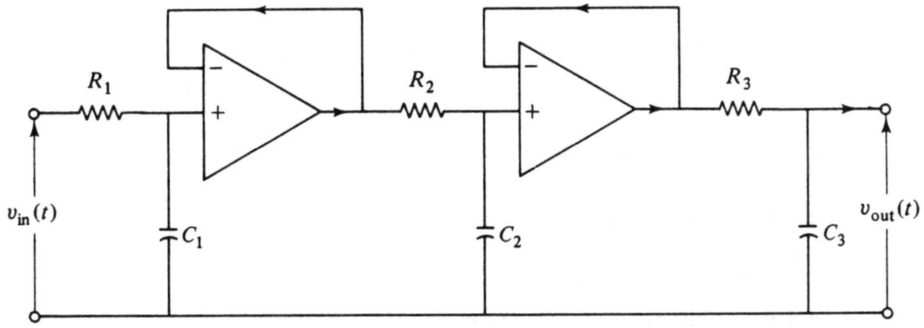

Figure P3-28

3-29. Derive Eq. (3.10-6) and show that the 3-dB frequencies are indeed $\omega_c + (\omega_0/2)$ and $\omega_c - (\omega_0/2)$ when Q is much larger than unity.

3-30. Show that the transfer function for the SAB band-pass circuit of Fig. 3.10-5(a) is given (for infinite op-amp gain) by

$$H(s) = \frac{-(1+b)R_2C_1s}{R_1R_2C_1C_2s^2 + [R_1(C_1+C_2) - bR_2C_1]s + 1},$$

where $b = R_b/R_a$.

3-31. Show that the transfer function for the SAB circuit of Fig. 3.10-5(b), with R_4 disconnected, has the general form of Eq. (3.10-9). (Assume that the op-amp has infinite input impedance, zero output impedance, and infinite gain.)

REFERENCES

[3-1] Ralph J. Schwarz and Bernard Friedland, *Linear Systems*, McGraw-Hill, New York, 1965, p. 12.

[3-2] H. W. Bode, *Network Analysis and Feedback Amplifier Design*, Van Nostrand Reinhold, New York, 1945.

[3-3] S. Ramo, J. R. Whinnery, and T. Van Duzer, *Fields and Waves in Communication Electronics*, Wiley, New York, 1967, p. 8.

[3-4] E. A. Guillemin, *The Mathematics of Circuit Analysis*, Wiley, New York, 1949.

[3-5] L. A. Zadeh, "Frequency Analysis of Variable Networks," *Proc. IRE*, **38,** pp. 291–299, March 1950.

[3-6] Raymond E. A. C. Paley and Norbert Wiener, "Fourier Transforms in the Complex Domain," *American Mathematical Society Colloquium Publication 19*, New York, 1934.

[3-7] A. N. Papoulis, *The Fourier Integral and Its Applications*, McGraw-Hill, New York, 1962, pp. 215–217.

[3-8] H. A. WHEELER, "The Interpretation of Amplitude and Phase Distortion in Terms of Paired Echos," *Proc. IRE*, **27,** pp. 359–385, June 1939.

[3-9] M. E. VAN VALKENBURG, *Introduction to Modern Network Synthesis*, Wiley, New York, 1960.

[3-10] B. O. PEIRCE, *A Short Table of Integrals*, Ginn, Boston, 1929, p. 75.

[3-11] L. WEINBERG, *Network Analysis and Synthesis*, McGraw-Hill, New York, 1962, Chap. 11.

[3-12] D. A. CALAHAN, *Modern Network Synthesis*, Vol. 1: Approximation, Hayden, New York, 1964.

[3-13] Y. J. LUBKIN, *Filter Systems and Design*, Addison-Wesley, Reading, Mass., 1970.

[3-14] JOHN L. HILBURN and DAVID E. JOHNSON, *Manual of Active Filter Design*, 2nd ed. McGraw-Hill, New York, 1983.

[3-15] ROBERT R. SHEPARD, "Active Filters: Part 12, Shortcuts to Network Design," *Electronics*, **42,** No. 17, pp. 82–91, Aug. 18, 1969.

[3-16] P. E. FLEISCHER, "Sensitivity Minimization in a Single Amplifier Biquad Circuit," *IEEE Trans. Circuits Sys.*, **CAS23,** No. 1, pp. 46–55, Jan. 1976.

[3-17] J. J. FRIEND, C. A. HARRIS, and D. HILBERMAN, "STAR: An Active Biquadratic Filter Section," *IEEE Trans. Circuits Sys.*, **CAS22,** No. 2, pp. 115–121, Feb. 1975.

[3-18] L. P. HUELSMAN, *Active Filters: Lumped, Distributed, Integrated, Digital and Parametric*, McGraw-Hill, New York, 1970.

[3-19] S. K. MITRA, ed., *Active Inductorless Filters*, IEEE Press, New York, 1971.

[3-20] R. W. NEWCOMB, *Active Integrated Circuit Synthesis*, Prentice-Hall, Englewood Cliffs, N.J., 1968.

[3-21] W. R. BENNETT, *Introduction to Signal Transmission*, McGraw-Hill, New York, 1970.

CHAPTER 4

Sampling and Quantization

4.1 INTRODUCTION

When one observes a typical signal, such as the output of a microphone picking up someone's speech, on an oscilloscope, one sees a very complicated looking random wave apparently able to take on an infinity of different values and shapes. However, closer examination of the waveform reveals that there are limits to its variability. One finds, for instance, that the wave does not have any abrupt breaks; that is, it is continuous. Also, on still closer examination (which may involve expansion of the time scale) one can see that all slopes and curvatures are finite. Thus, although quite random and undisciplined in first appearance, the wave is seen to be subject to some important constraints.

These constraints are imposed in part by the limitations of the physical apparatus (e.g., the human vocal tract) that generates the signal. These constraints translate into a limit on the lower and upper frequencies in the signal. For speech this ranges from about 100 to about 4000 Hz; for television signals it ranges from 0 to about 4 MHz, but it is always finite.

One important consequence of finite signal bandwidth is that signals can be represented by samples taken at discrete instants called *sampling instants*. As we shall show in the following sections, such sampling introduces essentially no loss, and it is possible to find interpolation functions that accurately reconstruct the original signal from the samples. The fact that band-limited signals can be accurately represented by discrete samples is of major importance in signal theory. It means that the undenumerable infinity of a continuous signal representation can be reduced to a countable infinity. Countability implies that the signal can be represented by a vector, and therefore the tools of vector-space mathematics can be utilized in the analysis of signals and of signal processing.

From the practical point of view, sampling makes it possible to deal with a signal only at isolated instants of time. This permits, for instance, time sharing of transmission facilities by a procedure referred to as time-division multiplexing. Perhaps even more importantly the waveform samples can be digitized (i.e., converted to numbers having only a finite number of digits). Although such a conversion is generally approximate, it can be made as exact as desired by using a large enough number of digits. By sampling and digitizing, signals can be converted into a series of numbers that can be handled by a digital computer or any other digital circuit. In fact, this is the only way in which signals can be processed by digital computers. Hence sampling of signals is an essential preprocessing step in all applications of digital systems to signal processing.

In the following sections we shall consider some of the consequences of the bandwidth constraint on signals. We derive several versions of the sampling theorem and show how the signal can be reconstructed from the samples. We study time-division multiplexing (TDM) of several signals and some of the problems, such as intersymbol interference, that have to be solved if TDM is to be practical. We then consider a number of digitizing schemes.

Before getting into a detailed discussion of sampling, we must briefly consider a general property of the frequency limitation. We saw in Sec. 2.9 that the Fourier spectrum of a time-limited signal such as the square pulse extends to infinite frequencies. By the duality principle, we conclude that a signal having a finite bandwidth must extend to infinity in time. Apparently, a signal cannot be finite in both time and frequency. This observation is more precisely enunciated in the following rule ([4-1], p. 70):

A strictly band-limited signal cannot be simultaneously time limited, and vice versa.

By strictly band limited we mean

$$X(f) = 0, \qquad \text{for } |f| > W, \tag{4.1-1}$$

and, similarly, strictly time limited means

$$x(t) = 0, \qquad \text{for } |t| > T. \tag{4.1-2}$$

The proof of this statement rests on the Paley–Wiener condition [Eq. (3.7-1)]. According to this criterion, if Eq. (4.1-2) holds, $x(t - T)$ can be regarded as the output of a causal system, and therefore $e^{j2\pi fT}X(f)$ cannot vanish for any finite (or infinite) frequency interval. The converse follows by duality.

Although the strict time and frequency limitations given in Eqs. (4.1-1) and (4.1-2) cannot be imposed simultaneously, we can deal with signals that are *essentially* time and frequency limited. This means, for instance, that the spectrum of a time-limited signal has only very small magnitude outside some frequency band. (For a good discussion of this subject, see [4-2].)

4.2 IMPLICATIONS OF THE FREQUENCY LIMITATION

We now consider a strictly band-limited signal, that is, one that satisfies Eq. (4.1-1). We assume also that $|X(f)|$ remains finite throughout the interval $-W < f < W$. It then follows that $x(t)$ and all its time derivatives exist, since the nth derivative of $x(t)$ is

$$\frac{d^n x(t)}{dt^n} = (2\pi j)^n \int_{-W}^{W} f^n X(f) e^{j2\pi ft}\, df, \tag{4.2-1}$$

and by hypothesis the integral on the right exists for any finite n. Formally, $x(t)$ can therefore be expanded in a Taylor series about any (finite) particular value of $t = t_0$:

$$\begin{aligned} x(t) &= x(t_0) + \frac{dx(t_0)}{dt}(t - t_0) + \frac{1}{2!}\frac{d^2x(t_0)}{dt^2}(t - t_0)^2 + \cdots \\ &= \sum_{n=0}^{\infty} \frac{x^{(n)}(t_0)}{n!}(t - t_0)^n. \end{aligned} \tag{4.2-2}$$

This series converges for all t. To show this, we calculate the remainder in Taylor's formula

$$x(t) = \sum_{n=0}^{N} \frac{x^{(n)}(t_0)}{n!}(t - t_0)^n + R_N(t).$$

As is shown in any standard calculus text,

$$R_N(t) = \frac{x^{(N+1)}(\xi)(t - t_0)^{N+1}}{(N + 1)!},$$

where ξ lies between t and t_0. Assume that the maximum absolute value of $X(f)$ in the interval $-W \le f \le W$ is X_m; such a maximum exists because we required $X(f)$ to be finite. Then $|x^{(N+1)}(\xi)|$ is bounded from above according to

$$|x^{(N+1)}(\xi)| \le 2WX_m(2\pi W)^{N+1},$$

and therefore

$$|R_N(t)| \le \frac{2WX_m(2\pi W|t - t_0|)^{N+1}}{(N + 1)!}. \tag{4.2-3}$$

It is easily shown (for instance, by examining the ratio R_{N+1}/R_N) that $\lim_{N\to\infty} R_N(t) = 0$. Therefore, by the definition of analyticity ([4-3], p. 128), $x(t)$ is analytic at t. This means that if $x(t)$ and all derivatives are known at a particular value of time t_0, $x(t)$ can be completely predicted for all future times. However, such predictability is inconsistent with the requirement that an information-bearing signal must be unpredictable to the receiver. Thus, in a sense, we find that a strictly band-limited signal cannot be used to convey information.

The paradox here again lies in the requirement that $x(t)$ be strictly band limited. This requirement is violated if the signal contains even a tiny amount of random noise. As shown in Chapter 9, the spectrum of *white* noise extends over a very broad band (theoretically infinite) and therefore the combination of signal and noise is not strictly band limited in the sense used here. The practical effect of noise is to make the measurement of $x(t_0)$ and the derivatives of $x(t)$ uncertain, and therefore the distance over which $x(t)$ can be accurately extrapolated becomes, in fact, quite small.

Sampling Theorem

The Taylor series expansion can be thought of as a means of expressing a continuous waveform in terms of a countable set of measurements. Although the Taylor series is not a particularly useful way to do this, the fact that it converges for band-limited signals suggests that there may be other expansions that would be more useful.

Probably the most useful of these is the sampling theorem. In a somewhat general form this theorem states that a signal whose Fourier transform vanishes for all frequencies above W Hz is completely specified by samples taken at the rate of at least $2W$ samples per second.

The general principle was demonstrated by Nyquist [4-4] by considering a section of length T of a signal $x(t)$ having a bandwidth of W and regarding it as one period of a periodic signal of period T. The T-second section can be expanded in a Fourier series. The fundamental frequency in this expansion is $2\pi/T$ radians per second. Because $x(t)$ has a bandwidth of W, it is argued, albeit not quite correctly, that the Fourier expansion has a maximum frequency of $2\pi W = 2\pi n_{max}/T$ radians per second. Then if WT is assumed to be an integer, we find that $n_{max} = WT$.

This argument is not completely correct, because as we have seen earlier a time function of finite duration has an infinite bandwidth. It therefore cannot be exactly represented by a finite Fourier series. Hence the Fourier series

$$x_s(t) = c_0 + \sum_{n=1}^{WT} \left[a_n \cos \frac{2\pi nt}{T} + b_n \sin \frac{2\pi nt}{T} \right] \tag{4.2-4}$$

can be only approximately equal to the T-second section of $x(t)$. The approximation can, however, be very good, especially in regions away from the end points of the T-second interval and if WT is very large.

Suppose we know the values of $x_s(t) \simeq x(t)$ at the points $0 \le t_0 < t_1 < t_2 < \cdots < t_{2TW} \le T$. Then Eq. (4.2-4) can be regarded as a set of $2WT + 1$ simultaneous equations in the unknowns c_0, a_n, and b_n. These equations can be solved if they are all independent (i.e., if the coefficient matrix is nonsingular). [The elements of this matrix are in the form cos $(2\pi nt_i/T)$, sin $(2\pi nt_i/T)$, or 1, depending on whether they multiply a_n's, b_n's, or c_0.] There are many ways of choosing the sample times t_i, $i = 0, 1, 2, \ldots, 2WT$, to make the coefficient matrix nonsingular; in fact, almost any random choice will do. Once c_0, a_n, and b_n are known, $x_s(t)$ is, in principle, specified.

We see that $2WT + 1$ sample values are required, or if $WT >> 1$, approximately $2WT$ values. We say that the signal has $2WT$ *degrees of freedom*. Clearly, if the sample values are uniformly spaced, the sampling rate is $2W$ samples per second. There is, however, no need for the samples to be uniformly spaced. Also note that we can differentiate Eq. (4.2-4) once or several times and generate different equations in which some derivative of $x(t)$ appears on the left. Thus some of the $2WT$ samples can be of $x(t)$, others of $\dot{x}(t)$, and so on.

Beyond having a theorem which states that $x(t)$ is specified by a countable number of samples, one would like some simple formula for interpolating $x(t)$ for values of t other than the sampling times. Although Eq. (4.2-4) is potentially such a formula, at least when $2WT >> 1$, it is not particularly simple. For one thing, it does not provide an explicit connection between the sample values and the function. Also, it provides no information about how well a small number of terms of the summation approximates the function. Finally, it is not exact.

The commonly used version of the sampling theorem is the uniform sampling theorem, also referred to as the Whittaker–Shannon sampling theorem.

Uniform Sampling Theorem

If $X(f)$, the Fourier transform of $x(t)$, is band limited such that $X(f) = 0$ for $|f| \geq W$, then $x(t)$ is completely specified by instantaneous samples uniformly spaced in time, with a uniform intersample period T_s such that $T_s \leq 1/(2W)$. At times other than the sampling instants, $x(t)$ is given by the interpolation formula

$$x(t) = \sum_{n=-\infty}^{\infty} x(nT_s) \operatorname{sinc}\left(\frac{t - nT_s}{T_s}\right). \tag{4.2-5}$$

As we shall show, the summation on the right is an exact representation of $x(t)$. Note that the sample values $x(nT_s)$ appear explicitly in this formula. Because of the orthogonality of the sinc functions, only one term of the summation is nonzero at the sampling instants. Although interpolation to other instants requires theoretically an infinite number of terms, the fact that the sinc function goes to zero for large arguments makes it possible to approximate $x(t)$ quite well by a relatively small number of samples in the neighborhood of t. This is shown in Fig. 4.2-1. Note that the contributions from samples other than the ones in the immediate neighborhood of a particular point t tend to cancel.

The maximum sampling interval is $T_s = 1/(2W)$. This interval is known as the Nyquist interval, and the corresponding sampling rate as the Nyquist rate. In practice, a shorter interval is often used to account for the fact that signals are never exactly band limited. A commonly used value is $T_s = 0.7/(2W)$.

To prove the sampling theorem, suppose that $X(f) = 0$ for $|f| \geq W$. It is then possible to regard $X(f)$ as a single cycle of a spectrum that repeats periodically with a period (in frequency) of $2W$ (see Fig. 4.2-2). This can be compactly expressed by the identity

$$X(f) = \operatorname{rect}\left(\frac{f}{2W}\right) \sum_{n=-\infty}^{\infty} X(f - 2nW). \tag{4.2-6}$$

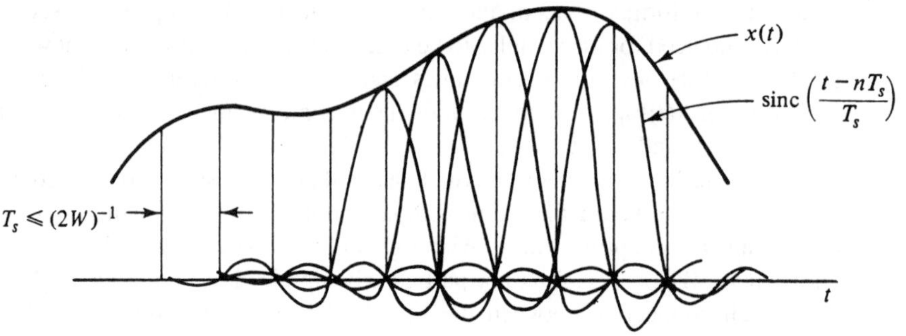

Figure 4.2-1. Between-sample interpolating property of the sinc functions.

It is shown in Chapter 2 [Eq. (2.12-20)] that the inverse Fourier transform of the summation in Eq. (4.2-6) is the "comb"

$$\frac{1}{2W} \sum_{n=-\infty}^{\infty} x\left(\frac{n}{2W}\right) \delta\left(t - \frac{n}{2W}\right).$$

Therefore, if we inversely Fourier-transform Eq. (4.2-6), we get

$$\begin{aligned} x(t) &= 2W \operatorname{sinc} 2Wt * \left[\frac{1}{2W} \sum_{n=-\infty}^{\infty} x\left(\frac{n}{2W}\right) \delta\left(t - \frac{n}{2W}\right)\right] \\ &= \sum_{n=-\infty}^{\infty} x\left(\frac{n}{2W}\right)\left[\operatorname{sinc} 2Wt * \delta\left(t - \frac{n}{2W}\right)\right], \end{aligned} \tag{4.2-7}$$

and this is identical with Eq. (4.2-5) if T_s is set equal to $1/(2W)$.

It is worth examining Eq. (4.2-7) more carefully, since it actually describes the ideal sampling and recovery process in some detail. Thus, consider the comb expression

$$\sum_{n=-\infty}^{\infty} x\left(\frac{n}{2W}\right) \delta\left(t - \frac{n}{2W}\right).$$

This can also be written in the form of the product $x(t) \sum_{n=-\infty}^{\infty} \delta[t - (n/2W)]$ and can therefore be thought of as the operation of a delta-function sampler operating on the original signal $x(t)$ (see Fig. 4.2-3). Although delta-function samplers are not practical, a real sampler consisting of a switch (electronic or otherwise) that periodically connects

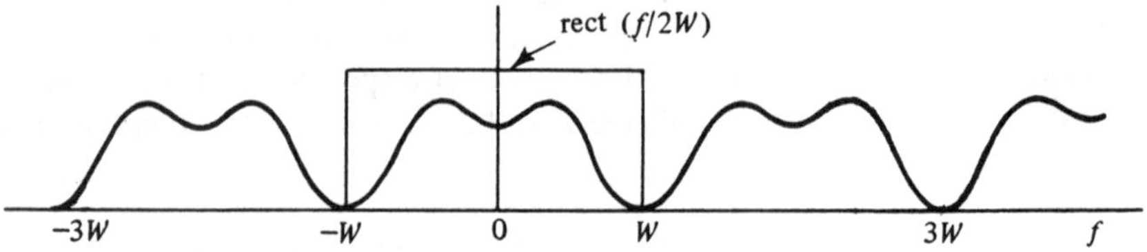

Figure 4.2-2. Periodic spectrum.

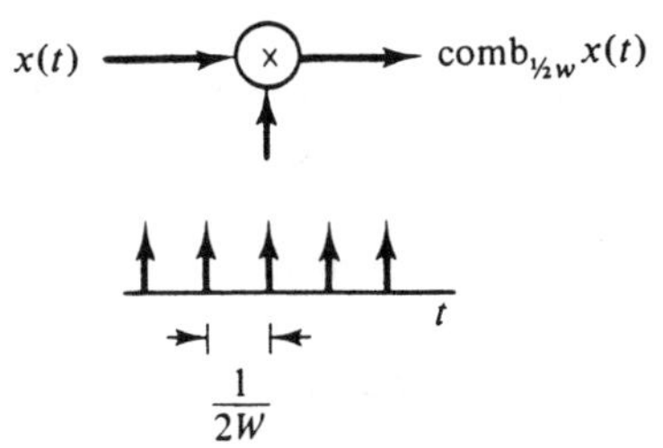

Figure 4.2-3. Ideal sampler.

the signal to the output can approximate such samplers. The approximation can be made very good by making the "on" time of the switch very short. This, in effect, replaces the delta-function sequence by a sequence of narrow pulses with a very short duty cycle. Since $T_s\, x(t) \sum_{n=-\infty}^{\infty} \delta(t - nT_s)$ and $\sum_{n=-\infty}^{\infty} X[f - (n/T_s)]$ are Fourier pairs, we see that the sampling operation results in a periodic spectrum with period $1/T_s$. Thus the periodic spectrum that we introduced somewhat artificially in the proof of the sampling theorem actually exists in the system: it is the spectrum at the output of the sampler. It then follows that multiplication of the spectrum by rect $(f/2W)$ or the equivalent, convolution of the sampled signal by $2W$ sinc $2Wt$, can be regarded as the operation of a reconstruction filter that converts the sampler output back to the original signal. By Eqs. (4.2-6) and (4.2-7), this filter is an ideal low-pass filter (in the sense discussed in Chapter 3) with passband W (see Fig. 4.2-4).

We now have another reason the sampling interval must be less than $1/(2W)$. The generation of the periodic spectrum will take place no matter what the sampling interval is; but if the sampling interval exceeds $1/(2W)$, the period of the spectrum is less than $2W$, and therefore there will be spectral overlap. This is shown in Fig. 4.2-5.

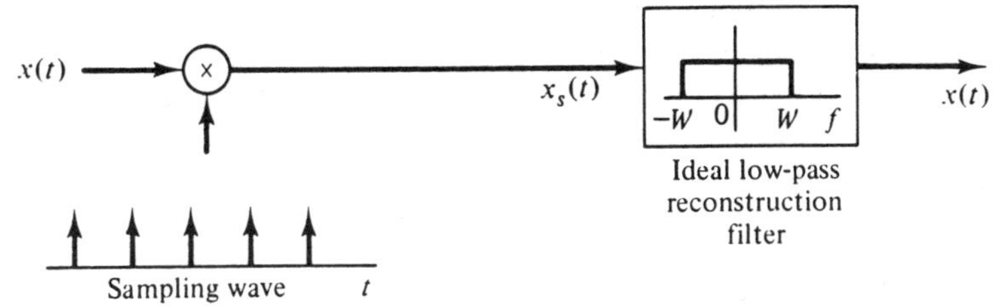

Figure 4.2-4. Complete sampling and reconstruction system.

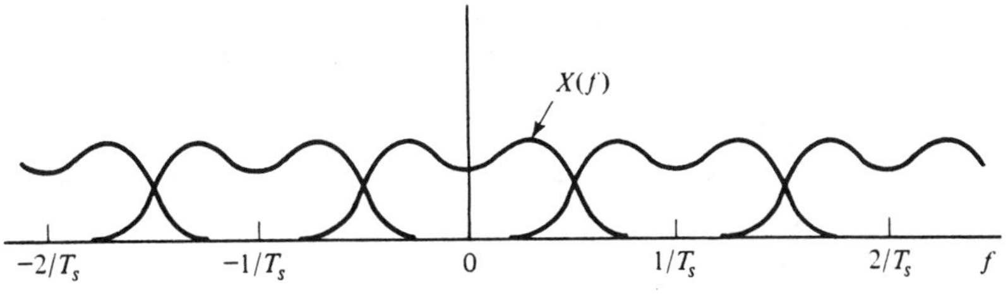

Figure 4.2-5. Spectral overlap.

Note that this overlap, also called *spectral folding* or *aliasing*, causes distortions in the original spectrum. Frequency components at the high-frequency edge of the spectrum are reflected down to lower frequencies. For example, if a signal containing the typical 60-Hz line-frequency noise is sampled at 100 samples/sec, corresponding to a folding frequency of 50 Hz, a nonzero spectral component will appear at 40 Hz even though the original signal had no energy at 40 Hz.

The effect is illustrated in another way in Fig. 4.2-6, which shows a sine wave sampled at a rate slightly larger than half of its frequency (actually considerably less than the Nyquist rate). The signal reconstruction from the sample values looks like a low-frequency sine wave; it has taken an "alias."

Since aliasing distortion causes high-frequency spectral components to appear at lower frequencies, it seriously affects the intelligibility of speech signals. The kind of spectral reversal produced by aliasing is in fact the basis for a form of speech scrambler used to preserve confidentiality in telephone conversations. For this reason, a low-pass filter is frequently used *ahead* of the sampler to prevent aliasing. The cutoff frequency of this low-pass filter is set at something like 35% or 40% of the sampling frequency to make certain that there are no significant spectral components at half the sampling frequency or above. The elimination of the high-frequency components in the signal obviously degrades the fidelity of transmission to some extent, but the loss of intelligibility is much less than it would be if aliasing were permitted. As an example, in time-division multiplexing of telephone conversations (discussed in Sec. 4.4) the telephone signals are passed through a low-pass filter with cutoff frequency of 3.2 kHz before being sampled at an 8-kHz sampling frequency.

Although the mapping from $x(t)$ to the sample values $x(nT_s)$ is unique, the inverse mapping from the sequence of samples to the continuous time function is not. In fact, the sequence of samples can yield any function $y(t) = x(t) + z(t)$, where $x(t)$ is given by Eq. (4.2-5) and $z(t)$ is an arbitrary function that is zero at all the sampling instants (see Fig. 4.2-7). Note that $z(t)$ has a minimum frequency of $W = 1/2T_s$ Hz. This phenomenon was first noted by E. T. Whittaker in 1915 [4-5], and it is clear that the function $x(t)$ given by Eq. (4.2-5) is only the *lowest-frequency* function that can be

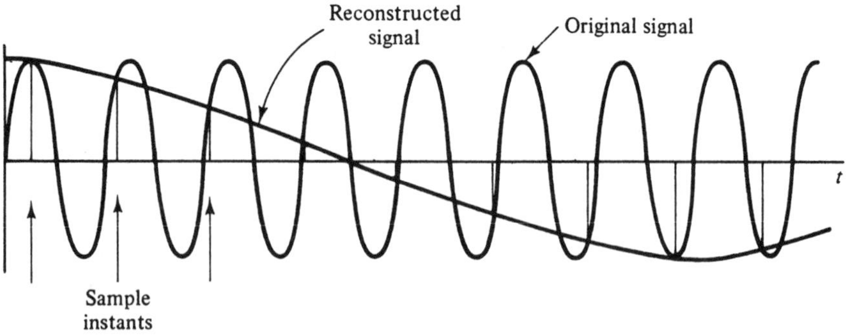

Figure 4.2-6. Showing the effect of sampling below the Nyquist rate.

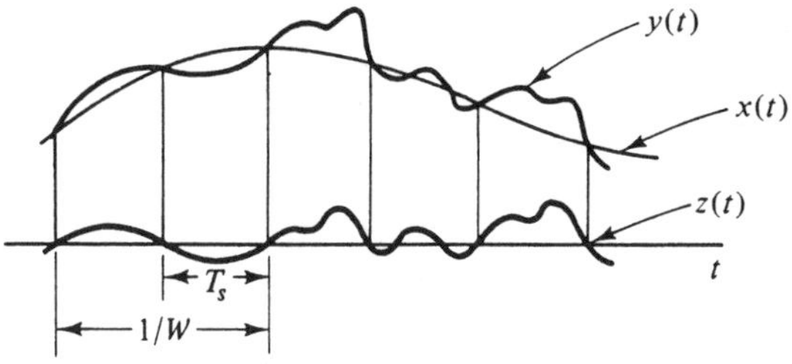

Figure 4.2-7. Illustration of nonuniqueness of inverse mapping of sample values.

passed through the set of sample points. The statement of the theorem guarantees that this function has a bandwidth of *less than* W, whereas the function $y(t)$ formed by adding a nonzero $z(t)$ to $x(t)$ must have a bandwidth of *at least* W. Thus there is no inconsistency.

Sampling of Band-Pass Signals

For band-pass signals, we assume that $X(f)$ is nonzero only in the bands $|f - f_0| < W/2$, $|f + f_0| < W/2$, as shown in Fig. 4.2-8. Although such a signal can be regarded as having a spectrum extending from $-f_0 - W/2$ to $f_0 + W/2$, one can reduce the sampling rate greatly by taking advantage of the fact that the effective bandwidth is much smaller.

As in previous discussions of band-pass signals, we find it convenient to deal with the *complex analytic form* $\xi(t)$ of the signal $x(t)$ (see Sec. 2.13) for which Re $\xi(t) = x(t)$, Im $\xi(t) = \hat{x}(t)$. The spectrum of $\xi(t)$ exists only for positive frequencies and is given there by $\Xi(f) = 2X(f)$.

We can then use the same procedure as in the baseband case. The band-limited spectrum is regarded as one cycle of a periodic spectrum and is written in the form

$$\Xi(f) = \text{rect}\left(\frac{f - f_0}{W}\right) \sum_{n=-\infty}^{\infty} \Xi(f - nW). \tag{4.2-8}$$

Fourier-transforming and using Eq. (2.12-20), we obtain

$$\begin{aligned} \xi(t) &= \sum_{n=-\infty}^{\infty} \xi\left(\frac{n}{W}\right)\left[\text{sinc}\,(Wt)e^{j2\pi f_0 t} * \delta\left(t - \frac{n}{W}\right)\right] \\ &= \sum_{n=-\infty}^{\infty} \xi\left(\frac{n}{W}\right) \text{sinc}\,(Wt - n)e^{j2\pi f_0[t-(n/W)]}. \end{aligned} \tag{4.2-9}$$

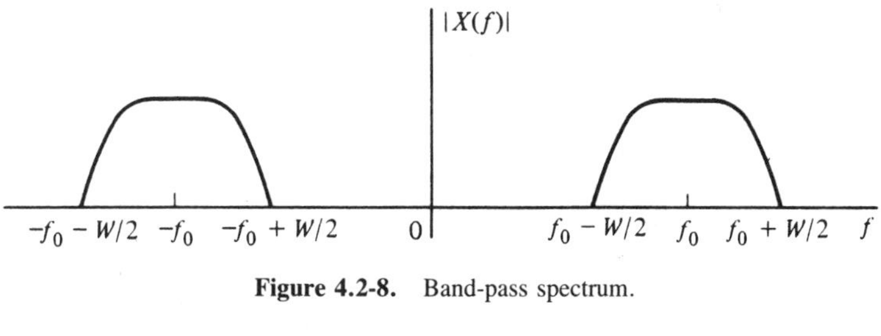

Figure 4.2-8. Band-pass spectrum.

To get back the original signal, take the real part of Eq. (4.2-9):

$$x(t) = \text{Re} \sum_{n=-\infty}^{\infty} \left[x\left(\frac{n}{W}\right) + j\hat{x}\left(\frac{n}{W}\right) \right] \text{sinc}\,(Wt - n) e^{j2\pi f_0[t-(n/W)]}$$

$$= \sum_{n=-\infty}^{\infty} \text{sinc}\,(Wt - n) \left[x\left(\frac{n}{W}\right) \cos \omega_0 \left(t - \frac{n}{W}\right) \right. \tag{4.2-10}$$

$$\left. - \hat{x}\left(\frac{n}{W}\right) \sin \omega_0 \left(t - \frac{n}{W}\right) \right]$$

$$= \sum_{n=-\infty}^{\infty} \text{sinc}\,(Wt - n) \left| \xi\left(\frac{n}{W}\right) \right| \cos \left\{ \omega_0 \left(t - \frac{n}{W}\right) \right. \tag{4.2-11}$$

$$\left. + \arg \left[\xi\left(\frac{n}{W}\right) \right] \right\}.$$

Note that the Nyquist sampling interval is now $1/W$; that is, it is twice as long as for baseband sampling. However, at each point it is necessary to make a measurement of a complex quantity, which means either the real and imaginary part, or the amplitude and phase. Thus at each sampling point there are two measurements, and therefore the minimum number of uniform samples in time T is $2WT$, which is the same as for the baseband signal of bandwidth W.

Here again it is possible, and usually desirable, to sample the signal at a higher rate to account for the fact that the spectrum is not precisely band limited.

4.3 PRACTICAL ASPECTS OF SAMPLING

In practice, sampling is performed by high-speed switching circuits. Typical circuits utilize field-effect transistors (FETs) as the switching element, together with suitable drivers to apply the sampling pulses. The effect is to connect the input to the output during the time that the sampling signal is "high" and to disconnect the output when it is "low." Since the switching pulses have finite width, this circuit approximates the delta-function sampling pulses by short rectangular pulses. An equivalent circuit employing a mechanical switch and the resulting output signal are shown in Fig. 4.3-1.

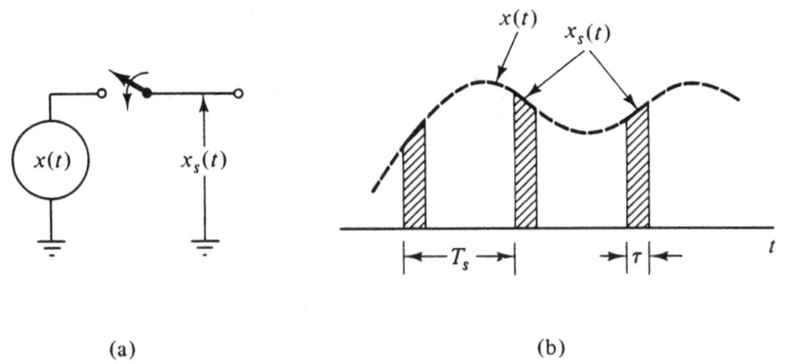

Figure 4.3-1. (a) Elementary sampling circuit and (b) sampled wave.

The effect of the finite width of the sampling pulses is easily investigated. The sampled wave $x_s(t)$ can be written as

$$x_s(t) = x(t) \cdot g(t), \tag{4.3-1}$$

where $g(t)$ is the sampling function given by

$$g(t) = \sum_{n=-\infty}^{\infty} \text{rect}\left(\frac{t - nT_s}{\tau}\right). \tag{4.3-2}$$

The Fourier series expansion of the sampling function was derived in Chapter 2 [Eq. (2.4-20)] and is given by

$$\begin{aligned} g(t) &= \sum_{n=-\infty}^{\infty} d \, \text{sinc}\,(nd) e^{j2\pi nt/T_s} \\ &= d[1 + 2 \sum_{n=1}^{\infty} \text{sinc}\,(nd) \cos n\omega_s t], \end{aligned} \tag{4.3-3}$$

where $d = \tau/T_s$ is the duty cycle and $\omega_s = 2\pi/T_s$ is the sampling frequency. Substituting Eq. (4.3-3) into Eq. (4.3-1) gives

$$x_s(t) = d[x(t) + 2 \sum_{n=1}^{\infty} \text{sinc}\,(nd) x(t) \cos n\omega_s t], \tag{4.3-4}$$

and therefore the Fourier spectrum of the sampled signal is

$$X_s(\omega) = dX(\omega) + d \sum_{n=1}^{\infty} \text{sinc}\,(nd)[X(\omega - n\omega_s) + X(\omega + n\omega_s)]. \tag{4.3-5}$$

This spectrum is illustrated in Fig. 4.3-2 under the assumption that $x(t)$ is sampled at a sufficiently high frequency so that there is no aliasing, that is, such that $X(\omega)$ does not extend beyond $\pm\omega_s/2$.

Note that the finite width of the sampling pulses causes each lobe of the spectrum to be multiplied by d sinc (nd). It does not alter the shape of the central lobe of the spectrum. An ideal low-pass filter with a passband of $\omega_s/2$ can therefore be used to recover the spectrum of the original signal. Thus the finite width of the sampling pulses is seen to have no important effects on the sampling process.

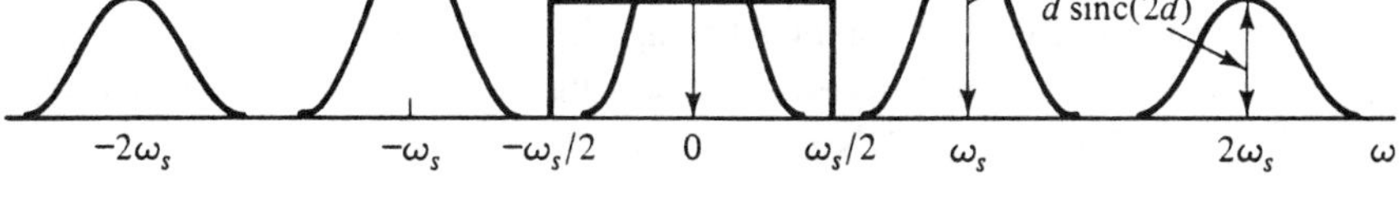

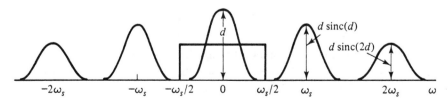

Figure 4.3-2. Effect of finite-width sampling.

Shape of the Sampling Pulses

Electronic switching circuits typically will not produce perfect rectangular sampling pulses; thus we consider arbitrary pulse shapes. Again, let

$$x_s(t) = x(t) \cdot g(t) \tag{4.3-6}$$

but with

$$g(t) = \sum_{n=-\infty}^{\infty} p(t - nT_s), \tag{4.3-7}$$

where $p(t)$ is some arbitrary pulse shape. Fourier transforming gives

$$\begin{aligned} X_s(f) &= X(f) * \frac{1}{T_s} \sum_{n=-\infty}^{\infty} P\left(\frac{n}{T_s}\right)\delta\left(f - \frac{n}{T_s}\right) \\ &= \frac{1}{T_s} \sum_{n=-\infty}^{\infty} X\left(f - \frac{n}{T_s}\right) P\left(\frac{n}{T_s}\right) \\ &= \left(\frac{1}{T_s}\right) X(f) \cdot P(0) + \frac{1}{T_s} \sum_{\substack{n=-\infty \\ n \neq 0}}^{\infty} X\left(f - \frac{n}{T_s}\right) P\left(\frac{n}{T_s}\right). \end{aligned} \tag{4.3-8}$$

The Fourier transform of $g(t)$ was obtained by using the Poisson sum formula [Eq. (2.12-20)]. We see that the lobes of the sampled spectrum are now multiplied by $P(n/T_s)$; however, since the central lobe is multiplied by the constant $P(0)$, sampling with arbitrary pulses that multiply the original signal is essentially the same as ideal sampling.

Instantaneous Sampling

Certain practical sampling circuits such as the sample-and-hold (S/H) circuit, to be discussed shortly, sample the signal over a very short time but deliver pulses that are stretched out in time. One reason for doing this is that the transmission of very narrow pulses requires an excessive bandwidth in all the circuits making up the transmission channels; for example, for delta-function pulses this bandwidth would be infinite. The stretching out of the pulses may therefore be intentional, right in the sampler, or it may simply be a consequence of the filtering action of the circuits that handle the pulses after they are generated. If we suppose that the impulse response of the stretching circuit is $h(t)$, then the sampled output is

$$x_s(t) = \left[x(t) \cdot \sum_{n=-\infty}^{\infty} \delta(t - nT_s)\right] * h(t) = \sum_{n=-\infty}^{\infty} x(nT_s)h(t - nT_s). \tag{4.3-9}$$

Such a sampled wave for $h(t)$ having the form of a square pulse is shown in Fig. 4.3-3.[†] Note the difference between this figure and Fig. 4.3-1, which represents a type of sampling sometimes referred to as *natural sampling* [4-6].

[†] As shown in Fig. 4.3-3, this is not causal. Why?

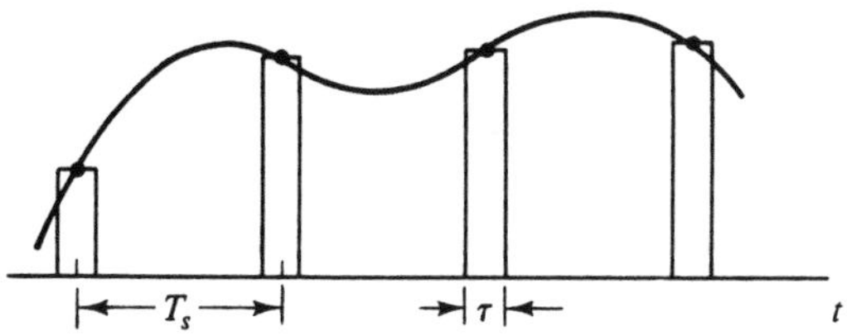

Figure 4.3-3. Instantaneous sampling.

The Fourier spectrum of the sampled signal of Eq. (4.3-9) is given by

$$\begin{aligned} X_s(f) &= \left[X(f) * \frac{1}{T_s} \sum_{n=-\infty}^{\infty} \delta\left(t - \frac{n}{T_s}\right)\right] H(f) \\ &= \frac{1}{T_s} \sum_{n=-\infty}^{\infty} X\left(f - \frac{n}{T_s}\right) H(f), \end{aligned} \tag{4.3-10}$$

where $H(f)$ is the Fourier transform of $h(t)$. As before, only the central lobe of this spectrum is passed by the reconstruction filter, but this is now given by $X(f)H(f)$. Thus instantaneous sampling alters the spectrum of the signal, and to recover the original spectrum it would be necessary to use a compensating or equalization filter having the transfer function $1/H(f)$.

In practical circuits, sampling cannot be truly instantaneous; in fact, the output of the sampler is generally a weighted average of the input during a short but finite sampling period. If $h(t)$ is the impulse response of such a weighting circuit, the sampled output and its spectrum are also given by Eqs. (4.3-9) and (4.3-10). Both effects (i.e., the weighting over a finite time and the subsequent holding or stretching) may occur together, and the resulting output will again be given by Eqs. (4.3-9) and (4.3-10), but with a more complicated $h(t)$. Thus Eqs. (4.3-9) and (4.3-10) represent a general form of *filtered sampling* characterized by the fact that the signal is convolved with the sampling function rather than being multiplied as in natural sampling.

Sample-and-Hold Circuit

A special case of filtered sampling occurs in the very frequently used sample-and-hold (S/H) circuit, shown in ideal form in Fig. 4.3-4. During the time that the sampling pulse is "high" the switch is on, and the capacitance C charges up to the value of the input signal. During the remainder of the sampling period the switch is off, and the signal is held on the capacitor. Ideally, the switch is "on" for only an instant during which the capacitor charges up to the exact value of the input. Then the S/H output is the staircase signal shown in Fig. 4.3-4(b). This is described by

$$x_s(t) = \sum_{n=-\infty}^{\infty} x(nT_s) \operatorname{rect}\left[\frac{t - (n + 1/2)T_s}{T_s}\right], \tag{4.3-11}$$

which is in the form of Eq. (4.3-9) with

$$h(t) = \operatorname{rect}\left[\frac{t - (T_s/2)}{T_s}\right]. \tag{4.3-12}$$

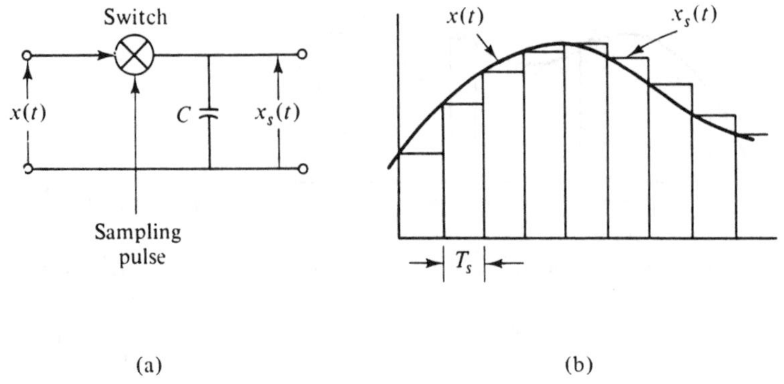

Figure 4.3-4. (a) Sample-and-hold (S/H) circuit. (b) S/H output.

The hold time of the S/H circuit need not extend over the entire sampling period T_s; in fact, it is frequently much shorter to permit sampling pulses from several signals to be interleaved. This is discussed more fully later. Clearly, if the hold time is $T \leq T_s$, the transfer function of the S/H circuit is

$$H(f) = T\ \text{sinc}\ (Tf)e^{-j\pi Tf}. \tag{4.3-13}$$

Practical S/H circuits differ from the ideal in that the switch has to be "on" for a finite time. Since the charging current is limited by the combined switch resistance and source output impedance, the capacitor voltage can reach only some fraction of the input during the sample part of the cycle. There is also a certain amount of leakage from the capacitor during the hold period because of finite load resistance. The size of the capacitor used in any given application is, in fact, generally chosen as a best compromise between minimizing the charging time constant and maximizing the leakage time constant.

Nonideal Reconstruction Filters

Practical filters don't have an infinitely sharp frequency cutoff characteristic, and therefore a certain amount of signal energy from the adjacent lobes of the sampled spectrum is always passed. This is shown in Fig. 4.3-5; the part shown shaded represents undesirable leakage of signal from adjacent spectral lobes. The effect is a high-frequency hiss or

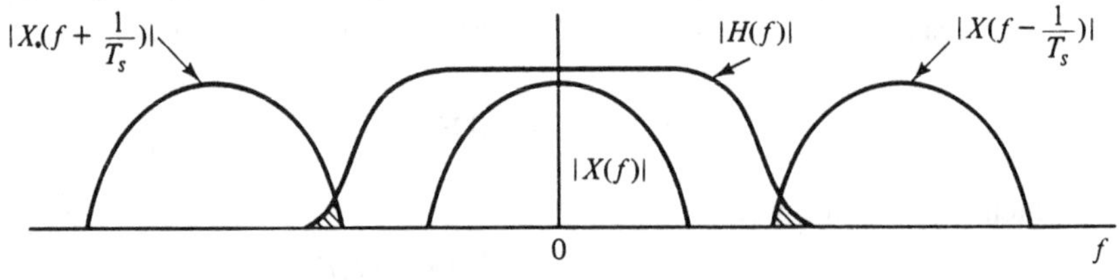

Figure 4.3-5. Reconstruction filter with a passband that is too large.

whistle in the reconstruction. It can be eliminated by narrowing the bandwidth of the filter or, more effectively, by using a higher sampling frequency to provide more separation between the spectral lobes. Note that the effect discussed here is different from aliasing, where significant portions of the spectral lobes overlap. The distortion caused by aliasing can only be eliminated by using a higher sampling frequency or by reducing the signal bandwidth before sampling, while the high-frequency noise considered here can at least in theory also be eliminated by using a sharper cutoff reconstruction filter.

4.4 TIME-DIVISION MULTIPLEXING

There are many applications of the principle of sampling in communications systems. One of the most important is time-division multiplexing (TDM), which is commonly used to simultaneously transmit several different signals over a single channel. Each of the signals is sampled at a rate in excess of the Nyquist rate. The samples are interleaved, and a single composite signal consisting of all the interleaved pulses is transmitted over the channel. At the receiving end the interleaved samples are separated by a synchronous switch or *demultiplexer*, and then each signal is reconstructed from the appropriate set of samples.

The samples can be coded into pulses suitable for transmission in a number of different ways. The most obvious is probably pulse amplitude modulation (PAM), where the signal is converted into a train of pulses whose amplitude is proportional to the amplitude of the signal at the sample points. However, it is also possible to keep the pulse amplitudes constant and to vary their width or their position relative to some reference point in proportion to the values of the signal sample. This results in pulse duration modulation (PDM) and pulse position modulation (PPM), respectively. Other schemes include pulse code modulation (PCM), where the pulses represent the signal in some form of digital code. For the purpose of our discussion of TDM, we assume that PAM is used. We consider the various other types of pulse modulation in the next two sections.

Time-division multiplexing is widely used in telephony, telemetry, radio, and data processing. Data about temperature, magnetic intensity, and vehicle attitude, measured on various instruments aboard a spacecraft, are almost always transmitted this way. All the instrument outputs are sampled at regular intervals, and the samples are impressed on a single radio-frequency carrier that is beamed to earth. In telephony, both frequency-division multiplexing (FDM) and TDM are jointly used to permit many hundreds of different conversations to utilize the same microwave link. When a computer is used to monitor several different experiments, the outputs of various instruments are multiplexed so that the computer can deal with them sequentially, one at a time.

The PAM–TDM system is illustrated in Fig. 4.4-1, where we show an elementary switching circuit that acts to multiplex the N signals $s_1(t)$, $s_2(t)$, . . . , $s_N(t)$. Instead of the mechanical commutator shown in the figure, practical multiplexers are generally all electronic and are available in the form of integrated circuits.

Figure 4.4-1 shows how the signals are sampled and how the samples are interleaved

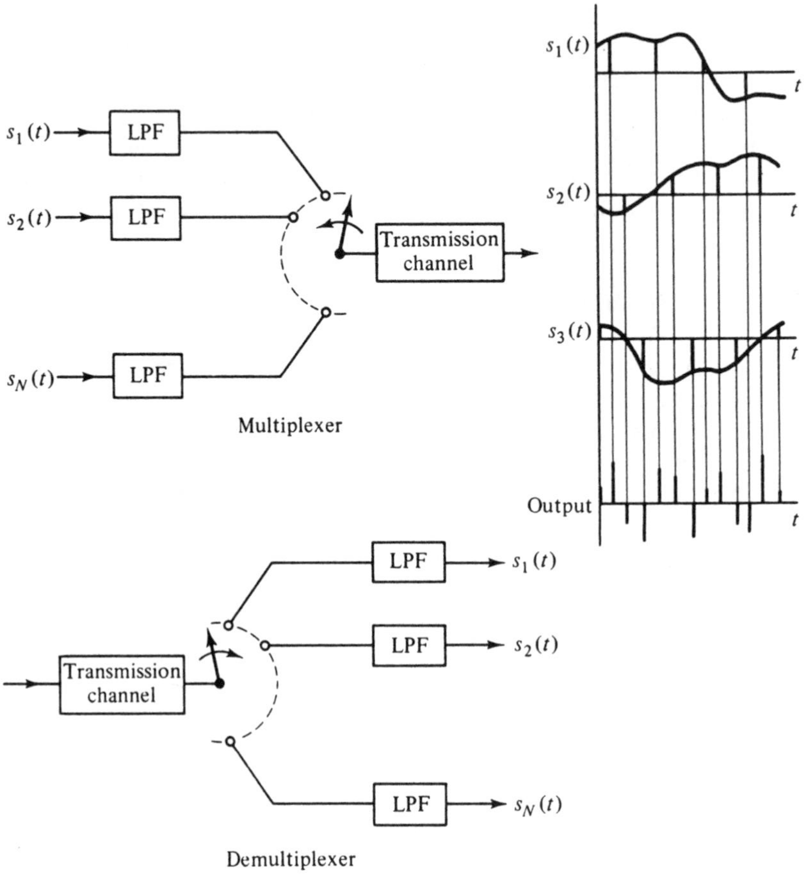

Figure 4.4-1. Time-division multiplexing.

before being transmitted. At the receiver a demultiplexer separates the interleaved samples and sends them on to appropriate reconstruction filters. It is obvious that the multiplexer and demultiplexer must operate in precise synchronism. Synchronization is, in fact, one of the major problems, especially with high-speed data systems and where the transmitter and receiver are physically far apart. In such systems, synchronization is generally achieved by utilizing very stable local *clock* oscillators. These, in turn, are kept in synchronism by transmitting an occasional special synchronizing signal. It is also possible to derive timing information from the signal pulses themselves by averaging over long periods of time. The synchronization problem and the various methods used in practice to solve it are discussed in more detail in the literature ([4-7] and [4-8]).

For channels involving a radio link, the pulses are themselves modulated onto a carrier using AM or FM modulation, discussed in Chapters 6 and 7. For this purpose it is often undesirable to have both positive and negative pulses. Also, synchronization may be simplified if the pulses never go to zero (i.e., so that a positive pulse is always present at each pulse time). Both of these objectives can be easily achieved by adding a dc level larger than the largest expected signal amplitude to all signals before sampling. The resulting pulse train is referred to as unipolar PAM, while the kind of system shown in Fig. 4.4-1 is bipolar. For very long-distance communication, where transmitter power may be at a premium, bipolar PAM has the advantage over unipolar PAM of requiring less signal power for the same signal-to-noise ratio at the receiver.[†]

An important consideration in TDM is the bandwidth required for the channel. An estimate of this is easily obtained if it is assumed that the individual signals $s_1(t)$, . . . , $s_N(t)$ all have the same bandwidth W. (See Prob. 4-12 for an example of unequal bandwidths.) Then the sampling rate for each signal must be at least $2W$ samples/sec; but to provide for the less-than-perfect cutoff characteristics of practical low-pass filters the sampling rate should be $2W_1$, where W_1 exceeds W by a small amount called a *guard band*. For instance, if the $\{s_i(t)\}$ are telephone conversations, $W = 3.2$ kHz and the sampling rate might be 8 kHz, providing a 1.6-kHz guard band. The number of revolutions per second at which the rotary switch in Fig. 4.4-1 rotates is equal to the sampling rate. If there are N signals, the length of each sampling pulse cannot exceed $1/N$ of a revolution or $1/(2W_1N)$ sec.

Suppose for the moment that the pulses are square (as in Fig. 4.3-3). The Fourier spectrum for a square pulse of length τ is $\tau \operatorname{sinc}(\tau f)$. The bandwidth may be defined in terms of the first zero crossing of the sinc function and is therefore given by

$$B = \frac{1}{\tau}, \tag{4.4-1}$$

and for $\tau = 1/(2W_1N)$ the channel bandwidth is

$$B = 2W_1N \text{ Hz}. \tag{4.4-2}$$

This is the bandwidth for pulses with maximum duty cycle, shorter pulses requiring a larger bandwidth.

At the expense of somewhat poorer separation between adjacent pulses, one can define B as the 3-dB bandwidth, which for square pulses is approximately $1/(2\tau)$. This gives

$$B_{3\text{ dB}} = \frac{1}{2\tau} = W_1N \text{ Hz}. \tag{4.4-3}$$

The bandwidth requirements appear to be more or less what one would expect: proportional to the signal bandwidth and to the number of signals to be multiplexed. This result is very similar to that obtained for frequency-division multiplexing (FDM), where different signals are sent on different frequency carriers (see Sec. 6.10). Thus, insofar as required channel bandwidth is concerned, these two techniques are approximately equivalent.

[†] This is discussed in more detail in Chapter 10.

Intersymbol Interference

The question of whether the bandwidth should be calculated according to Eq. (4.4-2) or (4.4-3), or some other way, can be answered much more precisely by considering the spillover from one pulse into the adjacent time slot and setting some upper limit on the permissible amount. Such spillover is called *intersymbol interference*, and in a TDM system it results in cross talk (i.e., a signal in one channel can be heard in an adjacent channel). Cross talk is generally highly objectionable. The amount of interference clearly depends on the channel bandwidth, because if the bandwidth is too small, pulses that

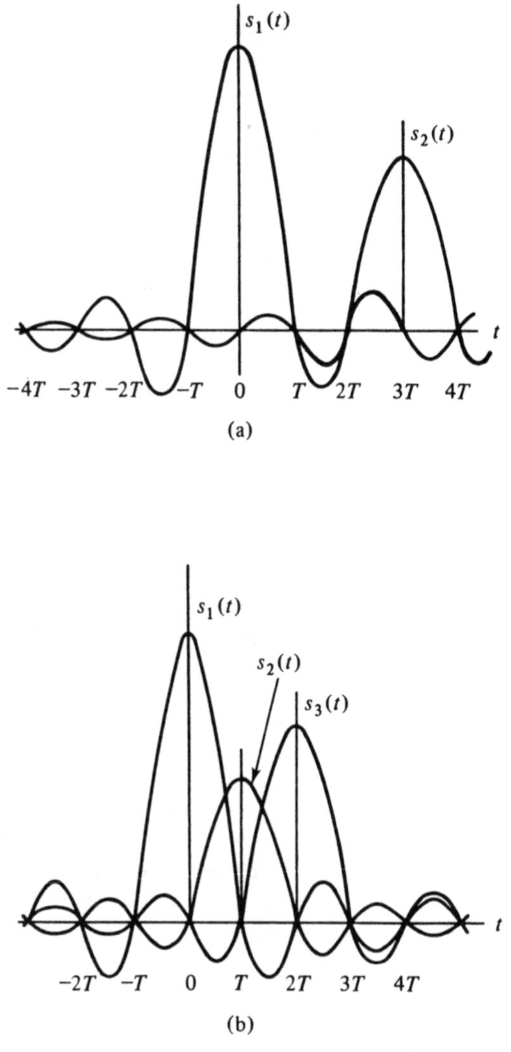

Figure 4.4-2. Zero intersymbol interference with sinc pulses. (a) Two sinc pulses whose maxima are separated by an integral number of sample points. (b) Three sinc pulses as close together as possible without interfering with each other.

are well separated at the point of origin will be spread out by the time they reach the receiver. However, additional factors are the pulse shape and the method used by the receiver to detect the pulses.

Thus, suppose that the receiver samples the signal using instantaneous sampling. Also, suppose that the pulse shape at the receiver has its maximum at the sampling point and goes through zero at all adjacent sampling points (see Fig. 4.4-2); then there will be no intersymbol interference, even though the received pulses persist over several time slots. In fact, this result is obtained if the received pulses have the form sinc $[(t - nT)/T]$, where $n = 0, \pm 1, \pm 2, \ldots$ and where T is the time between samples. This shape is obtained by using delta-function pulses at the transmitter and giving the remainder of the channel the characteristic of an ideal low-pass filter with a passband $W = 1/(2T)$ (see Fig. 4.4-3). Since by our previous discussion $T = 1/(2W_1N)$, this argument gives a channel bandwidth of W_1N, that is, as in Eq. (4.4-3).

In practice, a somewhat larger bandwidth has to be used because the ideal low-pass filter characteristic is difficult to realize and also because any inexactness in the sampling instants would again result in intersymbol interference. In fact, since the sinc function decays very slowly, it is even possible (although not likely) that the sum of pulse tails at times other than integral multiples of T could diverge. The object is to retain the property that the pulse goes through zero at adjacent sample points, but in addition the tails should also be small so that small jitter in sampling time at the receiver does not cause large intersymbol interference. It was shown by Nyquist [4-4] that this objective can be obtained quite easily (albeit at the expense of increasing the channel bandwidth) by using a channel filter characteristic having a more gradual falloff characteristic than the rect $(\cdot)$ transfer function shown in Fig. 4.4-3. Nyquist's result has been extended by Gibby and Smith [4-9], whose conclusions we briefly outline here.

Suppose that the channel output signal in Fig. 4.4-3 is $h(t)$. For zero intersymbol interference from transmitted pulses spaced T seconds apart, we must have

$$h(mT) \equiv h_m = \begin{cases} h_0, & \text{for } m = 0, \\ 0, & \text{for } m \neq 0. \end{cases} \tag{4.4-4}$$

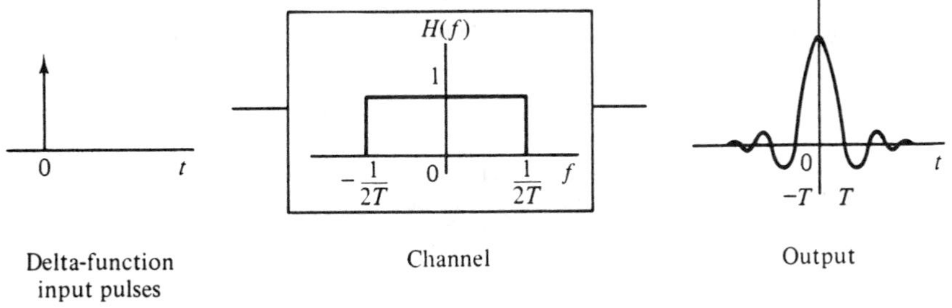

Figure 4.4-3. Output signal of a channel having the ideal low-pass filter characteristic when the input is an impulse.

To translate this constraint into a requirement on the channel transfer function $H(f)$, we consider the periodic (in frequency) function $\Sigma_{n=-\infty}^{\infty} H[f - (n/T)]$. By use of the Poisson sum formula [actually the dual of Eq. (2.12-18)], we can write

$$\sum_{n=-\infty}^{\infty} H\left(f - \frac{n}{T}\right) = T \sum_{m=-\infty}^{\infty} h_m e^{-j2\pi mTf}. \tag{4.4-5}$$

Hence, substituting the constraint equation (4.4-4) into Eq. (4.4-5), we find that $H(f)$ must satisfy

$$\sum_{n=-\infty}^{\infty} H\left(f - \frac{n}{T}\right) = Th_0. \tag{4.4-6}$$

In other words, the sum of $H(f)$ and all its replicas shifted by intervals of n/T in frequency must be a constant. In general, $H(f)$ is complex, and therefore Eq. (4.4-6) should be written in the form

$$\begin{aligned} \sum_{n=-\infty}^{\infty} \operatorname{Re} H\left(f - \frac{n}{T}\right) &= Th_0, \\ \sum_{n=-\infty}^{\infty} \operatorname{Im} H\left(f - \frac{n}{T}\right) &= 0. \end{aligned} \tag{4.4-7}$$

We may now choose any function $H(f)$ having a gradual roll-off and satisfying Eq. (4.4-6) or (4.4-7). The example considered by Nyquist was the *raised cosine*

$$H(f) = \begin{cases} \frac{1}{2}(1 + \cos \pi f T), & |f| < \frac{1}{T}, \\ 0, & \text{otherwise}. \end{cases} \tag{4.4-8}$$

This satisfies Eq. (4.4-5), as can be seen in Fig. 4.4-4. The corresponding impulse response is

$$h(t) = \frac{1}{T}\left[\operatorname{sinc}\frac{2t}{T} + \frac{1}{2}\operatorname{sinc}\left(\frac{2t}{T} + 1\right) + \frac{1}{2}\operatorname{sinc}\left(\frac{2t}{T} - 1\right)\right]. \tag{4.4-9}$$

This function is shown in Fig. 4.4-5(b). Note that $h(t)$ is exactly zero at all adjacent sample points, but it also is quite small at all other $|t| > T$. Practical filters whose transfer functions satisfy Eq. (4.4-7) are relatively easy to construct.

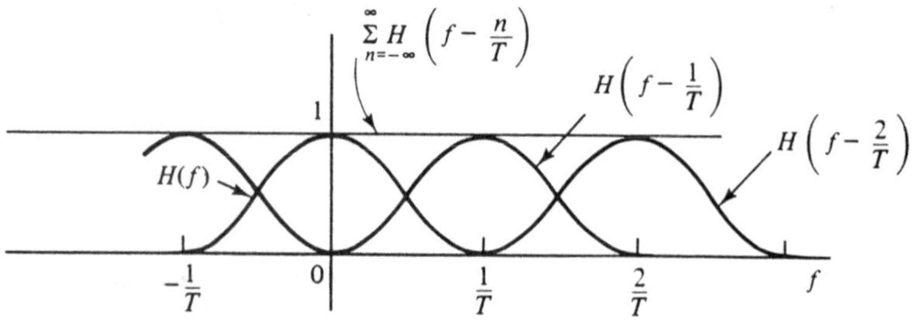

Figure 4.4-4. The sum of raised cosine functions shifted by intervals of n/T is constant.

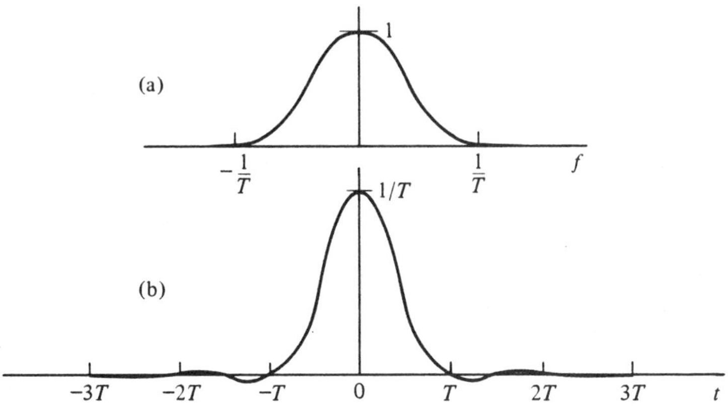

Figure 4.4-5. (a) Raised cosine transfer function and (b) its inverse Fourier transform.

Observe that the bandwidth for the raised cosine spectrum is $1/T = 2W_1N$, as given in Eq. (4.4-2).

In practice, the received pulse shape is determined not only by shaping filters at the transmitter but also by characteristics of the transmission medium. The latter may not always be the same; for instance, in telephony different trunk lines generally have different transmission loss functions. To prevent intersymbol interference from this source, adaptive equalizing filters are inserted at the transmission line terminals ([4-8] and [4-10]).

In Sec. 10.6 we will return to intersymbol interference in the context of digital modulation methods.

4.5 PULSE DURATION AND PULSE POSITION MODULATION

Besides PAM, two additional analog methods for impressing the sample value on a pulse-train carrier are pulse duration modulation† (PDM) and pulse position modulation (PPM). Typical pulse shapes for the three techniques are shown in Fig. 4.5-1. If $s(t)$ is the signal that modulates square pulses, mathematical expressions for the three forms of pulse modulation are as follows:

$$\text{PAM:} \qquad x(t) = \sum_{n=-\infty}^{\infty} [s(nT) + K] \operatorname{rect}\left[\frac{t - nT}{\tau}\right], \tag{4.5-1}$$

where $K \geq |s(t)|_{\max}$ is a constant added to the signal to prevent the pulses from becoming negative, τ is the pulse width, and T is the pulse period.

$$\text{PDM:} \qquad x(t) = \sum_{n=-\infty}^{\infty} \operatorname{rect}\left[\frac{t - nT}{\tau}\right], \qquad \tau = \tau(s) = as(nT) + K. \tag{4.5-2}$$

† PDM is similar to the halftone process used for printing photographs in books and newspapers.

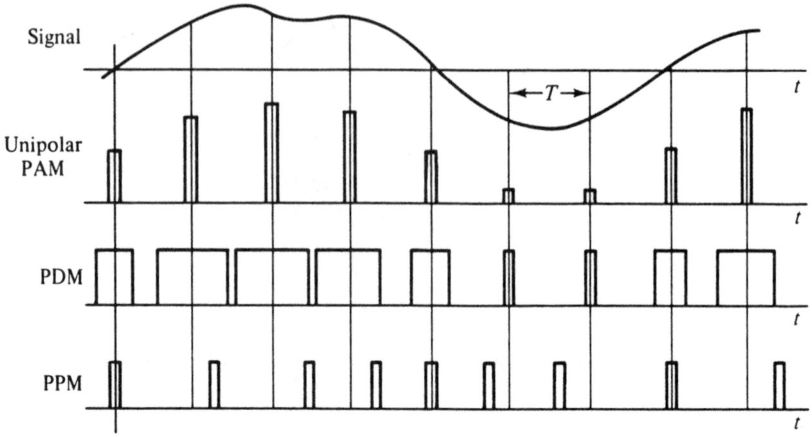

Figure 4.5-1. Analog pulse modulation methods.

Here the pulse width τ is made proportional to $s(nT)$,[†] and it is necessary to choose a and K such that $0 < \tau < T$; that is,

$$K > a|s(t)|_{max},$$
$$K + a|s(t)|_{\max} < T.$$

$$\text{PPM:} \quad x(t) = \sum_{n=-\infty}^{\infty} \text{rect}\left[\frac{t - nT - \alpha}{\tau}\right], \qquad (4.5\text{-}3)$$
$$\alpha = as(nT).$$

The pulse position is proportional to $s(nT)$, and the maximum shift must be less than $\frac{1}{2}(T - \tau)$; that is,

$$a\,|s(t)|_{max} < \tfrac{1}{2}(T - \tau).$$

A number of variations of these pulse trains are frequently encountered. Bipolar PAM has already been mentioned. The PDM train shown is symmetric around the sampling point, but frequently the leading edge is fixed and only the trailing edge is modulated. The PPM pulse train shown is bipolar in the sense that negative signal levels result in pulses that precede the sample instant. In practice, unipolar PPM, where the pulse always trails the sampling instant, may be easier to generate.

The main advantage of PDM and PPM over PAM is the fixed amplitude of the pulses, which makes them relatively immune from additive noise. Also, amplifiers and repeaters need not be linear; in fact, limiting or clipping amplifiers that eliminate all amplitude variation and transmit only the rise and fall times of the pulses are commonly used.

[†] Although proportional modulation is simplest, other monotonic functions $f[s(nT)]$ could be used as well.

Greater noise immunity could be achieved if the pulses were square, since additive disturbances would leave the rise and fall times unchanged. In practice, the pulses cannot be square because of finite channel bandwidths, and therefore they are affected by noise to some extent. This is shown in Fig. 4.5-2. We assume that the receiver determines the pulse arrival time by observing the instant at which the trailing edge of the pulse passes a certain threshold (usually set at half of the pulse amplitude). Such threshold circuits are referred to as *slicing circuits*, and the threshold is called the *slicing level*. Note that for a square pulse the arrival time is unaffected by the noise as long as the noise amplitude is less than the slicing level. However, for a pulse with a finite slope, there is an error that is proportional to the reciprocal of the slope of the trailing edge. It is easily demonstrated that the slope is proportional to the bandwidth. For example, consider a unit amplitude Gaussian pulse,

$$s(t) = e^{-\pi B^2 t^2}, \tag{4.5-4}$$

with Fourier transform

$$S(f) = \frac{1}{B} e^{-\pi f^2/B^2}. \tag{4.5-5}$$

We can loosely define the bandwidth by B. The slope is given by

$$s'(t) = \frac{ds(t)}{dt} = -2\pi B^2 t e^{-\pi B^2 t^2}, \tag{4.5-6}$$

and if t is evaluated, say, for a slicing level of $e^{-1/2}$, we get

$$t_0 = \pm \frac{1}{\sqrt{2\pi}\,B}, \tag{4.5-7}$$

and

$$|s'(t_0)| = \sqrt{2\pi}\, B e^{-1/2}. \tag{4.5-8}$$

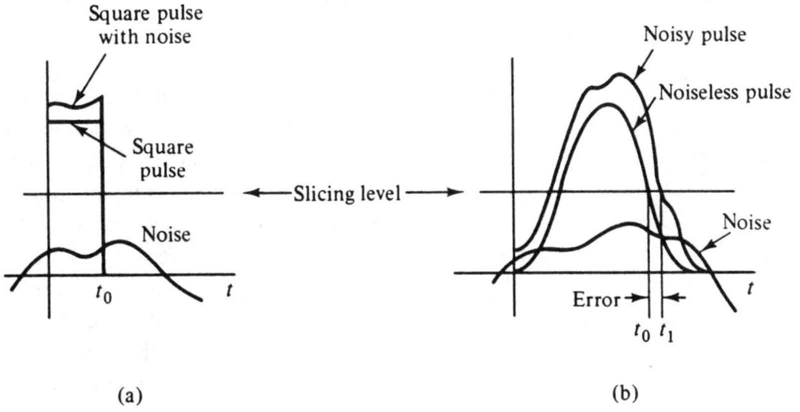

Figure 4.5-2. Noise effect in pulse demodulation. (a) Square pulse. (b) Pulse with finite bandwidth.

Thus, for a particular noise amplitude, the error will be inversely proportional to B, or the mean-square error is inversely proportional to B^2.

There is an inconsistency in this argument because we have assumed tacitly that the noise will remain unchanged as the bandwidth of the channel is increased. Actually, if we increase the bandwidth to make the pulses more nearly square, we generally also increase the noise power. The analysis of this problem is therefore more complicated. The general problem of the relation between bandwidth and signal-to-noise ratio appears again in our discussion of FM (Chapter 7) and is discussed in more detail in Chapter 9.

Generation of PDM and PPM

A simple circuit that will generate both PDM and PPM is shown in Fig. 4.5-3. A sawtooth or triangular wave whose fundamental frequency is the desired sampling rate is added to the signal, and the sum is passed into a slicing circuit. The slicing circuit is simply a comparator that puts out a "high" signal when $v_1 > v_2$ and a "low" signal otherwise. The slicing level is the value of the voltage v_2. The use of a triangular wave results in symmetrical PDM, whereas a sawtooth wave generates either leading-edge or trailing-edge modulation. The capacitor in the output differentiates the comparator output voltage, producing a positive spike when this voltage goes positive and a negative spike when it goes negative. The diode eliminates the negative spikes.

As shown in Fig. 4.5-3, the sampling produced by this circuit is nonuniform, because, in effect, sampling occurs whenever the combined wave crosses the slicing threshold in the positive direction and not when the signal level matches the level of the biasing waves. This kind of sampling is referred to as *natural* sampling and gives rise to some distortion ([4-11], [4-12]). Observe that the term *natural sampling* has a different meaning for PDM and PPM than for PAM (see Fig. 4.3-1).

Uniform sampling can be achieved with this circuit if the input is first passed through a sample-and-hold circuit. The S/H circuit performs the actual sampling, while the combination of triangle wave and slicing circuit determines the length or location of the respective PDM or PPM pulse.

Because of the complicated way in which signal information is transferred to pulse length or pulse position, the computation of the Fourier spectrum of PDM and PPM is not straightforward. Much of the early work on this problem is due to W. R. Bennett, and a fairly complete treatment can be found in [4-12] and [4-6]. Simplified analyses leading to approximate forms of the spectrum can be found in ([4-13] and [4-14]). One of the most important properties of the spectrum of PDM is that, if the pulse-repetition frequency is much higher than the highest signal frequency and if the modulation index a in Eq. (4.5-2) is small, then the lowest harmonic contains the signal. This can be understood by observing that under these conditions the width of, and therefore the energy in, each pulse is proportional to the signal. As a consequence, the signal can be recovered by simply passing the PDM wave through a low-pass filter that responds to the slowly changing average value of the pulses but rejects frequency components at the pulse-repetition frequency.

The Fourier spectrum of PPM can be shown to contain a term proportional to the

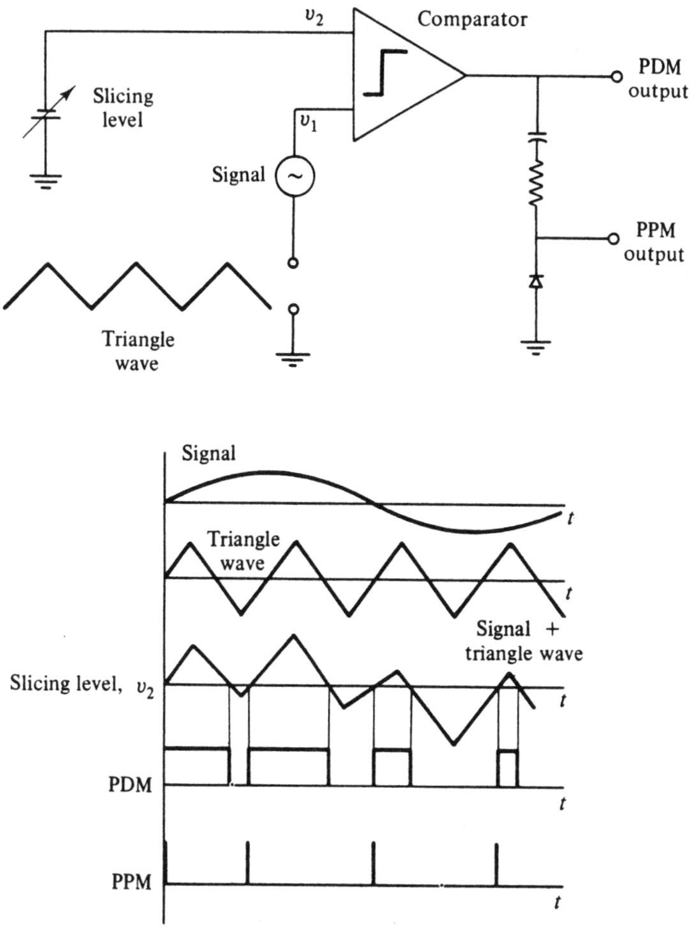

Figure 4.5-3. Generation of PDM and PPM.

derivative of the baseband signal. Hence PPM can be demodulated by passing the pulses through a low-pass filter that passes the signal and rejects harmonic components at the pulse-repetition frequency, followed by an integrator. In practice, PPM signals are often demodulated by first converting them to PDM. This can be done by using each pulse of the PPM signal to trigger a bistable flip-flop and resetting the flip-flop with pulses from a clock oscillator running at a constant pulse-repetition rate equal to the average rate of the PPM pulse stream. The resulting PDM is then demodulated by use of a low-pass filter.

An important feature of the Fourier spectrum of PDM and PPM is that the harmonics generally spread into the baseband even if the baseband signal bandwidth is considerably less than one-half of the average pulse-repetition frequency. This gives rise to a form of distortion not present in PAM; however, it can be minimized by using a low value of modulation index.

4.6 QUANTIZATION

Samples of a signal are quantized so that they may be stored or transmitted *symbolically*. By symbolic transmission we mean that one member of a finite set of symbols is sent for each signal sample. We often think of these symbols as integers or, more commonly, as integers expressed in binary form. In an ordinary quantization problem, we might assign the symbols to be the integers from 0 to 63, and we might express these in the binary forms 000000 to 111111. The reason for this choice is that a simple quantizer can be thought of as rounding an analog number to the nearest one of a set of levels, and the integer denotes which level in order this was. If the analog number is closest to level 17, for instance, the quantizer represents the number by integer 17, or as binary 010001. Whatever these outputs may be, it is important to realize that the quantizer's purpose is to convert a signal into a stream of symbols. Overall, this is being done in two steps: (1) a sampler discretizes the signal's time axis, and (2) a quantizer discretizes the signal amplitudes.

Quantization is not a method of transmission, but of conversion to a new form. The new form may be stored or it may be transmitted using one of the digital transmission methods of Chapter 10. The idea of quantization poses some interesting new questions. How many symbols are needed to perfectly represent an analog signal? If infinitely many are needed, what accuracy is obtainable from a certain fixed number? These are questions of information theory, and we shall look into them in Secs. 11.1 to 11.4.

The most profound question regarding quantization is why do it at all. Assuming we are free to choose an analog or digital format, why choose a digital one? A full discussion of this appears in Sec. 10.1, but we can summarize it here. For its part, advanced communications theory shows that, as the bandwidth available for transmission grows and the power available declines, digital transmission eventually becomes more efficient. From a commercial point of view, signals in digital form may be easier to switch and transmit, especially if the network must handle a variety of conflicting signal types. Finally, in long-distance networks like telephone systems, it is often necessary to use repeaters at regular intervals to amplify a signal that would otherwise be lost because of the curvature of the earth or some other source of loss. Here the advantage of digital transmission can be overwhelming.

Digital symbols have the advantage in a system with repeaters because it is possible to amplify and regenerate a symbol exactly. Analog signals are inevitably distorted at least slightly during regeneration, and the small distortions contributed by each repeater may build up to serious proportions. An example of this is given by the transmission of a telephone call between New York and San Francisco, cities that lie 5000 km apart. An earth-bound line at microwave frequencies needs about 100 repeaters to bridge this distance. Suppose that the signal is kept to power S and that each of the 100 transmission links contributes noise of power N_0. Thus the signal-to-noise ratio (SNR) in each link is S/N_0. We shall find in Chapter 6 and later chapters that the noise powers in the links add, while the signal power does not, so that the overall SNR is $S/100N_0$. Restated, the maintenance of a given SNR across the whole transmission requires an SNR 20 dB higher than this in each link, a hundredfold increase in transmitter power. Quite a different law governs a digital system. We shall find in Chapters 10 and 11 that the important parameter now is the probability of symbol error. If this is p in each link, it will be about $100p$ in a 100-link system. An overall probability of p can be obtained by designing each link to have $p/100$. We shall see in Chapter 10 that a phase-shift digital modulation needs *only 2 dB* added power to effect this change at a commercially useful error rate.

The abbreviation PCM, which stands for *pulse code modulation*, is a commonly used term for quantization. Actually, quantization need not produce pulses and it is not a method of modulation as the term is used in this book. The literal meaning of PCM relates to the early history of quantization. Today the term generally refers to one of the more straightforward quantization methods.

Pulse code modulation appears to have been first invented in 1926 by Paul M. Rainey (U.S. Patent No. 1,608,527, Nov. 30, 1926). Rainey's patent dealt with the transmission of a facsimile signal over telegraph channels by a process of sampling, quantizing, and coding. PCM was reinvented by A. H. Reeves (French Patent No. 853183, Oct. 23, 1939). Reeves's patent contains specific electronic circuits and proposed 32 quantizing levels as suitable for constant-volume speech. During World War II, PCM was invented once more at Bell Laboratories, this time for the purpose of providing secret telephoning by means of an enciphering method developed by Vernam in World War I [4-16]. Vernam's method was developed for use with binary telegraphy and consists of adding a random key k to the message M, modulo 2; that is, $0 + 0 = 0$, $0 + 1 = 1$, $1 + 0 = 1$, $1 + 1 = 0$. The resulting telegraphic message was completely scrambled and unintelligible but could be easily unscrambled at the receiving end by adding the key again, since $(M + k + k) \bmod 2 = M$. The application of this method to speech converted into a binary code by PCM is in principle straightforward. It was only the first of many other investigations into the use of PCM to permit various discrete transmission and detection systems to be used for analog signals.

Basic Linear PCM

We shall now describe the simplest form of quantization, in which an analog sample is converted to symbols that represent uniformly spaced levels. This is called *linear PCM*.

Figure 4.6-1 shows how an eight-level linear PCM system works. The output of this quantizer is a 3-bit binary word, since 3 bits will specify which of the eight levels is selected; engineers refer to this scheme as a 3-bit quantizer or a 3-bit PCM. The input range of the quantizer is divided into six identical *bins*, plus regions at the top and bottom that extend to $+\infty$ and $-\infty$. A sample lying in one of the bins results in an output equal to the binary code assigned to the bin. Henceforth, the real-number value of the sample will be the *reproducer level* located at the midpoint of the bin. The

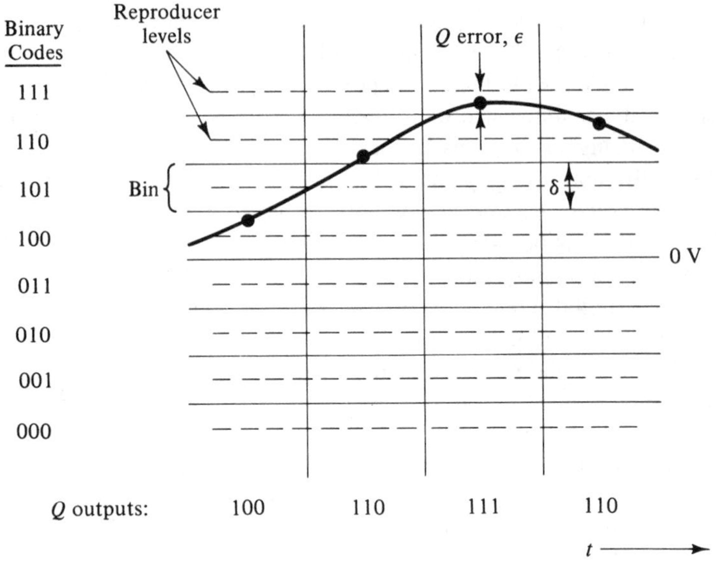

Figure 4.6-1. Three-bit linear PCM.

difference between the sample and the reproducer level becomes the *quantizer noise*. The positions of the reproducer levels and the bins define the quantizer.

We have seen that $2W$ samples per second are required to reconstruct perfectly a signal of bandwidth W. If the samples are quantized with a b-bit quantizer, an overall rate of

$$2bW \text{ bits/sec} \tag{4.6-1}$$

is needed. There will still be some error in the reconstruction from these bits because of the quantizer noise (to which may be added distortion from an imperfect reconstruction filter). We can reduce this noise by augmenting b, that is, by increasing the fineness of the bins, but the noise cannot be removed without an infinite fineness. It is characteristic of systems that carry digitized analog signals that a signal is reproduced with a certain irreducible noise, but that the noise does not increase as the signal passes through the system. This is because the signal is carried symbolically by bits, which can be transmitted virtually error free.

Quantizing noise in PCM is of three types. The first can be reduced to a desired specification by setting b; the other two can be eliminated by proper design and enough bits.

Random Noise

The first and most important kind derives from the small, random amounts by which the samples miss the reproducer levels. One of these errors is indicated in Figure 4.6-1. If the range of the signal is much bigger than the size δ of the bins, the location of a sample within a bin is essentially random. Note that the choice of a bin itself is not random with a typical real-life signal (the bin is probably close to the bin of the previous sample), but the error ε within a bin is nearly random and is uniformly distributed across the range $(-\delta/2, \delta/2)$. Random noise has a white, hissing sound.

Now we shall borrow a little probability theory from Chapter 8 and perform a simple calculation of the SNR prior to reconstruction due to random-type noise. The average energy in the uniformly distributed ε is

$$E[\varepsilon^2] = \int_{-\delta/2}^{\delta/2} \varepsilon^2 \frac{1}{\delta}\, d\varepsilon = \frac{\delta^2}{12}. \tag{4.6-2}$$

The average energy in the samples themselves comes from their square rms value, A_{rms}^2. This means that the SNR is

$$\text{SNR} = \frac{A_{\text{rms}}^2}{\delta^2/12}. \tag{4.6-3}$$

In writing this we have made the implicit assumption that the samples always lie within the total range of the quantizer. It is quite important that this be true, since the noise caused by samples that fall well into the top or bottom bins, called *overload noise*, is easily heard or seen. A standard assumption in audio work is that overload noise is sufficiently reduced if the quantizer range is set to $(-4A_{\text{rms}}, 4A_{\text{rms}})$. Making this assumption also allows us to eliminate A_{rms} from Eq. (4.6-3), since it must now be true that

$$\delta = \frac{8A_{\text{rms}}}{2^b - 2}$$

for a b-bit quantizer. Equation (4.6-3) now becomes $\text{SNR} = (\frac{3}{16})2^{2b}$. Usually, engineers express this in decibel form, which gives

$$\begin{aligned} \text{SNR}_{\text{dB}} &= 10[b \log_{10} 4 + \log_{10} (\tfrac{3}{16})] \\ &= 6b - 7.3 \qquad \text{(the 6-dB rule)}. \end{aligned} \tag{4.6-4}$$

Equation (4.6-4) gives a fundamental relationship between the bit size of linear PCM and its random-noise SNR; specifically, it says that each additional bit adds 6 dB to the quantizer SNR. For this reason, Eq. (4.6-4) is called the *6-dB rule*. Its significance extends far beyond linear PCM since almost all known quantization schemes obey a 6-dB rule, with only the 7.3 constant varying among schemes. This constant depends as well on how the quantizer range relates to A_{rms}, but all such relations lead to a 6-dB slope per bit.

Granular Noise

This type of noise occurs when the signal amplitude spans only a few quantizer bins. Granular noise is random noise that has ceased to be random because the bins are too few or the signal is too small. Now the quantizing error depends on which bin contains the sample. To the ear the error is a harsh, granular sound resembling gravel being poured into a barrel.

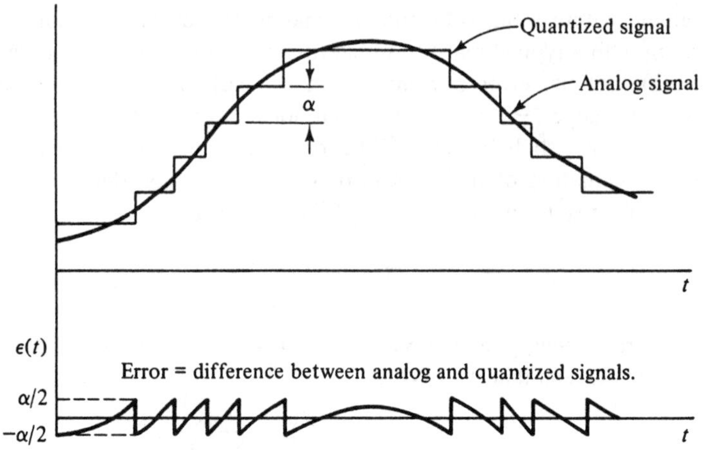

Figure 4.6-2. Granular quantizing noise.

Figure 4.6-2 depicts noise of the granular type. In plotting the noise we have assumed that the sampling speed is sufficient and that the reconstruction filter does not affect the noise; actually, the filter will round off the corners of the noise some, but the unwelcome granular character will still be present. Noise of the random type tends to consist of pulses of small, varying amplitude, unlike the constant pulses shown here. Granular noise can be removed by increasing the bit rate until all noise is of the random type. The method of companding, which we shall take up later, is also effective.

Hunting Noise

This type of noise is also visible in Figure 4.6-2. It is caused by the tendency of the quantizing error to oscillate between positive and negative values; particularly with constant signals or signals near zero, the bin selection may hunt between two adjacent bins. Hunting is audible as a singing tone at half the sampling rate and may be visible as a crosshatching in an image. The effect is obscured and submerged as the signal variation grows. In a voice-transmission system, it can be removed by filtering out the tone or by redesigning the bins adjacent to zero so that hunting does not occur during silences.

We shall now look at two everyday examples of PCM.

Example 1: The US/Canadian Digital Telephone System

Digital telephone lines in North America carry signals that have been sampled at 8 kHz and digitized by 8-bit PCM. The PCM is actually an instantaneous companding scheme, which we shall discuss shortly, and the eighth PCM bit is sometimes borrowed for network housekeeping or parity checking. The nominal analog bandwidth of the telephone channel is 300 to 3200 Hz, with the range from 3200 to 4000 Hz serving as a guard band to make pre-alias and reconstruction filtering easier. The total bit rate per channel is thus 64 kb/sec. Digital channels in other parts of the world often use a 9600-Hz sampling rate combined with the same 8-bit quantization.

By the 6-dB rule for linear PCM, a telephone channel should have an SNR of about 40 dB. (This is reduced somewhat by the instantaneous companding scheme to 30 to 35 dB). As a telephone call is passed through an extended network, it is important to properly maintain signal amplitudes as the call enters and leaves digital parts of the network; otherwise, the signal either overloads the quantizers or suffers from granular noise. A properly treated signal in an all-digital network will suffer no degradation at all, and it will in fact be impossible to judge the distance of the call from its noise level. In a malfunctioning network, a perceptive listener can easily identify all three types of noise that we have discussed.

Example 2: The Compact Disc

This is a read-only digital recording medium that is intended to replace ordinary phonograph records and tapes. Bits are recorded as micron-sized depressions in a glass surface that is read by coherent laser light. The digitization scheme is 16-bit linear PCM with a 44.1-kHz sampling rate, which gives a signal bandwidth of about 0 to 20 kHz and an SNR, by the 6-dB rule, in the neighborhood of 90 dB. The excellent clarity of sound from a compact disc can be traced to two sources. First, by placing the digital conversion at the recording microphone, essentially all unintentional distortion can be removed from the recording–playback chain. Second, the very high SNR of the disc has the effect of increasing the dynamic range of the recording. It does this by allowing a very quiet passage at, say, 50 dB below the loudest passage to be recorded with an SNR still as high as 40 dB. A vinyl analog disc after a few playings would play this passage at perhaps only 10-dB SNR, an unacceptable level, and consequently the quiet passage must be artificially increased before it is recorded. Dynamic range can be improved in this way by any means that improves the SNR of the medium, but the first type of improvement, the removal of record–playback processing noise, is only possible with digital recording.

Compact discs have several other interesting features, including run-length coding to remove long runs of 0's or 1's and error-correction coding to replace bits lost in imperfections in the glass disc. We shall discuss these kinds of coding in Chapter 11.

Encoding and Decoding Circuits

Conversion of the continuous signal into a digital code is usually done by an analog-to-digital converter (ADC). At the receiver a digital-to-analog converter (DAC) reverses the process. There are a large number of different types of converters, and we shall describe here only a few representative ones. The reader is referred to [4-17] for more details.

Since several commonly used ADC circuits employ a DAC, we consider the latter first. One of the most popular techniques for D/A conversion is the R-$2R$ ladder method. This consists of a resistive ladder network having series values of R and shunt values of $2R$ as shown in Fig. 4.6-3. The digital signal to be converted is stored in an auxiliary register not shown in the figure so that all its bits are simultaneously and continuously available. The switches shown at the bottom of the ladder are electronic gating devices

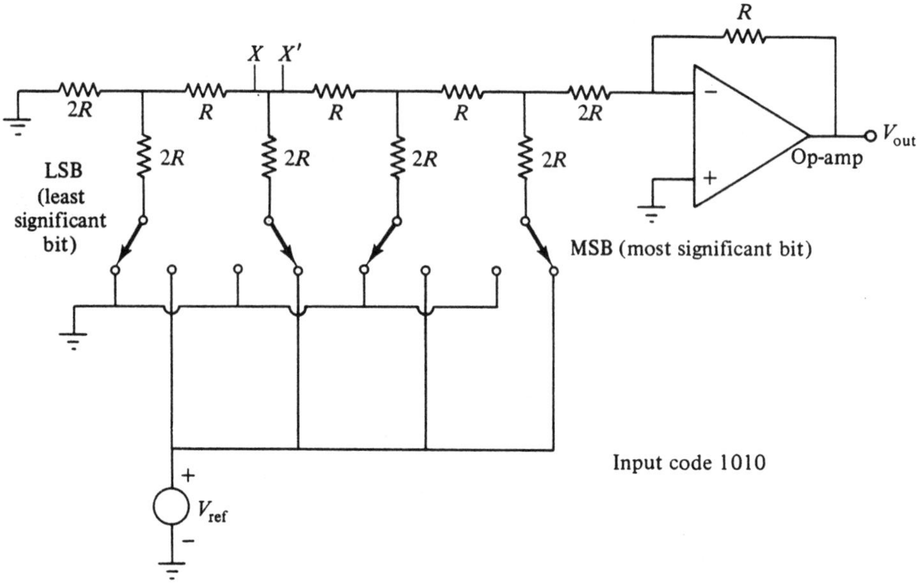

Figure 4.6-3. D/A converter using R-$2R$ ladder.

such as transistors or FETs that are activated by the digital input. If a particular bit of the input is a 1, the corresponding switch connects its shunt resistor to the positive reference voltage; if a bit is a 0, the corresponding shunt resistor is grounded. In Fig. 4.6-3, the binary number 1010 is applied to the converter input. Observe that the most significant bit (MSB) is applied to the switch nearest to the operational amplifier.

The operation of the network is based on the binary division of current as it flows down the ladder. This can be seen by examining the points X and X' in the ladder. As explained in Chapter 3, the operational amplifier input is a *virtual ground*; that is, the effect of the large gain and negative feedback is to permit essentially no voltage to exist at this point. Also, we can assume that the internal impedance of the reference voltage is zero. For these reasons, we find that the resistance looking to the left from point X is $R + (2R/2) = 2R$. Similarly, the resistance looking to the right from X is R. At point X' the resistance looking to the right is $2R$, and looking to the left it is R. This is true at any point on the ladder. If a shunt resistor is switched to the reference voltage source, the source sees a resistance of $2R$ in series with two $2R$ resistances in parallel (i.e., $3R$), and therefore the current in the shunt branches is either $V_{ref}/3R$ or zero. At the junction the current divides equally, with half flowing to the left and half flowing to the right. The right-hand current goes to the next junction, where it again divides in half, and so on to the right end of the ladder. Thus the current from the source feeding into the junction marked with X and X' will be halved three times before reaching the operational amplifier, and the signal produced by it will be proportional to $\frac{1}{8}$. Similarly, the current flowing upward in the last branch to the right is halved once, and therefore its contribution to the total is $\frac{1}{2}$. The currents from all the "on" switches add in the output because of superposition. Hence, with the switch setting shown, the analog output is proportional to $\frac{1}{2} + \frac{1}{8}$, which is exactly the value of the binary number 1010 if the binary point is assumed to be to the left of the number (i.e., 0.1010). The ladder can be extended at will to accommodate as many bits as desired. An important advantage of the circuit is that it uses only two resistance values so that matching and temperature tracking are very simple. The accuracy of the circuit can therefore be made very high.

One of the most widely used A/D converters is the successive approximation type, because it combines high resolution and high speed. This converter operates with a fixed conversion time per bit, independent of the value of the analog input. The system is illustrated in Fig. 4.6-4 and operates by comparing the input voltage with the DAC output. For the sake of concreteness, we consider an 8-bit converter.

The input signal is assumed to be held in a *hold* circuit and is therefore constant during the entire conversion process. At the start of the conversion cycle the logic circuit generates the digital number 10000000 (i.e., the most significant bit is 1, and all other bits are 0) and applies it to the D/A converter. The resulting analog output will be one-half of the reference voltage, V_{ref}, and this is compared to the input signal. If the input signal is larger, the logic circuit next applies 11000000 to the D/A converter, resulting in an output of $\frac{3}{4}V_{ref}$. Otherwise, the logic circuit turns off the MSB and applies 01000000, giving a D/A output of $\frac{1}{4}V_{ref}$. This process repeats down to the least significant bit (LSB), after which the output register contains the digital output. The way in which the input signal is approximated and the resulting digital output are shown in Fig. 4.6-4(b).

The operation described thus far is for unipolar signals; that is, the range of input signals is between zero and V_{ref}. The circuit is, however, easily converted to bipolar operation by connecting the two resistors R shown in the figure. If one assumes that both the signal and the reference are zero-impedance sources, then the input to the comparator will be $\frac{1}{2}(V_{in} + V_{ref})$. This varies between 0 for $V_{in} = -V_{ref}$ to V_{ref} for $V_{in} = V_{ref}$.

Analog-to-digital conversion can also be accomplished by applying the analog signal to a precision integrator and turning on a clock until the integrator output reaches a certain fixed level. The clock pulses can be used to step up a counter so that the digital output is the number appearing in the counter when the clock is stopped. Still another, somewhat similar approach is to have the analog signal control the frequency of a voltage-controlled oscillator, whose output is applied to a counter for a certain fixed time. Both methods have the disadvantage that the precision of conversion depends on the accuracy of analog devices such as an integrator or a voltage-controlled oscillator.

With all A/D conversion methods, it is best to use a sample-and-hold circuit before the ADC to keep the analog signal fixed during the conversion time. Otherwise, the output will be the digital equivalent of a not too clearly defined weighted average of the analog signal existing during the conversion period.

In PCM, after each input sample is digitized it must be converted into a sequential string of pulses for transmission. This can be done by use of a simple binary multiplexer

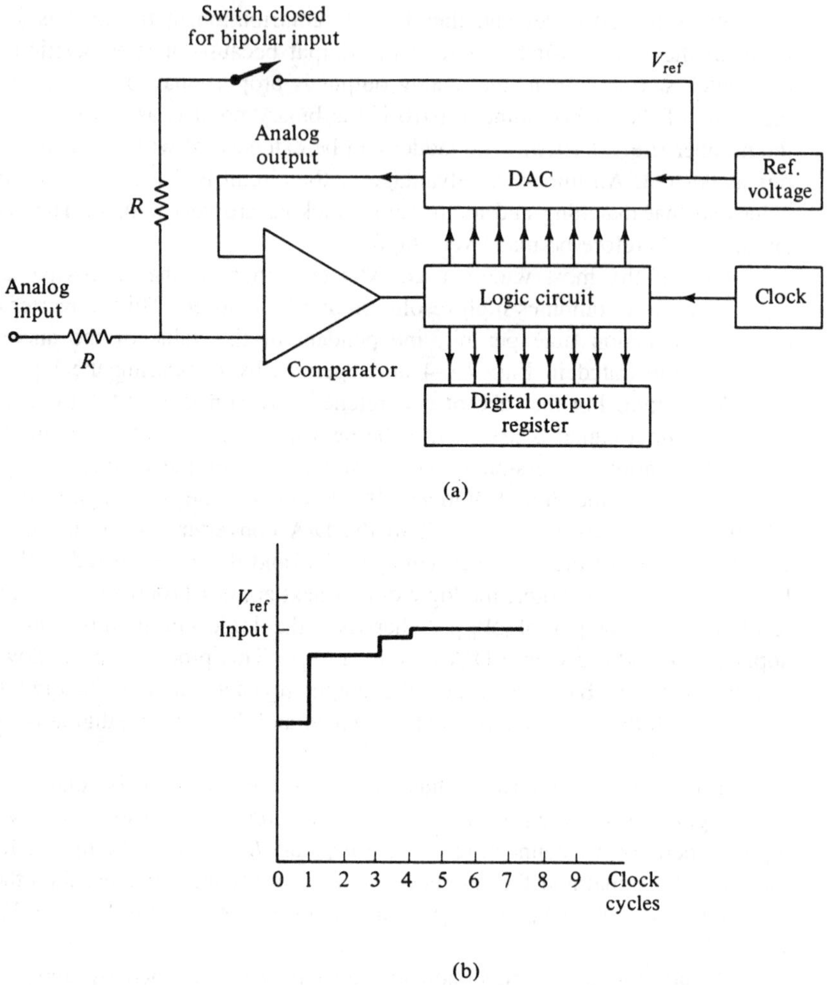

Figure 4.6-4. Analog-to-digital converter using the successive approximation method. (a) Block diagram. (b) Output for a typical input signal.

connected to the A/D converter output register. At the receiver a demultiplexer reverses the process and loads the group of pulses corresponding to each input sample into the D/A input register.

All the A/D and D/A systems described here are for binary codes, but the principle is easily adapted to higher-level codes. In the D/A converter the switches at the bottom of the ladder could be replaced by a precision operational amplifier that would apply one of several levels to the shunt resistances. The binary output of the A/D converter can always be converted to a multilevel output by feeding each group of m bits to a small DAC.

Both D/A and A/D converters are available today in large-scale-integrated-circuit (LSI) packages at moderate cost. Thus the complexity, lack of reliability, and expense of early coding systems are no longer factors against the use of PCM.

More Complex Quantization Schemes

There are a number of sophisticated versions of simple linear PCM. These often take advantage of or defend against some peculiarity of the signal. The result is a better average SNR than would be obtained with linear PCM, but the penalty is often a higher cost conversion circuit. All the schemes here obey the 6-dB rule except delta modulation.

A simple improvement to linear PCM is to make the quantizer bins of different sizes. If the probability distribution of the samples is reliably known, design procedures in fact exist that produce the least mean-square error set of bins for a b-bit quantizer. More bins of smaller size should be concentrated, for instance, at amplitudes the signal is most likely to visit. This kind of optimal quantizer is called a *Max quantizer*; the design procedure appears in [4-15].

Another motivation for variable bin sizes is presented by a signal whose short-term power level keeps changing. Voice signals, for example, consist of a sequence of randomly appearing loud and soft sounds. It is pointless to compute a probability distribution for these samples because at the local level they appear to be a long sequence of samples at one average power, followed by another sequence at quite a different power. A low-level sample train tends to produce objectionable granular noise; but if the quantizer range is compressed, loud sounds will overload the new range. An effective solution to this is the technique of *companding*, an acronym of compression and expanding. In this method a number of small bins are crowded near the zero point of the quantizer range, and larger bins are spaced out through the large-amplitude parts of the range. The result is that the SNR is somewhat poorer for loud signals, but granular noise with small signals is removed.

One can implement this type of companding either by scaling all the bins or by passing a signal first through a nonlinear amplifier and then to a linear PCM converter. Some of these nonlinear companding characteristics appear in Figure 4.6-5, which gives

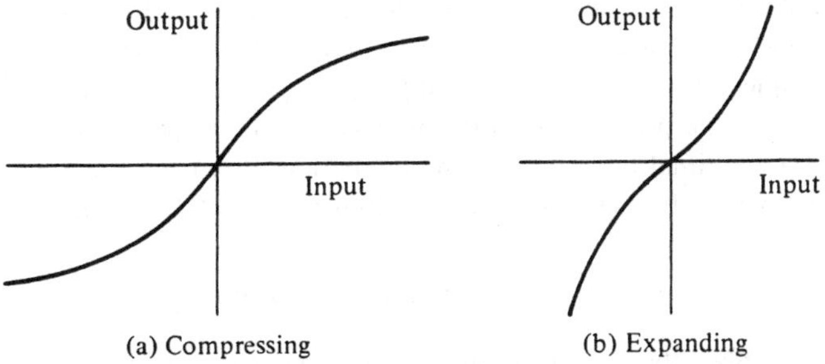

Figure 4.6-5. Companding. If the compressor is used at the A-to-D conversion, the expander must be used at the D-to-A output.

the input–output plots of such an amplifier. All digital telephone channels employ companding, and the North American standard for this is the *μ-law* compander defined by

$$|v_{\text{out}}| = \frac{\log(1 + \mu|v_{\text{in}}|)}{\log(1 + \mu)}, \qquad \mu = 100. \tag{4.6-5}$$

It is assumed that the sign is taken care of separately and that $|v_{\text{in}}| < 1$; the denominator assures that $|v_{\text{out}}| \leq 1$. At the digital-to-analog end of the channel, the inverse nonlinearity must be carried out. Companding usually degrades the quantizing of a stationary (in the sense of Chapter 9) signal with a well-defined probability distribution; the proper quantizer for this is the optimal Max quantizer.

The full name for a quantizer that uses the nonlinear characteristic of Eq. (4.6-5) is an *instantaneous compander*, after the fact that the amplitude scaling is done immediately when the sample appears. Other companding schemes expand or contract a set of linear bins over time in response to a changed signal power. The *Jayant adaptive quantizer*, for instance, expands its bin spacing if it detects the presence of the signal in the outer bins; it can easily determine this by looking at the second bit in the output code. Other companded quantizers compute the signal power, transmit it, and scale accordingly once each block of data. All these schemes are discussed in [4-15].

Another basic technique of quantization is *differential* PCM. It depends on a different signal characteristic, this time the correlation among the samples. Everyday signals like speech or a picture scan line are heavily correlated because they have a low-pass signal spectrum. This means that the present sample is likely to be close in amplitude to the previous one and to the signals before that; the extreme case is a dc signal, for which the previous sample exactly predicts the present one. A differential PCM scheme takes advantage of correlation by predicting the upcoming sample from previous samples and then PCM-coding the *difference* between this guess and the actual sample. Differential PCM will find increasing use in the near future for commercial voice communication.

Differential PCM works well for correlated-sample signals because the difference signal is much smaller than the original signal. A b-bit quantizer produces a proportionately smaller error with this small input. Miraculously, it can be shown that this small error is the error that appears in the much larger reconstructed signal! A typical differential PCM circuit is shown in Figure 4.6-6. The prediction is made by the FIR digital filter in the dashed box (see Chapter 5), and Q denotes an ordinary quantizer with difference input $d(k)$. The quantizing of $d(k)$ produces an error $\varepsilon(k)$, so the reproduced value of the sample $f(k)$ is the predictor output $\tilde{f}(k)$ plus $d(k) + \varepsilon(k)$. This reproduced value is also sent to the predictor to form the basis of future predictions.

The optimal tap gains for the prediction filter can be computed from the statistical correlations of the samples; this is discussed in [4-15]. The 6-dB rule for differential PCM is given by

$$\text{SNR}_{\text{dB}} = 6b - 7.3 + \text{SNI}_{\text{dB}}. \tag{4.6-6}$$

The term SNI (the signal-to-noise improvement ratio) can again be calculated from the signal correlation. It is perhaps 13 to 25 dB for telephone speech, depending on the present speech sound, and 25 to 35 dB for wideband music. Consequently, the SNR

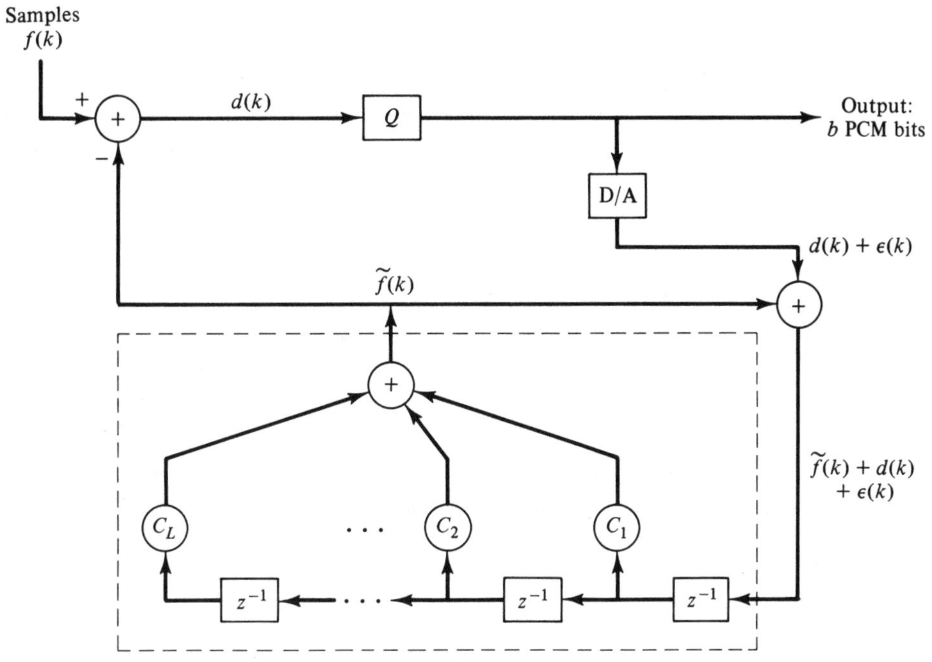

Figure 4.6-6. Typical differential PCM circuit. $\tilde{f}(k)$ is a prediction of $f(k)$. Digital filter notation is from Chapter 5.

performance of differential PCM is much higher than it is for ordinary PCM. Comparison of (4.6-6) to (4.6-4) shows that differential PCM for voice saves 3 to 4 bits per sample over an ordinary scheme. Commercial differential PCM schemes in fact operate at 24 to 32 kb/sec instead of the 64 kb/sec standard PCM rate.

A method for converting analog signals to a string of binary digits that requires much simpler circuitry than PCM is *delta modulation* (DM) [4-18]. A simple DM system is shown in Fig. 4.6-7. When the output $v(t)$ of the integrator is less than $s(t)$, the comparator output is positive, and therefore the modulator output consists of positive pulses. Otherwise, it consists of negative pulses. If $v(t) < s(t)$, the positive pulses entering the integrating network cause $v(t)$ to increase. The integrator output therefore follows $s(t)$. The sequence of positive and negative pulses constitutes the digital output code. At the receiver, an integrating circuit similar to the one used in the transmitter reconstitutes the analog signal from the pulses, and the low-pass filter removes some of the quantizing noise.

The integrating circuit may be simply an *RC* low-pass filter with a time constant much larger than the period of the lowest signal frequency. Thus the DM encoder consists of only a few very simple components. However, because of the simple coding scheme used, a relatively large pulse repetition rate and correspondingly large bandwidth are needed for acceptable fidelity. There are a number of variations of the simple scheme

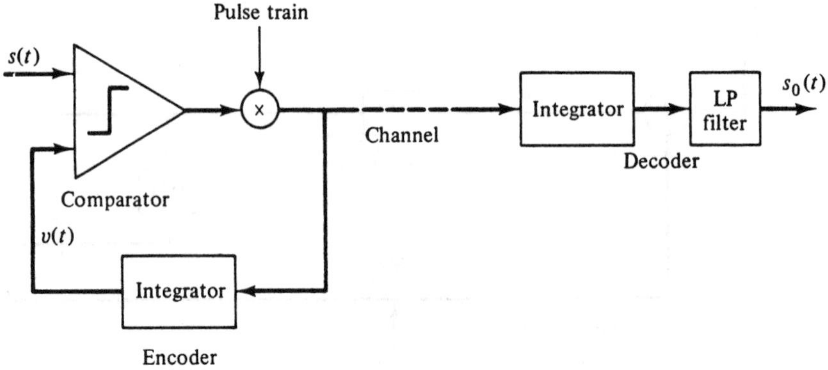

Figure 4.6-7. Delta-modulation system.

described here that have somewhat better performance characteristics at the expense of a more complicated implementation. For details see, for instance, [4-15].

4.7 SUMMARY

In this chapter we have dealt with some of the signal properties arising from the fact that signals are generally band limited. Although signals cannot be strictly confined to a finite time and frequency interval, in practice one can define reasonable confinement intervals. In particular, frequency confinement was shown to permit representation of a continuous signal in terms of a sequence of discrete samples. We introduced the sampling theorem, which says that the minimum required number of samples for signal reconstruction is $2W$ per second, where W is the signal bandwidth. From these samples the signal can, in theory, be exactly reproduced for all values of time, and we discussed some of the reconstruction methods used in practice. Sampling underlies a variety of pulse-transmission systems used in practice; we discussed PAM, PDM, PPM, and PCM systems. PCM employs an additional quantization of signal amplitudes at each sampling instant, which permits analog signals to be transmitted in symbolic digital form. We discussed a number of PCM schemes. Some of the theory underlying these will appear in Chapter 11.

PROBLEMS

4-1. Use the identity

$$\sum_{n=-\infty}^{\infty} \delta(t + nT) = \frac{1}{T} \sum_{n=-\infty}^{\infty} e^{jn\omega_0 t}, \qquad \omega_0 \equiv \frac{2\pi}{T},$$

and the fact that, for a linear system, the responses are given by

Input	Response
$\delta(t + nT)$	$h(t + nT)$
$e^{jn\omega_0 t}$	$H(n\omega_0)e^{jn\omega_0 t}$

to prove that

$$\sum_{n=-\infty}^{\infty} h(nT) = \frac{1}{T}\sum_{n=-\infty}^{\infty} H(n\omega_0).$$

This result is one version of *Poisson's summation formula*.

4-2. **(a)** Show that Eq. (4.3-5) can be written as

$$X_s(\omega) = d \sum_{n=-\infty}^{\infty} \text{sinc}\,(nd)X(\omega - n\omega_s),$$

where $X(\omega)$ is the spectrum of $x(t)$ and $X_s(\omega)$ is the spectrum of the pulse-sampled signal.

(b) Compute $X_s(\omega)$ when $x(t) = 1 + m \cos \omega_m t$, and identify the term that represents the spectrum of the sampling pulses.

4-3. Consider the sampling theorem with $T_s = 1/(2W)$. Then Eq. (4.2-5) takes the form

$$x(t) = \sum_{n=-\infty}^{\infty} x\left(\frac{n}{2W}\right) \phi_n(t),$$

where $\phi_n(t) \equiv \text{sinc}\; 2W[t - n/(2W)]$. Show that the set of functions $\phi_n(t)$, $n = \pm 1, \pm 2, \ldots$, are orthogonal over the interval $-\infty < t < \infty$. *Hint*: Parseval's theorem

$$\int_{-\infty}^{\infty} v(t)u^*(t)\,dt = \int_{-\infty}^{\infty} V(f)U^*(f)\,df$$

shows that if two functions are orthogonal in one domain their transforms will be orthogonal in the other domain.

4-4. Let the energy, E, in $x(t)$ be given by

$$E = \int_{-\infty}^{\infty} |x(t)|^2\,dt.$$

Show that if $x(t)$ is strictly band limited to B Hz then

$$E = \frac{1}{2B}\sum_{n=-\infty}^{\infty} \left| x\left(\frac{n}{2B}\right)\right|^2.$$

4-5. Assume a more general form of the "sampling" theorem in which $x(t)$ has the representation

$$x(t) = \sum_n x_n\phi_n(t),$$

where x_n are the "samples" and the set $\{\phi_n(t)\}$ is orthonormal over $a < t < b$. Show that

$$E \equiv \int_a^b |x(t)|^2\,dt = \sum_n |x_n|^2.$$

4-6. The output, y, of a square-law device is $y = x^2$ if x is the input. How can $y(t)$ be written in terms of the sampled values of $x(t)$? At what rate must $x(t)$ be sampled to enable the reconstruction of $y(t)$? Assume that $x(t)$ is band limited to B Hz.

4-7. (Frequency sampling theorem) Show that if $x(t) = 0$ for $|t| > T$ then

$$X(f) = \sum_{n=-\infty}^{\infty} X\left(\frac{n}{2T}\right) \operatorname{sinc} 2T\left(f - \frac{n}{2T}\right).$$

4-8. In the sample-and-hold circuit of Fig. 4.3-4(a), show that the sampler/capacitor system is represented by a transfer function

$$H(\omega) = \frac{1 - e^{-j\omega T_s}}{j\omega}$$

if the hold time is equal to the sampling interval. Derive Eq. (4.3-13) from this result.

4-9. N baseband signals, each limited to W Hz, are time multiplexed. Assume the minimum adequate sampling rate. What is the minimum required bandwidth required to transmit the multiplexed signal on a PAM system?

4-10. Thirty baseband channels, each band limited to 3.2 kHz, are sampled and multiplexed at an 8-kHz rate.

(a) What is the required bandwidth for transmission of the multiplexed samples on a PAM system?

(b) Explain the function of the "excess" sampling rate. What problems are encountered when the theoretical minimum rate is used?

4-11. It is suggested that the S/H circuit described by Eq. (4.3-13) be used as a low-pass filter to reconstruct the signal from its samples. Assume that the hold time is equal to the sampling period. Let the normalized response of the S/H circuit be

$$H(f) = \operatorname{sinc}\left(\frac{f}{f_s}\right) e^{-j2\pi(f/f_s)}, \qquad f_s \equiv \frac{1}{T_s}.$$

Consider a baseband signal with highest frequency B Hz.

(a) Let $f_s = 2B$. How effective is the filter in keeping out the higher lobes in the periodic spectrum?

(b) Let $f_s = 6B$. Indicate how the performance of the filter has improved by computing the response for the wanted (baseband) spectrum and the unwanted (first-order) spectrum.

4-12. A PAM telemetry system multiplexes four signals: $s(t)$, $i = 1, \ldots, 4$. Two of the signals $s_1(t)$, $s_2(t)$ have $B = 80$ Hz, while $s_3(t)$, $s_4(t)$ have $B = 1000$ Hz. The sampling rate for $s_3(t)$ and $s_4(t)$ is $f_{sm} = 2400$ samples/sec. Assume that other sampling rates can be derived from f_{sm} by dividing by powers of 2. Design a multiplexing system that does a preliminary multiplexing of $s_1(t)$ and $s_2(t)$ into a single sequence, $s_5(t)$, and a final multiplexing of $s_5(t)$, $s_3(t)$, and $s_4(t)$. What is the required minimum sampling rate for the two-step multiplexing? How does it compare with the required sampling rate if all signals were simultaneously multiplexed? Sketch a block diagram for a two-step demultiplexer. How many more signals of 80-Hz bandwidth could this system accommodate without an increase in sampling rate?

4-13. Show that the impulse response of the raised cosine filter [Eq. (4.4-9)] can be written in a somewhat more revealing form as

$$h(t) = \frac{\sin(2\pi t/T)}{2\pi t(1 - 4t^2/T^2)}.$$

This function has its first nulls at $t = \pm T$ and is very small *outside* the range $|t| < T$. Hence intersymbol interference is reduced.

4-14. The circuit shown in Fig. P4-14 (adapted from [4-12], p. 249) is a simplified PPM generator. The signal $s(t)$ is applied to the base of the transistor, in series with a periodic sweep signal described by

$$V(t) = \frac{2V_0}{T}(t - nT), \qquad nT - \frac{T}{2} < t < nT + \frac{T}{2}, \qquad n = 0, \pm 1, \ldots .$$

Assume that collector current flows when the base–emitter voltage exceeds zero.

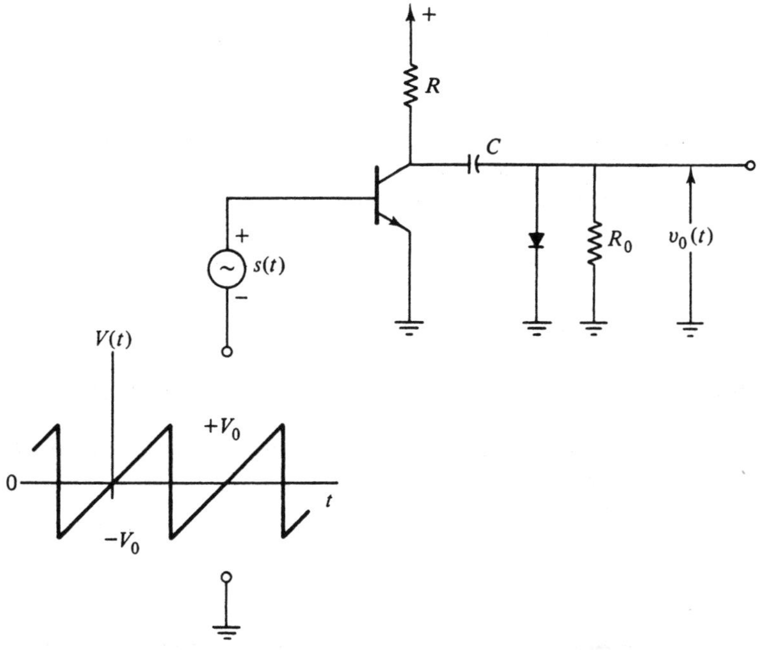

Figure P4-14

(a) Explain how this circuit works.

(b) Let $g(t)$ describe the output pulse when the latter is produced at $t = 0$. Let $s(t)$ be described by

$$s(t) = A\cos(\omega_c t + \theta),$$

where A, ω_c, and θ are constant. Show that the output signal $v_0(t)$ is given by

$$v_0(t) = \sum_{n=-\infty}^{\infty} g(t - t_n),$$

where t_n represents the time at which the nth pulse occurs and satisfies

$$A \cos(\omega_c t_n + \theta) + \frac{2V_0}{T}(t_n - nT) = 0.$$

(See [4-12] for a discussion of this problem.)

4-15. To reduce intersymbol interference due to pulse distortion by the channel, *equalizing* filters in the form of tapped delay lines are inserted between the A/D converter at the sender and the receiver. Figure P4-15 shows a transversal equalizer consisting of $2N + 1$ taps on a delay line, $2N + 1$ adjustable gain elements, and $2N$ delay elements.

(a) Show that the output of such a filter can be written as

$$\begin{aligned} x_{\text{eq}}(t) &= \sum_{i=1}^{2N+1} G_i x[t - (i-1)\tau] \\ &= \sum_{i=-N}^{N} F_i x[t - (i+N)\tau], \end{aligned}$$

where $F_i \equiv G_{i+N+1}$.

(b) Suppose that $x(t)$ has a peak of 1 at $t = 0$ and intersymbol interference on either side. Let $x_{\text{eq}}(t)$ be sampled at $t_k = (k + N)\tau$, $k = \ldots, -2, -1, 0, 1, \ldots$. How should the gains $\{F_i\}_{i=-N}^{N}$ be adjusted so that

$$x_{\text{eq}}(t_k) = \begin{cases} 1, & k = 0, \\ 0, & k \neq 0. \end{cases}$$

For a discussion of this problem see [4-10].

4-16. Suppose it is known that samples are going to be uniformly distributed between 0 V and 1 V.

(a) Derive the best location for the bins and reproducer levels if 1-bit quantization is tc be used.

(b) Repeat for a 3-bit quantizer.

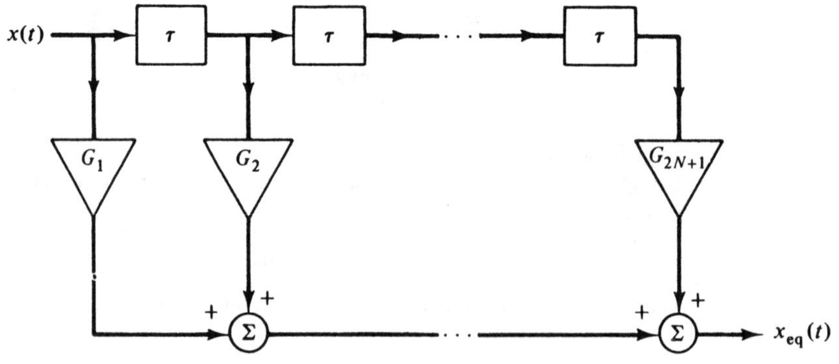

Figure P4-15

4-17. Find the SNR of the two quantizers in Prob. 4-16, and compare to the answer given by the 6-dB rule, Eq. (4.6-4).

4-18. Suppose it is desired to digitize sonar signals whose bandwidth is 300 Hz. Design a system to do this with the following specifications:
(i) Sample 20% above the Nyquist rate to allow for imperfect sampling.
(ii) Linear PCM, using the standard assumption on range, is to achieve 40 dB SNR.

4-19. **(a)** Rederive the 6-dB rule, but this time assume that the signal is uniformly distributed between two values. Assume as usual that linear PCM is employed with many small bins.
(b) Repeat for a signal known to be a sinusoid.

4-20. Figure 4.6-6 gave the circuit for a DPCM transmitter that converts samples to bits. Design the receiver circuit that re-creates the approximated samples. Keep in mind that the original samples are unavailable at the receiver.

REFERENCES

[4-1] A. Papoulis, *The Fourier Integral and Its Applications*, McGraw-Hill, New York, 1962.

[4-2] D. Slepian, "On Bandwidth," *Proc. IEEE*, **64,** No. 3, pp. 292–300, March 1976.

[4-3] R. C. Buck, *Advanced Calculus*, McGraw-Hill, New York, 1965.

[4-4] H. Nyquist, "Certain Topics in Telegraph Transmission Theory," *Trans. AIEE*, **47,** pp. 617–644, April 1928.

[4-5] E. T. Whittaker, "On the Functions Which Are Represented by the Expansions of the Interpolation Theory," *Proc. Roy. Soc. Edinburgh*, **35,** pp. 181–194, 1915.

[4-6] H. E. Rowe, *Signals and Noise in Communication Systems*, Van Nostrand Reinhold, New York, 1965, p. 225.

[4-7] W. R. Bennett and J. R. Davey, *Data Transmission*, McGraw-Hill, New York, 1965, Chap. 14.

[4-8] R. W. Lucky, J. Salz, and E. J. Weldon, *Principles of Data Communications*, McGraw-Hill, New York, 1968.

[4-9] R. A. Gibby and J. W. Smith, "Some Extensions of Nyquist's Telegraph Transmission Theory," *Bell System Tech. J.*, **44,** pp. 1487–1510, Sept. 1965.

[4-10] R. W. Lucky, "Techniques for Adaptive Equalization of Digital Communications Systems," *Bell System Tech. J.*, **45,** pp. 255–286, 1966.

[4-11] W. R. Bennett, "Statistics of Regenerative Digital Transmission," *Bell System Tech. J.*, **37,** pp. 1501–1542, Nov. 1958.

[4-12] M. Schwarz, W. R. Bennett, and S. Stein, *Communication Systems and Techniques*, McGraw-Hill, New York, 1966, Chap. 6.

[4-13] A. B. Carlson, *Communication Systems*, McGraw-Hill, New York, 1968, pp. 293–294.

[4-14] A. B. Carlson, *Communication Systems*, 2nd ed., McGraw-Hill, New York, 1975, pp. 312–314.

[4-15] N. S. Jayant and P. Noll, *Digital Coding of Waveforms: Principles and Applications to Speech and Video*, Prentice-Hall, Englewood Cliffs, N.J., 1984.

[4-16] G. S. Vernan, "Cipher Printing Telegraph Systems for Secret Wire and Radio Telegraphic Communications," *AIEE Trans.*, **65,** pp. 295–301, Feb. 1926.

[4-17] H. Schmidt, *Analog/Digital Conversion*, Van Nostrand Reinhold, New York, 1970.

[4-18] E. M. Deloraine, S. Van Mierlo, and B. Derjavich, French Patent No. 932,140, Aug. 1946, U.S. Patent No. 2,629,857, Feb. 24, 1953.

CHAPTER 5

Discrete System Theory

5.1 INTRODUCTION

The linear system theory presented in Chapter 3 dealt essentially with continuous-time signals such as the waveforms generated in speech or television. These signals are defined for every instant in a given interval and can, in principle, take on any value. In Chapter 4 we showed that band-limited signals can be specified by instantaneous samples taken at discrete times. Sampling converts continuous signals into a sequence of numbers. These numbers can be further processed and transmitted, as is done, for example, in pulse code modulation. Such a sequence of numbers is called a discrete-time signal, and the systems that deal with them are discrete-time systems or, more briefly, discrete systems. If, in addition, the signal levels are discrete (i.e., quantized), the discrete signal is said to be digital.

The sampling of continuous waveforms is only one means of generating discrete signals. Another obvious source of discrete signals is the digital computer, or, in fact, any of the vast variety of digital circuits that are in use today. These devices generate streams of numbers that may not have any very clear relation to analog signals. Much present-day communication equipment and circuitry must be designed to handle such number streams.

5.2 DISCRETE LINEAR SYSTEMS

The definition of linearity given in Chapter 2 and its relation to superposition carries over directly to the discrete case. Generally, the input to a discrete linear system is a *sequence* of numbers. Then, if y_n is the response to an input sequence u_n and z_n is the response to an input sequence v_n, the response to $a_1u_n + a_2v_n$ is $a_1y_n + a_2z_n$.

If the input to a discrete linear system consists of the sequence $x_1, x_2, \ldots, x_n$, the output consists of a weighted sum of the inputs:

$$\begin{aligned} y_1 &= h_{11}x_1 + h_{12}x_2 + \cdots + h_{1k}x_k + \cdots \\ y_2 &= h_{21}x_1 + h_{22}x_2 + \cdots + h_{2k}x_k + \cdots \\ &\;\;\vdots \\ y_n &= h_{n1}x_1 + h_{n2}x_2 + \cdots + h_{nk}x_k + \cdots. \\ &\;\;\vdots \end{aligned} \tag{5.2-1}$$

The set of coefficients $\{h_{nk}\}$ characterizes the system.

Equation (5.2-1) can be written in several equivalent and more convenient ways. Allowing for input strings that may be, at least in theory, infinitely long, we can write

$$y_n = \sum_{k=-\infty}^{\infty} h_{nk}x_k, \qquad n = \ldots, -1, 0, +1, \ldots. \tag{5.2-2}$$

This form is analogous to the superposition integral

$$y(t) = \int_{-\infty}^{\infty} h(t, \tau)x(\tau)\, d\tau \tag{5.2-3}$$

if we identify the indexes n and k with the time variables t and τ, respectively, and the constants h_{nk} with the impulse response $h(t, \tau)$. In fact, consider the special input consisting of the discrete impulse function:

$$x_k = \begin{cases} 1, & k = i, \\ 0, & \text{otherwise.} \end{cases}$$

Then the response is just h_{ni}. Hence, for any k, h_{nk} lends itself to the following interpretation: h_{nk} is the response at "time" n to a unit function applied at "time" k. We put quotes around the word *time* because the indexes do not necessarily have to refer to real time, although in communication systems they often do.

Because of the close analogy between Eqs. (5.2-2) and (5.2-3), the former is referred to as a *discrete superposition summation* or just (*discrete*) *superposition*. For a time-discrete system where the subscripts n, k do refer to time, *causality* requires that y_n cannot depend on future values of x_k (i.e., values of x_k such that $k > n$). Also, the input sequence is typically of finite duration and can be assumed to begin at $k = 0$ or $k = 1$. The choice of time origin is arbitrary. Hence Eq. (5.2-2) for *causal* systems with inputs that begin at $k = 1$ takes the form

$$y_n = \sum_{k=1}^{n} h_{nk}x_k, \tag{5.2-4}$$

which is in direct analogy to the continuous case where

$$y(t) = \int_0^t h(t, \tau)x(\tau)\, d\tau. \tag{5.2-5}$$

Another convenient description of the system of equations in Eq. (5.2-1) results from using vector notation. In this representation the input and output sequences are represented by column vectors **x** and **y**, where **x** has the K elements $x_1, \ldots, x_K$ and **y** has, say, the N elements $y_1, \ldots, y_N$. Then, through the use of basic matrix multiplication, Eq. (5.2-1) is converted to

$$\mathbf{y} = \mathbf{Hx}, \tag{5.2-6}$$

where $\mathbf{y} = (y_1, \ldots, y_N)^T$, $\mathbf{x} = (x_1, \ldots, x_K)^T$, the T denotes transpose, and **H** is given by

$$\mathbf{H} = \begin{bmatrix} h_{11} & h_{12} & h_{13} & \cdots & h_{1K} \\ h_{21} & h_{22} & h_{23} & \cdots & h_{2K} \\ \cdot & & & & \cdot \\ \cdot & & & & \cdot \\ \cdot & & & & \cdot \\ h_{N1} & & \cdots & & h_{NK} \end{bmatrix}. \tag{5.2-7}$$

We see that a discrete linear system can be characterized by a weighting matrix **H** instead of the impulse or step-function response. The matrix **H** is referred to as the *transmission matrix* of a system. It can contain a finite or infinite number of elements. The matrix for a causal system is triangular (Prob. 5-1).

We said earlier that discrete systems need not necessarily be indexed by real time. Often the discrete input data are already collected on magnetic tape or disk, and outputs can be computed using stored "future" values of the input. For this reason, concepts such as causality or even time invariance play a less important role in discrete system theory than in continuous-time systems. On the other hand, when $\mathbf{y} = \mathbf{Hx}$ is used to model a real-time sampled-data system as shown in Fig. 5.2-1, then causality and time invariance are notions of principal importance.

We have already mentioned causality. The concept of time invariance in discrete

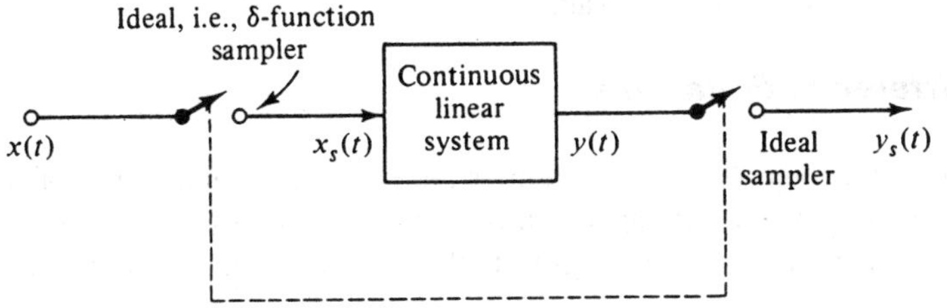

Figure 5.2-1. Real-time ideal sampled-data system.

systems is a direct carry-over from the continuous case. For the latter, a system is said to be time invariant if $h(t, \tau) = h(t - \tau)$. In a discrete, linear, time-invariant (DLTI) system, the impulse response satisfies

$$h_{nk} = h_{n-k} \tag{5.2-8}$$

an Eq. (5.2-2) reduces to

$$y_n = \sum_{k=-\infty}^{\infty} h_{n-k}x_k. \tag{5.2-9}$$

Equation (5.2-9) is referred to as a *discrete convolution*.† Although it may not be immediately apparent, the transmission matrix for a DLTI system has far fewer *distinct* entries than is the case for a noninvariant system (Prob. 5-2). The transmission matrix for a causal DLTI system has the form

$$\mathbf{H} = \begin{bmatrix} h_0 & & & & & & \\ & h_0 & & & \mathbf{0} & & \\ h_1 & & h_0 & & & & \\ & h_1 & & h_0 & & & \\ h_2 & & h_1 & & h_0 & & \\ & h_2 & & h_1 & & h_0 & \\ & & h_2 & & h_1 & & h_0 \end{bmatrix} \tag{5.2-10}$$

Note that h_{nk} is a function of only one subscript (i.e., $n - k \equiv m$), and hence the elements along diagonals are all equal.

For DLTI systems that are also causal, with inputs that begin at $k = 1$, the expression equivalent to Eq. (5.2-4) is

$$y_n = \sum_{k=1}^{n} h_{n-k}x_k. \tag{5.2-11}$$

If in Eq. (5.2-11) we let $m \equiv n - k$, we obtain

$$y_n = \sum_{m=0}^{n-1} h_m x_{n-m}. \tag{5.2-12}$$

It is convenient to denote a discrete convolution as $y_n = h_n * x_n$; the subscript n on both h and x is standard usage.

Difference Equations

The description of DLTI systems by the discrete impulse response h_m is not the only one in use. A powerful approach is a representation by linear, constant-coefficient difference equations. This is analogous to the representation of continuous systems by differen-

† It is assumed in discrete convolution that the x_k's and h_k's are equally and uniformly spaced along the time axis, i.e., $x_k = x(k\Delta)$ and $h_k = h(k\Delta)$. It is sometimes convenient to take $\Delta = 1$.

tial equations. The coefficients of the difference equation can be related to the discrete impulse response, and this also is analogous to the relation that exists between the impulse response and the coefficients of the differential equation describing a continuous system. We confine ourselves to a brief discussion of the difference-equation method.

A Kth-order difference equation can be written as

$$\sum_{k=0}^{K} a_k y_{n-k} = \sum_{k=0}^{M} b_k x_{n-k}, \qquad a_0 = 1, \tag{5.2-13}$$

where the $\{a_k\}$ and $\{b_k\}$ are coefficients that characterize the system and $a_K \neq 0$ if the system is Kth order. The order of the equation is K because the present value of y (i.e., y_n) depends on the previous K values $y_{n-1}, \ldots, y_{n-K}$. For most systems of engineering interest, $M < K$.

There is a somewhat subtle notational paradox in Eq. (5.2-13). The subscripts on the a_k and b_k identify the coefficients and imply their position in the equation but do *not* refer to time. On the other hand, the subscripts on the input and output variables x_i and y_i, respectively, often *do* refer to time; therefore, it has become conventional to write Eq. (5.2-13) as

$$\sum_{k=0}^{K} a_k y(n-k) = \sum_{k=0}^{M} b_k x(n-k), \qquad a_0 = 1, \tag{5.2-14}$$

which clearly differentiates between the two meanings. When $y(\cdot)$ and $x(\cdot)$ are written as functions of integers, the implication is that the sampling is uniform in time. It is convenient to write Eq. (5.2-14) as

$$y(n) = -\sum_{k=1}^{K} a_k y(n-k) + \sum_{k=0}^{M} b_k x(n-k), \tag{5.2-15}$$

which directly shows the dependence of $y(n)$ on its past values and on the input and its past values.†

As an example, consider the first-order difference equation

$$y(n) = -5y(n-1) + x(n) \tag{5.2-16}$$

with initial condition $y(-1) = 0$. Let $x(n) = n$. The solution can be obtained recursively as follows:

$$\begin{aligned} y(0) &= -5y(-1) + x(0) = 0 \\ y(1) &= -5y(0) + x(1) = 1 \\ y(2) &= -5y(1) + x(2) = -3 \\ y(3) &= -5y(2) + x(3) = 18 \\ &\;\vdots \end{aligned} \tag{5.2-17}$$

† The notation $y(n-k)$, $x(k)$, . . . , rather than y_{n-k}, x_k, . . . , will be adopted for the remainder of this chapter.

Another approach is to compute the solution directly as the sum of the driven solution $y_d(n)$ and the transient solution $y_t(n)$. For $y_t(n)$ we write

$$y_t(n) + 5y_t(n - 1) = 0 \tag{5.2-18}$$

and assume a solution of the form $A\gamma^n$, where A and γ are unknown. Then, substituting $A\gamma^n$ directly into Eq. (5.2-18) enables us to write

$$A\gamma^{n-1}(\gamma + 5) = 0,$$

for which the only nontrivial solution is

$$y_t(n) = A(-5)^n. \tag{5.2-19}$$

To compute $y_d(n)$ we *assume* a solution $y_d(n) = Bn + C$. When this result is substituted into Eq. (5.2-16), we obtain

$$Bn + C = -5[B(n - 1) + C] + n. \tag{5.2-20}$$

By matching coefficients of like powers of n, we get

$$B = \tfrac{1}{6}, \qquad C = \tfrac{5}{36}.$$

The complete solution, with A still unknown, is therefore

$$y(n) = \tfrac{1}{6}n + \tfrac{5}{36} + A(-5)^n. \tag{5.2-21}$$

When the initial condition $y(-1) = 0$ is inserted in Eq. (5.2-21), we get

$$A = -\tfrac{5}{36}.$$

The completely determined solution is finally

$$y(n) = \frac{6n + 5}{36} - \frac{5}{36}(-5)^n, \tag{5.2-22}$$

which has an advantage over Eq. (5.2-17) in that $y(n)$ can be computed for any n without first recursively computing $y(n - 1)$, $y(n - 2)$, ... A very simple filter that is represented by Eq. (5.2-16) is shown in Fig. 5.2-2.

We have already said that invariant systems require far fewer descriptors than

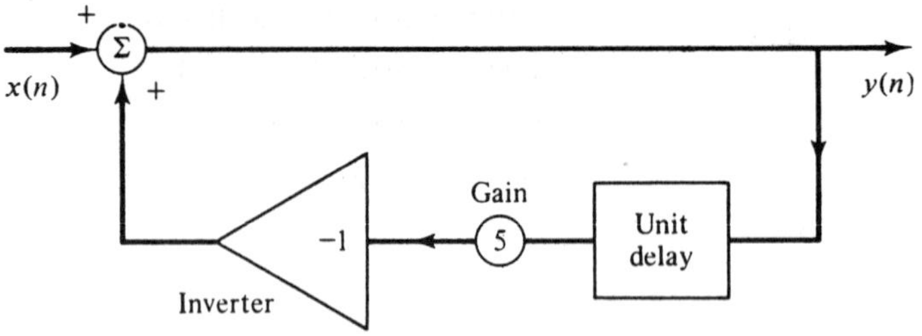

Figure 5.2-2. First-order system described by the difference equation $y(n) = -5y(n - 1) + x(n)$.

other systems. But that isn't the only advantage offered by the property of invariance; by direct analogy to the continuous case, input–output calculations for DLTI systems can be conveniently done with the discrete Fourier transform. With the aid of the latter, convolution is replaced by multiplication, and the transform technique offers a real savings in the amount of labor involved provided that a convenient way of computing the discrete Fourier transform exists. Fortunately, such an algorithm does exist and is known as the fast Fourier transform. It is discussed in Sec. 5.6. First, however, we shall discuss the theory behind the discrete Fourier transform.

5.3 THE DISCRETE FOURIER TRANSFORM

In the following discussion we consider discrete signals resulting from sampling of a continuous-time function $x(t)$ at a uniform sampling rate $1/\Delta t$. This time function will in general have a Fourier transform, and we can define a discrete Fourier transform (DFT) for the sampled time function that approximates the continuous Fourier transform (CFT) in the same way as the sequence $x(1) \cdots x(N)$ approximates the function $x(t)$. Since digital computers can only work with discrete signals, Fourier transforms obtained with a digital computer are always discrete Fourier transforms.

To develop the discrete Fourier transform, consider the expression for the CFT [Eq. (2.8-7)]:

$$X(f) = \int_{-\infty}^{\infty} x(t)e^{-j2\pi ft}\, dt. \tag{5.3-1}$$

For discrete signals, the time variable t becomes $i\,\Delta t$, and the function $x(t)$ is replaced by $x(i)$. Also, because we are interested only in discrete frequencies, the frequency variable f is replaced by $k\,\Delta f$, and $X(f)$ becomes $X(k)$. In practice, the number of time or frequency samples that can be handled is a finite number N, determined by the size of the computer. If the signal is sampled at the Nyquist rate, then Δt has the value

$$\Delta t = \frac{1}{2W}, \tag{5.3-2}$$

where W is the bandwidth. However, with N frequency components separated by a frequency increment Δf, the total frequency range that can be considered is $N\,\Delta f$, so that

$$2W = N\,\Delta f. \tag{5.3-3}$$

Combining Eqs. (5.3-2) and (5.3-3), we see that

$$\Delta t\,\Delta f = \frac{1}{N}. \tag{5.3-4}$$

The infinite limits of integration must be replaced by finite limits, and the integral itself is replaced by a sum. The infinitesimal dt becomes $\Delta t = 1/(2W) = T/N$, where T is the length of time over which the N samples of $x(t)$ have been obtained. Then, finally, Eq. (5.3-1) is transformed to

$$X(k) = \frac{T}{N}\sum_{i=1}^{N} x(i)e^{-j2\pi ki/N}, \qquad j = \sqrt{-1}, \tag{5.3-5}$$

and by direct analogy the inverse transform is

$$x(i) = \frac{2W}{N}\sum_{k=1}^{N} X(k)e^{j2\pi ki/N}. \tag{5.3-6}$$

The validity of these formulas is easily demonstrated by inserting the definition for $X(k)$ from Eq. (5.3-5) into Eq. (5.3-6). This gives

$$\begin{aligned} x(m) &= \frac{2W}{N}\sum_{k=1}^{N}\frac{T}{N}\sum_{i=1}^{N} x(i)e^{-j2\pi ki/N}e^{j2\pi km/N} \\ &= \frac{2WT}{N^2}\sum_{k=1}^{N}\sum_{i=1}^{N} x(i)e^{j2\pi k(m-i)/N}. \end{aligned} \tag{5.3-7}$$

Note the use of the index m to prevent confusion with the index i used in the second summation; this is equivalent to the use of a dummy variable τ in the demonstration of the Fourier identity in Sec. 2.8. Since both sums in Eq. (5.3-7) are finite, their order can be interchanged, giving

$$x(m) = \frac{2WT}{N^2}\sum_{i=1}^{N} x(i)\sum_{k=1}^{N} e^{j2\pi k(m-i)/N}. \tag{5.3-8}$$

The second summation has the form of a geometric series:

$$\sum_{k=1}^{N} G^k = \frac{G(1 - G^N)}{1 - G}, \tag{5.3-9}$$

where $G \equiv e^{j2\pi(m-i)/N}$. Since $m - i$ is an integer, $G^N = 1$, and therefore the series sums to zero unless $G = 1$ [i.e., when $m = i \bmod (N)$]. In this case it has the form $\Sigma_{i=1}^{N}(1)$ and sums to N. Thus Eq. (5.3-8) becomes

$$x(m) = \frac{2WT}{N}x(m). \tag{5.3-10}$$

However, it is easily shown that $2WT = N$. This follows from the fact that $\Delta t = 1/(2W)$ and $T = N\,\Delta t$. Thus Eq. (5.3-10) is an identity.

The discrete Fourier pair is often given in the form

$$X(k) = \sum_{i=1}^{N} x(i)U_N^{-ik}, \tag{5.3-11}$$

$$x(i) = \frac{1}{N}\sum_{k=1}^{N} X(k)U_N^{ik}, \tag{5.3-12}$$

where $U_N \equiv e^{j2\pi/N}$. This form eliminates all reference to T and W since these constants are only scale factors and are essentially irrelevant to the basic operation. The validity of this pair is as easily demonstrated as that of the previous one; in fact, the discussion subsequent to Eq. (5.3-10) can be omitted.

Because the series of Eq. (5.3-9) sums to N for all $m = i \bmod (N)$, we see that the inverse transform of the sequence $\{X(k)\}$ gives not only $x(i)$ but also all $x(i \pm cN)$, where c is any integer. Similarly, the discrete Fourier transform of $\{x(i)\}$ results in $X(k \pm cN)$. Hence we have

$$x(i) = x(i + N) = x(i + 2N) \cdots, \qquad i = 1, \ldots, N, \tag{5.3-13}$$

$$X(k) = X(k + N) = X(k + 2N) \cdots, \qquad k = 1, \ldots, N. \tag{5.3-14}$$

Thus both the transform and its inverse yield periodic sequences. In most applications one is interested in only a single period. For the time series, the period is $T = N\,\Delta t$, and in frequency the period is $f_s = N\,\Delta f = 2W$.

When the $\{x(i)\}$ sequence is real, the real part of $X(k)$ is symmetric about the folding frequency $f_s/2$. This is easily demonstrated by computing, say, $[X(N/2) + 1]$ and $X[(N/2) - 1]$ using Eq. (5.3-11) (assume N is even). We have

$$X\left(\frac{N}{2} + 1\right) = \sum_{i=1}^{N} x(i)e^{-j\pi i - j2\pi i/N} = \sum_{i=1}^{N} x(i)(-1)^i e^{-j2\pi i/N}$$

$$X\left(\frac{N}{2} - 1\right) = \sum_{i=1}^{N} x(i)e^{-j\pi i + j2\pi i/N} = \sum_{i=1}^{N} x(i)(-1)^i e^{j2\pi i/N}.$$

Hence $X[(N/2) + 1] = X^*[(N/2) - 1]$, where the superscript asterisk denotes complex conjugate and

$$\text{Re}\left[X\left(\frac{N}{2} + 1\right)\right] = \sum_{i=1}^{N} x(i)(-1)^i \cos\frac{2\pi i}{N} = \text{Re}\left[X\left(\frac{N}{2} - 1\right)\right]. \tag{5.3-15}$$

Thus the real part is symmetric, and the imaginary part of $X(k)$ is antisymmetric about $N/2$. But since $X(k)$ is a periodic sequence with period N, we also have that $X[(N/2) + k] = X[(-N/2) + k] = X^*[(N/2) - k]$ so that $X(k)$ has Hermitian symmetry about $k = 0$. Furthermore, we see from Eq. (5.3-11) that, when $k = N$, $X(N)$ is proportional to the dc value of the $\{x(i)\}$. Therefore, we associate $k = N$ with dc. The lowest positive frequency above dc is $k = 1$; the nearest negative frequency to dc is $N - 1$. Hence we identify the Fourier coefficients $X(k)$ for $1 \le k \le N/2$ as positive-frequency components, while those for $N/2 < k < N$ are negative-frequency components. The maximum fre-

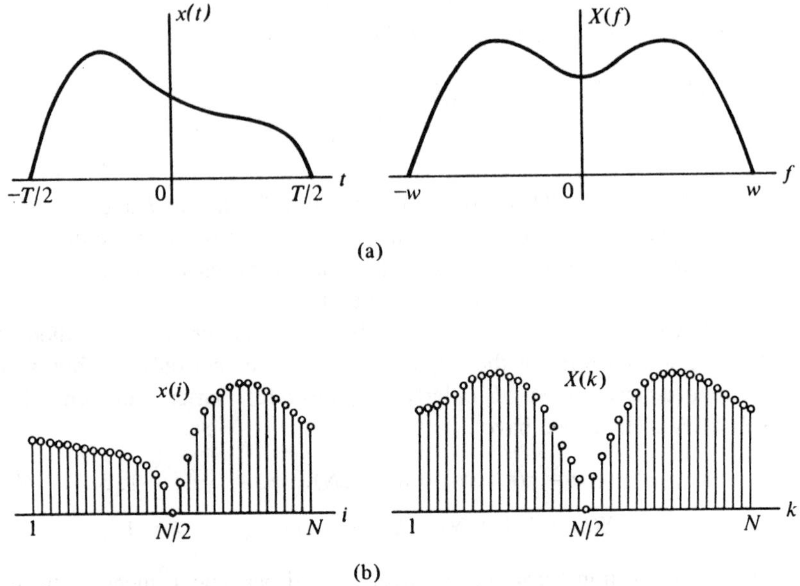

Figure 5.3-1. (a) Continuous signal and its Fourier transform. (b) Discrete signal and its Fourier transform.

quency corresponds to $k = +N/2$. For the time series $x(i)$, we similarly identify $N/2 < i < N$ as negative time and $1 \le i \le N/2$ as positive time. See Fig. 5.3-1.

5.4 PROPERTIES OF THE DISCRETE FOURIER TRANSFORM

Most of the elementary properties of the continuous Fourier transform presented in Sec. 2.9 have their counterpart in the discrete Fourier transform. We consider some of them here.

Parseval's Theorem

This takes the form

$$\sum_{i=1}^{N} |x(i)|^2 = \frac{1}{N}\sum_{k=1}^{N} |X(k)|^2 \tag{5.4-1}$$

and is proved by writing

$$\begin{aligned}\sum_{i=1}^{N} |x(i)|^2 &= \sum_{i=1}^{N} x^*(i)\cdot\frac{1}{N}\sum_{k=1}^{N} X(k)e^{j2\pi ik/N} \\ &= \frac{1}{N}\sum_{k=1}^{N} X(k)\sum_{i=1}^{N} x^*(i)e^{j2\pi ik/N} = \frac{1}{N}\sum_{k=1}^{N} |X(k)|^2.\end{aligned} \tag{5.4-2}$$

Linearity

If $X(k)$ and $Y(k)$ are the DFT of $x(i)$ and $y(i)$, respectively, the DFT of $ax(i) + by(i)$ is $aX(k) + bY(k)$.

Shifting Property

Let $X(k)$ be the DFT of $x(i)$, and consider the DFT of $x(i + m)$, where m is an integer. This is given by

$$\sum_{i=1}^{N} x(i + m)e^{-j2\pi ki/N} = \sum_{l=1+m}^{N+m} x(l)e^{-j2\pi(l-m)k/N} \tag{5.4-3}$$

$$= e^{j2\pi mk/N} \sum_{l=1+m}^{N+m} x(l)e^{-j2\pi kl/N}$$

$$= e^{j2\pi mk/N}X(k). \tag{5.4-4}$$

In the first step we have substituted $l \equiv i + m$ so that instead of counting from 1 we start counting from $m + 1$. In the last step we make use of the fact that the $x(l)$ are periodic, that is, $x(l) = x(l + N)$, and so on. Hence, as long as we sum N successive terms of the form $x(l)e^{-j2\pi kl/N}$, we get the same $X(k)$ no matter where we start the summing process.

Convolution Multiplication

Discrete data are usually obtained by *uniform* sampling of continuous-time signals. When the system is shift- or time-invariant, we can use discrete convolution to compute the output:

$$y(m) = \sum_{i=1}^{N} x(i)h(m - i). \tag{5.4-5}$$

Taking the discrete Fourier transform of both sides yields

$$Y(k) = \sum_{m=1}^{N} y(m)e^{-j2\pi mk/N} = \sum_{m=1}^{N}\sum_{i=1}^{N} x(i)h(m - i)e^{-j2\pi mk/N}$$

$$= \sum_{i=1}^{N} x(i)e^{-j2\pi ik/N} \sum_{m=1}^{N} h(m - i)e^{-j2\pi(m-i)k/N} \tag{5.4-6}$$

or

$$Y(k) = X(k)H(k). \tag{5.4-7}$$

In the second step we exchange orders of summation and multiply by $e^{-j2\pi(ik-ik)/N} = 1$. The last step uses the periodicity of the $h(m)$, which implies that summing from $1 - i$ to $N - i$ is the same as summing from 1 to N. Just as the CFT enjoys the convolution-multiplication property for continuous waveforms, so does the DFT for discrete signals; this property is, in fact, one of the most important reasons for using the DFT.

Some of the other Fourier properties discussed in Chapter 2 also have their discrete counterparts. Duality is easily proved and yields an additional set of simple relations. The scaling property of the continuous Fourier transform is not used very much in the discrete case since scale information generally does not appear in the discrete formulation.

5.5 PITFALLS IN THE USE OF THE DFT

The discrete Fourier transform is in some ways conceptually simpler than the continuous transform because it deals with finite sets of numbers. Questions of convergence or continuity never arise. Also, the various Fourier relations are exact, and there is no need to consider concepts such as the delta function that are sometimes regarded as being of doubtful validity.

Problems arise when the discrete Fourier transform is used to approximate the continuous Fourier transform. The DFT is a mapping of one periodic number sequence into another, while the CFT usually maps a finite interval in time (or frequency) into an infinite interval in frequency (or time). The problems are then mainly due to truncation effects and sampling approximations, and some of them have already been discussed in Chapter 4.

Consider the Fourier transformation of a cosine wave of frequency f_0 Hz, as shown in Fig. 5.5-1. Line (a) shows that the cosine, which extends over the infinite interval $-\infty < t < \infty$, transforms into two delta functions at $f = \pm f_0$. If the cosine is observed over only a finite time period, the frequency function is the convolution of the delta functions with a sinc function as shown in line (b). Sampling of the time-limited cosine is equivalent to multiplication in the time domain by a comb of delta functions or a convolution in the frequency domain by a related comb of delta functions [see Eq. (2.12-24)]. The sampling comb and its transform are shown in line (c) and the sampled time function and its continuous Fourier transform in line (d). As discussed in Chapter 4, there will be aliasing or overlap in the spectra unless the sampling rate is large enough. Since the spectrum of the truncated cosine, or of any time function extending over a finite time interval, extends to infinity on the frequency axis, a certain amount of aliasing will always take place, even though it can be minimized by using a sufficiently large sampling rate.

In any case, it is clear that the spectrum is no longer exactly that of the truncated cosine wave, and therefore retransforming will result in a somewhat altered wave.

Sampling of the spectrum results in convolution of the time function with a delta-function comb so that the single cosine-pulse sequence is transformed into a periodic sequence. This is shown in line (e). Here again there will generally be some aliasing if the rate at which the spectrum is sampled is not high enough. Finally, if both the time and frequency function are reconstructed from N samples, we see that neither will in general be exact.

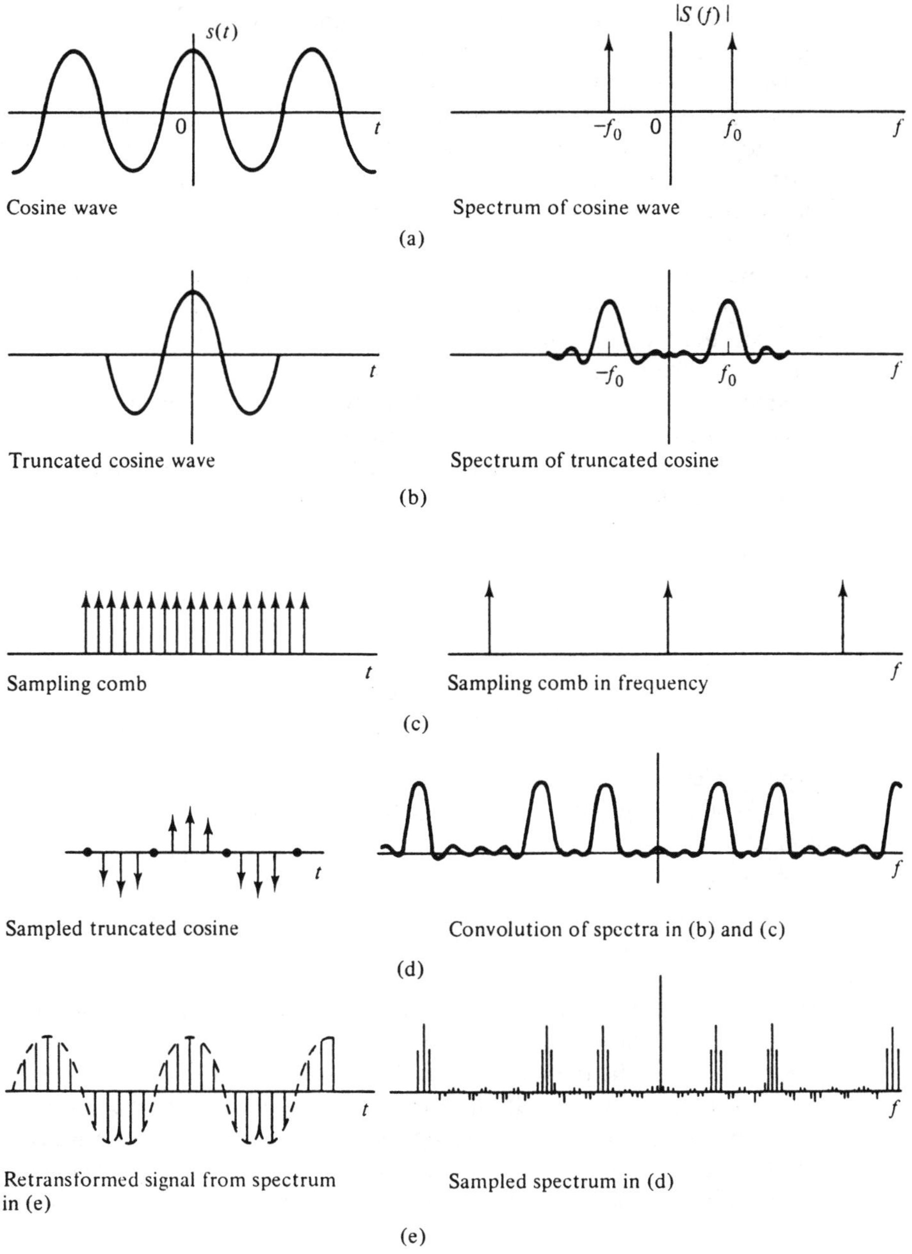

Figure 5.5-1. Effects of truncation and sampling. (After Bergland [5-1]).

In addition to the aliasing errors, truncation causes an effect referred to as *leakage*. This is illustrated in Fig. 5.5-1, line (b), where we show that the delta-function spectrum of the infinite cosine is converted into a sinc spectrum because of truncation. The word leakage refers to the fact that energy in the original spectral components at $f = \pm f_0$ leaks to other frequencies after truncation in time. The truncation shown in Fig. 5.5-1 is rather severe in order to clearly show the effect. However, if we wanted to compute the Fourier spectrum of a cosine, or any other function, we would always have to perform the transformation on a finite piece of the function because of the limitations in computer memory. Thus truncation is usually necessary (unless the function is naturally time limited), and it will result in the smearing effect shown.† Leakage can be reduced by using truncation functions that reduce the time function more gradually near the ends of the interval than the rect function illustrated in Fig. 5.5-1. Such functions are referred to as *data windows*, and much work has gone into the design of windows that reduce leakage without distorting the spectrum too much. A commonly used window is the cosine bell window, shown in Fig. 5.5-2. If the original time function is $x(t)$, then the truncated function is $x_T(t) = w(t)x(t)$, where

$$w(t) = \begin{cases} \dfrac{1}{2} - \dfrac{1}{2}\cos 2\pi \dfrac{t}{\alpha T}, & 0 < t \le \dfrac{\alpha T}{2}, \\ 1, & \dfrac{\alpha T}{2} < t < T - \dfrac{\alpha T}{2}, \\ \dfrac{1}{2} - \dfrac{1}{2}\cos 2\pi \dfrac{T-t}{\alpha T}, & T - \dfrac{\alpha T}{2} < t < T. \end{cases} \tag{5.5-1}$$

Typical values of α are 0.1 to 0.2.

Another frequently used window function is the generalized Hamming‡ window, defined by

$$w(t) = \begin{cases} \alpha - (1-\alpha)\cos\dfrac{2\pi t}{T}, & 0 \le t \le T, \\ 0, & \text{otherwise}. \end{cases} \tag{5.5-2}$$

This is the raised cosine pulse already encountered several times in this book. Note that for $\alpha \neq \frac{1}{2}$ there is a jump in $w(t)$ at $t = 0$ and $t = T$. When $\alpha = 0.54$, the window is called the (no-longer-generalized) Hamming window; when $\alpha = 0.50$, it is called the Hanning§ window [5-3]. An advantage of the Hanning window is that multiplication by $\frac{1}{2}$ is a particularly simple operation on the computer since it involves merely a right shift of the binary representation of the multiplicand. The Hamming window is optimum in the sense that of all raised cosine windows it has the smallest maximum side lobe level ([5-2], p. 99).

† There is no leakage in the DFT if the input spans exactly one fundamental period ([5-2], p. 92). However, in most applications the truncation of the input cannot be arranged to guarantee this.

‡ After Richard W. Hamming.

§ After the Austrian mathematician Julius von Hann.

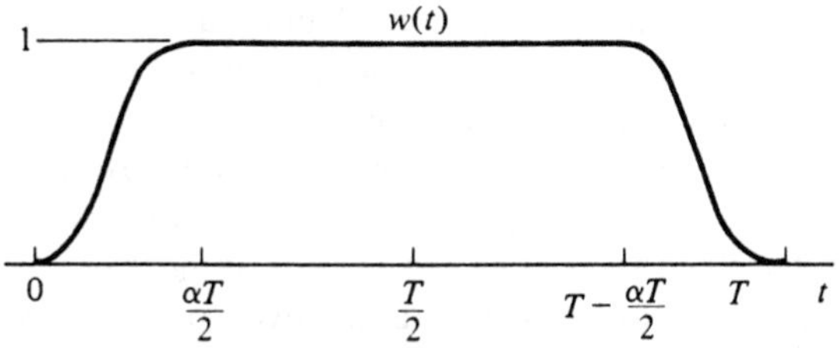

Figure 5.5-2. Data window with cosine roll-off.

Instead of multiplying the input signal by $w(t)$ one can, equivalently, convolve the spectrum of the input with the Fourier transform of the data window. The latter is called a *spectral window*. For the generalized Hamming window, the spectral window has the form (delay terms omitted)

$$W(f) = \alpha T \operatorname{sinc} fT + \frac{1-\alpha}{2} T \operatorname{sinc}(Tf - 1) + \frac{1-\alpha}{2} T \operatorname{sinc}(Tf + 1). \tag{5.5-3}$$

The DFT equivalent of Eq. (5.5-3) is

$$W(k) = \alpha \operatorname{sinc} k + \frac{1-\alpha}{2} \operatorname{sinc}(k - 1) + \frac{1-\alpha}{2} \operatorname{sinc}(k + 1), \tag{5.5-4}$$

and the discrete convolution of $W(k)$ with $X(k)$ results in replacing each point $X(k)$ in the original DFT by $\alpha X(k) + [(1 - \alpha)/2][X(k + 1) + X(k - 1)]$ (see Prob. 5-13). Thus the operation of the window function is nothing more than a weighted moving average of three adjacent spectral lines.

Data and spectral windows are discussed further in Sec. 9.5 where they are considered in relation to the problem of spectral estimation.

A third problem in the use of the DFT arises from the improper use of Eq. (5.4-7) to obtain the output of a linear system. The impulse response of such a system is generally defined only over a finite time interval T and is assumed to be zero outside this interval. Similarly, the input signal may exist over some time T_1. A typical input signal and impulse response are shown in Fig. 5.5-3. Convolution of these two signals results in a signal such as shown in Fig. 5.5-3(c), which typically extends over a time interval $T + T_1$.

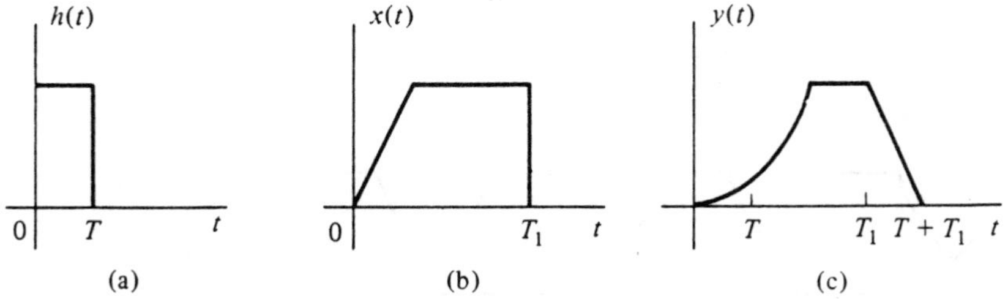

Figure 5.5-3. (a) Finite-duration impulse response. (b) Finite-duration signal. (c) Output.

Suppose that instead of actually performing the convolution, the output is obtained by multiplying the discrete Fourier transforms of $x(t)$ and $h(t)$, using Eq. (5.4-7). To get as short a sampling interval as possible, one might naively decide to obtain the N uniformly spaced samples from the longer of the two intervals T or T_1. However, since the DFT treats all functions as being periodic, this results in fact in the convolution of two periodic signals, such as the ones shown in Figs. 5.5-4(a) and (b). The output would be as shown in Fig. 5.5-4(c). Note that the result, even over one period, is quite different from that shown in Fig. 5.5-3.

A simple cure for this problem is to lengthen the sampling interval to be at least equal to $T + T_1$. Then, even though the signals are still periodic, the nonzero parts will be separated by intervals of zero signals such that the correct output is obtained in any one of the extended periods. See Fig. 5.5-5.

In practice, one frequently wants to convolve signals having widely differing periods. This is true, for instance, when the input is a random signal extending over indefinite time, and the impulse response can be assumed to have only a finite length. Such convolutions are most rapidly performed by using the fast Fourier transform (FFT), described next, to transform the impulse response and adjacent intervals of the signal, using Eq. (5.4-7), and then retransforming. For this operation to yield valid results, one must use overlapping intervals as indicated in Fig. 5.5-6. The length of each interval, containing N samples, is $2T$, where T is the time over which $h(t)$ is nonzero. Convolution

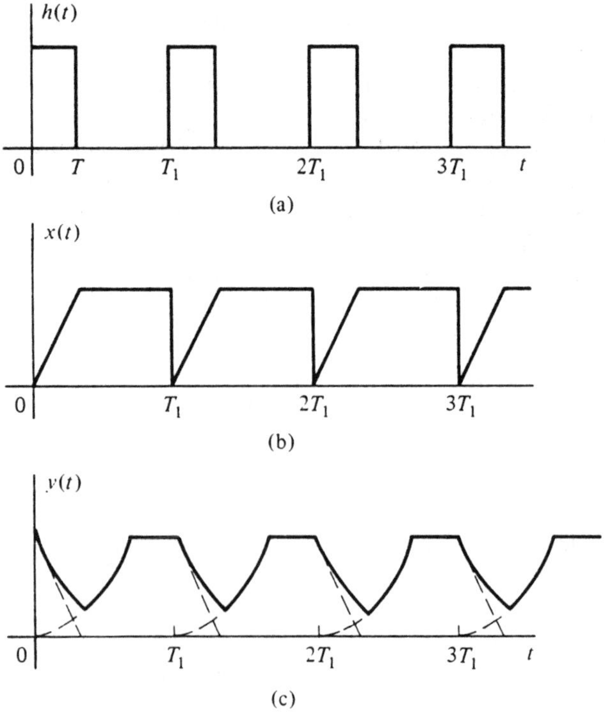

Figure 5.5-4. Periodicities induced by the DFT. (a) Impulse response. (b) Input. (c) Distorted output.

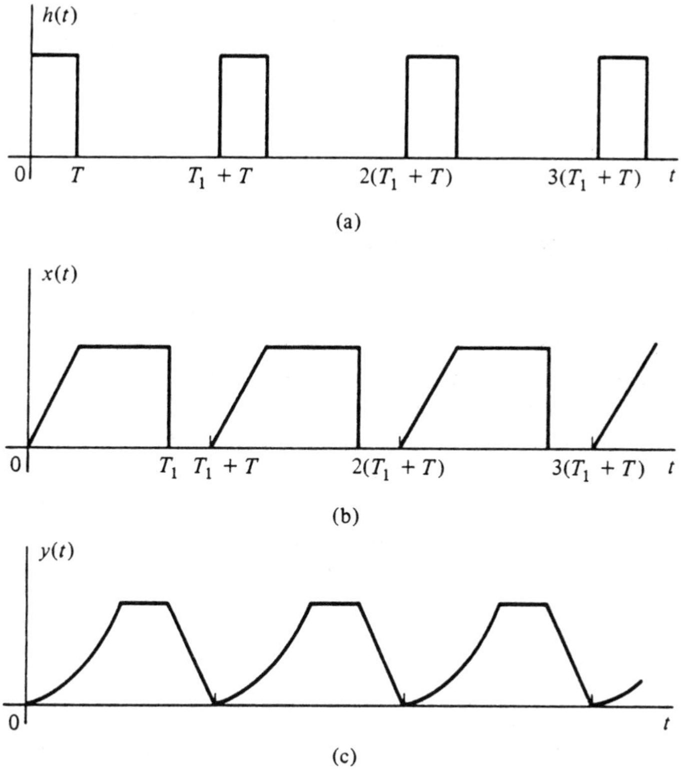

Figure 5.5-5. Periodicities induced by the DFT when a sufficiently large interval is used to produce the correct output. (a) Impulse response. (b) Input. (c) Undistorted output appears as a periodic signal.

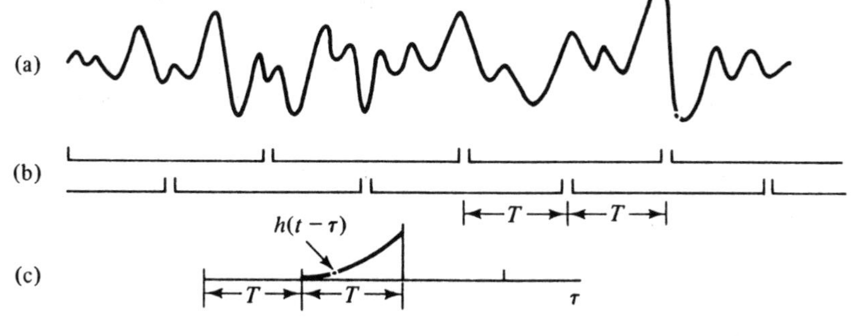

Figure 5.5-6. Use of overlapping intervals for processing signals of indefinite time duration with the DFT of a time-limited impulse response. (a) Input. (b) Overlapping intervals of length $2T$. (c) impulse response.

of this $h(t)$ with an equally long interval of signal results in incorrect results for the first $N/2$ spectral components but in valid results for the last $N/2$ components. We therefore retain only the last $N/2$ components from each set of N that is generated, and after each complete computation we advance the signal T seconds (i.e., $N/2$ points) and repeat the process. For a complete discussion of this procedure, see [5-1].

5.6 THE FAST FOURIER TRANSFORM

The discrete Fourier transform has in recent years become one of the most widely used computational tools because of the development in 1965 of the fast Fourier transform (FFT) algorithm by J. W. Cooley and J. W. Tukey [5-4]. This algorithm reduces the number of computations required to perform the DFT from something on the order of N^2 to $N/2 \log_2 N$. For $N = 1024$ the reduction factor is about $10^6/5000$, or 200. Thus, if a typical operation take 10 μs, the time required for the transformation is reduced from 10 sec to 50 ms. Machines are now available in which the FFT is hard-wired so that a 1024-point FFT may be obtained in less than 10 ms. This is fast enough so that fairly extensive signal processing can be done on line in many cases.

To explain the algorithm, consider the expression

$$X(k) = \sum_{i=0}^{N-1} x(i)U^{ik}. \tag{5.6-1}$$

This is essentially Eq. (5.3-11), with the indexing changed to run from 0 to $N - 1$ and with $U \equiv U_N^{-1}$. The algorithm is particularly simple if N is a power of 2, although this is not essential [5-5]. For the sake of our discussion, let $N = 8$. The indexes i and k can be expressed as binary numbers in the form

$$\begin{aligned} i &= (i_2, i_1, i_0), \\ k &= (k_2, k_1, k_0), \end{aligned} \tag{5.6-2}$$

where i_2, i_1, and so on, take on only the values 0 or 1. Only three binary digits are needed since $N = 8$; for example, if $i = 5$, it becomes 101 in binary form. The decimal equivalent of the number (i_2, i_1, i_0) is $4i_2 + 2i_1 + i_0$ and similarly for the k's.

With the binary number representation, Eq. (5.6-1) can be rewritten in the form

$$\begin{aligned} X(k_2, k_1, k_0) &= \sum_{i_0=0}^{1} \sum_{i_1=0}^{1} \sum_{i_2=0}^{1} x(i_2, i_1, i_0)U^{(4i_2+2i_1+i_0)(4k_2+2k_1+k_0)} \\ &= \sum_{i_0=0}^{1} \sum_{i_1=0}^{1} \sum_{i_2=0}^{1} x(i_2, i_1, i_0)U^{k_0(4i_2+2i_1+i_0)}U^{2k_1(2i_1+i_0)}U^{4k_2i_0}, \end{aligned} \tag{5.6-3}$$

where the second line is obtained from the first by multiplying out the exponent and observing that terms such as $U^{16k_2i_2}$ or $U^{8k_1i_2}$ can be dropped since they are equal to 1.

The equation can now be solved recursively as follows. We let

$$
\begin{aligned}
A_1(k_0, i_1, i_0) &\equiv \sum_{i_2=0}^{1} x(i_2, i_1, i_0)U^{k_0(4i_2+2i_1+i_0)}, \\
A_2(k_0, k_1, i_0) &\equiv \sum_{i_1=0}^{1} A_1(k_0, i_1, i_0)U^{2k_1(2i_1+i_0)}, \\
A_3(k_0, k_1, k_2) &\equiv \sum_{i_0=0}^{1} A_2(k_0, k_1, i_0)U^{4k_2 i_0}, \\
X(k) = X(k_2, k_1, k_0) &= A_3(k_0, k_1, k_2).
\end{aligned}
\tag{5.6-4}
$$

Note that each sum contains only two terms, each of which involves multiplication by some power of U. For each k there are therefore 6 complex multiplications and 3 complex additions, or since there are 8 k's, a total of 48 multiplications and 24 additions. Half of the multiplications can be eliminated immediately by noting that the exponent of U is zero for the 4 A_1's for which k_0 is zero in the first line,† and similarly for the 4 A_2's in the second line for which $k_1 = 0$ and for the 4 A_3's in the third line for which $k_2 = 0$. This follows since $U^0 = 1$. A further halving of the number of multiplications is possible if half of the additions are replaced by subtractions. This follows because $U^4 = -U^0$, $U^5 = -U^1$, and so on. This has the effect in the first line, for instance, of making $A_1(1, i_1, i_0) = [x(0, i_1, i_0) - x(1, i_1, i0)]U^{(2i_1+i_0)}$, and so on.

A signal flow diagram illustrating these factors is shown in Fig. 5.6-1 for the 8-point algorithm. In this diagram the junction of two arrows represents an addition, and a minus sign near one line at a junction means that this line should have its sign reversed before addition (i.e., it should be subtracted). Numbers such as U^0, U^1, placed next to lines mean multiplication by that amount. Note that there are 12 additions, 12 subtractions, and 12 multiplications. The numbers in this example indicate exactly each of $N/2 \log_2 N$ additions, subtractions, and multiplications. More generally this can be shown to be true whenever the number of points is a power of 2 ([5-6], Chap. 6). Actually 7 of the multiplications are by U^0 and could be omitted, leaving only 5 multiplications.‡ It is easily shown that for an N-point algorithm this remaining reduction in multiplications results in $(N/2) \log_2 N - (N - 1)$ multiplications.

Multiplications are in general an order of magnitude more time consuming than additions or subtractions. The amount of time required to perform an N-point FFT is therefore on the order of $(N/2) \log_2 N - (N - 1)$ times the time required to perform a single complex multiplication, or, since a complex multiplication is equivalent to four real multiplications, the time is $2N \log_2 N - 4(N - 1)$ real multiplication times. This contrasts with the roughly N^2 complex multiplications that would be needed if the DFT were directly evaluated using Eq. (5.3-11). Figure 5.6-2 shows the substantial time saving afforded by the FFT for large N [5-1].

Examination of the flow chart of Fig. 5.6-1 shows that the transform can be done "in place," that is, by writing all intermediate results over the original data sequence and writing the final answer over the intermediate results. Thus no storage is needed

† That is, $A_1(0, 0, 0)$, $A_1(0, 0, 1)$, $A_1(0, 1, 0)$, $A_1(0, 1, 1)$.

‡ The multiplications by U^0 are omitted in Fig. 5.6-1.

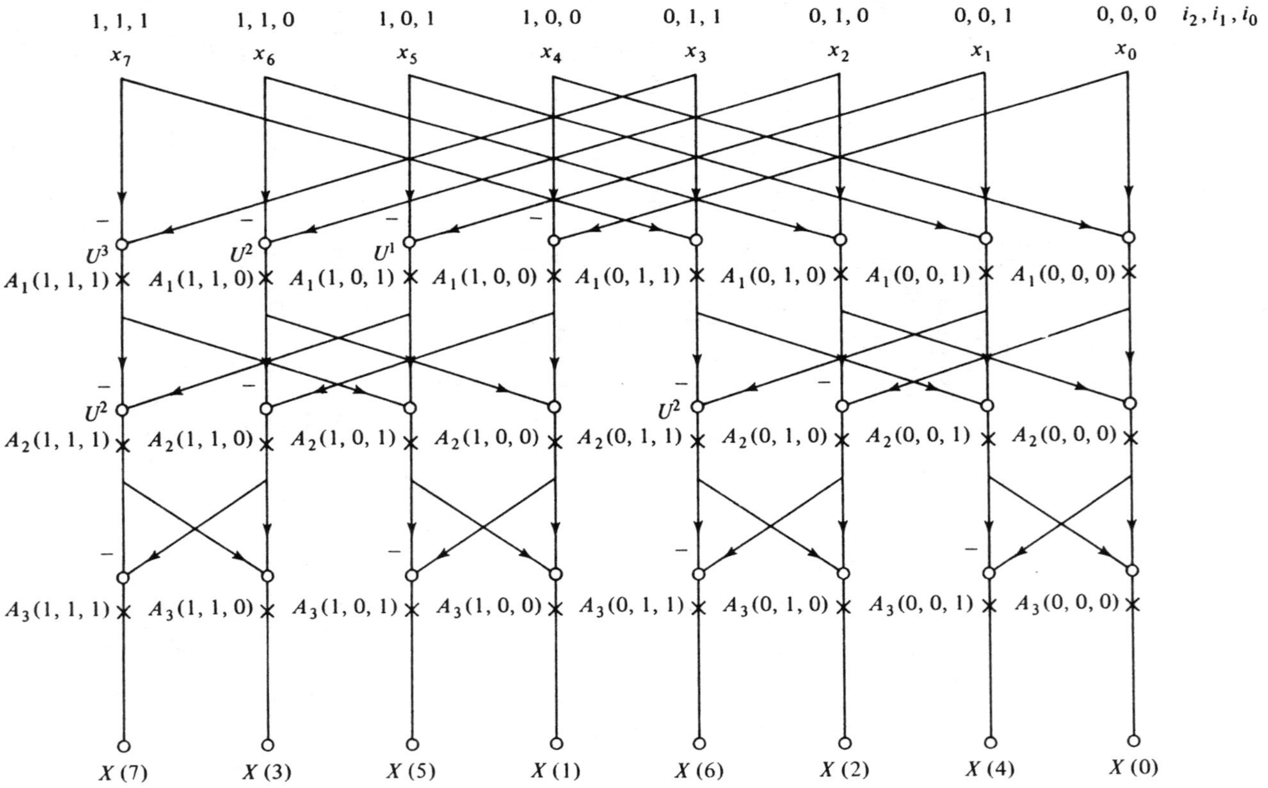

Figure 5.6-1. Signal flow diagram for the FFT algorithm. Nodes represent algebraic addition, and the numbers U^k next to some links represent multiplication by that amount. The values at the crosses are the A's computed at that level.

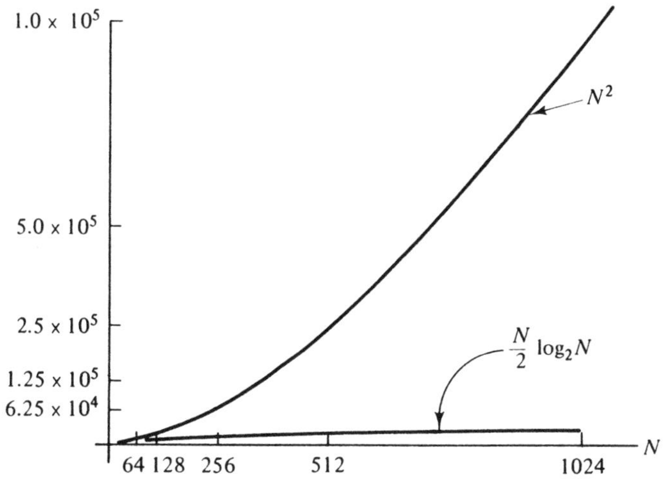

Figure 5.6-2. Required number of multiplications in a direct DFT calculation versus the number required in the FFT algorithm. (After Bergland [5-1]).

beyond that required for the original N complex numbers. This is an additional important advantage of the FFT. Also, since the FFT is essentially a series of nested 2-point DFTs, it is in principle possible to do a transform for N larger than the available storage space by doing several smaller FFTs and then combining the results. The algorithm is discussed in much more detail in some of the references ([5-4] to [5-7]). References [5-2] and [5-6] contain short FORTRAN programs that implement the algorithm.

5.7 THE Z TRANSFORM

While the DFT and FFT have their principal applications in the numerical *computation* of Fourier transforms by computer, the Z transform has its principal application as an *analytical tool* in the *analysis* of discrete-time systems. The Z transform and the DFT are closely related and share many properties. However, unlike the DFT, the frequency variable in the Z transform is a continuous complex variable, z, instead of a discrete real variable k. In a sense, one may regard the DFT as a weighted average of the input data, while the Z transform may be regarded as a *resolution* of the input data into signals of the form z^{-n}, where z is a complex variable. The one-sided Z transform of a discrete signal $x(n)$ is defined by

$$\mathcal{Z}_1[x(n)] = X_1(z) = \sum_{n=0}^{\infty} x(n)z^{-n}, \tag{5.7-1}$$

while the two-sided Z transform is defined by

$$\mathcal{Z}_2[x(n)] = X_2(z) = \sum_{n=-\infty}^{\infty} x(n)z^{-n}. \tag{5.7-2}$$

We shall consider only the one-sided Z transform here; therefore, expressions such as $X(z)$ or $Z[x(u)]$ all refer to one-sided Z transforms in the sequel.

The variable z is an arbitrary complex variable; but for the Z transform to exist, z must lie inside a region of convergence that depends on the particular sequence $x(n)$ being transformed. If $x(n)$ has nonzero values only over some finite range of values of n, the summation in Eq. (5.7-1) converges for all z. Also, if $x(n) = 0$ for $n < 0$ and is otherwise reasonably well behaved, the series $X_1(z)$ converges if $|z|$ is larger than some radius of convergence r.†

One main reason for using the Z transform is that it possesses the convolution-multiplication property of all Fourier-type transformations. This permits the replacement of the relatively complex procedure of convolution by the much simpler one of multiplication. Convolution of two discrete-time signals was discussed in Sec. 5.2. However, instead of assuming that $x(m)$ begins at $m = 1$, as we did in Eq. (5.2-11), we shall for convenience assume that $x(m) = 0$ for $m < 0$. In that case we write

$$y(n) = \sum_{m=0}^{n} x(m)h(n-m), \tag{5.7-3}$$

where $h(\cdot)$ is a causal response. Taking the Z transform of both sides of Eq. (5.7-3) gives

$$\begin{aligned} Y(z) &= \sum_{n=0}^{\infty} y(n)z^{-n} = \sum_{n=0}^{\infty}\sum_{m=0}^{\infty} x(m)h(n-m)z^{-n} \\ &= \sum_{m=0}^{\infty} x(m)z^{-m} \sum_{k=-m}^{\infty} h(k)z^{-k} \\ &= \sum_{m=0}^{\infty} x(m)z^{-m} \sum_{k=0}^{\infty} h(k)z^{-k} \\ &= X(z)H(z), \end{aligned} \tag{5.7-4}$$

where

$$H(z) = \sum_{k=0}^{\infty} h(k)z^{-k}. \tag{5.7-5}$$

In the second equality of the first line of this development we have changed the upper limit on the summation from n to infinity; this does not affect anything since $h(\cdot)$ is zero for negative arguments. In the second line we interchanged orders of summation and substituted the index $k \equiv n - m$. In the third line we again used the fact that $h(\cdot)$ is causal and is therefore zero for negative arguments.

The usefulness of the Z transform depends in part on the possibility of finding simple *closed-form expressions* for commonly used sequences $x(n)$ and $h(n)$. Then the

† For the one-sided Z transform, the conditions for convergence are (1) $|x(n)| < \infty$ for all finite n, and (2) $|x(n)| \leq kr^n$ if $n \geq N$, for some finite k, r, and N. The smallest value of r for which this inequality is satisfied is the radius of convergence.

resulting $Y(z)$ can be analyzed or inverted to study the properties of the response. In this respect the Z transform differs from the DFT, where closed-form expressions are mostly irrelevant. We shall consider the Z transforms of some common functions.

The Unit Step Function

Assuming that the step takes place at $n = 0$, we have that

$$x(n) = \begin{cases} 1, & \text{for } n = 0, 1, 2, \ldots, \\ 0, & \text{for } n < 0. \end{cases}$$

Then

$$X(z) = \sum_{n=0}^{\infty} z^{-n} = \frac{1}{1 - z^{-1}}. \tag{5.7-6}$$

Note that the summation is just a geometric series in z^{-1}. It converges if $|z| > 1$; therefore, the radius of convergence is 1.

The Delayed Unit Step

Suppose the step takes place at $n = k$; that is,

$$x(n) = \begin{cases} 1, & n < k, \\ 0, & n \geq k. \end{cases}$$

Then

$$X(z) = \sum_{n=k}^{\infty} z^{-n} = \frac{z^{-k}}{1 - z^{-1}}. \tag{5.7-7}$$

The radius of convergence is again 1. This example is a particular application of the *shifting property* of the Z transform:
If

$$X(z) = \mathscr{Z}[x(n)],$$

then

$$\mathscr{Z}[x(n - k)] = z^{-k}X(z) + z^{-k} \sum_{m=-k}^{m=-1} x(m)z^{-m}. \tag{5.7-8}$$

The proof is left as an exercise (Prob. 5-14). The second term vanishes if $x(m) = 0$ for negative m.

The Linearly Increasing Sequence

Here

$$x(n) = \begin{cases} n, & n \geq 0, \\ 0, & n < 0. \end{cases} \tag{5.7-9}$$

Then

$$X(z) = \sum_{n=0}^{\infty} nz^{-n}. \tag{5.7-10}$$

This summation, as well as others in which n appears to higher powers, is easily evaluated by differentiating Eq. (5.7-6) with respect to z, which results in $-\Sigma_{n=0}^{\infty} nz^{-n-1}$. Thus the $\mathcal{Z}$ transform desired here is

$$\mathcal{Z}[n] = -z\frac{d}{dz}\left(\frac{1}{1-z^{-1}}\right) = \frac{z^{-1}}{(1-z^{-1})^2}. \tag{5.7-11}$$

The Exponential Sequence

If

$$x(n) = \begin{cases} a^n, & n \geq 0, \\ 0, & n < 0. \end{cases}$$

then

$$\mathcal{Z}[a^n] = \sum_{n=0}^{\infty} a^n z^{-n} = \frac{1}{1-az^{-1}}. \tag{5.7-12}$$

The radius of convergence is now seen to be $|a|$. If $|a| < 1$, the sequence $x(n)$ converges to zero, and this permits a smaller radius of convergence for z. On the other hand, if $|a| > 1$, a^n diverges, and then a larger z magnitude is needed for convergence. Note that by making the identification $a = e^{-c}$ this result is seen to be analogous to the Fourier transform of the exponential function.

The Discrete Impulse

This is the discrete equivalent of the Dirac delta function discussed in Sec. 2.11. It is defined by

$$\delta_{mn} = \begin{cases} 1, & n = m, \\ 0, & n \neq m. \end{cases} \tag{5.7-13}$$

This function is also called the unit function or the Kronecker delta. It is sometimes written $\delta(n - m)$. If $m = 0$, $\delta(n)$ is 1 for $n = 0$, and it is zero otherwise. The corresponding Z transform is

$$\mathcal{Z}[\delta_{mn}] = z^{-m}, \tag{5.7-14}$$

and for the particular value $m = 0$

$$\mathcal{Z}[\delta_{0n}] = 1. \tag{5.7-15}$$

Again note the analogy to the continuous Fourier transform, especially Eq. (5.7-15). The discrete impulse has the advantage over the Dirac delta function of being finite and requiring no limiting arguments. This is another illustration of the relative analytic simplicity of discrete systems compared to continuous ones.

The Finite Square Pulse

As a final example, we consider the sequence

$$x(n) = \begin{cases} 1, & n = 0, 1, \ldots, m-1, \\ 0, & \text{otherwise}. \end{cases} \tag{5.7-16}$$

Then

$$X(z) = \sum_{n=0}^{m-1} z^{-n} = \frac{1 - z^{-m}}{1 - z^{-1}}.$$

This can be put into the more symmetric form

$$X(z) = z^{-(m-1)/2} \left[\frac{z^{m/2} - z^{-m/2}}{z^{1/2} - z^{-1/2}} \right], \tag{5.7-17}$$

where the factor $z^{-(m-1)/2}$ can be regarded as resulting from a shift of $(m - 1)/2$ sample points, so that the result in brackets is the Z transform of a symmetrical pulse. By using $z = e^{j\omega}$, Eq. (5.7-17) can, for small ω, be identified as an approximation to the sinc function; this demonstration is left as an exercise. Because of the finite number of terms, $X(z)$ converges everywhere.

5.8 THE INVERSE Z TRANSFORM

The one-sided Z transform can be computed by several methods, but the most general is the method of contour integration using the calculus of residues. Since

$$X(z) = \sum_{n=0}^{\infty} x(n) z^{-n},$$

the Z transform is recognized *to be a Laurent series* [5-8] about the point $z = 0$. Therefore, the coefficients are given by

$$x(n) = \frac{1}{2\pi j} \int_C X(z) z^{n-1}\, dz, \tag{5.8-1}$$

where C is any contour that encloses all the singularities of $X(z)$. The function $X(z)$ must be analytic on C. The residue theorem states that, if C is a closed contour within which an arbitrary function $F(z)$ is analytic except for a finite number of singular points $z_1, z_2, \ldots, z_N$ interior to C, then

$$\frac{1}{2\pi j}\int_C F(z)\,dz = \text{sum of residues of } F \text{ at } z_1, z_2, \ldots, z_N. \qquad (5.8\text{-}2)$$

To use the residue theorem to compute $x(n)$, we let $F(z) \equiv X(z)z^{n-1}$.

Since the Laurent expansion is unique, any other way of getting $X(z)$ into a form in which the coefficients can be recognized is acceptable as an inversion method. A simple method applicable to Z transforms that are ratios of polynomials in z is just to divide the denominator into the numerator. Still another method of performing the inversion is to use partial fractions to put the transform into a form where the inverses can be recognized. Some examples are given next.

Example

Consider

$$X(z) = \frac{z^2 - 9z}{z^2 - 6z + 5}.$$

1. *Partial fractions.*

$$X(z) = \frac{z^2 - 9z}{z^2 - 6z + 5} = \frac{z^2 - 9z}{(z-1)(z-5)} = \frac{2z}{z-1} - \frac{z}{z-5}$$
$$= \frac{2}{1 - z^{-1}} - \frac{1}{1 - 5z^{-1}}.$$

By Eq. (5.7-6), the first term corresponds to the step function sequence with a coefficient of 2. The second term is in the form of Eq. (5.7-12) with $a = 5$. Thus

$$x(n) = 2 - 5^n, \qquad n = 0, 1, 2, \ldots. \qquad (5.8\text{-}3)$$

2. *Residues.* The function $X(z)z^{n-1}$ has poles at $z = 1$ and $z = 5$. The residue at $z = 1$ is

$$\lim_{z\to 1} \frac{(z-1)(z^2 - 9z)z^{n-1}}{(z-1)(z-5)} = 2,$$

and the residue at 5 is

$$\lim_{z\to 5} \frac{(z-5)(z^2 - 9z)z^{n-1}}{(z-1)(z-5)} = -5^n.$$

Therefore, $x(n) = 2 - 5^n$ as before. Note that the function $X(z)z^{n-1}$ may have a pole at the origin for $n = 0$; in fact, this will generally happen if the numerator of $X(z)$ does not contain z as a factor. Then the residue for $z = 0$ must also be computed. As an example, if $X(z) = 10/(z-1)(z-5)$, the residues of $X(z)z^{n-1}$ for $n > 0$ are -2.5 and $5^n/2$; but for $n = 0$, there are three poles with residues 2, -2.5, and 0.5. Therefore, for this function $x(0) = 0$, $x(n) = 2.5(5^{n-1} - 1)$ for $n > 0$.

3. *Long division.* The function $X(z)$ is first converted into a function of z^{-1} by

dividing the numerator and denominator by the highest power of z. In our example we divide by z^2. This results in

$$X(z) = \frac{1 - 9z^{-1}}{1 - 6z^{-1} + 5z^{-2}}.$$

The denominator is then divided into the numerator as follows:

$$\begin{array}{r l}
 & 1 - 3z^{-1} - 23z^{-2} - 123z^{-3} \cdots \\
1 - 6z^{-1} + 5z^{-2} \big) & 1 - 9z^{-1} \\
 & 1 - 6z^{-1} + 5z^{-2} \\
\hline
 & -3z^{-1} - 5z^{-2} \\
 & -3z^{-1} + 18z^{-2} - 15z^{-3} \\
\hline
 & \qquad -23z^{-2} + 15z^{-3} \\
 & \qquad -23z^{-2} + 138z^{-3} - 115z^{-4} \\
\hline
 & \qquad\qquad -123z^{-3} + 115z^{-4}
\end{array}$$

Thus

$$x(0) = 1$$

$$x(1) = -3$$

$$x(2) = -23$$

$$x(3) = -123$$

$$\vdots$$

$$x(n) = 2 - 5^n,$$

as before. Note that the method of long division yields numerical values for the $x(n)$ from which a functional form has to be inferred.

Properties of the Z Transform

The properties of the Z transform are very similar to those of the DFT, already discussed in some detail. We shall therefore confine ourselves here to a listing of the most important properties without proof (Table 5.8-1). By way of example, consider the computation of the Z transform of na^n. Use property 5 with $k = 1$. The transform for a^n is $1/(1 - az^{-1})$, and therefore the desired transform is

$$z^{-1}\frac{d}{dz^{-1}}\left(\frac{1}{1 - az^{-1}}\right) = \frac{az^{-1}}{(1 - az^{-1})^2}.$$

The Z transform is discussed in considerable detail in books dealing with discrete signal processing. See particularly [5-9] to [5-12].

TABLE 5.8-1 PROPERTIES OF THE ONE-SIDED Z TRANSFORM

Property	Time sequence	Z Transform
1. Linearity	$ax(n) + by(n)$	$aX(z) + bY(z)$
2. Shifting	$x(n-k)$	$z^{-k}X(z) + z^{-k}\sum_{m=-k}^{m=-1} x(m)z^{-m}$
3. Scale change	$x(an)$ (a is an integer)	$X(z^{a^{-1}})$
4. Multiplication by a^n	$a^n x(n)$	$X(a^{-1}z)$
5. Multiplication by n^k	$n^k x(n)$	$\left(z^{-1}\frac{d}{dz^{-1}}\right)^k X(z)$
6. Forward difference	$\Delta x(n) = x(n+1) - x(n)$	$(z-1)X(z) - zx(0)$
7. Backward difference	$\nabla x(n) = x(n) - x(n-1)$	$(1 - z^{-1})X(z) - z^{-1}x(-1)$
8. Sums	$\sum_{k=-\infty}^{n} x(k)$	$\dfrac{X(z) + \sum_{k=-\infty}^{-1} x(k)}{1 - z^{-1}}$
9. Convolution	$y(n) = \sum_{m=0}^{\infty} h(n-m)x(m)$	$Y(z) = H(z)X(z)$
10. Product of two functions	$y(n) = x(n)\cdot g(n)$	$Y(z) = \frac{1}{2\pi j}\int_C X(\lambda)G(z\lambda^{-1})\frac{d\lambda}{\lambda}$ All poles of $X(\lambda)$ inside C, all poles of G outside C

5.9 RELATION BETWEEN THE Z TRANSFORM AND THE DFT

In Eq. (5.7-1), consider a sequence $x(n)$ having nonzero values for only a finite number of n's; that is,

$$x_N(n) = \begin{cases} x(n), & 0 \le n < N, \\ 0, & \text{otherwise.} \end{cases} \tag{5.9-1}$$

By Eq. (5.7-1), the Z transform of this truncated sequence is

$$X(z) = \sum_{n=0}^{N-1} x_N(n)z^{-n}. \tag{5.9-2}$$

The DFT of the sequence $x(n)$ is given in Eq. (5.3-11). With a minor change of notation, this can be written in the form†

$$\tilde{X}(k) = \sum_{n=0}^{N-1} x_N(n)U_N^{-kn}, \tag{5.9-3}$$

† The tilde over $\tilde{X}(k)$ is used to designate the DFT.

where $U_N = e^{j2\pi/N}$. Thus

$$\tilde{X}(k) = X(e^{j2\pi k/N}); \tag{5.9-4}$$

that is, the DFT is the Z transform evaluated for equally spaced values of z on the unit circle.

Since the sequence $x_N(n)$ and its DFT $\tilde{X}(k)$ uniquely define each other, it should be possible to obtain $X(z)$ for all values of z from the DFT. To obtain this relation, we start with Eq. (5.9-2), but for $x_N(n)$ we use the inverse DFT relation:

$$x_N(n) = \frac{1}{N} \sum_{k=0}^{N-1} \tilde{X}(k) U_N^{kn}. \tag{5.9-5}$$

Then Eq. (5.9-2) becomes

$$X(z) = \frac{1}{N} \sum_{n=0}^{N-1} \sum_{k=0}^{N-1} \tilde{X}(k)(U_N^k z^{-1})^n,$$

which, by a change in the order of summation, becomes

$$\begin{aligned} X(z) &= \sum_{k=0}^{N-1} \tilde{X}(k) \frac{1}{N} \sum_{n=0}^{N-1} (U_N^k z^{-1})^n \\ &= \sum_{k=0}^{N-1} \tilde{X}(k) \frac{1 - U_N^{kN} z^{-N}}{N(1 - U_N^k z^{-1})} \\ &= \sum_{k=0}^{N-1} \frac{\tilde{X}(k)(1 - z^{-N})}{(1 - U_N^k z^{-1})N} \end{aligned} \tag{5.9-6}$$

In going from the first line to the second we have summed the sum over n using the formula for a finite geometric series, and in the second step we used the fact that $U_N^{kN} = 1$.

Equation (5.9-6) is the desired relation and can be regarded as an interpolation formula by which $X(z)$ is expressed in terms of its values at N equally spaced points on the unit circle. In this respect it is somewhat similar to Eq. (4.2-5), which relates a function of time to its values at discrete-time points. The similarity can be made somewhat more evident by considering $X(e^{j\omega})$, that is, $X(z)$ evaluated on the unit circle. By Eq. (5.9-6), this is given by

$$X(e^{j\omega}) = \sum_{k=0}^{N-1} \frac{\tilde{X}(k)(1 - e^{-j\omega N})}{(1 - e^{-j[\omega - (2\pi k/N)]})N}. \tag{5.9-7}$$

By factoring a factor $e^{-j\omega N/2}$ out of the numerator and $e^{-j[\omega-(2\pi k/N)]/2}$ from the denominator, this can be put into the form

$$X(e^{j\omega}) = \sum_{k=0}^{N-1} \tilde{X}(k)\Phi\left(\omega - \frac{2\pi k}{N}\right). \tag{5.9-8}$$

where

$$\Phi(\omega) = \frac{\sin \omega N/2}{N \sin \omega/2} e^{-j\omega(N-1)/2} \tag{5.9-9}$$

is the interpolation function. Note that it is somewhat similar to the function sinc $[(t - n\tau_s)/\tau_s]$ appearing in Eq. (4.2-5). In particular, for $\omega = 2\pi l/N$ (i.e., on the DFT sample points) $\Phi[\omega - (2\pi k)/N]$ is zero for all $l \neq k$ and it is equal to 1 for $l = k$. Thus Eq. (5.9-7) is seen to give the exact value of $X(z)$ on the sample points. Equation (5.9-6) is the basis for one form of digital filter, as will be seen in Sec. 5.11.

5.10 DIGITAL FILTERS

Digital filters or digital signal processors are small special-purpose digital computers designed to implement an algorithm that converts an input sequence $x(n)$ into a desired output sequence $y(n)$. Such filters employ devices such as adders, multipliers, shifters, and delay elements rather than resistors, capacitors, or operational amplifiers. As a result they are generally unaffected by factors such as component accuracy, temperature stability, and long-term drift that afflict analog filter circuits. Also, many of the circuit restrictions imposed by physical limitations of analog devices can be removed or at least circumvented in a digital processor. On the other hand, digital filter designs have to take into account such things as finite word size, roundoff errors, aliasing, and other factors.

Of the large variety of possible digital processors the only ones considered in any detail here are the linear time-invariant (LTI) [also called linear shift-invariant (LSI)] systems. Linearity was discussed at the beginning of this chapter; time (or shift) invariance means that if the input sequence $x(n)$ produces the output sequence $y(n)$ then the input $x(n - n_0)$ produces the output $y(n - n_0)$ for all n_0. The input–output relation for LTI systems is the discrete convolution

$$y(n) = \sum_{m=-\infty}^{\infty} x(m)h(n - m), \tag{5.10-1}$$

where $h(n)$ is the discrete impulse response, that is, the sequence generated by the filter when the input is the discrete impulse $\delta(n)$, given by[†]

$$\delta(n) = \begin{cases} 1, & n = 0, \\ 0, & n \neq 0. \end{cases} \tag{5.10-2}$$

Recursive Structures

LTI systems are further subdivided into recursive and nonrecursive structures. A recursive system is one in which the current output value depends on preceding values of both the output and the input; that is,

$$y(n) = \mathcal{H}[y(n - 1), y(n - 2), \ldots, x(n), x(n - 1), \ldots], \tag{5.10-3}$$

[†] See Eq. (5.7-13).

where $\mathcal{H}$ is some linear operator. Recursive LTI systems are describable by linear constant-coefficient difference equations:

$$y(n) = \sum_{m=0}^{k} a_m x(n-m) - \sum_{m=1}^{k} b_m y(n-m). \tag{5.10-4}$$

They are therefore fairly direct analogs of ordinary continuous filters that are described by linear constant-coefficient differential equations, and their properties are similar. Their frequency response (actually the Z transform) has poles and zeros, and the response will be unstable if the poles lie outside the unit circle.

We can show this in a somewhat heuristic manner by writing the Z transform relationship between input and output:

$$Y(z) = H(z)X(z), \tag{5.10-5}$$

where $H(z)$ and $X(z)$ are transforms of the impulse response and input, respectively, and $Y(z)$ is the transform of the output. If the output sequence is computed from an inversion of a partial fraction expansion of $Y(z)$, such an expansion will contain terms generated by the poles of $H(z)$. These terms represent the transient response and will be of the form

$$a_1 z_1^k + a_2 z_2^k + \cdots + a_n z_n^k, \tag{5.10-6}$$

where the a_i are coefficients, the z_i are the poles of $H(z)$, and k is the discrete-time index. Clearly, if the output is to remain bounded for a bounded input, $|z_i|$ for $i = 1, \ldots, n$ should be less than unity. This in turn requires that *the poles be inside the unit circle*. Therefore, the poles of a stable system must satisfy

$$|z_i| < 1, \qquad \text{for } i = 1, \ldots, n. \tag{5.10-7}$$

Also, the amplitude and phase characteristics of $H(z)$ must satisfy some of the constraints in Chapter 3.

The location of the poles determines not only the stability but also the inherent character of the transient response of the system. For example, suppose that a typical term in the partial fraction expansion of $Y(z)$ is

$$\frac{a_i z}{z - z_i} = \frac{a_i}{1 - z_i z^{-1}}.$$

From Eq. (5.7-12) we know that this term gives rise to a sequence $x(n) = z_i^n$, $n = 0, 1, \ldots$. We already know that if $|z_i| < 1$ the transient response decays with increasing k and the system is stable. If z_i is real, the algebraic sign of z_i determines whether the sequence alternates or behaves monotonically. If the poles appear as complex conjugate pairs, that is, $z^{(1)} = \alpha + j\beta$ and $z^{(2)} = \alpha - j\beta$, stability requires that $(\alpha^2 + \beta^2)^{1/2} < 1$. Figure 5.10-1 shows some typical sequences and their corresponding pole locations.

Recursive filters are frequently referred to as IIR (infinite impulse response) systems to distinguish them from nonrecursive FIR (finite impulse response) systems whose impulse response is finite by design and equal to the number of delay elements employed in their construction. We discuss FIR filters later. Note that in practice the impulse

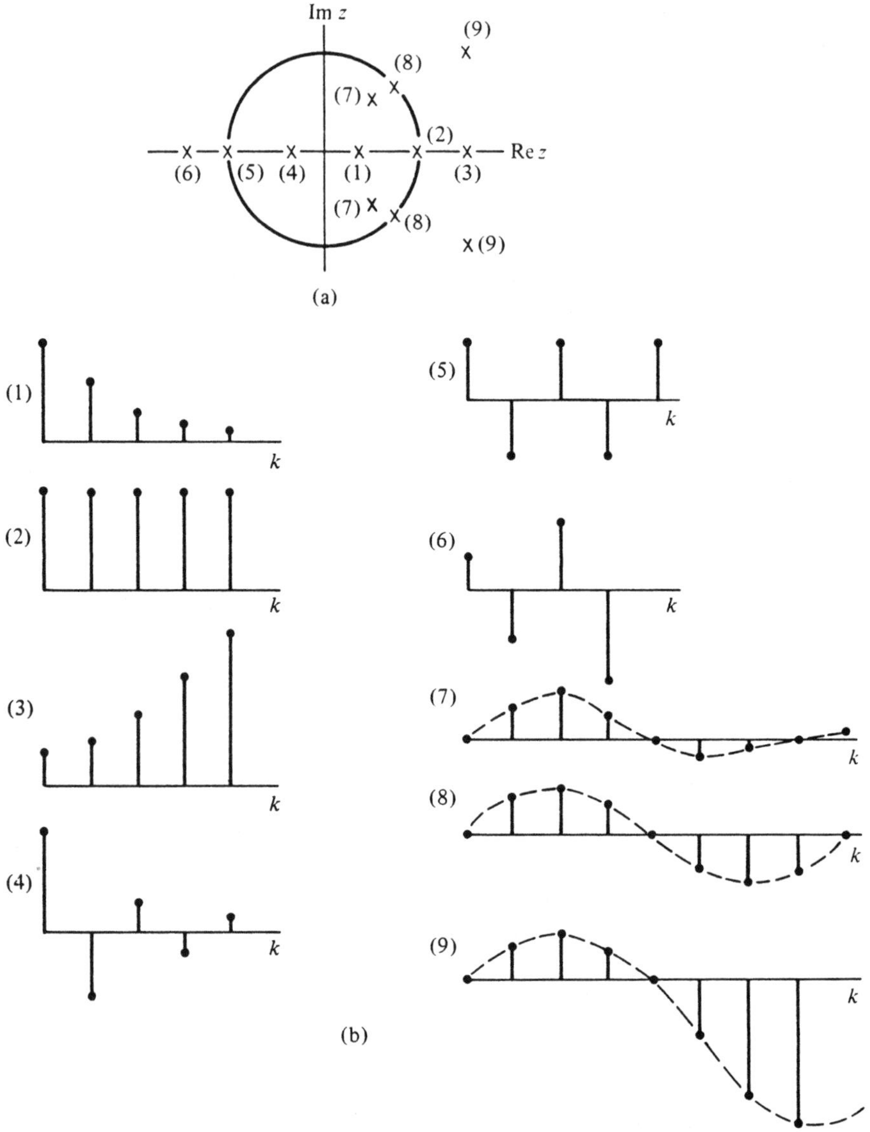

Figure 5.10-1. (a) Typical pole locations of $H(z)$. (b) The corresponding discrete-time sequences generated by such poles.

response time of stable IIR filters is actually also finite because of finite-word-size effects. That is, when the response has decayed to a value smaller than the smallest number that can be represented in the special-purpose digital computer that implements the filter, the response is effectively zero.

There are a number of standard realizations for IIR systems. A realization that

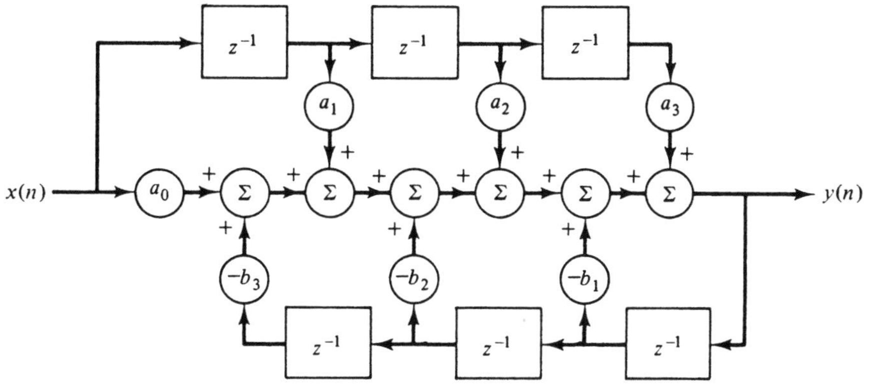

Figure 5.10-2. Example of a direct form 1 digital filter realization.

implements Eq. (5.10-4) directly is the *direct form 1* realization, in which separate delays are used to generate both the lagged inputs and outputs. An example of a third-order, direct form 1 realization is shown in Fig. 5.10-2. It is called a third-order system because it realizes a third-order difference equation in $y(n)$. In other words, the present value of $y(n)$ depends on $y(n-1)$, $y(n-2)$, and $y(n-3)$.

From the direct form 1 realization it is easy to derive $H(z)$. First, Eq. (5.10-4) is rewritten as

$$\sum_{m=0}^{k} a_m x(n-m) = \sum_{m=0}^{k} b_m y(n-m), \tag{5.10-8}$$

where $b_0 = 1$. The Z transform then yields

$$X(z) \sum_{m=0}^{k} a_m z^{-m} = Y(z) \sum_{m=0}^{k} b_m z^{-m},$$

or, equivalently,

$$H(z) = \frac{Y(z)}{X(z)} = \frac{\sum_{m=0}^{k} a_m z^{-m}}{\sum_{m=0}^{k} b_m z^{-m}}, \tag{5.10-9}$$

where the a_m and b_m are the gain parameters in the direct form 1 realization.

In Fig. 5.10-2, the blocks labeled z^{-1} are unit delays; in practice, these would be implemented by *shift registers*. The "gains" $a_1, a_2, \ldots, b_1, b_2, \ldots$ represent multiplication by constant numbers, and the blocks with the Σ label are adders. Thus the circuit employs shift-register delays, multipliers, and adders.

A somewhat different form requiring fewer delays can be developed from Eq. (5.10-9) by writing

$$
\begin{aligned}
H(z) &= \frac{\sum_{m=0}^{k} a_m z^{-m}}{\sum_{m=0}^{k} b_m z^{-m}} \\
&= \left(\frac{1}{\sum_{m=0}^{k} b_m z^{-m}}\right) \cdot \sum_{m=0}^{k} a_m z^{-m} \\
&= H_1(z) \cdot H_2(z).
\end{aligned}
\tag{5.10-10}
$$

$H_1(z)$ can be represented by a block diagram similar to the lower part of Fig. 5.10-2. $H_2(z)$ is simply a weighted sum of delayed versions of the input. Hence, for a third-order system as in Fig. 5.10-2, we obtain the block diagram of Fig. 5.10-3. Inspection of the circuit shows that the two sets of delay elements do the same thing and can therefore be replaced by a single set. The resulting canonical form, referred to as *direct form 2*, is shown in Fig. 5.10-4. Note that half as many delay elements are needed as in the filter shown in Fig. 5.10-2. In fact, since the filter shown is third order and contains just three delay elements, it has a minimum number of delay elements. However, this is not the only realization having this feature. For instance, it is always possible to write $H(z)$ as a product of factors:

$$
H(z) = H_1(z) H_2(z) \cdots H_k(z),
\tag{5.10-11}
$$

where the $H_i(z)$ are first- or second-order filters. The factored form suggests a cascade connection as shown in Fig. 5.10-5. Each element of this cascade may evidently be realized as a first- or second-order filter having the direct form 2 shown in Fig. 5.10-4, and then the cascade will have a minimum number of delay elements as well. Other

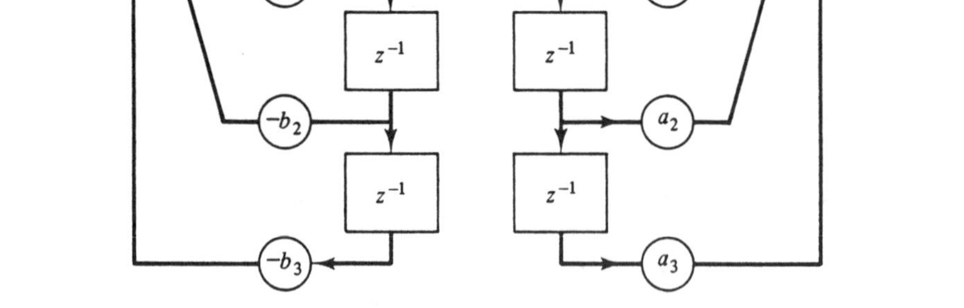

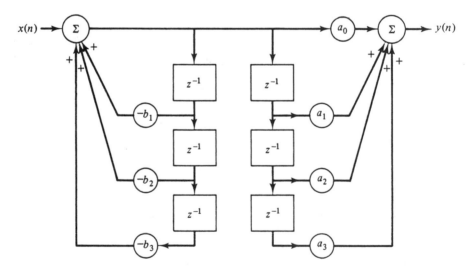

Figure 5.10-3. Realization of Eq. (5.10-10) for a third-order system.

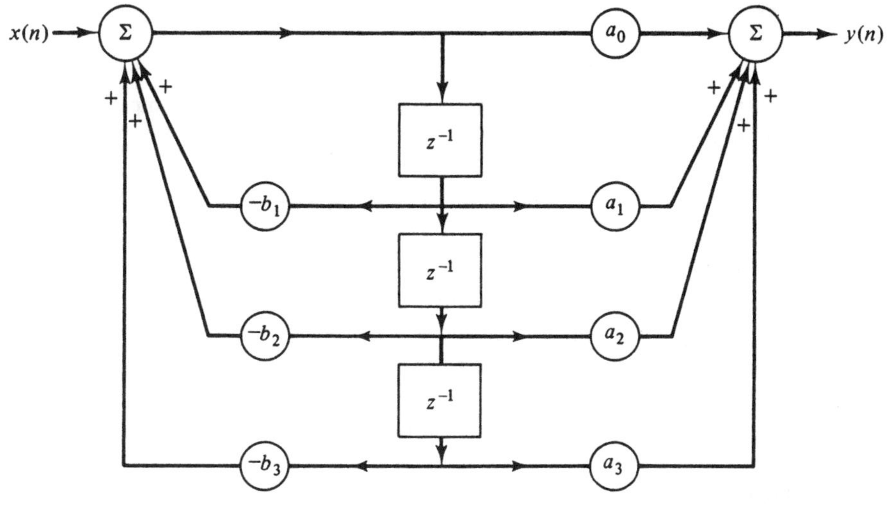

Figure 5.10-4. Direct form 2 realization.

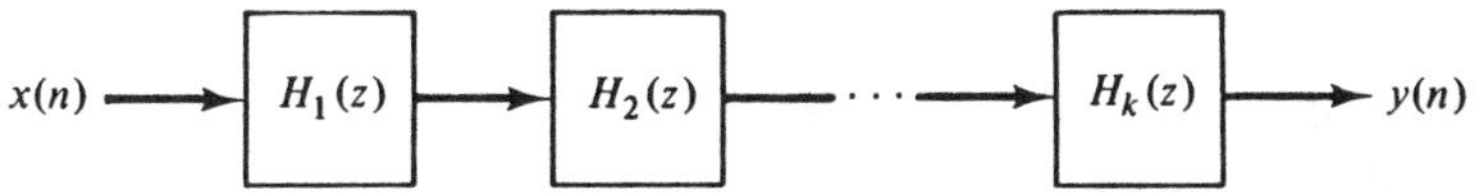

Figure 5.10-5. Cascade connection.

possibilities are parallel combinations [obtained by expressing $H(z)$ as a sum of terms] or any combination of series and parallel connections.

Although all these different forms are described by the same equations, they do not perform the same way in practice because of the effects of finite word size.† For instance, there is a severe coefficient sensitivity problem with the two direct forms shown in Figs. 5.10-2 and 5.10-4 when the poles of $H(z)$ are close together or near the unit circle. Conversely, it turns out that cascade connections in which poles and zeros are distributed among the various sections so that poles and zeros that are close to each other in the z plane are placed in the same section of the cascade tend to be less sensitive to word-size noise effects [5-13]. The general problem of how best to arrange a digital filter so as to minimize finite-word-size effects is clearly very important, but it is beyond the scope of this necessarily very brief treatment. Readers interested in more details are referred to one of a number of excellent references ([5-2], [5-6], [5-7]).

† This is a much less severe problem with analog filters where different configurations of the block diagram generally result in essentially the same output.

5.11 NONRECURSIVE (FIR) FILTERS

A widely used class of filters does not use output feedback. These filters are therefore nonrecursive and have a finite impulse response (FIR). One simple form of FIR structure is obtained by direct implementation of the discrete convolution:

$$y(n) = \sum_{m=-\infty}^{\infty} h(m)x(n-m). \tag{5.11-1}$$

Suppose that $h(m) = 0$ for $m < 0$, and also that, for $m \geq N$, $h(m)$ has decayed to a value small enough to be neglected. Because of finite word size there is always such an N if $h(m)$ is a stable impulse response, but in practice a smaller number may be used to reduce the length of the filter. Then Eq. (5.11-1) becomes

$$y(n) = \sum_{m=0}^{N-1} h(m)x(n-m) = \sum_{m=0}^{N-1} a_m x(n-m). \tag{5.11-2}$$

This is a difference equation just like Eq. (5.10-4), but it is nonrecursive since $y(n-m)$ does not appear on the right. The Z transform of Eq. (5.11-2) has all its poles at the origin; hence it cannot represent an unstable system. The length of the impulse response is exactly N samples.

The structure suggested by Eq. (5.11-2) is shown in Fig. 5.11-1 and is often referred to as a *transversal filter*. Another designation is *tapped delay-line filter*; this name comes from the fact that the string of delays shown in Fig. 5.11-1 acts like a discrete tapped delay line. In fact, analog versions of this filter that use a tapped delay line and continuously adjustable gains have been used [5-14]. In principle, the transversal filter can realize any arbitrary causal impulse response. Questions of stability or realizability do not arise. There are no difficulties in realizing impulse responses that have arbitrary jumps or that have radically different forms for different values of time n. Also, by defining the input sequence as the signal somewhere near the middle of the delay line (i.e., by considering the input only after a fixed delay), it is possible to get an impulse response with finite values for negative argument. The resulting "noncausality" is essentially mathematical; the filter is still causal in real-time operation.

Transversal filters are easily made time variable by varying the a's. There are simple algorithms for automatically adjusting the a's so that some overall performance

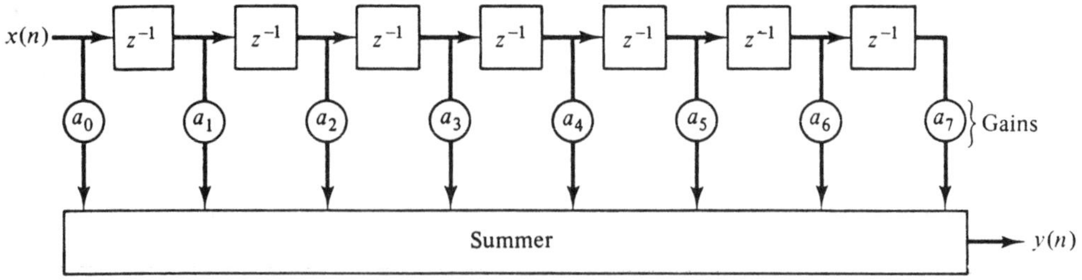

Figure 5.11-1. Transversal filter.

criterion is minimized. Such systems find considerable use in adaptive equalizers ([5-14], [5-15]).

The major disadvantage of the transversal filter is the large number of delay elements and multipliers that are required. For instance, an IIR filter can be designed to have a fourth-order Butterworth response with no more than eight delay elements and about the same number of multipliers and adders. The number of delay elements and multipliers needed for an FIR design tends to be much greater. This is particularly true for filters having a sharp cutoff characteristic, since they have a long impulse response. The advent of large-scale integration has, however, greatly lessened the disadvantages of complexity, and it is possible to construct special-purpose digital circuits for FIR filters that occupy only a single integrated circuit chip and are therefore physically no larger than a corresponding IIR filter. A simple digital circuit involving only two shift-register memories and a very minimum of additional circuitry has, in fact, been designed to implement the FIR algorithm ([5-6], p. 543). The arrangement for $N = 8$ is shown in Fig. 5.11-2. At the moment shown, the accumulator has been cleared and $a_7x(n - 7)$ is entered. During the first shift cycle a new value of $x(n)$ enters at the left, and $x(n - 7)$ is shifted off at the right. During the remaining seven shift cycles, both shift registers circulate, and it is left as an exercise to show that after a total of eight shifts $y(n) = \sum_{m=0}^{7} a_m x(n - m)$ is in the accumulator register and can be shifted out. After $y(n)$ is shifted out, a new sample of $x(n)$ enters, and a new sample of $y(n)$ leaves the system only once per complete shift cycle. Thus for an N-stage filter the shift-register speed must be N times the sampling frequency of the signal. With current technology it is possible to obtain shift-register memories with more than 1000 stages, switchable at rates in excess of 20 MHz in a single IC chip. Thus a 1000-point FIR filter capable of sampling rates of 20 kHz is quite feasible. Evidently, higher speeds can be attained at the expense of greater complexity by paralleling additional units.

The circuit shown in Fig. 5.11-2 using shift-register memories can, of course, also be implemented with random-access memories (i.e., with a small general-purpose computer, possibly involving a microprocessor). Similar simple circuits can be designed for IIR filters. (See [5-6], Chap. 11.)

FIR filters can be designed to have precise linear phase characteristics. As pointed out in Chapter 3, this means that, except for a delay corresponding to the slope of the phase versus frequency curve of the transfer function, the input is reproduced exactly at the output. All that is needed for a filter to have a linear phase response is that for $n = 0, 1, \ldots, (N/2 - 1)$

$$h(n) = h(N - 1 - n), \tag{5.11-3}$$

where, for the sake of simplicity, we assume N to be even. (A slightly different result holds for odd N.) To show that Eq. (5.11-3) implies linear phase, write the Z transform:

$$\begin{aligned} H(z) &= \sum_{n=0}^{(N/2)-1} h(n)z^{-n} + \sum_{N/2}^{N-1} h(n)z^{-n} \\ &= \sum_{n=0}^{(N/2)-1} [h(n)z^{-n} + h(n)z^{-(N-1-n)}], \end{aligned} \tag{5.11-4}$$

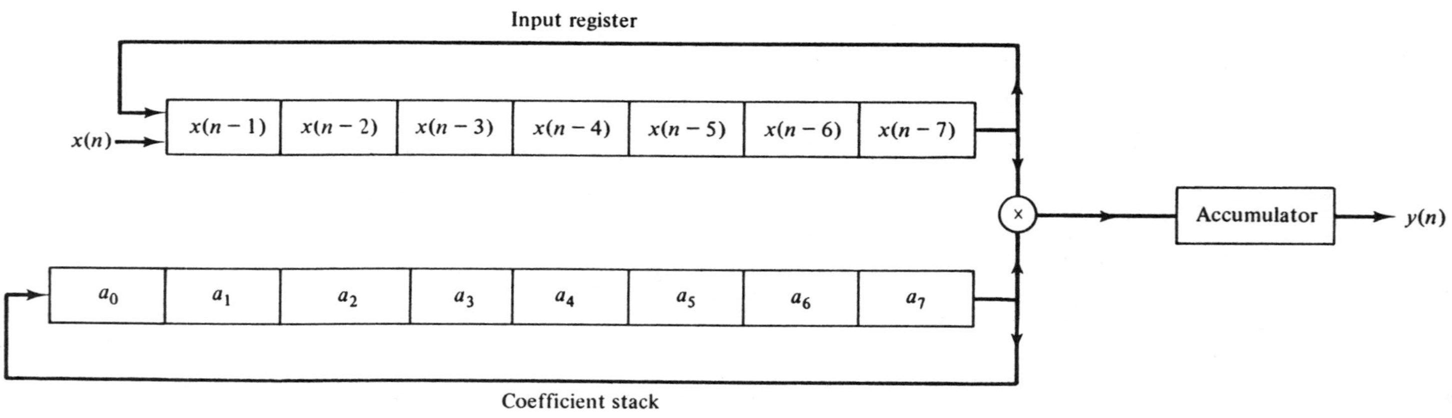

Figure 5.11-2. Digital realization of an FIR filter.

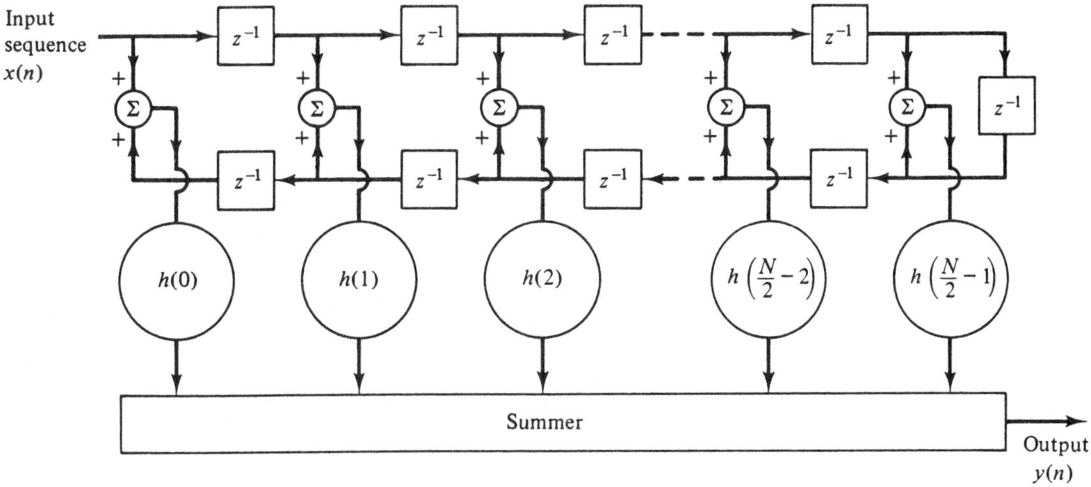

Figure 5.11-3. FIR filter having a linear phase characteristic.

where the second step follows if one makes the change of variable $n' = N - 1 - n$ in the second summation and then uses Eq. (5.11-3). Evaluating $H(z)$ for $z = e^{j\omega}$ gives

$$\begin{aligned} H(e^{j\omega}) &= e^{-j\omega(N-1)/2} \sum_{n=0}^{(N/2)-1} h(n)\{e^{j\omega[n-(N-1)/2]} + e^{-j\omega[n-(N-1)/2]}\} \\ &= 2e^{-j\omega(N-1)/2} \sum_{n=0}^{(N/2)-1} h(n) \cos \omega \left(n - \frac{N-1}{2}\right). \end{aligned} \tag{5.11-5}$$

Since the summation is real, phase shift is contributed only by the factor $\exp[-j\omega(N - 1)/2]$, and it is seen to be linear with slope $-(N - 1)/2$. An FIR filter structure suggested by Eq. (5.11-4) is given in Fig. 5.11-3 (for even N).

The single delay line filter shown in Fig. 5.11-1 is the FIR counterpart of one of the two direct forms for IIR filters shown in Fig. 5.10-1; and when the Z transform of the filter response is written as a product or sum, the resultant expressions suggest filter realizations by cascade or parallel structures here as well. It has been shown [5-16] that, as with IIR filters, the cascaded form of the FIR filter tends to be less sensitive to coefficient errors and finite-word-size effects than the direct form.

Frequency Sampling Structure

This is another form of nonrecursive filter structure. It is suggested by the relation between the Z transform and the discrete Fourier transform, developed in Sec. 5.9 [Eq. 5.9-6)]. If instead of $X(z)$ we write $H(z)$ in this equation, we obtain

$$H(z) = \frac{1 - z^{-N}}{N} \sum_{k=0}^{N-1} \frac{\tilde{H}(k)}{1 - z^{-1}e^{j2\pi k/N}}, \tag{5.11-6}$$

where the $\tilde{H}(k)$ are the DFT coefficients corresponding to the impulse response sequence $h(n)$ of the desired filter. This expression suggests the filter structure shown in Fig. 5.11-4.

Since the zeros of the function $1 - z^{-N}$ are the N roots of 1 (i.e., $e^{j2\pi n/N}$ for $n = 0, 1, \ldots, N - 1$), each of the poles in the parallel branches is in theory exactly canceled by a zero contributed by the series part of the circuit, and therefore the transfer function has no nonzero poles. (It does have N poles at the origin as the transfer functions of all FIR filters do, however.) In practice, because of finite-word-length effects, this cancellation is not perfect. Also, it is undesirable to have poles right on the unit circle since this results in stability problems. Therefore, the multipliers $e^{j2\pi k/N}$ shown in Fig. 5.11-4 are in practice replaced by $re^{j2\pi k/N}$, where r is a real

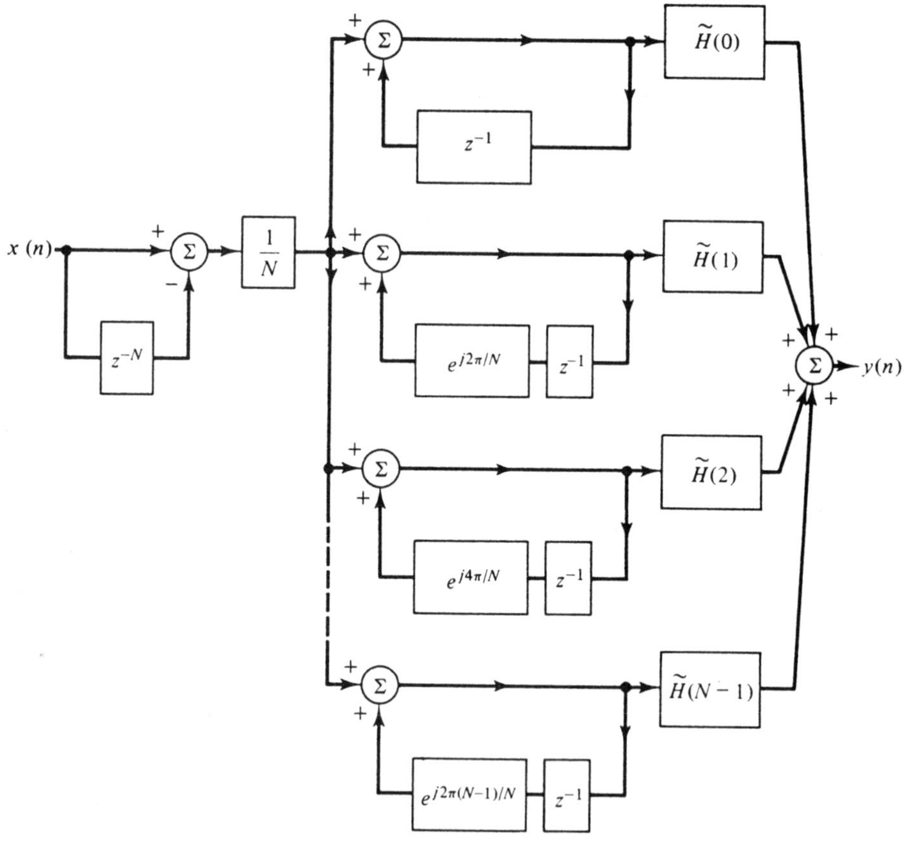

Figure 5.11-4. Frequency sampling filter.

number slightly smaller than unity. For both of these reasons the filter is not exactly an FIR filter. It is referred to as a frequency sampling filter because the basic coefficients $\tilde{H}(k)$ are the values of the filter frequency response $H(e^{j\omega})$ sampled at N equally spaced points around the unit circle.

The chief advantage of this filter is that it can result in a much simpler structure than the direct, or transversal, form, especially when the desired filter is a low-pass, high-pass, or band-pass filter with one or more stopbands. Since the frequency samples in the various stopbands are zero, the corresponding parallel branches can be omitted. Thus a narrowband filter may require only a small number of branches, in contrast to the transversal form where a narrowband filter with a sharp cutoff characteristic tends to be especially long. Another possible advantage is that, if several filters with the same number N of frequency samples are used in a filter bank, the $-z^{-N}$ feed-forward and the feedbacks in the individual branches need to be implemented only once, with different sections of the bank differing only in the multipliers $\tilde{H}(k)$.

Since the $\tilde{H}(k)$ and also the factors $re^{j2\pi k/N}$ are complex, the hardware implementation of the circuit shown in Fig. 5.11-4 requires complex multipliers and adders. By making use of some of the symmetry properties of the Z transforms of real impulse responses, the structure can be modified so that only real operations are needed. For details of this procedure, see [5-7, Sec. 4.5].

5.12 FAST CONVOLUTION

The convolution equation (5.10-1) is referred to as "slow" convolution to distinguish it from the fast convolution symbolized by the relation

$$y(n) = \mathcal{F}^{-1}\{\mathcal{F}[h(n)] \cdot \mathcal{F}[x(n)]\}. \tag{5.12-1}$$

The symbols $\mathcal{F}$ and $\mathcal{F}^{-1}$ refer here[†] to discrete Fourier transformation (DFT) and inverse DFT, respectively, and the reason Eq. (5.12-1) is called "fast" is that the FFT is used to perform the actual transformation. The block diagram of a filter employing fast convolution is shown in Fig. 5.12-1. The input sequence goes into a buffer memory holding N words. When this is full, it dumps the N values of $x(n)$ into the FFT processor and then commences to store the next N input samples. The FFT output can be manipulated in any arbitrary way by the digital processor. There are no realizability, phase, or stability constraints. The result is then inverse-transformed in the second FFT box and loaded into the output buffer from which the output samples can be taken at some constant rate. Note that even if the two FFT processors and the multiplication by $H(k)$ were instantaneous, there would be a minimum delay between input and output of N samples because of the requirement of filling the input buffer before applying the FFT. Thus the elimination of the realizability constraints is bought at the price of a delay,

[†] In other places in this book $\mathcal{F}$ refers to the *continuous* Fourier transform operator. The usage is made self-evident by the context of the discussion.

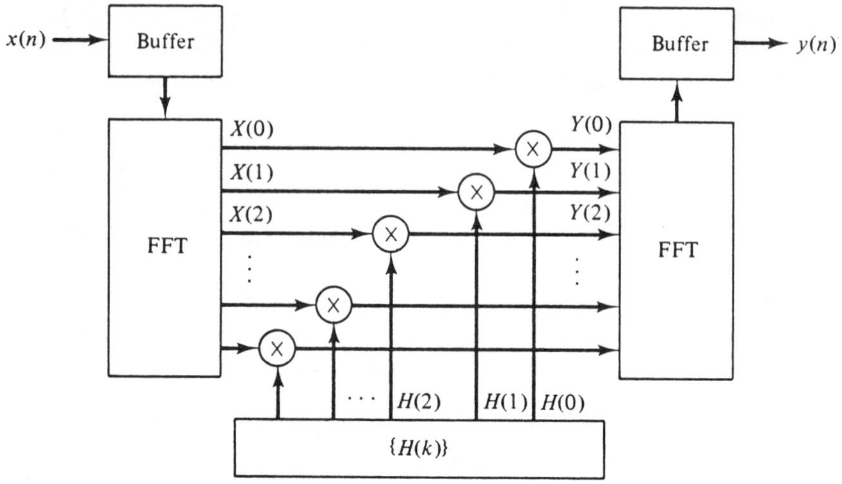

Figure 5.12-1. Digital filter employing fast convolution.

just as with the FIR filter. In other words, "there is no free lunch." However, because of the general availability of FFT processors either in hardware or in software, the fast convolution approach is widely used in practice.

Certain precautions must be observed in the application of the fast convolution technique. These have to do with sampling errors, aliasing errors, the fact that the DFT implicitly deals with harmonic sequences, and so on. All these considerations have already been discussed in Sec. 5.5, where the DFT was discussed in detail.

5.13 SUMMARY

In this chapter we have studied the properties of discrete linear systems (DLS), and we have shown that a DLS can be characterized by a transmission matrix **H** instead of the impulse or step-function responses that are commonly used for continuous systems. For systems that are time-invariant, we saw that a constant-coefficient *difference* equation can be used to relate output to input.

We investigated the properties of the DFT and found that the convolution-multiplication property, so useful in the continuous case, carried over to the discrete case as well. We saw that the DFT can be computed rapidly and efficiently with the FFT algorithm.

We found that the Z transform, which is essentially analogous to the Laplace transform in the continuous case, was a useful tool for the analysis of discrete filters. With respect to the latter, we studied both recursive and nonrecursive structures, and we showed that the former will be unstable if its Z-transform poles lie outside the unit circle. Nonrecursive, finite impulse response filters are fundamentally stable since the impulse response is N unit delays long.

We ended the chapter by considering the technique known as fast convolution, which uses the FFT algorithm for rapid signal processing.

PROBLEMS

5-1. Write, or at least indicate, the form of the most succinct transmission matrix **H** for the following cases:

(a) **H** noncausal, **x** extends from $k = -\infty$ to $k = +\infty$.

(b) **H** noncausal, **x** is nonzero for $1 \leq k \leq K$.

(c) **H** causal, **x** extends from $k = -\infty$ to $k = +\infty$.

(d) **H** causal, **x** nonzero for $1 \leq k \leq K$.

5-2. Consider a discrete system with a square transmission matrix **H.** How many independent elements are there in **H** when the system is

(a) Noncausal, non-time-invariant?

(b) Causal, non-time-invariant?

(c) Causal, time-invariant.

5-3. Use discrete convolution to compute the sequence $\{y_n\}$ when x_k and h_k are as follows:

k	0	1	2	3	4	5	6	7	8→
x_k	0	1	1	1	1	1	0	all zeros	
h_k	1	1	1	1	0	0	0	all zeros	

5-4. **(a)** Write a difference equation that describes the system in Fig. P5-4.

(b) Compute $y(n)$, by recursion, when $x(n) = n^2$.

(c) Compute $y(n)$ by solving the difference equation in part (a).

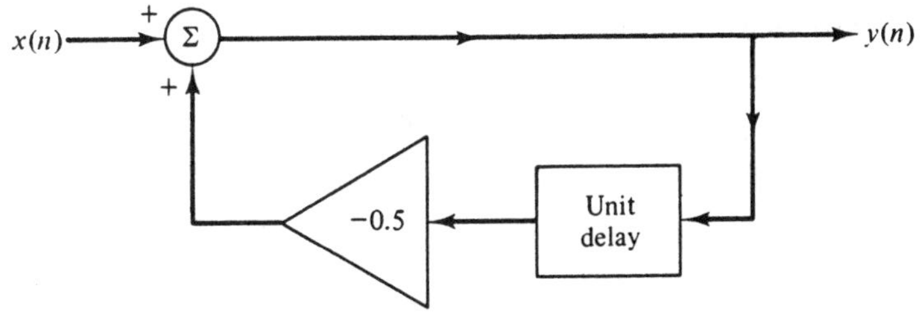

Figure P5-4

5-5. **(a)** Write a difference equation that describes the second-order system shown in Fig. P5-5.

(b) Let $y(-2) = y(-1) = 0$. Solve for $y(n)$ by recursion when

$$x(n) = \begin{cases} 1, & 0 \leq n \leq 3, \\ 0, & \text{otherwise,} \end{cases}$$

and $b_0 = 1$, $a_1 = \frac{1}{2}$, and $a_2 = 2$. Plot the output.

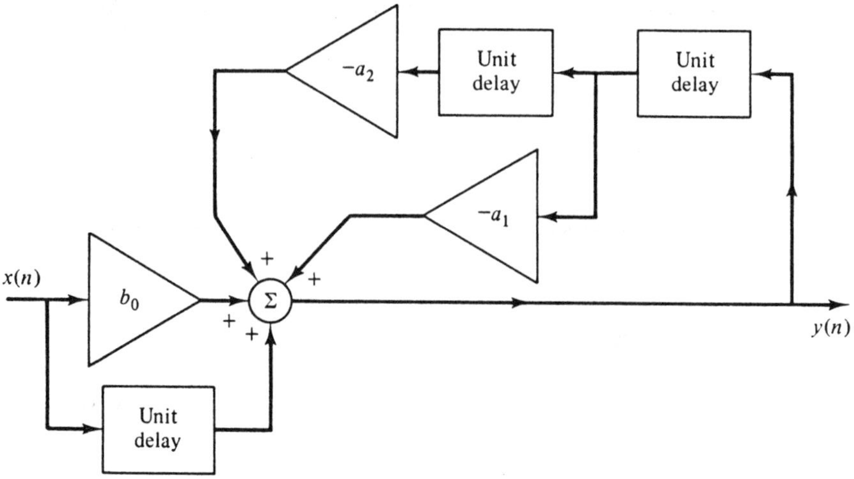

Figure P5-5

5-6. Demonstrate that Eqs. (5.3-11) and (5.3-12) form a valid discrete Fourier transform pair.

5-7. Demonstrate the following for arbitrary k:

(a) $X(-k) = X^*(k)$.

(b) $X\left(\frac{N}{2} - k\right) = X^*\left(\frac{N}{2} + k\right)$,

Here $X(\cdot)$ is the DFT and the sequence $\{x(i)\}$ is real. Assume N is even.

5-8. The discrete system counterpart of the derivative is the *difference*. For the sequence $\{x(i)\}$ the first forward difference [written $\Delta x(i)$] is the sequence $\{x(i + 1) - x(i)\}$. The backward difference [written $\nabla x(i)$] is $\{x(i) - x(i - 1)\}$.

(a) Find the DFT for the forward difference $\Delta x(i)$.

(b) Find the DFT for the backward difference $\nabla x(i)$.

5-9. Find the DFT for the product $x(i)y(i)$.

5-10. A waveform $x(t) = \cos(2\pi/T_0)t$ is sampled by the comb function

$$g(t) = \sum_{n=-\infty}^{\infty} \delta(t - nT).$$

(a) Sketch $x(t)g(t)$, assuming that $T \simeq T_0/10$.

(b) Assume a rectangular truncation function

$$p(t) = \text{rect}\left[\frac{t - (T_0 - T)/2}{T_0}\right],$$

and sketch $|P(f)|$.

(c) Sketch $h(t) \equiv x(t)g(t)p(t)$ and $H(f) = \mathcal{F}[h(t)]$.

(d) Sketch $H(f) \sum_{n=-\infty}^{\infty} \delta[f - (n/T_0)]$, and compare with $X(f)$. What does the inverse transform of the sampled spectrum look like?

5-11. Repeat Prob. 5-10, except choose a truncation interval that covers slightly more than one period. For example, choose

$$p(t) = \text{rect}\left[\frac{t - (T_0 - T/2)/2}{T_0 + T/2}\right].$$

5-12. Write expressions for each of the waveforms shown in Fig. 5.5-1, and carefully label all significant parameters such as zero crossing and sampling widths. Justify that neither the DFT nor the retransformed sampled wave need resemble the original spectrum and signal.

5-13. Show that convolving a DFT signal spectrum with the discrete generalized Hamming window is equivalent to replacing every $X(k)$ in the original sequence with the weighted average

$$\alpha X(k) + \frac{1 - \alpha}{2}[X(k + 1) + X(k - 1)].$$

5-14. Establish the shifting property of the Z transform given in Eq. (5.7-8).

5-15. The one-sided Z transform for a function $x(n)$ is given by

$$F(z) = \frac{1 + 2z}{z^2 + z}.$$

Determine the region of convergence, and compute $x(n)$ for all n.

5-16. Compute the discrete-time signal associated with

$$X(z) = \frac{1}{(z - 1)(z - 2)}$$

if $X(z)$ represents the one-sided Z transform. Describe the regions of convergence.

5-17. Starting with Eq. (5.9-7), prove that $X(e^{j\omega})$ can be written as in Eqs. (5.9-8) and (5.9-9). *Hint*: Use the fact that $(-1)^{2k} = 1$ for $k = 0, 1, 2, \ldots$.

5-18. Prove that

$$H(z) = \frac{\text{response to } z^n}{z^n},$$

where $H(z) \equiv \mathscr{Z}[h_k]$. State your assumptions.

5-19. Prove that the realizations in Fig. 5.10-3 or 5.10-4 are equivalent to the one shown in Fig. 5.10-2. *Hint*: In Fig. 5.10-3, define an auxiliary variable $w(n)$ as the signal in the direct (undelayed) link.

5-20. Prove that the Z transform of an FIR filter with impulse response $h(m)$ that satisfies

$$h(m) = \begin{cases} =0, & m < 0, m \geq N, \\ \neq 0, & \text{otherwise}, \end{cases}$$

has its poles at the origin. Why is stability therefore not a problem?

REFERENCES

[5-1] G. D. Bergland, "A Guided Tour of the Fast Fourier Transform," *IEEE Spectrum*, pp. 41–52, July 1969.

[5-2] Samuel D. Stearns, *Digital Signal Analysis*, Hayden, New York, 1975.

[5-3] R. B. Blackman and J. W. Tukey, *The Measurement of Power Spectra*, Dover, New York, 1958.

[5-4] J. W. Cooley and J. W. Tukey, "An Algorithm for the Machine Calculation of Complex Fourier Series," *Math. Comput.*, **19,** p. 297, April 1965.

[5-5] G.A.E. Subcommittee on Measurement Concepts, "What Is the Fast Fourier Transform?," *IEEE Trans. Audio Electroacoustics*, **AU-15,** pp. 45–55, June 1967.

[5-6] Lawrence R. Rabiner and Bernard Gold, *Theory and Application of Digital Signal Processing*, Prentice-Hall, Englewood Cliffs, N.J., 1975.

[5-7] Alan V. Oppenheim and Ronald W. Schafer, *Digital Signal Processing*, Prentice-Hall, Englewood Cliffs, N.J., 1975.

[5-8] R. V. Churchill, *Introduction to Complex Variables and Applications*, McGraw-Hill, New York, 1948, p. 102.

[5-9] Ralph J. Schwarz and Bernard Friedland, *Linear Systems*, McGraw-Hill, New York, 1965, Chap. 8.

[5-10] J. R. Ragazzini and G. F. Franklin, *Sampled Data Control Systems*, McGraw-Hill, New York, 1958, Chap. 4.

[5-11] H. Freeman, *Discrete Time Systems*, Wiley, New York, 1965.

[5-12] E. I. Jury, *Theory and Application of the Z-Transform Method*, Wiley, New York, 1964.

[5-13] L. B. Jackson, "Roundoff Noise Analysis for Fixed-Point Digital Filters Realized in Cascade or Parallel Form," *IEEE Trans. Audio Electroacoustics*, **AU-18,** pp. 107–122, June 1970.

[5-14] R. W. Lucky, "Automatic Equalization for Digital Communication," *Bell System Tech. J.*, **44,** pp. 547–588, April 1965.

[5-15] J. H. Chang and F. B. Tuteur, "A New Class of Adaptive Array Processors," *J. Acoust. Soc. Am.*, **49,** No. 3 (Part 1), pp. 639–649, March 1971.

[5-16] O. Hermann and H. W. Schuessler, "On the Accuracy Problem in the Design of Nonrecursive Digital Filters," *Arch. Elek. Ubertragung*, **24,** pp. 525–526, 1970.

CHAPTER 6

Amplitude Modulation Systems and Television

6.1 INTRODUCTION

It is no exaggeration to state that in the communication sciences modulation holds a central place. The terms amplitude modulation and frequency modulation are used by every layperson at one time or another, even if their only associations with these words are popular music for the former and classical music for the latter. Nevertheless, the word modulation has a distinct technical meaning and was defined by an appropriate committee of the IRE [6-1] as "the process . . . whereby some characteristic of a wave is varied in accordance with another wave." We shall be concerned with *controlled* modulation, that is, the *desired* and controlled shifting of the spectrum of a message wave that is usually baseband and contains information of interest to humans. The term baseband refers generally to a low-pass wave such as simple speech or, as in more complicated systems, a multiplex wave consisting of many low-pass waves.

There are two fundamental reasons we modulate. One has to do with the laws of electromagnetic propagation, which require that the size of the radiating element be a significant fraction of the wavelength of the signal to be transmitted. Thus the transmission of a 1000-Hz signal by a quarter-wave antenna would require a radiating element 75 kilometers long if we didn't modulate the 1000-Hz signal onto a high-frequency carrier. There exists a mismatch or gap between the frequencies that the human ear can detect and the frequencies at which electromagnetic energy can be efficiently radiated. Modulation bridges this gap.

The other main reason for modulating is the need for simultaneous transmission of different signals. The signals of interest to a human are primarily in a frequency band that spans from tens of hertz to several thousand hertz. If we didn't modulate,

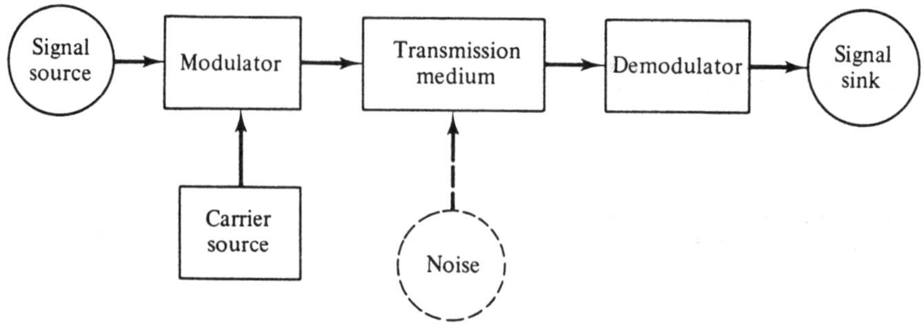

Figure 6.1-1. Ideal system that ignores all undesirable effects except channel noise.

we could only broadcast one baseband signal in any locality at a given time; simultaneous transmission of more than one signal would cause the signals to overlap without hope of separation. Through use of modulation, however, we can transmit many messages over the same medium and still ensure their separability at the receiver. It is the multichannel capability furnished by modulation that enables us to have many radio and television channels in the same locality at the same time. The same principles are in use when transmitting many telephone messages over the same cable. The separation of the messages by different carrier assignment is sometimes called frequency-division multiplexing. In the case of telephoning, signals are also separated in time by a process called time-division multiplexing (TDM), which is not a carrier-modulation technique. TDM was discussed in Sec. 4.4.

There are other, secondary reasons we modulate that have to do with obtaining noise reduction, limiting the size and weight of circuits and components, and other factors. Spurious modulation such as occurs to transmitted signals during electrical storms and modulation due to faulty practices and components are of course also commonplace, but we shall not discuss them further.

Since modulation involves the generation of new frequency components (i.e., shifting of spectral bands), we cannot modulate by using linear, time-invariant systems. Modulation is generally done by using time-varying linear systems or systems using one or more nonlinear elements.

In this chapter we shall discuss amplitude modulation (AM) and related modulation methods. AM is very easy to understand and is essentially a direct translation of the message spectrum to the carrier frequency. Closely related schemes are double sideband (DSB), single sideband (SSB), and vestigial sideband (VSB). All these modulation methods are closely related; either both sidebands, or one sideband, or fractions of both, with or without a carrier signal, are transmitted between source and receiver. At the receiver, we must demodulate the signal and recover the baseband message $s(t)$, also sometimes called the information.† A block diagram of an ideal communication system is shown in Fig. 6.1-1.

Perhaps the most widely used communication system of all and one that uses all

† Also sometimes loosely called *intelligence*.

these modulation methods is color television. Black/white and color TV are therefore discussed at length at the end of this chapter.

6.2 AMPLITUDE MODULATION

An ordinary AM wave can be constructed by adding to the baseband signal $s(t)$ a constant and multiplying the sum by a sine wave. Our set of messages $\{s(t)\}$ will be assumed normalized and band-limited so that the set is described by $\{s(t):|_{\max} \leq 1; S(f) = 0, |f| > W\}$. W is the bandwidth of the signal. The normalization $|s(t)|_{\max} \leq 1$ enables us to write for the AM signal

$$x(t) = A_c[1 + ms(t)] \cos (2\pi f_c t + \phi), \qquad (6.2\text{-}1)$$

where $x(t)$ is the AM signal, m is the modulation index, f_c is the carrier frequency, A_c is the carrier amplitude, and ϕ is a constant-phase term that can be set to zero by appropriate choice of the time origin. The number m should not exceed the value of unity; for $m \geq 1$, $ms(t) + 1$ can go negative if $s(t)$ takes its most negative value. This means that $x(t)$ has undergone a 180° phase reversal and will lead to distortion in the recovery of $s(t)$ *if the simplest AM detection, that is, envelope detection, is used.* When $m = 1$, we have 100% modulation; for $m > 1$, we have *overmodulation*. Figure 6.2-1 shows AM waveforms for various values of m and a sinusoidal modulating signal.

The Fourier transform of Eq. (6.2-1) gives the spectrum of $x(t)$. It is (with ϕ set to zero for convenience)

$$X(f) = \frac{A_c}{2}\delta(f - f_c) + \frac{A_c}{2}\delta(f + f_c) + \frac{mA_c}{2}S(f - f_c) + \frac{mA_c}{2}S(f + f_c). \qquad (6.2\text{-}2)$$

Figure 6.2-2(a) shows a possible spectrum for $s(t)$; Fig. 6.2-2(b) shows the spectrum of $x(t)$. Several interesting results are now apparent. The bandwidth of the AM wave is

$$B = 2W, \qquad (6.2\text{-}3)$$

that is, twice as great as the bandwidth of the original signal. Also, the AM wave consists of a *pair* of sidebands each of width W. Foldover distortion occurs when the carrier frequency is not high enough (Fig. 6.2-3). The foldover phenomenon also occurs in sampling theory, that is, when the sampling rate is too low (Sec. 4.2). For example, in Fig. 4.2-5 it can be seen that the high-frequency components of a displaced spectrum fold into the lower frequencies of its neighbor. For no foldover distortion in AM, we must have

$$f_c > W.$$

We can now describe some of the other modulation schemes that are closely related to AM. If the carrier component in Eq. (6.2-2) is absent, we obtain a wave of the form

$$x(t) = Bs(t) \cos 2\pi f_c t, \qquad (6.2\text{-}4)$$

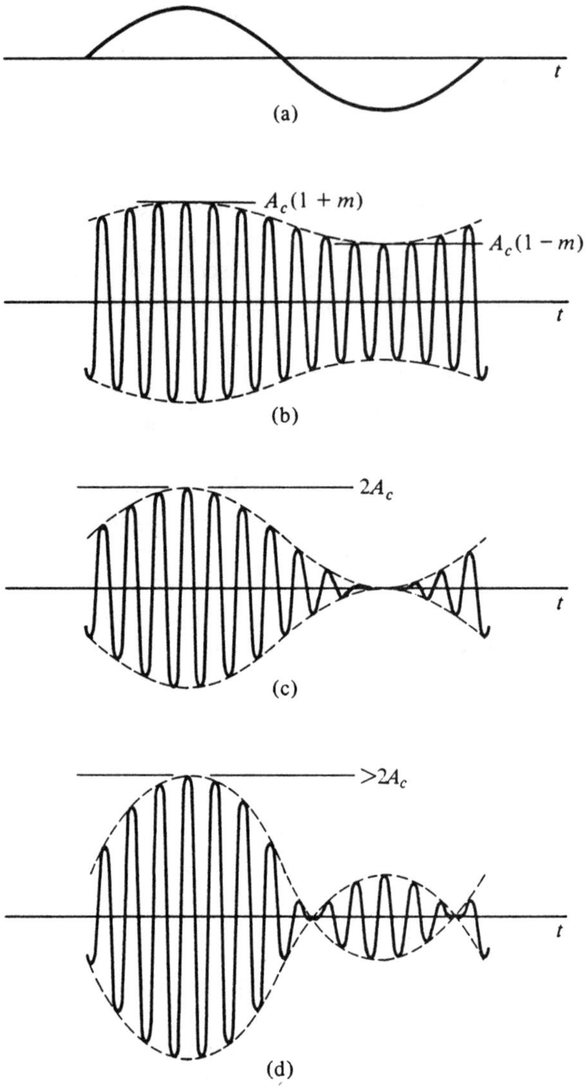

Figure 6.2-1. AM waveform. (a) $s(t)$. (b) $x(t)$ for $m < 1$. (c) $x(t)$ for $m = 1$. (d) $x(t)$ for $m > 1$. Note 180° phase reversal in carrier.

where m has been dropped. The wave in Eq. (6.2-4) is known as double-sideband suppressed carrier (DSBSC) or just double sideband (DSB) for short. Its spectrum is identical with the spectrum in Fig. 6.2-2 except for the missing carrier. If, in addition, we remove either of the sidebands, we obtain single-sideband modulation (SSB). In either case, a small carrier component (sometimes called a vestigial carrier) may be transmitted to synchronize a *local* oscillator to the carrier frequency to simplify demodulation. If a large portion of one sideband is sent with a small portion of the other sideband, we eliminate the abrupt frequency cutoff required in SSB and still obtain many of its

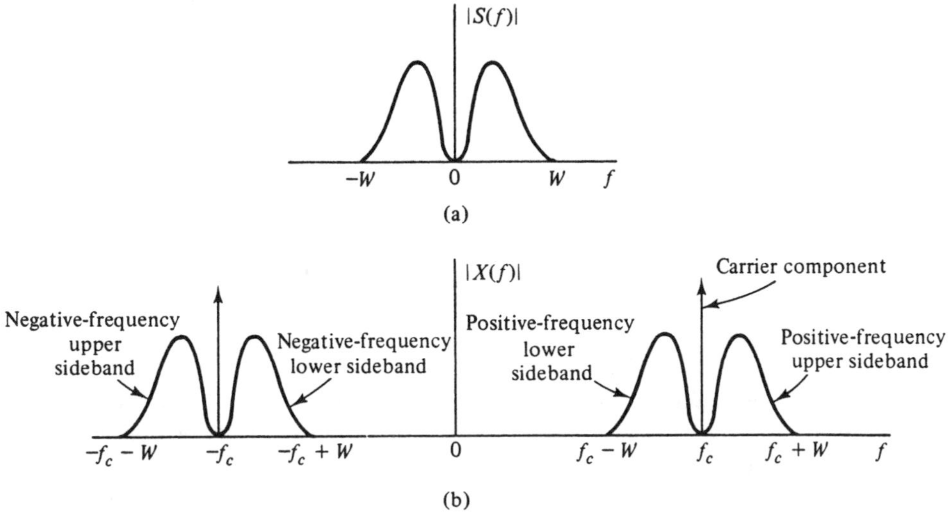

Figure 6.2-2. (a) Spectrum of $s(t)$. (b) Spectrum of the AM curve.

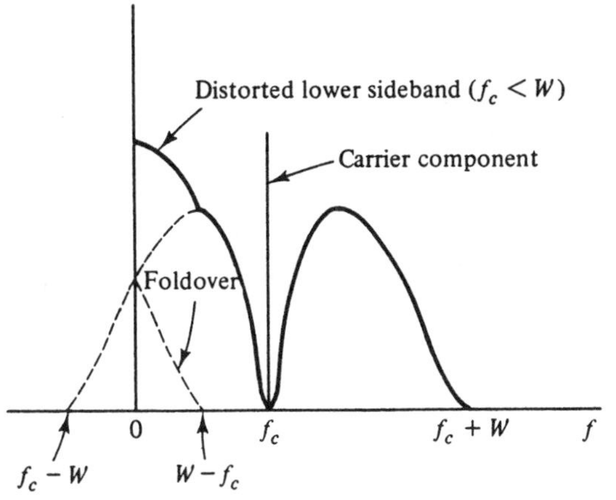

Figure 6.2-3. Sideband foldover if $f_c < W$. The low frequencies of the lower sideband (which contain the high-frequency information in the baseband) add to the lower frequencies and produce distortion.

advantages. This scheme is called vestigial sideband (VSB) and is in use in TV broadcasting.

6.3 MODULATORS

The generation of an AM wave can be conceptually described as in Fig. 6.3-1. The key operation is the ideal product between the carrier signal and the biased signal $1 + ms(t)$. The product of two signals can be obtained in a number of ways. One way is to use one signal to control the gain of an active device and use the other signal as input. Another way is to use one or more nonlinear elements that furnish the desired product but also, frequently, generate additional terms that need to be removed by filtering.

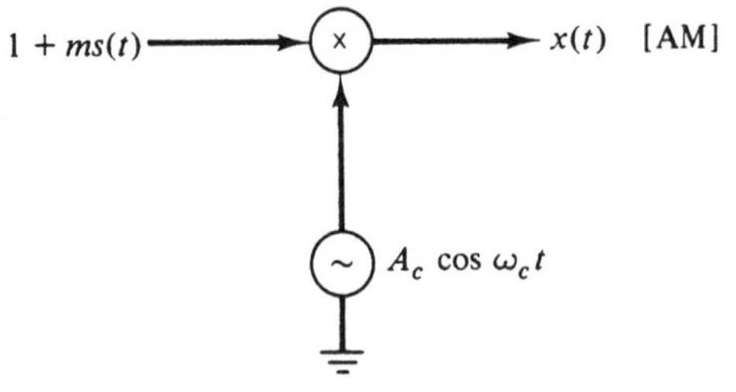

Figure 6.3-1. Ideal generation of an AM wave.

Multipliers

An analog multiplier is a three-port device that produces at the output port a signal that is proportional to the product of the two input signals present at its input ports. The variable transconductance multiplier (VTM) uses one signal to control the gain (transconductance) of an active device, which then amplifies the instantaneous value of the other signal in proportion to the instantaneous value of the control signal. The active device has the configuration of an emitter-coupled differential amplifer except that one input is used to control the current in the common emitter. A practical realization of a VTM is the monolithic unit shown in Fig. 6.3-2 (No. AD 530, Analog Devices,

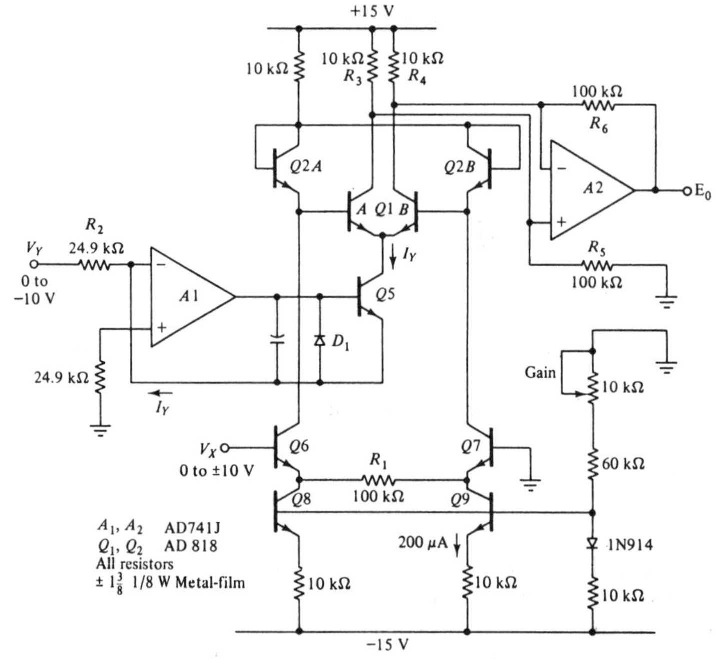

Figure 6.3-2. Practical two-quadrant variable transconductance multiplier. (Adapted from Sheingold [6-3], by permission.)

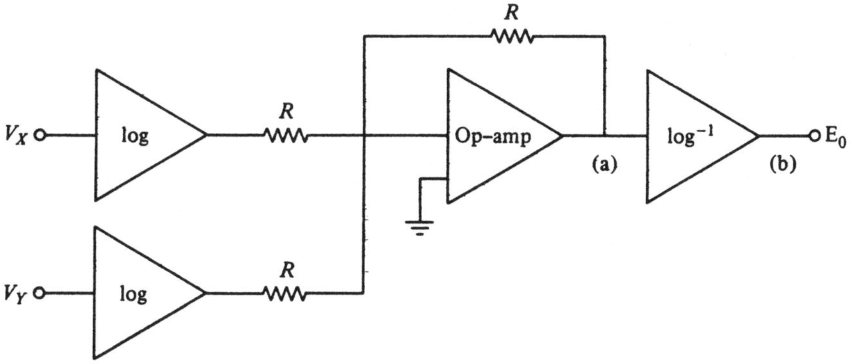

Figure 6.3-3. Logarithmic multiplier. At (a) we have log V_X + log V_Y. At (b) we have $E_0 = KV_XV_Y$, where K is a scale factor.

Inc.). A discussion of this circuit is given in [6-2] and [6-3]. It is not too difficult to show that the output signal E_0 is given by

$$E_0 = KV_XV_Y, \tag{6.3-1}$$

where K is a constant and V_X and V_Y are the two input signals to be multiplied.

Another type of multiplier in common use is the logarithmic multiplier ([6-3], p. 233), which uses logarithmic amplifiers to multiply together two positive signals. This operation is called one-quadrant operation. A bipolar signal can be multiplied by another bipolar signal if these signals are first added to positive dc levels that exceed the magnitude of the largest negative excursion of the signals. Figure 6.3-3 shows a logarithmic amplifier. The purpose of the op-amp is to facilitate the addition of the signals generated by the logarithmic amplifiers and to provide isolation from the log^{-1} circuit.

Square-Law Modulators

Semiconductor diodes and certain other types of nonlinear devices have a volt–ampere characteristic not unlike that shown in Fig. 6.3-4. When such a device is connected to

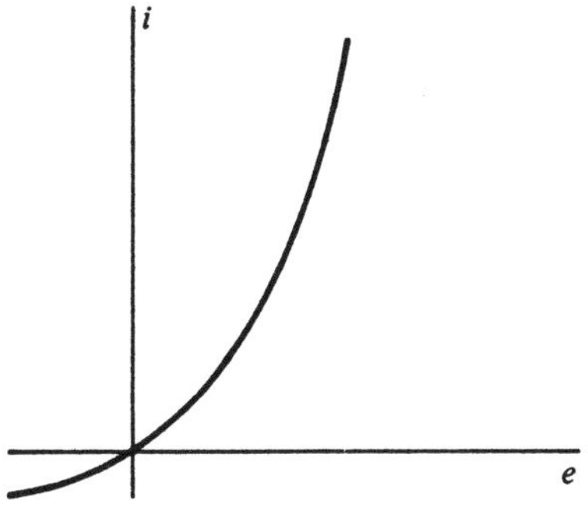

Figure 6.3-4. Volt–ampere characteristic of P-N diodes and certain other nonlinear devices.

a resistor, the voltage developed across the resistor can be described as a power series of the form

$$e_0 = \sum_{l=1}^{\infty} a_l e_i^l, \tag{6.3-2}$$

where e_0 is taken as the output voltage and e_i is the voltage impressed across the device. If $|e_i|$ is small enough, we can generally approximate the transfer characteristic by retaining only the first two terms. Then we say that we have a square-law device. For such a device,

$$e_0 = a_1 e_i + a_2 e_i^2. \tag{6.3-3}$$

Let the input consist of the sum of modulating signal plus carrier; that is,

$$e_i(t) = s(t) + \cos 2\pi f_c t.$$

The output is then

$$\begin{aligned} e_0(t) &= a_1 s(t) + a_2 s^2(t) + a_2 \cos^2 2\pi f_c t \\ &\quad + a_1\left[1 + \frac{2a_2}{a_1} s(t)\right] \cos 2\pi f_c t, \end{aligned} \tag{6.3-4}$$

which can be written as

$$\begin{aligned} e_0(t) &= A_c s(t) + a_2 s^2(t) + \frac{a_2}{2}(1 + \cos 4\pi f_c t) \\ &\quad + \underbrace{A_c[1 + ms(t)] \cos 2\pi f_c t}_{\text{AM wave}}, \end{aligned} \tag{6.3-5}$$

where $A_c \equiv a_1$ and $m \equiv 2a_2/a_1$. The unscaled spectrum of $e_0(t)$ is shown in Fig. 6.3-5. A band-pass filter of bandwidth $2W$ centered at f_c will isolate the AM wave. The filter bandwidth can, of course, be larger provided it does not admit the other spectra. Note that for modulation with a square-law device we must have

$$f_c > 3W \qquad \text{(square-law modulation)}. \tag{6.3-6}$$

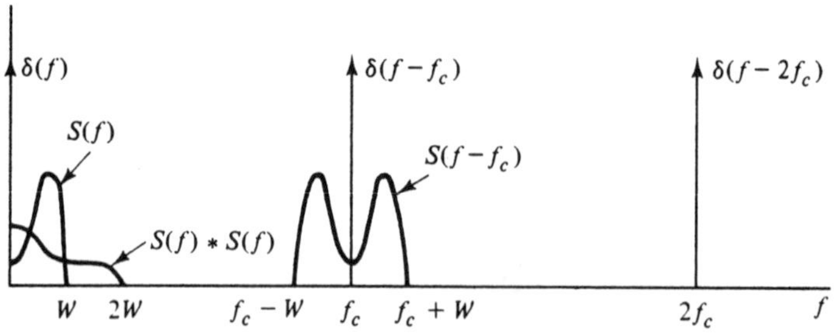

Figure 6.3-5. Spectrum of $e_0(t)$.

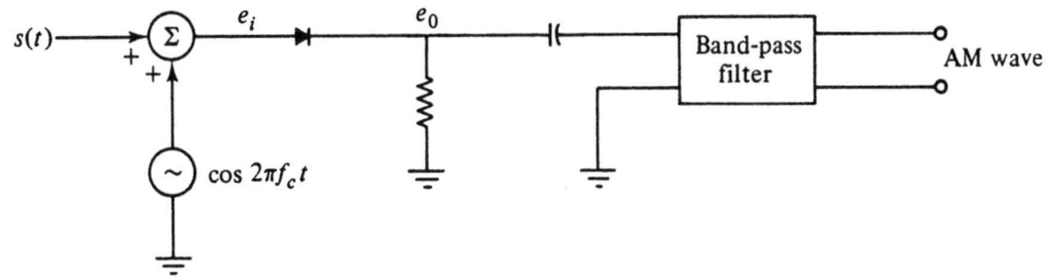

Figure 6.3-6. Square-law diode modulator.

A square-law diode modulator is shown in Fig. 6.3-6. The voltage e_i is assumed to be very small in order for the square-law model to be valid.

Switch-Type Modulators

The square-law transfer characteristic assumed for diodes and other nonlinear devices, such as nonlinear resistors and reactors, is only an approximation and, in practice, it is possible to get distortion terms that fall within the passband and that cannot be removed by filtering. This problem can, to some extent, be overcome by switch-type modulators.

In a switch-type modulator, the modulating signal is switched either in polarity or on–off at the carrier rate. A polarity switch is shown in Fig. 6.3-7. The signal flow paths for the two carrier polarities are shown in Fig. 6.3-8. We assume that the amplitude of the switching signal is much greater than $|s(t)|_{\max}$.

The output signal $e(t)$ can be written as

$$e(t) = s(t)g(t), \tag{6.3-7}$$

where

$$g(t) = \begin{cases} 1, & \text{for } \cos 2\pi f_c t > 0, \\ -1, & \text{for } \cos 2\pi f_c t < 0. \end{cases} \tag{6.3-8}$$

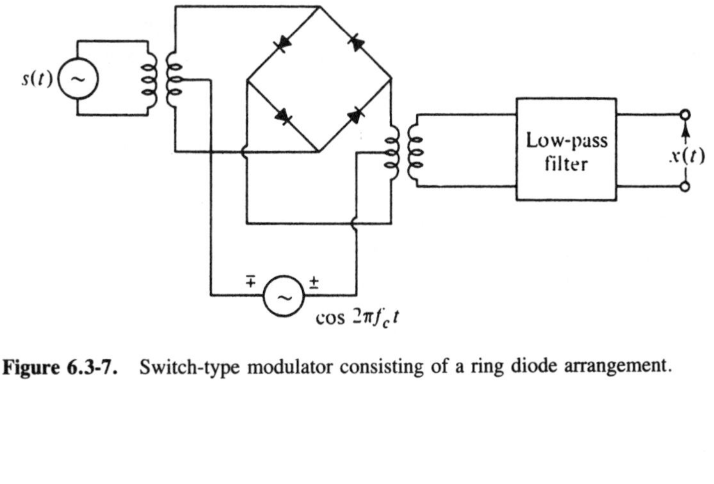

Figure 6.3-7. Switch-type modulator consisting of a ring diode arrangement.

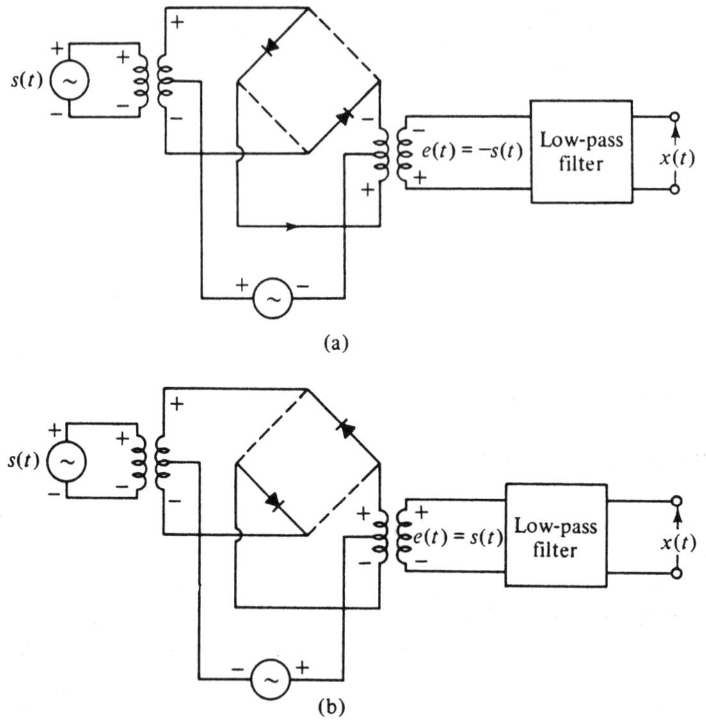

Figure 6.3-8. (a) Odd signal parity during first half of carrier cycle. (b) Even signal parity during second half of carrier cycle.

That is, $g(t)$ is a square wave of unit amplitude and fundamental frequency f_c. We note that this (idealized) system is linear since

$$\begin{aligned} g(t)[s_1(t) + s_2(t)] &= g(t)s_1(t) + g(t)s_2(t) \\ &= e_1(t) + e_2(t). \end{aligned} \tag{6.3-9}$$

However, the system is *time varying* since the output depends on the dynamical state of the system at time t. By using the Fourier series expansion of the square wave, the output signal $e(t)$ can be written as

$$e(t) = s(t)\left[\frac{4}{\pi} \cdot \sum_{n=0}^{\infty} (-1)^n(2n+1)^{-1} \cos 2\pi(2n+1)f_ct\right], \tag{6.3-10}$$

and the positive frequency spectrum is as shown in Fig. 6.3-9. To separate the unwanted sidebands from the signal of interest, we require that $3f_c - W > f_c + W$ or $f_c > W$. A filter centered at f_c with bandwidth $2W$ can be used to separate the modulated signal from the other terms. The output $x(t)$ has the form

$$x(t) = Bs(t) \cos 2\pi f_ct, \tag{6.3-11}$$

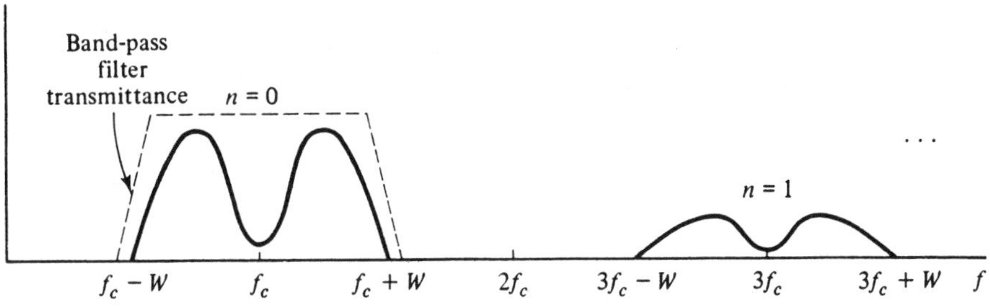

Figure 6.3-9. Spectrum of the output of a ring modulator.

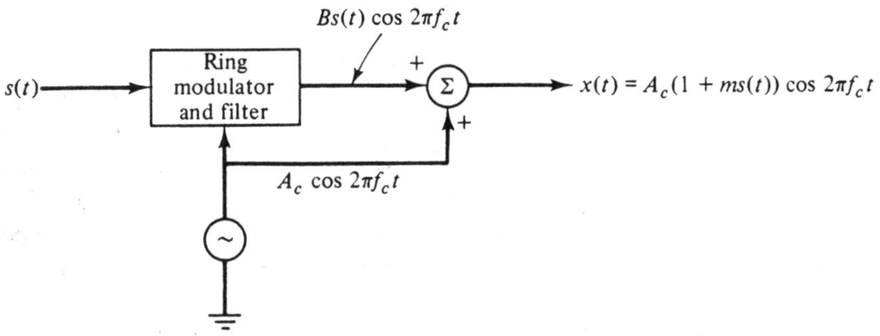

Figure 6.3-10. Generation of AM wave from DSB wave.

which is identical with Eq. (6.2-4). Thus we see that balanced switching produces a DSB wave that can be converted to an AM wave by the addition of a signal A_c cos $2\pi f_c t$ with A_c sufficiently large (Fig. 6.3-10).

Balanced Modulators

The ring-diode modulator is but one example of a type of modulator known as a *balanced modulator*. The output of such a modulator is typically DSB. When the modulating element is a nonlinear device with transfer characteristic given by Eq. (6.3-3) with $a_1 \neq 0$, it is possible to produce a DSB wave and simulate an ideal multiplier through the use of the balanced configuration shown in Fig. 6.3-11.

The signal e_1 is proportional to $a_1(s + \xi) + a_2(s^2 + 2s\xi + \xi^2)$, while the signal e_2 is proportional to $a_1(-s + \xi) + a_2(s^2 - 2s\xi + \xi^2)$. The difference is proportional to $2a_1 s + 4a_2 s\xi$, and the $2a_1 s$ term is removed by the filter. The output is a pure DSB wave and simulates the product of $s(t)$ and the carrier $\xi(t) = \cos 2\pi f_c t$. If the constant $a_1 = 0$, there would be no need for band-pass filtering. However, very few nonlinear devices satisfy the equation $y = x^2$. Note that if the two nonlinear devices in Fig. 6.3-11 are not matched potentially troublesome terms such as $s^2(t)$ remain after summation.

A balanced modulator circuit available as an integrated-circuit package is shown

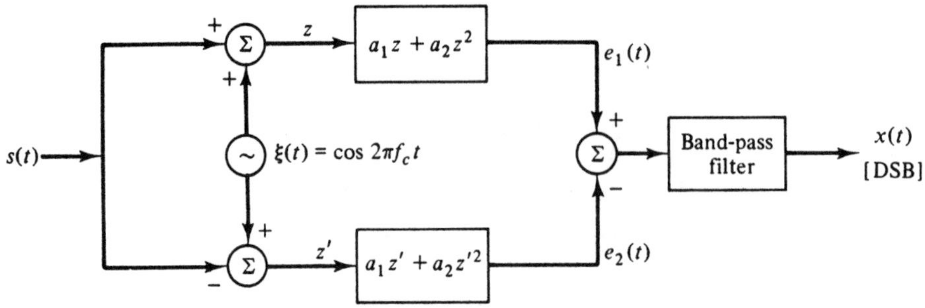

Figure 6.3-11. Use of a balanced modulator to generate DSB.

in Fig. 6.3-12(a). In this circuit the carrier signal is applied to the bases of the four transistor switches Q_1, Q_2, Q_3, and Q_4, and it acts to switch on either pair Q_1, Q_4 or pair Q_2, Q_3. The modulating signal is applied to the emitters of the same four transistors by the two transistors Q_5 and Q_6. If transistors Q_5, Q_6 are operating in the linear region, then their collector voltages are proportional to their base voltages. When line 8 is positive, the Q_3 and Q_2 switches are closed, and the positive input appears at $V_0(+)$ and the negative input at $V_0(-)$. During the next half-cycle of the carrier, the positive signal input goes to $V_0(-)$ and the negative signal input to $V_0(+)$. Thus the signal is in effect multiplied by a balanced square wave. Typical input and output wave shapes for this circuit are shown in Fig. 6.3-12(b). Note that this circuit differs only in some minor details from the multiplier circuit shown in Fig. 6.3-2. It acts as a switching modulator if the carrier-signal amplitude is sufficiently large.

Class C Amplifier Modulation

The modulators discussed so far are low level (i.e., low-power modulation). For AM, high-level modulation (i.e., high-power modulation) is achieved with the aid of tuned class C amplifers.

A class C amplifier is an amplifier for which the operating point is chosen so that the output current or voltage is zero for more than one-half of an input sinusoidal signal cycle. A simplified version of a class C amplifier is shown in Fig. 6.3-13.

With the carrier signal off, the values of V_{BB} and R_B are so chosen that the transistor is well beyond cutoff (for an *n-p-n* transistor this requires that $V_{BE} < 0.5$ V for germanium and < 0.7 V for silicon). The transistor goes into conduction only near the positive peak portion of the carrier cycle. The transistor is thus switched on and off by the carrier at the carrier frequency, f_c. In the absence of $s(t)$, the output is a series of pulses that is filtered by the tuned *RLC* circuit (sometimes called a tank circuit) to produce a constant amplitude collector current at the frequency f_c. When the collector supply voltage is varied by $s(t)$, the envelope of the collector current follows the variations in $s(t)$ and an AM wave is produced. Class C amplifiers are widely used for the generation of high-level AM signals. A simplified block diagram for an AM broadcast system is

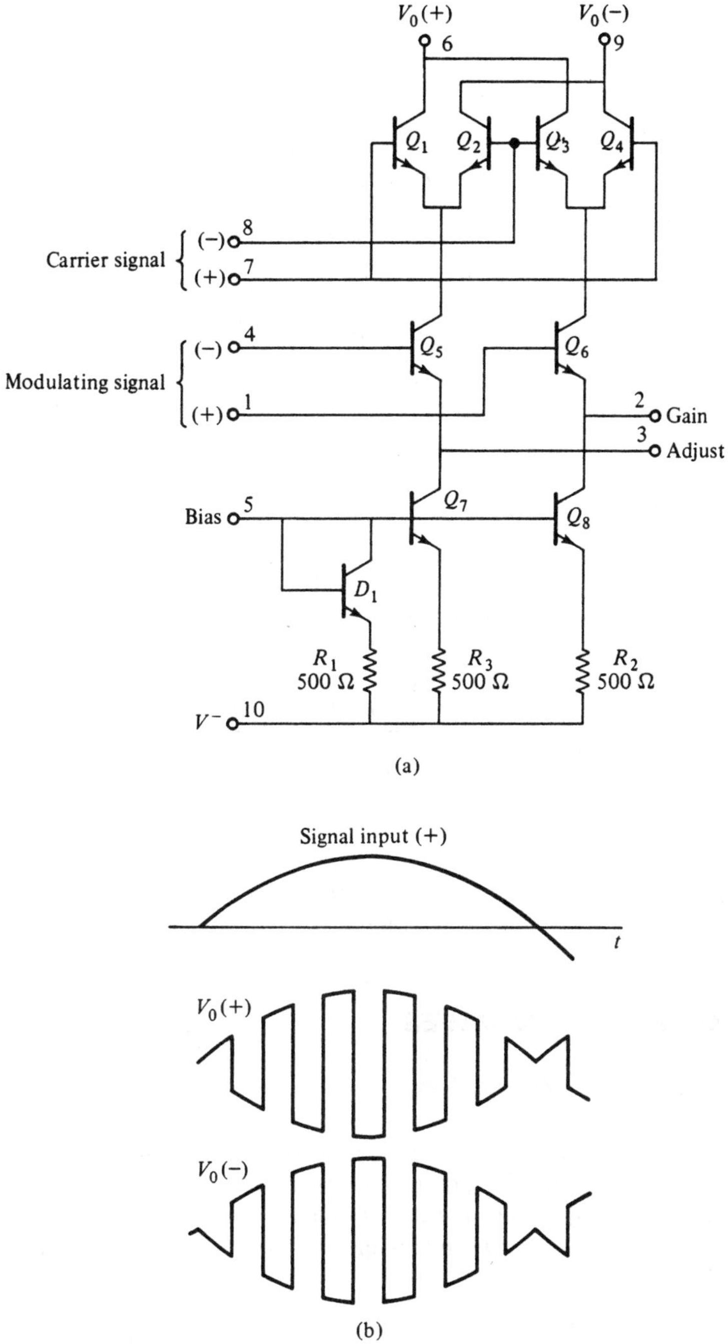

Figure 6.3-12. (a) Integrated circuit for balanced modulator–demodulator (Signetics 5596). (b) Typical input and output wave shapes.

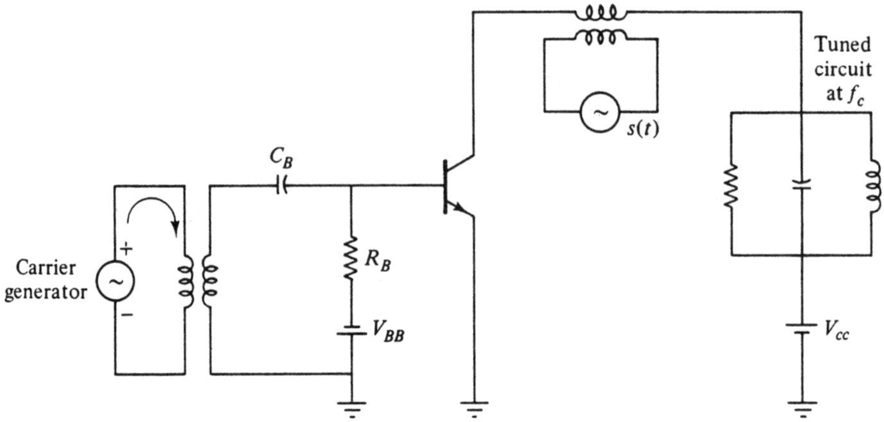

Figure 6.3-13. Simplified circuit of a class C amplifier.

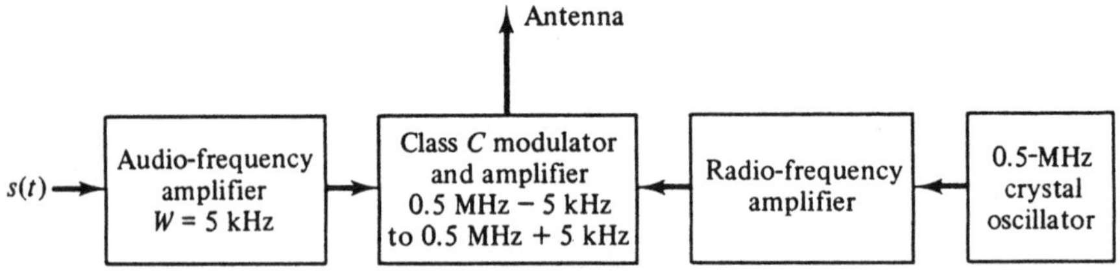

Figure 6.3-14. AM high-level broadcast system tuned to 500 kHz.

shown in Fig. 6.3-14. The level of $s(t)$ in class C modulation is generally much higher than in the circuits described earlier.

6.4 DETECTION OF AM WAVES

Synchronous Detection

From a conceptual point of view, synchronous or homodyne detection is perhaps the simplest, although it is almost never used to detect AM waves. Nevertheless, we shall illustrate how synchronous detection might be used to detect AM waves. With $x(t) = A_c[1 + ms(t)] \cos 2\pi f_c t$ representing the incoming AM wave, the product of $x(t)$ with a carrier gives

$$x(t) \cos 2\pi f_c t = \tfrac{1}{2} A_c[1 + ms(t)] \cos (4\pi f_c t) + \tfrac{1}{2} A_c + \tfrac{1}{2} A_c ms(t). \qquad (6.4\text{-}1)$$

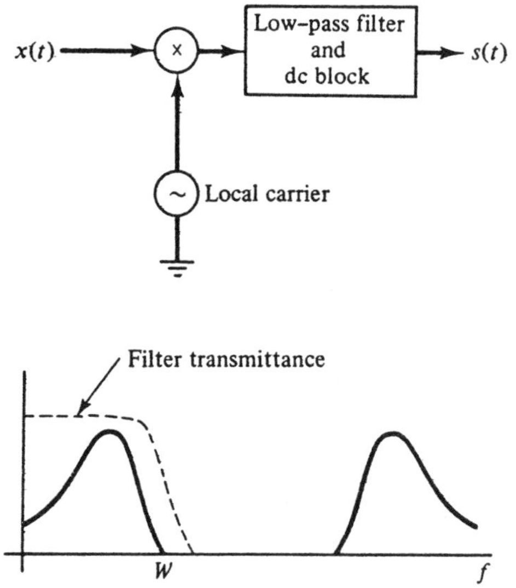

Figure 6.4-1. Synchronous detection of AM waves.

The last term is the desired signal. A block diagram for the synchronous detector is shown in Fig. 6.4-1. In principle, the generation of the carrier signal at the receiver can be done with a crystal oscillator† tuned to precisely the same frequency as the incoming carrier, by separation of the incoming carrier by narrowband filtering or by tuning a local oscillator. Practical methods for obtaining a carrier wave at the receiver will be discussed later. With few exceptions, some form of synchronous detection is generally required to demodulate DSB, SSB, VSB, or some other derived form of AM. However, AM itself is ordinarily detected with an *envelope detector*.

Homodyne techniques (i.e., when signals of the same frequencies are multiplied) are a special case of the more general *heterodyne* principle, which is widely used in AM receivers. Heterodyning is the process of generating new frequencies by multiplying carrier signals of different frequencies. Heterodyning leads to a frequency shift of the carrier signals; that is, $\cos 2\pi f_1 t \cdot \cos 2\pi f_2 t = \frac{1}{2}[\cos 2\pi(f_1 + f_2) + \cos 2\pi(f_2 - f_1)t]$. One of the two signals thus produced is then filtered out. If the sum frequency is rejected, we say that the signal has been down-converted. If the difference frequency is rejected, the signal is said to be up-converted. Thus homodyning is down conversion to baseband (or up conversion to double the frequency). A block diagram of a heterodyne system is shown in Fig. 6.4-2. To enable separation of the up-converted signal from the down-converted signal without sideband foldover, we require that (assume $f_2 > f_1$)

$$\text{(i) } f_2 - f_1 > W,$$

$$\text{(ii) } f_1 > W, \tag{6.4-2}$$

$$\text{(iii) } f_2 > 2W.$$

† However, there will be, in general, a phase difference between the local demodulating carrier and the incoming carrier signal. This could result in significant loss of detected power.

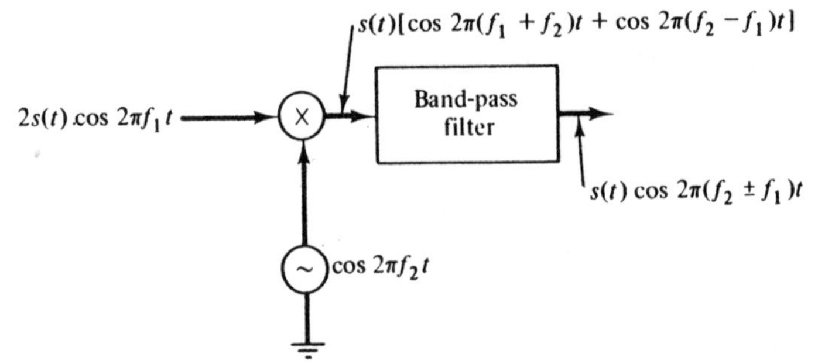

Figure 6.4-2. Heterodyning of frequency conversion.

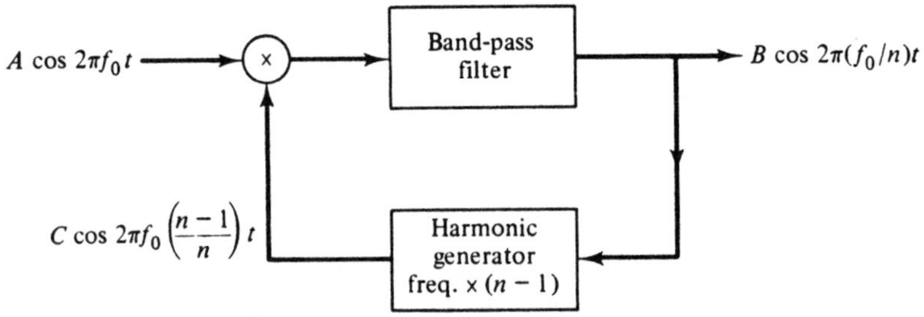

Figure 6.4-3. Regenerative frequency divider.

The last condition follows from the first two. The product of the two signals can be achieved by the techniques discussed in Sec. 6.3.

An interesting modification of a heterodyne system, discussed in ([6-4], p. 191), is the regenerative frequency divider, which takes an input frequency f_0 and produces an output frequency f_0/n. The system is shown in Fig. 6.4-3. We leave it to the reader (Prob. 6-12) to show that if the band-pass filter rejects the upper frequency $(2n - 1)f_0/n$ then the output is indeed f_0/n.

Frequency conversion of the form f/N, where N is a multiple of 2, can be done with digital circuitry. For example, a four-stage ripple counter ([6-5], p. 301) can be used to produce the frequencies $f_0/2$, $f_0/4$, $f_0/8$, and $f_0/16$ from an input pulse sequence of frequency f_0. Although the intermediate signals are square waves, appropriate band-pass filtering will generate sinusoids of the same frequency.

The Frequency Synthesizer

An application of both the heterodyning principle and of frequency division can be found in the frequency synthesizer, an instrument used to generate very precisely controllable frequencies.

The basic idea is illustrated in the circuit shown in Fig. 6.4-4. A crystal-controlled

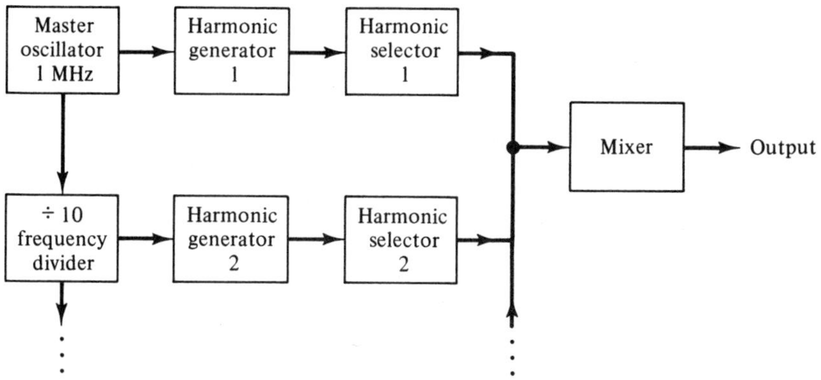

Figure 6.4-4. Frequency synthesizer.

master oscillator generates a fixed frequency of, say, 1 MHz. This signal is passed through a nonlinear device to produce a signal rich in harmonics of 1 MHz. An adjustable narrowband filter is used to select one of these harmonics.

The master-oscillator signal is also passed through a series of frequency dividers, each one of which generates an output signal whose frequency is precisely one-tenth of the input frequency. In this way, signals whose frequency is exactly 0.1 MHz, 0.01 MHz, and so on, are generated. Only one of these subharmonic generators is shown in the figure. The output of each subharmonic generator is again passed through a harmonic generator and a harmonic selector and finally into a mixer. A narrowband filter at the output of the mixer selects either the sum or the difference frequency of the mixer output. The output frequencies made available by the two-stage system shown in the figure therefore range from $1.0 - 0.9 = 0.1$ MHz to $9 + 0.9 = 9.9$ MHz in 0.1-MHz steps. Further subdivision of the frequency output is provided by additional stages of heterodyning with the harmonics of additional subharmonic generators.

Detection with Rectifiers

A nonlinear operation on the AM signal can produce a signal that, after filtering, furnishes the message $s(t)$. The linear half-wave and linear full-wave rectifiers shown in Fig. 6.4-5 are devices useful for such operations. The linear full-wave rectifier has transfer characteristic $e_0 = |e_i|$. Hence, denoting an AM wave by

$$x(t) = r(t) \cos 2\pi f_c t,$$

where $r(t) \equiv A_c[1 + ms(t)] \geq 0$, we can write for the output of the full-wave rectifier

$$z(t) = |x(t)| = r(t)|\cos 2\pi f_c t|. \tag{6.4-3}$$

The dc value of $|\cos 2\pi f_c t|$ is $2/\pi$. The complete Fourier series for $|\cos 2\pi f_c t|$ is

$$|\cos 2\pi f_c t| = \frac{2}{\pi} - \frac{4}{\pi} \sum_{n=1}^{\infty} \frac{(-)^n}{4n^2 - 1} \cos 2\pi(2n)f_c t. \tag{6.4-4}$$

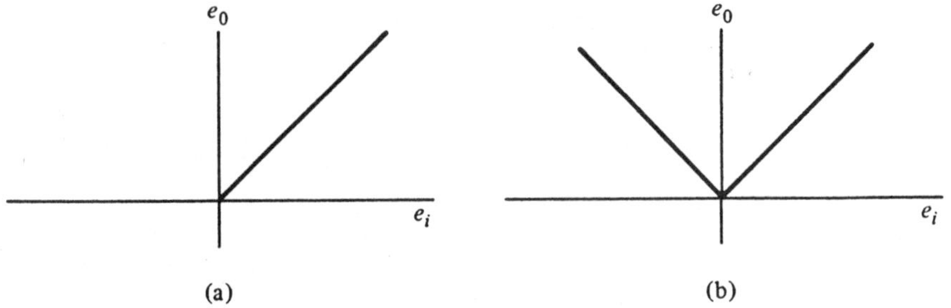

Figure 6.4-5. (a) Half-wave and (b) full-wave rectifier transfer characteristics.

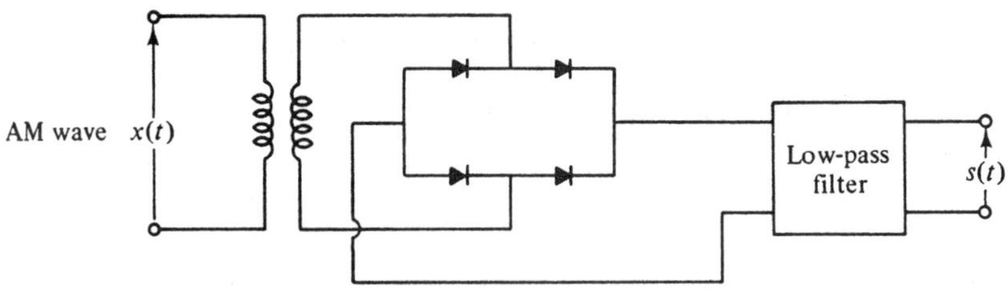

Figure 6.4-6. Detection of an AM wave with full-wave rectifier.

If we substitute Eq. (6.4-4) into Eq. (6.4-3), we obtain

$$z(t) = \frac{2A_c}{\pi}[1 + ms(t)] - \frac{4A_c}{\pi}[1 + ms(t)] \cdot \sum_{n=1}^{\infty} \frac{(-)^n}{4n^2 - 1} \cos 2\pi(2n)f_c t. \tag{6.4-5}$$

It is clear that if $f_c > W$, an ideal low-pass filter will enable the separation of the term $2A_c[1 + ms(t)]/\pi$ from the other terms. A dc blocking capacitor will then remove the dc bias and leave us with $s(t)$. If $s(t)$ contains a dc term, that too will be removed. In TV, the loss of the dc component in $s(t)$ can be a problem, and dc restoration circuitry is required. A full-wave rectifier detector for AM signals is shown in Fig. 6.4-6.

The Envelope Detector

The simplest and most economical circuit for detecting AM waves is the half-wave rectifier. Despite the fact that, as we shall see in subsequent sections, both DSB and SSB have important advantages over AM, the popularity of AM over DSB and SSB is due mostly to the fact that AM waves can be detected by rectifiers, whereas both DSB and SSB require some form of synchronous detection.

An AM envelope detector with capacitive filtering is shown in Fig. 6.4-7. A

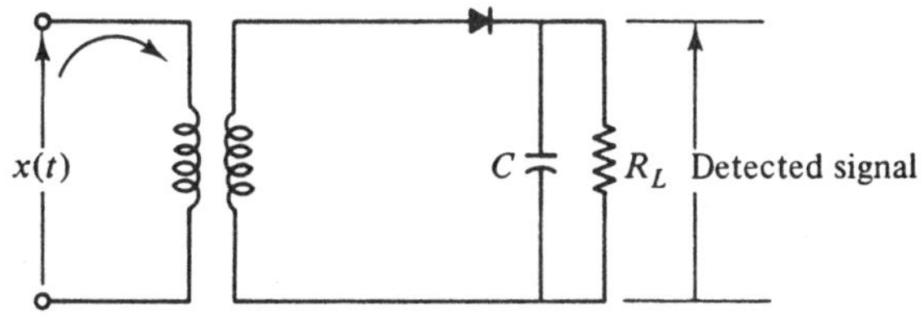

Figure 6.4-7. AM demodulation with an envelope detector.

thorough analysis of this circuit is given in [6-6, p. 606]. If the capacitor were not there, the output would simply be a half-wave-rectified version of the input (Fig. 6.4-8). However, with the capacitor filter in place, the output is essentially the input envelope with a small ripple superimposed. While the diode is conducting, the capacitor charges to the input voltage with a time constant given by the product of C and the resistance resulting from the parallel combination of R_L and the diode forward resistance R_f. Since R_f is assumed very small, for all practical purposes the input is impressed directly across the load. When the diode is reverse biased, the capacitor discharges through R_L with a time constant equal to R_LC. The value of R_LC should be large enough to smooth the waveform (i.e., prevent excessive ripple), but not so large as to prevent the load voltage from following the changes in the *envelope*. Figure 6.4-9 shows a closeup view of the output voltage variation in an envelope detector. The computations of the cut-in and cut-out times can be done graphically ([6-6, p. 609]). Figure 6.4-9 greatly exaggerates the rate of change of the envelope with respect to the carrier. At a carrier frequency of 1000 kHz and a highest audio modulating frequency of 20 kHz, there should be 50 carrier cycles for each audio cycle. Only when the audio frequency is an order (or orders) of magnitude lower in frequency than the carrier frequency does it make sense to speak of an envelope.

It should be obvious that envelope detection is fundamentally simpler and easier

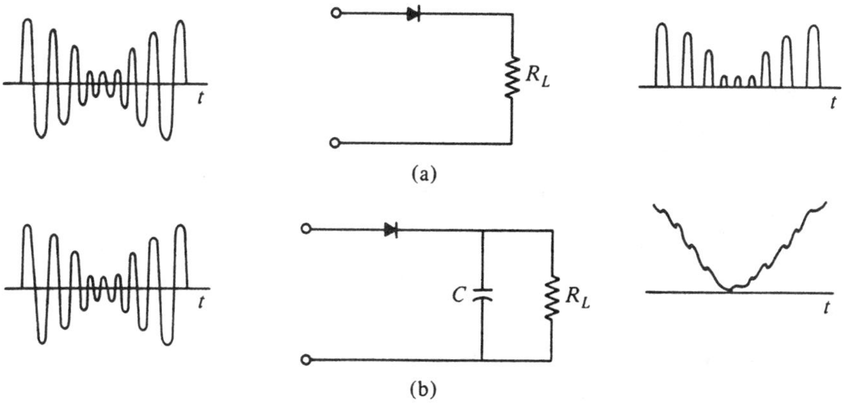

Figure 6.4-8. (a) Output without filtering; (b) output with filtering.

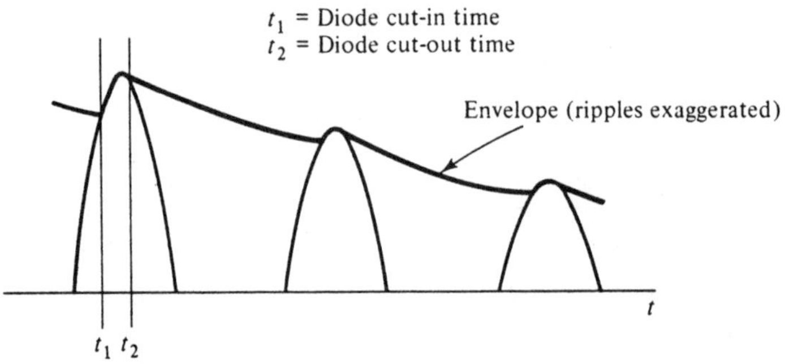

Figure 6.4-9. Envelope variations in envelope detector.

than homodyne detection. However, we can establish a rough equivalence between the two by looking at envelope detection from the following, approximate, point of view. The carrier waveform operates the diode as an on–off switch with switching function

$$g(t) \simeq \frac{1}{2} + \frac{2}{\pi} \sum_{n=0}^{\infty} (2n + 1)^{-1}(-1)^n \cos 2\pi(2n + 1)f_c t. \tag{6.4-6}$$

If we ignore the loading effect of the *RC* low-pass filter at the output of the diode, the voltage developed across the output terminals is proportional to

$$e_0(t) = g(t)x(t). \tag{6.4-7}$$

The product of $x(t)$ with the $n = 0$ component gives rise to a term that contains just $s(t)$. The addition of a low-pass filter removes all components but $s(t)$. Hence an approximate equivalent representation of the envelope detector can be constructed as in Fig. 6-4-10. When Fig. 6.4-10 is compared with Fig. 6.4-1, which shows synchronous detection, the relation between the two methods becomes more apparent. Thus the large carrier signal that is always present in AM can be regarded as a way of including the waveform needed for homodyne demodulation in the transmission. We discuss later other methods for including such synchronizing signals in the transmission. These are generally more energy efficient, but also require more complicated circuitry than the simple envelope detector.

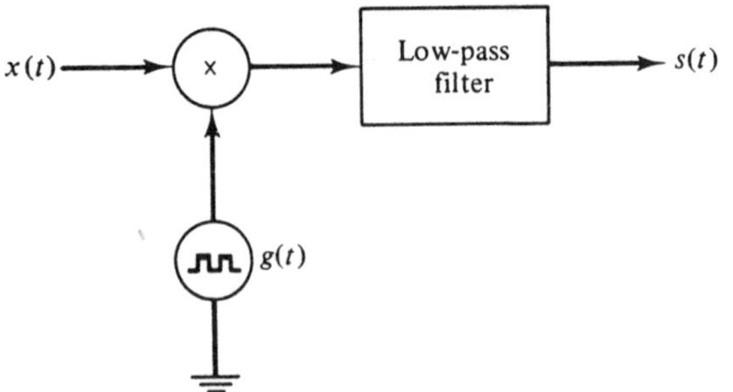

Figure 6.4-10. Equivalent representation of envelope detector that ignores loading of *RC* low-pass filter.

6.5 THE SUPERHETERODYNE RECEIVER

There are several different types of AM receivers, but by far the most widely used is the superheterodyne (superhet) receiver shown in Fig. 6.5-1. The superheterodyne principle is used in low-cost as well as high-quality receivers.

The superhet receiver must basically carry out two functions: the amplification of the band containing the selected signal and the detection of the information $s(t)$. Superheterodyne operation refers to the frequency conversion from the variable RF (radio-frequency) to the fixed IF (intermediate-frequency) signal that is ultimately detected with an envelope detector.

In a typical situation, the signals appearing at the antenna are a superposition of a large number of radio, TV, other man-made, as well as non-man-made signals. Assume that the RF amplifier is tuned to the frequency f_c, and consider at point A, a specific AM wave of the form

$$e_A(t) = r(t) \cos 2\pi f_c t, \tag{6.5-1}$$

where

$$r(t) = a[1 + ms(t)], \qquad a \text{ is a constant.}$$

At point B, the signal is

$$e_B(t) = a_1 r(t)[\cos 2\pi f_{\text{IF}} t + \cos 2\pi(2f_c + 2f_{\text{IF}})t]. \tag{6.5-2}$$

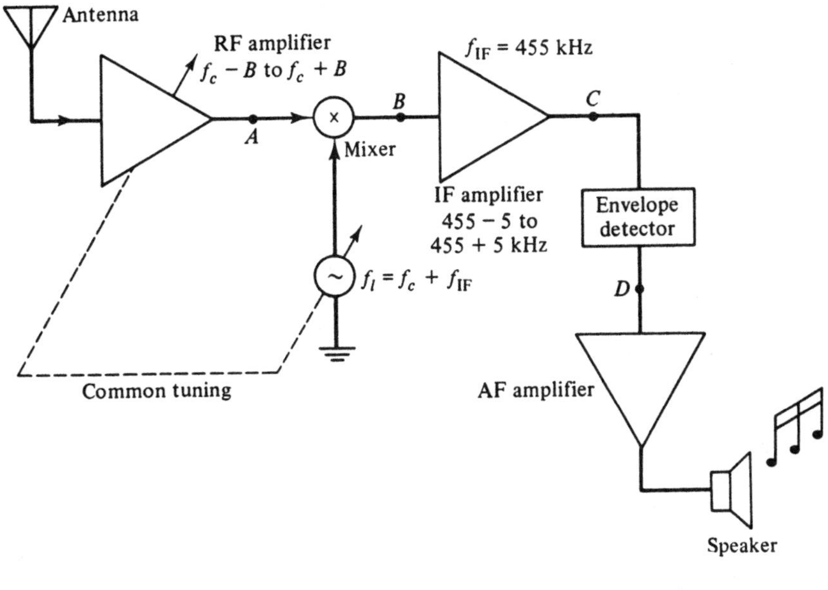

Figure 6.5-1. Principal components of the superheterodyne receiver.

Note that the common tuning of the variable-frequency local oscillator with the RF amplifier is an ingenious way of moving the RF band past a fixed IF band-pass window. This results in much better band-pass characteristics for the IF stages, since the design of these units can be optimized for a fixed frequency. For example, the IF stage can be designed for a fixed relative bandwidth ($\Delta f/f_{\mathrm{IF}}$) of 10/445 instead of having to handle a variable relative bandwidth, which is a significant complication.

The second term in the brackets in Eq. (6.5-2) represents the sum frequency and is rejected by the IF amplifier. At C, the signal has the form

$$e_C(t) = a_2 r(t) \cos 2\pi f_{\mathrm{IF}} t. \tag{6.5-3}$$

The envelope detector generates the signal

$$e_D = a_3 r(t), \tag{6.5-4}$$

and, finally, the AF (audio-frequency) amplifier amplifies $s(t)$ and drives the speaker.

It is possible to use the detected signal for automatic volume control (AVC). If the envelope variations are smoothed out completely, the output of the AVC circuit will be a dc signal proportional to the carrier amplitude. This signal can be used to control the gain of the IF stage so that if the carrier fades the gain increases, thereby tending to keep the volume constant.

The bandwidth, $2B$, of the RF amplifier should be no less than $2W$ but may be larger since *adjacent-channel selectivity* is furnished by the IF stage. The partial selectivity of the RF stage furnishes *image-channel* rejection. This is important because otherwise image channels can be detected as well as the desired ones. To explain this phenomenon, note that without the selectivity of the RF stage there are in fact two channels that will pass the IF stage and hence be detected. If the local oscillator frequency f_i is set at[†]

$$f_i = f_c + f_{\mathrm{IF}},$$

then the RF frequency

$$f_{\mathrm{RF}} = f_c$$

and its image

$$f_{\mathrm{RF}}^{(i)} = f_c + 2f_{\mathrm{IF}}$$

will fall within the band of the IF stage. The image is rejected by the RF stage; hence the name image-channel rejection.

The frequency band for standard AM broadcasting is 550 to 1600 kHz. The IF frequency is fixed at 455 kHz, and there is a 10-kHz channel bandwidth per station ([6-7], p. 21.49). For perfect audible reproduction of speech and music, a frequency range from 30 to 15,000 kHz is desirable. Broadcast AM obviously falls far short of this, but high-quality broadcasting of AM signals provides a ± 2-dB response over the

[†] The choice of $f_i = f_c + f_{\mathrm{IF}}$ leads to a smaller tuning range for the local oscillator than $f_i = f_c - f_{\mathrm{IF}}$. See Prob. 6-11.

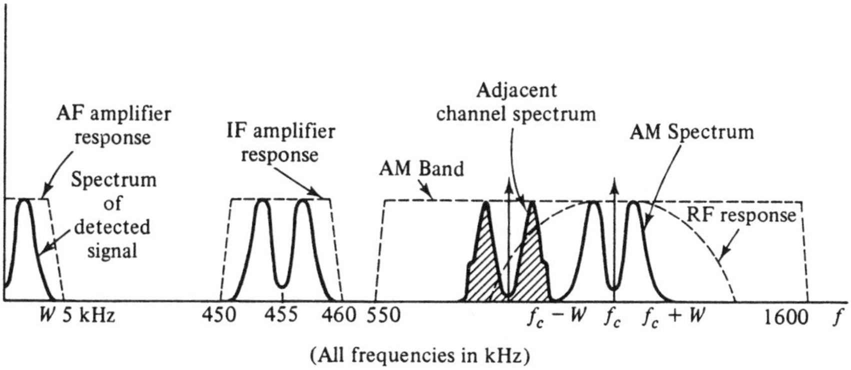

Figure 6.5-2. Spectra of AM signals in the superheterodyne receiver. The adjacent station may pass the RF amplifier but will not pass the IF amplifier.

frequency range from 30 to 4500 Hz. Figure 6.5-2 shows the locations in the spectrum of various signals.

Most AM and AM/FM superheterodyne receivers now come with most of the active components located on a single integrated-circuit chip [6-8].

*6.6 THE SUPERHETERODYNE PRINCIPLE IN SPECTRUM ANALYSIS

Practical spectrum analyzers are frequently of a type shown in Fig. 6.6-1. This configuration is closely related to the superheterodyne receiver. The amplifier/filter is centered at a *fixed* frequency f_0 and has a *fixed* bandwidth, Δ.

The device marked VCO is a voltage-controlled oscillator. Its frequency is proportional to the applied voltage. To understand the operation of this system, we write $x(t)$ as a Fourier series:

$$x(t) = \sum_{n=1}^{\infty} 2|c_n| \cos (2\pi f_n t + \theta_n), \tag{6.6-1}$$

where $f_n = (1/T_s)n$, T_s = period of the signal, and θ_n is the phase associated with the nth harmonic. Assume that the dc term in $x(t)$ has been filtered out. The signal $e(t)$ is given by

$$\begin{aligned} e(t) &= \sum_{n=1}^{\infty} |c_n| \cos \{2\pi[(f_x - f_n)t + f_0 t] - \theta_n\} \\ &+ \sum_{n=1}^{\infty} |c_n| \cos \{2\pi[(f_x + f_n)t + f_0 t] + \theta_n\}. \end{aligned} \tag{6.6-2}$$

The narrowband filter effectively passes only the component whose frequencies are in the range $[f_0 - (\Delta/2), f_0 + (\Delta/2)]$; this means that $f_n = f_x$ and that $|c_n|$ is detected and

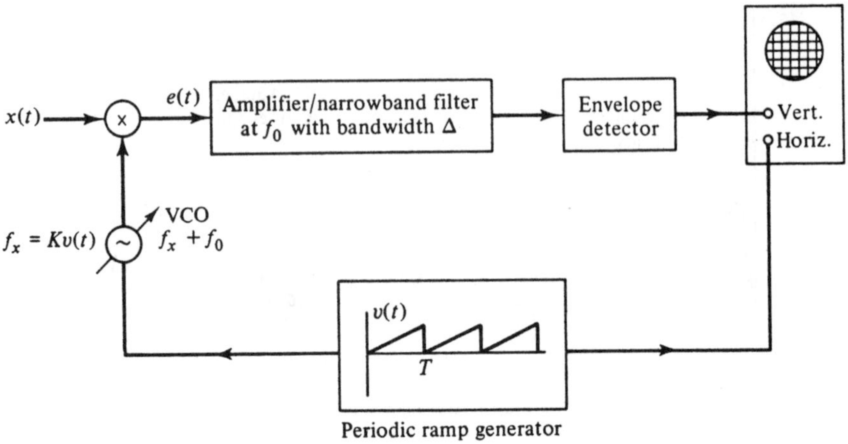

Figure 6.6-1. Scanning spectrum analyzer.

displayed on an oscilloscope whose horizontal deflection is derived from the voltage $v(t)$, which is proportional to f_x.† In the absence of an RF image-rejection filter, images may be a problem and the sweep *range* must be restricted to less than $2f_0$.

If at time $t = 0$ the frequency of the VCO is f_0 and if at time $t = T$ the frequency is $f_M = f_0 + K_1 v(T)$, then the total band over which the spectrum is examined is $f_M - f_0 = K_1 v(T) = K_2 T$ (the K's are constants of proportionality). If Δ is the width of the narrowband filter, then the number of frequency resolution cells is

$$N = \frac{f_M - f_0}{\Delta}. \tag{6.6-3}$$

A large N implies a high-resolution spectrum analyzer. During the sweep time T, the entire band $f_M - f_0$ must be examined. The time allowed per resolution cell is T/N. To get a meaningful response from the filter, its response time τ_c must satisfy

$$\tau_c \ll \frac{T}{N}.$$

At the same time, we know that $\tau_c \simeq \Delta^{-1}$ (see Fig. 3.6-3 with B replaced by Δ). Hence

$$\tau_c \simeq \frac{1}{\Delta} \ll \frac{T\Delta}{f_m - f_0} \tag{6.6-4}$$

or

$$T \gg \frac{f_M - f_0}{\Delta^2} = \frac{N^2}{f_M - f_0}. \tag{6.6-5}$$

† Only one harmonic component falls within the passband at any one time if $T_s^{-1} > \Delta$.

Thus for high resolution we require long sweep times. For $f_M - f_0 = 10{,}000$ Hz and $N = 100$, $T \gg 1$ sec. Since $T_s^{-1} > \Delta$, we must also satisfy $T \gg (f_m - f_0)T_s^2 < NT_s$. In commercial spectrum analyzers the frequency sweep and resolution adjustments are frequently mechanically coupled to implement these inequalities.

6.7 DOUBLE SIDEBAND (DSB)

Before considering the problem of detecting DSB, we might ask why we should consider DSB in the first place if AM is so easily detected with an envelope detector. The answer lies in the fact that DSB is more efficient than AM. Since the carrier is not sent, there is considerable saving in power. To see how much power is saved, consider the time-averaged transmitted power in an AM wave, which is

$$P_{AM} = \lim_{T\to\infty} \frac{1}{T}\int_{-T/2}^{T/2} x^2(t)\,dt \equiv \langle x^2(t)\rangle \tag{6.7-1a}$$

$$= \frac{A_c^2}{2}[1 + m^2\langle s^2(t)\rangle], \tag{6.7-1b}$$

where we have used the fact that the carrier frequency $f_c \gg W$ the message bandwidth, and the $\langle\ \ \rangle$ represent the time-averaging operation indicated in Eq. (6.7-1a). The AM power can be decomposed as

$$P_{AM} = P_c + 2P_{SB}, \tag{6.7-2}$$

where $P_c = A_c^2/2$ is the carrier power and $P_{SB} = m^2A_c^2\langle s(t)^2\rangle/4$ is the signal-associated power in a sideband. Since $m|s(t)| \le 1$, the maximum value of P_{SB} is $A_c^2/4$. Therefore, *at least 50% of the transmitted power is in the carrier wave*, which, by itself, carries no information. Thus DSB is indeed more efficient than AM; however, DSB cannot be detected with an envelope detector, which would tend to detect $|s(t)|$ rather than $s(t)$. The detection circuitry for DSB is more complicated and requires additional electronics. As stated earlier, detection of DSB requires some form of homodyne detection. For this to work properly, the local carrier must have the right frequency and phase. If the carrier has the wrong phase, then the detected signal has the form (Fig. 6.7-1) $\tilde{s}(t) =$

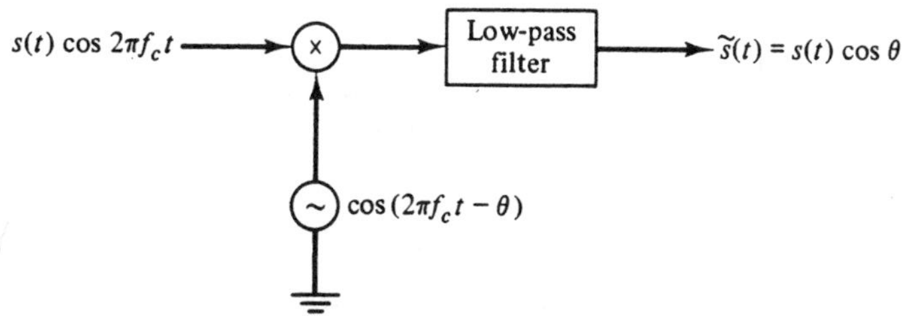

Figure 6.7-1. Detection of DSB with a carrier with the wrong phase.

$s(t)\cos\theta$. Thus, for a phase error of only 45°, the detected power is reduced by half. As $\theta \rightarrow \pi/2$, the detected signal vanishes.

The effect of demodulating with the *wrong frequency* is much worse than demodulating with the wrong phase and leads to serious distortion. To see this, consider a modulating tone of the form $s(t) = 2A_m \cos 2\pi f_m t$ that DSB-modulates a carrier $\cos 2\pi f_c t$. The DSB wave is then

$$x(t) = A_m\{\cos[2\pi(f_c + f_m)t] + \cos[2\pi(f_c - f_m)t]\}. \tag{6.7-3}$$

Now assume that the local carrier in Fig. 6.7-1 furnishes a signal $\cos[2\pi(f_c + \Delta)t]$, where Δ is the error frequency. The output of the low-pass filter is then proportional to

$$\begin{aligned}\tilde{s}(t) &= \cos 2\pi(f_m + \Delta)t + \cos 2\pi(f_m - \Delta)t \\ &= 2\cos 2\pi f_m t \cos 2\pi\Delta t.\end{aligned} \tag{6.7-4}$$

If Δ is a very low frequency, the detected signal will exhibit beats instead of a steady tone. For somewhat larger values of Δ, the sum and difference frequency terms result in a completely distorted and unintelligible output.

A small amount of carrier, when added to the DSB signal, can be used to generate a carrier of the same phase and frequency at the receiver, which can then be used in homodyne detection. Adding a small amount of carrier to a DSB wave does not convert it to an AM wave, and it cannot be detected with an envelope detector. Figure 6.7-2 shows a DSB detector in which the carrier is generated from the sidebands with the aid of a square-law device. The narrowband filter centered at $2f_c$ produces a sine wave of frequency $2f_c$, and a 2:1 frequency converter produces the locally generated carrier frequency.

Another scheme, proposed by Costas [6-9], uses what amounts to a phase-locked loop to generate a constant-frequency carrier signal from the sidebands. The system is shown in Fig. 6.7-3. Assume that the local oscillator is at frequency $f < f_c$ and the input signal is $x(t) = s(t)\cos 2\pi f_c t$. After low-pass filtering the multiplier output signal is proportional to

$$e(t) = \frac{s^2(t)}{4}\sin[4\pi(f_c - f)t - 2\theta], \tag{6.7-5}$$

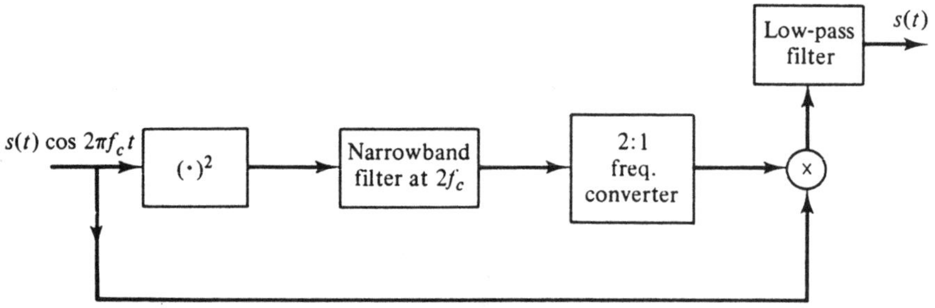

Figure 6.7-2. DSB detector using a square-law device.

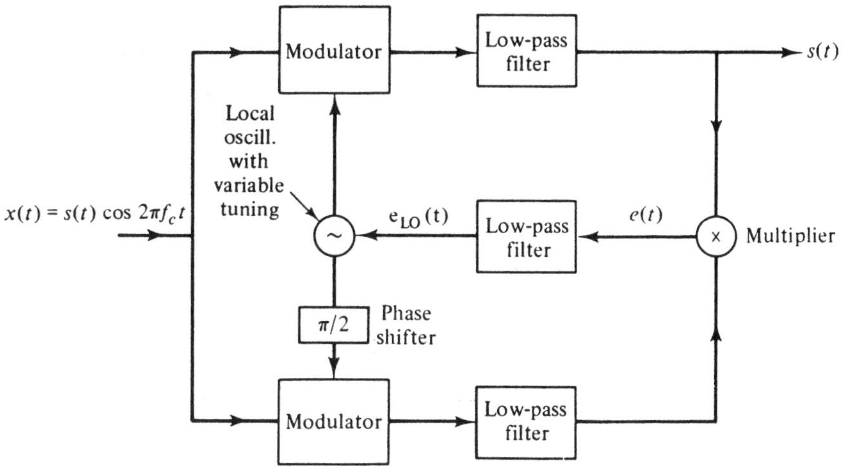

Figure 6.7-3. Costas's scheme for DSB detection.

where θ is the phase difference between the local oscillator and the incoming wave when both are at the same frequency. If $f_c - f$ is small compared to frequencies at which there are significant components in $s(t)$, the amplitude variations in $e(t)$ can be filtered out, and the voltage driving the oscillator becomes essentially proportional to

$$e_{LO}(t) = K \sin [4\pi(f_c - f)t - 2\theta]. \tag{6.7-6}$$

The local oscillator will adjust its frequency until $e_{LO}(t) \to 0$. This happens not only when $f_c = f$ but when $\theta = 0$. Thus the loop drives not only to the correct frequency but the correct phase as well. Because phase-locked loops are so important in any AM-coherent detection scheme, as well as FM, we shall discuss them in greater detail in Sec. 6.11.

Demodulation with a carrier of the correct phase is essential in the case of DSB *quadrature-carrier multiplexing*. This is a bandwidth-conservation scheme that enables two separate DSB signals to occupy the same band while still enabling their separation at the receiver. The central idea is that the transmitted signals are uniquely encoded by separating the phases of the two carriers (both of the same frequency) by $\pi/2$ radians. Figure 6.7-4 shows the system. The output of channel 1 is $s_1(t)$, while the output of channel 2 is $s_2(t)$. It is quite important to maintain equal gain for the two sidebands to avoid interchannel interference.

If the local carrier has the wrong phase, then we can expect interference between the channels, in addition to loss of desired signal power. Thus channel 1 would give

$$\hat{s}_1(t) = s_1(t) \underbrace{\cos \theta}_{\text{loss of power}} - \underbrace{s_2(t) \sin \theta}_{\text{interference}} \tag{6.7-7}$$

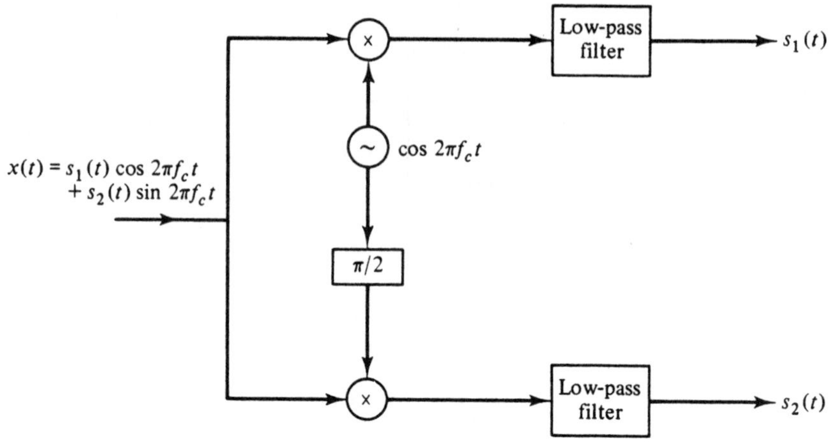

Figure 6.7-4. Detection in quadrature two-channel multiplexing.

while channel 2 would give

$$\hat{s}_2(t) = s_1(t)\sin\theta + s_2(t)\cos\theta. \tag{6.7-8}$$

Quadrature-carrier multiplexing is used in color TV (Sec. 6.15). There, synchronization pulses are transmitted to keep the local oscillator at the right frequency and phase.

6.8 Single Sideband (SSB)

Since the information in DSB is in fact duplicated in the two sidebands, the transmission bandwidth of the modulated wave can be reduced by 50% (i.e., made equal to W) by sending only one sideband rather than two. In SSB systems one of the sidebands of the DSB signal is removed, usually by direct filtering, and the remaining sideband is sent by itself or with a low-level carrier. Thus SSB is more "efficient" than DSB from the point of view of bandwidth conservation.

Removing one of the sidebands by filtering is not easy to do when there are significant signal components at low frequencies. In such a case, vestigial sideband (VSB) is preferred. However, in the case of speech, articulation tests show that components below 200 Hz are not necessary for intelligibility, and natural sounding speech does not require the transmission of components below 100 Hz. Even for orchestral music, reasonably good reproduction is obtained when no frequency components below 80 Hz are transmitted. An SSB generator is shown in Fig. 6.8-1. The type of signal spectrum that is conveniently filtered with a sideband filter is shown in Fig. 6.8-2. The absence of important low-frequency components in the spectrum is important because realizable sideband filters have finite transition regions in the frequency domain. Abrupt cutoff is not possible (Sec. 3.8).

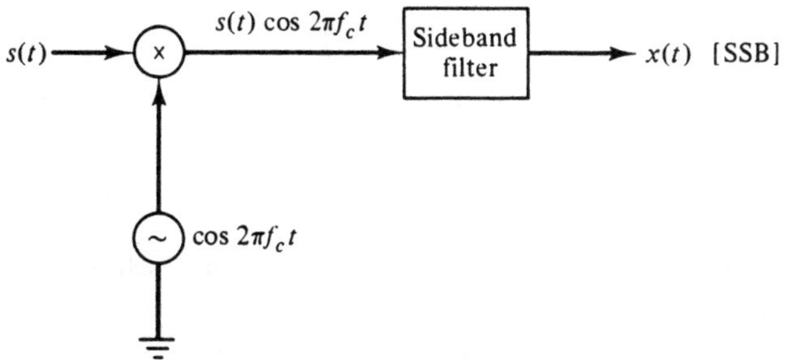

Figure 6.8-1. Generation of an SSB wave.

An alternative to sideband filtering is suggested by the fact that the SSB signal can be expressed by

$$x(t) = \frac{1}{2}[s(t)\cos 2\pi f_c t \pm \hat{s}(t)\sin 2\pi f_c t], \tag{6.8-1}$$

where $\hat{s}(t)$ is the Hilbert transform of $s(t)$ (see Sec. 2-13). This follows from the definition of SSB according to which an upper-sideband SSB wave is given by

$$\begin{aligned} x_{\text{SSB}}(t) &= \frac{1}{2}\int_{f_c}^{\infty} S(f-f_c)e^{j2\pi ft}\,df \\ &+ \frac{1}{2}\int_{-\infty}^{-f_c} S(f+f_c)e^{j2\pi ft}\,df. \end{aligned} \tag{6.8-2}$$

Now let $u = f - f_c$ in the first integral and $v = f + f_c$ in the second integral. Since u and v are dummy variables, we replace them by the common dummy variable f. The result is

$$\begin{aligned} x_{\text{SSB}}(t) &= \frac{1}{2}e^{j2\pi f_c t}\int_{0}^{\infty} S(f)e^{j2\pi ft}\,df + \frac{1}{2}e^{-j2\pi f_c t}\int_{-\infty}^{0} S(f)e^{j2\pi ft}\,df \\ &= \frac{e^{j2\pi f_c t}}{4}\int_{-\infty}^{\infty} S(f)e^{j2\pi ft}\,df + \frac{e^{-j2\pi f_c t}}{4}\int_{-\infty}^{\infty} S(f)e^{j2\pi ft}\,df \\ &+ \frac{e^{j2\pi f_c t}}{4}\int_{-\infty}^{\infty} S(f)\,\text{sgn}\,(f)e^{j2\pi ft}\,df + \frac{e^{-j2\pi f_c t}}{4}\int_{-\infty}^{\infty} S(f)\,\text{sgn}\,(-f)e^{j2\pi ft}\,df \\ &= \frac{1}{2}[s(t)\cos 2\pi f_c t - \hat{s}(t)\sin 2\pi f_c t]. \end{aligned} \tag{6.8-3}$$

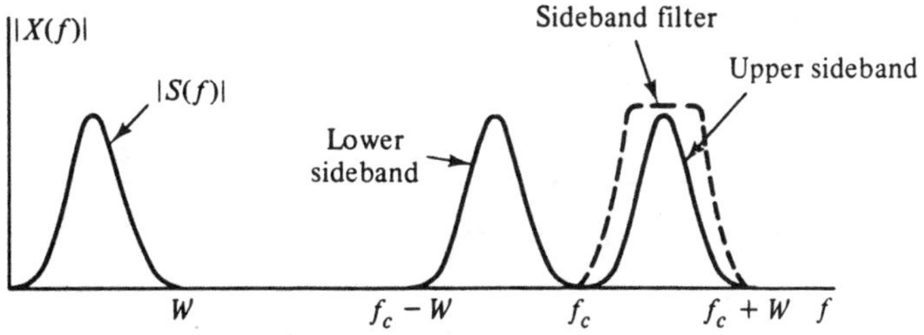

Figure 6.8-2. Sideband filtering of signals with negligible energy at low frequencies.

We use the notation introduced in Sec. 2.13,

$$\hat{s}(t) = -j \int_{-\infty}^{\infty} S(f) \operatorname{sgn}(f) e^{j2\pi ft}\, df, \tag{6.8-4}$$

and the fact that sgn $(f) = -\operatorname{sgn}(-f)$. Equation (6.8-3) is identical with Eq. (6.8-1) when the minus sign is used there.

The system suggested by this result is shown in Fig. 6.8-3. In this scheme, the central element is the quadrature filter, which introduces a 90° phase shift in every frequency component in the signal band. If we ignore the introduction of the delay elements in Fig. 6.8-3, which enable a causal system to approximate a noncausal one, we see that the signal obtained at A' is the Hilbert transform of $s(t)$. Thus at A' we have

$$\hat{s}(t) = \int_{-\infty}^{\infty} H(f) S(f) e^{j2\pi ft}\, df, \tag{6.8-5}$$

where

$$H(f) = \begin{cases} -j, & f > 0, \\ 0, & f = 0, \\ j, & f < 0. \end{cases} \tag{6.8-6}$$

At B the signal is $s(t) \cos 2\pi f_c t$, and at B' it is $\hat{s}(t) \sin 2\pi f_c t$. Hence, depending on the sign at the summing junction, the SSB wave is described by Eq. (6.8-1). The *difference* of the signals at B and B' produces the upper-sideband (USSB) wave, and the *sum* the lower-sideband (LSSB) wave.

Filters that shift the phase of a signal by 90° over a wide band of frequencies (such as the audio band) are difficult to realize, and therefore the scheme shown in Fig. 6.8-3 is not practical. It is possible, however, to design two filters whose phase

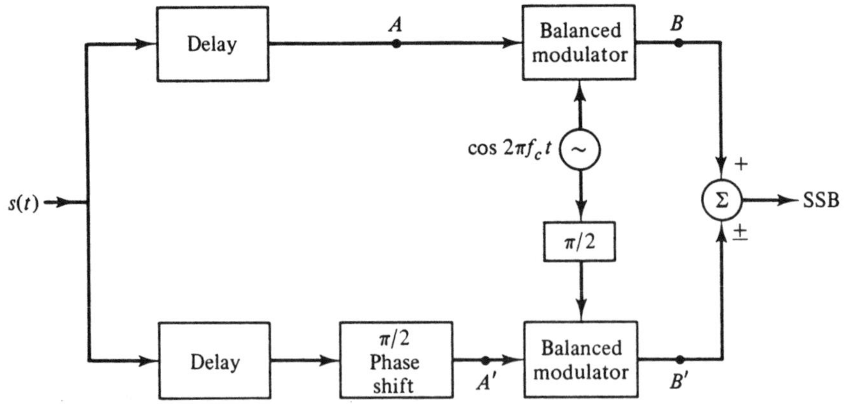

Figure 6.8-3. Generation of SSB without sideband filtering, through the use of phase shifting.

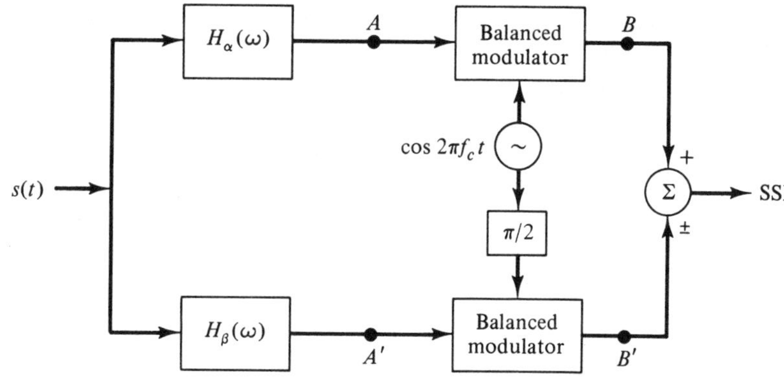

Figure 6.8-4. Practical SSB modulator using filters whose phase-shift differs by 90°.

shift differs by approximately 90° over a fairly wide band [6-10]. A practical method for SSB modulation making use of such a pair of filters is shown in Fig. 6.8-4. It is easily shown (see Prob. 6-16) that if the phase of the filters $H_\alpha(\omega)$ and $H_\beta(\omega)$ in that figure differs by 90° the output is SSB.

A technique for detecting an SSB wave is shown in Fig. 6.8-5. At A, the signal is given by [we ignore the factor of $\frac{1}{2}$ in Eq. (6.8-1)]

$$[s(t)\cos 2\pi f_c t \pm \hat{s}(t)\sin 2\pi f_c t]\cos 2\pi f_c t = \frac{s(t)}{2}(1 + \cos 4\pi f_c t) \pm \frac{\hat{s}(t)}{2}\sin 4\pi f_c t.$$

Hence if the low-pass filter does not pass the high-frequency terms, the detected signal is proportional to $s(t)$. If the carrier has the wrong phase, there will be a term proportional to $\hat{s}(t)$. For pulse-type signals, this results in serious distortion. For example, the Hilbert transform of a pulse of the form rect $[(t - T)/T]$ is shown in Fig. 2.13-5. The "horns" lead to high peak voltages. Sudden high voltages lead to flashover, energy losses by corona discharge, and destruction of insulation. For these reasons SSB would seem a doubtful choice for transmitting pulse-type signals in which abrupt signal changes occur frequently.

Since human hearing is largely insensitive to the relative phase between the spectral components, detecting an SSB wave with a demodulating signal of the wrong phase

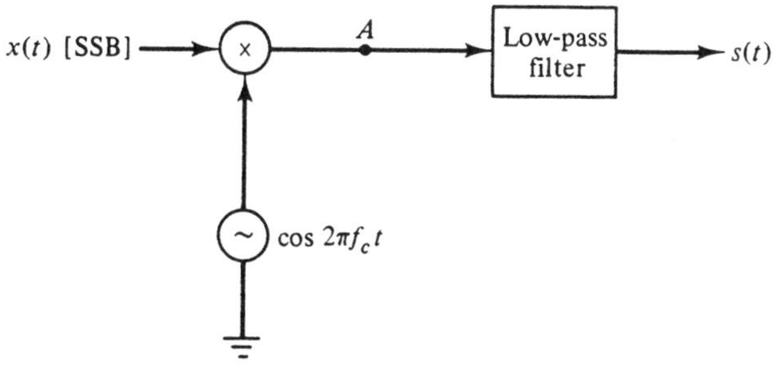

Figure 6.8-5. Homodyne detection of an SSB wave.

does not generally lead to audible distortion. But if the frequency of the demodulating signal differs from the carrier frequency, all sine-wave components in the demodulated signal are shifted away from their correct frequency by the amount of the frequency error. Thus, if the modulating signal is a pure tone of the form $s(t) = A_m \cos 2\pi f_m t$, the upper-sideband SSB wave is proportional to

$$x(t) = A_m \cos 2\pi(f_c + f_m)t. \tag{6.8-7}$$

If this wave is now detected in a homodyne system in which the local carrier is at frequency $f_c + \Delta$, where Δ is the frequency error, the demodulated signal is proportional to

$$\tilde{s}(t) = A_m \cos 2\pi(f_m - \Delta)t. \tag{6.8-8}$$

The effect of the frequency error on a composite signal such as speech or music is to destroy the harmonic relationships existing in such signals. This causes speech to take on a characteristic ''Donald Duck'' quality. Music demodulated with an incorrect carrier frequency acquires a bell or gonglike quality that is generally highly objectionable (unless one wants to create special effects). Thus, even though small frequency errors in the demodulating signal will not destroy intelligibility (as is the case with DSB), accurate tuning of SSB demodulators is important. An automatic SSB demodulator that makes use of the destruction of the harmonic relationships in speech by carrier offset error has recently been suggested in the literature [6-11].

An SSB wave can be rewritten as an envelope on a carrier with a time-dependent phase $\theta(t)$ added to the linear phase $2\pi f_c t$. Thus we can write

$$\begin{aligned} x_{\text{SSB}}(t) &= s(t) \cos 2\pi f_c t \pm \hat{s}(t) \sin 2\pi f_c t \\ &= R(t) \cos [2\pi f_c t + \theta(t)], \end{aligned} \tag{6.8-9}$$

where

$$R(t) = \sqrt{s^2(t) + \hat{s}^2(t)}, \tag{6.8-10a}$$

$$\theta(t) = \mp \tan^{-1} \frac{\hat{s}(t)}{s(t)}. \tag{6.8-10b}$$

Equation (6.8-9) indicates that, if we detect an SSB signal with an envelope detector, serious nonlinear distortion occurs. If $s(t)$ is a simple rect $(\cdot)$ pulse, the detected signal will have the form shown in Fig. 6.8-6. In a real system, the peaks may be large but

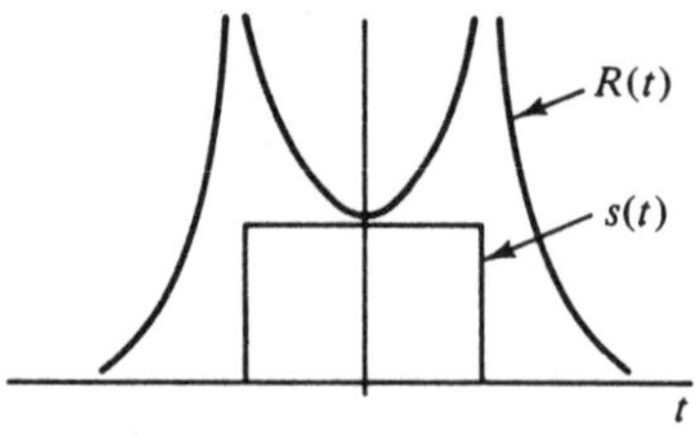

Figure 6.8-6. Result of detecting SSB with an envelope detector.

finite, and the signal will not extend to $t = \pm\infty$. The peaks are called the *horns*, and the rise and fall at the beginning and end of the signal are called *smears*.

If an upper-sideband, tone-modulated SSB wave is envelope-detected, the detected signal is proportional to

$$R(t) = \sqrt{\cos^2 2\pi f_m t + \sin^2 2\pi f_m t}$$
$$= 1.$$

Hence no tone, at any frequency, is heard. The addition of a large carrier signal, in phase with the transmitted carrier, can reduce the distortion and enable SSB detection with an envelope detector. Thus

$$x_{SSB}(t) + A \cos 2\pi f_c t = R(t) \cos [2\pi f_c t + \theta(t)], \tag{6.8-11}$$

where $R(t)$ is now given by

$$R(t) = \sqrt{[s(t) + A]^2 + \hat{s}^2(t)} \tag{6.8-12}$$
$$\simeq A + s(t), \tag{6.8-13}$$

if the higher-order terms in the power series expansion of the square root are neglected. The constant bias A can be removed, and $s(t)$ can be recovered.

6.9 VESTIGIAL SIDEBAND (VSB)

In the case of speech and music, the generation of SSB by sideband filtering is feasible because of the absence of unimportant message components at low frequencies. Such is not the case in TV, where, as we shall see in Sec. 6.12, there are important components at very low frequencies. To maintain bandwidth conservation while eliminating the need for the critical filtering of a sideband, a compromise technique known as VSB is used. In VSB, most of one sideband is passed along with a vestige of the other sideband. VSB is widely used in TV, facsimile, and certain data-signal-transmission systems. The typical bandwidth required to transmit a VSB wave is about 1.25 that of SSB.

The VSB-modulated signal is produced by passing the DSB signal through a filter that removes part of one of the sidebands. Such a system and the spectra at various points in it are shown in Fig. 6.9-1. The filter transfer function shown in this figure removes part of the lower sideband, but the same result would be obtained if the upper sideband were partially removed. The transmission channel is not shown in Fig. 6.9-1(a), but it would come anywhere between points B and C (i.e., between the DSB modulator and the homodyne demodulator).

The main point to notice in this system is the spectrum at D (i.e., after demodulation but before the low-pass filter). Observe that the spectrum near zero frequency is the sum of the left and right partial spectra that have been shifted to $f_c = 0$ by the demodulation

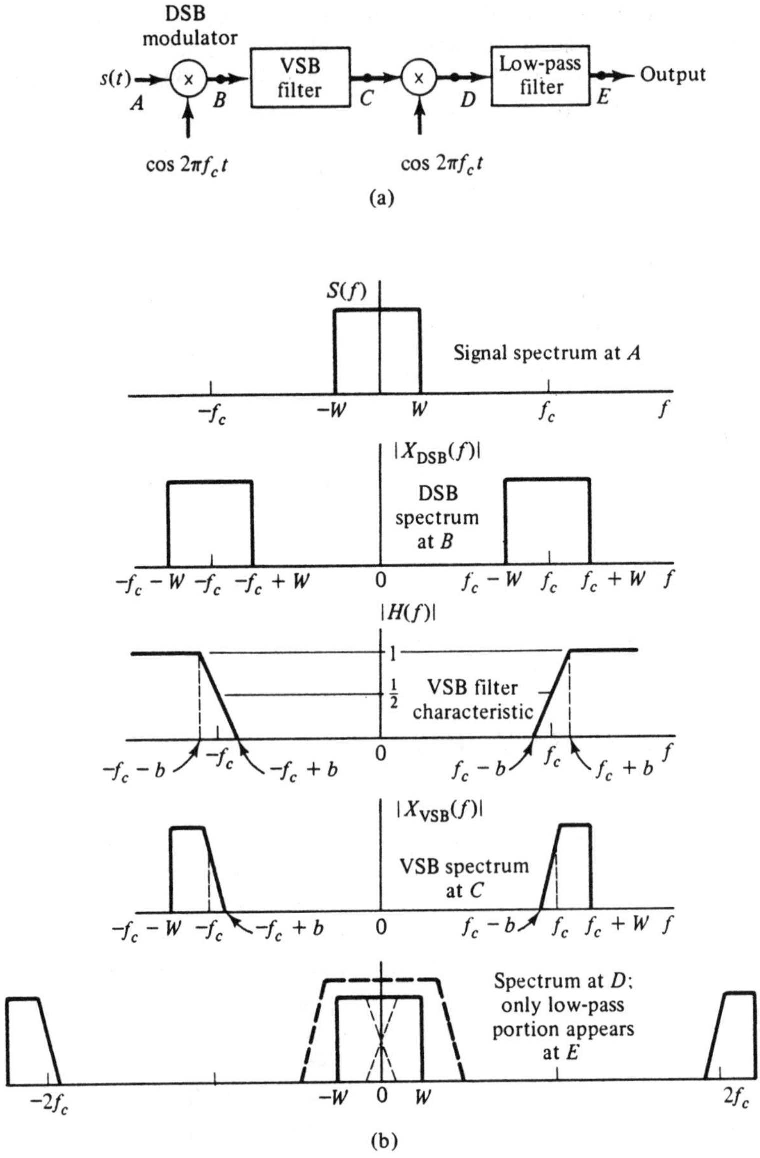

Figure 6.9-1. (a) VSB system. (b) Spectra at various points in a VSB system.

process. The sum spectrum should have the same shape as the original signal spectrum at point A if there is to be no distortion. Analytically we have

Spectrum at A is proportional to $S(f)$,
Spectrum at B is proportional to $\frac{1}{2}[S(f+f_c)+S(f-f_c)]$,
Spectrum at C is proportional to $H(f)[S(f+f_c)+S(f-f_c)]$,
Spectrum at D is proportional to $\frac{1}{2}H(f+f_c)[S(f+2f_c)+S(f)]$
$+\frac{1}{2}H(f-f_c)[S(f-2f_c)+S(f)]$.

The central lobe of the spectrum at D should be $S(f)$; hence we have

$$S(f)=\tfrac{1}{2}[H(f+f_c)\,S(f)+H(f-f_c)S(f)]$$

or

$$H(f+f_c)+H(f-f_c)=2. \tag{6.9-1}$$

The number 2 on the right in Eq. (6.9-1) is arbitrary and can be replaced by any constant; in fact, Eq. (6.9-1) is often written in the form

$$H(f+f_c)+H(f-f_c)=2H(f_c). \tag{6.9-2}$$

The most important property of $H(f)$ is that it has odd symmetry in the neighborhood of $f=f_c$ and the 50% response level at f_c. The filter can be anywhere in the channel between the modulator and demodulator; in fact, part of it can be physically in the transmitter while the remainder is in the receiver as long as the relation in Eq. (6.9-2) is satisfied by the overall transfer function. Also, if the VSB filter is not exactly symmetrical as required in Eq. (6.9-2), further equalization of the spectrum is in principle possible in the low-pass filter (between D and E).

The analytical form of the VSB signal is easily obtained by inverse-Fourier-transforming the spectrum at C. This results in

$$\begin{aligned} x(t) &= \int_{-\infty}^{\infty} H(f)[S(f+f_c)+S(f-f_c)]e^{j2\pi ft}\,df \\ &= \int_{-\infty}^{\infty} H(f-f_c)S(f)e^{j2\pi(f-f_c)t}\,df + \int_{-\infty}^{\infty} H(f+f_c)S(f)e^{j2\pi(f+f_c)t}\,df. \end{aligned} \tag{6.9-3}$$

The symmetry requirement on $H(f)$ can be expressed in the form

$$\begin{aligned} H(f-f_c) &= H(f_c)-Y(f) \\ H(f+f_c) &= H(f_c)+Y(f)' \end{aligned} \qquad (-W\le f\le W), \tag{6.9-4}$$

where $Y(f)$ is a function having odd symmetry about $f=0$; that is,

$$Y(-f)=-Y(f) \tag{6.9-5}$$

(see Fig. 6.9-2). If we use these results in Eq. (6.9-3) and also, for convenience, let $H(f_c)=\frac{1}{2}$, we get

$$x(t)=s(t)\cos 2\pi f_ct-q(t)\sin 2\pi f_ct, \tag{6.9-6}$$

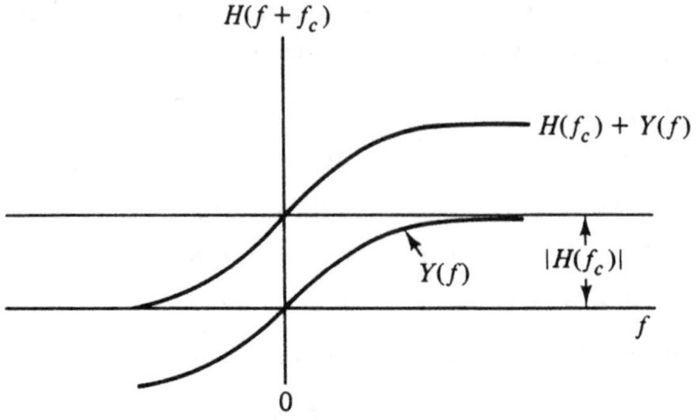

Figure 6.9-2. Transmittance of the VSB filter (upper sideband) showing the constant past $H(f_c)$ and the past having odd symmetry.

where

$$q(t) = -2j \int_{-\infty}^{\infty} Y(f)S(f)e^{j2\pi ft}\, df. \tag{6.9-7}$$

It goes without saying that the bandwidth W of $s(t)$ satisfies $W < f_c$. If the function $Y(f)$ in Fig. 6.9-2 is replaced by $-Y(f)$, the lower sideband is retained; hence, changing the minus sign in Eq. (6.9-6) to plus results in VSB with lower sideband retained. As the transition region becomes smaller, $Y(f)$ approaches sgn (f), $q(t) \to \hat{s}(t)$, and the signal approaches SSB. If $Y(f) \to 0$, $\hat{q}(t) \to 0$, and the signal approaches DSB. For true VSB operation, the frequency transition region of $H(f)$ should satisfy $0 < 2b < 2W$ (see Fig. 6.9-1). Equation (6.9-6) implies that a VSB or SSB signal can be generated from a DSB signal by injecting a quadrature component obtained from the modulating signal by a filtering operation. The quadrature component partially or totally cancels one of the sidebands. Several other ways of looking at VSB are given in [6-12].

As in the case of SSB, VSB should be detected by homodyne means. To facilitate detection at the receiver, a small pilot carrier can be transmitted along with the VSB

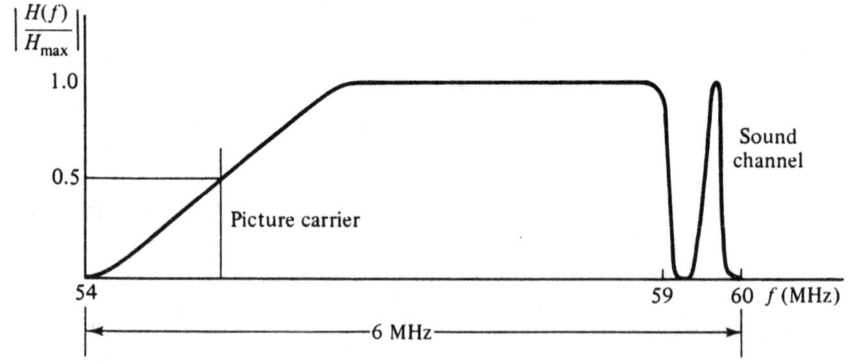

Figure 6.9-3. Frequency response of a TV receiver tuned to the 54- to 60-MHz channel.

signal to synchronize a local oscillator.† A phase-locked loop can be used for this purpose. In TV systems, the transmitted signal is not quite VSB because there is no attempt to rigidly control the transition region in the spectrum of the signal. The actual VSB shaping is done at the receiver with a VSB filter. A typical frequency response for a VSB filter for TV is shown in Fig. 6.9-3.

6.10 FREQUENCY-DIVISION MULTIPLEXING (FDM)

FDM, like TDM discussed in Sec. 4.4, is a scheme for transmitting, simultaneously, several messages over the same transmission link while enabling separation of the messages at the receiver. FDM is used in telephone systems, telemetry, stereo broadcasting, and communication networks. The Bell Telephone System multiplexes as many as 3600 4-kHz separate messages on its L4 carrier system. In a sense, the broadcast signals in a given geographical area form an FDM system because all of them occupy the same transmission medium at the same time and yet can be individually detected in the home receiver.

Figure 6.10-1 shows a block diagram of an FDM-AM system for several messages $\{s_i'(t),\ i = 1, \ldots, n\}$. The low-pass filter cuts off at W Hz, thereby ensuring that no message occupies a band wider than W Hz. The frequency-truncated message, $s_i(t)$, modulates a carrier signal at frequency f_i. The initial carrier signal is called the subcarrier and serves to uniquely encode the different messages $\{s_i(t)\}$.

For AM, the spacing between adjacent subcarriers, $\Delta f \equiv f_i - f_{i-1}$, must be at least as great as $2W$. The composite signal $s_b(t)$, consisting of the sum of all the modulated signals $\{x_i(t)\}$, is still treated as baseband because it has not yet modulated a high-frequency carrier wave. For AM encoding, the composite signal is

$$s_b(t) = \sum_{i=1}^{N} x_i(t) = \sum_{i=1}^{N} A_i[1 + m_i s_i(t)] \cos(2\pi f_i t + \theta_i), \tag{6.10-1}$$

and its spectrum is shown in Fig. 6.10-2.

Modulation of a carrier by $s_b(t)$ may not be required in certain situations. Even when $s_b(t)$ modulates a carrier, the nature of the modulation may be completely different from the initial modulation of the signals $\{s_i(t)\}$. For example, the generation of $s_b(t)$ might be done with SSB encoding, as is often the case since baseband bandwidth is conserved and adjacent carrier frequencies need be separated only by W Hz. The final modulation of the carrier might be by wideband frequency modulation (FM) to take advantage of the noise-immunity properties of this type of modulation. The designation AM/FM FDM is used to describe a frequency-division multiplexing scheme whereby

† VSB can be detected by envelope detection if a large amount of carrier is added. In fact, this is how it is normally done in commercial TV.

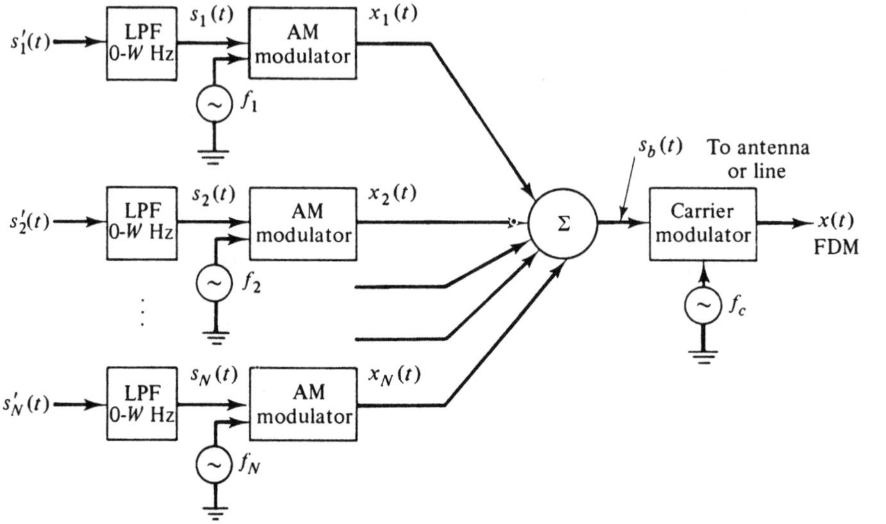

(a)

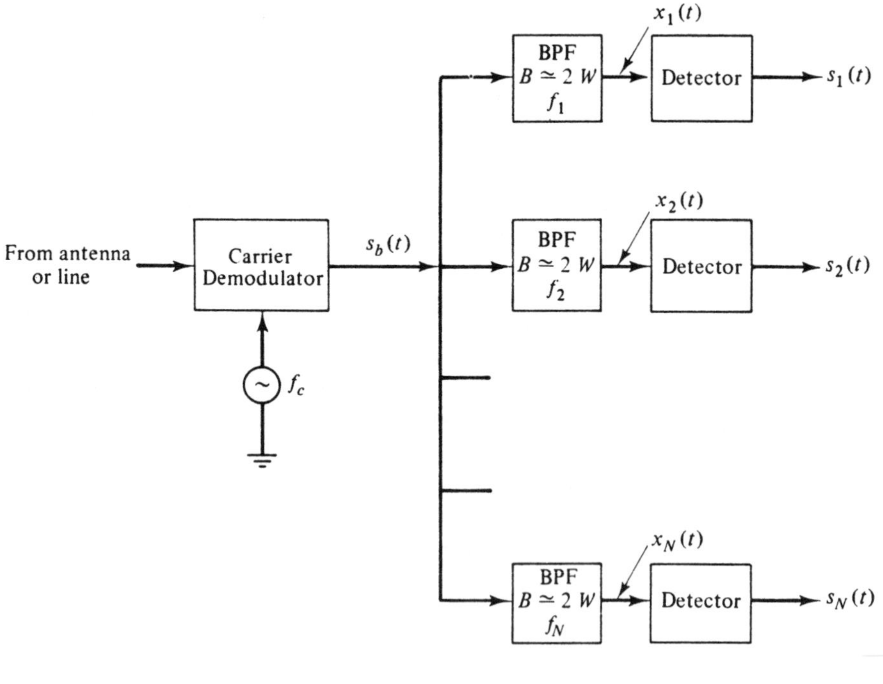

(b)

Figure 6.10-1. FDM system. (a) Transmitter. (b) Receiver.

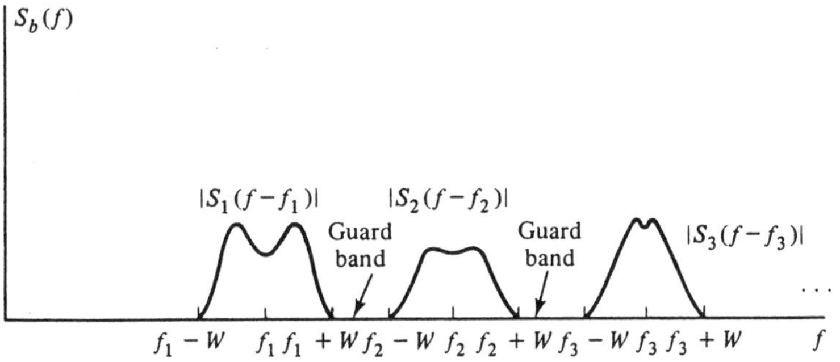

Figure 6.10-2. Spectrum of the baseband signal $s_b(t)$ in FDM.

$s_b(t)$ is generated by AM, while $x(t)$ is generated by FM. An example of an FM/FM FDM system for space telemetry is given in Chapter 7.

In FDM systems, as well as many other communication systems, *cross talk* is an important problem and results when a signal $s_i(t)$ modulates a carrier at frequency f_j assigned to a different signal $s_j(t)$. This phenomenon results from nonlinearities in the system, and great care, including the use of feedback techniques, is taken to reduce nonlinear effects.† Another problem is the spilling over of significant spectral components from one band into a band reserved for another carrier signal. This can be caused by nonlinearities also. To reduce this kind of distortion, guard bands, as shown in Fig. 6.10-2, are used to separate adjacent-channel spectra.

Compatible Stereo

An important example of an FDM system is to be found in the generation of stereophonic signals that are compatible with monophonic receivers. Let $s_L(t)$ and $s_R(t)$ denote the left and right message signals, respectively. The generation of the composite baseband $s_b(t)$ is shown in simplified form in Fig. 6.10-3(a).

The sum signal, $s_L(t) + s_R(t)$, can be separated by a low-pass filter and is available for monophonic reception. For stereophonic systems, $s_L(t) + s_R(t)$ is separated from the DSB signal by a low-pass filter. The DSB signal is separated from the monophonic signal by a 23- to 53-kHz band-pass filter and coherently detected by a locally generated (from the 19-kHz pilot) carrier at 38 kHz. The two signals are then added and subtracted to furnish $s_L(t)$ and $s_R(t)$ separately. Figure 6.10-4 is a block diagram of the system.‡

† Another important source of cross talk is coupling between physically separated but adjacent, in the electromagnetic sense, transmission media.

‡ We shall see that in FM the detected noise spectrum increases *quadratically* with frequency. This means that, for example, in DSB-FM FDM stereo, the $s_L - s_R$ signal sits in a band from 23 to 53 kHz where the noise is more pronounced than in the band from 0 to 15 kHz. This accounts for the "noisyness" of stereo compared with mono reception.

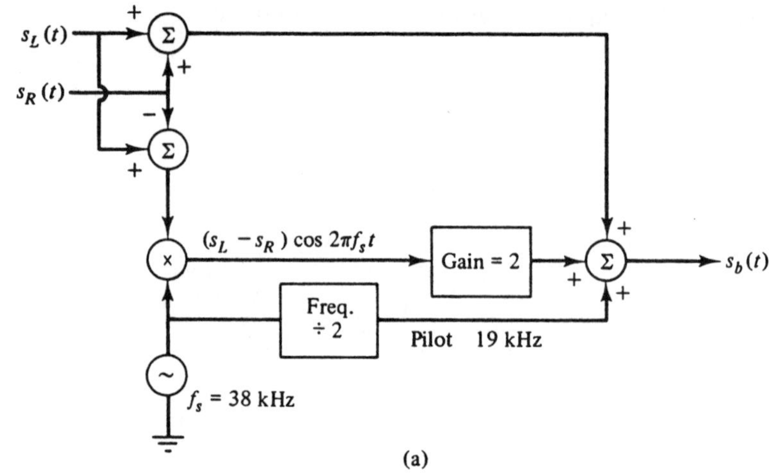

(a)

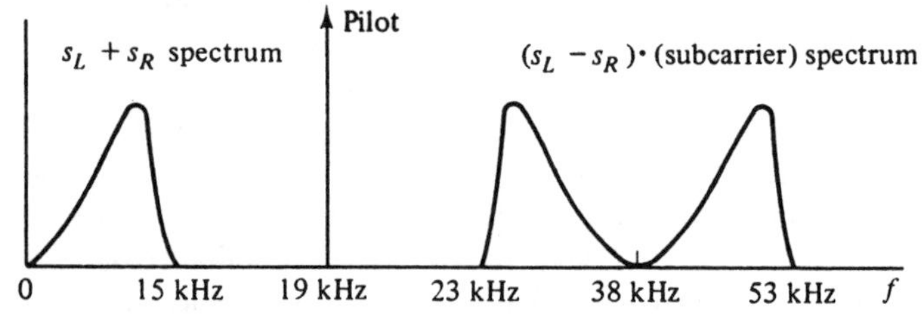

(b)

Figure 6.10-3. (a) Generation of stereo baseband signal. (b) Spectrum of baseband signal.

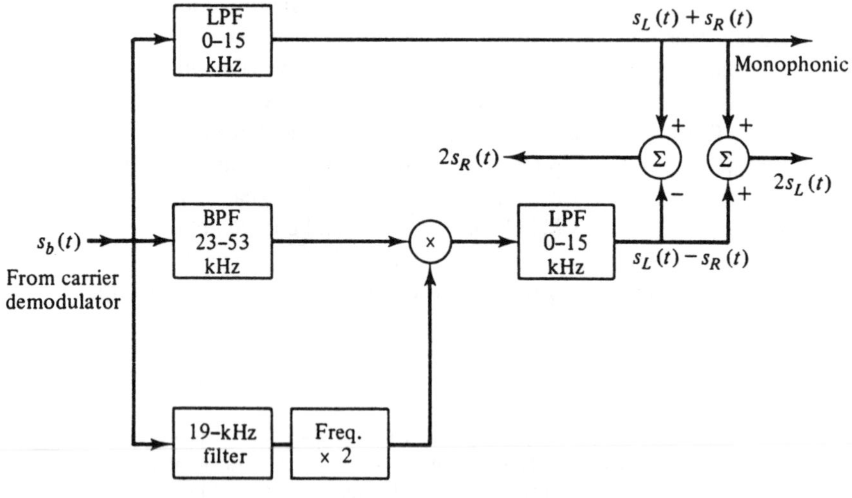

Figure 6.10-4. Detection of monophonic-compatible stereophonic signals.

*6.11 PHASE-LOCKED LOOP

In the homodyne systems discussed in earlier sections, the local oscillator at the receiver was required to track the frequency and phase of the received carrier. This tracking can be done with a phase-locked loop (PLL), which, in simplest terms, consists of a multiplier, a linear filter/amplifier, and a voltage-controlled oscillator (VCO). Fundamentally, the phase-locked loop is an extremely narrowband filter that extracts a carrier-frequency component from the modulated signal and removes the sideband information. For example, a PLL would extract the component $A \cos \omega_c t$ from $A[1 + ms(t)] \cos \omega_c t$. Our discussion of such a loop will closely follow the analysis furnished by Viterbi [6-13], who discusses many of the sophisticated problems associated with the operation of a PLL.

Figure 6.11-1 shows the basic PLL system. We shall assume that all initial conditions are zero and that $e(t) = 0$ when $x(t) = 0$. The quiescent frequency of the VCO will be denoted by ω_0 (radians per second). Use of ω_0 instead of f_0 eliminates the need for writing 2π. The signals at various points in the system are denoted

$$x(t) = \sqrt{2}\, A \sin \theta(t), \tag{6.11-1}$$

$$y(t) = \sqrt{2}\, K_1 \cos \theta'(t), \tag{6.11-2}$$

$$z(t) = x(t)y(t), \tag{6.11-3}$$

$$e(t) = \int_0^t z(\tau)h(t - \tau)\, d\tau, \tag{6.11-4}$$

and the *instantaneous* radian frequency, ω, of the VCO output signal is

$$\omega \equiv \dot{\theta}'(t) = \omega_0 + K_2 e(t). \tag{6.11-5}$$

The constant K_2 has units of radians per second per volt. The signal $z(t)$ is

$$\begin{aligned} z(t) &= x(t)y(t) \\ &= 2AK_1 \sin \theta(t) \cos \theta'(t) \\ &= K_1 A\{\sin [\theta(t) - \theta'(t)] + \sin [\theta(t) + \theta'(t)]\}, \end{aligned} \tag{6.11-6}$$

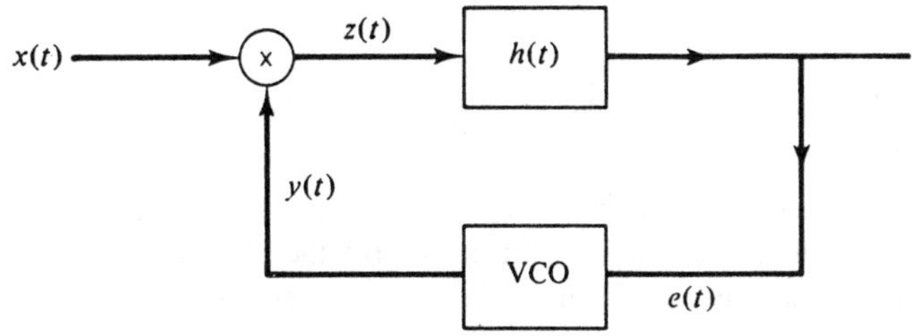

Figure 6.11-1. Phase-locked loop.

so that $e(t)$ is given by, from Eq. (6.11-4),

$$e(t) = \int_0^t K_1 A \sin[\theta(\tau) - \theta'(\tau)]h(t-\tau)\,d\tau + \int_0^t K_1 A \sin[\theta(\tau) + \theta'(\tau)]h(t-\tau)\,d\tau. \tag{6.11-7}$$

Note that $\theta(t)$ and $\theta'(t)$ are close to $\omega_0 t$. The filter $h(t)$ is essentially low pass so that the second integral, which involves a term of frequency near $2\omega_0$, is zero. The basic equation characterizing the PLL is obtained from Eq. (6.11-5); that is,

$$\dot{\theta}'(t) = \omega_0 + AK_2K_1 \int_0^t \sin[\theta(\tau) - \theta'(\tau)]h(t-\tau)\,d\tau. \tag{6.11-8}$$

To reduce Eq. (6.11-8) to a more convenient form, we make the following definitions:

$$\begin{aligned} K &\equiv K_1K_2 = \text{loop gain}, \\ \phi(t) &\equiv \theta(t) - \theta'(t) = \text{instantaneous phase error}, \\ \xi(t) &\equiv \theta(t) - \omega_0 t = \text{deviation of } \theta(t) \text{ about a linear phase } \omega_0 t, \\ \xi'(t) &= \theta'(t) - \omega_0 t = \text{deviation of } \theta'(t) \text{ about a linear phase } \omega_0 t. \end{aligned} \tag{6.11-9}$$

Equation (6.11-8) can now be written as

$$\dot{\phi}(t) = \dot{\xi}(t) - AK \int_0^t h(t-\tau)\sin\phi(\tau)\,d\tau. \tag{6.11-10}$$

This is a nonlinear integrodifferential equation that is quite difficult to solve. Viterbi discusses the general solution [6-13]. We shall only discuss the noise-free linear approximation to Eq. (6.11-10).

The Linear Approximation under Noise-Free Conditions

If the loop is near phase lock, then by definition $\phi(\tau) \simeq 0$, $\sin\phi(\tau) \simeq \phi(\tau)$, and Eq. (6.11-10) reduces to

$$\dot{\phi}(t) = \dot{\xi}(t) - AK \int_0^t h(t-\tau)\phi(\tau)\,d\tau. \tag{6.11-11}$$

This expression approximates the PLL by a simple linear feedback loop. Such loops are studied in elementary feedback control theory and are most conveniently analyzed by use of Laplace transform methods. We assume that the reader is familiar with the general procedure. We define the Laplace transforms of $\phi(t)$, $\xi(t)$, $h(t)$, and $\xi'(t)$ by

$$\tilde{\phi}(s) = \mathscr{L}[\phi(t)] \equiv \int_0^\infty \phi(t)e^{-st}\,dt,$$

$$\tilde{\xi}(s) = \mathscr{L}[\xi(t)],$$

$$H(s) = \mathscr{L}[h(t)],$$

$$\tilde{\xi}'(s) = \mathscr{L}[\xi'(t)].$$

The Laplace transform of Eq. (6.11-11) [for $\dot{\phi}(0) = \dot{\xi}(0) = 0$] is

$$s\tilde{\phi}(s) + AKH(s)\tilde{\phi}(s) = s\tilde{\xi}(s) \tag{6.11-12}$$

or

$$\tilde{\phi}(s) = \frac{1}{1 + AKH(s)/s}\tilde{\xi}(s) \equiv T(s)\tilde{\xi}(s). \tag{6.11-13}$$

A block diagram of the linear model in terms of Laplace transform operators is shown in Fig. 6.11-2.

Let us now assume that $H(s)$ is the transfer function associated with an ideal integrator in parallel with a direct connection. Then $H(s) = 1 + Bs^{-1}$, where B is a gain constant. The loop will be stable if the gain constants are adjusted so that all the poles of $T(s)$ are in the left-half s plane. Let $x(t)$ be a constant-frequency sinusoid $2\sqrt{A}\cos(\omega t + \theta_0)$ applied at $t = 0$. Then

$$\begin{aligned} \xi(t) &= (\omega - \omega_0)t + \theta_0, \\ \tilde{\xi}(s) &= \frac{\omega - \omega_0}{s^2} + \frac{\theta_0}{s}, \end{aligned} \tag{6.11-14}$$

and use of Eq. (6.11-13) gives

$$\tilde{\phi}(s) = \frac{s^2}{s^2 + AKs + ABK}\left(\frac{\omega - \omega_0}{s^2} + \frac{\theta_0}{s}\right). \tag{6.11-15}$$

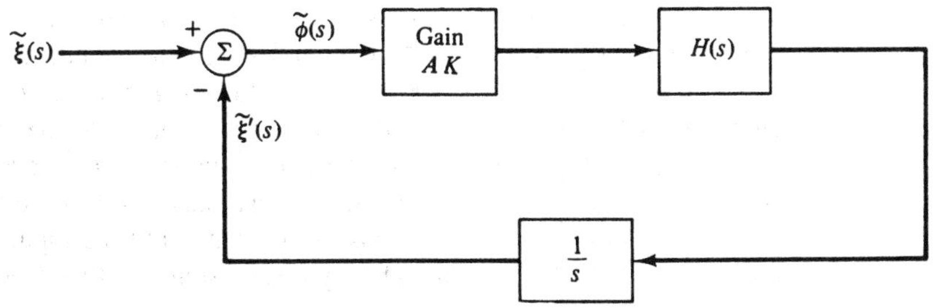

Figure 6.11-2. Block diagram of the linear model of the PLL.

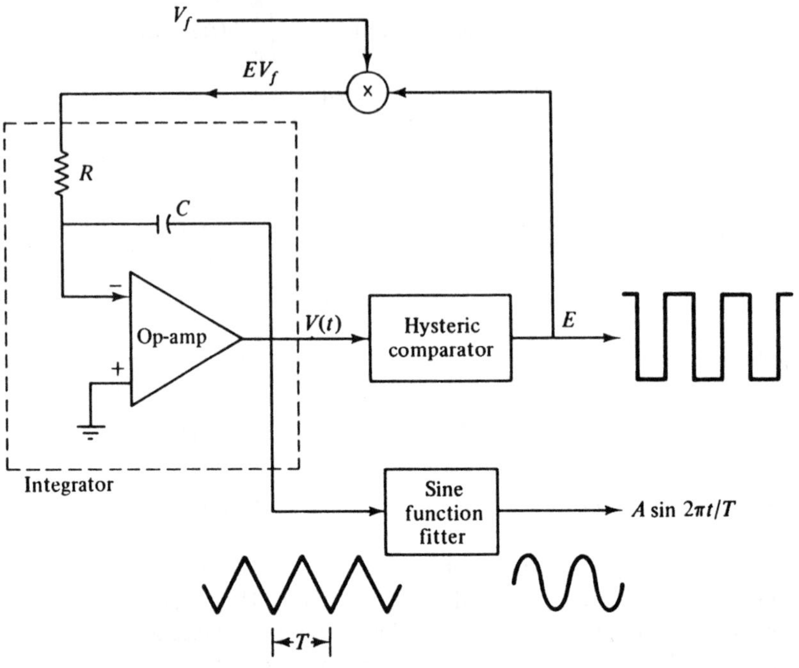

Figure 6.11-3. Simplified VCO. (Adapted from Sheingold [6-3], by permission.)

The final value theorem then gives

$$\lim_{t\to\infty} \phi(t) = \lim_{s\to 0} s\tilde{\phi}(s) = 0. \tag{6.11-16}$$

Hence the error in tracking the instantaneous phase goes to zero, and the loop eventually locks on the correct frequency and phase.

*The Voltage-Controlled Oscillator

We have seen that the VCO is an important component of a phase-locked loop. It has many other uses (e.g., in spectral analyzers and FM stereo demodulators). A simple VCO circuit is shown in Fig. 6.11-3 ([6-3], p. 73). Let us first ignore the multiplier in the feedback path and assume a direct feedback of the signal E to the integrator. The output of the hysteric comparator is one of the two stable states E^+ or E^-. It switches to E^+ when the input exceeds V^+ and remains in that state until the input falls below V^-, whereupon it switches to E^-. It remains in that state until the input again exceeds V^+. Many devices, including operational amplifiers using positive feedback, can be made to develop hysteresis.†

† For example, a Schmitt trigger is a circuit that has this property [6-14, p. 389].

Assume that the output has just gone into the E^+ state. The output of the integrator, starting from the time when $V = V^+$, is

$$\begin{aligned} V(t) &= -\frac{1}{RC}\int_0^t E^+\,d\tau + V^+ \\ &= V^+ - \frac{E^+}{RC}t, \qquad E^+ > 0. \end{aligned} \tag{6.11-17}$$

When $V(t) = V^-$, E switches to E^-. The elapsed time is therefore

$$\Delta t_1 = RC\frac{V^+ - V^-}{E^+}. \tag{6.11-18}$$

Assume now that $E^- < 0$. When E switches to E^-, the output of the integrator rises linearly until $V = V^+$. The elapsed time is

$$\Delta t_2 = RC\frac{V^+ - V^-}{|E^-|}. \tag{6.11-19}$$

The period T is $\Delta t_1 + \Delta t_2$, and the frequency is given by

$$f = \frac{1}{T} = \frac{E^+}{[1 - (E^+/E^-)](V^+ - V^-)RC}. \tag{6.11-20}$$

If $V^+ = -V^- = V_0$ and $E^+ = -E^- = E_0$, then

$$f = \frac{E_0/V_0}{4RC}. \tag{6.11-21}$$

Returning to Fig. 6.11-3, we see that the feedback signal is proportional to V_f, the frequency-control voltage. The frequency is then a linear function of V_f given by

$$f = \frac{V_f E_0/V_0}{4RC}. \tag{6.11-22}$$

The output of the integrator is a symmetrical triangular wave. It can be filtered or shaped into a sine wave by a function fitter. The same can be done with the comparator output.

6.12 TELEVISION (TV)

Television is the transmission of visual images by electrical signals. The signals can be transmitted by radiation, coaxial cables, telephone wires, or esoteric means such as laser beams. The image can be defined by a brightness function, $B(x, y, t)$, which depends on three variables: the x and y coordinates of the scene and time. An electrical signal is a function of only one variable—time. For an electrical signal to absorb all the variations in $B(\cdot)$, the image must be sequentially *scanned*. The process of scanning is central in TV and requires synchronization signals to keep track of what portion of the electrical signal is associated with what portion of the image. Each image is broken down in a predetermined systematic sequence of waveforms and can be reconstructed at the receiver to produce a faithful replica of the original scene. This is in contrast to motion picture photography in which the whole scene is recorded at once as an image on photographic film. The former is an example of serial processing, while the latter might be called parallel processing. A TV system is shown in Fig. 6.12-1. The essential components are the camera; the deflection, synchronization (sync), and blanking generator; the video, sync, and blanking mixer/amplifier; and the carrier modulator. The TV camera is a device that contains optics capable of producing an image on a target consisting of a large number of photosensitive elements. An electron beam, produced by an electron gun and deflected by the signals from the deflection generator, scans the charge pattern on the photosensitive surface and produces a signal current proportional to the charge sensed at that instant; this is called the video signal. During retrace, the electron beam is turned off by the blanking signals. The process of scanning produces a one-to-one correspondence between the (x, y) position of the scanned image brightness and the temporal values of the video signal. There are several different types of TV camera devices, operating on variations of this principle, and they are variously called orthicon, image orthicon, vidicon, plumbicon, and so on. The signal is combined with sync and blanking pulses in a mixer/amplifier and then the composite video signal is RF-modulated with a carrier in the transmitter terminal.

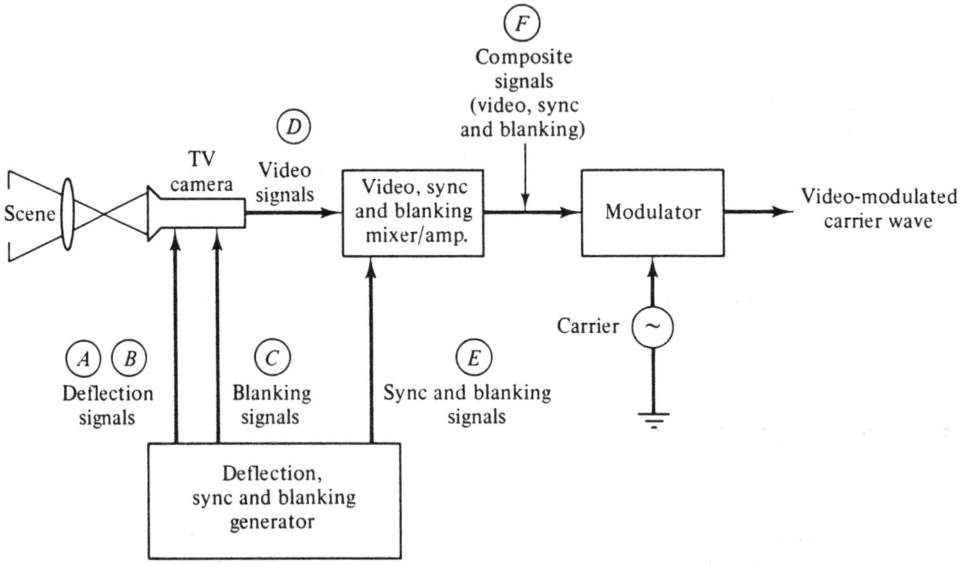

Note: circled letters refer to waveforms in Fig. 6.12-3(b)

Figure 6.12-1. Generation of TV signals.

Camera Tubes: The Image Orthicon

There are a number of television camera tubes in use, and they broadly fall into two classes: photoemissive and photoconductive. The image orthicon [6-7] is a photoemissive tube that combines high sensitivity with wide dynamic range. Figure 6.12-2 shows the structure of the image orthicon. The principle of operation is easily understood. The optical system generates a focused image on the *photocathode*, which is photosensitive material that emits electrons from its rear surface in proportion to the amount of light impinging on its front surface. The electron image, that is, the electron emission distribution $E(x, y)$, is proportional to the incident image brightness $B(x, y)$.

The emitted electrons are attracted to the target mosaic (TM), which is several volts positive with respect to the photocathode. The impinging primary electrons give rise to secondary emission at the TM, and these secondary electrons are collected by a fine mesh screen located adjacent to the TM. A positive charge residue is left on the TM that is proportional to the brightness of the optical image. This charge configuration is essentially static during the frame interval.

The scanning electron beam furnishes the electrons to neutralize the positive charges on the rear face of the TM. If a point is highly positive (corresponding to a high-intensity image point), many electrons will be collected and the return-beam current will be low. If a point is only slightly positive, the return-beam current will be high. The variations imposed on the current constitute the *video signal*.

The electrons absorbed on the rear surface of the TM diffuse throughout the surface and neutralize the existing positive charge. The target is thus prepared for the next frame.

The return-beam current is amplified by an electron multiplier target structure surrounding the electron gun. The amplified-beam current leaves the tube through an output electrode and generates a voltage signal across a coupling resistor.

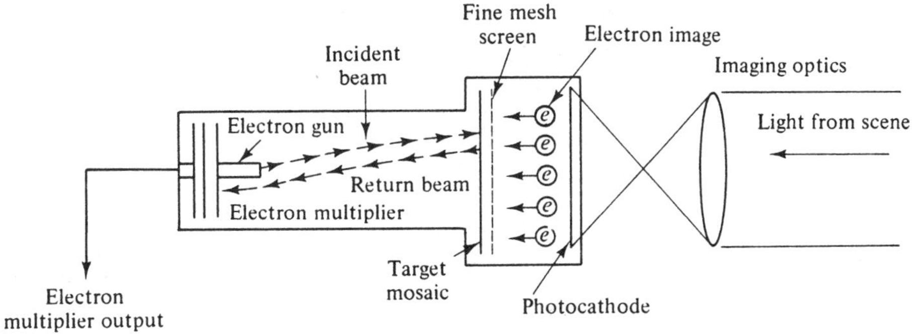

Figure 6.12-2. Image orthicon TV camera.

Waveforms

The scanning of the photosensitive target in the camera tube by the electron gun is, in many respects, analogous to the reading of a printed page by a human reader. The target is scanned in line-by-line fashion, from top to bottom, as shown in Fig. 6.12-3(a). The scanning pattern is commonly called a *raster* scan. A complete frame (i.e., a complete single picture of the image field) consists of a 525-line raster. However, in standard TV broadcasting practice, this is decomposed into two 262.5-line rasters that are *interlaced*. We have ignored the interlacing in Fig. 6.12-3(a). Figure 6.12-3(b) shows the important waveforms in a TV system [6-15]. The actual TV synchronization signals are considerably more complicated than the waves shown here, which have been simplified for ease of presentation. Hanson shows actual TV broadcast synchronization signals in Chapter 7 of Reference [6-16]. The first horizontal line in Fig. 6.12-3(a), from 1 to 2, is generated by a sawtooth pulse of the X-axis deflection waveform shown in curve A of Fig. 6.12-3(b). This signal, also called the fast deflection, is controlled by the timing element in the deflection, sync, and blanking generator shown in Fig. 6.12-1. The timing element produces a set of horizontal and vertical sync pulses (curve H) that mark the beginning of the horizontal and vertical deflection waveforms. When the sweep 1 to 2 is over, the beam is retraced to position 3. This retrace or *flyback* happens in a very brief time, corresponding to 2 to 3 in curve A. During the vertical (4 to 1) retrace, blanking signals (curve C) turn off the electron-beam signal to avoid spurious and meaningless video outputs. The combined sync/blanking waveform is shown in curve E. The vertical deflection signal is much slower than the horizontal deflection signal, there being 262.5 horizontal deflection pulses per vertical deflection pulse. Vertical deflection pulses are shown in curve B.

Assume that the camera focuses on a blackboard containing a chalk drawing of

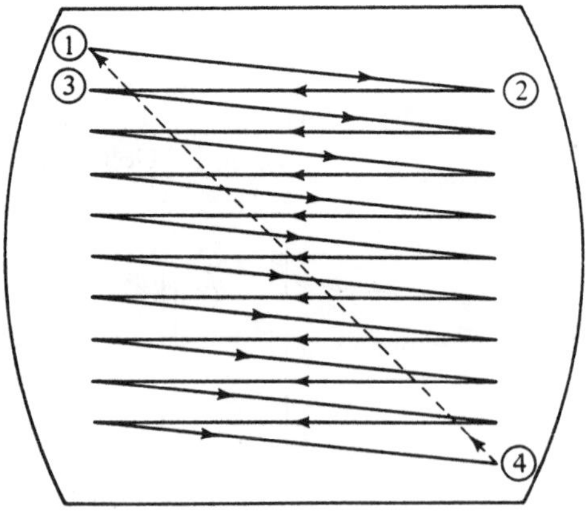

The numbers in (a) are associated with the corresponding numbers in (b).

(a)

Figure 6.12-3. (a) TV raster screen.

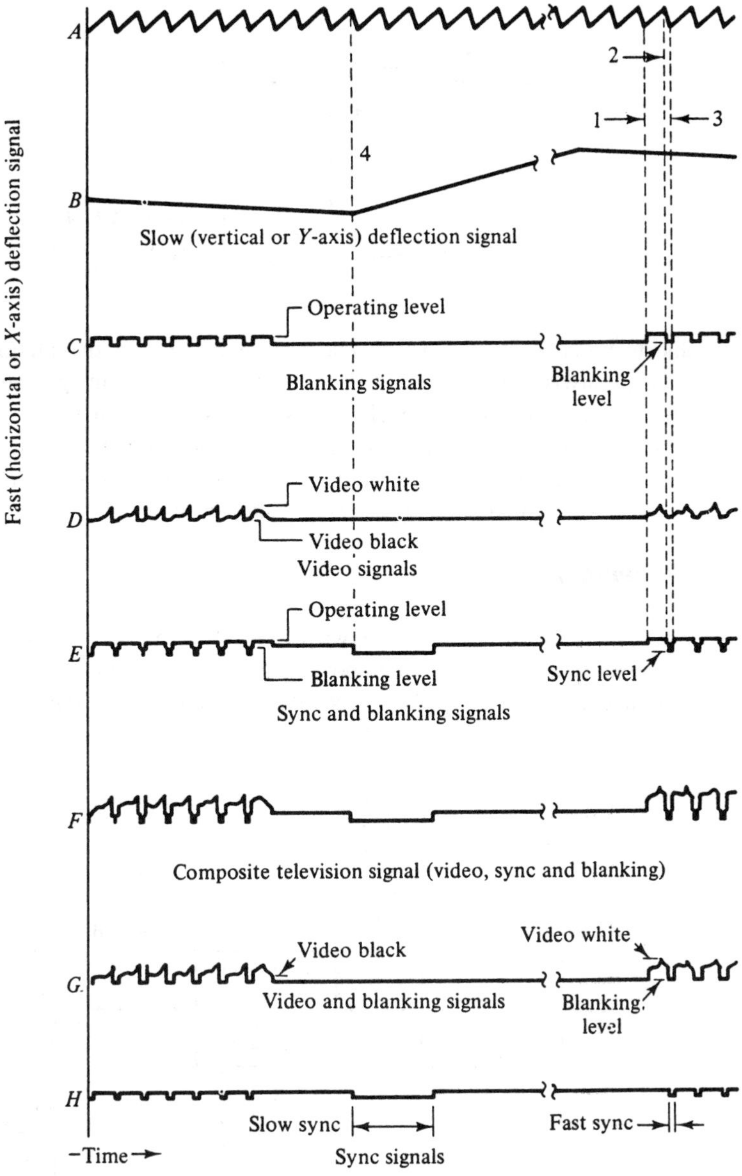

(b)

Figure 6.12-3 (continued). (b) TV waveforms. (Courtesy of Hughes Aircraft Co.)

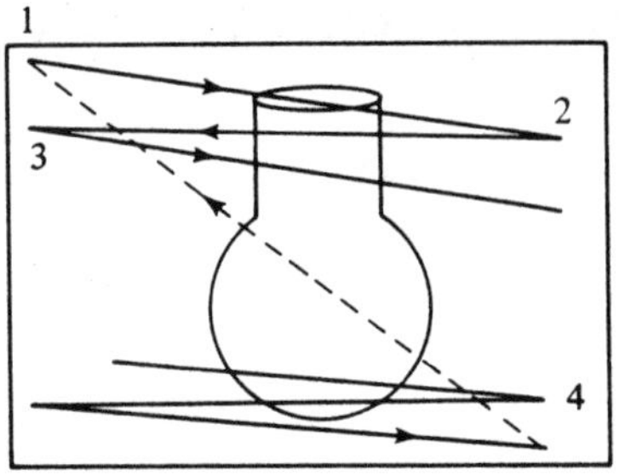

Figure 6.12-4. Scanning the charge pattern on the photosensitive surface in the camera tube.

the object shown in Fig. 6.12-4. The charge pattern on the photosensitive element would be pointwise proportional to the brightness of the image, and the line-by-line scan would generate a series of elemental signals as shown in curve D of Fig. 6.12-3(b). Each elemental signal is associated with a single horizontal sweep, and the ensemble of these signals represents, for a single picture, an orderly dissection of the image into a waveform from which we can reconstruct the scene.

The Receiver

A TV receiver block diagram is shown in Fig. 6.12-5. The RF signal is demodulated at the receiver terminal, and the composite signal (i.e., video, sync, and blanking signal) is transmitted to a video amplifier and sync separator. The sync signals are removed

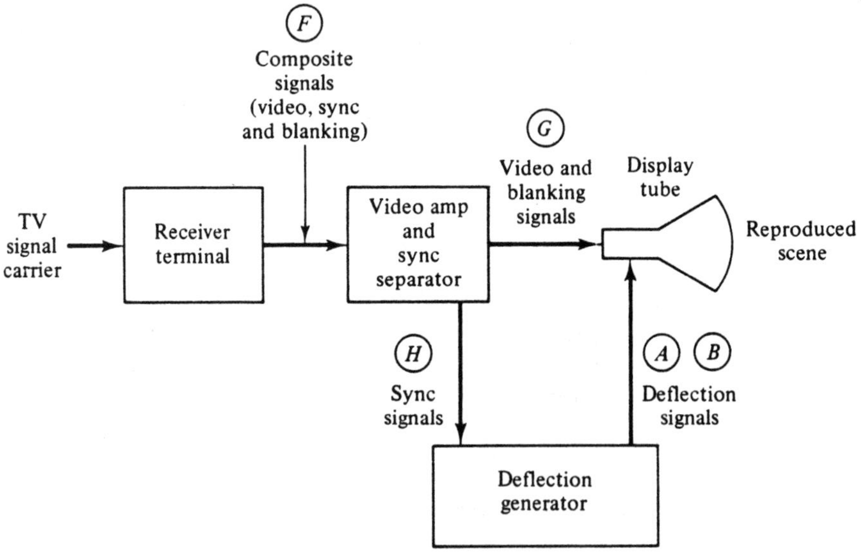

Note: Circled letters refer to waveforms in Fig. 6.12.3(b)

Figure 6.12-5. Reception of TV signals.

from the composite video and used to synchronize a deflection generator, while the video plus blanking signal modulates the electron beam. Without the x-y deflection signals from the deflection generator, the electron beam would produce a single flickering spot of light on the phosphor screen of the display tube. The deflection signals move the electron beam in a systematic manner to reconstitute the image.

The composite video signal is shown in curve F in Fig. 6.12-3(b). This is the modulating signal for the RF waveform. The video and blanking signals that drive the receiver display tube are shown in curve G. In our discussion we have associated a large positive signal with video white and a small positive signal with video black. In actual broadcast TV, maximum signal denotes a video black, while video white goes with low-level signals. The FCC standard TV wave is shown in Fig. 6.12-6.

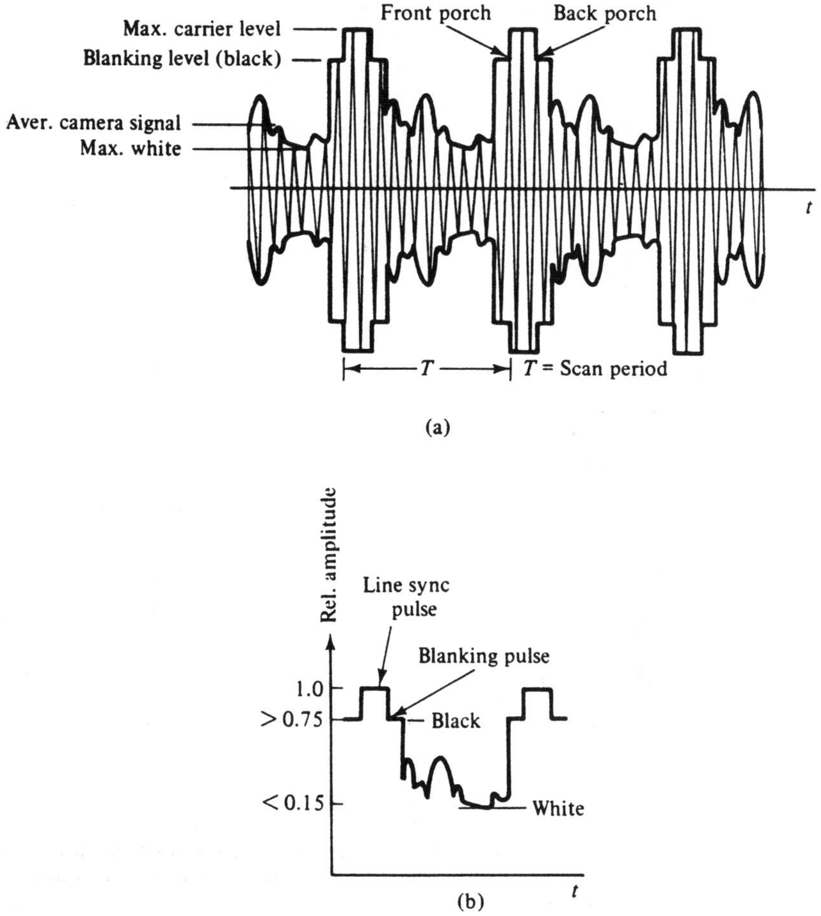

Figure 6.12-6. TV wave as specified by the FCC. (a) Carrier wave. (b) Details of composite video.

The removal of the sync pulses from the detected video can be done with a clipping circuit, as shown in Fig. 6.12-7. The clipping should be done at the blanking level to ensure that the camera signal does not affect synchronization. The separation of the horizontal sync pulses from the vertical sync pulses cannot be done by amplitude discrimination; instead, separation by waveform content is used. The horizontal sync pulses are enhanced by a differentiating network, while the vertical sync pulses, having a much lower repetition frequency, are enhanced by an integrating circuit. A simplified block diagram of a system to separate the two sync signals is shown in Fig. 6.12-8. The actual circuits are more involved [6-16, Chap. 7].

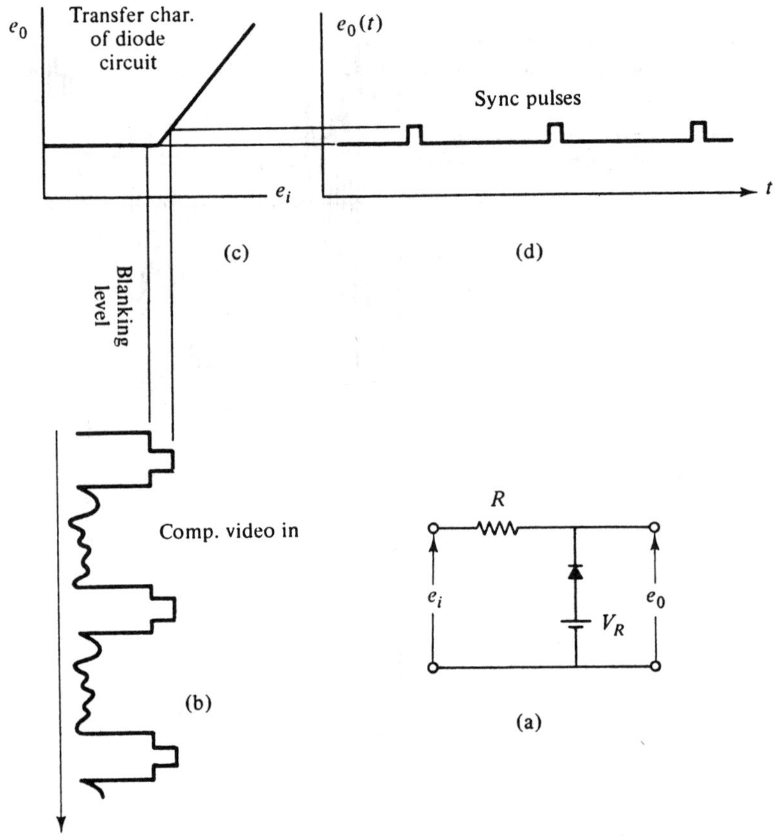

Figure 6.12-7. Removal of sync pulses at receiver for synchronization. (a) Diode clipper. (b) Composite video input signal. (c) Transfer characteristic of diode circuit. (d) Separated sync pulses.

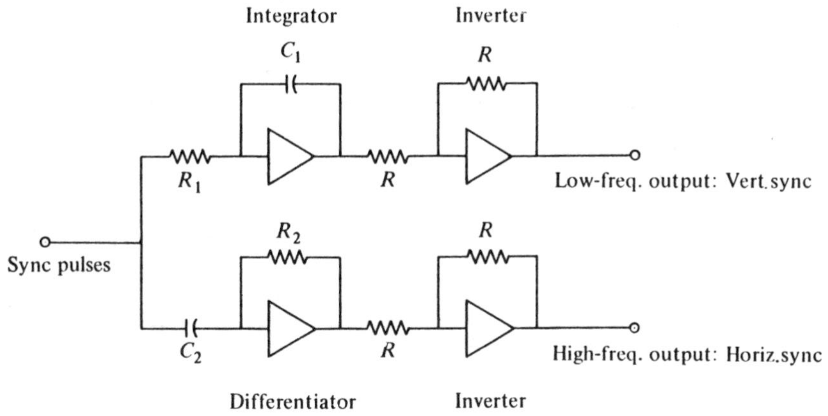

Figure 6.12-8. System for separating the vertical from the horizontal sync pulses.

6.13 BANDWIDTH CONSIDERATIONS FOR TV

In U.S. broadcast TV, a single TV frame consists of 525 lines of which about 490 are actually active in synthesizing the image. The inactive lines travel from the bottom to the top during the vertical retrace. The number 525 was chosen because it is composed of simple odd factors ($525 = 3 \times 5 \times 5 \times 7$) that simplify the task of frequency division. To reduce flicker in the reproduced image, each full frame actually consists of two interlaced 262.5-line fields. Sixty fields are transmitted per second, two fields to a frame. The line-scanning frequency is 525×30 Hz $= 15.75$ kHz, and the image (frame) repetition frequency is 30 Hz. The aspect ratio (ratio of raster width to raster height) is $\frac{4}{3}$.

It is found that 60 fields (or 30 frames) per second produces a flicker-free effect on the human eye because of the phenomenon of persistence of vision. The number of picture elements scanned per second is roughly equal to $30 \times 525 \times 525 \simeq 8.3 \times 10^6$. To calculate the maximum required frequency, it is customary to assume that the picture elements are arranged as alternate black and white squares along the scanning line. There are therefore two elements per cycle of the wave. The first sinusoidal harmonic of this square wave has frequency $\simeq 8.3 \times 10^6/2 = 4.15$ MHz. The actual maximum frequency in the standard video signal is around 4.2 MHz. The spectrum of the transmitted TV signal is shown in Fig. 6.13-1. One sideband is transmitted plus 25% of the other. The VSB shaping, as explained in Sec. 6-9, is done at the receiver by a VSB filter. The response spectrum of the RF receiver is shown in Fig. 6.13-2.

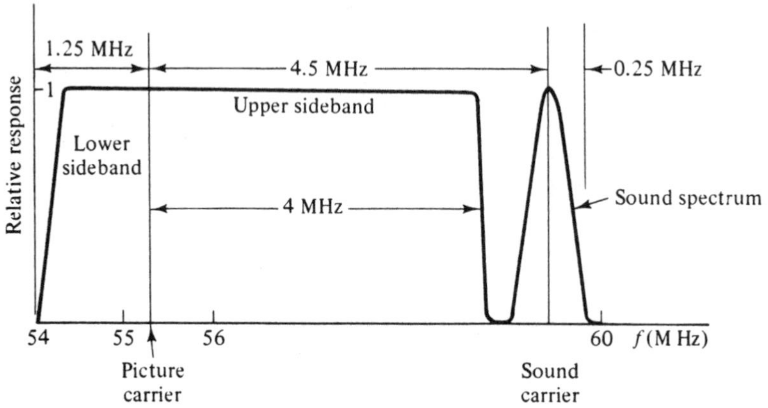

Figure 6.13-1. Spectrum of TV modulated wave (typical channel).

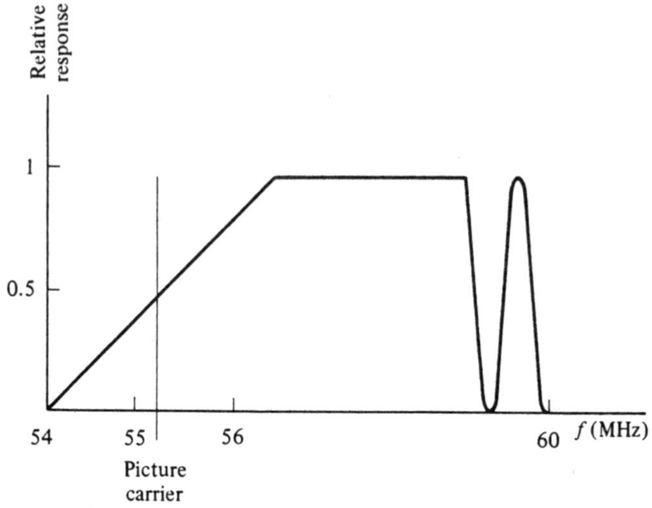

Figure 6.13-2. Frequency response of VSB filter at receiver.

6.14 STRUCTURE OF THE SPECTRUM OF A TV WAVE

A black/white TV frame is essentially a two-dimensional signal and can be analyzed by using a two-dimensional extension of the Fourier series. (Cf. Sec. 2.14 ff.) Following Bennett [6-17, p. 111], we model the scanning as a doubly infinite array of image fields. If the retrace time in a raster scan is ignored, the electrical output associated with the model (Fig. 6.14-1) is the same as in actual scanning.

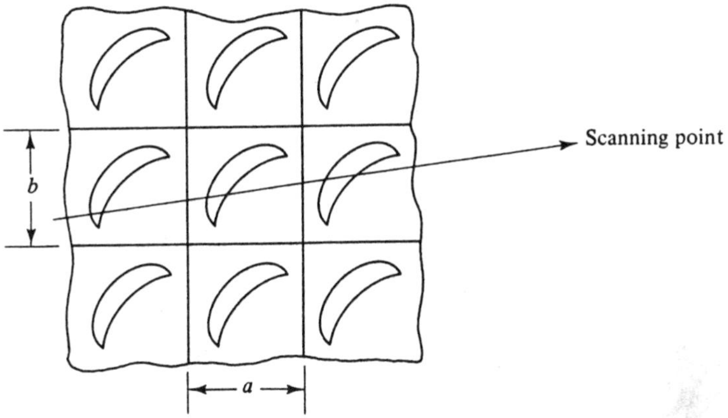

Figure 6.14-1. Model for scanning process using a doubly infinite array of image fields.

We let $B(x, y)$ denote the brightness of the still (i.e., nonchanging) image. Then

$$B(x, y) = \sum_{n=-\infty}^{\infty} \sum_{m=-\infty}^{\infty} b_{mn} \exp\left[j2\pi\left(\frac{m}{a}x + \frac{n}{b}y\right)\right], \tag{6.14-1}$$

where

$$b_{mn} = \frac{1}{ab}\int_{-a/2}^{a/2} dx \int_{-b/2}^{b/2} dy\, B(x, y) \exp\left[-j2\pi\left(\frac{m}{a}x + \frac{n}{b}y\right)\right]. \tag{6.14-2}$$

The positions x and y are related to the time variable t according to

$$x = V_x t, \qquad y = V_y t, \tag{6.14-3}$$

where V_x and V_y are the scan velocities in the x and y directions, respectively. The electrical signal $e(t)$ is proportional to $B(V_x t, V_y t)$, so we can write (K being a constant of proportionality)

$$e(t) = K \sum_{n=-\infty}^{\infty} \sum_{m=-\infty}^{\infty} b_{mn} \exp\left[j2\pi\left(m\frac{V_x}{a}t + n\frac{V_y}{b}t\right)\right]. \tag{6.14-4}$$

The quantities a/V_x and b/V_y are recognized as the times required for a complete scan in the x and y directions, respectively. Hence V_x/a represents the line-scanning frequency (i.e., 15.75 kHz) and V_y/b represents the image repetition frequency (i.e., 30 Hz). Thus

$$e(t) = K \sum_{n=-\infty}^{\infty} \sum_{m=-\infty}^{\infty} b_{mn} \exp\left[j2\pi(m \times 15.75 \times 10^3 + n \times 30)t\right]. \tag{6.14-5}$$

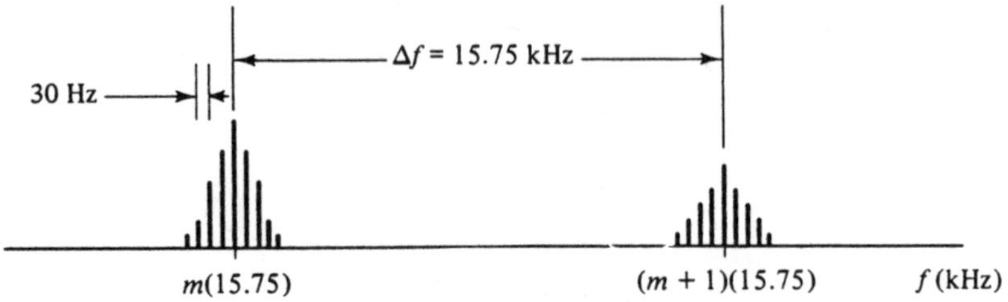

Figure 6.14-2. Spectrum of the scanned video signal.

The line-scanning harmonics are spaced 15.75 kHz apart, and clustered about each is an array of satellites spaced 30 Hz apart. The spectrum is shown in Fig. 6.14-2. When there is motion in the scene being televised, there will be some smearing of the lines in the spectrum. In practice, wide gaps appear between the line-scanning harmonics. The existence of these gaps allows for the transmission of color TV signals in the same 6-MHz band allocated to black/white TV, thus enabling compatible color TV.

6.15 COLOR TELEVISION

Compatible color TV (CCTV) is a system of color TV, developed in 1954, that fits into the existing monochrome channel assignments and permits either color or black/white reception depending on the type of receiver available.

It is well known that arbitrary color images can be reproduced by using only three primary colors: red, green, and blue. This principle is used not only in color TV but also in color photography and color printing. The three primary colors, red (R), green (G), and blue (B), when added together in equal amounts produce the colors illustrated in Fig. 6.15-1. Other colors can be produced by varying the intensities of the primary colors.

A color TV camera contains separating optics that resolve the light from the

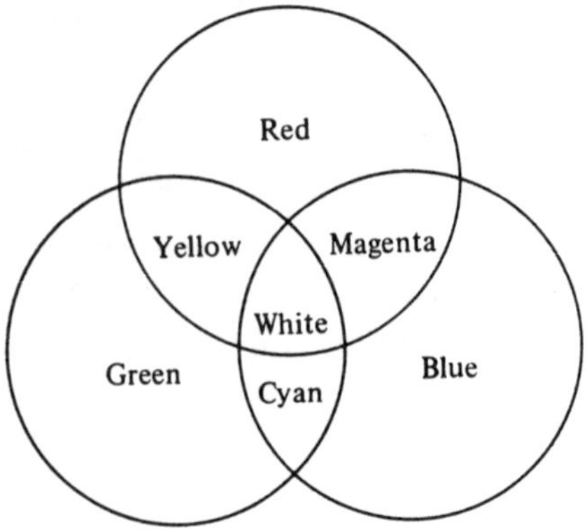

Figure 6.15-1. Additive color mixing.

scene into the three primary colors R, G, B. A set of three camera tubes then produces three video signals, one for each primary color, $s_R(t)$, $s_G(t)$, and $s_B(t)$. These three signals could be transmitted individually and used subsequently to reconstruct a color image at the receiver. However, this method is not used because additional bandwidth would be required, and the system would not be compatible with existing monochrome receivers.

Before discussing what is actually transmitted, we stress that any three linear, independent equations involving $s_R(t)$, $s_G(t)$, and $s_B(t)$ can be used to compute the individual signals $s_R(t)$, $s_G(t)$, and $s_B(t)$. For example, we could transmit the three independent signals $f_1(t)$, $f_2(t)$, and $f_3(t)$ given by

$$\begin{aligned} f_1(t) &= \alpha_{11}s_R(t) + \alpha_{12}s_G(t) + \alpha_{13}s_B(t), \\ f_2(t) &= \alpha_{21}s_R(t) + \alpha_{22}s_G(t) + \alpha_{23}s_B(t), \\ f_3(t) &= \alpha_{31}s_R(t) + \alpha_{32}s_G(t) + \alpha_{33}s_B(t), \end{aligned} \tag{6.15-1}$$

This system of equations is conveniently written in matrix form as

$$\mathbf{f} = \mathbf{As}, \tag{6.15-2}$$

where $\mathbf{f} = (f_1, f_2, f_3)^T$, $\mathbf{A} = [\alpha_{ij}]$ is a nonsingular 3×3 matrix, $\mathbf{s} = (s_R, s_G, s_B)^T$, and T denotes transpose. A receiver need only matrix the signals $\mathbf{f}$ according to $\mathbf{s} = \mathbf{A}^{-1}\mathbf{f}$ to recover the primary color signals. This is in fact what is actually done.

A monochrome receiver (i.e., one that shows black, white, and gray) requires a brightness or *luminance* signal, derived from the three color signals, that closely matches the luminance of a conventional, monochrome, video signal. It turns out that the required mix is given by[†]

$$f_Y(t) = 0.30s_R(t) + 0.59s_G(t) + 0.11s_B(t). \tag{6.15-3}$$

The luminance produced by this signal is "compatible" with the black/white image produced by a conventional black/white TV system. Brightness levels are additive. Thus if $s_R(t) = s_G(t) = s_B(t) = 1$ (which corresponds to maximum or saturated signals), $f_Y(t) = 1$, which corresponds to the sensation of white. If blue is absent, the monochrome brightness is 0.89. However, if green is absent, the brightness falls to 0.41.

We still require two additional, independent signals for transmission of color information. These are the *chrominance* signals given by

$$\begin{aligned} f_I(t) &= 0.60s_R(t) - 0.28s_G(t) - 0.32s_B(t), \\ f_Q(t) &= 0.21s_R(t) - 0.52s_G(t) + 0.31s_B(t). \end{aligned} \tag{6.15-4}$$

In all the above, we have assumed that $|s_R(t)| \le 1$, $|s_G(t)| \le 1$, $|s_B(t) \le 1$.

The chrominance is that portion of the composite color signal that represents the color information in the televised scene. The reasons the chrominance signals have the particular forms given in Eq. (6.15-4) has to do with efficient utilization of certain

[†] This is an empirical result.

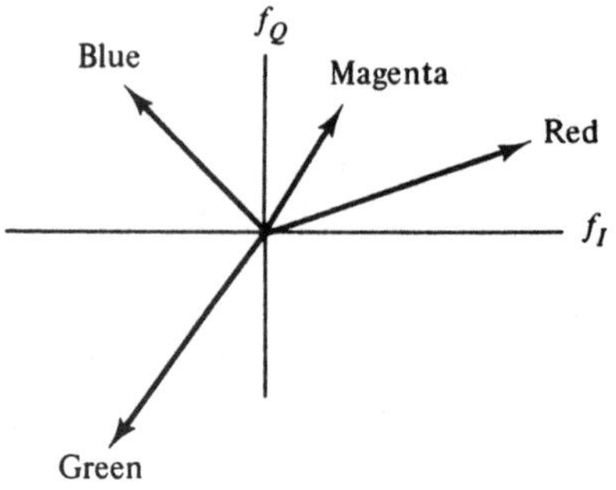

Figure 6.15-2. $f_I - f_Q$ color plane and the three saturated primaries. Also shown in a nonsaturated magenta.

features of human color vision. The specifics are discussed in [6-16, p. 254]. However, in principle, other independent equations could serve the purpose as well.

The terms *hue* and *saturation* refer to important concepts in the accurate reconstruction of color. Hue is the attribute of colors that permits them to be classed as red, yellow, green, or blue or as intermediate between any contiguous pair of these colors. Saturation refers to chromatic purity, the freedom from dilution with white. It is a measure of the degree of difference from a gray having the same brightness. The purest color is said to be 100% saturated. A strongly saturated blue, for example, is often erroneously called a "dark" blue. A nonsaturated blue might be called light or pastel blue. However, hue and saturation are separate notions. Both hue and saturation are easily encoded by the *color vector* defined by

$$\mathbf{f}_c(t) = (f_I(t), f_Q(t)). \qquad 6.15\text{-}5)$$

When the color vector is applied to the three saturated primary colors, we obtain

$$\begin{aligned} [\mathbf{f}_c]_{\text{red}} &= (0.6, 0.21) = 0.63 \angle 19°, \\ [\mathbf{f}_c]_{\text{green}} &= (-0.28, -0.52) = 0.59 \angle 242°, \\ [\mathbf{f}_c]_{\text{blue}} &= (-0.32, 0.31) = 0.45 \angle 136°. \end{aligned} \qquad (6.15\text{-}6)$$

These are shown in Fig. 6.15-2. The magnitude of $\mathbf{f}_c(t)$, $|\mathbf{f}_c(t)|$, is proportional to the saturation, and its angle, arg $[\mathbf{f}_c(t)]$, is a measure of the hue. For example, a partially saturated magenta (red-blue) may have $s_G = 0$, $s_R = s_B = \frac{1}{2}$. The color vector is then

$$\mathbf{f}_c = (0.14, 0.26) = 0.3 \angle 62°.$$

Subcarrier Determination and Frequency Interlacing

The luminance signal (also known as the Y signal) produces the monochrome image and is therefore assigned the entire 4.2-MHz bandwidth. The I-Q chrominance signals are separated from the luminance by modulating a color subcarrier whose frequency f_{cc} is halfway between the 227th and 228th harmonic of the line-scanning frequency, f_s. If f_s were 15.75 kHz, this would give

$$f_{cc} = 227.5 \times 15.75 = 3.583125 \text{ MHz}. \qquad (6.15\text{-}7)$$

However, the actual value of f_{cc} is slightly lower (3.579545 MHz to be exact) in order to avoid an objectionable beat frequency with the sound carrier, which lies at 4.5 MHz above the picture carrier ([6-13)], p. 246). To avoid the beat note and still maintain the relation $f_{cc} = 227.5f_s$, the line-scanning frequency is changed to 15.73426 kHz, and the field repetition frequency is reduced to 59.94 Hz. These slight changes require no changes in existing monochrome circuitry.

The choice of $f_{cc} \simeq 3.6$ MHz represents a reasonable compromise between two factors. On the one hand, the color subcarrier should be as high as possible to minimize its effect on monochrome images. On the other hand, the satisfactory transmission of the chrominance information requires a band of at least 0.6 MHz *above* the subcarrier. The upper limit of the video passband is 4.2 MHz; hence, $4.2 - 0.6 = 3.6 \text{ MHz} \simeq f_{cc}$.

So far we have considered the coarse determination of f_{cc}. The reason f_{cc} was chosen an odd multiple of half the line frequency will now be explained. In Sec. 6-14, it was pointed out that the spectrum of the monochrome signal consists of harmonics of f_s about which are clustered the harmonics of the frame rate (30 Hz). Motion in the televised scene slightly modifies the 30-Hz spacing, but this is of little consequence in our analysis. There are wide gaps between the harmonics of f_s where in fact additional signals can be added. The location of the color subcarrier midway between the 227th and 228th harmonics of f_s takes advantage of this fact since the line-scanning harmonics associated with the chrominance signals are located at

$$f_s\left(\frac{455}{2} \pm n\right), \qquad n = 0, 1, 2, \ldots . \tag{6.15-8}$$

The important point to note here is that these harmonics always fall in the relatively empty zones between the harmonics of the luminance signal. This is a reason compatible TV is feasible. The process of placing the chrominance carrier and its sidebands in spectral zones where there is little or no luminance information is known as frequency interlacing or *frequency interleaving*. Figure 6.15-3 illustrates the idea.

The spectral separation of chrominance and luminance information allows, at least in principle, the individual recovery of these signals. In theory, a crenulated or comb-type filter could be used to admit only the chrominance and not the luminance, and vice versa. However, this is not how the separation is done in practice. In fact, rather

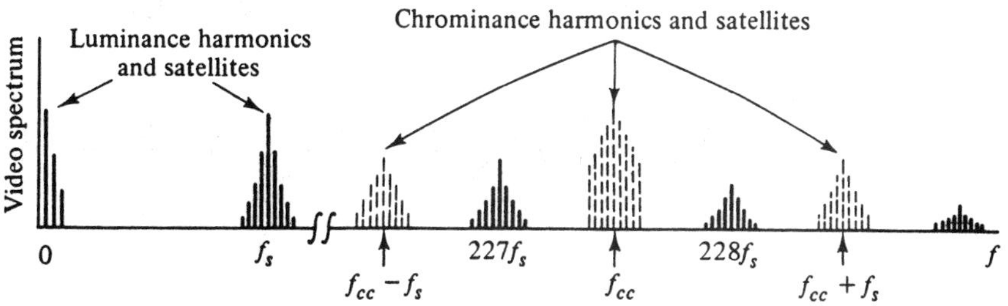

Figure 6.15-3. Interleaving of the chrominance harmonics in the luminance.

surprisingly, there is no need to separate the chrominance sidebands from the luminance signal, either in monochrome or color receivers. The effects of the chrominance sidebands on the luminance signal essentially go unnoticed by the viewer.

The meaningful luminance signal, $f_Y(t)$, and not the luminance induced by the subcarrier, dominates the chrominance because the color subcarrier frequency is exactly an odd multiple of one-half the line scanning frequency. The actual separation of chrominance from luminance is done by the time–space integration properties of the combined eye/TV-screen optical system. We shall consider the details of this phenomenon next.

Additional Properties of the Color Subcarrier

The sinusoidal variations of the subcarrier produce a flicker in the luminance that potentially could be disturbing. What are the properties of the flicker? First, since $f_{cc} \simeq 3.6$ MHz, the flickering is over a picture area no greater than a few picture resolution elements wide (there are roughly 250,000 picture resolution elements per frame, assuming 60 half-frames per second). So, at worst, the sinusoidal flicker is perceived more as a kind of noise than as spatial cross talk, which would, in fact, be much more disturbing.

Second, the fact that the color subcarrier is exactly an odd multiple of one-half the line-scanning frequency produces brightness variations that are consecutively reversed both in time and space. These variations are averaged partly by the time-integration property of the phosphor on the screen, partly by the phenomenon of persistence of vision associated with the human eye/brain system, and partly by the limited resolution of human vision, which behaves, to some extent, like a *spatial* low-pass filter.

Just how does the particular choice of f_{cc} furnish out-of-phase variations? For the sake of discussion, consider a five-line-per-field, one field-per-second, TV system. Three successive fields are shown in Fig. 6.15-4. The color subcarrier frequency is chosen,

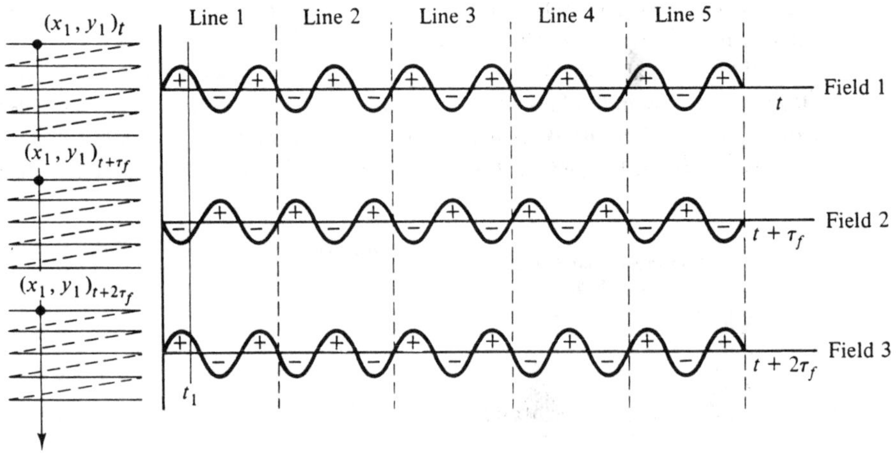

Figure 6.15-4. Effect of interlacing with a color subcarrier of frequency $f_{cc} = (2n + 1)f_s/2$, where n is an integer. Here $n = 1$.

for convenience of illustration, to be the third multiple of one-half the line-scanning frequency; that is, $f_{cc} = 3f_s/2$. Hence there are 7.5 cycles of the subcarrier per field. The result of using such a low value of f_{cc} is to give the reader a misleading impression of the rate of averaging. In actual television, the high-frequency variations in the x direction, generated by the subcarrier and its harmonics, work to the viewer's advantage.

First, we note that the chrominance-induced brightness variations at any two adjacent points in the y direction are always out of phase and therefore tend to cancel in a *spatial* average. Second, the luminances at t and $t + \tau_f$ (τ_f is the field time) at the *same point* are always out of phase and therefore tend to cancel in a *time* average.

In Fig. 6.15-4, the start of the arrow on the extreme left indicates a particular point in the field at time t and its direction is advancing field time. At 3.6 MHz, the variations in both the time and space averages should not be noticeable to the viewer. The effective response of the eye is controlled less by the instantaneous stimulation provided by one scan than by the integrated, *average* stimulation furnished by several scans.

The principles behind color-subcarrier cancellations are exactly the same when $f_{cc} = 445f_s/2$. The argument also extends to the sidebands of the color subcarrier; each sideband is an odd multiple of $f_s/2$. Figure 6.15-5 illustrates the effect of the frequency interlacing technique on the net luminance signal.

In practice, the cancellation of the subcarrier and its sidebands is not quite so good as the previous discussion might indicate. An important factor that contributes to less-than-perfect cancellation is the nonlinearity in the receiver circuitry and the screen. For example, positive excursions in the subcarrier signal might not promote the same response as negative excursions. Another factor is that the storage of information by the eye, from one frame to the next, may not be perfect ([6-18], p. 211).

Generation of the Composite Video

The three signals f_Y, f_I, and f_Q, when suitably combined, constitute the composite color video signal. Methods for separating the three signals from each other must be found if recovery of s_R, s_G, and s_B is to be possible at the receiver. We have already discussed how the color subcarrier enables the separation of f_Y from (f_I, f_Q). We shall discuss now how the two chrominance signals are separated from each other.

Certain properties of human vision enable the bandwidths of the I and Q signals to be significantly less than the 4.2 MHz allocated to the luminance. In particular, tests show that human vision does not detect color in *very small* objects and that the color of *small* objects can be satisfactorily reproduced by utilizing only two primary colors. The result is that if the nominal bandwidths of f_I, f_Q are 0.6 and 1.6 MHz, respectively, satisfactory color reproduction is possible.

The separation of the I and Q signals is done by modulating with a variation of DSB quadrature-carrier two-channel multiplexing, discussed in Sec. 6.7. Figure 6.15-6 illustrates the technique.

The Q signal, being band limited to 0.6 MHz, can be modulated as DSB; the I signal cannot. Hence the 0- to 0.6-MHz portion of the I signal is sent DSB, while the

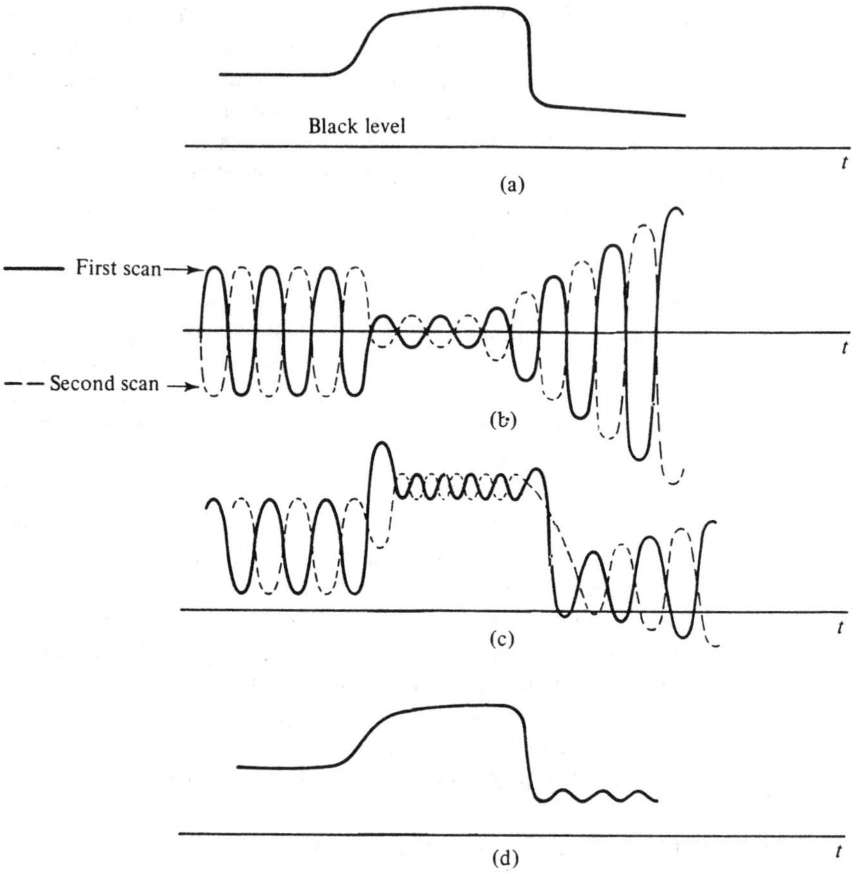

Figure 6.15-5. Effect of frequency interlacing. (a) Meaningful luminance signal. (b) Subcarrier signal. (c) Sum of (a) + (b). (d) Averaged luminance after two scans.

0.6- to 1.6-MHz portion is sent LSSB (lower-sideband SSB). The net result is a VSB signal that can be written as

$$x_I(t) = f_I(t) \cos \omega_{cc}t + \hat{f}_{IH}(t) \sin \omega_{cc}t,$$

where $\hat{f}_{IH}(t)$ is the quadrature component needed to produce VSB modulation, analogous to the signal $q(t)$ in Eq. (6.9-6). Equivalently, for the purpose of analysis, $x_I(t)$ can be decomposed into a sum of DSB and LSSB signals as

$$x_I(t) = \underbrace{[f_I(t) - f_{IH}(t)] \cos \omega_{cc}t}_{\text{DSB}} + \underbrace{f_{IH} \cos \omega_{cc}t + \hat{f}_{IH} \sin \omega_{cc}t}_{\text{LSSB}} \tag{6.15-9}$$

and

$$x_Q(t) = \underbrace{f_Q(t) \sin \omega_{cc}t.}_{\text{DSB}} \tag{6.15-10}$$

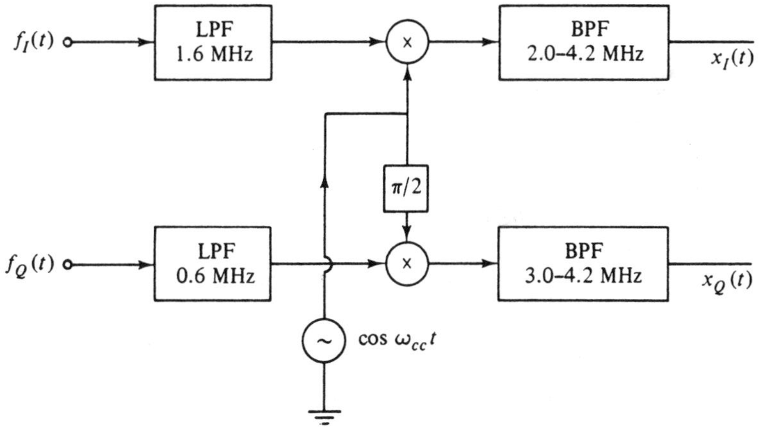

Figure 6.15-6. Separation of the chrominance signals by quadrature-carrier modulation.

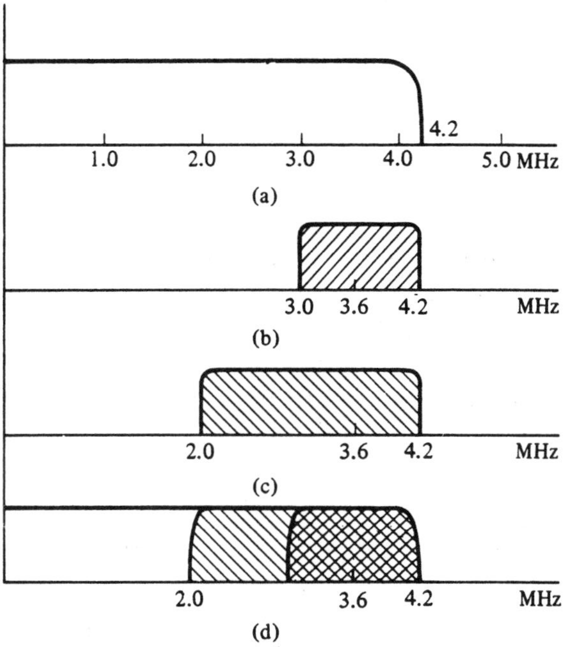

Figure 6.15-7. Utilization of the video spectrum of the color baseband signal. (a) *Y* or luminance component. (b) *Q* chrominance component. (c) *I* chrominance component. (d) Composite spectrum (am plitude of spectra not drawn to scale).

Figure 6.15-7 shows the video spectra for compatible color TV. In Eq. (6.15-9), $f_{IH}(t)$ represents the high-frequency (0.6 to 1.6 MHz) portion of $f_I(t)$, and $\hat{f}_{IH}$ is the Hilbert transform of f_{IH}. The reader may have noticed that the LSSB portion of Eq. (6.15-9) has in fact twice the amplitude of the DSB part.† This can be achieved with a band-pass filter that furnishes twice as much gain in the 2- to 3.0-MHz zone than in the

† Recall that SSB is described by $x(t) = \frac{1}{2}[s(t) \cos \omega_{cc} t \pm \hat{s}(t) \sin \omega_{cc} t]$.

3.0- to 4.2-MHz zone. Equivalently, a VSB filter that has a step-type characteristic can be put into the receiver that compensates for the attenuation of one sideband in the I signal. The filter should give a relative gain of 2 for all frequencies in the I signal between 2 and 3.0 MHz ([6-18], p. 244).

A simplified version of the generation of the composite baseband signal $s_b(t)$ is shown in Fig. 6.15-8.

The color burst signal is a short sample of carrier that is added for frequency and phase synchronization of the local oscillator at the receiver. It is added to the trailing edge (*back porch*) of the horizontal blanking pulse. The horizontal sync signal is derived

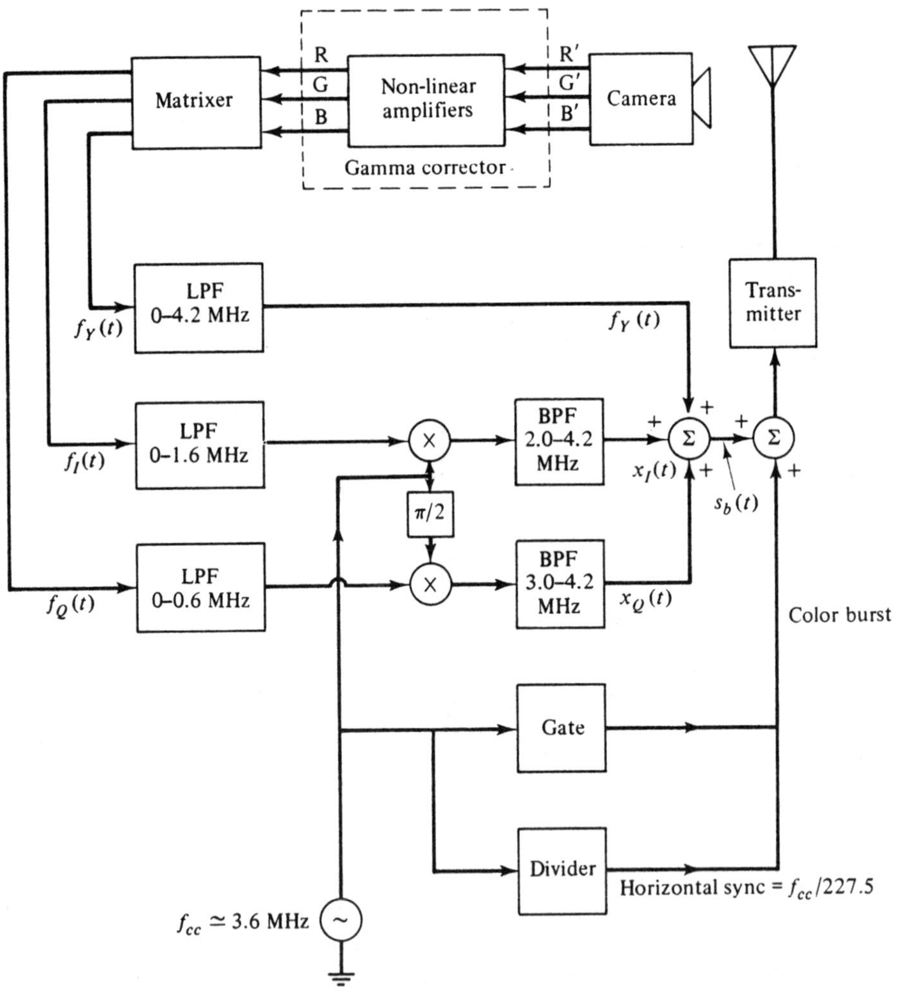

Figure 6.15-8. Simplified block diagram showing the generation of $s_b(t)$. The gamma correctors are needed to establish a linear characteristic between input and output signal.

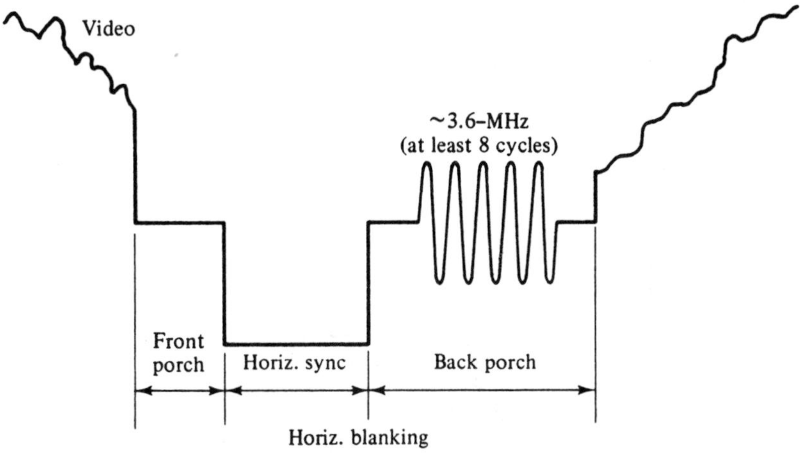

Figure 6.15-9. Horizontal blanking pulse for color TV showing the color burst.

from the subcarrier signal by countdown circuitry that divides the 3.57954-MHz frequency down to the 15.734-kHz horizontal frequency. The vertical sync signal is derived from the horizontal frequency. The color burst is shown in Fig. 6.15-9.

Demodulation

Ignoring sync pulses, the baseband signal for color TV is

$$s_b(t) = f_I(t) \cos \omega_{cc}t + f_Q(t) \sin \omega_{cc}t + \hat{f}_{IH}(t) \sin \omega_{cc}t + f_Y(t). \tag{6.15-11}$$

At the receiver, demodulation of the carrier wave is done as in a black/white monochrome receiver. Demultiplexing of the baseband signal, $s_b(t)$, is done after envelope detection. The demultiplexing of $s_b(t)$ into the three primary color signals is shown in Fig. 6.15-10.

The separation of the luminance signal is done by frequency interlacing. For all practical purposes, therefore, the signal at A is $f_Y(t)$. At B' there is a signal proportional to

$$\begin{aligned} e_{B'}(t) = f_I(t) + [f_Q(t) + \hat{f}_{IH}(t)] \sin 2\omega_{cc}t + 2f_{YH}(t) \cos \omega_{cc}t \\ + f_I(t) \cos 2\omega_{cc}t, \end{aligned} \tag{6.15-12}$$

where $f_{YH}(t)$ is the high-frequency (2.0 to 4.2 MHz) component of the luminance signal. None of the double-frequency terms pass through the low-pass filter; $f_{YH}(t) \cos \omega_{cc}t$ is centered at odd harmonics of one-half the line frequency so that it is separated, or made "invisible," by interlacing. The only effective signal at B is therefore $f_I(t)$.

At C', there is a signal proportional to

$$\begin{aligned} e_{C'}(t) = f_Q(t) - [f_Q(t) + \hat{f}_{IH}(t)] \cos 2\omega_{cc}t + f_I(t) \sin 2\omega_{cc}t \\ + 2f_{YH}(t) \sin \omega_{cc}t + \hat{f}_{IH}(t). \end{aligned} \tag{6.15-13}$$

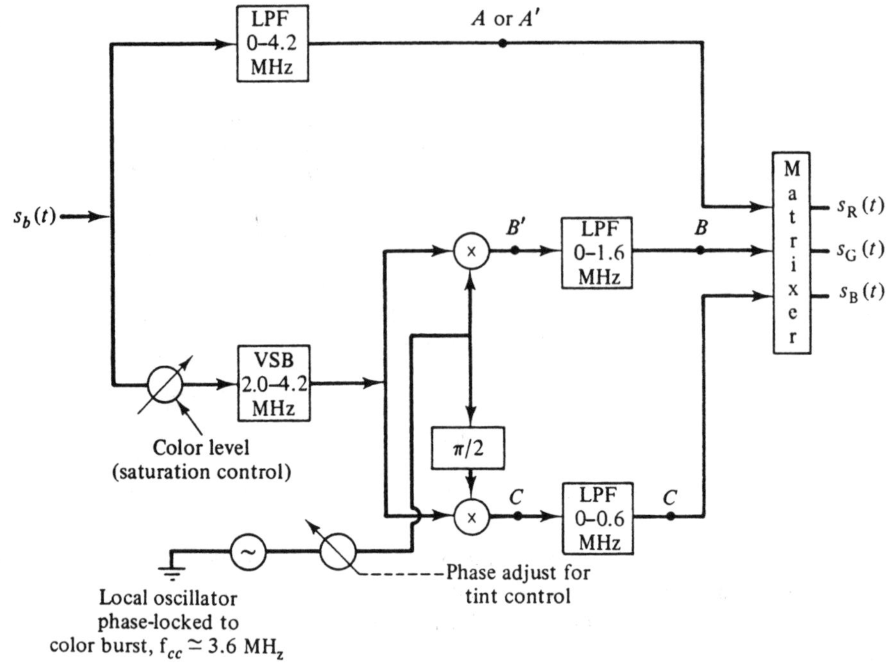

Figure 6.15-10. Demultiplexing of $s_b(t)$ and restoration of primary color signals.

None of the double-frequency components is passed by the low-pass filter. The terms involving $\sin \omega_{cc}t$ are made invisible through interlacing; $\hat{f}_{IH}(t)$ has components only in 0.6 to 1.6 MHz. Hence it is rejected by the filter, and the only effective signal at C is proportional to $f_Q(t)$.

A conventional sync separator is used to produce the pulses necessary for control of the deflection circuitry. The sync pulses also open a gate circuit, which in turn admits the eight-cycle subcarrier burst. The burst is amplified and compared with the output of a local oscillator. An error voltage, proportional to the difference between burst and LO signal, is used to lock the frequency and phase of the local oscillator to the burst. Tint control is obtained by varying the phase of the LO in the color circuit. We leave it as an exercise to show that this rotates the color vector shown in Fig. 6.15-2.

Matrixing of f_Y, f_I, and f_Q produces the three primary color signals:

$$\begin{aligned} s_R(t) &= f_Y(t) - 0.96f_I(t) + 0.26f_Q(t), \\ s_G(t) &= f_Y(t) - 0.28f_I(t) - 0.64f_Q(t), \\ s_B(t) &= f_Y(t) - 1.10f_I(t) + 1.7f_Q(t). \end{aligned} \tag{6.15-14}$$

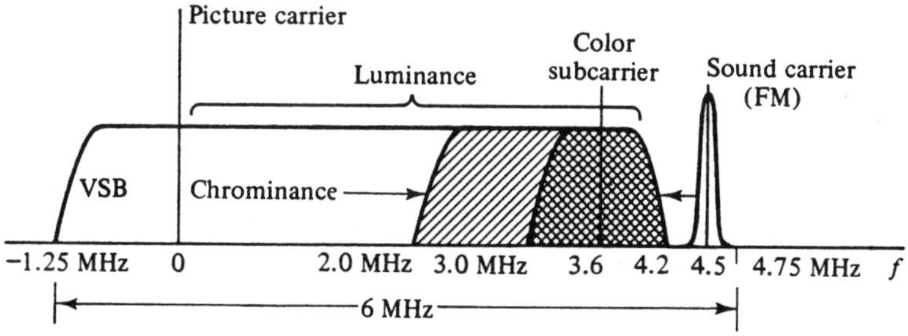

Figure 6.15-11. Standard broadcast spectrum.

We leave it as an exercise (Prob. 6-33) to determine what happens when the chrominance signals are absent (i.e., a color receiver receiving monochrome images).

The TV sound signal modulates a carrier that is 4.5 MHz above the picture carrier. The sound carrier is *frequency modulated,* a process we shall discuss in Chapter 7. The complete TV channel spectrum is shown in Fig. 6.15-11.

6.16 SUMMARY

In this chapter we have examined the principles of amplitude modulation and their application to various systems. We saw that AM and its derivative schemes basically involve the controlled shifting of the signal spectrum to various points along the frequency scale. We saw that the fundamental operation of shifting the baseband spectrum from the origin to the frequency f_c basically requires multiplication by a carrier wave of the form $\cos 2\pi f_c t$ (or $\sin 2\pi f_c t$).

Although what is commonly called ordinary AM (i.e., DSB modulation with transmitted carrier) was seen to be inferior (with respect to power conservation and/or bandwidth requirements) to DSBSC and SSB, the ease with which ordinary AM can be detected more than makes up for its deficiencies and accounts for its widespread use in commercial broadcasting. We studied frequency-division multiplexing (FDM) and showed how FDM can be used in stereo broadcasting that is compatible with monophonic systems.

Finally, we investigated how TV systems manage to transmit the information in visual imagery by electrical signals. We saw that the process of scanning enables the mapping of the two-dimensional image brightness function into one-dimensional electrical signals. We showed that color TV can be made compatible with existing monochrome systems at the cost of increased complexity and the incorporation of almost every AM scheme in existence.

AM systems are discussed in a number of books. Good treatments of the subject can be found in [6-19], [6-20], [6-17], [6-21], and [6-22], but there are others.

PROBLEMS

6-1. Consider an AM wave

$$x(t) = A_c[1 + ms(t)] \cos 2\pi f_c t,$$

where $s(t)$ is a real modulating signal of bandwidth W.

(a) Under what conditions does

$$x_+(t) = \frac{A_c}{2}[1 + ms(t)]e^{j2\pi f_c t}$$

represent the positive-frequency portion of $x(t)$?

(b) Assuming that the conditions in part (a) are met, show that

$$x_+(t) = \tfrac{1}{2}[x(t) + j\hat{x}(t)],$$

where $\hat{x}(t)$ is the Hilbert transform of $x(t)$.

6-2. Assume that in the AM wave

$$x(t) = A_c[1 + ms(t)] \cos 2\pi f_c t$$

the message bandwidth $W < f_c$. Decompose $s(t)$ into $s_+(t)$ and $s_-(t)$, which represent the signals reconstructed from the positive and negative portions of the spectrum, respectively; for example,

$$s_+(t) = \int_0^\infty S(f)e^{j2\pi ft}\,df.$$

Sketch the spectrum of $x(t)$, and identify which of the terms $s_\pm e^{\pm j2\pi f_c t}$ produce which portions of the spectrum.

6-3. Consider an AM wave modulated by a periodic signal of the form

$$s(t) = \sum_{n=-N}^{N} c_n e^{j\omega_n t}, \qquad \omega_n = \frac{2\pi n}{T}.$$

(a) What condition must be satisfied to prevent foldover distortion?

(b) Show that the modulated wave can be written as

$$x(t) = A_c(1 + mc_0) \cos \omega_c t + A_c m \sum_{n=1}^{N} |c_n| \{\cos [(\omega_c + \omega_n)t + \theta_n] + \cos [(\omega_c - \omega_n)t - \theta_n]\}.$$

6-4. A linear half-wave rectifier is a device with transfer characteristic

$$y(t) = \begin{cases} x(t), & x(t) > 0, \\ 0, & \text{otherwise.} \end{cases}$$

Show that if $x(t)$ is a standard AM wave the output of the rectifier contains a term proportional to $s(t)$. Design a system that will recover $s(t)$.

6-5. Consider the modulator shown in Fig. P6-5. The diode D has infinite backward resistance and forward resistance R_f. Assume piecewise linearity for the diode operating characteristic and $s(t) \ll A_c$.

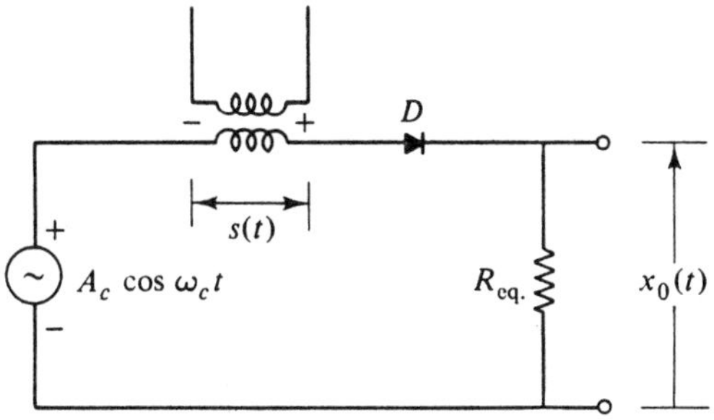

Figure P6-5

(a) Compute the output $x_0(t)$ by replacing the nonlinear circuit by an equivalent linear time-varying circuit. Show that the output can be written as

$$x_0(t) = [A_c \cos \omega_c t + s(t)]g(t),$$

where $g(t)$ is an appropriate switching function. Determine $g(t)$.

(b) What is the required filtering if an AM wave is desired? Assume that the signal bandwidth, W, satisfies $W < f_c$.

6-6. Because of a nonlinearity in the system, a DSB generator produces a modulated signal

$$x(t) = s(t) \cos \omega_c t + a_1[s(t) \cos \omega_c t]^2.$$

What relation between carrier frequency f_c and signal bandwidth W must be satisfied to enable the removal of the error term $[s(t) \cos \omega_c t]^2$?

6-7. A DSB wave is modulated by the following signal:

$$s(t) = \sum_{n=1}^{5} n^{-1} \cos n\omega_m t, \qquad \omega_m < \omega_c.$$

(a) What constraint must be applied to prevent sideband foldover?

(b) Assuming that the constraint in part (a) is satisfied, sketch the positive-frequency spectrum of the DSB wave.

6-8. Consider an AM wave modulated by a tone at frequency f_n:

$$x(t) = A_c[1 + m \cos \omega_n t] \cos \omega_c t.$$

Because of inadequate filtering, the upper sideband associated with $x(t)$ is totally attenuated. Assuming that $m \ll 1$, compute the resultant envelope.

6-9. Consider the analog modulator shown in Fig. P6-9. The half-square-law devices have transmittance

$$e_{out} = \begin{cases} Ke_{in}^2, & e_{in} > 0, \\ 0, & e_{in} \le 0. \end{cases}$$

Take $e_1(t) = A_1[1 + ms(t)]$ and $e_2(t) = A_2 \cos \omega_c t$. Assume that $e_1 + e_2$ as well as $e_1 - e_2$ are greater than zero. Show that the output is an AM wave with modulation index strictly less than unity.

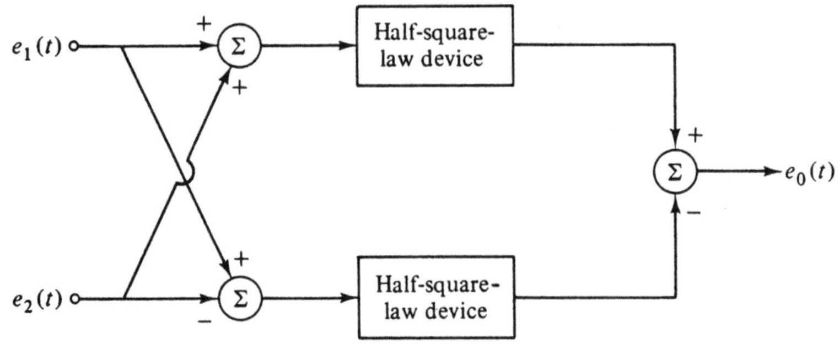

Figure P6-9

6-10. It is possible to increase the modulation of an AM wave before transmission by subtracting excess carrier from the modulated signal. Given an AM wave with m strictly less than 1,

$$x(t) = A[1 + ms(t)] \cos \omega_c t,$$

how much carrier must be removed before unity modulation is achieved? This technique is known as *carrier cancellation*.

6-11. Design a superheterodyne receiver for the following parameters: 50 channels, message bandwidth = 7.5 kHz, RF tuning range to begin at 600 kHz. Let f_{IF} = 500 kHz. What are reasonable values of the IF and AF bandwidths? What is the advantage of taking the local oscillator frequency as $f_{LO} = f_c + f_{IF}$ instead of $f_c - f_{IF}$? Sketch a block diagram of the receiver.

6-12. Show that in the regenerative frequency divider of Fig. 6.4-3 the output frequency is f_0/n if the filter rejects the frequency $(2n - 1)f_0/n$.

6-13. An equivalent circuit of the last IF stage and envelope detector in an AM demodulator is shown in Fig. P6-13.

(a) Explain qualitatively how this circuit works.

(b) Take $i_I(t) = I \cos \omega_{IF} t$; show that $v_0(t)$ is proportional to I. By extension, if $i_I(t) = g(t) \cos \omega_{IF} t$ with $g(t) \geq 0$ and "slowly varying," then $v_0(t)$ is proportional to $g(t)$.

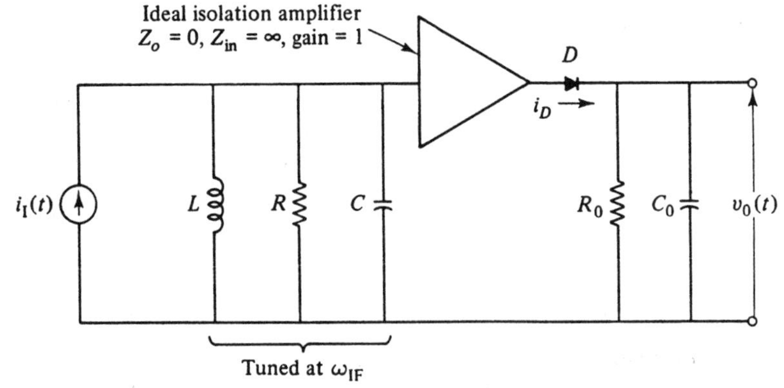

Figure P6-13

6-14. (DSB and SSB) Consider a square wave of amplitude A and period T. Let $\omega_m \equiv 2\pi/T$. Assume that the square wave is made band limited by passing it through a low-pass filter with cutoff frequency $f_{co} = Nf_m$. Call the output of the low-pass filter $s(t)$.

(a) Obtain an expression for the DSBSC wave

$$x(t) = s(t) \cos \omega_c t$$

in terms of the Fourier components of $s(t)$. Assume that $\omega_c > N\omega_m$.

(b) Show that the upper-sideband SSB signal can be written as

$$x(t) = \frac{A}{\pi} \sum_{n=1,3,5\ldots}^{N} \frac{\sin (n\pi/2)}{n} \cos (\omega_c + n\omega_m)t.$$

(c) Expand $x(t)$ in part (b) into the form

$$x(t) = \alpha(t) \cos \omega_c t - \beta(t) \sin \omega_c t,$$

and consider the SSB wave at $t = T/4$. What can be said about the peak power in $x(t)$ as $N \to \infty$? What possible disadvantage can you foresee in the use of SSB modulation?

6-15. A method proposed by D. K. Weaver ("A Third Method of Generation and Detection of SSB Signals," *Proc. IRE*, **44,** pp. 1703–1705, Dec. 1956) for generating SSB is shown in Fig. P6-15. The method is applicable to signals with finite energy gap near zero frequency. Let the signal be given by

$$s(t) = \sum_{n=1}^{N} c_n \cos (\omega_n t + \phi_n),$$

where $\omega_L \leq \omega_n \leq \omega_L + 2\pi W$ for all n in $n = 1, \ldots, N$. Let $\omega_0 = \omega_L + 2\pi W$ and $\omega_c \gg \omega_0 + \omega_L + 2\pi W$.

(a) Explain how this system produces an SSB wave by sketching the spectrum at each point. What is the required cutoff frequency for the LPF? Why is it necessary for the signal to have a low-frequency gap?

(b) What advantage does this method have over the phase-shifting method discussed in Sec. 6.8?

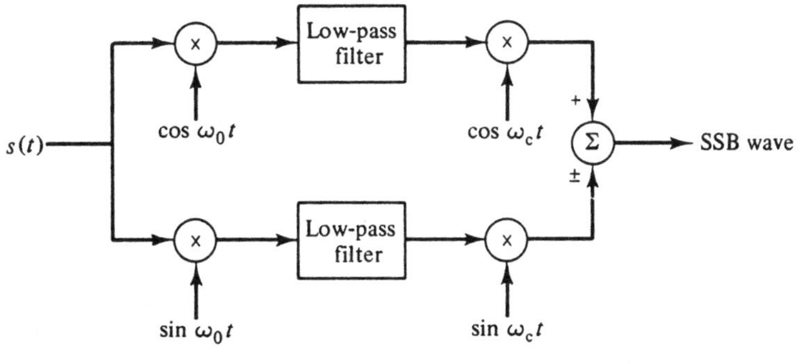

Figure P6-15

6-16. Show that the circuit of Figure 6.8-4, in which the signal at A' lags the signal at point A by 90° over a large frequency band, produces a SSB modulated output.

6-17. Consider the modulated wave

$$x(t) = A_m s(t) \cos \omega_c t + A_m q(t) \sin \omega_c t.$$

(a) What is the envelope of the resulting signal if an in-phase carrier term $A_c \cos \omega_c t$ is added to $x(t)$?

(b) Let $q(t) = \hat{s}(t)$, where $\hat{s}(t)$ is the Hilbert transform of $s(t)$. Under what conditions can $x(t) + A_c \cos \omega_c t$ be reasonably detected with an envelope detector? (The resulting scheme is called compatible single-sideband or CSSB.)

(c) Describe $q(t)$ when $x(t)$ represents, in turn, a (1) DSBSC, (2) SSB, and (3) VSB wave.

6-18. Consider the following expression for an AM-type modulated wave:

$$x(t) = s(t) \cos \omega_c t - q(t) \sin \omega_c t,$$

where $S(f)$ and $Q(f)$, which represent the Fourier transforms of $s(t)$ and $q(t)$, respectively, are zero for $|f| \geq f_c$. Let $x(t)$ appear as an input to a system with impulse response $h(t)$ and transmittance $H(f) = \mathcal{F}[h(t)]$. Show that the output $y(t) = x(t) * h(t)$ can be written as

$$y(t) = r(t) \cos \omega_c t - p(t) \sin \omega_c t,$$

where

$$\mathrm{r}(t) = h_1(t) * s(t) - h_2(t) * q(t)$$

and

$$p(t) = h_2(t) * s(t) + h_1(t) * q(t).$$

The relations between $h_1(t)$, $h_2(t)$, and $h(t)$ are implied by

$$h_c(t) = h_1(t) + jh_2(t),$$

where

$$h_c(t) \equiv \mathcal{F}^{-1}[H_c(f)] = \mathcal{F}^{-1}\{1/2 \cdot [1 + \operatorname{sgn}(f + f_c)]H(f + f_c)\}.$$

For a discussion of this problem, see [6-17, p. 193].

6-19. Use the results obtained in Prob. 6-18 to design a filter with transmittance $H(f)$ that will convert:

(a) A DSB wave into a SSB wave.

(b) A VSB wave into a SSB wave.

6-20. Message scrambling:

(a) Analyze the system shown in Fig. P6-20 and show that a reversal of the message spectrum takes place. The resulting "message" represents scrambled speech.

(b) Design a system that will work as an unscrambler. *Hint*: Consider an identical system.

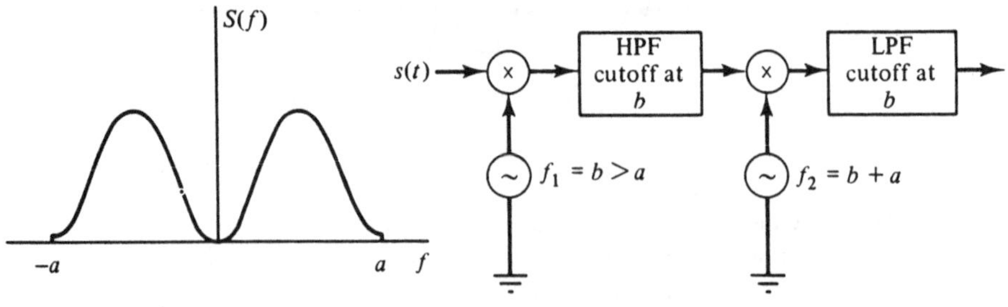

Figure P6-20

6-21. The system shown in Fig. P6-21 is proposed for securing the confidentiality of $s(t)$ in a DSB system. Let $H(f) = H_0 e^{j\phi(f)}$, where $\phi(f)$ is the phase-scrambling function and is real. Let $H(f) \simeq 0$ for $f > f_0$, where $W \leq f_0 < 2f_c - W$. Let $S(f) = 0$ for $f > W$. Explain how the system works. (In optics this procedure is known as coding through a random phase mask.)

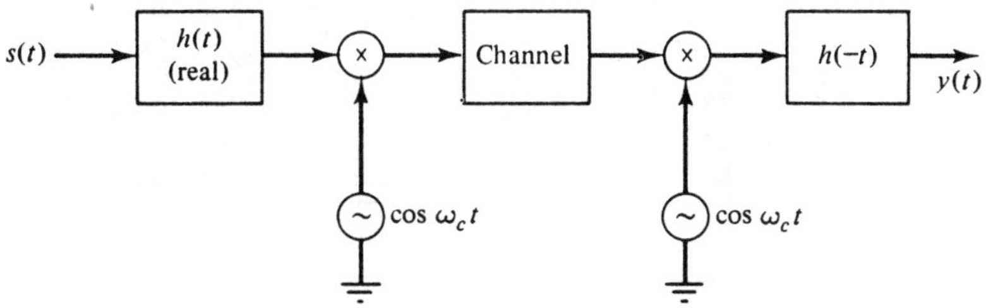

Figure P6-21

6-22. Design an FDM system that accommodates ten 12-kHz channels. Two of the 12 kHz should be reserved for guard bands, and the remaining spectrum should accommodate a 10-kHz signal. Choose an appropriate set of subcarriers. What kind of subcarrier modulation do you recommend? What is the bandwidth of the carrier wave if DSB modulation is chosen?

6-23. Compute the Fourier coefficients b_{mn} in Eq. (6.14-5) for a brightness distribution $B(x, y)$ as shown in Fig. P6-23.

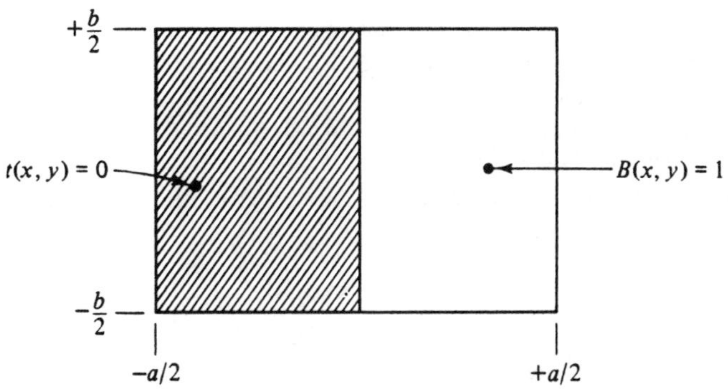

Figure P6-23

6-24. Repeat Prob. 6-21, except consider a brightness distribution consisting of a Ronchi ruling (Fig. P6-24).

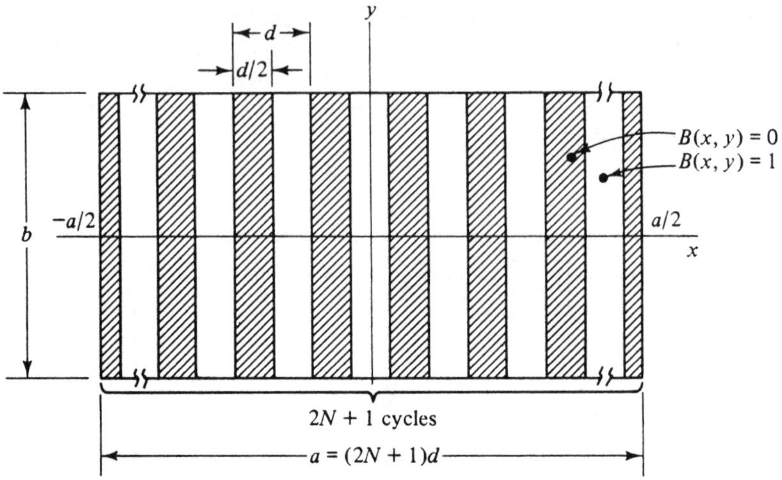

Figure P6-24

6-25. In Sec. 6.14, the scanning process was modeled as a *point* moving across the video field. In actual practice, a finite aperture must be used. Define the aperture by a transmittance function $h(\xi, \eta)$, where (ξ, η) are rectangular coordinates defined with respect to an origin at the center of the aperture and

$$h(\xi, \eta) = \frac{\text{transmitted brightness}}{\text{incident brightness}} \qquad \text{at } (\xi, \eta).$$

(a) Show that if the aperture is centered at (x, y) the video current is proportional to

$$I_h = \int_{-\infty}^{\infty}\int_{-\infty}^{\infty} h(\xi, \eta)B(x + \xi, y + \eta)\, d\xi\, d\eta,$$

where B is the incident brightness.

(b) Write an expression for $h(\xi, \eta)$ when the aperture is a rectangular hole of width w and height h.

(c) Compute the current when the aperture in part (b) is applied to the brightness distribution in Fig. P6-23.

6-26. Use the Fourier series expansion of $B(x, y)$ in Eq. (6.14-1) to show that the current I_h of Prob. 6-25 can be written as

$$I_h = \sum_{n=-\infty}^{\infty}\sum_{m=-\infty}^{\infty} H_{mn}^{*}c_{mn} \exp\left[j2\pi\left(\frac{m}{a}x + \frac{n}{b}y\right)\right],$$

where H_{mn} is given by

$$H_{mn} = \int_{-\infty}^{\infty}\int_{-\infty}^{\infty} h(\xi, \eta) \exp\left[-j2\pi\left(\frac{m}{a}\xi + \frac{n}{b}\eta\right)\right] d\xi\, d\eta$$

and $h(\xi, \eta)$ is a real function. What effect does the filter $\{H_{mn}\}$ have on the spectrum of the video signal?

6.27. The aspect ratio $\mathscr{A}$ is the ratio of relative image width W to image height L in the TV image. Show that if the resolution, in resolvable lines per unit distance, is to be the same in the horizontal and vertical directions then

$$\mathscr{A} \equiv \frac{W}{L} = \frac{n_H}{n_V},$$

where n_V, n_H are the number of rows and columns of the finest (i.e., most detailed) checkerboard pattern that can be resolved (Fig. P6-27).

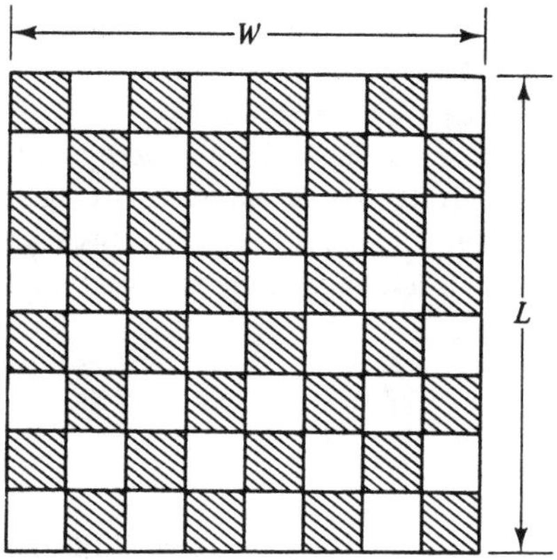

Figure P6-27

6-28. (Continuation of Prob. 6-27) If a sinusoidal signal of maximum video frequency W is applied to a TV monitor, a sequence of dark and light zones results, much like a checkerboard row. The *effective* line-scan time, T_E, is the difference between the total line-scan time, T_L, and the horizontal retrace time, T_{HR}.

(a) Show that $n_H = 2WT_E = 2W(T_L - T_{HR})$.

(b) The "utilization" factor or Kell factor, K, is a number that relates n_V to the useful number of scan lines per frame. If N = total number of scan lines and N_{VR} = number of scan lines blanked out during the vertical retrace, then

$$n_V = K(N - N_{VR}).$$

If the electron scanning beam were perfectly aligned with the rows of the checkerboard pattern and suffered no randomness in position, K would be 1. In practice, K is less than 1 ($K \simeq 0.7$). Use this fact and the results from Prob. 6-27 and this problem to show that the bandwidth W can be written as

$$W = 0.35\mathscr{A}\frac{N}{T_L}\frac{1 - N_{VR}/N}{1 - T_{HR}/T_L}.$$

6-29. (Continuation of Prob. 6-28) The frame time T_F is the time required to produce a complete image by scanning. Show that T_F is given by

$$T_F = \frac{0.714\mathscr{A}n_V^2}{W(1 - N_{VR}/N)(1 - T_{HR}/T_L)}.$$

6-30. A Mars–Earth facsimile transmission system has the following parameters: horizontal sweep = 300°, vertical sweep = 60°, number of horizontal lines = 9150, number of gray levels = 64, carrier frequency = 2.2 GHz (2.2×10^9 Hz), and maximum signaling rate = 4000 bits/sec. Estimate the frame time T_f if both N_{VR} and T_{HR} are negligible and can be taken as zero.

6-31. In Fig. 6.15-5, explain why the average luminance signal obtained after averaging the composite signal (luminance plus subcarrier signal) after two scans is slightly different from the original luminance signal. Why is the average luminance "bumpy" in regions where the composite signal overshoots the black level?

6-32. Consider a five-line-per-field, one-field-per-second TV system in which the color subcarrier f_{cc} is chosen to be the *first* multiple of one-half the line-scanning frequency f_s.
- **(a)** How many cycles of subcarrier are there per field?
- **(b)** How many cycles of subcarrier are there per line?
- **(c)** Draw a time-phase diagram as in Fig. 6.15-4 to verify that the subcarrier-induced luminance variations tend to cancel in time and space averaging.

6-33. Explain what happens when a color TV monitor receives a black/white picture? The ability of a color monitor to reproduce a black/white image is termed *reverse compatibility*.

REFERENCES

[6-1] *IRE Dictionary of Electronics Terms and Symbols*, The IEEE Inc., New York, 1961, p. 92.

[6-2] B. Gilbert, "A New Wide-Band Amplifier Technique," *IEEE J. Solid State Circuits*, **SC-3,** pp. 353–365, Dec. 1968.

[6-3] D. H. Sheingold, ed., *Non-Linear Circuits Handbook*, Analog Devices, Inc., Norwood, Mass., 1974.

[6-4] A. B. Carlson, *Communication Systems: An Introduction to Signals and Noise in Electrical Communications*, McGraw-Hill, New York, 1968.

[6-5] V. H. Grinich and H. G. Jackson, *Introduction to Integrated Circuits*, McGraw-Hill, New York, 1975.

[6-6] J. Millman and C. C. Halkias, *Electronic Devices and Circuits*, McGraw-Hill, New York, 1967.

[6-7] K. Henney, *Radio Engineering Handbook*, McGraw-Hill, New York, 1950.

[6-8] W. Deil and R. J. McFadyen, "Single-Slice Superhet," *IEEE Spectrum*, **17,** No. 3, pp. 54–57, March, 1977.

[6-9] J. P. Costas, "Synchronous Communication," *Proc. IRE*, **44,** pp. 1713–1718, Dec. 1956.

[6-10] S. Darlington, "Realization of Constant Phase Difference," *Bell Syst. Tech. J.*, **29,** pp. 94–104, Jan. 1950.

[6-11] D. Starer and A. Nehorai, "Adaptive SSB Carrier Offset Determination," submitted to *IEEE Trans. Acoust. Speech and Signal Processing*.

[6-12] F. S. Hill, Jr., "On Time Domain Representations for Vestigial Sideband Signals," *Proc. IEEE*, **62,** pp. 1032–1033, July 1974.

[6-13] A. J. Viterbi, *Principles of Coherent Communications*, McGraw-Hill, New York, 1966.

[6-14] J. Millman and H. Taub, *Pulse, Digital and Switching Waveforms*, McGraw-Hill, New York, 1965.

[6-15] J. W. Sandberg, "Assembling and Displaying Slow-Scan TV Pictures," Hughes Customer Application Notes 91–11–009, Hughes Aircraft Co., Oceanside, Calif., April 1973.

[6-16] L. H. Hansen, *Introduction to Solid-State Television Systems*, Prentice-Hall, Englewood Cliffs, N.J., 1969.

[6-17] W. R. Bennett, *Introduction to Signal Transmission*, McGraw-Hill, New York, 1970.

[6-18] J. W. Wentworth, *Color Television Engineering*, McGraw-Hill, New York, 1955.

[6-19] A. B. Carlson, *Communication Systems: An Introduction to Signals and Noise in Electrical Communications*, 3rd ed., McGraw-Hill, New York, 1986.

[6-20] M. Schwartz, W. R. Bennett, and S. Stein, *Communication Systems and Techniques*, McGraw-Hill, New York, 1966.

[6-21] P. F. Panter, *Modulation, Noise and Spectral Analysis*, McGraw-Hill, New York, 1965.

[6-22] H. E. Rowe, *Signals and Noise in Communication Systems*, Van Nostrand Reinhold, New York, 1965.

CHAPTER 7

Angle Modulation

7.1 INTRODUCTION

Angle modulation encompasses *phase modulation* (PM) and *frequency modulation* (FM) and refers to the process by which the quantity $\theta(t)$ in the expression

$$x(t) = A \cos \theta(t) \tag{7.1-1}$$

is controlled by a message $s(t)$. The amplitude A is intended to be constant, and any variations in A constitute a form of noise. As always, we shall assume that the class of modulating signals $\{s(t)\}$ is band limited to W Hz. Although both FM and PM will be discussed in this chapter, the emphasis will be on FM since it is by far the most important angle-modulation process in use.

Historically, FM was first correctly analyzed by John R. Carson [7-1], who discussed the relatively wide bandwidths required. E. H. Armstrong [7-2] was among the first to recognize the noise-suppression properties of FM and designed an FM modulator based on PM. Commercial broadcast FM is radiated in the band extending from 88 to 108 MHz. A single channel is nominally 200 kHz wide.

Tests have shown that an interfering audio signal will create objectionable interference if its level is as high as 30 to 40 dB below the desired signal ([7-3], pp. 21–65). Thus the feasible service areas are the zones in which the desired component of the resulting audio is at least 35 dB above the interference. In the case of AM (amplitude modulation, *not* angle modulation), the power ratio of desired AM to interference must be at least 35 dB if the interfering wave is on the same carrier frequency. In the case of FM systems, however, the ratio of desired FM to interference need only be 6 dB to meet the same audio criterion. A property of FM that makes this possible is that the information in FM signals is in the zero crossings of the wave and not in its amplitude where the primary effect of the interference is manifested. The amplitude variations induced by the interference are removed in the FM receiver by a limiter circuit.

The main advantages of FM over AM are as follows:

1. Improved signal-to-noise ratio. Tests have shown as much as a 25-dB increase in this ratio over AM with respect to automobile ignition, X-ray generation, and other man-made interference.
2. A smaller geographical interference area when two nearby FM transmitters are operating simultaneously on the same frequency.
3. Less radiated power required for the same signal-to-noise ratio.
4. More efficient use of transmitting equipment.
5. The existence of uniform and well-defined service areas for a given transmitter since the FM signal-to-noise ratio remains high until the field intensity reaches a low value (threshold effect).

Against these important advantages, FM also suffers some serious drawbacks. An FM wave typically requires a large bandwidth, up to 20 times the amount required for AM. FM systems are generally more complicated than corresponding AM ones. FM modulation is also strongly nonlinear; this means that superposition does not hold and that the analysis of FM waves is more difficult than the analysis of AM. In fact, we shall see that FM analysis uses more approximations and is less rigorous than is the analysis for AM.

7.2 DEFINITIONS

The instantaneous radian frequency is defined by

$$\omega = \lim_{\Delta t \to 0} \frac{\Delta\theta}{\Delta t} = \dot{\theta}(t) \qquad \text{(radians per second)}, \tag{7.2-1}$$

or equivalently,

$$f = \frac{1}{2\pi}\dot{\theta}(t) \qquad \text{(hertz)}.$$

The fact that frequency, which is a measure of the number of cycles per second, changes continuously with time should not be considered paradoxical. The same concept holds in mechanics, where velocity can change continuously with time despite the fact that it is a measure of the number of meters traversed per second.

In Eq. (7.2-1), let $\theta(t) = \omega_c t + \theta_0$, where ω_c and θ_0 are constants. We then obtain

$$x(t) = A\cos(\omega_c t + \theta_0). \tag{7.2-2}$$

This type of wave is called an unmodulated carrier and conveys no information. The instantaneous radian frequency is a constant given by

$$\dot{\theta}(t) = \frac{d}{dt}(\omega_c t + \theta_0) = \omega_c,$$

(i.e., the carrier frequency). The instantaneous phase is a linear function of time and increases in direct proportion to t, with slope ω_c.

A phase-modulated wave is described by the expression

$$x_{PM}(t) = A \cos [\omega_c t + \theta_0 + K's(t)], \tag{7.2-3}$$

where K' is a constant with units of radians per volt if $s(t)$, the modulating signal, has units of volts. If we normalize $s(t)$ to satisfy $|s(t)| \leq 1$, we find that the peak phase deviation from the unmodulated phase $\omega_c t + \theta_0$ is just K'.

Let the modulating signal be a pure tone of radian frequency ω_m; that is,

$$s(t) = A_m \cos \omega_m t; \tag{7.2-4}$$

then

$$x_{PM}(t) = A \cos [\omega_c t + A_m K' \cos \omega_m t], \tag{7.2-5}$$

where θ_0 has, without loss of generality, been set equal to zero. The constant

$$K_d \equiv A_m K' \tag{7.2-6}$$

is called the *phase deviation* and is a measure of the maximum shift of the phase of $x_{PM}(t)$ from $\omega_c t$. The phase deviation depends on the *amplitude* of the tone.

An FM wave is described by

$$x(t) = A \cos \left[\omega_c t + \theta_0 + K'' \int_{t_0}^{t} s(\lambda)\, d\lambda \right], \tag{7.2-7}$$

where K'' is a constant with units of radians per volt-second. By choosing t_0 appropriately, we can cancel θ_0. As in the case of PM, we set θ_0 equal to zero.

The instantaneous phase is[†]

$$\theta(t) = \omega_c t + K'' \int^{t} s(\lambda)\, d\lambda, \tag{7.2-8}$$

and the instantaneous frequency f [we omit writing $f(t)$, which could easily be misread as "function of time"] is given by

$$f = \frac{\dot{\theta}(t)}{2\pi} = f_c + Ks(t), \qquad f_c \equiv \frac{\omega_c}{2\pi}, \qquad K \equiv \frac{K''}{2\pi}. \tag{7.2-9}$$

[†] The arbitrary nature of the lower limit is indicated by leaving it off altogether.

From Eq. (7.2-8), we see that if we first integrate $s(t)$ and then allow it to phase-modulate a carrier we have achieved FM modulation indirectly. In fact, this method of producing an FM wave was first used by Armstrong and is called *indirect* FM.

Let us again consider a modulating signal consisting of a pure tone, as in Eq. (7.2-4). Then the FM wave takes the form

$$x_{FM}(t) = A \cos\left(\omega_c t + \frac{A_m K''}{\omega_m} \sin \omega_m t\right). \tag{7.2-10}$$

The maximum frequency deviation from the frequency of the unmodulated carrier is

$$\Delta\omega \equiv \left|\frac{d}{dt}\left(\frac{A_m K''}{\omega_m} \sin \omega_m t\right)\right|_{\max} = A_m K''. \tag{7.2-11}$$

The quantity $\Delta\omega$ or $\Delta f = \Delta\omega/2\pi$ is appropriately called the *frequency deviation*. The frequency deviation divided by the modulating frequency is the maximum phase difference between the FM wave and the unmodulated carrier; it is called the *modulation index* and is frequently denoted by β:

$$\beta \equiv \frac{\Delta\omega}{\omega_m} = \frac{\Delta f}{f_m} \qquad \text{(radians)}. \tag{7.2-12}$$

Equations (7.2-5) and (7.2-10) are very much alike. In the steady state, for a fixed ω_m and K' and K'' adjusted so that $K' = K''/\omega_m$, there would be no way of distinguishing the PM wave from the FM wave from a single trace. Only by varying the modulating frequency ω_m, while keeping the amplitude constant, could one distinguish between the two waves. No change in peak phase deviation would be observed in the PM case, whereas an inverse relation between ω_m and peak phase deviation could be observed in the FM case.

Both the PM and FM waves can, under conditions of sinusoidal modulation, be described by a wave of the form

$$x(t) = A \cos[\omega_c t + \xi \sin(\omega_m t + \gamma)], \tag{7.2-13}$$

where $\xi = \beta$ and $\gamma = 0$ for FM, and $\xi = K_d$ and $\gamma = \pi/2$ for PM.

Angle-modulated waves do not have the appearance of AM waves. For example, consider a modulating signal consisting of a linear ramp; that is, $s(t) = btu(t)$, where b is a constant and $u(t)$ is the unit step. For $t > 0$, the PM and FM waves are described as follows:

PM: $\quad x_{PM}(t) = A \cos(\omega_c t + K'bt), \qquad$ inst. frequency $\dot{\theta} = \omega_c + K'b,$

FM: $\quad x_{FM}(t) = A \cos\left(\omega_c t + \frac{K''b}{2} t^2\right), \qquad \dot{\theta} = \omega_c + K''bt.$

These results enable us to draw the waveforms in Fig. 7.2-1 for the *periodic-ramp-modulating* signal.

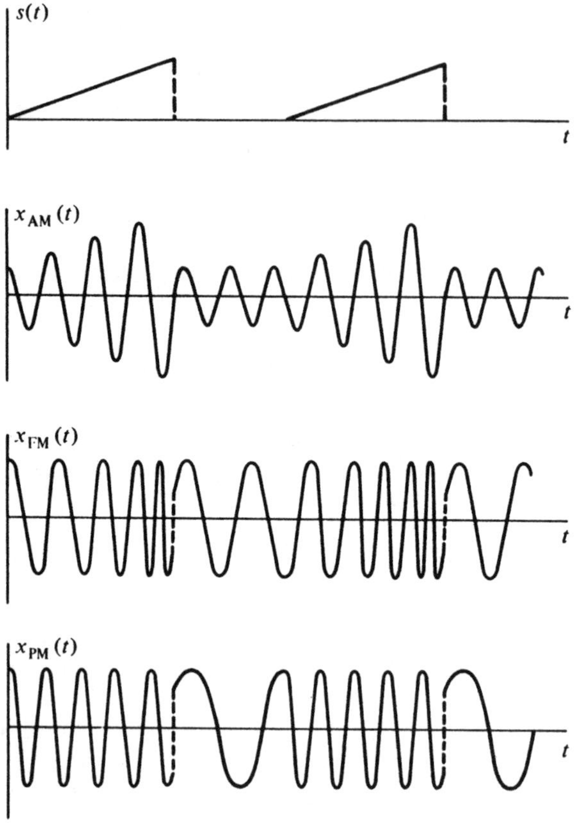

Figure 7.2-1. Carrier waveforms for AM, FM, and PM when the modulation is a periodic ramp as shown. The PM wave switches discontinuously between two frequencies separated by an amount $K'b/2\pi$ Hz.

7.3 FOURIER SPECTRUM OF ANGLE-MODULATED SIGNALS

We consider initially an angle-modulated signal of the form

$$x(t) = A\cos(\omega_c t + \xi \sin \omega_m t). \tag{7.3-1}$$

that is, the modulation is a pure sinusoid. An equivalent expression is

$$\begin{aligned} x(t) &= \operatorname{Re} Ae^{j(\omega_c t + \xi \sin \omega_m t)} \\ &= \operatorname{Re} Ae^{j\omega_c t} e^{j\xi \sin \omega_m t}. \end{aligned} \tag{7.3-2}$$

Thus the desired spectrum is the spectrum of the factor $e^{j\xi \sin \omega_m t}$ shifted to the carrier frequency ω_c.

The function $e^{j\xi \sin \omega_m t}$ is clearly a periodic function with period $T = 2\pi/\omega_m$. It therefore has a Fourier-series representation:

$$e^{j\xi \sin \omega_m t} = \sum_{n=-\infty}^{\infty} c_n e^{jn\omega_m t}, \tag{7.3-3}$$

where, by Eq. (2.4-15),

$$c_n = \frac{1}{T}\int_{-T/2}^{T/2} e^{j\xi \sin \omega_m t} e^{-jn\omega_m t}\, dt$$
$$= \frac{1}{T}\int_{-T/2}^{T/2} e^{j(\xi \sin \omega_m t - n\omega_m t)}\, dt. \tag{7.3-4}$$

By making the change of variable $\lambda = \omega_m t$, we get

$$c_n = \frac{1}{2\pi}\int_{-\pi}^{\pi} e^{j(\xi \sin\lambda - n\lambda)}\, d\lambda \equiv J_n(\xi), \tag{7.3-5}$$

where $J_n(\xi)$ is the nth-order Bessel function of the first kind. Equation (7.3-5) is one of the defining equations for $J_n(\xi)$ [see also Eq. (2.16-6) for the case of $n = 0$]. Equivalently,

$$c_n = J_n(\xi) = \frac{1}{\pi}\int_0^{\pi} \cos(\xi \sin\lambda - n\lambda)\, d\lambda, \tag{7.3-6}$$

which follows directly from Eq. (7.3-5) by converting the exponential function into trigonometric form and observing that both $\sin(\cdot)$ and its argument are odd functions; hence, the sine term vanishes. Equation (7.3-6) shows the c_n to be real. Substituting back into Eq. (7.3-3) results in

$$e^{j\xi \sin \omega_m t} = \sum_{n=-\infty}^{\infty} J_n(\xi) e^{jn\omega_m t}. \tag{7.3-7}$$

This equation is known as the Bessel–Jacobi identity, and it is the desired Fourier expansion. By multiplying both sides by $\exp(j\omega_c t)$ and equating real and imaginary parts, we obtain

$$\cos(\omega_c t + \xi \sin \omega_m t) = \sum_{n=-\infty}^{\infty} J_n(\xi) \cos(\omega_c t + n\omega_m t), \tag{7.3-8}$$

$$\sin(\omega_c t + \xi \sin \omega_m t) = \sum_{n=-\infty}^{\infty} J_n(\xi) \sin(\omega_c t + n\omega_m t). \tag{7.3-9}$$

We see that angle modulation with a sinusoid results in a line spectrum with lines at $\omega_c \pm n\omega_m$, and with the height of the lines given by $J_n(\xi)$. See Fig. 7.3-1.

The precise shape of the spectrum depends on the behavior of $J_n(\xi)$. Many properties of the Bessel functions can be inferred from the defining equations (7.3-5 and 7.3-6) (see Prob. 7-4). Other properties are more easily obtained from the series form of the Bessel function. This can be obtained from the Bessel–Jacobi identity as follows: Write

$$\sin \omega_m t = \frac{e^{j\omega_m t} - e^{-j\omega_m t}}{2j} = \frac{1}{2j}(z - z^{-1}), \tag{7.3-10}$$

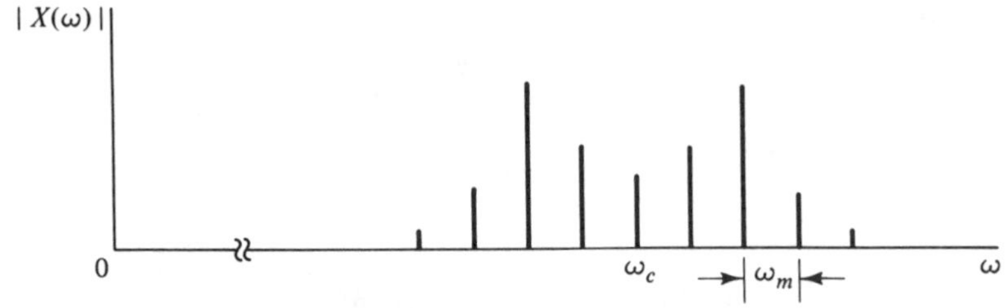

Figure 7.3-1. Line spectrum of an FM signal modulated by a single sinusoid

where $z = e^{j\omega_m t}$. Then $e^{j\xi \sin \omega_m t} = e^{(\xi/2)(z-z^{-1})} = e^{\xi z/2} e^{-\xi z^{-1}/2}$, and by using the Taylor series expansion for the exponential function

$$e^Y = \sum_{l=0}^{\infty} \frac{Y^l}{l!}$$

one gets

$$e^{j\xi \sin \omega_m t} = \sum_{l=0}^{\infty} \sum_{i=0}^{\infty} \frac{\xi^{i+l} z^{i-l}}{2^{i+l} i! l!}. \tag{7.3-11}$$

The substitution $n = i - l$ results in

$$e^{j\xi \sin \omega_m t} = \sum_{n=-\infty}^{\infty} z^n \sum_{l=0}^{\infty} \frac{(\xi/2)^{n+2l}(-1)^l}{(n+l)! l!} \tag{7.3-12}$$

If, finally, we replace z by $e^{j\omega_m t}$ and use the Bessel–Jacobi identity (Eq. 7.3-7), we see that

$$J_n(\xi) = \sum_{l=0}^{\infty} \frac{(\xi/2)^{n+2l}(-1)^l}{(n+l)! l!} \tag{7.3-13}$$

Properties of the Bessel Function

1. $J_0(0) = 1$.
2. $J_n(0) = 0$ if n is a nonzero integer.
3. $J_n(\xi) = J_n(-\xi)$ for even n.
4. $J_n(\xi) = -J_n(-\xi)$ for odd n.
5. $J_n(\xi) = J_{-n}(\xi)$ for even n.
6. $J_n(\xi) = -J_{-n}(\xi)$ for odd n.
7. $J_n(\xi) \to 0$ for $n \gg \xi$.
8. For small ξ, $J_n(\xi) \sim (\xi/2)^n/n!$.

The general shape of $J_n(\xi)$ as a function of ξ is shown in Fig. 7.3-2, and the variation with n for fixed ξ in Fig. 7.3-3. Bessel functions behave somewhat like damped sinusoids, but the zeros are not evenly spaced. The first few zeros for $J_0(\xi)$ and $J_1(\xi)$ are given in Table 7.3-1.

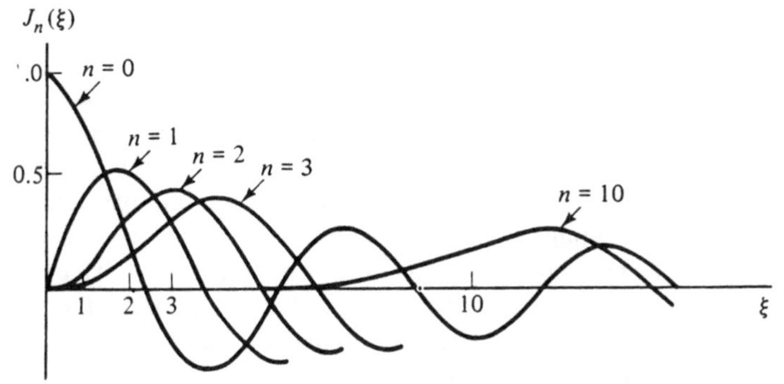

Figure 7.3-2. Variations of $J_n(\xi)$ with ξ.

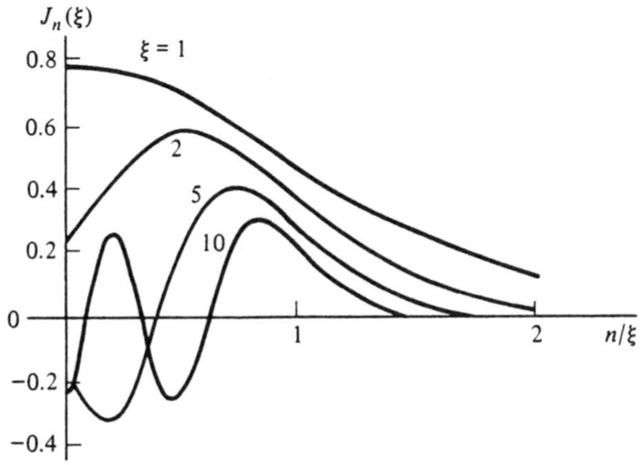

Figure 7.3-3. Variations of $J_n(\xi)$ versus n with ξ as a parameter.

TABLE 7.3-1 ZEROS OF BESSEL FUNCTIONS

$J_0(\xi_n) = 0$		$J_1(\lambda_n) = 0$	
Roots ξ_n	$J_1(\xi_n)$	Roots ξ_n	$J_0(\xi_n)$
2.4048	0.5191	0.0000	1.0000
5.5201	−0.3403	3.8317	−0.4028
8.6537	0.2715	7.0156	0.3001
11.7915	−0.2325	10.1735	−0.2497
14.9309	0.2065	13.3237	0.2184
18.0711	−0.1877	16.4706	−0.1965
21.2116	0.1733	19.6159	0.1801

Simple Bandwidth Calculation

Bessel function property 7 determines the effective bandwidth of an angle-modulated signal; that is, $J_n(\xi)$ vanishes rapidly for $n \gg \xi$. This is true particularly for large n and ξ, where n need not be very much larger than ξ for J_n to become negligibly small (cf. Fig. 7.3-2). This property has the effect of limiting the number of lines in the line spectrum to something like 2ξ (ξ lines on either side of the carrier).

Consider now an FM signal. From Eq. (7.2-12) we have $\xi = \beta = \Delta f/f_m$, where Δf is the frequency deviation. The effective bandwidth of the FM spectrum is the number of significant lines multiplied by the frequency spacing between them (i.e., nf_m). But if the significant number of lines is approximately equal to ξ (or β), the bandwidth is approximately equal to

$$\mathrm{BW} \approx 2\beta f_m = 2\Delta f \tag{7.3-14}$$

(i.e., twice the frequency deviation). This is what one would expect intuitively. The bandwidth generated by sweeping a carrier from the frequency $f_c - \Delta f$ to $f_c + \Delta f$ is $2\Delta f$. The approximation is especially good for large β (i.e., if a low-frequency modulation sweeps the carrier back and forth slowly). [Observe from Eq. (7.2-12) that β varies inversely with f_m.] Here we get many closely spaced lines, and the spectrum extends essentially from $f_c - \Delta f$ to $f_c + \Delta f$. For a high-frequency modulating signal, β is small, the lines are farther apart, and some of them will be outside the band $2\Delta f$. This is particularly true in the case of *narrowband* FM (NBFM), where $\beta \ll 1$ and where the bandwidth has to be at least $2f_m$. (NBFM is considered in more detail in Sec. 7.5.) Figure 7.3-4 shows several typical FM spectra and illustrates some of these points.

Composite Signal-Modulating Functions

The results that we have obtained so far are for a modulating function consisting of a single sine wave. More interesting modulating functions, such as those resulting from speech or music, can be represented by a Fourier series (i.e., by a sum of sine waves). To begin with, we therefore consider a signal consisting of two sine waves:

$$s(t) = A_1 \cos \omega_1 t + A_2 \cos \omega_2 t. \tag{7.3-15}$$

There are now two modulating indexes, $\beta_1 = A_1K''/\omega_1$ and $\beta_2 = A_2K''/\omega_2$, and the complex form of an FM wave can be written as

$$e^{j(\omega_c t+\beta_1 \sin \omega_1 t+\beta_2 \sin \omega_2 t)} = \sum_{n=-\infty}^{\infty} \sum_{k=-\infty}^{\infty} J_n(\beta_1)J_k(\beta_2)e^{j(\omega_c t+n\omega_1 t+k\omega_2 t)}. \tag{7.3-16}$$

Equating real and imaginary parts results in

$$\cos(\omega_c t + \beta_1 \sin \omega_1 t + \beta_2 \sin \omega_2 t)$$
$$= \sum_{n=-\infty}^{\infty} \sum_{k=-\infty}^{\infty} J_n(\beta_1)J_k(\beta_2) \cos(\omega_c + n\omega_1 + k\omega_2)t \tag{7.3-17}$$

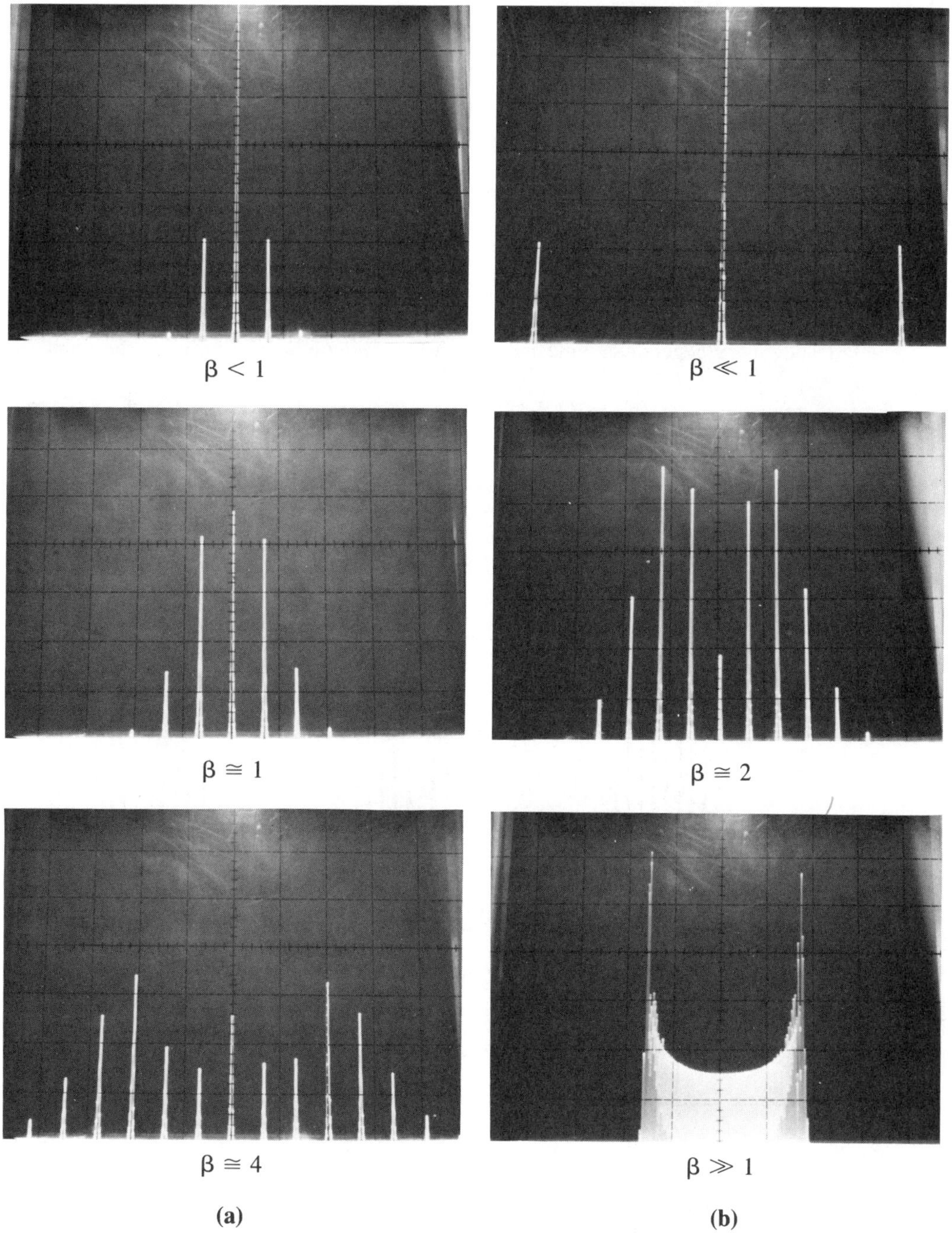

Figure 7.3-4. Spectrum of an FM wave as β increases due to (a) increasing Δf, fixed f_m; (b) decreasing f_m, fixed Δf.

and

$$\sin(\omega_c t + \beta_1 \sin\beta_1 t + \beta_2 \sin\omega_2 t)$$
$$= \sum_{n=-\infty}^{\infty} \sum_{k=-\infty}^{\infty} J_n(\beta_1)J_k(\beta_2)\sin(\omega_c + n\omega_1 + k\omega_2)t. \qquad (7.3\text{-}18)$$

If $\beta_2 = 0$, all k components vanish except for $k = 0$. The result is Eqs. (7.3-8) and (7.3.9). The frequencies $\omega_c + n\omega_1$, $n = 0, \pm1, \pm2, \ldots$, are generated by the modulating tone $A_1 \cos \omega_1 t$. Similarly, when $\beta_1 = 0$, we obtain only the frequencies $\omega_c + k\omega_2$, $k = 0, \pm1, \pm2, \ldots$, which are generated by $A_2 \cos \omega_2 t$. However, when both modulating tones are active, we obtain all possible combinations of $\omega_c + n\omega_1 + k\omega_2$. The new frequencies $n\omega_1 + k\omega_2$ ($n, k \neq 0$) clearly show how nonlinear FM and, more generally, angle modulation are. Figure 7.3-5 illustrates the point for $f_2 > f_1$.

Equations (7.3-16) to (7.3-18) can be generalized to a signal with M Fourier components. Thus if

$$s(t) = \sum_{i=1}^{M} A_i \cos(\omega_i t + \psi_i),$$

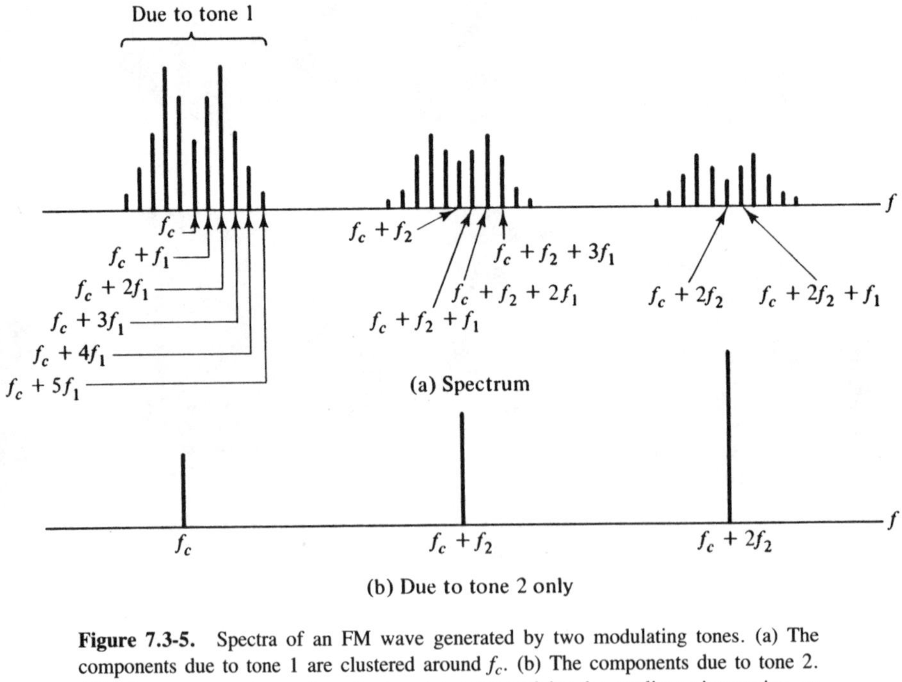

Figure 7.3-5. Spectra of an FM wave generated by two modulating tones. (a) The components due to tone 1 are clustered around f_c. (b) The components due to tone 2. All other components are beat frequencies generated by the nonlinear interaction on tones 1 and 2 in the FM modulation process.

the FM wave takes the form

$$x_{FM}(t) = A_c \cos\left[\omega_c t + \sum_{i=1}^{M} \beta_i \sin(\omega_i t + \psi_i)\right], \tag{7.3-19}$$

which can be written as

$$\begin{aligned} x_{FM}(t) &= \text{Re}\,(\exp\{j[\omega_c t + \sum_{i=1}^{M} \beta_i \sin(\omega_i t + \psi_i)]\}) \\ &= \text{Re}\left(\sum_{k_1=-\infty}^{\infty}\sum_{k_2=-\infty}^{\infty}\cdots\sum_{k_M=-\infty}^{\infty} J_{k_1}(\beta_1)\cdots J_{k_M}(\beta_M)\right. \\ &\quad \left.\cdot \exp\left\{j\left[\omega_c t + \sum_{i=1}^{M} k_i(\omega_i t + \psi_i)\right]\right\}\right) \\ &= \sum_{k_1}\cdots\sum_{k_M}\left[\prod_{i=1}^{M} J_{k_i}(\beta_i)\right]\cos\left[\omega_c t + \sum_{l=1}^{M} k_i(\omega_i t + \psi_i)\right]. \end{aligned} \tag{7.3-20}$$

This expression is unwieldy and not very informative. To see what happens in a particular case, consider a square-wave modulating function. This has the Fourier-series expansion [cf. Eq. (2.4-24)]

$$s(t) = \frac{4}{\pi}\left[\cos\frac{2\pi t}{T} - \tfrac{1}{3}\cos\frac{6\pi t}{T} + \tfrac{1}{5}\cos\frac{10\pi t}{T}\cdots\right] \tag{7.3-21}$$

We observe first that, since the higher-frequency terms of the square wave are all *harmonics* of the fundamental frequency, the combination of frequencies indicated in Eqs. (7.3-17) and (7.3-19) does *not* in fact generate new frequencies in the FM signal. All that the higher-frequency terms of $s(t)$ do is to change the magnitude of the harmonics of the FM signal produced by the lowest frequency in $s(t)$. Also, we note that, since the amplitudes of the higher harmonics in $s(t)$ are inversely proportional to the harmonic orders, the β's for the higher frequencies of $s(t)$ decrease as the square of the harmonic order. Thus the number of spectrum lines in the FM signal affected by the higher-frequency terms in the Fourier expansion of $s(t)$ rapidly decreases with harmonic order. We see therefore that the main effect of modulation with a square-wave signal is to alter the shape of the spectrum of the FM wave; it has relatively little effect on the bandwidth, which is determined largely by the harmonic with the largest β. Although this analysis is specific to the square wave, it suggests that the situation may not be too different for other kinds of composite signals. We consider next some of the empirical rules that are used to determine FM bandwidths in more general cases.

7.4 GENERAL BANDWIDTH CONSIDERATIONS FOR FM

We have seen that even a tone-modulated FM wave has a line spectrum that extends over an infinite band. However, in this respect FM signals are no different from many other kinds of signals; for instance, we found the same thing in the Fourier analysis of a square pulse in Chapter 2. In all these cases we have to decide what we mean by bandwidth. This issue was addressed rigorously by D. Slepian [4-2], but for our purposes we determine the bandwidth by the spread of *significant* components in the spectrum. Estimates of the spread of these significant components can be obtained a follows.

We have already observed that for single-tone modulation with large β, the total bandwidth is approximately $2\beta f_m$ = twice the frequency deviation Δf. However, a shortcoming of this result is that if $\beta \ll 1$, as is the case in NBFM, Δf may be less than f_m, and we would have no modulation at all if we assign a bandwidth $2\Delta f$. To retain some modulation, we must keep at least the first pair of sidebands at $f_c \pm f_m$. Thus, to ensure that we have modulation by the tone for any β, we could require that the bandwidth be given by

$$B_{\text{CR}} = 2\Delta f + 2f_m = 2f_m(1 + \beta) \tag{7.4-1}$$

$$\begin{aligned} &\simeq 2f_m\beta = 2\Delta f, \qquad \beta \gg 1 \\ &\simeq 2f_m, \qquad \beta \ll 1 \text{ (NBFM).} \end{aligned} \tag{7.4-2}$$

Equation (7.4-1) is a handy rule of thumb and a special case of a bandwidth formula called *Carson's rule* (the subscript CR is for Carson's rule).

For high-quality FM transmission or for modulation with high-frequency signals, the "$n > \beta$" rule for determining significant components may not suffice, and we then need to be more specific about what we mean by significant Fourier components. Suppose that we define as significant all components for which $|J_n(\beta)| \geq \epsilon$. This criterion requires that significant sidebands be at least $100\epsilon\%$ of the amplitude of the unmodulated carrier. Let L be the largest integer, for β fixed, for which this criterion is met; that is,

$$|J_L(\beta)| \geq \epsilon, \qquad |J_{L+1}(\beta)| < \epsilon.$$

Then L depends on β and ϵ, and the required bandwidth is

$$B_\epsilon = 2L(\beta, \epsilon)f_m. \tag{7.4-3}$$

Figure 7.4-1 is a plot of $L(\beta, \epsilon)$ versus β for the 1% ($\epsilon = 0.01$) and 10% sideband ($\epsilon = 0.1$) criteria. The 10% criterion may lead to slight distortion, while 1% is overly conservative. The dashed line is a reasonable compromise ([7-5], p. 240).

To obtain an idea of the relation among bandwidth, number of significant components, and β, let us consider commercial FM for which the FCC requires that Δf be no greater than 75 kHz and the modulating tones range from, say, 30 Hz to 15 kHz. The computation of β, B_{CR}, $L(\beta, \epsilon)$, and B_ϵ for $\epsilon = 0.01$ and 0.1 is summarized in Table 7.4-1. This table demonstrates several interesting facts concerning FM bandwidth requirements. First, by assuming that all tones produce the maximum frequency deviation, we see that the largest modulating frequency requires the largest bandwidth and that the largest β's are associated with the smallest bandwidths. Second, for large values of β, the bandwidth is simply twice the frequency deviation, regardless of which criterion is chosen. Third, Carson's rule for estimating bandwidth corresponds closely to the 10% sideband bandwidth, although it is somewhat more conservative than this criterion. The FCC channel-width allocation for FM stations is 200 kHz. Although this bandwidth

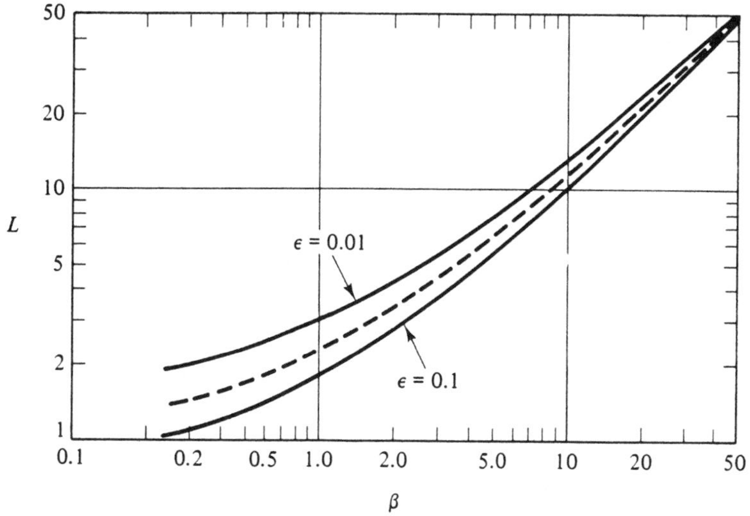

Figure 7.4-1. Significant sideband pairs L as a function of β. (Adapted from Carlson ([7-5], p. 240), by permission.)

pertains to modulation by a signal $s(t)$ that is not a pure tone, we shall see that its determination is based on the same principles as in the case of tone modulation.

We have assumed so far that all tones produce the maximum frequency deviation of 75 kHz. But if the various tone amplitudes do not produce the same frequency deviation, then it is not simply the highest frequency that requires the greatest bandwidth. For example, let a 15-kHz tone have an amplitude that is only 40% of the maximum. Then the frequency deviation is 30 kHz and $\beta = 2$. If we use the 1.0% criterion, then $L(2, 0.01) = 6$, giving $B_{0.01} = 150$ kHz. Likewise, the 10% criterion gives $L(2, 0.1) = 3$, which furnishes $B_{0.1} = 90$ kHz. These bandwidths are much lower than the required bandwidths for full-amplitude tone modulation. Thus the bandwidth is seen to depend on frequency deviation *and* modulation frequency. This being the case, we might ask the following: Given an ensemble of tones, $\{s_i(t)\}$, which single modulating

TABLE 7.4-1 BANDWIDTHS FOR VARIOUS TONES, $\Delta f = 75$ KHZ

f_m (kHz)	β	$L(\beta, 0.01)$	$L(\beta, 0.1)$	$B_{0.01}$ (kHz)	$B_{0.1}$ (kHz)	B_{CR} (kHz)
15	5	8	6	240	180	180
10	7.5	11	8	220	160	170
3.75	20	22	20	165	150	157
1	75	75	75	150	150	152
0.05	1500	1500	1500	150	150	150

tone will require the greatest transmission bandwidth? An approximate analytical approach to this problem is furnished by Carson's formula, Eq. (7.4-1):

$$\begin{aligned} B_{CR} &= 2f_m(1 + \beta) \\ &= 2(KA_m + f_m). \end{aligned} \tag{7.4-4}$$

Clearly, the greatest bandwidth is determined by the tone $s^*(t)$, with amplitude A_m^* and frequency f_m^*, which satisfies

$$\max_{s_i} 2(KA_{m_i} + f_{m_i}) = 2(KA_m^* + f_m^*), \tag{7.4-5}$$

where A_{m_i} and f_{m_i} are the amplitude and frequency, respectively, of $s_i(t)$. In the *worst-case* analysis, we let $A_m = 1$ and $f_m = W$. The resulting bandwidth, B_{max}, is given by

$$B_{max} = 2([\Delta f]_{max} + W), \tag{7.4-6}$$

where $[\Delta f]_{max}$ is the maximum allowed frequency deviation and W is the highest frequency in the baseband.

The use of Carson's rule to compute B_{max} is somewhat arbitrary. For example, Carlson ([7-5], p. 241) uses the relation

$$L(\beta, \bar{\epsilon}) \simeq \beta + C, \tag{7.4-7}$$

which is an approximate fit to the dashed line in Fig. 7.4-1. $\bar{\epsilon}$ is a compromise between the 0.1 and 0.01 sideband criteria (the dashed line in Fig. 7.4-1), and C is a constant representing the extrapolated ordinate when $\beta = 0$. The constant C lies between 1 and 2, but setting $C = 2$ furnishes a formula that estimates the required bandwidth somewhat more conservatively than Carson's rule of thumb. Use of Eq. (7.4-7) in Eq. (7.4-3) furnishes the formula

$$B_{\bar{\epsilon}} = 2(\Delta f + 2f_m). \tag{7.4-8}$$

The worst-case bandwidth involving a signal of maximum amplitude at the highest baseband frequency is

$$[B_{\bar{\epsilon}}]_{max} = 2[(\Delta f)_{max} + 2W]. \tag{7.4-9}$$

For $(\Delta f)_{max} = 75$ kHz and $W = 15$ kHz, this formula gives a bandwidth of 210 kHz, which is quite close to the 200 kHz assigned to commercial FM channels in the United States.

Modulation by a Composite Signal

The exact determination of the bandwidth for an arbitrary modulating signal is difficult to do without resorting to a numerical evaluation of an expression of the type given in Eq. (7.3-20). However, for some signals, particularly those that have no significant components above some highest frequency, say W, such as speech or music, the required transmission bandwidth can be estimated from other considerations. In particular, the bandwidth formulas developed earlier in this section can be extended to furnish useful estimates of bandwidth, although the strong nonlinearity of FM does not permit the results to be as rigorous as we might like. The bandwidth estimation formulas are sometimes empirically adjusted to furnish estimates that are more in line with the bandwidths that are required in practice.

In the analysis of the FM spectrum generated by a square-wave modulation, we found that the bandwidth was determined largely by the harmonic with the largest β. In general, the harmonic structure of the modulating signal is more complicated, but similar considerations can be used. If the important frequencies in a composite signal are not harmonically related, we can use the fact that, when a carrier is modulated by two tones f_1 and f_2, sidebands occur at frequencies $f_c + kf_2 + nf_1$, where n and k are positive and negative integers. For $f_2 > f_1$, the beat-frequency sidebands show up as satellites around the frequencies $f_c + kf_2$, and the amplitudes of these components are proportional to the product $J_n(\beta_1)J_k(\beta_2)$, which becomes negligible when $J_k(\beta_2)$ becomes negligible. To a first approximation, then, we can say that in the case of a composite signal the bandwidth is determined by a full-amplitude tone (i.e., one that produces maximum frequency deviation) at the highest significant frequency, W, in the baseband. If β^* denotes the ratio of maximum frequency deviation to W, then the bandwidth can be estimated from Eq. 7.4-3, which furnishes

$$B_\epsilon = 2L(\beta^*, \epsilon)W. \tag{7.4-10}$$

Carson's rule is more convenient to use and furnishes

$$B_{CR} = 2W(1 + \beta^*), \tag{7.4-11}$$

and the "conservative" linear approximation, i.e., Eq. (7.5-10), furnishes

$$[B_\epsilon]_{max} = 2W(2 + \beta^*). \tag{7.4-12}$$

Example

In commercial FM, the FCC allows $[\Delta f]_{max} = 75$ kHz. Let the bandwidth of $s(t)$ be $W = 15$ kHz. Then $\beta^* = 5$. Carson's rule gives $B_{CR} = 180$ kHz, while Eq. (7.4-12) gives 210 kHz. The latter figure is closer, on the conservative side, to the 200 kHz nominally assigned.

Variations from the above formulas are also in use. For example, a formula that is used in FM telephony is

$$B = 2W(1 + \alpha\beta^*),$$

where α is a constant that depends on the quality of transmission. For commercial telephony $\alpha = 1$, which, for $[\Delta f]_{max} = 15$ kHz and $W = 3$ kHz (this bandwidth enables voice identification), furnishes $\beta^* = 5$. The bandwidth is then 36 kHz. For higher-fidelity communication larger values of α are necessary.

7.5 NARROWBAND ANGLE MODULATION

The analysis of angle-modulated waves is somewhat simplified if the phase deviation, K_d, in PM or the modulation index, β, in FM is considerably smaller than 1 radian. Consider a PM wave modulated by a single-frequency tone at radian frequency ω_m. Then

$$\begin{aligned} x_{\mathrm{PM}}(t) &= \cos(\omega_c t + K_d \cos \omega_m t) \\ &= \cos \omega_c t \cos(K_d \cos \omega_m t) - \sin \omega_c t \sin(K_d \cos \omega_m t) \\ &= \cos \omega_c t \left(1 - \frac{K_d^2 \cos^2 \omega_m t}{2!} + \frac{K_d^4 \cos^4 \omega_m t}{4!} - \cdots\right) \\ &\quad - \sin \omega_c t \left(K_d \cos \omega_m t - \frac{K_d^3 \cos^3 \omega_m t}{3!} + \cdots\right), \end{aligned} \tag{7.5-1}$$

where we have replaced $\cos(K_d \cos \omega_m t)$ and $\sin(K_d \cos \omega_m t)$ by their Maclaurin series expansion.

If $K_d < 0.2$ radian, then, to a good approximation, we can omit all powers above the first and obtain

$$\begin{aligned} x_{\mathrm{PM}}(t) &= \cos \omega_c t - K_d \sin \omega_c t \cos \omega_m t \\ &= \cos \omega_c t - \tfrac{1}{2} K_d [\sin(\omega_c + \omega_m)t + \sin(\omega_c - \omega_m)t]. \end{aligned} \tag{7.5-2}$$

The spectrum of such a wave is shown in Fig. 7.5-1 and superficially resembles AM in that the only significant frequencies are a pair of symmetrical sidebands at $f = f_c - f_m$ and $f = f_c + f_m$, and a carrier frequency f_c.

However, the sidebands of narrowband PM sum to a component in quadrature with the carrier, while in AM the sidebands sum to a component in phase with the carrier. This is easily seen from a phasor diagram.† The complex form of Eq. (7.5-2) is

$$\begin{aligned} \tilde{x}_{\mathrm{PM}}(t) &= e^{j\omega_c t}\left[1 + \frac{K_d}{2} e^{j(\omega_m t + \pi/2)} + \frac{K_d}{2} e^{j(-\omega_m t + \pi/2)}\right] \\ &= e^{j\omega_c t} X_{\mathrm{PM}}(t), \end{aligned} \tag{7.5-3}$$

where the $X_{\mathrm{PM}}(t)$ are the three terms in square brackets. The real wave $x_{\mathrm{PM}}(t)$ is reconstructed from

$$x_{\mathrm{PM}}(t) = \mathrm{Re}\,[e^{j\omega_c t} X_{\mathrm{PM}}(t)]. \tag{7.5-4}$$

The factor $\exp(j\omega_c t)$ is the carrier phasor and rotates in the counterclockwise direction at an angular velocity ω_c. Without loss of generality, we can suppress its effect by

† For the reader not familiar with this notion, the phasor is basically a vector representation of the complex or "analytic" form of a signal $x(t)$. For example, $x(t) = A \cos(\omega_c t + \theta)$ can be represented by the complex signal $\tilde{x}(t) = A \exp[+j(\omega_c t + \theta)]$. The phasor would then be interpreted as having length A and angle $\omega_c t + \theta$ from the real axis. It would be drawn as a vector whose projection on the real axis is $x(t)$.

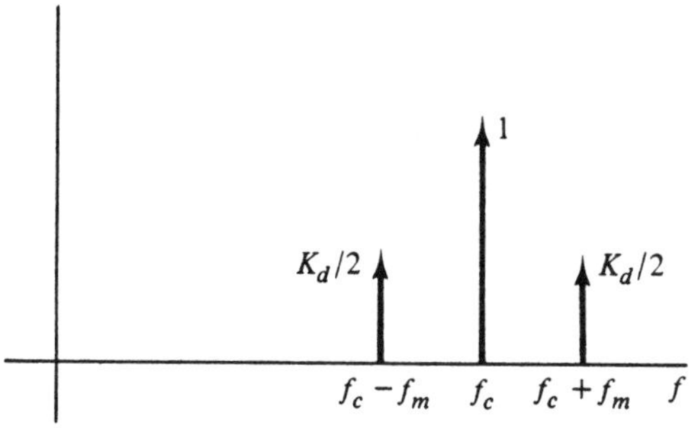

Figure 7.5-1. Amplitude spectrum of narrowband PM.

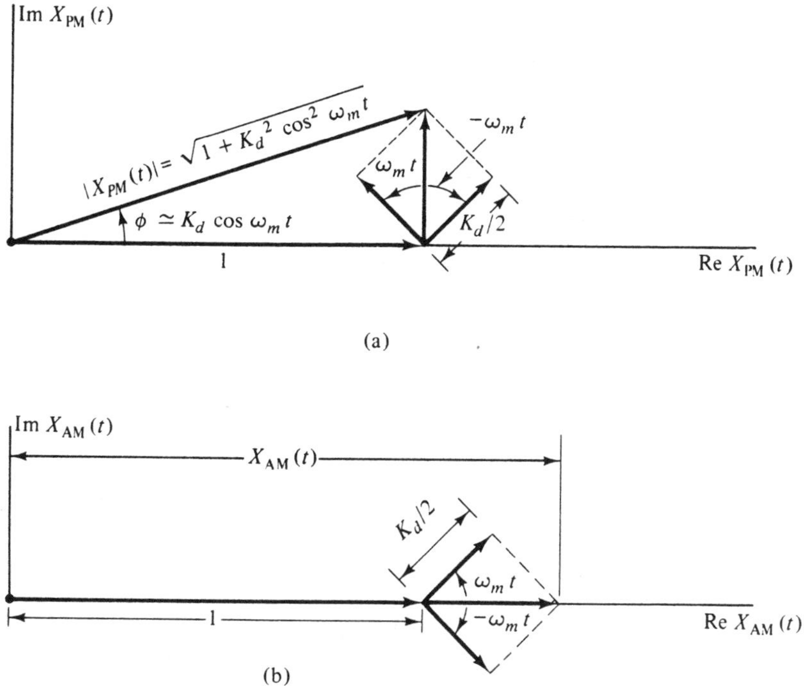

Figure 7.5-2. Phasors for narrowband PM and AM. (a) PM. (b) AM.

considering all rotations relative to it. The important quantity is then the resultant $X_{PM}(t)$. It is shown in Fig. 7.5-2(a). The phasor diagram for AM is shown in Fig. 7.5-2(b). Note that in PM the amplitude of the wave hardly changes† but the phase does, while in AM there is no phase change but a considerable amplitude change.

The same results can be obtained for narrowband FM (NBFM). In this case

$$x_{FM}(t) = \cos(\omega_c t + \beta \sin \omega_m t),$$

† The small change in amplitude results from neglecting all the higher power terms in Eq. (7.5-1). The exact expression would show no amplitude change since by definition there is none in angle modulation.

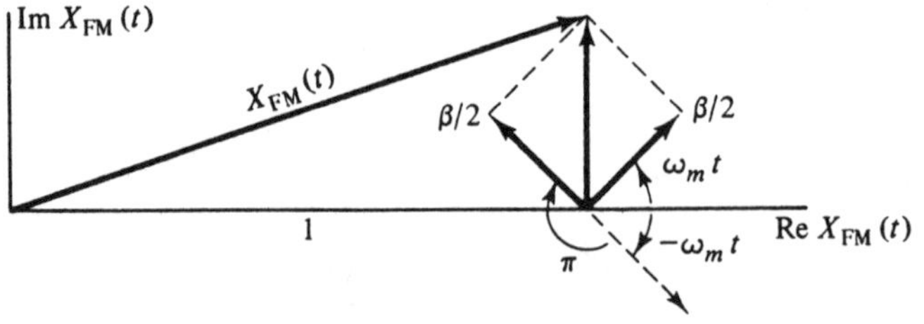

Figure 7.5-3. Phasor diagram for narrowband FM. As in the case of PM, the sidebands sum to a component in quadrature with the carrier.

which, for $\beta \ll 1$ radian (i.e., $\beta \leq 0.2$ radian), gives

$$x_{FM}(t) \simeq \cos \omega_c t - \frac{\beta}{2}[\cos(\omega_c - \omega_m)t - \cos(\omega_c + \omega_m)t]. \tag{7.5-5}$$

Equation (7.5-5) was derived using techniques similar to those leading up to Eq. (7.5-2). The phasor diagram is obtained from the complex wave

$$\tilde{x}_{FM}(t) = e^{j\omega_c t} X_{FM}(t),$$

where

$$X_{FM}(t) = 1 + \frac{\beta}{2} e^{-j(\omega_m t - \pi)} + \frac{\beta}{2} e^{j\omega_m t}. \tag{7.5-6}$$

Figure 7.5-3 shows how the sidebands sum to a component in quadrature with the carrier.

The spectrum of narrowband angle modulation is thus similar to AM except for the phase quadrature of the sidebands with respect to the carrier. This suggests a method of generating narrowband angle modulation in terms of AM circuitry with which we are already familiar. We recall that to generate AM we could add a carrier to a DSB output produced by a balanced modulator [Fig. 7.5-4(a)]. If we phase-shift the carrier by 90° before addition, we produce a PM wave [Fig. 7.5-4(b)]. If $s(t)$ is integrated before modulation, we produce an FM wave [Fig. 7.5-4(c)]. These systems apply to narrowband angle modulation only, but they form the basis for the Armstrong method for generating wide-band FM considered in the next section.

7.6 INDIRECT GENERATION OF FM: THE ARMSTRONG METHOD

E. H. Armstrong was the first to demonstrate the feasibility and merits of FM by using an indirect technique to generate FM. The indirect FM (IFM) method consists of three important steps:

1. Integrate the modulating signal $s(t)$ to produce the signal

$$x(t) = \int^t s(\lambda)\, d\lambda.$$

2. Phase-modulate a carrier with $z(t)$ to produce NBFM.
3. Use a system of frequency multipliers to convert NBFM to wideband FM.

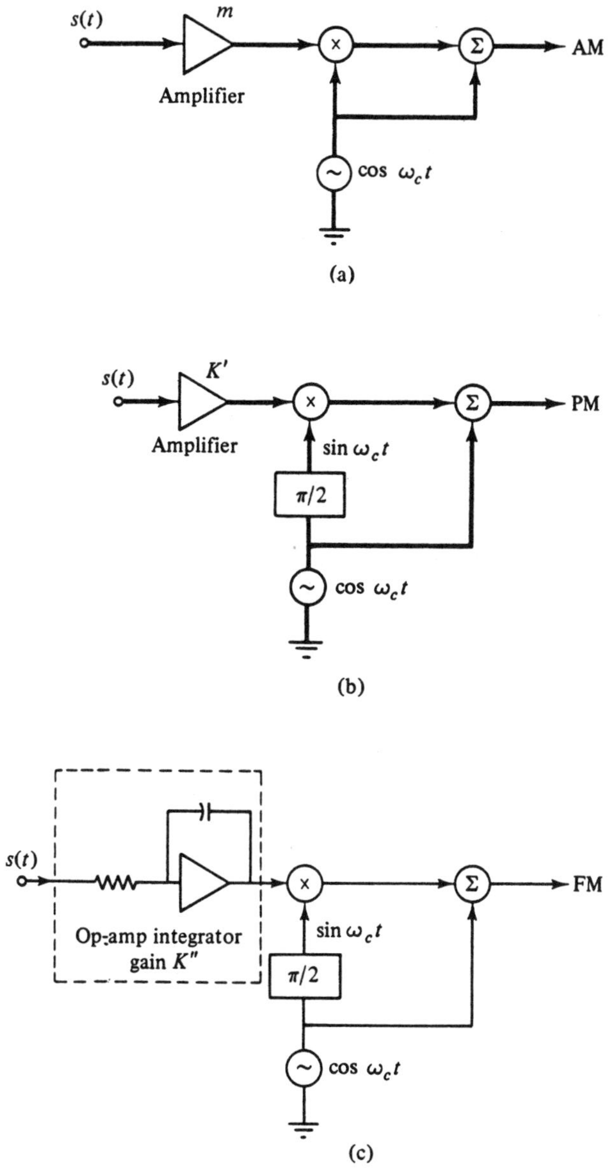

Figure 7.5-4. (a) Generation of AM. (b) Generation of NBPM. (c) Generation of NBFM.

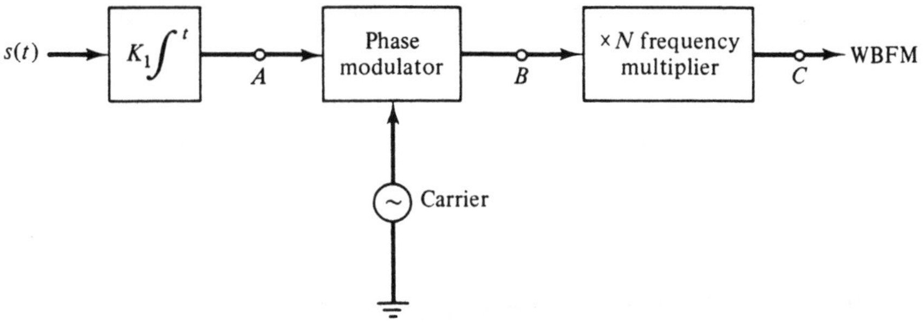

Figure 7.6-1. Method of generating indirect FM.

A block diagram of an IFM system is shown in Fig. 7.6-1. At point A the output is

$$\xi(t) = K_1 \int^t s(\lambda)\, d\lambda = K_1 z(t), \tag{7.6-1}$$

and at point B the output is proportional to

$$\begin{aligned} x_1(t) &= \cos\,[\omega_{c_1} t + K_1'' \int^t x(\lambda)\, d\lambda] \\ &= \cos\,(\omega_{c_1} t + \beta_1 \sin \omega_m t) \end{aligned} \tag{7.6-2}$$

if sine-wave modulation is assumed. The phase deviation $\beta_1 \equiv (\Delta f)_1/f_m$ is typically much less than 0.5 radian, so $x_1(t)$ can be considered, within reason, an NBFM wave. The frequency multiplier produces a signal with frequency ω given by

$$\omega = N\dot{\theta}_1(t), \tag{7.6-3}$$

where $\theta_1(t) \equiv \omega_{c_1} t + \beta_1 \sin \omega_m t$. Hence

$$\omega = N\omega_{c_1} + N\beta_1 \omega_m \cos \omega_m t.$$

The new phase is proportional to the integral of $N\dot{\theta}_1(t)$, so at C we have

$$x(t) = \cos\,(N\omega_{c_1} t + N\beta_1 \sin \omega_m t). \tag{7.6-4}$$

By writing $\omega_c = N\omega_{c_1}$ and $\beta = N\beta_1$, we can rewrite Eq. (7.6-4) in standard form:

$$x(t) = \cos\,(\omega_c t + \beta \sin \omega_m t),$$

where β is now a value typical of wideband FM.

Example

Let f_m range from 100 to 10,000 Hz, and let the maximum frequency deviation, Δf, at the output be 50 kHz. Then

$$\beta_{\min} = \frac{50 \times 10^3}{10^4} = 5,$$

$$\beta_{\max} = \frac{50 \times 10^3}{10^2} = 500.$$

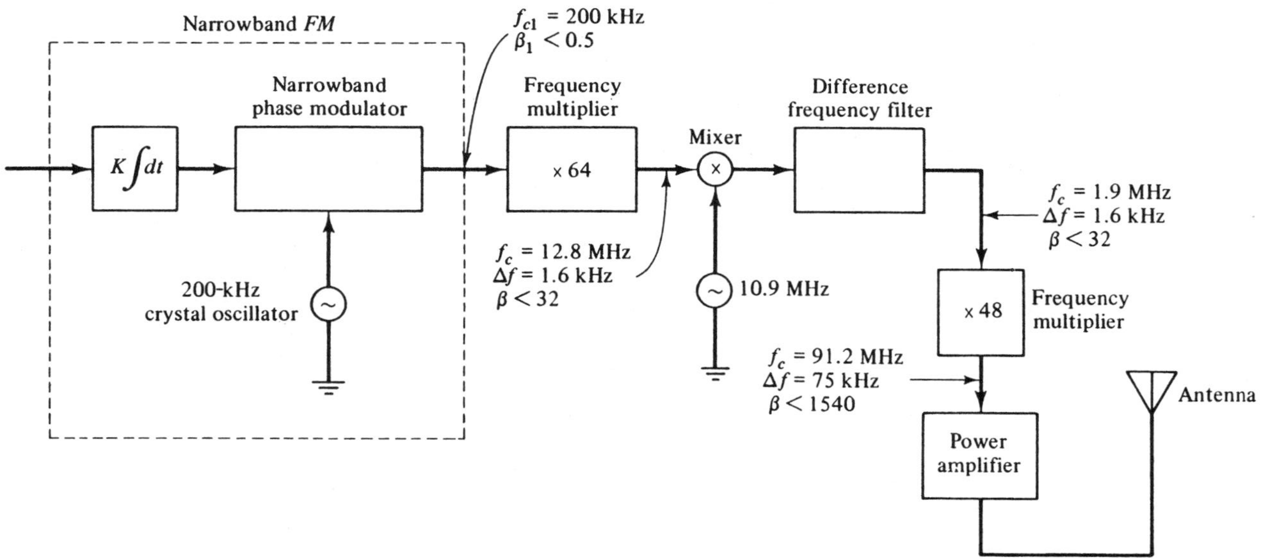

Figure 7.6-2. Armstrong-type indirect FM transmitter. (Adapted from M. Schwartz, *Information Transmission, Modulation and Noise*, McGraw-Hill Book Company, New York, 1959, p. 134, by permission.)

If $[\beta_1]_{max} = 0.5$, then the required frequency multiplication is

$$N = \frac{\beta_{max}}{[\beta_1]_{max}} = \frac{500}{0.5} = 1000.$$

The maximum allowed frequency deviation at the *input* is

$$\Delta f_1 = \frac{50 \times 10^3}{1000} = 50 \text{ Hz}$$

if the maximum specified Δf at the output is 50 kHz. If the initial carrier frequency f_{c_1} were, say, 200 kHz, then the final frequency would be $f_c = 200$ MHz. This figure is too high for standard FM broadcasting, and frequency converters are used to reduce f_c to the desired band. For example, if we heterodyne $x(t)$ in Eq. (7.6-4) with a carrier wave of frequency, f_{LO}, the modulation index $N\beta_1$ remains unaffected but the wave is shifted to the new carrier frequency, $Nf_c - f_{LO}$.

A block diagram of an Armstrong-type indirect transmitter is shown in Fig. 7.6-2. The largest modulation index furnished by the NBFM system is determined by the lowest modulating frequency ($\simeq$ 50 Hz), and this determines the required frequency multiplication. For example, if $\beta_1 \simeq 0.5$ for a 50-Hz tone, then the 75-kHz frequency deviation requires a $\beta = 1500$. The required multiplication is $1500/0.5 = 3000$. The largest tone frequency determines the bandwidth. The value of β_1 for the 15-kHz signal is approximately 1.7×10^{-3}. The extremely small initial values of β_1 are required to prevent distortion due to amplitude-modulation effects that can occur in the generation of NBFM by the method shown in Fig. 7.4-4.

The indirect method of producing FM enjoys good frequency stability, which constitutes a distinct advantage over the direct method (Sec. 7.7). However, the repeated stages of frequency multiplication as well as the extra heterodyning require considerable circuit complexity, which must be judged a disadvantage. The signal integrator must operate over a frequency range on the order of 1000 to 1. If the integration is done with *RC* elements in an operational amplifier configuration, the integrating capacitor must have high Q and small leakage over the frequency range of the signal.

7.7 GENERATION OF FM BY DIRECT METHODS

In the direct generation of FM, all that is essentially required is a device whose output frequency varies linearly with the level of the applied signal. The voltage-controlled oscillator is such a device, and a particular design of a VCO was discussed in Chapter 6. Another technique for achieving the same result is to use a tuned circuit oscillator in which one of the resonant circuit elements is a variable reactance. A useful device in this respect is the varactor, which is a diode whose barrier capacitance C depends on the applied voltage. The capacitance depends inversely on the width of the space-charge layer, which in turn depends on the externally applied voltage. The capacitance versus applied voltage relation for varactor diodes can be approximated, over a limited region of operation, by a linear function. A description of varactors (also called varicaps

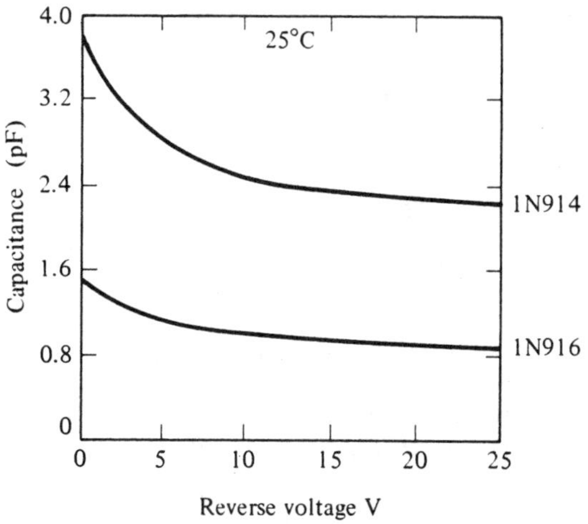

Figure 7.7-1. Capacitance variation with applied voltage for silicon diodes. (Courtesy of Fairchild Semiconductor Corporation.)

or voltacaps) is furnished in [7-6, p. 137]. Figure 7.7-1 shows typical capacitance versus applied voltage variations for varactor diodes.

Figure 7.7-2 shows, in simplified form, a tuned circuit oscillator in which the resonant frequency is controlled by a varactor diode. The resistors R_1, R_2, and R_e furnish the quiescent self-bias to start the oscillations, and the dynamic self-bias is obtained from the R_2C'' combination due to the flow of base current. The polarities of the primary

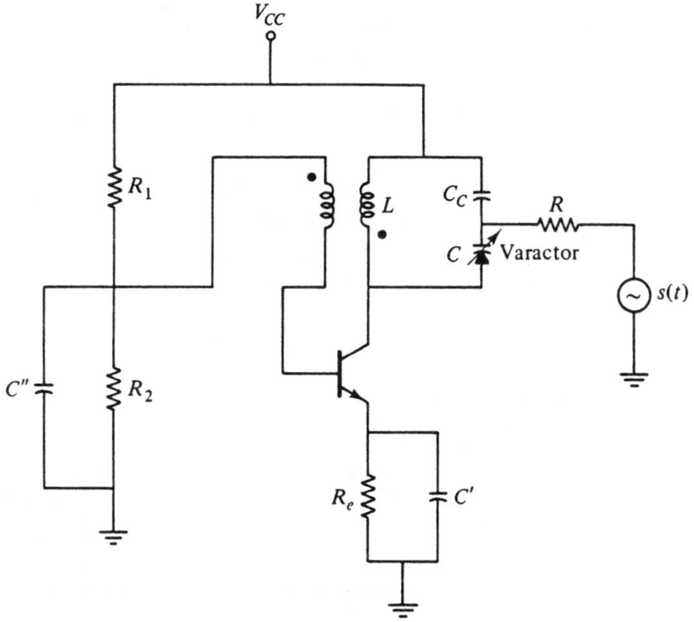

Figure 7.7-2. Tuned circuit oscillator with a voltage-controlled varactor in the tuned circuit.

and secondary windings are reversed, which furnishes a 180° phase shift, in the collector-to-base feedback; the additional 180° phase shift between base and collector gives a net loop phase shift of zero. The capacitance C_C represents a high impedance to $s(t)$; but if $C_C \gg C$, the instantaneous resonant frequency is unaffected by C_C and is given by (winding losses are disregarded)

$$\omega = [LC]^{-1/2}. \tag{7.7-1}$$

Now assume that when a signal $s(t)$ is applied the capacitance can be described by

$$C = C_0\left[1 - \frac{\Delta C}{C_0} s(t)\right], \qquad |s(t)| < 1, \tag{7.7-2}$$

where C_0 is the capacitance when $s(t) = 0$ and ΔC represents the maximum deviation in capacitance from C_0. If we assume that $\Delta C/C_0 \ll 1$, then the instantaneous oscillator frequency may be written as

$$\omega \simeq \omega_c\left[1 + \frac{\Delta C}{2C_0} s(t)\right], \tag{7.7-3}$$

where $\omega_c = [LC_0]^{1/2}$. The output signal can be written as

$$x(t) = \cos\left[\omega_c t + K'' \int^t s(\lambda)\, d\lambda\right], \tag{7.7-4}$$

which is the standard form for an FM wave, with $K'' \equiv (\omega_c \Delta C)/2C_0$. We can obtain some idea of how large ΔC must be from Eq. (7.7-3). For example, if $f_c = 30$ MHz and the maximum frequency deviation is $\Delta f = 25$ kHz, then

$$\frac{\Delta C}{2C_0} = \frac{\Delta f}{f_c} = \frac{25}{30} \times 10^{-3} = 0.833 \times 10^{-3}.$$

A block diagram for an illustrative direct FM system is shown in Fig. 7.7-3. The main advantages of direct FM are the reduced need for frequency multiplication and heterodyne frequency conversion. On the minus side, the high carrier frequencies must be stabilized against even small drifts since, in the example considered, a drift of only 0.08% corre-

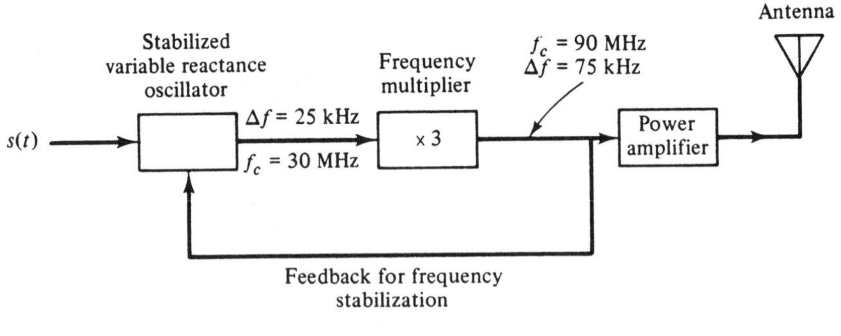

Figure 7.7-3. Simplified diagram of a direct FM system.

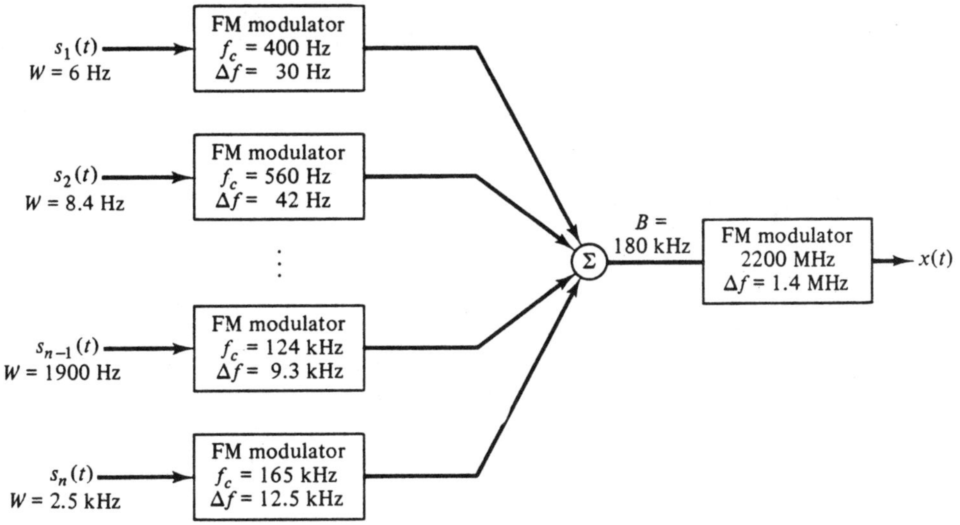

Figure 7.7-4. FM/FM system for space telemetry (proportional-bandwidth subcarrier channels). For each channel, $\Delta f/W \simeq 5$ and $\Delta f/f_c = 7.5\%$. The nth carrier is given by $f_n = f_{n-1}(W_n/W_{n-1})$. (From M. Schwartz ([7-7], p. 256), by permission.)

sponds to the maximum allowed frequency deviation. For this reason, feedback is used to stabilize the oscillator frequency. A frequency drift error signal can be detected with an appropriate discriminator and used to control the oscillator.

Some typical direct FM systems for use in space telemetry are given in [7-7, pp. 255–256]. An example of such a system, which uses multiplexing of several channels, is shown in Fig. 7.7-4.

7.8 FM SIGNALS IN LINEAR NETWORKS

Before discussing the problem of FM demodulation, it is worthwhile to consider the problem of transmitting an FM wave through a linear network. Formally at least, the solution is straightforward. If we denote the network transmittance by $|H(\omega)|$ exp $[-j\gamma(\omega)]$, where $|H(\omega)|$ is the magnitude of the transmittance and $\gamma(\omega)$ is the phase, the response of the network to

$$x(t) = \cos(\omega_c t + \beta \sin \omega_m t) \tag{7.8-1}$$

is

$$y(t) = \sum_{n=-\infty}^{\infty} |H(\omega_c + n\omega_m)| J_n(\beta) \cos[(\omega_c + n\omega_m)t - \gamma(\omega_c + n\omega_m)], \tag{7.8-2}$$

where we have used Eq. (7.3-8); that is,

$$x(t) = \sum_{n=-\infty}^{\infty} J_n(\beta) \cos(\omega_c + n\omega_m)t. \tag{7.8-3}$$

Equation (7.8-2) gives the solution that is associated with an FM wave modulated by a single tone. Even in this most elementary case, the expression is very complicated and furnishes little insight. For arbitrary modulating signals, an expression such as Eq. (7.3-20) would be needed, and the meaning would be even less clear.

The Quasi-Static Method

An alternative to the direct but possibly tedious FM-response computations described by the previous equations is the quasi-static method (QSM). The underlying assumption in the QSM is that the steady-state solution evaluated at the instantaneous frequency is a good first approximation that can be systematically improved by the addition of higher-order terms. Unfortunately, convergence of the resultant series is not guaranteed, and estimation of the remaining error at any particular stage of the calculation is not easy ([7-8], p. 229). However, we shall not be concerned with convergence problems in the discussion that follows.

We write

$$x(t) = \cos\left[\omega_c t + K'' \int^t s(\lambda)\, d\lambda\right] = \text{Re}\, z(t), \tag{7.8-4}$$

where

$$z(t) = e^{j[\omega_c t + \mu(t)]} \tag{7.8-5}$$

and

$$\mu(t) \equiv K'' \int^t s(\lambda)\, d\lambda. \tag{7.8-6}$$

Let the transmittance of the network (Fig. 7.8-1) be $H(\omega)$. The response $y(t)$ is given by

$$y(t) = \text{Re}\,\eta(t) = \text{Re} \int_{-\infty}^{\infty} Z(\omega) H(\omega) e^{j\omega t} \frac{d\omega}{2\pi}, \tag{7.8-7}$$

where the capital letters indicate Fourier transforms of the lowercase functions; that is, $Z(\omega) = \mathcal{F}[z(t)]$, and so on. We describe the transmittance by

$$H(\omega) = A(\omega) e^{-j\gamma(\omega)}, \tag{7.8-8}$$

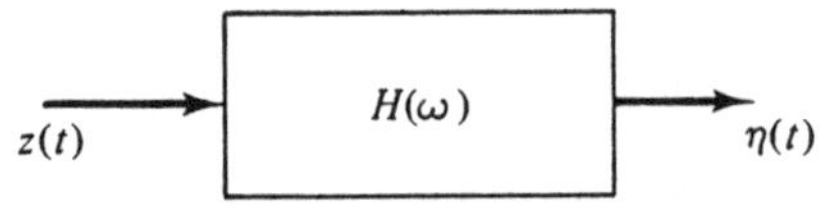

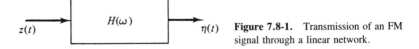

Figure 7.8-1. Transmission of an FM signal through a linear network.

where $A(\omega)$ and $\gamma(\omega)$ are real. Now assume that $H(\omega)$ can be expanded in a Taylor series about $\omega = \omega_c$. Then

$$H(\omega) = H(\omega_c) + H'(\omega_c)(\omega - \omega_c) + H''(\omega_c)\frac{(\omega - \omega_c)^2}{2!} + \cdots + H^{(n)}(\omega_c)\frac{(\omega - \omega_c)^n}{n!} + \cdots, \tag{7.8-9}$$

where

$$H^{(i)}(\omega) \equiv \frac{d^i}{d\omega^i}[H(\omega)]_{\omega=\omega_c}. \tag{7.8-10}$$

If we substitute Eq. (7.8-9) into Eq. (7.8-7) and consider $\eta(t)$ rather than Re $\eta(t)$, we obtain

$$\eta(t) = \sum_{n=0}^{\infty} \frac{1}{n!} H^{(n)}(\omega_c) \int_{-\infty}^{\infty} Z(\omega)(\omega - \omega_c)^n e^{j\omega t} \frac{d\omega}{2\pi} \tag{7.8-11}$$

$$= \sum_{n=0}^{\infty} \frac{1}{n!} H^{(n)}(\omega_c) e^{j\omega_c t} \int_{-\infty}^{\infty} Z(\omega)(\omega - \omega_c)^n e^{j(\omega-\omega_c)t} \frac{d\omega}{2\pi}. \tag{7.8-12}$$

Observe that

$$\frac{1}{j}\frac{d}{dt}\left[\int_{-\infty}^{\infty} Z(\omega) e^{j(\omega-\omega_c)t} \frac{d\omega}{2\pi}\right] = \int_{-\infty}^{\infty} Z(\omega)(\omega - \omega_c) e^{j(\omega-\omega_c)t} \frac{d\omega}{2\pi} \tag{7.8-13}$$

and, more generally, that

$$\frac{1}{j^n}\frac{d^n}{dt^n}\int_{-\infty}^{\infty} Z(\omega) e^{j(\omega-\omega_c)t} \frac{d\omega}{2\pi} = \int_{-\infty}^{\infty} Z(\omega)(\omega - \omega_c)^n e^{j(\omega-\omega_c)t} \frac{d\omega}{2\omega}. \tag{7.8-14}$$

Since

$$z(t) = \int_{-\infty}^{\infty} Z(\omega) e^{j\omega t} \frac{d\omega}{2\pi},$$

it follows that

$$z(t)e^{-j\omega_c t} = \int_{-\infty}^{\infty} Z(\omega) e^{j(\omega-\omega_c)t} \frac{d\omega}{2\pi},$$

so that, from Eq. (7.8-14),

$$\frac{1}{j^n}\frac{d^n}{dt^n}[e^{-j\omega_c t} z(t)] = \int_{-\infty}^{\infty} Z(\omega)(\omega - \omega_c)^n e^{j(\omega-\omega_c)t} \frac{d\omega}{2\pi}. \tag{7.8-15}$$

If we use Eqs. (7.8-15) and (7.8-5) in Eq. (7.8-12), we obtain

$$\begin{aligned}\eta(t) &= \sum_{n=0}^{\infty} \frac{1}{n!} H^{(n)}(\omega_c) e^{j\omega_c t} \frac{1}{j^n} \frac{d^n}{dt^n} [e^{-j\omega_c t} z(t)] \\ &= z(t)\{H(\omega_c) + \dot{\mu}(t) H'(\omega_c) + \tfrac{1}{2} H''(\omega_c) \\ &\quad \cdot [\dot{\mu}(t)]^2 - j\ddot{\mu}(t)] + \cdots\}.\end{aligned} \tag{7.8-16}$$

Now consider the special case wher $\ddot{\mu}(t)$ and all higher time derivatives of the modulating signal can be ignored, as well as the higher-frequency derivatives (say above the second) of the transmittance $H(\omega)$. Then Eq. (7.18-16) can be approximated by

$$\eta(t) \simeq z(t)\{H(\omega_c) + \dot{\mu}(t)H'(\omega_c) + \tfrac{1}{2}H''(\omega_c)[\dot{\mu}(t)]^2\}. \tag{7.8-17}$$

The instantaneous frequency is $\omega_c + \dot{\mu}(t)$. An expansion for $H[\omega_c + \dot{\mu}(t)]$ about the carrier frequency ω_c gives

$$\begin{aligned}H[\omega_c + \dot{\mu}(t)] &= H(\omega_c) + H'(\omega_c)\dot{\mu}(t) \\ &\quad + \frac{H''(\omega_c)}{2} [\dot{\mu}(t)]^2 + \cdots .\end{aligned} \tag{7.8-18}$$

But Eq. (7.8-18), with higher terms omitted, is precisely the term in brackets in Eq. (7.8-17). Hence for "slowly" varying $\mu(t)$ (with time) and slowly varying $H(\omega)$ (with frequency), we can write

$$\eta(t) \simeq z(t)H[\omega_c + \dot{\mu}(t)]. \tag{7.8-19}$$

The interpretation of Eq. (7.8-19) is that, to a first approximation and under the conditions given, the complex response is simply the product of the complex input and the transmittance evaluated at the instantaneous frequency $\omega_c + \dot{\mu}(t)$.

Schwartz, Bennett, and Stein ([7-8], p. 232) give numerous references on the QSM. When the modulating signal changes abruptly, as in square-wave modulation, or the transmittance $H(\omega)$ contains many wiggles over the band of significant instantaneous frequencies, the QSM is generally not useful.

7.9 DEMODULATION OF FM SIGNALS

Slope Detection

Demodulation of an FM wave requires a device that produces an output signal whose amplitude is a linear function of the frequency of the input signal. The single-tuned circuit shown in Fig. 7.9-1 can, with the proper choice of parameters and proper choice of operating region, furnish the required transfer characteristic.

If the linear region of the amplitude response can be described by an equation of the form

$$|H(\omega)| = A(\omega - \omega_c) + B, \tag{7.9-1}$$

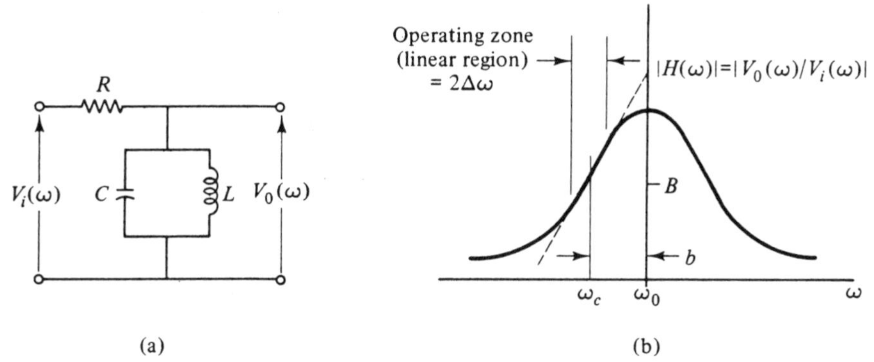

Figure 7.9-1. Use of a single-tuned circuit to demodulate an FM wave. (a) The circuit. (b) The amplitude of the transfer function.

then use of Eq. (7.8-19) furnishes

$$\eta(t) = [A\dot{\mu}(t) + B] \exp [\, j\{\omega_c t + \mu(t) - \gamma[\omega_c + \dot{\mu}(t)]\}], \tag{7.9-2}$$

where $\dot{\mu}(t) = Ks(t)$ and $\gamma(\omega)$ is the phase of $H(\omega)$.

The response $y(t)$ is simply the real part of $\eta(t)$. Thus

$$y(t) = [A\dot{\mu}(t) + B] \cos \{\omega_c t + \mu(t) - \gamma[\omega_c + \dot{\mu}(t)]\}. \tag{7.9-3}$$

If $y(t)$ is injected into an envelope detector and the dc term B is removed by a blocking capacitor, we recover the signal $s(t)$. Detection of FM waves in which the slope of a tuned circuit is used in the manner described previously is quite appropriately called *slope detection*. It is interesting to investigate the conditions required for slope detection with single-tuned circuits. To this end, consider Fig. 7.9-1, where $\omega_0 = (LC)^{-1/2}$, ω_c is the carrier frequency, $\omega_0 - \omega_c \equiv b$, and $\Delta\omega$ denotes the frequency deviation of the input signal frequency from the carrier frequency. The magnitude of the transfer function $H(\omega)$ is given by

$$|H(\omega)| = \frac{1}{\sqrt{1 + C^2R^2[(\omega^2 - \omega_0^2)/\omega]^2}}. \tag{7.9-4}$$

If we substitute $\omega = \omega_c + \Delta\omega$, $\omega_0 = \omega_c + b$ and assume that

$$\begin{aligned} &\text{(i) } \Delta\omega \ll b \ll \omega_c, \\ &\text{(ii) } b^2 \ll 2\omega_c\Delta\omega, \\ &\text{(iii) } b \ll [2CR]^{-1}, \end{aligned} \tag{7.9-5}$$

then Eq. (7.9-4) can be written in the form

$$|H(\omega)| = A(\omega - \omega_c) + B, \tag{7.9-6}$$

where $B \equiv 1 - 2[CR]^2b^2$ and $A \equiv [2CR]^2b$. We leave the details as an exercise for the reader.

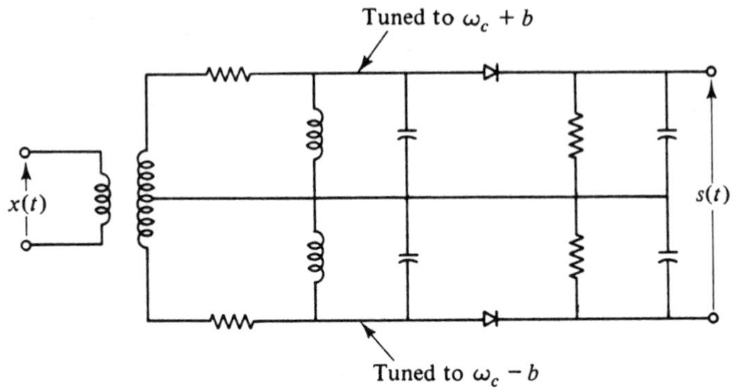

Figure 7.9-2. Balanced discriminator.

A much better system than a single-tuned circuit is the balanced discriminator shown in Fig. 7.9-2. The transfer function is shown in Fig. 7.9-3. One resonant circuit is tuned to $\omega_c + b$, while the other is tuned to $\omega_c - b$. The total transfer characteristic over the linear region is described by

$$\begin{aligned} H_T(\omega) &= [1 - 2(CR)^2b^2] + (2CR)^2b(\omega - \omega_c) \\ &\quad - \{[1 - 2(CR)^2b^2] - (2CR)^2b(\omega - \omega_c)\} \\ &= A(\omega - \omega_c), \end{aligned} \tag{7.9-7}$$

where $A \equiv 2(2CR)^2b$. The assumption of a linear region is only an approximation valid in the region around f_c. The response, from Eq. (7.9-2), is just

$$\eta(t) = A\dot{\mu}(t) \exp [j\{\omega_c t + \mu(t) - \gamma[\omega_c + \dot{\mu}(t)]\}], \tag{7.9-8}$$

and the envelope detector furnishes, without the use of a blocking capacitor, the signal $s(t)$.

Phase-Locked Loop (PLL)

The PLL was discussed in Chapter 6 in connection with synchronous detection. In that application, the PLL was used as an extremely narrowband filter to remove the carrier component of the AM signal, thereby enabling homodyne detection through mixing.

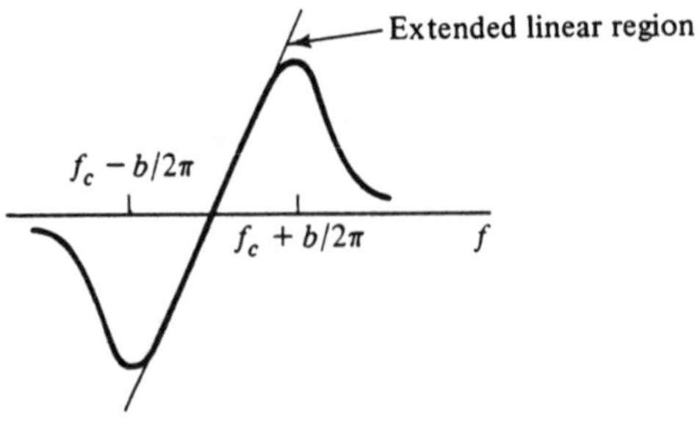

Figure 7.9-3. Equivalent transfer function of balanced discriminator.

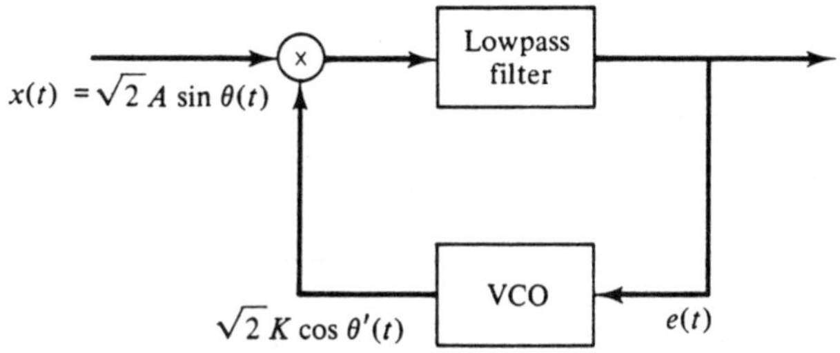

Figure 7.9-4. Use of a phase-locked loop to demodulate FM.

The PLL can also be used to demodulate FM signals. The operation of the PLL in this connection is easily explained by considering the circuit in Fig. 7.9-4. In Sec. 6-11, we showed that when the loop was near or at phase lock then $\theta(t) \simeq \theta'(t)$. If $x(t)$ is an FM wave and the loop is at or near phase lock, then

$$\theta'(t) \simeq \theta(t) = \omega_c t + K'' \int^t s(\lambda)\, d\lambda. \tag{7.9-9}$$

The VCO frequency, $\dot{\theta}'(t)$, is given by

$$\dot{\theta}'(t) = \omega_0 + K_v e(t), \tag{7.9-10}$$

where ω_0 is the frequency of the VCO when $e(t) = 0$ and K_v is a constant. The instantaneous input frequency is

$$\dot{\theta}(t) = \omega_c + K'' s(t). \tag{7.9-11}$$

Upon comparing Eq. (7.9-10) with Eq. (7.9-11) we see that, when $\dot{\theta}'(t) = \dot{\theta}(t)$,

$$e(t) = K_v^{-1}(\omega_c - \omega_0) + K_v^{-1}K'' s(t). \tag{7.9-12}$$

Hence, except for a removable dc term, we see that $e(t)$ is proportional to $s(t)$ and that demodulation has been achieved. The low-pass filter must reject the sum frequency signal produced by the multiplier but admit all significant components in the baseband signal $s(t)$. The PLL usually follows the IF stage in the receiver. The spurious amplitude variations in the FM signal should be removed before reaching the PLL to avoid distortion in the output. A limiter circuit is therefore often included in the IF stage to remove these variations.

Phase-Shift Discriminator (PSD)

The PSD was at one time widely used in FM receivers, but it has now been largely replaced by circuits such as the quadrature detector, which are more compatible with integrated circuitry. Figure 7.9-5(a) shows a PSD using a doubly tuned circuit in which there is an ac connection between the primary and secondary windings of the transformer. A simplified equivalent circuit for the ac is shown in Fig. 7.9-5(b). The capacitors labeled C_b are ac bypass capacitors. The resistor R_e, shown in dashed lines, is sometimes added to equalize the Q's of the two circuits; however, it is omitted in our discussion. How the PSD works can best be understood by considering the phase of E_2 with respect

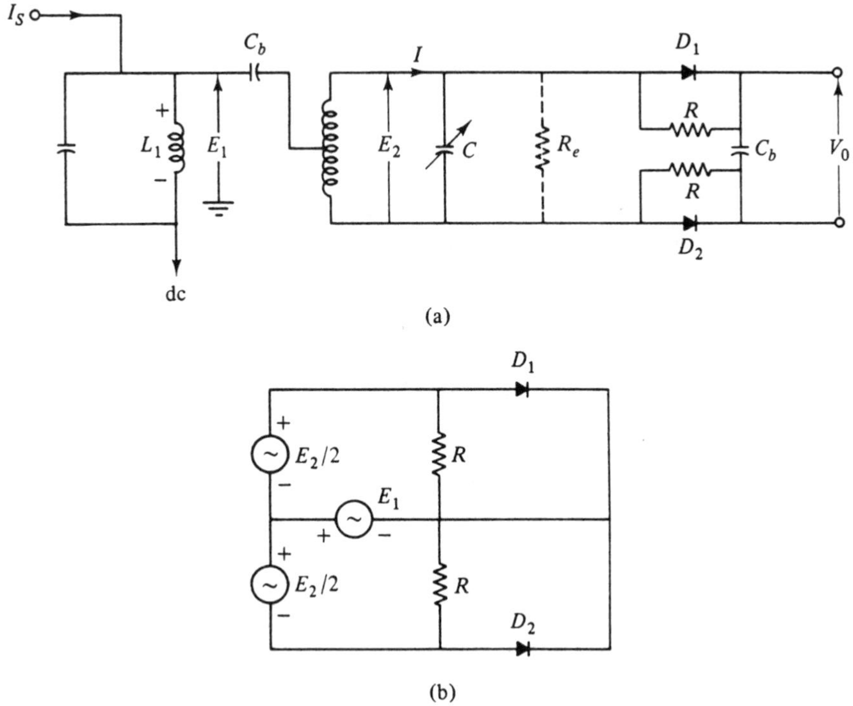

Figure 7.9-5. (a) Phase-shift discriminator. (b) Simplified ac equivalent circuit.

to E_1. The transformer together with the capacitor consists of a tuned circuit. At the resonant frequency ω_0 (which depends on the Thevenin's equivalent inductance) the circuit is resistive, and the current I is in phase with E_1. Hence E_2 lags E_1 by 90°. If $\omega > \omega_0$, the circuit is inductive, and the current I lags E_1 by some angle. The voltage E_2 always lags I by 90°; hence E_2 lags E_1 by an angle greater than 90°. When $\omega < \omega_0$, the circuit is capacitative, and the current leads E_1. Hence E_2 lags E_1 by an angle less than 90°. With these observations in mind we can now return to Fig. 7.9-5(b) and draw the phasor relationships for the voltages across D_1 and D_2. These are shown in Fig. 7.9-6. Under normal operation both tuned circuits are tuned to the carrier frequency ω_c; hence, $\omega_0 = \omega_c$.

The rectified output voltage, V_0, is proportional to the difference in the amplitudes (envelopes) of the ac, that is,

$$|E_{D_1}| - |E_{D_2}|,$$

and therefore is zero in the absence of modulation. The output voltage is not necessarily a linear function of the frequency deviation, and circuit parameters have to be carefully chosen to achieve good linearity. This is one of the drawbacks of the circuit. When tuning the PSD, the primary is adjusted for symmetry of response on either side of ω_c, and the secondary is adjusted to obtain zero response at ω_c ([7-3], p. 12.35).

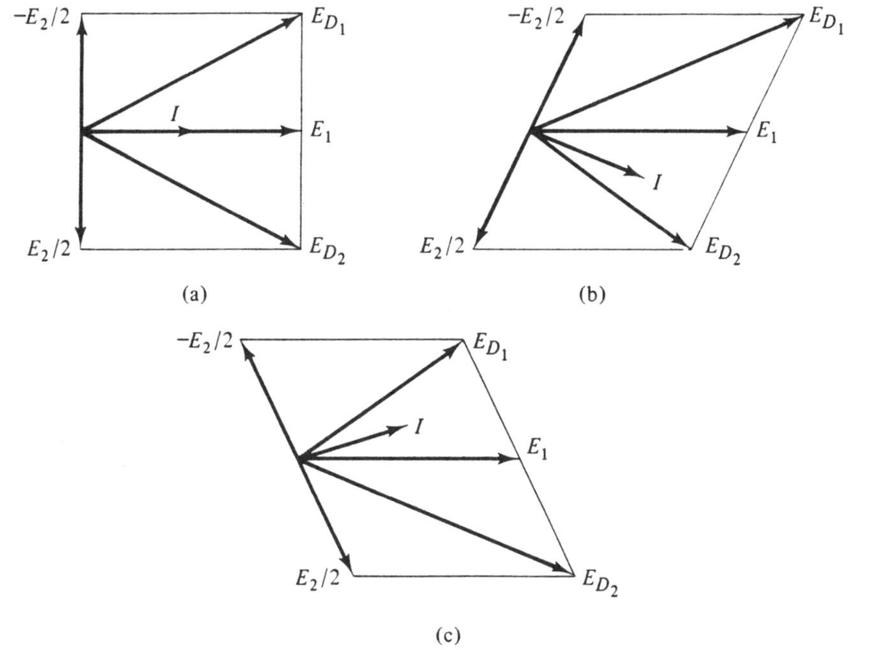

Figure 7.9-6. Phasor diagrams for phase-shift discriminator. (a) Resonance, $\omega = \omega_0$. (b) $\omega > \omega_0$. (c) $\omega < \omega_0$.

One reason for the superiority of FM over AM is its insensitivity to variations in signal amplitude since most man-made and natural noise sources affect amplitude rather than phase or frequency. To achieve this amplitude insensitivity, FM receivers generally incorporate a limiter. The limiter is usually part of the IF amplifier; in fact, it generally *is* the IF amplifier, which is designed to saturate for relatively small inputs. Clearly, the more stages of amplification or limiting that the IF amplifier has, the smaller the input needed to saturate it. The number of stages of limiting is therefore one factor that distinguishes a high-quality receiver from one of lesser quality.

One way of gaining some of the advantages of the limiter without actually using one is the ratio detector shown in Figure 7.9-7. This is a variant of the circuit shown in Fig. 7.9-5. The main difference between the two circuits is that one diode is reversed and that a large capacitor C_e is placed across the output resistors. The effect of this capacitor is to maintain a constant voltage across the transformer secondary so that the output signal is substantially unaffected by variations in input amplitude. The circuit therefore acts like a combination of PSD and limiter. It is generally true that a circuit designed to perform several functions simultaneously performs none of them too well, and the ratio detector is no exception. However, it was for many years commonly used in inexpensive FM radios. Like the PSD, this circuit has now been largely superseded by large-scale integrated (LSI) circuit chips.

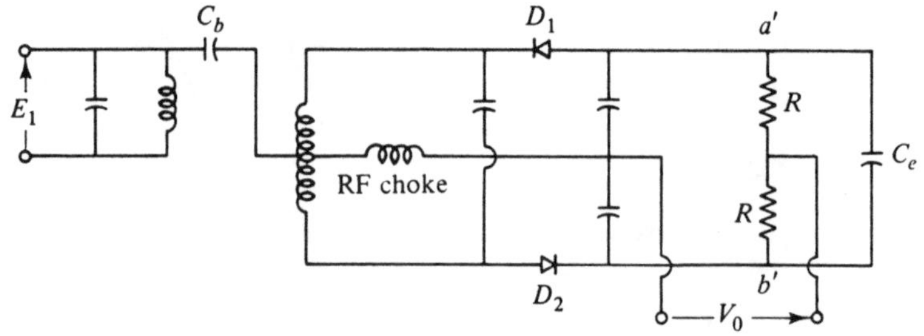

Figure 7.9-7. Ratio detector.

Quadrature Detector

The quadrature detector is shown in Fig. 7.9-8. The input to the demodulator is typically a 10.7-MHz IF signal; the output, at E, is the detected signal. The resonant circuit is normally tuned to the carrier frequency of the IF signal, ω_0, and the Q of the circuit is chosen to satisfy

$$2Q \ll \frac{\omega_0}{\Delta\omega}, \qquad \Delta\omega \equiv \omega - \omega_0, \tag{7.9-13}$$

where $\Delta\omega$ is the frequency deviation. To understand how the circuit works, we first compute the transfer function $V_2(\omega)/V_1(\omega)$, where $V_2(\omega) = \mathcal{F}[v_2(t)]$ and $V_1(\omega) = \mathcal{F}[v_1(t)]$. This is

$$\frac{V_2(\omega)}{V_1(\omega)} = H(\omega) = \frac{1}{1 + (L_c/L)[1 - (\omega/\omega_0)^2] + j\omega(L_c/R)}. \tag{7.9-14}$$

If we recall that the Q of a parallel RLC circuit is given by

$$Q = \frac{R}{\omega_0 L}, \tag{7.9-15}$$

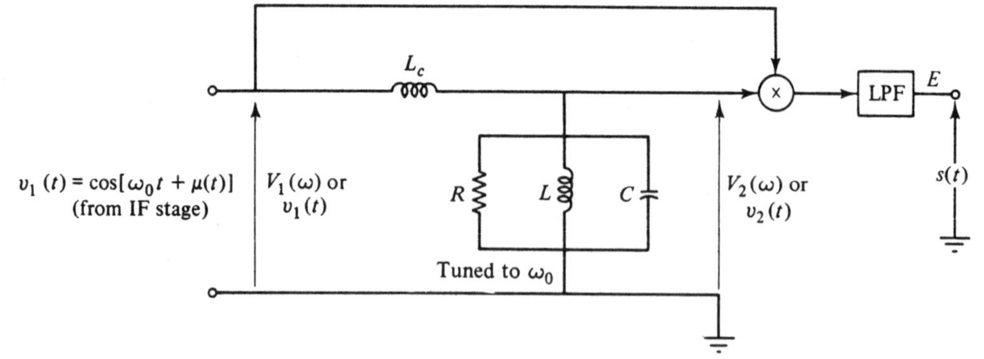

Figure 7.9-8 Quadrature detector.

then use of Eq. (7.9-13) enables us to write, to a very good approximation,

$$H(\omega) \simeq A \exp\left\{j\left[\frac{\pi}{2} - 2Q\left(\frac{\omega - \omega_0}{\omega_0}\right)\right]\right\}, \tag{7.9-16}$$

where A is $R/L_c\omega_0$. We are assuming that we are operating near resonance and that the reactance of L_c is large.† We now apply the quasi-static method and write the input signal as

$$z(t) = e^{j[\omega_0 t + \mu(t)]}, \qquad [v_1(t) = \text{Re } z(t)], \tag{7.9-17}$$

where

$$\mu(t) = K'' \int^t s(\lambda)\, d\lambda. \tag{7.9-18}$$

The complex output signal, from Eqs. (7.8-17) and (7.9-16), is

$$\begin{aligned} \eta(t) &= z(t) \exp\left[j\left(\frac{\pi}{2} - 2Q\frac{\dot{\mu}}{\omega_0}\right)\right] \\ &= \exp\left\{j\left[\frac{\pi}{2} + \mu(t) + \omega_0 t - 2Q\frac{\dot{\mu}}{\omega_0}\right]\right\}, \end{aligned} \tag{7.9-19}$$

where gain constants have been ignored. The output voltage is

$$\begin{aligned} v_2(t) &= \text{Re } [\eta(t)] \\ &= -\sin\left[\omega_0 t + \mu(t) - 2Q\frac{\dot{\mu}}{\omega_0}\right]. \end{aligned} \tag{7.9-20}$$

The signal obtained at point E after the balanced modulator is proportional to

$$\begin{aligned} v_1(t)v_2(t) &= \sin\left(2Q\frac{\dot{\mu}}{\omega_0}\right) \\ &\simeq 2Q\left(\frac{K''}{\omega_0}\right)s(t) \qquad \text{(sum frequency term assumed filtered out).} \end{aligned} \tag{7.9-21}$$

The last line follows from Eq. (7.9-13) and the facts that $\dot{\mu}(t) = K''s(t)$ and $|s(t)| \leq 1$. From Eq. (7.9-21) we see that the signal $s(t)$ has been recovered from the modulated carrier. The two key steps in this scheme are the $\pi/2$ phase shifting of the input signal and the taking of the product of the phase-shifted signal with the original. The quadrature detector is used in the RCA CA3089 E monolithic FM-IF chip [7-9], and a somewhat similar circuit is used in the Signetics ULN2111 chip [7-10].

† More specifically, we are assuming that $R/(L_c\omega_0) \ll 1$.

Zero-Crossing Detector

Another type of FM detector takes advantage of the fact that a measure of the instantaneous frequency is the number of zero crossings per unit time. It is in fact not difficult to show that the message can be reconstructed from knowledge of the zero crossings alone. Consider the FM wave

$$x(t) = A \cos \left[\omega_c t + K'' \int^t s(\lambda)\, d\lambda \right] = A \cos \theta(t). \tag{7.9-22}$$

Let t_1 and t_2 be the times associated with two adjacent zero crossings. Then

$$\theta(t_2) - \theta(t_1) = \pi = \omega_c(t_2 - t_1) + K'' \int_{t_1}^{t_2} s(\lambda)\, d\lambda. \tag{7.9-23}$$

The bandwidth, W, of the message $s(t)$ is assumed much less than the bandwidth, B, of the modulated wave. Hence $s(t)$ is essentially constant over the interval $[t_1, t_2]$, and $s(\lambda)$ can be taken outside the integral and given any argument t in $[t_1, t_2]$. Under these circumstances we obtain

$$[\omega_c + K''s(t)](t_2 - t_1) = \pi. \tag{7.9-24}$$

The term $\omega_c + K''s(t)$ is simply the derivative of the instantaneous phase. Hence it represents the instantaneous frequency ω_i or f_i, which is

$$\omega_i = \omega_c + K''s(t) \simeq \frac{\pi}{t_2 - t_1} \tag{7.9-25}$$

or

$$f_i = f_c + Ks(t) = \frac{1}{2(t_2 - t_1)}, \qquad K \equiv \frac{K''}{2\pi}. \tag{7.9-26}$$

If we count the number of zero crossings in an interval T that is large compared to f_c^{-1} but small compared to B^{-1}, we can assume that $s(t)$ is reasonably constant over T while ensuring that we have a reasonable number of zero crossings. If n_T denotes the number of zero crossings in T, then the spacings between adjacent zero crossings will not depart significantly from $t_2 - t_1$, and

$$n_T \simeq \frac{T}{t_2 - t_1} \tag{7.9-27}$$

so that

$$f_i = f_c + Ks(t) \simeq \frac{n_T}{2T}. \tag{7.9-28}$$

Hence n_T can be used to recover $s(t)$. A number of practical systems are available that work on this principle ([7-11], p. 618). For example, one system uses a monostable multivibrator that is triggered on the positive-sloping edge of a hard-limited FM wave and produces a pulse of short duration every time it is triggered. A subsequent low-pass filter with time constant T serves as an integrator and does the averaging indicated in Eq. (7.9-28). A balancing branch can be used to eliminate the f_c term in Eq. (7.9-28), allowing for $s(t)$ to be obtained directly.

Summary of Detectors

It might be useful at this point to briefly point out some of the merits of the various FM demodulators. The single-tuned circuit shown in Fig. 7.9-1 was presented mainly to show how a simple circuit might work as an FM demodulator; in practice, it does not have a sufficiently wide range of linearity to make it useful. The balanced slope detector shown in Fig. 7.9-2 is an improvement over the single-tuned circuit, but it is difficult to tune to give a linear response and is rarely used in practice.

The PLL is a fairly complicated circuit for a job that can be done more easily by simpler circuits. Its linearity depends on the linearity of the internal VCO, but there are designs for which excellent linearity can be guaranteed. A possible drawback of the PLL is that most of them are designed for bandwidths less than the 15 kHz required for demodulation of broadcast FM; however, this is not a fundamental limitation, and there are high-quality receivers that use a PLL as a demodulator.

The PSD and ratio detectors were at one time very popular, but are now largely obsolete.

The quadrature detector offers very high linearity, especially if, instead of the simple resonant circuit shown in Fig. 7.9-8, a doubly tuned resonant circuit is used. It is commonly used in high-quality receivers. Except for the tuned circuit, the rest of the detector, including the IF amplifier, is usually mounted on a single chip (e.g., RCA CA 3089E).

The zero-crossing detector is capable of better than 0.1% linearity over frequency deviations approaching the carrier frequency. Commercial units are available that handle carrier frequencies as high as 100 MHz, although above 10 MHz, divide-by-two and divide-by-ten counters are used to relax rather strenuous circuit requirements (e.g., recovery time of the pulse generator must be less than 10 ns). In general, this type of detector is best suited when exceptional linearity over very large frequency deviations is required. It is less useful when the frequency deviation is a small fraction of the carrier frequency.

*7.10 INTERFERENCE IN FM

When an interfering wave $y(t) = B \cos \omega_i t$ is superimposed on an FM wave, the instantaneous phase, as well as the amplitude, is disturbed. Presence of a signal exerts a masking effect on the interference, which could lead to less severe requirements on the interference rejection ability of the system. For this reason, the study of the effect of interference on an unmodulated carrier is important. Let $x(t) = A \cos \omega_c t$ represent the unmodulated carrier. The total signal (i.e., interference plus carrier) is then

$$z(t) = A \cos\omega_c t + B \cos \omega_i t. \tag{7.10-1}$$

If $\cos \omega_i t$ is written as $\cos (\omega_i - \omega_c + \omega_c)t$ and expanded according to the formula

$$\cos (a + b) = \cos a \cos b - \sin a \sin b,$$

we obtain a useful representation for $z(t)$ in terms of envelope and phase:

$$z(t) = R(t) \cos [\underbrace{\omega_c t + \phi(t)}_{\theta(t)}], \tag{7.10-2}$$

where

$$R(t) = [(A + B \cos \omega_D t)^2 + B^2 \sin^2 \omega_D t]^{1/2}, \tag{7.10-3}$$

$$\phi(t) = \tan^{-1} \frac{B \sin \omega_D t}{A + B \cos \omega_D t}, \tag{7.10-4}$$

and

$$\omega_D \equiv \omega_i - \omega_c. \tag{7.10-5}$$

The quantity ω_D is the difference between the carrier frequency and the frequency of the interfering wave.† The phasor diagram for the resultant wave is shown in Fig. 7.10-1.

The most interesting case in practice is when $B/A < 1$. With $r \equiv B/A$, we can rewrite Eqs. (7.10-3) and (7.10-4) as

$$R(t) = A(1 + r^2 + 2r \cos \omega_D t)^{1/2}, \tag{7.10-6}$$

$$\phi(t) = \tan^{-1} \frac{r \sin \omega_D t}{1 + r \cos \omega_D t}. \tag{7.10-7}$$

For the case $r \ll 1$, $R(t)$ and $\phi(t)$ can be simplified to

$$R(t) \simeq 1 + r \cos \omega_D t, \qquad \phi(t) \simeq r \sin \omega_D t$$

so that

$$z(t) \simeq A(1 + r \cos \omega_D t) \cos (\omega_c t + r \sin \omega_D t). \tag{7.10-8}$$

To a first approximation, then, the effect of a small-amplitude interference wave is to amplitude- *and* phase-modulate the carrier such that the resulting modulation index is r.

An ideal PM detector is insensitive to $R(t)$ and furnishes an output proportional to $\phi(t)$. Likewise, an ideal FM detector delivers an output proportional to $\dot{\phi}(t)$. The computation of $\dot{\phi}(t)$ is straightforward if we recall that

$$\frac{d}{dt}(\tan^{-1} u) = \frac{1}{1 + u^2} \frac{du}{dt}.$$

† The interfering wave can also be regarded as a component at frequency ω_i from a noise signal.

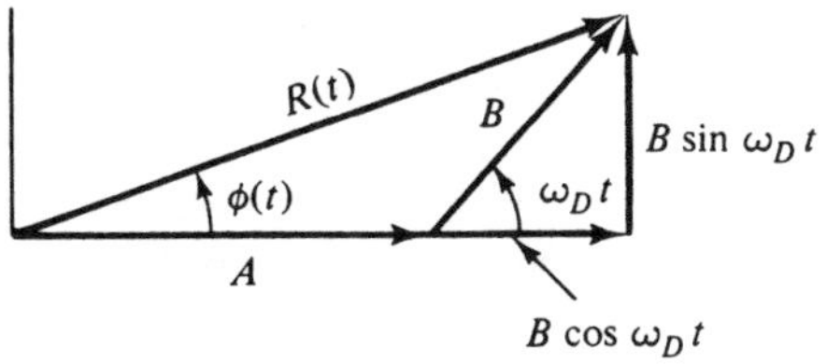

Figure 7.10-1. Phasor diagram showing the effect of interference.

The result is

$$\dot{\phi}(t) = r\omega_D \frac{r + \cos \omega_D t}{1 + 2r \cos \omega_D t + r^2} \equiv \omega_e \tag{7.10-9a}$$

and

$$\dot{\theta}(t) = \omega_c + r\omega_D \frac{r + \cos \omega_D t}{1 + 2r \cos \omega_D t + r^2} \qquad [\theta = \omega_c t + \phi(t)]. \tag{7.10-9b}$$

The quantity $\dot{\phi}(t)$ represents the frequency modulation induced by the interfering wave and represents the (error) departure of the instantaneous frequency from ω_c. Since $\dot{\phi}(t)$ is an error term, we denote it by ω_e and refer to it as *error frequency*.

An examination of Eq. (7.10-9a) shows that

$$\omega_e = \begin{cases} \dfrac{-r}{1-r}\,\omega_D, & \omega_D t = (2n+1)\pi\,, \\[2ex] \dfrac{r^2}{1+r^2}\,\omega_D, & \omega_D t = (2n+1)\dfrac{\pi}{2}\,, \\[2ex] \dfrac{r}{1+r}\,\omega_D, & \omega_D t = 2n\pi, \end{cases} \tag{7.10-10}$$

where $n = 0, \pm 1, \ldots$. Because of interference, one obtains the signal shown in Fig. 7.10-2 instead of detecting a null, which an ideal detector would furnish in the absence of a modulating signal $s(t)$.

The analysis of more complicated cases, such as interference between FM waves

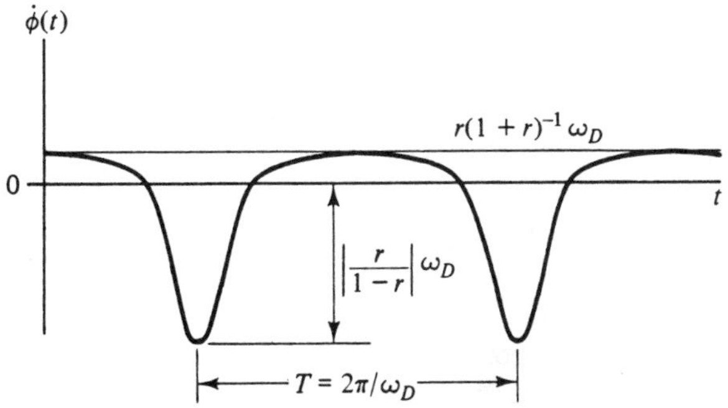

Figure 7.10-2. Detected instantaneous frequency associated with the interference of two carrier-type signals.

in common or adjacent channels, follows the same principles, although the algebra gets more involved. In common channel interference, two messages from different sources appear in the demodulated signal. This phenomenon is known as *cross talk* and is generally very disturbing to the listener. Corrington [7-12] and Panter [7-13] give extensive discussions of various types of interference and show how the distortion in the detected signal varies with the parameter r.

7.11 PREEMPHASIS AND DEEMPHASIS FILTERING (PDF)

An examination of the results obtained in Section 7.10 shows that, other things being equal, the interference increases with increasing value of ω_D [see, e.g., Eq. (7.10-9a) or (7.10-10)]. For many signals of common interest, such as speech and music, most of the energy is generally located in the lower frequency range. The small-amplitude high-frequency components of the baseband do not produce the full frequency deviation that is allowed. Hence the assigned bandwidth is not uniformly filled with signal energy, there being less at high frequencies. Unfortunately, as we shall see in Chapter 9, the receiver noise density increases as the square of frequency. The net result can be an intolerably low ratio of signal-to-noise power at the higher signal frequencies. To offset this undesirable phenomenon, an ingenious scheme known as preemphasis and deemphasis filtering is employed. The central idea is easily explained. At the transmitter, where signal levels are high and noise is, to some extent, under control, the modulating signal, $s(t)$, is purposely altered with the aim of enhancing the higher frequencies relative to the lower ones. This tailoring of the signal produces distortion, and the receiver must therefore invert the process to furnish an undistorted version of $s(t)$. The initial alteration of the signal that produces high-frequency enhancement is known as preemphasis filtering; the inverse process is deemphasis filtering. The deemphasis filter enhances the low frequencies relative to the higher ones. The net gain is that the noise now adds to a *strong* high-frequency signal and is suppressed in the deemphasis filtering. By the simple expedient of boosting the most noise-susceptible portion of the signal *before* noise becomes a problem and then inverting the process, an increase in signal-to-noise ratio is manifest at the receiver.

The technique of PDF is not confined to FM systems. However, it is widely used in FM because it is effective and the PDF circuitry is relatively simple. The deemphasis filter should be located in the receiver *after* the discriminator. Preemphasis filtering is done at the transmitter before modulation. A typical preemphasis network is shown in Fig. 7.11-1. The transfer function for the PE filter is

$$H_{\text{PE}}(\omega) = \hat{K}\frac{1 + j\omega\tau_1}{1 + j\omega\tau_2} \simeq \hat{K}(1 + j\omega\tau_1), \qquad \text{for } \omega < \omega_2 \equiv \frac{1}{\tau_2}, \tag{7.11-1}$$

where $\hat{K} \equiv R_2(R_1 + R_2)^{-1}$, $\tau_1 = CR_1$, and $\tau_2 = CR_1R_2(R_1 + R_2)^{-1} \simeq CR_2$. A typical value for τ_1 is 75 μs. This means that components above 2.1 kHz are "emphasized." The value of τ_2 is relatively unimportant provided only that $(2\pi\tau_2)^{-1}$ is at least as

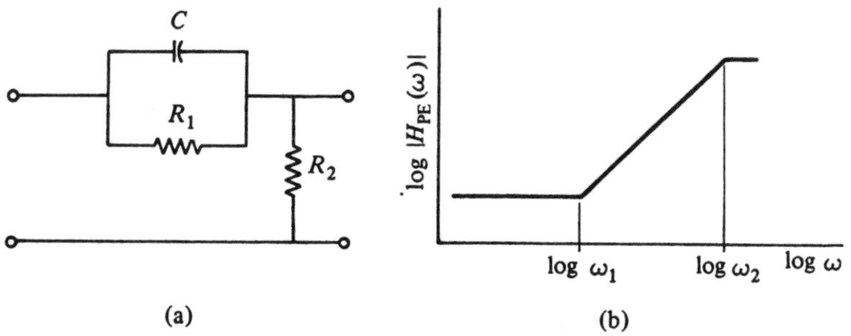

Figure 7.11-1. (a) Preemphasis filter. (b) Asymptotic frequency response.

great as the highest audio frequency (in hertz) for which preemphasis is desired; for quality reception this might be around 15 kHz.

The deemphasis filter must perform the inverse of the preemphasis filter to avoid a net distortion of $s(t)$. A deemphasis network is shown in Fig. 7.11-2. The transfer function for the DE network is

$$H_{DE}(\omega) = \frac{1}{1 + j\omega\tau_1}, \tag{7.11-2}$$

where $\tau_1 = R_1C$. The break frequency is 2.1 kHz when $\tau_1 = 75$ μs.

We note that

$$H_{PE}(\omega)H_{DE}(\omega) \simeq \hat{K}, \qquad \text{for } \omega < \omega_2, \tag{7.11-3}$$

which is the requirement for no distortion.

An interesting observation can be made with reference to Eq. (7.11-1). In the "active" region where $\omega_1 < \omega < \omega_2$, the transfer function of the preemphasis filter is approximately

$$H_{PE}(\omega) \simeq j\hat{K}\omega\tau_1. \tag{7.11-4}$$

Thus if $S(\omega)$ represents the message spectrum, the response of the filter is seen to be the derivative of the signal. But frequency-modulating a carrier with the *derivative* of

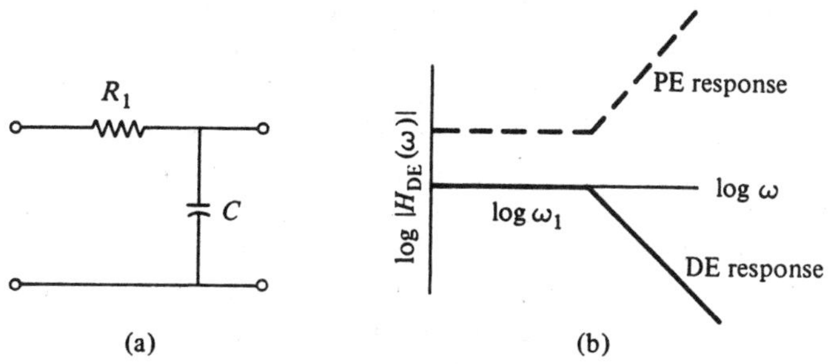

Figure 7.11-2. (a) Deemphasis filter. (b) Asymptotic frequency response.

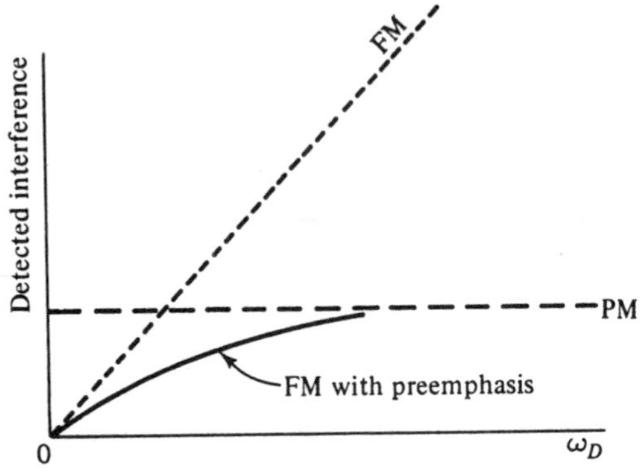

Figure 7.11-3. Detected interface for FM, PM, and FM with preemphasis.

a signal is equivalent to *phase-modulating* with the signal. Hence preemphasized FM is a mixture of FM and PM. In this connection it is interesting to compare the interference susceptibility of FM with that of PM. Equation (7.10-7) states that a PM detector produces an error signal $\phi_e(t)$ whose peak value is independent of the interfering frequency ω_D. For $r \ll 1$, the peak detected error is

$$\phi_e(t) = r, \tag{7.11-5}$$

while in the case of FM, the peak detected error is proportional to

$$\dot{\phi}_e = \omega_D r.$$

Hence we deduce that FM is superior to PM at low values of ω_D and that PM is superior to FM at large values of ω_D. Suitably designed FM with PDF incorporates the best features of both FM and PM. At low values of ω_D, the PDF should be inactive so that the detected interference is as in FM. At higher values of ω_D, the response of the preemphasis filter is essentially as in PM. The net result is shown in Fig. 7.11-3.

7.12 THE FM RECEIVER

A block diagram of an FM receiver is shown in Fig. 7.12-1. Most FM receivers are of the superheterodyne type. The tuning controls of the local oscillator and RF amplifier are mechanically coupled so that the output of the mixer is a constant-carrier-frequency (10.7 MHz) FM signal. The frequency of 10.7 MHz is slightly larger than one-half the FM broadcasting band of 20 MHz. This means that all image frequencies lie outside the broadcasting band. For example, if the local oscillator frequency is set at $f_{LO} = f_c + 10.7$ MHz, then the FM signals at f_c and $f_c + 2(10.7)$ MHz will be admitted by the IF stage. The frequency $f_c + 2(10.7)$ MHz is the image frequency. However, *there is no signal* in the FM broadcast band at $f_c + 2(10.7)$ MHz, as is clear from the following: When f_c has its lowest value of 88 MHz, the image frequency is 109.4 MHz, which is outside the broadcast band. The possibility still remains, however, that image interference

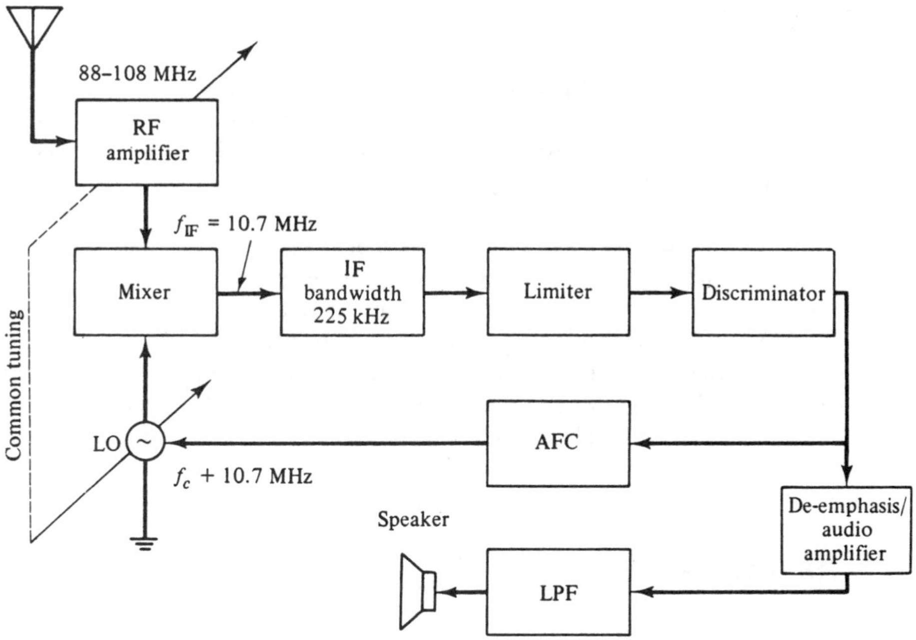

Figure 7.12-1. Block diagram of an FM receiver.

may be encountered from other services outside the 88- to 108-MHz band. One such service is the airport-aircraft communication channel, which operates above the FM band. The ability to reject image frequencies is a criterion of quality for an FM receiver, as is its ability to reject adjacent channels in a given geographical area (these are usually spaced no less than 400 kHz above and below the station). The latter is described by the *selectivity* of the receiver. Selectivity can be tested by tuning to a given station and injecting a second signal into an adjacent channel. The strength of the second signal is then raised until it becomes audible; the stronger the required signal, the more selective the receiver.

Other important criteria are harmonic distortion (HD) and intermodulation distortion (IMD). HD refers to the levels of undesired signals, produced at harmonics of signal frequencies, that are generated by nonlinearities in the system. HD depends on power level and generally increases as the audio power level increases. IMD refers to the levels of undesired signals produced at frequencies that are not harmonics of the signal frequencies. A power series expansion of a nonlinear characteristic would show that undesired harmonic and *inharmonic* (discordant) frequencies are produced, which are likely to fall in the middle or upper portion of the audio range. This makes excessive IMD very objectionable, and considerable efforts are made by manufacturers to keep IMD small in quality receivers. Many other figures of merit are used in describing the quality of FM receivers. Descriptions of these are given in the literature [7-13].

7.13 SUMMARY

We began the chapter by giving a mathematical description of angle modulation and deriving a Fourier resolution for FM and PM signals modulated by a single, constant-frequency tone. We then used these mathematical tools to derive approximate bandwidths for FM waveforms modulated by arbitrary signals.

The quasi-static method for dealing with the problem of transmission of FM signals through linear networks was then discussed and applied to the slope detector.

Several types of FM detectors were discussed and analyzed, including some suitable for manufacture as integrated circuits because they lack inductances.

We ended the chapter by considering FM interference and preemphasis and deemphasis filtering. References [7-5], [7-7], and [7-8] are useful for further reading, as are [7-12] and [7-13], in which some of the more complicated mathematical and statistical problems associated with FM interference are discussed. We shall return to signal-to-noise considerations in Chapter 9.

PROBLEMS

7-1. A real wave with arbitrary angle modulation $\phi(t)$ can be written as

$$x(t) = A\cos[\omega_c t + \phi(t)].$$

(a) Show that $x(t)$ may be written as

$$x(t) = \text{Re } z(t),$$

where $z(t) = u(t)e^{j\omega_c t}$ and $u(t) = Ae^{j\phi(t)}$

(b) Explain why $z(t)$ cannot be regarded as a complex analytic signal even if $\phi(t)$ is band limited.

(c) It is known from practice that for phase and frequency modulation the spectrum of $u(t)$ ultimately drops off rapidly with increasing $|\omega|$. Explain therefore why, if ω_c is large enough, we can write

$$z(t) \simeq \xi(t) \equiv x(t) + j\hat{x}(t),$$

where $\hat{x}(t)$ is the Hilbert transform of $x(t)$.

7-2. Define the average power in $x(t)$ over an interval T to be

$$P_{\text{ave}} = \frac{1}{T}\int_{-T/2}^{T/2} x^2(t)\,dt.$$

What should be the size relationships between T, ω_c, and the bandwidth of $\exp[j2\phi(t)]$ in order for the average power in $x(t) = A\cos[\omega_c t + \phi(t)]$ to register $A^2/2$?

7-3. At first glance it might seem simpler to define FM by adding a signal-dependent frequency to ω_c; that is,

$$x(t) = \cos\underbrace{[\omega_c + \omega(t)]t.}_{\phi(t)}$$

(a) Compute the instantaneous frequency $\omega_i \equiv \dot{\theta}(t)$ for the general case and when $\omega(t)$ is sinusoidal.

(b) Demonstrate the inadequacy of this definition by considering what happens when $t \to \infty$ and $\omega(t)$ is sinusoidal.

7-4. This problem deals with Bessel-function properties.

(a) By use of Eq. (7.3-5), show that $J_n(0) = \text{sinc}\ (n)$ and hence that properties 1 and 2 are true.

(b) By use of the change of variable $\lambda' = \pi - \lambda$ in Eq. (7.3-5) or (7.3-6), show that properties 3 to 6 are true.

(c) By sketching the function $\xi \sin \lambda - n\lambda$ for different ξ and n, convince yourself that property 7 is correct.

(d) Use the series expansion for the Bessel function to demonstrate property 8.

7-5. Compute the instantaneous phase and frequency for both PM and FM when the modulating signal, $s(t)$, is shown in Fig. P7-5.

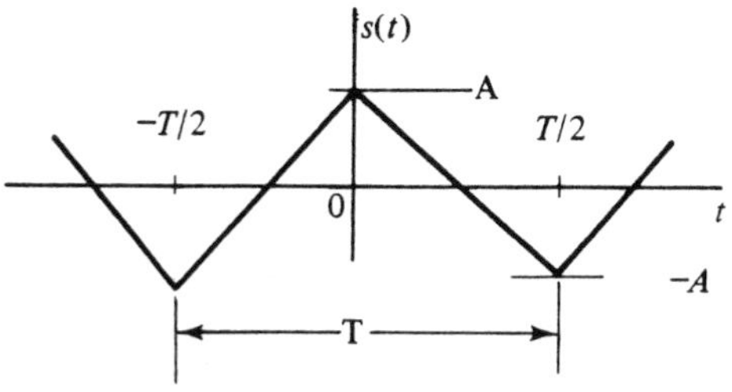

Figure P7-5

7-6. The spectra of AM and narrowband PM, although similar, have important differences. Use phasor diagrams as in Fig. 7.5-2 to compute the maximum variations in the amplitudes of the waves in the two cases. Which is greater?

7-7. Repeat Prob. 7-6 when FM is compared with AM instead of PM.

7-8. Verify that the three systems shown in Fig. 7.5-4 do in fact generate AM, NBPM, and NBFM when $s(t)$ is a sinusoidal modulating signal. What polarities should be shown at the summation points?

7-9. A 50-MHz carrier is to be frequency-modulated by a 20-kHz sine wave.

(a) The frequency deviation is 20 Hz. What is a minimum appropriate bandwidth?

(b) The frequency deviation is increased to 1.0 MHz. What is the appropriate bandwidth? Why is the approximation $B_{CR} \simeq 2\Delta f$ inadequate for part (a)?

7-10. Consider an FM wave

$$\begin{aligned} x(t) &= \text{Re}\ x_c(t) \\ &= \text{Re}\ A_c \exp\{j[\omega_c t + \phi(t)]\}, \end{aligned}$$

where $x_c(t)$ is the complex form of the wave and Re means "real part of." Let the modulating signal as well as $\phi(t)$ be periodic with period $T_m = 2\pi/\omega_m$. Show that $x_c(t)$ may be written as

$$x_c(t) = A_c \sum_{n=-\infty}^{\infty} c_n e^{j(\omega_c + n\omega_m)t},$$

where

$$c_n = \frac{1}{T_m} \int_{-T_m/2}^{T_m/2} e^{j\phi(t)} e^{-jn\omega_m t} \, dt.$$

7-11. Let the instantaneous frequency $\omega_i = \omega_c + \dot{\phi}$ be as shown in Fig. P7-11. Use the results of Prob. 7-10 to show that

$$x(t) = A_c \sum_{n=-\infty}^{\infty} \frac{2\beta}{\pi(\beta^2 - n^2)} \sin\left[(\beta - n)\frac{\pi}{2}\right] \cos(\omega_c + n\omega_m)t,$$

where $\beta \equiv \Delta\omega/\omega_m$.

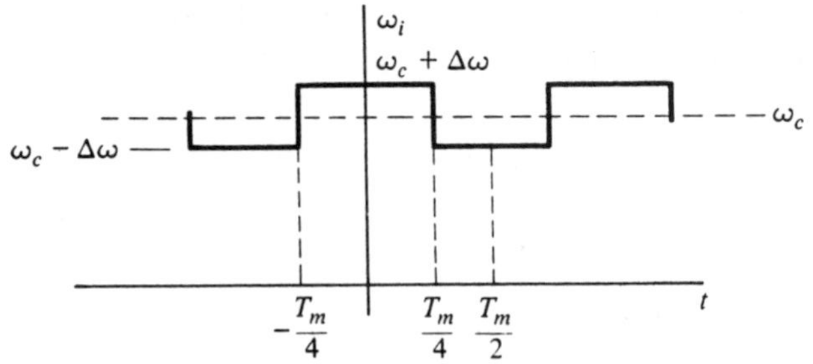

Figure P7-11

7-12. Assume that there exists an FM broadcasting system for which $\Delta f = 50$ kHz. Compute a table similar to Table 7.4-1 by considering sinusoidal modulating frequencies that range from 20 to 0.05 kHz. Which tones require the largest bandwidths?

7-13. Assume that for a particular FM system a sinusoidal tone at frequency f_m generates a frequency deviation given by

$$\Delta f = 75e^{-[(f_m - 500)/1000]^2} \qquad \text{(kHz)} \quad (f_m \text{ in Hz}).$$

Plot the required bandwidth, as determined by Carson's rule, for sinusoidal tones with frequencies varying over the audio range.

7-14. Plot the amplitude-frequency spectra of an FM wave for $\beta = 0.1$, 1, 5, and 10. Make two plots in which (a) β increases because Δf increases but f_m is held fixed and (b) β increases because f_m decreases but Δf is held fixed. Verify that as β becomes large the bandwidth approaches $2\Delta f$. Choose any reasonable increments.

7-15. Consider the phase-modulated wave

$$x(t) = \cos[\omega_c t + K' s(t)], \qquad |s(t)| \leq 1.$$

(a) Explain why K' must be restricted to $K' < \pi$.

(b) Discuss the required bandwidth for PM with a sinusoidal modulating signal. What happens when the phase deviation K_d changes? What happens when the modulating frequency f_m changes?

7-16. It is desired to design an Armstrong-type indirect FM transmitter. Let the modulating frequencies, f_m, range from $f_m = 50$ to $f_m = 5000$ Hz. The maximum frequency deviation at the output is to be no greater than 75 kHz. The initial NBFM $[\beta_1]_{max}$ is restricted to 0.2. Compute the required frequency multiplication and the largest frequency deviation at the input.

7-17. An FM signal is to be demodulated with a slope detector followed by an envelope detector. Assume that the (one-sided) transfer function associated with the slope detector is described by

$$H(\omega) = k_1(\omega - \omega_c)^2 + k_2(\omega - \omega_c) + k_3,$$

where, for simplicity, the $\{k_i\}$ are real constants in appropriate units.

(a) Compute the complex response $\eta(t)$ to the complex signal

$$z(t) = \exp\{j[\omega_c t + \mu(t)]\}.$$

(b) Let the modulation be a periodic ramp, that is, $s(t) = At$, $0 \le t \le T$, and so on. Compute the per-cycle power ratio of signal to distortion in the envelope-detected output.

7-18. Design a single-tuned *RLC* network (Fig. 7.9-1) to slope-detect an FM signal with carrier frequency of 50 MHz and $\Delta f = 75$ kHz.

7-19. It can be shown that demodulation of an FM wave is possible by direct differentiation followed by envelope detection and dc blocking. Explain how delay lines might be used to approximate the derivative of the wave by realizing the quantity

$$\dot{x}(t) \simeq \frac{x(t) - x(t - \tau)}{\tau}.$$

Draw the system, and suggest how small τ must be in order for the right side to be a good approximation of the derivative.

7-20. Consider an FM wave $x(t) = A \cos \theta(t)$, where $\theta(t) = \omega_c t + K'' \int^t s(\lambda)\, d\lambda$. Let t_1, t_2 $(t_2 > t_1)$ denote the time of two adjacent zeros of $x(t)$. Assume that

$$\int_{t_1}^{t_2} s(\lambda)\, d\lambda \simeq s(t)(t_2 - t_1), \qquad \text{where } t_1 \le t \le t_2.$$

Show that

$$K''s(t) = \frac{\pi}{T_{AC}} - \omega_c,$$

where $T_{AC} = t_2 - t_1$. This result implies that $s(t)$ can be detected by counting the zero crossings in $x(t)$. What property of FM waves accounts for this?

7-21. (Generalization of Prob. 7-20) Let n_z denote the number of zero crossings in time T_z. Show that if T_z satisfies

$$f_c^{-1} < T_z \ll W^{-1},$$

where W is the bandwidth of $s(t)$, then

$$Ks(t) \simeq \frac{n_z}{2T_z} - f_c.$$

Hence the signal can be detected by using a zero-crossing counter. Why must the inequalities on T_z be satisfied for this method to work?

7-22. (Multipath transmission in FM) Consider two FM waves coming from the same source but reaching the receiver by different paths. Let the direct wave be given by

$$x_1(t) = \cos \underbrace{[\omega_c(t - t_1) + \beta \sin \omega_m(t - t_1)]}_{\psi_1(t)},$$

and let the reflected (indirect) wave be given by

$$x_2(t) = \rho \cos \underbrace{[\omega_c(t - t_2) + \beta \sin \omega_m(t - t_2)]}_{\psi_2(t)},$$

where $\rho \le 1$ and $t_1(t_2)$ is the time for the direct (reflected) wave to reach the receiver. Draw a phasor diagram for this situation and show that the resultant wave can be written as

$$x_R(t) = A(t) \cos \phi(t),$$

where

$$A \equiv \sqrt{1 + \rho^2 + 2\rho \cos \psi(t)}\,,$$

$$\psi(t) \equiv \psi_2(t) - \psi_1(t),$$

$$\phi(t) \equiv \psi_1(t) + \tan^{-1} \frac{\rho \sin \psi(t)}{1 + \rho \cos \psi(t)}.$$

7-23. (Problem 7-22 continued) Consider the resultant wave $x_R(t)$ in Prob. 7-22. What is the detected signal if $x_R(t)$ is sent to a limiter followed by a balanced discriminator? (A balanced discriminator removes the dc term ω_c).

REFERENCES

[7-1] J. R. Carson, "Notes on the Theory of Modulation," *Proc. IRE*, **10,** pp. 57–64, Feb. 1922.

[7-2] E. H. Armstrong, "Method of Reducing the Effect of Atmospheric Disturbances," *Proc. IRE*, **16,** pp. 15–26, Jan. 1928.

[7-3] K. Henney, ed., *Radio Engineering Handbook*, 5th ed., McGraw-Hill, New York, 1959.

[7-4] M. Abramowitz and I. A. Stegun, *Handbook of Mathematical Functions*, Dover, New York, 1965.

[7-5] A. B. Carlson, *Communication Systems: An Introduction to Signals and Noise in Electrical Communications*, McGraw-Hill, New York, 1968.

[7-6] J. Millman and C. C. Halkias, *Electronic Devices and Circuits*, McGraw-Hill, New York, 1967.

[7-7] M. Schwartz, *Information Transmission, Modulation and Noise*, 2nd ed., McGraw-Hill, New York, 1970.

[7-8] M. Schwartz, W. R. Bennett, and S. Stein, *Communication Systems and Techniques*, McGraw-Hill, New York, 1966.

[7-9] *RCA Linear Integrated Circuits*, RCA Corporation, Somerville, N.J., 1974, pp. 427–432.

[7-10] *Signetics Digital, Linear, MOS Databook*, Signetics Corporation, Sunnyvale, Calif., 1972, pp. 6–128 to 6–133.

[7-11] K. K. Clarke and D. T. Hess, *Communication Circuits: Analysis and Design*, Addison--Wesley, Reading, Mass., 1971.

[7-12] M. S. Corrington, "Frequency Modulation Caused by Common and Adjacent Interference," *RCA Review*, **7**, pp. 552–560, Dec. 1946.

[7-13] P. F. Panter, *Modulation, Noise and Spectral Analysis*, McGraw-Hill, New York, 1965.

[7-14] *Understanding High Fidelity*, 2nd ed., Pioneer Electronic Corporation, Tokyo, 1975.

CHAPTER 8

Probability and Random Variables

8.1 INTRODUCTION

In this chapter we present some of the basic ideas of the theory of probability and random variables. We assume that the reader has had some prior exposure to set theory and is somewhat familiar with the notion of sets and set operations. The material discussed here is fairly standard and is covered in much greater depth in other places (e.g., [8-1]–[8-8]).

Probability as a Measure of Frequency of Occurrence

One approach to defining the probability of an event E is to perform an experiment n times. The number of times that E appears is denoted by n_E. Then the probability of E occurring is defined according to

$$P(E) = \lim_{n\to\infty} \frac{n_E}{n}. \tag{8.1-1}$$

Clearly, since $n_E \leq n$, we must have $0 \leq P(E) \leq 1$. One difficulty with this approach is that we can never perform the experiment an infinite number of times, so we can only estimate $P(E)$ from a finite number of trials. Second, we *postulate* that n_E/n approaches a limit in some sense as n goes to infinity. But consider flipping a fair coin 1000 times. The likelihood of getting exactly 500 heads is very small; in fact, if we flipped the coin 10,000 times, the likelihood of getting exactly 5000 heads is even smaller. As $n \to \infty$, the event of observing exactly $n/2$ heads becomes vanishingly small. Yet our intuition demands that $P(\text{head}) = \frac{1}{2}$ for a fair coin. Suppose we choose a $\delta > 0$; then we shall find experimentally that if the coin is truly fair the number of times that

$$\left|\frac{n_E}{n} - \frac{1}{2}\right| > \delta \tag{8.1-2}$$

becomes very small as n becomes large. Thus although it is very unlikely that at any stage of this experiment, especially when n is large, n_E/n is exactly $\frac{1}{2}$, this ratio will nevertheless hover around $\frac{1}{2}$, and the number of times it will make significant excursions away from the vicinity of $\frac{1}{2}$ in the sense of Eq. (8.1-2) becomes very small indeed.

Despite these problems with the frequency definition of probability, the relative frequency concept is essential in applying probability theory to the physical world.

Probability as the Ratio of Favorable to Total Outcomes

In this approach, which is not experimental, the probability of an event is computed a priori by counting the number of ways N_E that E can occur and forming the ratio N_E/N, where N is the number of all possible outcomes, (i.e., the number of all alternatives to E plus N_E). An important notion here is that all outcomes are equally likely. Since "equally likely" is really a way of saying equally probable, the reasoning is somewhat circular. Suppose we throw a pair of unbiased dice and ask what is the probability of getting a seven? We partition the outcome into 36 equally likely outcomes as shown in Table 8.1-1, where each entry is a possible outcome. The total number of outcomes is 36 if we keep the dice distinct. The number of ways of getting a seven is $N_7 = 6$. Hence

$$P(\text{getting a seven}) = \tfrac{6}{36} = \tfrac{1}{6}.$$

TABLE 8.1-1

		First die					
		1	2	3	4	5	6
Second die	1	2	3	4	5	6	7
	2	3	4	5	6	7	8
	3	4	5	6	7	8	9
	4	5	6	7	8	9	10
	5	6	7	8	9	10	11
	6	7	8	9	10	11	12

† *A priori* means relating to reasoning from self-evident propositions or presupposed by experience. *A posteriori* means relating to reasoning from observed facts.

Example

Throw a fair coin twice (note that since no physical experimentation is involved, there is no problem in postulating an ideal "fair coin"). The outcomes are HH, HT, TH, TT. The probability of getting at least one tail is computed as follows: With E denoting the event of getting at least one tail, the event E is the set of outcomes

$$E = \{HT, TH, TT\}.$$

The number of elements in E is $N_E = 3$; the number of all outcomes, N, is 4. Hence

$$P(\text{at least one } T) = \frac{N_e}{N} = \frac{3}{4}.$$

Axiomatic Approach

This is the approach followed in most modern textbooks on the subject. To develop it, we first define certain notions.

The *sample space* or *sample description space* S of a random event or experiment with random outcomes is the space of descriptions of all possible elementary outcomes of the experiment.

Examples

1. The experiment consists of throwing a coin twice. Then $S = \{HT, TH, HH, TT\}$.
2. The experiment consists of choosing a person at random and counting the hair on his (or her) head. Then $S = \{0, 1, \ldots\}$ (i.e., the set of nonnegative integers). An event E is a subset of S written $E \subset S$. Since $S \subset S$, S is also an event and is called the *certain event*. Hence S is both the sample space and the certain event.
3. The experiment consists of throwing a coin twice. Then $S = \{HH, HT, TH, TT\}$. The event, E, of getting at least one tail is

$$E = \{HT, TH, TT\} \subset S.$$

The empty set (i.e., the set containing no descriptions in S) is the impossible event, denoted by $\varnothing$. If S consists of n discrete elementary events (i.e., basic descriptions s_i, $i = 1, \ldots, n$), then there are 2^n distinct events that can be defined over S. For example, with $S = \{H, T\}$ (one throw of a coin), the events are

$$\{H\}, \quad \{T\}, \quad S, \quad \varnothing.$$

Exercise. Show that $m = 2^n$ events can be defined over S.

Before giving the axiomatic definition of probability, we remind the reader that the union (sum) of two sets is the set of all elements that are in at least one of the two sets. The intersection (product) of two sets is the set of all elements that appear only in both sets. A symbolic representation of union, $\cup$, and intersection, $\cap$, is obtained

† The intersection of two sets, $A \cap B$, will be written as AB.

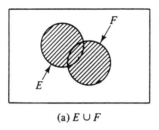

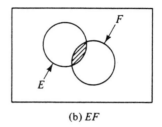

(a) $E \cup F$ (b) EF

Figure 8.1-1

with the use of *Venn diagrams*. In Fig. 8.1-1(a) the union of the sets E and F is indicated by the shaded area. In Fig. 8.1-1(b), the intersection of E and F is shown by the shaded area. Another concept that we need is that of a *sigma* (σ) *field* $\mathcal{F}$. In general, a *field* $\mathcal{M}$ is a class of sets such that:

1. If $A \in \mathcal{M}$, then $A^c \in \mathcal{M}$ (A^c is the complement of A; it is the set of all elements not in A).
2. If $A \in \mathcal{M}$ and $B \in \mathcal{M}$, then $A \cup B \in \mathcal{M}$, $AB \in \mathcal{M}$, and $\varnothing \in \mathcal{M}$.
3. Generalization: If $A_1, \ldots, A_n$ belong to $\mathcal{M}$, so do the union $A_1 \cup A_2 \cup \cdots \cup A_n$ and the intersection $A_1 A_2 \cdots A_n$.

A sigma field $\mathcal{F}$ is a field that is closed under any countable set of unions, intersections, and combinations thereof. Thus, if $A_1, \ldots, A_n, \ldots$ belong to $\mathcal{F}$, so do

$$\bigcup_{i=1}^{\infty} A_i \quad \text{and} \quad \prod_{i=1}^{\infty} A_i.$$

If S has a countable number of elements, then every subset of S is an event, and therefore the set of all subsets of S forms a sigma field. However, when S is not countable (i.e., say, when $S = R =$ the real line), then not every subset of S can be constructed from countable unions and intersections. Such subsets are said to be nonprobabilizable and cannot be considered as events. These sets cannot be assigned a probability measure without violating the axioms in Sec. 8.2. However, such sets do not normally occur in problems of engineering interest.

Thus events consist of points $s \in S$. Every event is a subset of S. Moreover, every event is an element of a sigma field and we speak of the sigma field or *sigma algebra* of events. If A is an event, then $A \subset S$ and $A \in \mathcal{F}$. If the sample description space is the real line R, then events are probabilizable subsets of R, and the class of all such subsets is also a sigma field. This sigma field is often called the *Borel sigma field* or simply *Borel field* and denoted by the symbol $\mathcal{B}$. Events on the real line are sets of numbers called Borel sets.

8.2 AXIOMATIC DEFINITION OF PROBABILITY

Probability is a function $P(\cdot)$ that assigns to every $E \subset S$ a number $P(E)$ called the probability of E such that

$$\text{(i)}\ P(E) \geq 0, \tag{8.2-1}$$

$$\text{(ii)}\ P(S) = 1, \tag{8.2-2}$$

$$\text{(iii)}\ P(E \cup F) = P(E) + P(F) \text{ if } EF = \varnothing. \tag{8.2-3}$$

These three axioms are sufficient to establish most elementary results. A fourth axiom,

$$P\left[\bigcup_{i=1}^{\infty} E_i\right] = \sum_{i=1}^{\infty} P(E_i) \text{ if } E_iE_j = \varnothing, \qquad \text{all } i \neq j,$$

must be included to enable one to deal rigorously with limits and countable unions. Note that E and F must be *events*; that is, they must be subsets of S in $\mathcal{F}$. A *probability space* $\mathcal{H}$ refers to the triplet

$$(S, \mathcal{F}, P), \tag{8.2-4}$$

where S is the sample description space of a random phenomenon, $\mathcal{F}$ is the sigma field of events, and P is the probability measure whose domain is S and whose range is $[0, 1]$.

Exercises

1. Show that $P(\varnothing) = 0$.
2. Show that with $E \in \mathcal{F}$, $F \in \mathcal{F}$,

$$P(EF^c) = P(E) - P(EF).$$

3. Show that $P(E) = 1 - P(E^c)$.
4. Show that $P(E \cup F) = P(E) + P(F) - P(EF)$.

Examples

1. The experiment consists of throwing a coin once. Hence

$$S = \{H, T\}.$$

The sigma field of events consists of the following sets: $\{H\}$, $\{T\}$, S, $\varnothing$. With $P(H)$ assumed $\frac{1}{2} = P(T)$ we have

$$P(H) = P(T) = \tfrac{1}{2}, \qquad P(S) = 1, \qquad P(\varnothing) = 0.$$

2. The experiment consists of throwing a die once. The outcome is the number of dots n_i appearing on the upface of the die. The set S is given by $S = \{1, 2, 3, 4, 5, 6\}$. The sigma field of events consists of 2^6 elements. Some are

$$\varnothing, \quad S, \quad \{1\}, \quad \{1, 2\}, \quad \{1, 2, 3\}, \quad \{1, 4, 6\}, \quad \{1, 2, 4, 5\}.$$

We assign

$$P(i) = \tfrac{1}{6}, \quad i = 1, \ldots, 6 \quad \text{(an assumption)}.$$

All probabilities can now be computed from the basic axioms and the assumed probabilities for the elementary events. Thus, with $A = \{1\}$ and $B = \{2, 3\}$, we obtain $P(A) = \frac{1}{6}$. Also $P(A \cup B) = P(A) + P(B)$, since $AB = \varnothing$. Furthermore, $P(B) = P(2) + P(3) = \frac{2}{6}$, so

$$P(A \cup B) = \tfrac{1}{6} + \tfrac{2}{6} = \tfrac{1}{2}.$$

3. The experiment consists of picking at random a numbered ball from 12 balls numbered 1 to 12 from an urn:

$$S = \{1, \ldots, 12\}.$$

Let

$$A = \{1, \ldots, 6\}, \qquad B = \{3, \ldots, 9\},$$

$$A \cup B = \{1, \ldots, 9\}, \qquad AB = \{3, 4, 5, 6\}, \qquad AB^c = \{1, 2\},$$

$$B^c = \{1, 2, 10, 11, 12\}, \qquad A^c = \{7, \ldots, 12\}, \qquad A^cB^c = \{10, 11, 12\},$$

$$(AB)^c = \{1, 2, 7, 8, 9, 10, 11, 12\}.$$

Hence

$$P(A) = P(1) + P(2) + \cdots + P(6),$$

$$P(B) = P(3) + \cdots + P(9),$$

$$P(AB) = P(3) + \cdots + P(6).$$

If $P(1) = \cdots = P(12) = \frac{1}{12}$, then $P(A) = \frac{1}{2}$, $P(B) = \frac{7}{12}$, $P(AB) = \frac{4}{12}$, and so on.

8.3 JOINT AND CONDITIONAL PROBABILITIES; INDEPENDENCE

Consider a probability space $(S, \mathcal{F}, P)$. Let A and B be two events defined over this space. Then AB is an event. The joint probability of the event A and B is $P(AB)$. The frequency interpretation is as follows:

$$P(A) \simeq \frac{n_A}{n},$$

$$P(B) \simeq \frac{n_B}{n}, \qquad \text{for } n \text{ large}, \tag{8.3-1}$$

$$P(AB) \simeq \frac{n_{AB}}{n},$$

where n_{AB} is the number of times *both A and B* occur. The conditional probability of A given B is defined by

$$P(A|B) = \frac{P(AB)}{P(B)}, \qquad P(B) > 0. \tag{8.3-2}$$

Frequency interpretation. From Eq. (8.3-1), we have

$$P(A|B) \simeq \frac{n_{AB}/n}{n_B/n} = \frac{n_{AB}}{n_B}, \qquad n_b > 0, \tag{8.3-3}$$

that is, the ratio of the number of times that A and B both occur to the number of times that at least B occurs.

From Eq. (8.3-2), we have $P(AB) = P(A|B)P(B)$. Two events A, B are said to be independent if

$$P(AB) = P(A)P(B). \tag{8.3-4}$$

Interpretation. If $P(A|B) = P(A)$, then observing B has no effect on the probability of A. Hence we say A and B are independent. If $A_1, \ldots, A_n$ are independent it follows that

$$P(A_1 \ldots A_n) = \prod_{i=1}^{n} P(A_i). \tag{8.3-5}$$

8.4 TOTAL PROBABILITY AND BAYES'S THEOREM

Let $A_1, \ldots, A_n$ be n mutually exclusive events. Let $\cup_{i=1}^{n} A_i = S$. With B denoting any event defined over the same probability space, we have

$$P(B) = P(B|A_1)P(A_1) + \cdots + P(B|A_n)P(A_n). \tag{8.4-1}$$

Proof: Since $A_iA_j = \varnothing$, $i \neq j$, we have

$$BS = B = B(A_1 \cup A_2 \cup \cdots \cup A_n) = \bigcup_{i=1}^{n} BA_i. \tag{8.4-2}$$

But $(BA_i)(BA_j) = \varnothing$ since $A_iA_j = \varnothing$. Hence

$$P(BA_1 \cup \cdots \cup BA_n) = P(BA_1) + \cdots + P(BA_n).$$

Using

$$P(BA_i) = P(B|A_i)P(A_i),$$

we obtain

$$P(B) = P\left[\bigcup_{i=1}^{n} (BA_i)\right] = \sum_{i=1}^{n} P(B|A_i)P(A_i). \tag{8.4-3}$$

$P(B)$ in Eq. (8.4-3) is sometimes called the unconditional or average probability of the event B.

Bayes's Theorem

A formula that is widely used in statistical communication theory, pattern recognition, and statistical inference is Bayes's theorem, which can easily be developed from the results of the preceding discussion. With $P(A_i) > 0$, $i = 1, \ldots, n$, $P(B) > 0$, $\cup_{i=1}^{n} A_i = S$, and $A_iA_j = \varnothing$ for $i \neq j$, we obtain

$$P(A_j|B) = \frac{P(A_j B)}{P(B)} = \frac{P(B|A_j)P(A_j)}{\Sigma_{i=1}^{n} P(B|A_i)P(A_i)}. \tag{8.4-4}$$

The probability $P(A_j|B)$ is sometimes called the *a posteriori* probability of A_j given B. Equation (8.4-4) enables the computation of the *a posteriori* probability in terms of the *a priori* conditional probabilities $P(B|A_i)$ and the causal probabilities $P(A_i)$.

Examples

1. In a binary communication system a zero or one is transmitted. Let X denote the transmitted symbol ($X = 0$ or $X = 1$) and Y the received symbol ($Y = 0$ or $Y = 1$). Let $P(X = 0) \equiv P_0$ and $P(X = 1) \equiv P_1 = 1 - P_0$, respectively. Due to noise in the channel, a zero can be received as a one with probability β, and a one can be received as a zero also with probability β. A one is observed. What is the probability that a one was transmitted?

Solution. The structure of the channel is shown in Fig. 8.4-1. Hence

$$\begin{aligned} P(X = 1|Y = 1) &= \frac{P(X = 1, Y = 1)}{P(Y = 1)} \\ &= \frac{P(Y = 1|X = 1)P(X = 1)}{P(Y = 1|X = 1)P(X = 1) + P(Y = 1|X = 0)P(X = 0)} \\ &= \frac{P_1(1 - \beta)}{P_1(1 - \beta) + P_0\beta}. \end{aligned}$$

If $P_0 = P_1 = \frac{1}{2}$, the *a posteriori* probability $P(X = 1|Y = 1)$ depends on β, as shown in Fig. 8.4-2. The channel is said to be noiseless if $\beta = 1$ or $\beta = 0$.

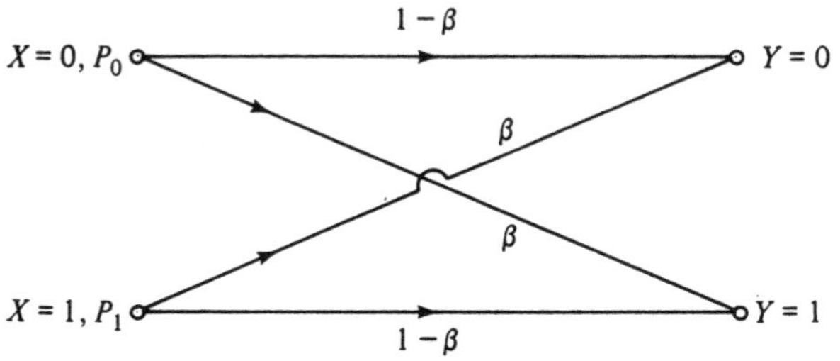

Figure 8.4-1

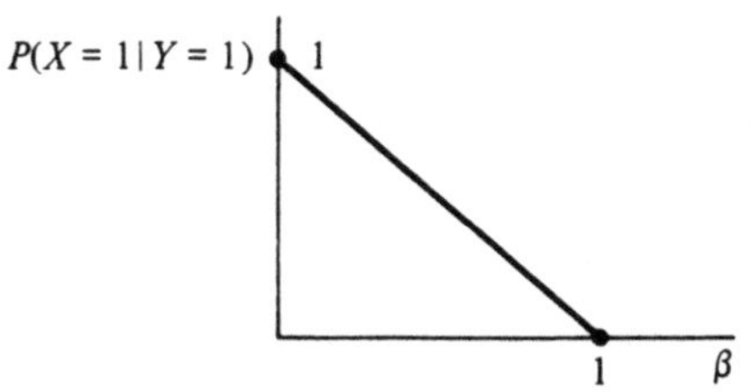

Figure 8.4-2

2. ([8-1], p. 119) Suppose there exists a (fictitious) test for cancer with the following properties. Let

$$\begin{aligned} A &= \text{event that the test states that tested person has cancer,} \\ B &= \text{event that person has cancer,} \\ A^c &= \text{event that test states person is free from cancer,} \\ B^c &= \text{event that person is free from cancer.} \end{aligned}$$

It is known that $P(A|B) = P(A^c|B^c) = 0.95$ and $P(B) = 0.005$. Is the test a good test?

Solution. To answer this question, we need to know the likelihood that a person actually has cancer if the test so states [i.e., $P(B|A)$]. Hence

$$\begin{aligned} P(B|A) &= \frac{P(B)P(A|B)}{P(A|B)P(B) + P(A|B^c)P(B^c)} = \frac{(0.005)(0.95)}{(0.95)(0.005) + (0.05)(0.995)} \\ &= 0.087. \end{aligned}$$

Hence in only 8.7% of the cases where the tests are positive will the person actually have cancer. This test has a very high false-positive rate and in this sense cannot be regarded as a good test.

8.5 BERNOULLI TRIALS

Consider the very simple experiment that consists of a single trial with a binary outcome X: a success $\{X = s\}$ with probability p or a failure $\{X = f\}$ with probability $q = 1 - p$. Thus $P(s) = p$ and $P(f) = q$, and the sample description space is $S = \{s, f\}$.

Suppose we do the experiment twice. The new sample description space S_2, written $S_2 = S \times S$, is the set of all ordered 2-tuples

$$S_2 = \{ss,\ sf,\ fs,\ ff\}.$$

The product $S \times S$ is called the Cartesian product. If we do n independent trials, the sample space is

$$S_n = \underbrace{S \times S \times \cdots \times S}_{n \text{ times}}$$

and contains 2^n elementary outcomes each of which is an ordered n-tuple. Thus

$$S_n = \{a_1, \ldots, a_M\}, \qquad \text{where } M = 2^n,$$

and $a_i = z_{i_1} z_{i_2} \ldots z_{i_n}$, where $z_{i_j} = s$ or f. Since each outcome z_{i_j} is independent of any other outcome, the joint probability is $P(z_{i_1}, \ldots, z_{i_n}) = P(z_{i_1})P(z_{i_2}) \ldots P(z_{i_n})$. Thus the probability of a given ordered set of k successes and $n - k$ failures is simply $p^k q^{n-k}$. For example, suppose we throw a coin three times with $p = P(H)$ and $q = P(T)$. The probability of the event $\{HTH\}$ is $pqp = p^2 q$. The probability of the event $\{THH\}$ is also $p^2 q$. The different events leading to two heads and one tail are

$$E_1 = \{HHT\},$$
$$E_2 = \{HTH\},$$
$$E_3 = \{THH\}.$$

If F denotes the event of getting two heads and one tail, then $F = E_1 \cup E_2 \cup E_3$. Since $E_iE_j = \varnothing$, $i \neq j$, we obtain $P(F) = P(E_1) + P(E_2) + P(E_3) = 3p^2q$.

Very frequently we are interested in this kind of result, that is, the probability of getting k successes in n tries without specification of order. The probability law that applies to this case is the *binomial law* $b(k; n, p)$, which is the probability of getting k successes in n *independent* tries with individual Bernoulli-trial success probability p. The probability law is given by

$$b(k; n, p) = \binom{n}{k} p^k q^{n-k}, \tag{8.5-1}$$

where

$$\binom{n}{k} \equiv \frac{n!}{k!(n-k)!} \tag{8.5-2}$$

is the binomial coefficient. The coefficient $\binom{n}{k}$ is the number of ways of getting unordered subsets of size k that may be formed from the members of a set of size n. How do we obtain this coefficient? Consider an urn containing n distinguishable balls. Suppose we draw k balls without replacement. From these k balls we can form $k!$ distinguishable (ordered) subsets. The total number of ordered subsets must equal $k!x_k$, where x_k is the different subsets of size k we can form from n. But $k!x_k$ must equal $n(n-1)\cdots(n-k+1)$ since this is the number of distinguishable subsets we can draw from the urn directly. Hence

$$x_k k! = n(n-1)\cdots(n-k+1) = \frac{n!}{(n-k)!} \tag{8.5-3}$$

or

$$x_k = \frac{n!}{k!(n-k)!} = \binom{n}{k}. \tag{8.5-4}$$

Examples

1. Suppose $n = 4$; that is, there are four balls numbered 1 to 4 in the urn. The number of distinguishable, ordered, samples of size 2 that can be drawn without replacement is 12; that is, {1, 2}, {1, 3}, {1, 4}, {2, 1}, {2, 3}, {2, 4}, {3, 1}, {3, 2}, {3, 4}, {4, 1}, {4, 2}, {4, 3}. The number of distinguishable unordered sets is 6; that is,

{1, 2}	{1, 3}	{1, 4}	{2, 4}	{3, 4}	{2, 3}
{2, 1}	{3, 1}	{4, 1}	{4, 2}	{4, 3}	{3, 2}

From the formula we obtain this result directly:

$$\binom{n}{k} = \frac{4!}{2!2!} = 6.$$

2. Ten independent, binary pulses per second arrive at a receiver. The error probability (i.e., a zero received as a one, or vice versa) is 0.001. What is the probability of at least one error per second?

$$P(\text{at least one error/sec}) = 1 - P(\text{no errors/sec})$$
$$= 1 - \binom{10}{0}(0.001)^0(0.999)^{10} = 1 - (0.999)^{10} \simeq 0.01.$$

Observation

$$P(S) = \sum_{k=0}^{n} b(k;\, n,\, p) = 1. \tag{8.5-5}$$

Why?

8.6 FURTHER DISCUSSION OF THE BINOMIAL LAW

We shall write down some self-evident formulas for further use. The probability $B(k;\, n,\, p)$ of not more than k successes in n tries is given by

$$B(k;\, n,\, p) = \sum_{i=0}^{k} b(i;\, n,\, p) = \sum_{i=0}^{k} \binom{n}{i} p^i q^{n-i}.$$

The probability of more than k successes in n tries is

$$\sum_{i=k+1}^{n} b(i;\, n,\, p) = 1 - B(k;\, n,\, p). \tag{8.6-1}$$

The probability of more than k successes but no more than j successes is

$$\sum_{i=k+1}^{j} b(i;\, n,\, p). \tag{8.6-2}$$

8.7 THE POISSON LAW

Suppose that $n \gg 1$, $p \ll 1$, but that np remains finite, say $np = a$. Recall that $q = 1 - p$. Hence

$$\binom{n}{k} p^k (1-p)^{n-k} \simeq \frac{1}{k!} a^k \left(1 - \frac{a}{n}\right)^{n-k},$$

where $n(n-1)\cdots(n-k+1) \simeq n^k$ because the binomial probability will have significant values around $k = np \ll n$. Hence in the limit as $n \to \infty$, $p \to 0$, $np = a$, and $k \ll n$, we obtain

$$b(k; n, p) \simeq \frac{1}{k!} a^k \left(1 - \frac{a}{n}\right)^{n-k} = \frac{a^k}{k!} e^{-a} \qquad (\text{as } n \to \infty). \tag{8.7-1}$$

Examples

1. A computer contains 10,000 components. Each component fails independently from the others, and the yearly failure probability per component is 10^{-4}. What is the probability that the computer will be working at the end of the year? Assume that the computer fails if one or more components fail.

Solution

$$p = 10^{-4}, \qquad n = 10{,}000, \qquad k = 0, \qquad np = 1.$$

Hence

$$b(0; 10{,}000, 10^{-4}) = \frac{1^0}{0!} e^{-1} = \frac{1}{e} = 0.368.$$

2. Suppose that n independent points are placed at random in an interval $(0, T)$. Let $0 < t_1 < t_2 < T$ and $t_2 - t_1 = \Delta t$. Let $\Delta t/T \ll 1$ and $n \gg 1$. What is the probability of observing exactly k points in Δt?

Solution. Consider a single point occurring in $[0, T]$. The probability of the point appearing in Δt is $\Delta t/T$. Let $p = \Delta t/T$. Every other point has the same probability of being in Δt. Hence the probability of finding k points in Δt is the binomial law

$$P(k \text{ points in } \Delta t) = \binom{n}{k} p^k q^{n-k}. \tag{8.7-2}$$

With $n \gg 1$ and $\Delta t/T \ll 1$, we use the approximation in Eq. (8.7-1) to give

$$b(k; n, p) \simeq \left(\frac{n\Delta t}{T}\right)^k \frac{e^{-n\Delta t/T}}{k!} \tag{8.7-3}$$

$$= (\lambda \Delta t)^k \frac{e^{-\lambda \Delta t}}{k!}, \tag{8.7-4}$$

where $\lambda \equiv n/T$ is the "average" number of points per unit interval.†

Equations (8.7-1) and (8.7-4) are examples of the *Poisson probability law*. The Poisson law with parameter $a(a > 0)$ is defined by‡

$$P(k \text{ events}) = e^{-a} \frac{a^k}{k!}, \tag{8.7-5}$$

† λ is often called the *Poisson rule parameter*.

‡ We use the term *event* here rather than "success."

where $k = 0, 1, 2, \ldots$. With $a = \lambda \Delta t$, where λ is the average number of events per unit time, and Δt the interval $(t, t + \Delta t)$, the probability of k events in Δt is

$$P[k \text{ events in } (t, t + \Delta t)] = e^{-\lambda \Delta t} \frac{(\lambda \Delta t)^k}{k!}. \tag{8.7-6}$$

In Eq. (8.7-6) we assumed that λ was independent of t. If λ depends on t, the probability of k events in $(t, t + \Delta t)$ is

$$\exp\left[-\int_t^{t+\Delta t} \lambda(\xi)\, d\xi\right] \frac{1}{k!} \left[\int_t^{t+\Delta t} \lambda(\xi)\, d\xi\right]^k. \tag{8.7-7}$$

8.8 RANDOM VARIABLES

Many random phenomena have sample description spaces that are sets of real numbers: the voltage $v_n(t)$, at time t, across a noisy resistor; the arrival time of the next customer at a movie theatre; the number of photons in a light pulse; the brightness level at a particular point on the TV screen; the number of times a light bulb will switch on before failing; the lifetime of a given living person; the number of hairs on the head of a person picked at random in New York City. In all these cases the sample description space is a subset of R, the real line. Sometimes this subset consists of a discrete set of countable points; other times the sample description space consists of an uncountable continuum of points. Nevertheless, whatever the nature of the subset (i.e., discrete, continuous, or mixed), it is often convenient to choose R as the sample description space. The set R is convenient even when the sample description space of a particular random phenomenon can be written more "compactly," as in the case of the number of hairs on a person's head where a compact sample space might be the set of integers $S = \{0, 1, \ldots\}$. By letting R be the sample description space, we simply add a lot of descriptions whose probability is zero. Although the set R may contain subsets of probability zero, the advantage of considering R is that we can generate a unified theory that will cover all random phenomena with real-valued numerical outcomes.

When R is the sample description space, every event is a subset of R. However, not every subset of R is necessarily an event. There exist certain sets that are not countable unions or intersections of intervals or points in R, and these sets are said to be nonprobabilizable. However, as pointed out in Section 8.1, such sets are of no importance in engineering, and we shall forget about them. All sets of engineering importance tend to be countable unions and intersections of intervals of the form (a, b), $(a, b]$, $[a, b)$, and $[a, b]$,† where a and b can be finite or infinite numbers. As already stated, such sets are called Borel sets, and the family of such sets is called the Borel field of events.

When the sample description space S is not numerical, we might want to generate a sample space from S that is numerical. For example, we might want to encode written

† $(a, b) = (a < x < b)$, $[a, b) = (a \le x < b)$, and so on.

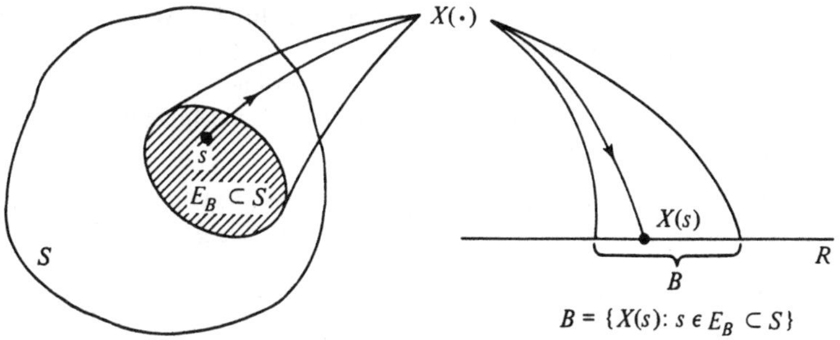

Figure 8.8.1. Symbolic representation of the function X(·).

text into signals for purposes of transmission or storage in a computer. Also, in many problems in communication theory we shall want to operate on random phenomena to generate new random phenomena by means of mathematical operations. To achieve these ends, we introduce and apply the notion of a random variable.

Let S be a sample description space (numerical or otherwise), and let its elements be $\{s\}$. A random variable $X(\cdot)$ is a *function* that assigns to every $s \in S$ a *number* $X(s)$. We also require that for every Borel set of numbers B the set $\{s\text{: } X(s) \in B\}$ is an event, that is, it belongs to $\mathcal{F}$ and is, therefore, in the domain of the probability function $P(\cdot)$.† The domain of $X(\cdot)$ is S, and its range is a subset of R. (See Fig. 8.8-1).

Under the mapping X we have in effect generated a new probability space $(R, \mathcal{B}, P_X)$, where R is the real line, $\mathcal{B}$ is the Borel sigma algebra of all subsets of R generated by countable unions and intersections of the form $(-\infty, x]$, and P_X is a set function assigning a number $P_X(B) \geq 0$ to each $B \in \mathcal{B}$. For example, we can speak of the event $(-\infty, x]$ and the probability $P_X[(-\infty, x]]$.

In communication theory we are rarely concerned with the functional form of X, and we rarely specify the set S, although it is always implied. Our interest is not in generating the numbers $\{X(s)\}$ from the elements of S. Primarily, we want to compute the probability that X will take values in a certain prespecified set. Some examples follow.‡

Examples

1. A bus arrives at random in $[0, T]$; let t denote the time of arrival. The sample description space S is $S = \{t\text{: } t \in [0, T]\}$. An r.v. X is defined by

$$X = \begin{cases} 1, & t \in \left[\dfrac{T}{4}, \dfrac{T}{2}\right], \\ 0, & \text{otherwise.} \end{cases}$$

† A mathematical definition of an r.v. $X(\cdot)$ is that X be a function on S to R such that the *inverse images* under X of all Borel sets in R are events.

‡ The notation r.v. will frequently be used for "random variable."

Assume that the arrival time is uniform over $[0, T]$. We can now compute the probability $P[X = 1]$ or $P[X = 0]$ or $P[X \leq 5]$.

2. An urn contains three balls colored white, black, and red. The experiment consists of choosing a ball at random from the urn. The sample description space is $S = \{W, B, R\}$. The random variable X is defined by

$$X = \begin{cases} \pi & s = W \text{ or } B, \\ 0, & s = R. \end{cases}$$

We can compute the probability $P(X \leq x_1)$, where x_1 is any number.

8.9 PROBABILITY DISTRIBUTION FUNCTION

In the first example we gave in the previous section, the basic set of events is $[R, \{0\}, \{1\}, \varnothing]$ for which the probabilities are $P(R) = 1$, $P(X = 0) = \frac{3}{4}$, $P(X = 1) = \frac{1}{4}$, and $P(\varnothing) = 0$. From these probabilities, we can infer any other probabilities, such as, for example, $P(X \leq 0.5)$. In many cases it is awkward to write down $P(\cdot)$ for every event. For this reason we introduce the notion of the probability distribution function (PDF). The PDF[†] is a pointwise function of x that contains all the information necessary to compute $P(E)$ for any E in the Borel set of events. The PDF $F_X(x)$ is defined by[‡]

$$F_X(x) = P(X \leq x) = P_X[(-\infty, x]]. \tag{8.9-1}$$

For the present we shall denote random variables by capital letters (e.g., X, Y, and Z) and the values they can take by lowercase letters (x, y, and z).

Properties of $F_X(x)$

1. $F_X(\infty) = 1$, $F_X(-\infty) = 0$.
2. $x_1 \leq x_2 \rightarrow F_X(x_1) \leq F_X(x_2)$; that is, $F_X(x)$ is a nondecreasing function of x.
3. $F_X(x)$ is continuous from the right; that is,

$$F_X(x) = \lim_{\epsilon \to 0} F_X(x + \epsilon), \qquad \epsilon > 0.$$

Proof of Property 2[§]

Consider the event $\{x_1 < X \leq x_2\}$ with $x_2 > x_1$. We write

$$0 \leq P(x_1 < X \leq x_2) \leq 1.$$

† PDF should not be confused with pdf, which will stand for probability density function.

‡ An expression such as $P(X \leq x)$ is shorthand for $P[s \in S\colon X(s) \leq x]$. Thus it is a probability on the original sample description space. On the other hand, $P_X[(-\infty, x]]$ is a probability on R. Note that, while $P_X(\cdot)$ is a set function, $P_X[(-\infty, x]]$ depends only on x. This makes it of great utility in calculations.

§ Property 1 follows from the fact that we require that $P(X = \infty) = P(X = -\infty) = 0$. The proof of Property 3 can be found in [8-3, p. 226].

But $\{X \le x_2\} = \{X \le x_1\} \cup \{x_1 < X \le x_2\}$, and

$$\{X \le x_1\}\{x_1 < X \le x_2\} = \emptyset.$$

Hence, by Eq. (8.2-3)

$$F_X(x_2) = F_X(x_1) + P(x_1 < X \le x_2)$$

or

$$P(x_1 < X \le x_2) = F_X(x_2) - F_X(x_1) \ge 0, \qquad \text{for } x_2 > x_1. \tag{8.9-2}$$

Exercise: Compute the probabilities of the events $\{X \ge a\}$, $\{a \le X < b\}$, $\{a \le X \le b\}$, $\{a < X \le b\}$, and $\{a < X < b\}$ in terms of $F_X(x)$ and $P(X = x)$.

Example

Toss a coin once; $S = \{H, T\}$ with $P(H) = p$ and $P(T) = q$; define $X(\cdot)$ by $X(H) = \pi$ and $X(T) = 0$.

Computation of $F_X(x)$

1. $x < 0$: The event $\{X \le x\} = \emptyset$, and $F_X(x) = 0$.
2. $0 \le x < \pi$: The event $\{X \le x\}$ is equivalent to the event $\{T\}$ since

$$X(H) = \pi > x,$$
$$X(T) = 0 \le x.$$

Hence $F_X(x) = q$.

3. $x \ge \pi$: The event $\{X \le x\}$ is the certain event since

$$X(H) = \pi \le x,$$
$$X(T) = 0 \le x.$$

The solution is drawn in Fig. 8.9-1.

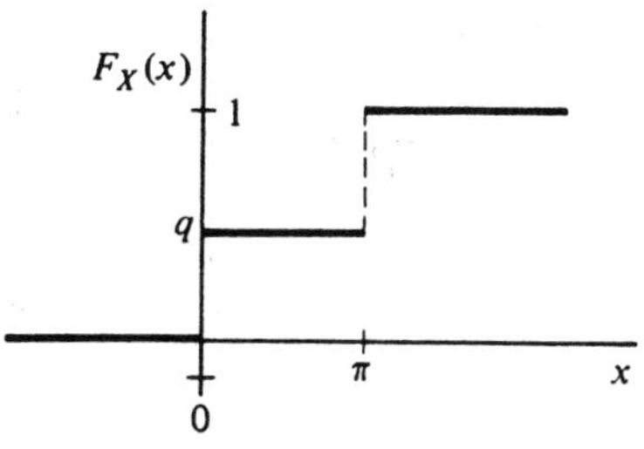

Figure 8.9-1. Probability distribution function associated with the coin-tossing experiment.

8.10 PROBABILITY DENSITY FUNCTION

The pdf, if it exists, is given by

$$f(x) = \frac{dF(x)}{dx}, \tag{8.10-1}$$

where $F(x) \equiv F_X(x)^\dagger$ since we are dealing only with a single random variable.

Properties

If $f(x)$ exists, then

$$\text{(i)} \int_{-\infty}^{\infty} f(\xi)\, d\xi = F(\infty) - F(-\infty) = 1, \tag{8.10-2}$$

$$\text{(ii)} \ F(x) = \int_{-\infty}^{x} f(\xi)\, d\xi = P(X \leq x), \tag{8.10-3}$$

$$\begin{aligned}\text{(iii)} \ F(x_2) - F(x_1) &= \int_{-\infty}^{x_2} f(\xi)\, d\xi - \int_{-\infty}^{x_1} f(\xi)\, d\xi \\ &= \int_{x_1}^{x_2} f(\xi)\, d\xi.\end{aligned} \tag{8.10-4}$$

Interpretation of f(x)

$$P(x < X \leq x + \Delta x) = F(x + \Delta x) - F(x).$$

If $F(x)$ is continuous in its first derivative, then, for sufficiently small Δx,

$$F(x + \Delta x) - F(x) = \int_{x}^{x+\Delta x} f(\xi)\, d\xi \simeq f(x)\, \Delta x.$$

Hence for small Δx

$$P(x < X \leq x + \Delta x) \simeq f(x)\, \Delta x. \tag{8.10-5}$$

Example

The univariate normal (Gaussian) pdf is

$$f(x) = \frac{1}{\sqrt{2\pi\sigma^2}} \exp\left[-\frac{1}{2}\left(\frac{x - \mu}{\sigma}\right)^2 \right]. \tag{8.10-6}$$

† For the present we shall dispense with the subscript X on $F(\cdot)$ unless we deal with more than one r.v.

TABLE 8.10-1

$$\text{erf}(x) = \frac{1}{\sqrt{2\pi}} \int_0^x \exp(-\tfrac{1}{2}t^2)\, dt$$

x	erf(x)	x	erf(x)
0.05	0.01994	2.05	0.47981
0.10	0.03983	2.10	0.48213
0.15	0.05962	2.15	0.48421
0.20	0.07926	2.20	0.48609
0.25	0.09871	2.25	0.48777
0.30	0.11791	2.30	0.48927
0.35	0.13683	2.35	0.49060
0.40	0.15542	2.40	0.49179
0.45	0.17364	2.45	0.49285
0.50	0.19146	2.50	0.49378
0.55	0.20884	2.55	0.49460
0.60	0.22575	2.60	0.49533
0.65	0.24215	2.65	0.49596
0.70	0.25803	2.70	0.49652
0.75	0.27337	2.75	0.49701
0.80	0.28814	2.80	0.49743
0.85	0.30233	2.85	0.49780
0.90	0.31594	2.90	0.49812
0.95	0.32894	2.95	0.49840
1.00	0.34134	3.00	0.49864
1.05	0.35314	3.05	0.49884
1.10	0.36433	3.10	0.49902
1.15	0.37492	3.15	0.49917
1.20	0.38492	3.20	0.49930
1.25	0.39434	3.25	0.49941
1.30	0.40319	3.30	0.49951
1.35	0.41149	3.35	0.49958
1.40	0.41924	3.40	0.49965
1.45	0.42646	3.45	0.49971
1.50	0.43319	3.50	0.49976
1.55	0.43942	3.55	0.49980
1.60	0.44519	3.60	0.49983
1.65	0.45052	3.65	0.49986
1.70	0.45543	3.70	0.49988
1.75	0.45993	3.75	0.49990
1.80	0.46406	3.80	0.49992
1.85	0.46783	3.85	0.49993
1.90	0.47127	3.90	0.49994
1.95	0.47440	3.95	0.49995
2.00	0.47724	4.00	0.49996

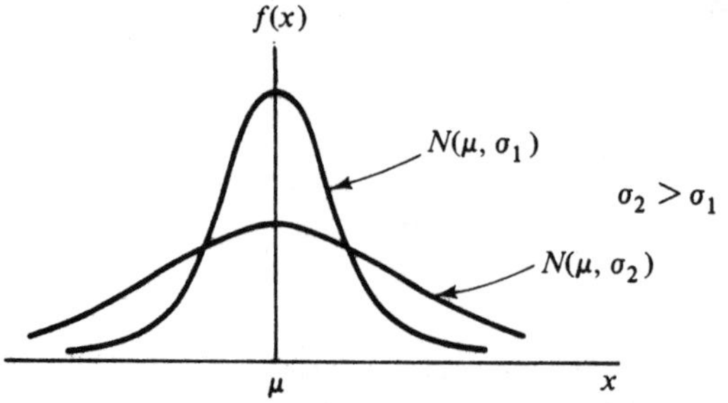

Figure 8.10-1. Univariate normal pdf.

There are two independent parameters: $\sigma \equiv$ standard deviation (σ^2 = variance) and $\mu \equiv$ the mean (Fig. 8.10-1). Consider a normal random variable, X, with mean μ and variance σ^2; we usually denote this by X: $N(\mu, \sigma^2)$. We have

$$P(a < X \leq b) = \frac{1}{\sqrt{2\pi\sigma^2}} \int_a^b \exp\left[-\left(\frac{1}{2}\right)\left(\frac{x-\mu}{\sigma}\right)^2\right] dx. \tag{8.10-7}$$

With $\beta = (x - \mu)/\sigma$, $d\beta = (1/\sigma)\, dx$, $b' = (b - \mu)/\sigma$, and $a' = (a - \mu)/\sigma$, we obtain

$$\begin{aligned} P(a < X \leq b) &= \frac{1}{\sqrt{2\pi}} \int_{a'}^{b'} \exp\left(-\frac{1}{2}\beta^2\right) d\beta \\ &= \frac{1}{\sqrt{2\pi}} \int_0^{b'} \exp\left(-\frac{1}{2}\beta^2\right) d\beta - \frac{1}{\sqrt{2\pi}} \int_0^{a'} \exp\left(-\frac{1}{2}\beta^2\right) d\beta \\ &\equiv \Phi(b') - \Phi(a'). \end{aligned} \tag{8.10-8}$$

The function $\Phi(x)$ is tabulated (see Table 8.10-1). It is sometimes referred to as the error function of x, abbreviated erf(x). Hence, if X is normal,

$$P(a < X \leq b) = \text{erf}\left(\frac{b-\mu}{\sigma}\right) - \text{erf}\left(\frac{a-\mu}{\sigma}\right). \tag{8.10-9}$$

8.11 CONTINUOUS, DISCRETE, AND MIXED RANDOM VARIABLES

If $F(x)$ is continuous for every x and its derivative exists everywhere except at a countable set of points, then we say that X is a continuous random variable. At points x where $F'(x)$ exists, the pdf is $f(x) = F'(x)$. At points where $F'(x)$ does not exist, we can assign any positive number to $f(x)$; $f(x)$ will then be defined for every x, and we are free to use the following important formulas:

$$F(x) = \int_{-\infty}^{x} f(\xi)\, d\xi, \tag{8.11-1}$$

$$P(x_1 < X \leq x_2) = \int_{x_1}^{x_2} f(\xi)\, d\xi, \tag{8.11-2}$$

and

$$P(B) = \int_{\text{all } \xi:\xi \in B} f(\xi)\, d\xi, \tag{8.11-3}$$

where, in Eq. (8.11-3), B is any event (i.e., any Borel set of real numbers). Equation (8.11-3) is an example of the pdf for a continuous r.v.

A *discrete random variable* has a staircase type of distribution function (Fig. 8.11-1). A probability measure for discrete r.v.'s is the probability *frequency* function, which is also known as the probability *mass* function $P(x_i)$. It is defined by

$$P(x_i) \equiv P(X = x_i) = F(x_i) - F(x_i^-), \tag{8.11-4}$$

where x_i^- is the point taken on the left of the jump at x_i.

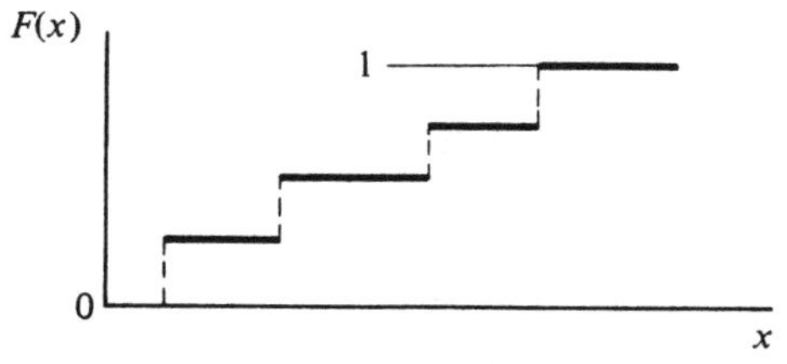

Figure 8.11-1. Probability distribution function for a discrete r.v.

Frequency Interpretation

For large n, $P(x_i) \simeq n_i/n$. The pdf for this case consists of impulse functions; that is,

$$f(x) = \frac{dF(x)}{dx} = \sum_i P(x_i)\delta(x - x_i) \tag{8.11-5}$$

and

$$P(x_1 < X \leq x_2) = \int_{x_1^+}^{x_2^+} f(\xi)\, d\xi. \tag{8.11-6}$$

The upper limit in the integration includes the impulse at $x = x_2$, while the lower limit avoids the integration of the impulse at $x = x_1$.

A mixed r.v. has a discontinuous probability distribution function, but not of the staircase type (Fig. 8.11-2).

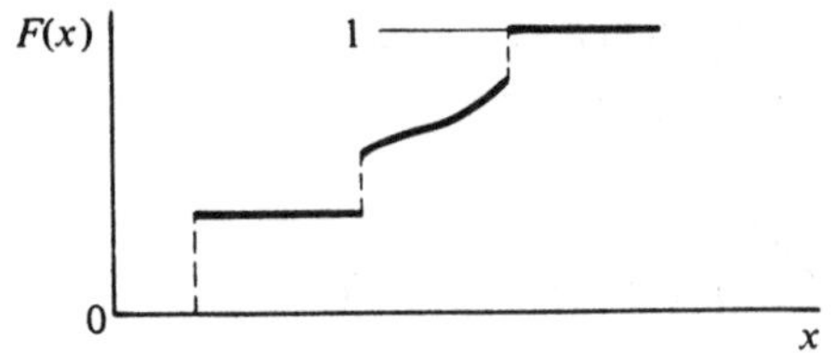

Figure 8.11-2. Probability distribution function for a mixed r.v.

8.12 CONDITIONAL AND JOINT DISTRIBUTIONS

Consider the event C consisting of all outcomes $s \in S$ such that $X(s) \leq x$ *and* $s \in B \subset S$.† The event C is then the set product of the two events $\{s: X(s) \leq x\}$ and $\{s: s \in B\}$. We define the *conditional distribution function of X given the event B as*

$$F(x|B) = \frac{P(C)}{P(B)} = \frac{P(X \leq x, B)}{P(B)}, \tag{8.12-1}$$

where $P(X \leq x, B)$ is the probability of the joint event $\{X \leq x\} \cap B$ with $P(B) \neq 0$. It is not difficult to prove ([8-2], p. 105) that $F(x|B)$ has all the properties of an ordinary distribution; that is, $x_1 \leq x_2 \rightarrow F(x_1|B) \leq F(x_2|B)$, $F(-\infty|B) = 0$, $F(\infty|B) = 1$, and so on. The conditional pdf is simply

$$f(x|B) = \frac{d}{dx}[F(x|B)]. \tag{8.12-2}$$

Generally, the event B will be expressed in terms of X.

Example

Let $B = \{X \leq 10\}$. We wish to compute $F(x|B)$.

1. For $x \geq 10$, the event $\{X \leq 10\}$ is a subset of the event $\{X \leq x\}$. Hence $P(X \leq 10, X \leq x) = P(X \leq 10)$, and use of Eq. (8.12-1) gives

$$F(x|B) = \frac{P(X \leq x, X \leq 10)}{P(X \leq 10)} = 1.$$

2. For $x \leq 10$, the event $\{X \leq x\}$ is a subset of the event $\{X \leq 10\}$. Hence $P(X \leq 10, X \leq x) = P(X \leq x)$ and

$$F(x|B) = \frac{P(X \leq x)}{P(X \leq 10)}.$$

The result is shown in Fig. 8.12-1. We leave it as an exercise for the reader (Prob. 8-20) to compute $F(x|B)$ when $B = \{b < X \leq a\}$.

† B is a more convenient symbol than E_B.

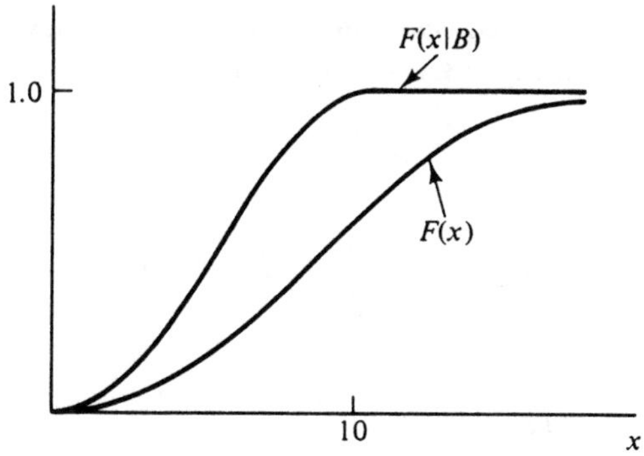

Figure 8.12-1

Joint Distribution

Suppose we are given two random variables X, Y defined on the same underlying probability space $(S, \mathscr{F}, P)$. The event $\{X \leq x, Y \leq y\} = \{X \leq x\} \cap \{Y \leq y\}$ consists of all outcomes $s \in S$ such that $X(s) \leq x$ *and* $Y(s) \leq y$.† The *joint distribution function* of X and Y is defined by

$$F_{XY}(x, y) = P(X \leq x, Y \leq y). \tag{8.12-3}$$

Since $\{X \leq \infty\}$ and $\{Y \leq \infty\}$ are certain events, we obtain

$$\{X \leq x, Y \leq \infty\} = \{X \leq x\}, \tag{8.12-4}$$

$$\{X \leq \infty, Y \leq y\} = \{Y \leq y\}, \tag{8.12-5}$$

so that

$$F_{XY}(x, \infty) = F_X(x), \tag{8.12-6}$$

$$F_{XY}(\infty, y) = F_Y(y). \tag{8.12-7}$$

The joint pdf, if it exists, is given by

$$f_{XY}(x, y) = \frac{\partial^2}{\partial x\, \partial y}[F_{XY}(x, y)]. \tag{8.12-8}$$

By twice integrating Eq. (8.12-8), we obtain

$$F_{XY}(x, y) = \int_{-\infty}^{x} d\xi \int_{-\infty}^{y} d\eta\, f_{XY}(\xi, \eta). \tag{8-12.9}$$

† What we mean when we say that X and Y are *defined on the same probability space* is merely that both X and Y share the same sample description space S as well as the same sigma field of events $\mathscr{F}$ and probability set measure P.

The functions $F_X(x)$ and $F_Y(y)$ are called *marginal* distributions if they are derived from a multivariate distribution as in Eqs. (8.12-6) and (8.12-7). Thus

$$F_X(x) = F_{XY}(x, \infty) = \int_{-\infty}^{x} d\xi \int_{-\infty}^{\infty} dy\, f_{XY}(\xi, y), \tag{8.12-10}$$

$$F_Y(y) = F_{XY}(\infty, y) = \int_{-\infty}^{y} d\eta \int_{-\infty}^{\infty} dx\, f_{XY}(x, \eta). \tag{8.12-11}$$

Since the marginal densities are given by

$$f_X(x) = F'_X(x),$$

$$f_Y(y) = F'_Y(y),$$

we obtain, by differentiating Eqs. (8.12-10) and (8.12-11),

$$f_X(x) = \int_{-\infty}^{\infty} f_{XY}(x, y)\, dy, \tag{8.12-12}$$

$$f_Y(y) = \int_{-\infty}^{\infty} f_{XY}(x, y)\, dx. \tag{8.12-13}$$

For discrete random variables we obtain equivalent results. Given the joint probability mass function $P_{XY}(X = x_i, Y = y_k)$ for all x_i, y_k, we compute the marginal mass function from

$$P_X(X = x_i) = \sum_{\text{all } y_k} P_{XY}(X = x_i, Y = y_k), \tag{8.12-14}$$

$$P_Y(Y = y_k) = \sum_{\text{all } x_i} P_{XY}(X = x_i, Y = y_k). \tag{8.12-15}$$

If X, Y are independent r.v.'s, then

$$F_{XY}(x, y) = F_X(x)F_Y(y), \tag{8.12-16}$$

$$f_{XY}(x, y) = f_X(x)f_Y(y), \tag{8.12-17}$$

$$P_{XY}(X = x_k, Y = y_j) = P_X(X = x_k)P_Y(Y - y_j). \tag{8.12-18}$$

Example

$$\begin{aligned} f_{XY}(x, y) &= \frac{1}{2\pi\sigma^2} \exp\left[-\frac{1}{2\sigma^2}(x^2 + y^2)\right] \\ &= \frac{1}{\sqrt{2\pi\sigma^2}} \exp\left(-\frac{1}{2}\frac{x^2}{\sigma^2}\right) \frac{1}{\sqrt{2\pi\sigma^2}} \exp\left(-\frac{1}{2}\frac{y^2}{\sigma^2}\right). \end{aligned} \tag{8.12-19}$$

Therefore, X and Y are independent.

8.13 FUNCTIONS OF RANDOM VARIABLES

In later work we shall have to deal with random variables that have been acted upon by operators and shall have to know how to compute the statistics of the output. We shall give some examples.

Examples

1. Let $Z = \max(X, Y)$, X, Y independent. What is $F_Z(z)$?

$$F_Z(z) = P(Z \le z) = P[\max(X, Y) \le z].$$

But

$$\begin{aligned} P[\max(X, Y) \le z] &= P(X \le z, Y \le z) \\ &= P(X \le z)P(Y \le z) = F_X(z)F_Y(z). \end{aligned}$$

2. The square-law device (Fig. 8.13-1) yields

$$\begin{aligned} F_Y(y) &= P(Y \le y) = P(X^2 \le y) = P(-\sqrt{y} \le X \le \sqrt{y}) \\ &= F_X(\sqrt{y}) - F_X(-\sqrt{y}) + P(X = -\sqrt{y}). \end{aligned}$$

But $P(X = -\sqrt{y}) = 0$ if X is a continuous r.v. Hence

$$f_Y(y) = \frac{d}{dy}F_Y(y) = \begin{cases} \dfrac{1}{2\sqrt{y}} f_X(\sqrt{y}) + \dfrac{1}{2\sqrt{y}} f_X(-\sqrt{y}), & y > 0, \\ 0, & y < 0. \end{cases} \tag{8.13-1}$$

3. Let the phase X of a sine wave be uniformly distributed in $[-\pi, \pi]$; that is,

$$f_X(x) = \begin{cases} \dfrac{1}{2\pi}, & |x| < \pi, \\ 0, & \text{otherwise}. \end{cases}$$

Let $Y = \sin X$; what is $f_Y(y)$? To compute $F_Y(y)$ we observe from Fig. 8.13-2 that for $1 \ge y \ge 0$ the event $\{\sin X \le y\}$ is given by $\{\pi - \sin^{-1} y < X \le \pi\} \cup \{-\pi < X \le \sin^{-1} y\}$. Since these are disjoint, we have

$$\begin{aligned} P(\sin X \le y) &= P(-\pi < X \le \sin^{-1} y) + P(\pi - \sin^{-1} y < X \le \pi) \\ &= F_X(\pi) - F_X(\pi - \sin^{-1} y) + F_X(\sin^{-1} y) - F_X(-\pi) = F_Y(y) \end{aligned}$$

so that

$$\begin{aligned} f_Y(y) = \frac{dF_Y(y)}{dy} &= f_X(\pi - \sin^{-1} y)\frac{1}{\sqrt{1 - y^2}} + f_X(\sin^{-1} y)\frac{1}{\sqrt{1 - y^2}} \\ &= \frac{1}{\pi}\frac{1}{\sqrt{1 - y^2}}, \qquad 0 \le y \le 1. \end{aligned} \tag{8.13-2}$$

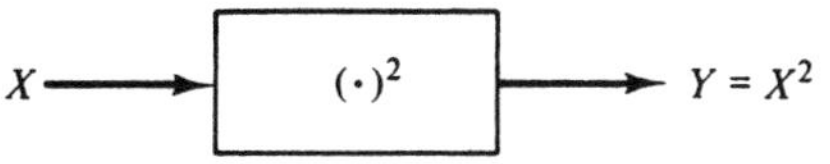

Figure 8.13-1

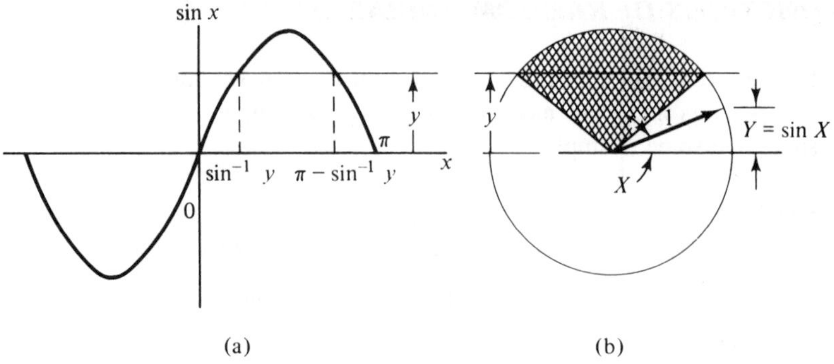

Figure 8.13-2. (a) The function $y = \sin x$. (b) The event $\{\sin X \le y\}$ is viewed as a unit vector that can rotate to any angle in the clear zone only.

Exercise. Repeat the calculation for $y < 0$, and show that the complete solution is

$$f_Y(y) = \begin{cases} \dfrac{1}{\pi} \dfrac{1}{\sqrt{1 - y^2}}, & |y| < 1, \\ 0, & \text{otherwise.} \end{cases} \tag{8.13-3}$$

4. Given $Z = X + Y$, X, Y independent r.v.'s, what is $f_Z(z)$?

Comment. This situation arises frequently in communication systems where, say, X represents a random signal and Y represents noise at a particular instant of time:

$$\begin{aligned} F_Z(z) = P(X + Y \le z) &= \iint_{x+y\le z} f_X(x) f_Y(y)\, dx\, dy \\ &= \int_{-\infty}^{\infty} dx\, f_X(x) \int_{-\infty}^{z-x} f_Y(y)\, dy \\ &= \int_{-\infty}^{\infty} F_Y(z - x) f_X(x)\, dx. \end{aligned} \tag{8.13-4}$$

Hence

$$\begin{aligned} f_Z(z) = \frac{d}{dz} F_Z(z) &= \int_{-\infty}^{\infty} f_Y(z - x) f_X(x)\, dx \\ &= f_Y(z) * f_X(z). \end{aligned} \tag{8.13-5}$$

Comment. When X, Y are independent, the pdf of the sum is the *convolution* of the pdf's. When X, Y are not independent, the solution is

$$f_Z(z) = \int_{-\infty}^{\infty} f_{XY}(x, z - x)\, dx. \tag{8.13-6}$$

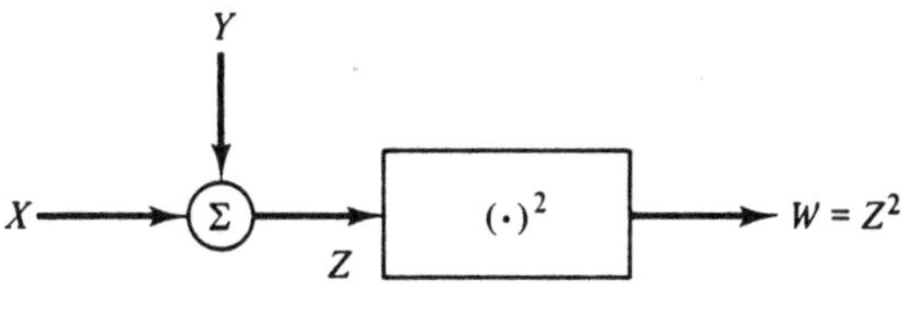

Figure 8.13-3

The general expression for evaluating the probability of any event E when X, Y are continuous r.v.'s is

$$P(E) = \iint_{\text{all } x,y \in E} f_{XY}(x, y)\, dx\, dy. \tag{8.13-7}$$

5. $X + Y$ injected into a square-law device [i.e., $W = Z^2 = (X + Y)^2$ (Fig. 8.13-3]. Let X and Y be independent, identically distributed r.v.'s, uniformly distributed in $(-\frac{1}{2}, \frac{1}{2})$. From Eq. (8.13-5) we compute $f_z(z) = (1 - |z|)\ \text{rect}\ (z/2)$. Now we use the result derived in Eq. (8.13-1) to obtain

$$f_W(w) = \begin{cases} \dfrac{1}{\sqrt{w}}(1 - \sqrt{w}), & 0 < w \le 1, \\ 0, & \text{otherwise.} \end{cases} \tag{8.13-8}$$

8.14 A GENERAL FORMULA FOR DETERMINING THE pdf OF A FUNCTION OF A SINGLE RANDOM VARIABLE

Suppose we are given a continuous r.v. X with its pdf $f_X(x)$. Consider the continuous transformation $y = g(x)$. What is the pdf of Y when X is applied to the system with transmittance $g(\cdot)$?

Solution. The event $\{y < Y \le y + dy\}$ can be written as a union of disjoint elementary events involving X. If the equation $y = g(x)$ has n real roots $x_1, \ldots, x_n$, then the disjoint elementary events have the form $E_i = \{x_i - |dx_i| < X < x_i\}$ if $g'(x_i)$ is negative or $E_i = \{x_i < X < x_i + |dx_i|\}$ if $g'(x_i)$ is positive. (See Fig. 8.14-1.) In either case, it follows from the definition of the pdf that $P(E_i) = f(x_i)|dx_i|$. Hence

$$\begin{aligned} P(y < Y < y + dy) &= f_Y(y)|dy| \\ &= \sum_{i=1} P(E_i) \\ &= \sum_{i=1}^{n} f_X(x_i)|dx_i|, \end{aligned} \tag{8.14-1}$$

or, equivalently, if we divide through by $|dy|$,

$$f_Y(y) = \sum_{i=1}^{n} f_X(x_i)\left|\frac{dx_i}{dy}\right|.$$

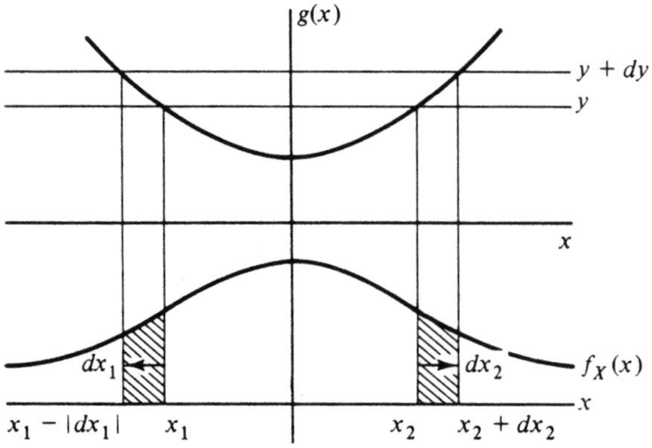

Figure 8.14-1

Since $y = g(x)$ and $dy/dx_i = g'(x_i)$, we obtain the important formula

$$f_Y(y) = \sum_{i=1}^{n} \frac{f_X(x_i)}{|g'(x_i)|}, \qquad x_i = x_i(y). \tag{8.14-2}$$

Equation (8.14-2) is a fundamental equation that is very useful in solving problems where the transmittance $g(x)$ has several roots. If, for a given y, the equation $y - g(x)$ has no real roots, then $f_Y(y) = 0$. Figure 8.14-1 illustrates the case when $n = 2$. From Fig. 8.14-1 we see that the event $\{y < Y \leq y + dy\}$ is identical to the event $\{x_1 - |dx_1| < X \leq x_1\} \cup \{x_2 < X \leq x_2 + |dx_2|\}$, and Eq. (8.14-1) follows with $n = 2$ in this case. The reader is cautioned that Eq. (8.14-2) is valid only when $g'(x_i) \neq 0$. When $g'(x_i) = 0$, indirect methods, such as illustrated in the first few examples in this section, must be used.

Examples

1. Let $g(\cdot)$ be the square-law device discussed earlier. Then $y - x^2 = 0$ has two roots (i.e., $x_1 = -\sqrt{y}$, $x_2 = +\sqrt{y}$ for $y > 0$). For $y < 0$, $y - x^2$ has no real roots; hence $f_Y(y) = 0$). Suppose X is a normal r.v. with mean zero and variance unity; then from Eq. (8.14-2),

$$f_Y(y) = \frac{1}{\sqrt{2\pi}} \frac{\exp[-\frac{1}{2}(\sqrt{y})^2]}{2\sqrt{y}} + \frac{1}{\sqrt{2\pi}} \frac{\exp[-\frac{1}{2}(-\sqrt{y})^2]}{2\sqrt{y}}, \qquad y > 0,$$

$$= \begin{cases} \dfrac{1}{\sqrt{2\pi y}} \exp(-\frac{1}{2}y), & y < 0, \\ 0, & \text{otherwise}. \end{cases} \tag{8.14-3}$$

2. Let $g(x_1, x_2) = x_1^2 + x_2^2$. With $Z_i = X_i^2$, $i = 1, 2$, we compute $f_{Z_i}(z)$ from Eq. (8.14-2) and compute the pdf of $W = Z_1 + Z_2$ from Eq. (8.13-5).

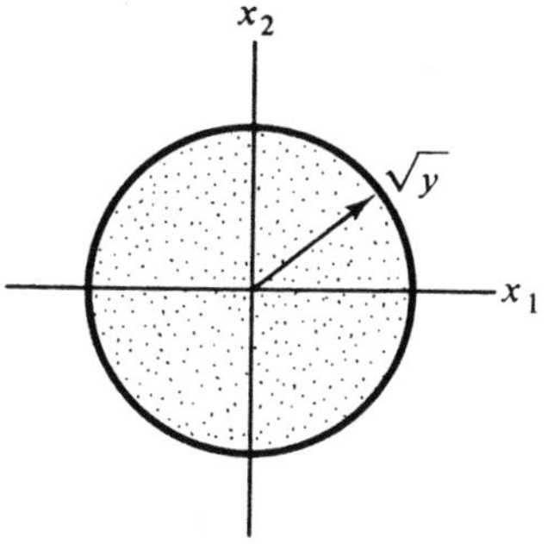

Figure 8.14-2

Comment. When X_1 and X_2 are normal, independent random variables, the computation of the pdf of $Y = X_1^2 + X_2^2$ is facilitated by an indirect approach. Thus

$$F_Y(y) = P(X_1^2 + X_2^2 \leq y) = \iint_{x_1^2+x_2^2 \leq y} f_{X_1X_2}(x_1, x_2)\, dx_1\, dx_2. \tag{8.14-4}$$

Note that the surface of integration is a circle about the origin (Fig. 8.14-2). We must compute

$$F_Y(y) = \iint_{x_1^2+x_2^2 \leq y} \frac{1}{2\pi\sigma^2} \exp\left[-\frac{1}{2}\left(\frac{x_1^2 + x_2^2}{\sigma^2}\right)\right] dx_1\, dx_2.$$

Let $x_1 = r \cos \theta$ and $x_2 = r \sin \theta$; then $r^2 = x_1^2 + x_2^2$ and $\theta = \tan^{-1} x_2/x_1$. Hence

$$F_Y(y) = \int_{\theta=0}^{2\pi} \int_0^{\sqrt{y}} \frac{r}{2\pi\sigma^2} \exp\left(-\frac{1}{2}\frac{r^2}{\sigma^2}\right) dr\, d\theta = 1 - e^{-y/2\sigma^2}, \qquad y \geq 0$$

$$= 0, \qquad y < 0 \text{ (note that } Y \text{ cannot take on negative values).}$$

The result may be conveniently written as

$$F_Y(y) = (1 - e^{-y/2\sigma^2})u(y), \tag{8.14-5}$$

where $u(y) = 1$ for $y \geq 0$ and $u(y) = 0$ for $y < 0$. The pdf is thus

$$f_Y(y) = \frac{dF_Y(y)}{dy} = \frac{1}{2\sigma^2} e^{-y/2\sigma^2} u(y), \tag{8.14-6}$$

which is known as the exponential probability law. Had we considered instead the pdf of $Y = \sqrt{X_1^2 + X_2^2}$, the surface of integration would have been a circular disk of radius y. For this case

$$F_Y(y) = \int_{\theta=0}^{2\pi} \int_0^{y} \frac{r}{2\pi\sigma^2} \exp\left(-\frac{1}{2}\frac{r^2}{\sigma^2}\right) dr\, d\theta$$

$$= \begin{cases} 1 - e^{-y^2/2\sigma^2}, & y \geq 0, \\ 0, & y < 0. \end{cases} \tag{8.14-7}$$

The pdf in this case is

$$f_Y(y) = \frac{y}{\sigma^2} e^{-y^2/2\sigma^2} u(y), \tag{8.14-8}$$

which is the *Rayleigh* density function. It is also known as the χ (chi) distribution with two degrees of freedom ([8-1], p. 181).

8.15 AVERAGES

The average value of X, also called the expected or mean value, is written $E(X)$ (read "the operator E acting on X") and is defined for a discrete r.v. by

$$\mu \equiv E(X) = \sum_i x_i P(X = x_i), \tag{8.15-1}$$

where E is the expectation operator and the summation over i really means "sum over all x_i such that $P(X = x_i) > 0$)." For a continuous r.v., we have

$$\mu \equiv E(X) = \int_{-\infty}^{\infty} x f_X(x)\, dx. \tag{8.15-2}$$

In more advanced work† it is shown that $E(X)$ can also be calculated according to

$$\begin{aligned} E(X) &= \lim_{\substack{\Delta x \to 0 \\ k \to \infty}} \sum_k x_k P(x_k < X \le x_k + \Delta x) \\ &= \lim_{\substack{\Delta x \to 0 \\ k \to \infty}} \sum_k x_k [F(x_k + \Delta x) - F(x_k)] \\ &= \int_R x\, dF. \end{aligned} \tag{8.15-3}$$

The definitions in Eqs. (8.15-1) and (8.15-2) extend to more general cases. Thus, with $g(X)$ denoting any function of X, we have

$$E[g(X)] = \int_{-\infty}^{\infty} g(x) f_X(x)\, dx, \qquad X \text{ continuous}, \tag{8.15-4}$$

$$E[g(X)] = \sum_i g(x_i) P(X = x_i), \qquad X \text{ discrete}. \tag{8.15-5}$$

The conditional expectation of X given $Y = y$, written $E(X|Y = y)$, is defined by

$$E(X|Y = y) = \int_{-\infty}^{\infty} x f_{X|Y}(x|y)\, dx, \qquad X \text{ continuous}, \tag{8.15-6}$$

$$E[X|Y = y_j] = \sum x_i P[X = x_i|Y = y_j], \qquad X, Y \text{ discrete}. \tag{8.15-7}$$

Note that the conditional expectations are functions of the particular value of y that the random variable Y has taken on.

† See, for example, [8-9, Chapter 6].

For several variables,

$$E[g(X, Y, Z)] = \int_{-\infty}^{\infty}\int_{-\infty}^{\infty}\int_{-\infty}^{\infty} g(x, y, z)\, f_{XYZ}(x, y, z)\, dx\, dy\, dz, \tag{8.15-8}$$

and

$$\begin{aligned} E(X + Y + Z) &= \int_{-\infty}^{\infty}\int_{-\infty}^{\infty}\int_{-\infty}^{\infty} (x + y + z) f_{XYZ}(x, y, z)\, dx\, dy\, dz \\ &= E(X) + E(Y) + E(Z). \end{aligned} \tag{8.15-9}$$

The proof of Eq. (8.15-9) is left as an exercise. This result can be generalized to

$$E\left[\sum_{i=1}^{N} X_i\right] = \sum_{i=1}^{N} E[X_i].$$

Note that *independence is not required*. The notation $E(X) = \bar{X}$ is also frequently used. We use it in what follows.

8.16 MOMENTS

We consider functions $g(\cdot)$ of X of the form X^r. We define the rth *moment* of X by

$$\zeta_r \equiv \overline{X^r} = \int_{-\infty}^{\infty} x^r f_X(x)\, dx \tag{8.16-1}$$

if X is continuous and

$$\zeta_r \equiv \overline{X^r} = \sum_i x_i^r P(X = x_i)$$

if X is discrete. In general, we shall omit writing expressions for the discrete case unless the extension from the continuous case is not obvious. We see from Eq. (8.16-1) that $\zeta_0 = 1$ and $\zeta_1 = \mu$. The rth *central moment* is more widely used. It is defined by

$$\begin{aligned} m_r &\equiv E[(X - \bar{X})^r] \\ &= \int_{-\infty}^{\infty} (x - \mu)^r f_X(x)\, dx. \end{aligned} \tag{8.16-2}$$

The most widely used moments are the first and second. The second central moment m_2 is also called the *variance* or dispersion. The symbol σ^2 is frequently used in place of m_2.

Exercise. Show that if $X{:}N(\mu,\sigma^2)$, the expected value of X is μ and the variance $m_2 = \sigma^2$. Also show that $m_3 = 0$ and $m_4 = 3\sigma^4$. Hence the symbols μ, σ^2 in the Gaussian pdf are aptly chosen.

Joint Moments

The ijth joint central moment of X and Y is defined by

$$m_{ij} = E[(X - \bar{X})^i (Y - \bar{Y})^j] = \int_{-\infty}^{\infty}\int_{-\infty}^{\infty} (x - \bar{X})^i (y - \bar{Y})^j f_{XY}(x, y)\, dx\, dy. \tag{8.16-3}$$

The most widely used joint central moment is m_{11}, which is called the *covariance*. The *correlation coefficient*, also called the *normalized covariance*, is defined by

$$\rho \equiv \frac{m_{11}}{\sqrt{m_{20}m_{02}}}. \tag{8.16-4}$$

Exercise. Show by direct substitution into Eq. (8.16-3) that $m_{11} = \rho\sigma_x\sigma_y$, where $f_{XY}(x, y)$ is given by

$$f_{XY}(x, y) = \frac{1}{2\pi\sigma_x\sigma_y\sqrt{1-\rho^2}} \cdot \exp\left\{-\frac{1}{2(1-\rho^2)}\left[\frac{(x-\bar{X})^2}{\sigma_x^2} - \frac{2\rho(x-\bar{X})(y-\bar{Y})}{\sigma_x\sigma_y} + \frac{(y-\bar{Y})^2}{\sigma_y^2}\right]\right\} \tag{8.16-5}$$

and $\sigma_x^2 = m_{20}$ and $\sigma_y^2 = m_{02}$. Equation (8.16-5) is the formula for two jointly normal random variables X, Y.

Exercise. Show that if X, Y are jointly Gaussian but uncorrelated random variables (i.e., $\rho = 0$), they are also independent.

Exercise. Show that any two statistically independent random variables are always uncorrelated. *Warning*: The converse is not always true.

8.17 MOMENT-GENERATING AND CHARACTERISTIC FUNCTIONS

In Section 8.16 we saw that the computation of moments requires a summation or integration for every moment that we want calculated. We shall now present a method for computing the moments from a single function which produces all moments by routine differentiation.

Definition. The moment-generating function $\theta(t)$ of a random variable X is given by

$$\theta(t) = E(e^{tX}). \tag{8.17-1}$$

When X is continuous, we obtain

$$\theta(t) = \int_{-\infty}^{\infty} e^{tx} f_X(x)\, dx, \tag{8.17-2}$$

and when X is discrete, we obtain

$$\theta(t) = \sum_i e^{tx_i} P(X = x_i). \tag{8.17-3}$$

Assume that X is continuous; then

$$e^{tX} = 1 + tX + \frac{(tX)^2}{2!} + \cdots + \frac{(tX)^n}{n!} + \cdots, \tag{8.17-4}$$

from which we obtain, assuming that $\zeta_r = \overline{X^r}$ exists,

$$E(e^{tX}) = 1 + t\mu + \frac{t^2}{2!}\zeta_2 + \cdots + \frac{t^n}{n!}\zeta^n + \cdots, \tag{8.17-5}$$

which is a linear combination of all the moments of X. To obtain any single moment, observe that (the differentiation is with respect to t)

$$\theta'(t)\Big|_{t=0} = E(Xe^{tX})_{t=0} = \mu, \tag{8.17-6}$$

$$\theta''(t)\Big|_{t=0} = E(X^2e^{tX})_{t=0} = \zeta_2, \tag{8.17-7}$$

$$\theta'''(t)\Big|_{t=0} = E(X^3e^{tX})_{t=0} = \zeta_3, \tag{8.17-8}$$

and so on. Hence the rth moment of X is obtained from

$$\zeta_r = \frac{d^r}{dt^r}[\theta(t)]\Big|_{t=0}. \tag{8.17-9}$$

Examples

1. Let $X = N(\mu, \sigma^2)$; then by direct evaluation of the integral

$$\theta(t) = \frac{1}{\sqrt{2\pi\sigma^2}} \int_{-\infty}^{\infty} \exp\left[-\frac{1}{2}\left(\frac{x-\mu}{\sigma}\right)^2\right] e^{tx}\,dx$$

we obtain

$$\theta(t) = \exp\,(t\mu + \tfrac{1}{2}t^2\sigma^2) \equiv e^{f(t)}, \tag{8.17-10}$$

where $f(t) = t\mu + \frac{1}{2}t^2\sigma^2$.

The first two moments are calculated to be

$$\theta'(t)\Big|_{t=0} = (\mu + t\sigma^2)e^{f(t)}\Big|_{t=0} = \zeta_1 = \mu, \tag{8.17-11}$$

$$\theta''(t)\Big|_{t=0} = [\sigma^2 + (\mu + t\sigma^2)^2]e^{f(t)}\Big|_{t=0} = \zeta_2 = \sigma^2 + \mu^2. \tag{8.17-12}$$

2. Let X be a discrete r.v. with a binomial distribution. The moment-generating function is

$$\theta(t) = \sum_{k=0}^{n} e^{tk} \binom{n}{k} p^k q^{n-k}$$
$$= \sum_{k=0}^{n} \binom{n}{k} (e^t p)^k q^{n-k}, \qquad p + q = 1. \tag{8.17-13}$$

By the binomial expansion theorem, this is simply $(pe^t + q)^n$. Hence for the binomial random variable

$$\theta(t) = (pe^t + q)^n, \tag{8.17-14}$$

$$\theta'(t) = npe^t(pe^t + q)^{n-1} = np, \qquad \text{at } t = 0, \tag{8.17-15}$$

so that $\mu = np$. Similarly, $\theta''(t)$ is given by

$$\theta''(t) = \frac{d}{dt}\theta'(t) = npe^t(pe^t + q)^{n-1} + n(n-1)p^2e^{2t}(pe^t + q)^{n-2}, \tag{8.17-16}$$

so that

$$\zeta_2 = npq + n^2p^2. \tag{8.17-17}$$

Note that the second central moment is $m_2 = \zeta_2 - \mu^2 = npq$. One could similarly define a *joint moment-generating function* according to

$$\theta_{X_1X_2\cdots X_n}(t_1, \ . \ . \ . \ , t_n) = E\left[\exp\left(\sum_{i=1}^{n} t_iX_i\right)\right]$$
$$= \sum_{k_1=0}^{\infty}\sum_{k_2=0}^{\infty}\cdots\sum_{k_n=0}^{\infty} \frac{t_1^{k_1}}{k_1!}\frac{t_2^{k_2}}{k_2!}\cdots\frac{t_n^{k_n}}{k_n!}\overline{X_1^{k_1}\cdots X_n^{k_n}}. \tag{8.17-18}$$

For example, with $n = 2$ we obtain

$$\theta_{X_1X_2}(t_1, t_2) = E[e^{t_1X_1+t_2X_2}], \tag{8.17-19}$$

so that the first and second moments can be obtained from

$$\overline{X_1} = \frac{\partial}{\partial t_1}\theta_{X_1X_2}(0, 0) \equiv \mu_1, \tag{8.17-20}$$

$$\overline{X_1^2} = \frac{\partial^2}{\partial t_1^2}\theta_{X_1X_2}(0, 0) = \mu_1^2 + m_{20}, \tag{8.17-21}$$

$$\overline{X_1X_2} = \frac{\partial^2}{\partial t_1\,\partial t_2}\theta_{X_1X_2}(0, 0) = m_{11} + \mu_1\mu_2, \tag{8.17-22}$$

$$\overline{X_2^2} = \frac{\partial^2}{\partial t_2^2}\theta_{X_1X_2}(0, 0) = \mu_2^2 + m_{02}, \tag{8.17-23}$$

$$\overline{X_2} = \frac{\partial}{\partial t_2}\theta_{X_1X_2}(0, 0) \equiv \mu_2, \tag{8.17-24}$$

where the arguments (0, 0) mean that all derivatives are evaluated at $t_1 = t_2 = 0$. Note that the variances and covariances can be expressed in terms of the preceding according to

$$m_{20} = \frac{\partial^2}{\partial t_1^2}[\theta_{X_1X_2}(0, 0)] - \left|\frac{\partial}{\partial t_1}[\theta_{X_1X_2}(0, 0)]\right|^2, \tag{8.17-25}$$

$$m_{11} = \frac{\partial^2}{\partial t_1\, \partial t_2}[\theta_{X_1X_2}(0, 0)] - \frac{\partial}{\partial t_1}[\theta_{X_1X_2}(0, 0)]\frac{\partial}{\partial t_2}[\theta_{X_1X_2}(0, 0)]. \tag{8.17-26}$$

We have seen that the pdf of the sum of independent random variables involves the convolution of their pdf's. Thus, if $Z = X_1 + \cdots + X_n$, where X_i, $i = 1, \ldots, n$, are independent random variables, the pdf of Z is furnished by

$$f_Z(z) = f_{X_1}(z) * f_{X_2}(z) * \cdots * f_{X_n}(z), \tag{8.17-27}$$

that is, the repeated convolution product.

The actual evaluation of Eq. (8.17-27) can be very tedious. However, we know from our studies of Fourier transforms that the Fourier transform of a convolution product is the product of the individual transforms. This property of the Fourier transform is also used conveniently in probability theory by introducing a function closely related to the moment-generating function called the *characteristic function*. The characteristic function $\Phi_X(\omega)$ for a continuous r.v. is defined conventionally by†

$$\Phi_X(\omega) = E(e^{j\omega X}) = \theta_X(j\omega) = \int_{-\infty}^{\infty} e^{j\omega x} f_X(x)\, dx, \tag{8.17-28}$$

which is seen to be the complex conjugate of the Fourier transform of $f_X(x)$. All the moments of a random variable *that exist* can be obtained from

$$E(X^k) = \frac{1}{j^k}\frac{d^k}{d\omega^k}\Phi_X(\omega)\bigg|_{\omega=0} \tag{8.17-29}$$

Equation (8.17-29) is obtained by differentiating Eq. (8.17-28) k times and multiplying by $(-j)^k$ at $\omega = 0$.

Consider the pdf of $Z = X_1 + X_2$, where X_1 and X_2 are independent r.v.'s with pdf's $f_{X_1}(x)$ and $f_{X_2}(x)$. Let us compute the characteristic function of Z. By definition

$$\begin{aligned}\Phi_Z(\omega) = E(e^{j\omega Z}) &= E[e^{j\omega(X_1+X_2)}] = E[e^{j\omega X_1}]E[e^{j\omega X_2}] \\ &= \Phi_{X_1}(\omega)\Phi_{X_2}(\omega).\end{aligned} \tag{8.17-30}$$

† The symbol $\Phi_X(\omega)$ has a totally different meaning from $\Phi(x)$ used in Sec. 8.10 to denote erf(x).

The same result is obviously obtained if $\Phi_Z(\omega)$ is computed directly from $f_Z(z)$:

$$\begin{aligned}\Phi_Z(\omega) &= \int_{-\infty}^{\infty} f_Z(z)e^{j\omega z}\,dz \\ &= \int_{-\infty}^{\infty} dz\, e^{j\omega z} \int_{-\infty}^{\infty} f_{X_1}(z-x)f_{X_2}(x)\,dx\end{aligned} \tag{8.17-31}$$

$$\begin{aligned}&= \int_{-\infty}^{\infty} dx\, f_{X_1}(x)e^{j\omega x} \int_{-\infty}^{\infty} du\, f_{X_2}(u)e^{j\omega u} \\ &= \Phi_{X_1}(\omega)\Phi_{X_2}(\omega).\end{aligned} \tag{8.17-32}$$

Thus the characteristic function of the sum of independent r.v.'s is the product of the characteristic functions. Note that by the Fourier inversion theorem the pdf of Z can be obtained directly from

$$f_Z(z) = \int_{-\infty}^{\infty} \Phi_Z(\omega)e^{-j\omega z}\frac{d\omega}{2\pi}. \tag{8.17-33}$$

For a discrete random variable X, the characteristic function is defined by

$$\Phi_X(\omega) = \sum_i e^{j\omega x_i}P(X = x_i). \tag{8.17-34}$$

Even though X is discrete, $\Phi_X(\omega)$ is a continuous function of ω. Furthermore, the magnitude of $\Phi_X(\omega)$ is bounded from above by unity since

$$\begin{aligned}|\Phi_X(\omega)| &= \left|\sum_i e^{j\omega x_i}P(X = x_i)\right| \leq \sum_i \left|e^{j\omega x_i}P(X = x_i)\right| \\ &= \sum_i P(X = x_i) = 1.\end{aligned} \tag{8.17-35}$$

Hence $|\Phi_X(\omega)| \leq 1$. A similar result holds when X is continuous. Equation (8.17-29) can be used to compute the moments whether X is discrete or continuous.

Examples

1. Let X_1 be $N(\mu_1, \sigma_1^2)$ and X_2 be $N(\mu_2, \sigma_2^2)$. Compute the pdf of $Z = X_1 + X_2$ if X_1 and X_2 are independent r.v.'s.

Solution. We can obtain $\Phi_{X_i}(\omega)$, $i = 1, 2$, by direct integration. However, since $\Phi_{X_i}(\omega) = \theta_{X_i}(j\omega)$, we use

$$\theta_{X_i}(t) = \exp(t\mu_i + \tfrac{1}{2}t^2\sigma_i^2),$$

from which we obtain

$$\begin{aligned}\Phi_{X_1}(\omega) &= \exp(j\omega\mu_1 - \tfrac{1}{2}\omega^2\sigma_1^2), \\ \Phi_{X_2}(\omega) &= \exp(j\omega\mu_2 - \tfrac{1}{2}\omega^2\sigma_2^2).\end{aligned} \tag{8.17-36}$$

Hence

$$\begin{aligned}\Phi_Z(\omega) &= \Phi_{X_1}(\omega)\Phi_{X_2}(\omega) \\ &= \exp(j\omega\mu - \tfrac{1}{2}\omega^2\sigma^2),\end{aligned} \tag{8.17-37}$$

where $\mu \equiv \mu_1 + \mu_2$ and $\sigma^2 \equiv \sigma_1^2 + \sigma_2^2$. The pdf of Z is

$$\begin{aligned}f_Z(z) &= \int_{-\infty}^{\infty} \Phi_Z(\omega)e^{-j\omega z}\frac{d\omega}{2\pi} \\ &= \frac{1}{\sqrt{2\pi\sigma^2}}\exp\left[-\frac{1}{2}\left(\frac{z-\mu}{\sigma}\right)^2\right].\end{aligned} \tag{8.17-38}$$

Thus the sum of two normal, independent r.v.'s is a normal r.v. whose mean is the sum of the means and whose variance is the sum of the variances. A similar result holds true for the sum of n normal, independent, random variables. The pdf of the sum is Gaussian with mean $\mu = \sum_{i=1}^n \mu_i$ and variance $\sigma^2 = \sum_{i=1}^n \sigma_i^2$.

2. Compute the first few moments of $Y = \sin\Theta$ if Θ is uniformly distributed in $[0, 2\pi]$.

Solution. We use the result in Eq. (8.15-4); that is, if $Y = g(X)$, then $\bar{Y} = \int_{-\infty}^{\infty} y f_Y(y)\,dy = \int_{-\infty}^{\infty} g(x) f_X(x)\,dx$. Hence

$$\begin{aligned}E(e^{j\omega Y}) &= \int_{-\infty}^{\infty} e^{j\omega y} f_Y(y)\,dy \\ &= \int_{-\infty}^{\infty} e^{j\omega\sin\theta} f_\Theta(\theta)\,d\theta \\ &= \frac{1}{2\pi}\int_{-\infty}^{\infty} e^{j\omega\sin\theta}\,d\theta \\ &= J_0(\omega),\end{aligned} \tag{8.17-39}$$

where $J_0(\omega)$ is the Bessel function of the first kind of order zero. A power series expansion of $J_0(\omega)$ gives

$$J_0(\omega) = 1 - \left(\frac{\omega}{2}\right)^2 + \frac{1}{2!2!}\left(\frac{\omega}{2}\right)^4 - \cdots. \tag{8.17-40}$$

Hence all the odd-order moments are zero. From Eq. (8.17-29), we compute

$$\overline{Y^2} = (-1)\frac{d^2}{d\omega^2}[J_0(\omega)]\bigg|_{\omega=0} = \tfrac{1}{2}, \tag{8.17-41}$$

$$\overline{Y^4} = (+1)\frac{d^4}{d\omega^4}[J_0(\omega)]\bigg|_{\omega=0} = \tfrac{3}{8}. \tag{8.17-42}$$

As in the case of joint moment-generating functions, we can define the joint characteristic function by

$$\Phi_{X_1\cdots X_n}(\omega_1, \ldots, \omega_n) = E[\exp(j\sum_{i=1}^{n}\omega_i X_i)]. \tag{8.17-43}$$

By the Fourier inversion property, the joint pdf is the inverse Fourier transform of $\Phi_{X_1\cdots X_n}(\omega_1, \ldots, \omega_n)$. Thus

$$f_{X_1 \cdots X_n}(X_1, \ldots, X_n) = \frac{1}{(2\pi)^n} \int_{-\infty}^{\infty} \cdots \int_{-\infty}^{\infty} \Phi_{X_1 \cdots X_n}(\omega_1, \ldots, \omega_n) \exp\left(-j \sum_{i=1}^{n} \omega_i x_i\right) \cdot d\omega_1 d\omega_2 \cdots d\omega_n. \tag{8.17-44}$$

The *rk*th joint moment can be obtained by differentiation. Thus with X, Y denoting any two random variables, we have

$$\zeta_{rk} \equiv E(X^r Y^k) = (-j)^{r+k} \left. \frac{\partial^{r+k}\Phi(\omega_1, \omega_2)}{\partial\omega_1^r \, \partial\omega_2^k} \right|_{\omega_1=\omega_2=0} \tag{8.17-45}$$

Exercise. Show that if X is $N(0, \sigma^2)$ then

$$\zeta_n = m_n = \begin{cases} 1 \cdot 3 \cdots (n-1)\sigma^n, & n \text{ even}, \\ 0, & n \text{ odd}. \end{cases} \tag{8.17-46}$$

Show also that $E|X|^n$ is given by

$$E|X|^n = \begin{cases} m_n, & n \text{ even}, \\ \dfrac{\sqrt{2}}{\pi} 2^{(n-1)/2} \left(\dfrac{n-1}{2}\right)! \sigma^n, & n \text{ odd}. \end{cases} \tag{8.17-47}$$

8.18 TWO FUNCTIONS OF TWO RANDOM VARIABLES

Suppose we are given two random variables X, Y defined on the same underlying probability space $(S, \mathscr{F}, P)$ and two real functions $w = g(x, y)$, $z = h(x, y)$. We form the two random variables $W = g(X, Y)$ and $Z = h(X, Y)$ and consider the computation of $f_{WZ}(w, z)$.

This problem is a direct extension of the problem considered in Sec. 8.14, and the solution has the same form as Eq. (8.14-2) except extended to two dimensions. We denote by (x_i, y_i), $i = 1, \ldots, n$, the n real solutions to the equations

$$g(x_i, y_i) = w, \; h(x_i, y_i) = z, \qquad i = 1, \ldots, n. \tag{8.18-1}$$

At each root we evaluate a normalizing factor called the *Jacobian*, denoted J_i at the ith root, and defined by

$$J_i = \left. \left| \frac{\partial w}{\partial x}\frac{\partial z}{\partial y} - \frac{\partial z}{\partial x}\frac{\partial w}{\partial y} \right| \right._{x=x_i, y=y_i} \tag{8.18-2}$$

The joint pdf of W and Z is given by

$$f_{WZ}(w, z) = \sum_{i=1}^{n} \frac{f_{XY}(x_i, y_i)}{|J_i|}, \tag{8.18-3}$$

where $x_i = x_i(w, z)$ and $y_i = y_i(w, z)$. The proof is given in several places, for example, in [8-3, p. 97]. Note that $|J_i|$ plays the same roles as $|g'(x_i)|$ in Eq. (8.14-2), which considered the single-variable transformation. We illustrate the application of Eq. (8.18-3) with an example.

Example

Let $W \equiv X$ and $Z \equiv X/Y$. Compute $f_{WZ}(w, z)$ and $f_Z(z)$ in terms of $f_{XY}(x, y)$.

Solution. The only solution to the set of equations

$$w = x$$

$$z = x/y$$

is $x = w$, $y = w/z$. Hence

$$|J_1| = \left|\frac{x}{y^2}\right| = \frac{z^2}{|w|}$$

and

$$f_{WZ}(w, z) = f_{XY}\left(w, \frac{w}{z}\right)\frac{|w|}{z^2}. \tag{8.18-4}$$

Suppose $f_{XY}(x, y)$ is given by

$$f_{XY}(x, y) = \frac{1}{2\pi\sigma_1\sigma_2\sqrt{1-\rho^2}}\exp\left\{-\left[\frac{1}{2(1-\rho^2)}\left(\frac{x^2}{\sigma_1^2} - \frac{2\rho xy}{\sigma_1\sigma_2} + \frac{y^2}{\sigma_2^2}\right)\right]\right\}; \tag{8.18-5}$$

then, from Eq. (8.18-4), $f_{WZ}(w, z)$ is given by

$$f_{WZ}(w, z) = \frac{|w|}{2\pi\sigma_1\sigma_2 z^2\sqrt{1-\rho^2}}\exp\left\{-\left[\frac{w^2}{2(1-\rho^2)}\left(\frac{1}{\sigma_1^2} - \frac{2\rho}{z\sigma_1\sigma_2} + \frac{1}{z^2\sigma_2^2}\right)\right]\right\}. \tag{8.18-6}$$

To obtain $f_Z(z)$, we integrate $f_{WZ}(w, z)\,dw$ over all values of w. Thus

$$f_Z(z) = \int_{-\infty}^{\infty} f_{WZ}(w, z)\,dw. \tag{8.18-7}$$

Despite the formidable appearance of $f_{WZ}(w, z)$, this is actually an easy integration. The result is

$$f_Z(z) = \frac{\sqrt{1-\rho^2}\,\sigma_1\sigma_2/\pi}{\sigma_2^2(z-\rho\sigma_1/\sigma_2)^2 + \sigma_1^2(1-\rho^2)}. \tag{8.18-8}$$

The distribution function of Z is given by

$$F_Z(z) = \int_{-\infty}^{z} f_Z(u)\,du. \tag{8.18-9}$$

Here, too, despite the formidable appearance of $f_Z(z)$, the integration is easy. We leave it as an exercise for the reader to show that†

$$F_Z(z) = \frac{1}{2} + \frac{1}{\pi}\tan^{-1}\frac{\sigma_2 z - \rho\sigma_1}{\sigma_1\sqrt{1-\rho^2}}. \tag{8.18-10}$$

† Alternatively, the integral can be looked up in tables of integrals such as those by B. O. Peirce and R. M. Foster, *A Short Table of Integrals*, Ginn, Boston, 1956, Eq. (70).

In the special case where X and Y are independent r.v.'s with common variance σ^2,

$$f_Z(z) = \frac{1}{\pi}\frac{1}{z^2 + 1} \tag{8.18-11}$$

and

$$F_Z(z) = \frac{1}{2} + \frac{1}{\pi}\tan^{-1} z. \tag{8.18-12}$$

Suppose we are interested in the probability that $X/Y < 0$. Then, with the aid of Eq. (8.18-10), we obtain

$$P\left(\frac{X}{Y} < 0\right) = P(Z < 0) = F_Z(0) = \frac{1}{2} - \frac{1}{\pi}\tan^{-1}\frac{\rho}{\sqrt{1-\rho^2}}. \tag{8.18-13}$$

Let

$$\alpha \equiv \tan^{-1}\frac{\rho}{\sqrt{1-\rho^2}}$$

(see Fig. 8.18-1). Then

$$P\left(\frac{X}{Y} < 0\right) = \frac{1}{2} - \frac{1}{\pi}\alpha \tag{8.18-14}$$

$$= \frac{\beta}{\pi}, \tag{8.18-15}$$

where $\beta = (\pi/2) - \alpha = \cos^{-1}\rho$. Considering the complexity of the joint Gaussian pdf, it is somewhat surprising that we end up with such simple expressions. The probability that $X/Y < 0$ will come up when we discuss Van Vleck's theorem in Chapter 9.

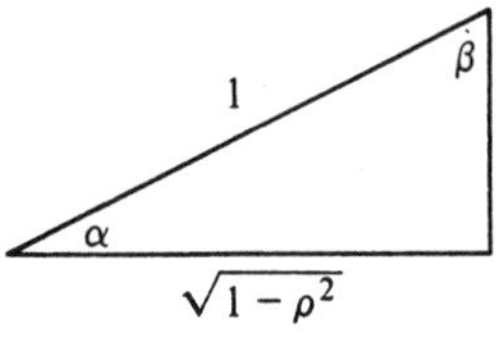

Figure 8.18-1

8.19 SUMMARY

In this chapter we gave a brief introduction to the basic principles of probability and random variables. Our emphasis has been on developing the tools needed for solving engineering-type probability problems. With this material well in hand, the reader should experience little difficulty in understanding the discussions on random signals and noise that appear in subsequent chapters.

It may not be immediately obvious to the reader that the point of view taken here stresses the systems configuration, which, in simplest form, amounts to the following: Given a random variable X and its probability law and a system function $g(\cdot)$, what is the probability law of $Y = g(X)$? In more advanced books, probability theory is discussed more in terms of abstract spaces, measure theory, and Lebesgue integration [8-9]. Fortunately, the vast majority of engineering problems can be solved without these advanced mathematical tools.

PROBLEMS

8-1. An urn contains three balls numbered 1, 2, 3. The experiment consists of drawing a ball at random, recording the number, and replacing the ball before the next ball is drawn. This is called sampling with replacement. What is the probability of drawing the same ball twice?

8-2. An experiment consists of drawing two balls *without* replacement from an urn containing six balls numbered 1 to 6. Describe the sample description space S. What is S if the first ball is replaced before the second is drawn?

8-3. The experiment consists of measuring the heights of each partner of a randomly chosen married couple. Describe S in convenient notation.

8-4. In Prob. 8-3, let E be the event that the man is shorter than the woman. Describe E in convenient notation.

8-5. An urn contains 10 balls numbered 1 to 10. Let E be the event of drawing a ball numbered no greater than 5. Let F be the event of drawing a ball numbered greater than 3 but less than 9. Evaluate E^c, F^c, EF, $E \cup F$, EF^c, E^cF, $E^c \cup F^c$, $EF^c \cup E^cF$, $EF \cup E^cF^c$, $(E \cup F)^c$, and $(EF)^c$. Express these events in words.

8-6. An experiment consists of drawing two balls at random, with replacement, from an urn containing five balls numbered 1 to 5. Three students Dim, Dense, and Smart were asked to compute the probability p that the sum of numbers appearing on the two draws equals 5. Dim computed $p = \frac{2}{15}$, arguing that there are 15 distinguishable unordered pairs and only 2 are favorable [i.e., (1, 4) and (2, 3)]. Dense computed $p = \frac{1}{9}$, arguing that there are 9 distinguishable sums (2 to 10) of which only 1 was favorable. Smart computed $p = \frac{4}{25}$, arguing that there were 25 distinguishable ordered outcomes of which 4 were favorable [i.e., (4, 1), (3, 2), (2, 3), and (1, 4)]. Why is $p = \frac{4}{25}$ the correct answer? Explain what is wrong with the reasoning of Dense and Dim.

8-7. Use the axioms given in Eqs. (8.2-1) to (8.2-3) to show the following ($E \in \mathcal{F}$, $F \in \mathcal{F}$):
(a) $P(\varnothing) = 0$.
(b) $P(EF^c) = P(E) - P(EF)$.
(c) $P(E) = 1 - P(E^c)$.
(d) $P(E \cup F) = P(E) + P(F) - P(EF)$.

8-8. Use the appropriate axiom and the results of Prob. 8-7 to show that the probability of $\{E$ or $F\}$ is given by

$$P(E \text{ or } F) = P(E) + P(F) - 2P(EF).$$

Hint: Write the event $\{E$ or $F\}$ as a union of two disjoint sets.

8-9. A fair die is tossed twice (a die is said to be fair if all outcomes 1, . . . , 6 are equally likely). Given that a 3 appears on the first toss, what is the probability of obtaining the sum 7 in two tosses?

8-10. A random-number generator generates integers from 1 to 9 (inclusive). All outcomes are equally likely; each integer is generated independently of any previous integer. Let Σ denote the sum of two consecutively generated integers; that is, $\Sigma = N_1 + N_2$. Given that Σ is odd, what is the conditional probability that Σ is 7? Given that $\Sigma > 10$, what is the conditional probability that at least one of the integers is > 7? Given that $N_1 > 8$, what is the conditional probability that Σ will be odd?

8-11. In the trinary communication channel shown in Fig. P8-11 a 3 is sent three times more frequently than a 1, and a 2 is sent two times more frequently than a 1. A 1 is observed; what is the conditional probability that a 1 was sent?

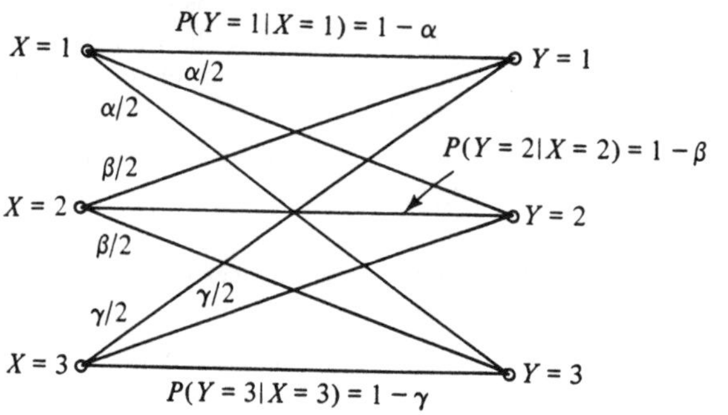

Figure P8-11

8-12. Derive the result in Eq. (8.5-5); that is,

$$P(S) = \sum_{k=0}^{n} b(k; n, p) = 1,$$

where S is the certain event and $b(\cdot)$ is the binominal probability law.

8-13. War-game strategists make a living by solving problems of the following type. There are 6 incoming ballistic missiles (BMs) against which are fired 12 antimissile missles (AMMs). The AMMs are fired so that two AMMs are directed against each BM. The single-shot-kill probability (SSKP) of an AMM is 0.8. The SSKP is simply the probability that an AMM destroys a BM. Assume that the AMMs do not interfere with each other and that an AMM can, at most, destroy only the BM against which it is fired. Compute the probability that (a) all BMs are destroyed, (b) at least one BM gets through to destroy the target, and (c) exactly one BM gets through.

8-14. Assume in Prob. 8-13 that the target was destroyed by the BMs. What is the conditional probability that only one BM got through?

8-15. An odd number of people want to play a game that requires two teams made up of even numbers of players. To decide who shall be left out to act as umpire, each of the N persons tosses a fair coin with the following stipulation: If there is one person whose outcome (be it heads or tails) is different from the rest of the group, that person will be the umpire. Assume that there are 11 players. What is the probability that a player will be "odd man out" on the first play?

8-16. In Prob. 8-15, derive a formula for the probability that the "odd man out" will occur on the nth play. *Hint*: Consider each play as an independent Bernoulli trial with success if an odd man out occurs and failure otherwise.

8-17. A smuggler, trying to pass himself off as a glass-bead importer, attempts to smuggle diamonds by mixing diamond beads among glass beads in the proportion of 1 diamond bead per 1000 glass beads. A harried customs inspector examines a sample of 100 beads. What is the probability that the smuggler will be caught?

8-18. Assume that a faulty receiver produces audible clicks to the great annoyance of the listener. The average number of clicks per second $\lambda(\tau)$ depends on the receiver temperature, which in turn depends on τ, the time from receiver turn-on, and is given by $\lambda(\tau) = 1 - e^{-\tau/10}$. Derive a formula for the probability of 0, 1, 2, . . . clicks during the first 10 sec of operation after turn-on.

8-19. In a restaurant known for its unusual service, the time X, in minutes, that a customer has to wait before he captures the attention of a waiter is specified by the following distribution function:

$$
\begin{aligned}
F_X(x) &= \left(\frac{x}{2}\right)^2, && \text{for } 0 \le x \le 1 \\
&= \frac{x}{4} && \text{for } 1 \le x \le 2 \\
&= \tfrac{1}{2} && \text{for } 2 \le x \le 10 \\
&= \frac{x}{20} && \text{for } 10 \le x \le 20 \\
&= 1 && \text{for } x \ge 20.
\end{aligned}
$$

(a) Sketch $F_X(x)$.
(b) Compute and sketch the pdf $f_X(x)$. Verify that the area under the pdf is indeed unity.
(c) What is the probability that the customer will have to wait (1) at least 10 minutes, (2) less than 5 minutes, (3) between 5 and 10 minutes, (4) exactly 1 minute?

8.20. Show that the conditioned distribution of X given the event $A = \{b < X \le a\}$ is

$$
F_X(x|A) = \begin{cases} 0, & x < b, \\ \dfrac{F_X(x) - F_X(b)}{F_X(a) - F_X(b)}, & b \le x < a, \\ 1, & x \ge a. \end{cases}
$$

8.21. In the following pdf's, compute the constant B required for proper normalization:

(a) Cauchy ($\alpha < \infty$, $\beta > 0$):

$$
f(x) = \frac{B}{1 + [(x - \alpha)/\beta]^2}, \qquad -\infty < x < \infty.
$$

(b) Maxwell ($\alpha > 0$):

$$
f(x) = \begin{cases} Bx^2 e^{-x^2/\alpha^2}, & x > 0, \\ 0, & \text{otherwise}. \end{cases}
$$

(c) Beta ($b > -1$, $c > -1$):

$$
f(x) = \begin{cases} Bx^b(1 - x)^c, & 0 \le x \le 1, \\ 0, & \text{otherwise}. \end{cases}
$$

(See formula 6.2.2 on p. 258 of [8-10].)

(d) Chi square ($\sigma > 0$, $n = 1, 2, \ldots$):

$$f(x) = \begin{cases} Bx^{(n/2)-1}\, e^{-x/2\sigma^2}, & x > 0 \\ 0, & \text{otherwise.} \end{cases}$$

8.22. A noisy resistor produces a voltage $v_n(t)$. At $t = t_1$, the noise level $X \equiv v_n(t_1)$ is known to be a Gaussian r.v. with pdf

$$f_X(x) = \frac{1}{\sqrt{2\pi\sigma^2}} \exp\left[-\frac{1}{2}\left(\frac{x}{\sigma}\right)^2\right].$$

Compute and plot the probability that $|X| > k\sigma$ for $k = 1, 2, \ldots$.

8.23. A noisy waveform is sampled at instants $t_1, t_2, \ldots, t_N$. The samples form a set of N random variables with joint pdf given by

$$f_{X_1 \cdots X_N}(x_1, \ldots x_N) = \frac{1}{(2\pi\sigma^2)^{N/2}} \exp\left(-\frac{1}{2\sigma^2}\sum_{i=1}^{N} x_i^2\right).$$

Are the $\{X_i\}$ independent of each other? Compute the probability that at least one sample has a value greater than 3σ.

8.24. (The general joint Gaussian pdf for two r.v.'s) Two random variables X_1, X_2 are said to be jointly Gaussian if their joint pdf is given by

$$f_{X_1X_2}(x_1, x_2) = \frac{1}{2\pi\sigma_1\sigma_2\sqrt{1-\rho^2}}$$

$$\exp\left\{-\frac{1}{2(1-\rho^2)}\left[\frac{(x_1-\mu_1)^2}{\sigma_1^2} - \frac{2\rho(x_1-\mu_1)(x_2-\mu_2)}{\sigma_1\sigma_2} + \frac{(x_2-\mu_2)^2}{\sigma_2^2}\right]\right\}.$$

(a) How many independent parameters characterize this pdf?
(b) Show that, if $\rho = 0$, X_1, X_2 are independent.
(c) Show that the marginal pdf's are Gaussian.
(d) Show that the loci of constant probability are ellipses.
(e) What is true about the principal axes of the ellipses of constant probability when $\rho = 0$?
(f) Assume that $\rho = 0$; under what circumstances do the ellipses degenerate into circles?

8.25. Let X be an r.v. with probability distribution function $F_X(x)$ and pdf $f_X(x)$. A new r.v. is generated by the transfer $Y = aX$. Compute $F_Y(y)$ when (a) $a > 0$ and (b) $a < 0$.

8.26. Let X be an r.v. with probability distribution function $F_X(x)$ and pdf $f_X(x)$. What is $F_Y(y)$ when Y is the output of an ideal half-wave rectifier; that is,

$$Y = \begin{cases} X, & X > 0, \\ 0, & \text{otherwise?} \end{cases}$$

8.27. Let X be an r.v. with probability distribution $F_X(x)$. Assume that $F_X(\cdot)$ is invertible; that is, $F_X^{-1}(x)$ exists and is well defined for each x. Show that the transformation $Y = F_X(X)$ produces a uniformly distributed r.v.

8.28. Let X be a discrete r.v. with binominal probability law; that is,

$$P(X = k) = \binom{n}{k} p^k q^{n-k}, \qquad k = 0, 1, \ldots, \qquad p + q = 1$$

Compute $E(X)$, σ_X^2.

8.29. Let X be a discrete r.v. with Poisson probability law with parameter $\lambda > 0$; that is,

$$P(X = k) = e^{-\lambda}\frac{\lambda^k}{k!}, \qquad k = 0, 1, \ldots .$$

Compute $E(X)$, σ_X^2.

8.30. (Chebyshev inequality) Let X be an r.v. with pdf $f(x)$ with mean and variance given by μ and σ^2, respectively.

(a) Show that

$$\sigma^2 \geq \int_{|x-\mu|>k\sigma} (x - \mu)^2 f(x)\, dx,$$

where k is an arbitrary constant, $k > 0$.

(b) Show that

$$\int_{|x-\mu|>k\sigma} (x - \mu)^2 f(x)\, dx \geq k^2\sigma^2 \int_{|x-\mu|>k\sigma} f(x)\, dx.$$

(c) Use parts (a) and (b) and the fact that

$$P(|X - \mu| \geq k\sigma) = \int_{|x-\mu|>k\sigma} f(x)\, dx$$

to show that

$$P(|X - \mu| \geq k\sigma) \leq \frac{1}{k^2}.$$

This important result is known as the *Chebyshev inequality*.

8.31. Let X be a uniform r.v. with pdf

$$f_X(x) = \begin{cases} (b - a)^{-1}, & a < x < b, \\ 0, & \text{otherwise}. \end{cases}$$

Compute the mean and variance of X.

8.32. Compute the mean and variance of X when X is an r.v. that has as pdf the Rayleigh law:

$$f_X(x) = \begin{cases} \dfrac{x}{\alpha^2} \exp\left[-\dfrac{1}{2}\left(\dfrac{x}{\alpha}\right)^2\right], & x > 0, \alpha > 0, \\ 0, & \text{otherwise}. \end{cases}$$

8.33. Show that the rth central moment of X

$$m_r = \int_{-\infty}^{\infty} (x - \mu)^r f_X(x)\, dx$$

can be expanded into a series

$$m_r = \sum_{i=0}^{r} (-1)^i \binom{r}{i} \mu^i \zeta_{r-i},$$

where ζ_{r-i} is defined in Eq. (8.16-1).

8.34. Let X, Y be two random variables. Show that $E(X + Y) = E(X) + E(Y)$. Extend this result to $E(Z)$, where $Z = \sum_{i=1}^{N} X_i$ and X_i, $i = 1, \ldots, N$, are r.v.'s.

8.35. (a) Show that, if $h_1(x) \le h_2(x)$ for all x, then

$$E[h_1(X)] \le E[h_2(X)].$$

(b) Show that for any $h(\cdot)$

$$|E[h(X)]| \le E|h(X)|.$$

8.36. Use the result of Prob. 8.33 or use a direct calculation to show that the variance of an r.v. X is given by

$$\sigma^2 = \zeta_2 - \mu^2.$$

8.37. The expectation of X is frequently estimated from the sample mean $\hat{\mu}$:

$$\hat{\mu} = \frac{1}{N}\sum_{i=1}^{N} X_i,$$

where the $\{X_i\}$ are N independent outcomes of X. Compute the mean and variance of $\hat{\mu}$. How fast does the variance of $\hat{\mu}$ go to zero with N?

8.38. (Regression) Two random variables X, Y are suspected of having a strong linear dependence on the basis of a scatter diagram. The scatter diagram is a plot of actual observations (x_i, y_i) on X, Y (Fig. P8-38). Assume that the dependence is modeled as $Y_p = \alpha + \beta X$.

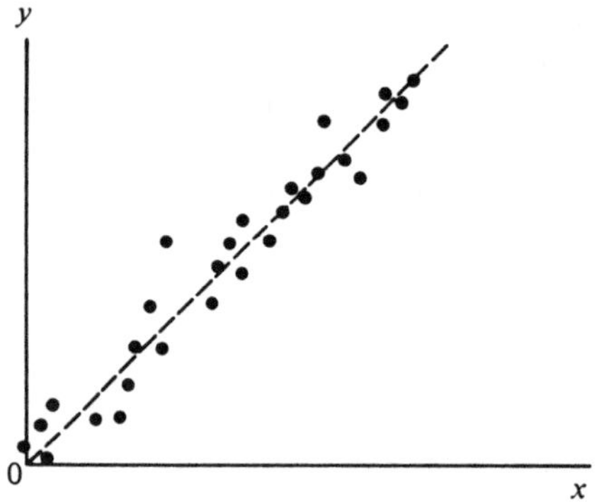

Figure P8-38

(a) Compute the mean-square error $\overline{\epsilon^2}$ between the predicted and actual value of Y; that is,

$$\overline{\epsilon^2} = E[Y - Y_p]^2.$$

(b) Compute

$$\frac{\partial \overline{\epsilon^2}}{\partial \alpha}, \qquad \frac{\partial \overline{\epsilon^2}}{\partial \beta}$$

to find the "best" α, β in the sense of minimizing $\overline{\epsilon^2}$.

(c) Show that the best predictor passes through $E(X)$, $E(Y)$.

8.39. Let X_1 and X_2 be independent, normal r.v.'s with mean and variance given by μ_1, σ_1^2 and μ_2, σ_1^2, respectively. Show that the mean and variance (μ, σ^2) of $X_1 + X_2$ is $\mu = \mu_1 + \mu_2$ and $\sigma^2 = \sigma_1^2 + \sigma_2^2$.

8.40. Let X be an r.v. that takes on values 0 and 1 with probability

$$P(X = 1) = p, \qquad P(X = 0) = q = 1 - p.$$

(a) Compute the moment-generating function.
(b) Compute the first few moments of X.
(c) Compute the mean and variance of an r.v. X using the moment-generating technique if $f_X(k) = e^{-\lambda}\lambda^k/k!$.

8.41. (Log-normal distribution) The r.v. X is said to be *log normal* if $Y = \log X$ is normal. Compute the mean and variance of X using moment-generating functions. *Hint*: $E(X) = E[e^{tY}]_{t=1}$. Similarly, $E(X^2) = E[e^{tY}]_{t=2}$, and so on.

8.42. Let Θ be uniformly distributed in $[-\pi/2, \pi/2]$. Compute the pdf of $Y = \sin \Theta$ using the method of characteristic functions. *Hint*: Use the approach that leads up to Eq. (8.17-39); that is, consider the transformation $y = \sin \theta$ in

$$\Phi_Y(\omega) = \int_{-\pi/2}^{\pi/2} e^{j\omega \sin\theta} \frac{d\theta}{\pi}$$

and compare with

$$\Phi_Y(\omega) = \int_{-\infty}^{\infty} e^{j\omega y} f_Y(y)\, dy.$$

8.43. Use the method of characteristic functions to compute the first four moments of X if the pdf of X is the exponential law; that is,

$$f_X(x) = \begin{cases} \lambda e^{-\lambda x}, & x > 0, \quad \lambda > 0, \\ 0, & \text{otherwise.} \end{cases}$$

8.44. Let X be a uniform r.v. over the interval $[-a, a]$. Let Y be a uniform r.v. over the interval $[0, 2a]$. X and Y are independent. Compute the characteristic function of $Z = X + Y$. Use this result to compute

$$\overline{Z} \quad \text{and} \quad \overline{[Z - \overline{Z}]^2}.$$

8.45. Consider the system in Fig. P8-45. Let X and Y be correlated, jointly normal r.v.'s with pdf's as given in Eq. (8.16-5). Use the method of joint characteristic functions to compute $\overline{W}$.

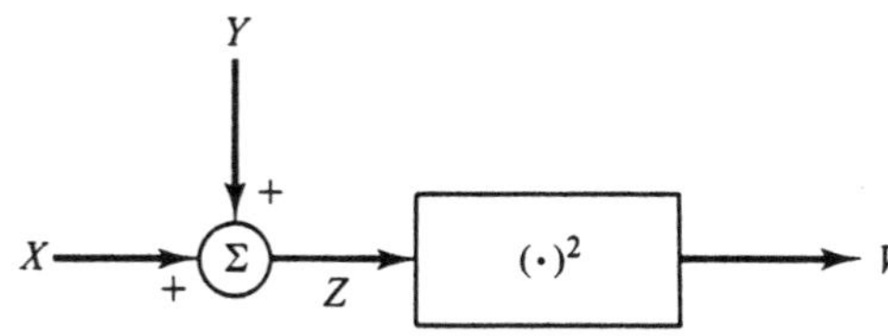

Figure P8-45

8.46. In communication theory, waveforms of the type

$$w(t) = x(t) \cos \omega t - y(t) \sin \omega t$$

appear quite often. At a particular instant of time, say $t = t_1$, $X \equiv x(t_1)$ and $Y \equiv y(t_1)$ are known to be Gaussian, uncorrelated r.v.'s. Compute the joint pdf of the *envelope* $Z \equiv (X^2 + Y^2)^{1/2}$ and *phase* $\phi \equiv \tan^{-1} Y/X$ of $w(t_1)$.

REFERENCES

[8-1] E. Parzen, *Modern Probability Theory and Its Applications*, Wiley, New York, 1960.

[8-2] A. Papoulis, *Probability, Random Variables, and Stochastic Processes*, McGraw-Hill, New York, 1965.

[8-3] H. Stark and J. W. Woods, *Probability, Random Processes and Estimation Theory for Engineers*, Prentice-Hall, Englewood Cliffs, N.J., 1986.

[8-4] C. W. Helstrom, *Probability and Stochastic Processes for Engineers*, Macmillan, New York, 1984.

[8-5] W. Feller, *An Introduction to Probability Theory and Its Applications*, 2 vols., Wiley, New York, 1950, 1966.

[8-6] W. F. Davenport, *Probability and Random Processes: An Introduction for Applied Scientists and Engineers*, McGraw-Hill, New York, 1970.

[8-7] M. O'Flynn, *Probability, Random Variables, and Random Processes*, Harper & Row, New York, 1982.

[8-8] P. Z. Peebles, Jr., *Probability, Random Variables, and Random Signal Principles*, McGraw-Hill, New York, 1980.

[8-9] R. M. Gray and L. D. Davisson, *Random Processes: A Mathematical Approach for Engineers*, Prentice-Hall, Englewood Cliffs, N.J., 1986.

[8-10] M. Abramowitz and I. A. Stegun, *Handbook of Mathematical Functions*, Dover, New York, 1965.

CHAPTER 9

Random Processes

9.1 INTRODUCTION

In communication systems, one frequently encounters waveforms that display irregular and unpredictable fluctuations in some characteristic of the wave. The waveforms associated with ordinary speech and music, the noise voltage across a resistor, electromagnetic emission from the sun and stars, the envelope of a TV signal, the instantaneous electron emission current in a tube, and telegraph and telephone signals all exhibit some inherent randomness that makes them "interesting." The study of waveforms of this type brings us to the subject of this chapter, which is *random* or *stochastic* processes.† A rigorous study of random processes is a formidable undertaking and well beyond our intention. Our aim is to present enough of the theory to enable the reader to make input–output computations for typical communication systems, when the signals are random. An important calculation of this type is the computation of the *signal-to-noise* ($\mathcal{S}/\mathcal{N}$) ratio for a communication system; communication engineers should understand how the parameters (i.e., gain, bandwidth, etc.) of the system influence the $\mathcal{S}/\mathcal{N}$ ratio.

Why study random processes? For one thing, because they include the most important waveforms in engineering, the ones that really contain information. If a waveform is completely predictable, if no characteristic of it (amplitude, phase, start time, spectrum, etc.) is unknown, then there is no point in sending it—it contains no information. A monochromatic sine wave of known amplitude, frequency, and phase is, despite its esthetic quality, extremely dull; its future values are as predictable as yesterday's weather.

† The word stochastic is derived from the Greek word *stochastiko*, which means "skillful in aiming or guessing at . . . ."

Whereas the unpredictability of some waveforms reflects their ability to impart information, the unpredictability of other waveforms is regarded only as noise and represents both an irritation and a challenge to designers. The thermal noise voltage observed across a resistor is an example of this type of waveform. This signal is generated by the thermal agitation of the electrons in the resistor; if we were interested in the kinetics of electron motion in that resistor, then, conceivably, the variations in the wave might give us useful information. However, in a communication-systems setting, the resistor noise voltage is a source of interference. Precisely how it limits our capability to enjoy, say, listening to music in an AM receiver is the concern of communication engineers.

The thermal noise voltage is an example of a signal with irregular fluctuations. Knowledge of the waveform for past values of time does not enable us to predict future values. However, not all random waveforms have irregular fluctuations. A sine wave with random phase remains a sine wave for all time, and knowledge of it for (say) $t < t_0$ specifies its values for $t > t_0$. Nevertheless, such a waveform is an example of a random process.

9.2 DEFINITION OF A RANDOM PROCESS

As we shall see, a random process is essentially nothing more than a collection of random variables related through a suitable indexing set, say T. If T is the set of integers, the random process is said to be of a discrete type. Otherwise, the random process is said to be continuous. However, this definition does not make clear how random variables are related to random waveforms. Suppose that we are performing an experiment with random outcomes whose sample description space is S. We already know that a random variable X is a function on S, which means that for every $s \in S$, $X(\cdot)$ assigns a number $X(s)$. Now let us assign to every $s \in S$ a function of two parameters, $x(t, s)$, where t denotes time and is an element of the indexing set T.† The family of functions $\{x(t, s), s \in S, t \in T\}$ then furnishes, depending on what is held fixed, four objects:

	$x(t,s)$	
	s fixed	s variable
t fixed	$x(t, s)$ is a number	$x(t, s)$ is a single random variable
t variable	$x(t, s)$ is a single function of time, also called a sample function	$x(t, s)$ is a family of sample functions or equivalently a collection of random variables (a random process)

† Most random signals of interest are one-dimensional time variations. However, if the indexing set is $R \times R$ with elements (x, y), we would have a *spatial* random process. Such processes occur in optics and image processing. Unless stated otherwise we shall assume *real* random processes; i.e., all sample functions are real.

This is the point of view taken by Papoulis [9-1] and clearly shows how random processes are related to random variables. If t is fixed, say at t_1, but s is variable, then $x(t_1, s) \equiv X_1$ is a random variable. For $t_2 \equiv t_1 + \Delta$ and s variable, we have $x(t_2, s) \equiv X_2$, a new random variable. Our process is generated by stepping the index parameters t. Note that both t and s must be varied; keeping t fixed by letting $x(t, \cdot)$ roam over S generates a random variable; varying t generates the random process.

As a somewhat more concrete example of a random process, consider the noise voltage $v(t)$ across a thermally agitated resistor. The value of this voltage at a particular instant of time, say $t = t_1$, constitutes a random variable $X_1 \equiv v(t_1)$. To obtain the statistics of X_1, we could take a very large number $(n \to \infty)$ of identical resistors and record their waveforms. These might look as in Fig. 9.2-1.

At $t = t_1$, the totality of observed levels represents essentially all the values that X_1 can take on. In particular, at $t = t_1$, we can obtain the relative frequency of occurrence of the event $\{x_1 < X_1 \leq x_1 + \Delta_1\}$ and estimate the pdf of X_1. Of course, this can be repeated for X_2 at t_2, X_3 at t_3, and so on. The *joint pdf* of X_1 and X_2 can similarly be obtained by considering the relative frequency of the event $\{x_1 < X_1 \leq x_1 + \Delta_1, x_2 < X_2 \leq x_2 + \Delta_2\}$. Proceeding in this way, we can obtain the nth-order joint pdf of $X_1, X_2, \ldots, X_n$.

In this example, the underlying experiment can be thought of as randomly picking a resistor. The sample description space S of the underlying experiment is the set of all n resistors. Once we have picked a particular resistor, say the ith, the resultant waveform $v_i(t) \equiv v(t, s_i)$ is a single function of time. On the other hand, the noise voltage at $t = t_i$ is a random variable defined on the probability space $(S, \mathscr{F}, \mathscr{P})$.

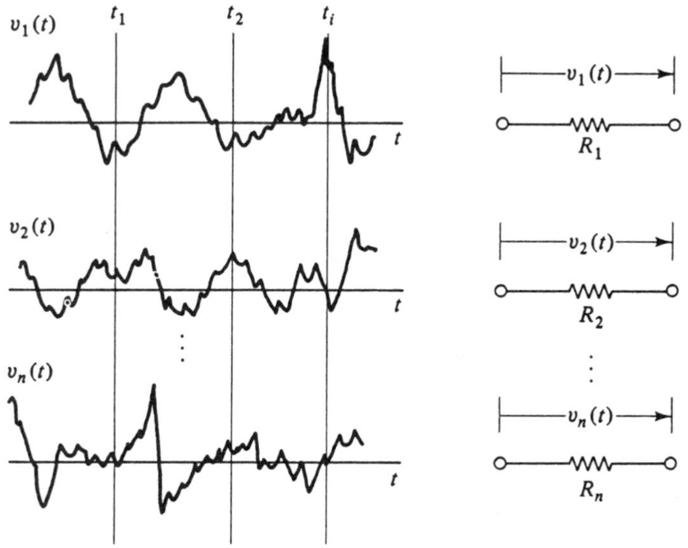

Figure 9.2-1. Noise voltage across resistors.

Notation

In denoting a random process, the functional dependence of waveforms and random variables on the outcomes $s \in S$ of the underlying experiment is usually suppressed. In general, the underlying experiment is not made explicit, although it is implied. The dependence on time, however, is made explicit. Some authors use bold or capital letters to differentiate a random process from an ordinary function of time. Others use lowercase symbols such as $x(t)$ to denote the sample functions of the process and reserve x_t for the random variable generated by fixing t [9-2]. Our own preference is to reserve the use of capital letters such as X, Y, Z for Fourier transforms of x, y, z. Hence, in general, we shall use a lowercase symbol $x(t)$ to denote the random variable at t or a sample function of the process. The specific interpretation of $x(t)$ will be inferred from the context. The use of lowercase letters is in distinct contrast to the notation used in Chapter 8, but the adoption of the former will enable us to dispense with introducing cumbersome or exotic symbols. On some occasions, specifically when we deal with vectors of random variables such as in the Gaussian process, we shall revert to capital notation to indicate that we are dealing with vectors instead of scalars. Also we shall occasionally use the notation $\{x(t)\}$ to indicate an ensemble of sample functions when it is necessary to do so. Finally, because of our desire to reserve capital letters primarily for Fourier transforms, we shall generally omit the use of capital subscripts on probability distribution or density functions.

Example

Consider the experiment of throwing a fair die. The sample description space S is $S = \{1, 2, 3, 4, 5, 6\}$. Let the random process be defined by

$$x(t) = kt, \qquad \text{if } s = k, \qquad k = 1, \ldots, 6. \tag{9.2-1}$$

The sample functions are shown in Fig. 9.2-2. At $t = 1$, $x(1)$ is an r.v. that takes on values $k = 1, \ldots, 6$ with probability $P[x(1) = k] = \frac{1}{6}$. All other values of $x(1)$ are

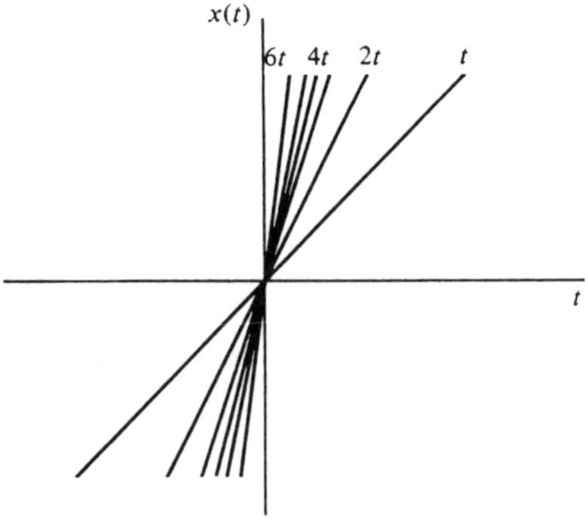

Figure 9.2-2. Sample functions of the process $x(t) = kt$.

associated with the impossible event. At $t = 2$, $x(2)$ is an r.v. that takes on values $2k$ for $k = 1, \ldots, 6$ with probability $P[x(2) = 2k] = \frac{1}{6}$.

The average value of $x(1)$ is

$$\overline{x(1)} = \tfrac{1}{6} \sum_{k=1}^{6} k = 3.5,$$

while

$$\overline{x(2)} = \tfrac{1}{6} \sum_{k=1}^{6} 2k = 7.$$

We see that the mean value of $x(t)$ depends on t. Consequently, this is an example of a *nonstationary* random process. Processes for which the mean and higher moments do not depend on t are said to be *stationary*. We shall say more about such processes later.

In the example considered, the sample functions were very regular, and knowing $x(t)$ for $t < t_1$ completely specifies $x(t)$ for all t. Regularity of sample functions is not the usual case, however, and most of our concern will be with functions that are highly irregular (i.e., noiselike). Knowledge of such functions for $t < t_1$ is insufficient to predict future values exactly.

9.3 STATISTICS OF A RANDOM PROCESS

Consider the random variable $x(t_1)$, where t_1 is a fixed value of t. With x_1 denoting a number, the probability of the event $\{x(t_1) \leq x_1\}$ is the probability distribution function (PDF) of $x(t_1)$; that is,

$$F(x_1, t_1) = P[x(t_1) \leq x_1]. \tag{9.3-1}$$

The pdf† is obtained by differentiating Eq. (9.3-1). Thus

$$f(x_1, t_1) = \frac{\partial F}{\partial x_1}. \tag{9.3-2}$$

$F(x_1, t_1)$ is called the first-order distribution of the process $x(t)$. The event $\{x(t_1) \leq x_1\}$ consists of all $s \in S$ such that $x(t_1, s) \leq x_1$. Similarly, given t_1 and t_2, $x(t_1)$ and $x(t_2)$ represent two random variables. Their joint distribution is the second-order distribution of $x(t)$. It is given by

$$F(x_1, x_2; t_1, t_2) = P[x(t_1) \leq x_1, x(t_2) \leq x_2]. \tag{9.3-3}$$

The associated second-order pdf is

$$f(x_1, x_2; t_1, t_2) = \frac{\partial^2 F}{\partial x_1 \, \partial x_2}. \tag{9.3-4}$$

† Recall that pdf stands for the probability density function.

The nth-order joint distribution function $F(x_1, \ldots, x_n; t_1, \ldots, t_n)$ and the nth-order pdf are direct extensions of the preceding. Thus

$$F(x_1, \ldots, x_n; t_1, \ldots, t_n) = P[x(t_1) \le x_1, \ldots, x(t_n) \le x_n], \tag{9.3-5}$$

$$f(x_1, \ldots, x_n; t_1, \ldots, t_n) = \frac{\partial^n F}{\partial x_1 \cdots \partial x_n}. \tag{9.3-6}$$

A random process is *completely characterized* if its nth-order statistics are known for all n. In other words, it is required that

$$F(x_1, \ldots, x_n; t_1, \ldots, t_n)$$

be known for every n and for every set and combination of the arguments $(x_1, \ldots, x_n)$ and $(t_1, \ldots, t_n)$. Fortunately, most problems of engineering importance require knowledge of only the first two orders.

Stationary Random Processes

A random process is said to be strict-sense stationary (sss) if

$$P[x(t_1) \le x_1, \ldots, x(t_n) \le x_n] = P[x(t_1 + \tau) \le x_1, \ldots, x(t_n + \tau) \le x_n] \tag{9.3-7}$$

for every combination of $\mathbf{t} = (t_1, \ldots, t_n)$, $\mathbf{x} = (x_1, \ldots x_n)$, τ, and n. The pdf of an sss process is independent of time shifts; this implies that sss waveforms cannot be switched on and off (i.e., they must be on for all time).

Averages of a Random Process

The average, expected value, or mean $\mu(t)$, of a random process is defined by

$$\mu(t) = E[x(t)] = \int_{-\infty}^{\infty} xf(x, t)\, dx. \tag{9.3-8}$$

For a stationary process, $\mu(t)$ is independent of time. The proof is easy:

$$\mu(t + \tau) = E[x(t + \tau)] = \int_{-\infty}^{\infty} xf(x, t + \tau)\, dx$$

$$= \int_{-\infty}^{\infty} xf(x, t)\, dx = \mu(t).$$

The *autocorrelation* function (also called the correlation or self-correlation) is given by

$$R(t_1, t_2) = E[x(t_1)x(t_2)] = \int_{-\infty}^{\infty}\int_{-\infty}^{\infty} x_1x_2 f(x_1, x_2; t_1, t_2)\, dx_1\, dx_2.$$

For a stationary process, this depends only on $\tau = t_2 - t_1$ (why?) and is written $R(\tau)$. The *autocovariance* or, simply, *covariance* is given by

$$\begin{aligned} r(t_1, t_2) &= E\{[x(t_1) - \mu(t_1)][x(t_2) - \mu(t_2)]\} \\ &= R(t_1, t_2) - \mu(t_2)\,\mu(t_1). \end{aligned} \tag{9.3-9}$$

The covariance of two random variables, say X_1 and X_2, is often written as $\text{cov}(X_1\ X_2)$.

For an sss process, the covariance, like the autocorrelation, depends only on $t_2 - t_1 = \tau$. We write it as $r(\tau)$. The nth joint moment of a random process is

$$E[x(t_1)x(t_2)\cdots x(t_n)] = \int_{-\infty}^{\infty}\int_{-\infty}^{\infty} x_1 x_2 \cdots x_n f(x_1, x_2, \ldots, x_n; t_1, t_2, \ldots, t_n)\,dx_1 \cdots dx_n. \tag{9.3-10}$$

Definition: A random process is said to be *wide-sense* or *weakly stationary* (wss) if

$$\text{(i) } \mu(t) = E[x(t)] \text{ is independent of } t, \tag{9.3-11a}$$

$$\text{(ii) } R(t_1, t_2) = R(t_2 - t_1) = R(\tau), \tag{9.3-11b}$$

$$\text{(iii) } R(0) = E[x^2(t)] < \infty. \tag{9.3-11c}$$

Such processes are very important in practice. Unless otherwise stated, we shall assume that a process is at least wide-sense stationary. Note that all sss processes are wss. The reverse is not true.

Properties of the Correlation Function

It is easily shown that $R(\tau)$ has the following properties:

$$\begin{aligned} &\text{(i) } R(\tau) = R(-\tau), \\ &\text{(ii) } R(0) \geq R(\tau), \\ &\text{(iii) } R(0) = \sigma^2 + \mu^2, \end{aligned} \tag{9.3-12}$$

where $\sigma^2 \equiv r(0)$ is the variance of the wss random process $x(t)$. Properties (i) and (iii) follow from the definition of $R(\tau)$ and property (ii) from the fact that

$$E[x(t_2) - x(t_1)]^2 \geq 0.$$

The correlation function is a measure of the statistical dependence of two random variables separated by a distance τ in time. If the process $x(t)$ contains no periodic or dc components, we generally expect $x(t_1)$ and $x(t_2)$ to be strongly dependent if $t_2 - t_1$ is small and essentially independent if $t_2 - t_1$ is large. For example, the outdoor temperature an hour from now is much more dependent on the present temperature than will be the temperature 365 days later. Such processes are said to satisfy a *strong mixing condition* [9-3]. For such processes, $\lim_{\tau\to\infty} E[x(t)x(t + \tau)] \to E[x(t)]E[x(t + \tau)] = \mu^2$. Most processes of engineering interest that do not have periodic components other than a dc component behave in this way.

9.4 EXAMPLES OF CORRELATION FUNCTION COMPUTATIONS

The Full-Random Binary Waveform: The "Telegraph" Signal

A typical sample function of this process is shown in Fig. 9.4-1. The average number of polarity switches (zero crossings) per unit time is λ. The probability of getting exactly k crossings in time τ is given by the Poisson law:

$$p(k) = e^{-\lambda\tau}\frac{(\lambda\tau)^k}{k!}. \tag{9.4-1}$$

Note that the events $\{k \text{ points in } (0, t)\}$, $k = 0, 1, \ldots$, are mutually exclusive. Let $x(t)$ denote the value of the waveform at t. The correlation function is given by

$$\begin{aligned} R(t_1, t_2) &= E[x(t_2)x(t_1)] \\ &= A^2P[x(t_1) = A, x(t_2) = A] \\ &\quad + A(-A)P[x(t_1) = A, x(t_2) = -A] \\ &\quad + (-A)(A)P[x(t_1) = -A, x(t_2) = A] \\ &\quad + (-A)^2P[x(t_1) = -A, x(t_2) = -A]. \end{aligned} \tag{9.4-2}$$

We assume that for *any* t_1, t_2 the events $\{x(t_1) = A, x(t_2) = A\}$ and $\{x(t_1) = -A, x(t_2) = -A\}$ are equally likely. The same assumption† holds for the events $\{x(t_1) = -A, x(t_2) = A\}$ and $\{x(t_1) = A, x(t_2) = -A\}$. We also assume that $P[x(t_1) = A] = P[x(t_2) = A] = \frac{1}{2}$. These assumptions make the waveform statistically symmetric above and below the axes. Hence

$$\begin{aligned} R(t_1, t_2) &= 2A^2\{P[x(t_1) = A, x(t_2) = A] - P[x(t_1) = -A, x(t_2) = A]\} \\ &= A^2\{P[x(t_2) = A|x(t_1) = A] - P[x(t_2) = -A|x(t_1) = A]\}. \end{aligned} \tag{9.4-3}$$

With $\tau \equiv t_2 - t_1$, we obtain, with the help of Eq. (9.4-1),

$$P[x(t_2) = A|x(t_1) = A] = \sum_{k\,\text{even}} e^{-\lambda\tau}\frac{(\lambda\tau)^k}{k!} \tag{9.4-4a}$$

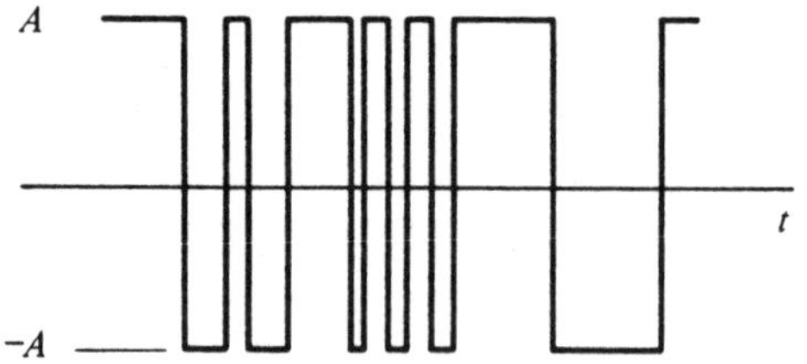

Figure 9.4-1. Sample function of the telegraph signal.

† What would happen if we went a step further and assumed that for any t_1, t_2 the four events $\{x(t_1) = \pm A, x(t_2) = \pm A\}$ are equally likely?

and

$$P[x(t_2) = -A|x(t_1) = A] = \sum_{k\,\text{odd}} e^{-\lambda\tau}\frac{(\lambda\tau)^k}{k!}. \qquad (9.4\text{-}4b)$$

A direct substitution of Eqs. (9.4-4) into Eq. (9.4-3) yields

$$R(t_1, t_2) = R(\tau) = A^2 e^{-2\lambda\tau}, \qquad \tau > 0. \qquad (9.4\text{-}5)$$

To obtain this result, we used the fact that

$$\sum_{k\,\text{even}} e^{-\lambda\tau}\frac{(\lambda\tau)^k}{k!} - \sum_{k\,\text{odd}} e^{-\lambda\tau}\frac{(\lambda\tau)^k}{k!} = e^{-\lambda\tau}\sum_{k=0}^{\infty}\frac{(\lambda\tau)^k}{k!}(-)^k$$
$$= e^{-2\lambda\tau}.$$

The complete solution that includes $\tau < 0$ (i.e., if $t_1 > t_2$) is

$$R(\tau) = A^2 e^{-2\lambda|\tau|}. \qquad (9.4\text{-}6)$$

This function is shown in Fig. 9.4-2.

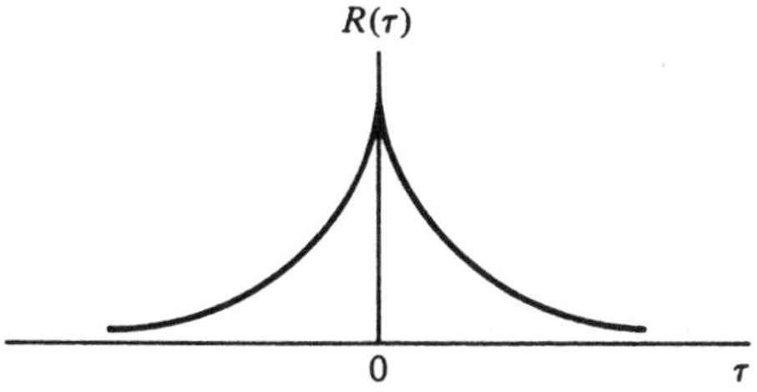

Figure 9.4-2. Correlation function of the telegraph signal.

Sine Wave with Random Phase

The random process $x(t)$ has sample functions $\{\cos(\omega_0 t + \theta)\}$, where θ denotes the values that a random variable Θ can take on and ω_0 is a constant. A sample function is shown in Fig. 9.4-3. If Θ is uniformly distributed in $[-\pi, \pi]$, the correlation function is given by

$$E[x(t_1)x(t_2)] = \frac{1}{2\pi}\int_{-\pi}^{\pi}\cos(\omega_0 t_1 + \theta)\cos(\omega_0 t_2 + \theta)\,d\theta$$
$$= \frac{1}{2}\cos\omega_0(t_2 - t_1). \qquad (9.4\text{-}7)$$

The process is wss, and the correlation function can be written in terms of $\tau \equiv t_2 - t_1$ as

$$R(\tau) = \tfrac{1}{2}\cos\omega_0\tau. \qquad (9.4\text{-}8)$$

Notice that the sample functions in this case are extremely regular; for a given element s of the sample space S, knowledge of $x(t)$ for $t < t_1$ specifies all future values.

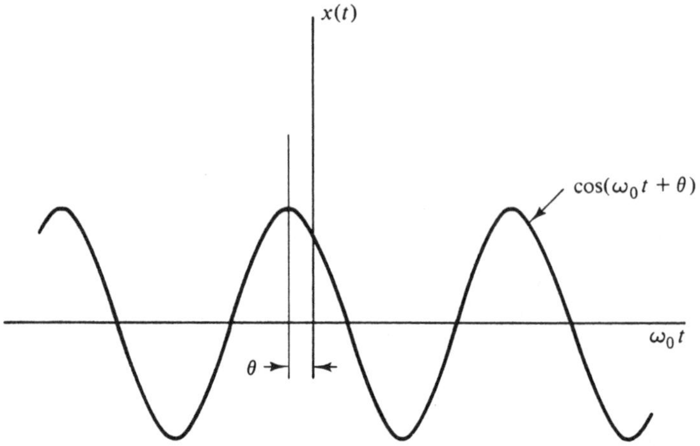

Figure 9.4-3. Sample function of the sine wave with random phase process.

Integrals of Random Processes

Given a random process $x(t)$, what meaning shall we give to the integral

$$\xi = \int_a^b x(t)\, dt? \tag{9.4-9}$$

Recall that a random process can be viewed as an ensemble of sample functions. If the integral in Eq. (9.4-9) converges, in the Riemann sense, for almost[†] every sample function $x(t, s)$, $s \in S$, then the numbers $\xi(s)$ define a random variable with sample description space $\{\xi(s): s \in S\}$. Integrals of random processes are important in the practical estimation of the low-order moments of certain types of processes called ergodic processes.

Time Averages and Ergodicity

Consider the limit, frequently called a *time average* of $x(t)$,

$$\hat{\mu} = \lim_{T\to\infty} \frac{1}{2T} \int_{-T}^{T} x(t)\, dt, \tag{9.4-10}$$

[†] Except, conceivably, for a "small" set of probability zero. If the integral does not converge for some $s \in S$, it is still possible to define $\xi(s)$ in terms of the limit of a sum of random variables, i.e., $\xi = \sum_{i=1}^{n} x(t_i)\, \Delta t_i$, which converges in the mean-square sense. We shall use this viewpoint on occasion.

where $x(t)$ denotes a stationary random process. Clearly, if $\mu = E[x(t)]$, then

$$\begin{aligned} E[\hat{\mu}] &= \lim_{T\to\infty} \frac{1}{2T} \int_{-T}^{T} E[x(t)]\, dt \\ &= \mu. \end{aligned} \tag{9.4-11}$$

In order to argue that $\hat{\mu}$ (an r.v.) is equal to μ, we must show that the variance of $\hat{\mu}$, written $\sigma_{\hat{\mu}}^2$, tends to zero as $T \to \infty$. If $E[\hat{\mu}] = \mu$ and $\sigma_{\hat{\mu}}^2 = 0$, then we can say that the time average equals the ensemble average and $x(t)$ has the property of *ergodicity of the mean*. The notion of ergodicity is extremely important in engineering because we do not usually have available a large number of sample functions from which to do an ensemble average. The notion of a random process consisting of a large ensemble of waveforms is useful as a model but does not reflect real-life situations too well. Usually, we have to estimate the statistics of a random process from an observation on a single function. Hence the question arises, To what extent does a single sample function represent the entire ensemble? This is something we cannot know with certainty. But if we can reasonably argue that the process is *ergodic*, then a single function is all we need.

A process $x(t)$ is said to be ergodic if all its statistics can be obtained by observing a single function $x(t, s)$ (s fixed) of the process. For an ergodic process all the moments can be determined from time averages performed on $x(t, s)$. In practice, however, most of the time we require only ergodicity of the first two moments; that is,

$$\hat{\mu} = \lim_{T\to\infty} \frac{1}{2T} \int_{-T}^{T} x(t)\, dt = \mu, \tag{9.4-12}$$

$$\hat{R}(\tau) = \lim_{T\to\infty} \frac{1}{2T} \int_{-T}^{T} x(t)x(t+\tau)\, dt = R(\tau). \tag{9.4-13}$$

We reiterate that in order for Eqs. (9.4-12) and (9.4-13) to have any meaning, the variances of $\hat{\mu}$ and $\hat{R}(\tau)$ must be zero as $T \to \infty$. Ergodic processes must be stationary, but the converse is not necessarily true (Probs. 9-5 and 9.6).

9.5 THE POWER SPECTRUM

Although the sample functions of many random processes have irregular shapes and cannot easily (if at all) be described by an equation, it still seems reasonable to talk about frequency content, bandwidth, average power, and the like. For example, the "grassy"-looking waveform shown in Fig. 9.5-1(a) is expected to have a higher bandwidth than the smoother-looking waveform shown in Fig. 9.5-1(b).

In earlier chapters we found that the analysis of nonperiodic, finite-energy signals

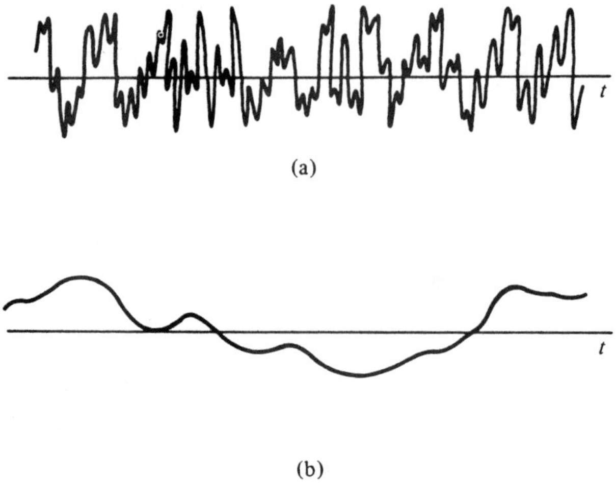

Figure 9.5-1. Sample function of (a) a high-frequency process and (b) a low-frequency process.

is conveniently done with the use of the Fourier transform. However, when $x(t)$ is a random process, the condition

$$\int_{-\infty}^{\infty} |x(t)|\, dt < \infty \tag{9.5-1}$$

is generally not satisfied,† and therefore the Fourier transform may not exist. Even integrals of the form

$$X_T(\omega) = \int_{-T}^{T} x(t)e^{-j\omega t}\, dt, \tag{9.5-2}$$

where T is finite, are not a useful frequency decomposition of a random process because $X_T(\omega)$, for fixed ω, is a random variable whose statistics are generally difficult to compute. In Sec. 9-4 we suggested that a way to view the integral of a random process was as a limit of a sum of many random variables. Hence computing the pdf of $X_T(\omega)$ would, typically, involve many repeated superposition-type integrations leading to extremely involved expressions.

For the reasons given, a sinusoidal resolution of a random process is generally not practical. On the other hand, a sinusoidal resolution of the *average power* in $x(t)$ is not only frequently possible but is extremely useful from an engineering point of view. Such a resolution is called the *power spectrum* or the *power spectral density*.

We already know that for a deterministic signal $x(t)$ the time-averaged power over an interval $(-T, T)$ is given by

$$P_T = \frac{1}{2T}\int_{-T}^{T} x^2(t)\, dt. \tag{9.5-3}$$

† For a wss process $x(t)$ is never "switched off," and therefore the integral of $|x(t)|$ grows without limit. It would be more precise to say that the r.v. $y = \int_{-T}^{T} |x(t)|\, dt$ has unbounded variance as $T \to \infty$. This is true even when $x(t)$, and not just its magnitude, is considered.

If $x(t)$ is a random process, however, the object, P_T, as given in Eq. (9.5-3) is a random variable. However, the quantity

$$E(P_T) = \frac{1}{2T}\int_{-T}^{T} \overline{x^2(t)}\, dt \tag{9.5-4}$$

is a number and plays the same role for a random process as P_T in Eq. (9.5-3) plays for a deterministic function of time. If $x(t)$ is a stationary random process, we might be interested in the average power over all time, which is

$$\lim_{T\to\infty} E(P_T) \equiv \overline{P} = \lim_{T\to\infty} \frac{1}{2T}\int_{-T}^{T} \overline{x^2(t)}\, dt. \tag{9.5-5}$$

This definition is only useful if the limit exists.

Equation (9.5-5) is a direct extension of familiar results. To determine the distribution of power in *frequency*, we proceed as follows: Define $x_T(t)$ by

$$x_T(t) = \begin{cases} x(t), & |t| < T, \\ 0, & |t| > T, \end{cases}$$

and let $X_T(f) = \mathscr{F}[x_T(t)]$. By Parseval's theorem,

$$E(P_T) = \int_{-\infty}^{\infty} \frac{\overline{x_T^2(t)}}{2T}\, dt = \int_{-\infty}^{\infty} \frac{\overline{|X_T(f)|^2}}{2T}\, df, \tag{9.5-6}$$

and, by Eq. (9.5-5), we obtain

$$\overline{P} = \int_{-\infty}^{\infty} \lim_{T\to\infty} \frac{\overline{|X_T(f)|^2}}{2T}\, df, \tag{9.5-7}$$

where we have interchanged the order of integration and limiting. The quantity

$$W(f) \equiv \lim_{T\to\infty} \frac{\overline{|X_T(f)|^2}}{2T} \tag{9.5-8}$$

is commonly taken as the definition of the power spectrum. Note that the *expectation* must be taken *before* the limiting operation to ensure convergence of $W(f)$. In terms of $W(f)$, the average power can be written as

$$\overline{P} = \int_{-\infty}^{\infty} W(f)\, df. \tag{9.5-9}$$

The relation between $W(f)$ and the correlation function $R(\tau)$ is furnished by the *Wiener–Khintchine theorem*. We shall consider this next.

Theorem: If, for a stationary process,

$$\int_{-\infty}^{\infty} |\tau R(\tau)|\, d\tau < \infty, \tag{9.5-10}$$

then

$$W(f) = \int_{-\infty}^{\infty} R(\tau)e^{-j2\pi f\tau}\,d\tau. \tag{9.5-11}$$

In words, if Eq. (9.5-10) is satisfied, then the power spectrum is the Fourier transform of the autocorrelation function. The Wiener–Khintchine theorem is of fundamental importance in the theory of stationary random processes.

Proof of the Wiener–Khintchine Theorem

From the definition of $X_T(f)$, we have

$$\begin{aligned}\overline{|X_T(f)|^2} &= \int_{-T}^{T} ds \int_{-T}^{T} dt\, \overline{x(t)x(s)}e^{-j2\pi f(s-t)} \\ &= \int_{-T}^{T} ds \int_{-T}^{T} dt\, R(s-t)e^{-j2\pi f(s-t)}.\end{aligned} \tag{9.5-12}$$

We have used the fact that $\overline{x(t)x(s)} \equiv R(s-t)$ and $\overline{|X_T(f)|^2} = \overline{X_T(f)X_T^*(f)}$. Note that the region of integration is the inside of a square of side T centered at the origin of the st plane. The evaluation of Eq. (9.5-12) can be simplified if we use the coordinate transformation $\alpha \equiv s - t$, $\beta \equiv s + t$. Since $\alpha = 0$ implies $s = t$ and $\beta = 0$ implies $s = -t$, the new region of integration is the inside of a rotated square or "diamond," shown in Fig. 9.5-2. For $\alpha < 0$, the integration with respect to β goes from $-(2T + \alpha)$ to $2T + \alpha$. However, for $\alpha > 0$, the integration on β goes from $-(2T - \alpha)$ to $2T - \alpha$. Hence the integration is broken up into two integrals as in

$$\begin{aligned}\overline{|X_T(f)|^2} = \tfrac{1}{2}\Bigg\{&\int_{-2T}^{0} d\alpha\, R(\alpha)e^{-j2\pi f\alpha} \int_{-2T-\alpha}^{2T+\alpha} d\beta \\ &+ \int_{0}^{2T} d\alpha\, R(\alpha)e^{-j2\pi f\alpha} \int_{-2T+\alpha}^{2T-\alpha} d\beta\Bigg\}.\end{aligned} \tag{9.5-13}$$

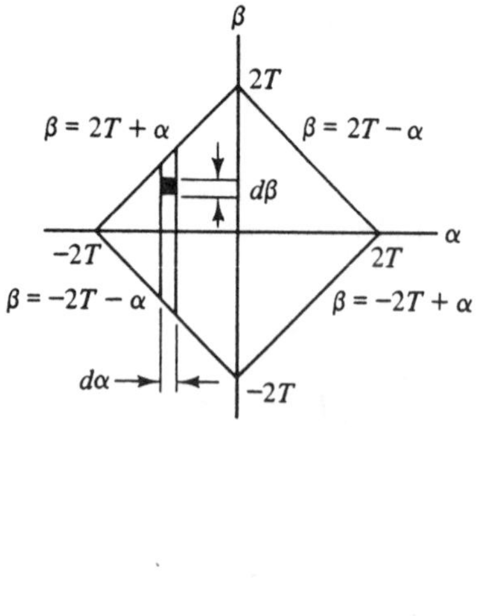

Figure 9.5-2. Region of integration for proof of Wiener–Khintchine theorem.

The factor of $\frac{1}{2}$ results from the fact that the Jacobian of the transformation is 2. The integration with respect to β gives

$$\overline{|X_T(f)|^2} = \int_{-2T}^{2T} R(\alpha)e^{-j2\pi f\alpha}\{2T - |\alpha|\}\, d\alpha. \tag{9.5-14}$$

Hence

$$\begin{aligned}\lim_{T\to\infty} \frac{\overline{|X_T(f)|^2}}{2T} &= \lim_{T\to\infty} \int_{-2T}^{2T} R(\alpha)e^{-j2\pi f\alpha}\, d\alpha \\ &- \lim_{T\to\infty} \int_{-2T}^{2T} \frac{|\alpha|R(\alpha)}{2T} e^{-j2\pi f\alpha}\, d\alpha.\end{aligned} \tag{9.5-15}$$

Because, by assumption,

$$\int_{-\infty}^{\infty} |\alpha R(\alpha)|\, d\alpha$$

is bounded, the limit of the second integral in Eq. (9.5-15) is zero. Thus

$$\begin{aligned}\lim_{T\to\infty} \frac{\overline{|X_T(f)|^2}}{2T} &= \int_{-\infty}^{\infty} R(\alpha)e^{-j2\pi f\alpha}\, d\alpha \\ &= W(f),\end{aligned} \tag{9.5-16}$$

which proves Eq. (9.5-11).

Properties of W(f)

From the definition of $W(f)$, we have that

$$W(f) \geq 0. \tag{9.5-17}$$

Also, since $R(\tau)$ is an even function of τ, we obtain

$$W(f) = \int_{-\infty}^{\infty} R(\tau) \cos 2\pi f\tau\, d\tau, \tag{9.5-18}$$

which shows that $W(f)$ is an *even* function of f. By the Fourier inversion property, it follows that

$$R(\tau) = \int_{-\infty}^{\infty} W(f)e^{j2\pi f\tau}\, df.$$

In particular, $R(\tau)$ evaluation at $\tau = 0$ gives the total power in $x(t)$:

$$R(0) = \int_{-\infty}^{\infty} W(f)\, df = \sigma^2 + \mu^2. \tag{9.5-19}$$

In words, the integral of $W(f)$ over $(-\infty, \infty)$ is the sum of the dc and ac power in the process $x(t)$. Finally, setting $f = 0$ in Eq. (9.5-18) shows that $W(0)$ is the area under the curve $R(\tau)$; that is,

$$W(0) = \int_{-\infty}^{\infty} R(\tau)\, d\tau. \tag{9.5-20}$$

Asymptotic Behavior of W(f)

The asymptotic behavior of $W(f)$ for large $|f|$ can be determined from the properties of $R(\tau)$ in the vicinity of $\tau = 0$. Formally, we obtain

$$\left.\frac{dR(\tau)}{d\tau}\right|_{\tau=0} \equiv R'(0) = \int_{-\infty}^{\infty} j2\pi f W(f)\, df, \tag{9.5-21}$$

$$R''(0) = -\int_{-\infty}^{\infty} (2\pi f)^2 W(f)\, df. \tag{9.5-22}$$

If $R'(\tau)$ is discontinuous at the origin, $R''(0)$ doesn't exist. Hence

$$\int_{-\infty}^{\infty} f^2 W(f)\, df$$

does not converge, which indicates that the asymptotic behavior of $W(f)$ is such that $W(f)$ decreases no more slowly than f^{-3}; that is,

$$W(f) \geq \frac{K}{f^3} \tag{9.5-23}$$

as $f \to \infty$ (K is a constant). On the other hand, if $R'(\tau)$ is continuous at the origin,

$$\int_{-\infty}^{\infty} f^2 W(f)\, df \tag{9.5-24}$$

exists, indicating that

$$W(f) < \frac{k}{f^3}, \qquad \text{as } f \to \infty. \tag{9.5-25}$$

(See Prob. 9-14.)

*Estimating W(f) from Measurements

The power spectrum is one of the most important quantities in electrical as well as other branches of engineering. It is important in determining the bandwidth of random signals and in the subsequent design of filters to pass or reject such signals. It is useful in distinguishing between an earthquake and a man-made explosion. It can be used to differentiate between normal and abnormal brain waves or normal and abnormal heart signals. It is even useful in economics where it has been used to analyze the behavior of stock-market fluctuations. For all these reasons, the measurement of $W(f)$ has been given prominent consideration in the literature, and entire books have been written on the subject. The trouble is that $W(f)$ is a somewhat complicated *average*, and, in general, it cannot be measured directly and certainly not exactly. It can only be estimated, albeit with a high degree of precision. We shall briefly consider some techniques for estimating $W(f)$. Let $x(t)$ be a stationary random process; then

$$W(f) = \lim_{T\to\infty} \frac{\overline{|X_T(f)|^2}}{2T} \tag{9.5-26}$$

is its power spectrum. The quantity

$$\mathcal{W}_T(f) = \frac{1}{2T}\left|\int_{-T}^{T} x(t)e^{-j2\pi ft}\,dt\right|^2 \tag{9.5-27}$$

would seem to be a reasonable estimate of $W(f)$. Unfortunately, for many processes, including the all-important Gaussian process, the quantity $\mathcal{W}_T(f)$ does not converge in a statistical sense (i.e., in the mean) to $W(f)$ even when $T \to \infty$ ([9-2], p. 107).

If $\mathcal{W}(f)$ is to a "good" estimate of $W(f)$, it should, on the average, closely approximate $W(f)$ and at the same time be *stable* (i.e., its variance should be low). The degree to which $\overline{\mathcal{W}(f)}$ approximates $W(f)$ is an indication of *fidelity* and is measured by a quantity called the *bias*, given by

$$\beta(f) = \overline{\mathcal{W}(f)} - W(f). \tag{9.5-28}$$

A small value of $|\beta(f)|$ is clearly desirable. However, $\beta(f)$ does not tell the whole story since it is a function of average values only. Equally important, therefore, is the stability of $\mathcal{W}(f)$. The stability of $\mathcal{W}(f)$ is a measure of its repeatability. A stable estimate is repeatable from record to record and is free from noiselike fluctuations unrelated to $W(f)$. The stability of $\mathcal{W}(f)$ is commonly given by the ratio

$$\frac{\text{var}\,[\mathcal{W}(f)]}{[\overline{\mathcal{W}(f)}]^2}, \tag{9.5-29}$$

where var $[\mathcal{W}(f)]$ is shorthand for $\overline{|\mathcal{W}(f) - \overline{\mathcal{W}(f)}|^2}$. If the ratio is close to unity, the estimate is very unstable.

There are several techniques available for extracting faithful and stable estimates from observations on $x(t)$. One approach is to generate $\mathcal{W}_T(f)$ as in Eq. (9.5-27) and then convolve it with a smoothing filter, $B(f)$. The smoothing filter $B(f)$ is sometimes called a *spectral window*, and the smoothing operation is described by

$$\mathcal{W}_B(f) = \int_{-\infty}^{\infty} \mathcal{W}_T(f - f')(B(f')\,df'. \tag{9.5-30}$$

A direct computation can be done to show that in the Gaussian case (and even somewhat more generally)

$$\frac{\text{var}\,[\mathcal{W}_T(f)]}{[\overline{\mathcal{W}_T(f)}]^2} \simeq 1, \qquad T \text{ large}. \tag{9.5-31}$$

However, when spectral smoothing is done,

$$\frac{\text{var}\,[\hat{W}_B(f)]}{[\hat{W}_B(f)]^2} \simeq \frac{E}{T}, \qquad T \text{ large}, \tag{9.5-32}$$

where E is the energy in the window; that is,

$$E \equiv \int_{-\infty}^{\infty} |B(f)|^2\, df. \tag{9.5-33}$$

Clearly, the number E/T can be made as small as desirable by choice of $B(f)$, but the increased stability is usually achieved at the expense of lowered fidelity. A very stable estimate may grossly fail to give a faithful rendition of $W(f)$ and hence be useless. In general, a compromise must be reached between fidelity and stability. The choice of $B(f)$, or, equivalently, its inverse Fourier transform $b(\tau)$ (called a *lag* or *data* window) is determined from the criterion of goodness set up for the estimator.

An example of smoothing by the method of Eq. (9.5-30) (i.e., direct realization of the convolution) is shown in Fig. 9.5-3. The example is taken from optics, and the random process is the *overlapping circular-grain model*.

The overlapping circular-grain model is used as an idealized model of the distribution of silver grains in uniformly exposed photographic film. The "random" part of this random process is the location of the circular, constant-radii grains. In reality, the grain sizes and shapes would also be random. Figure 9.5-3(a) shows a computer printout of the unsmoothed power spectrum estimate, $\hat{W}_T(f)$. What should be clear from this figure is the rapid fluctuation in intensity for small changes in frequency. This is symptomatic of an unstable spectrum.

Figure 9.5-3(b) shows $\hat{W}_B(f)$, computed as in Eq. (9.5-30) with the convolution done by digital computer. The particular spectral window, $B(f)$, that was used is given by

$$B(f) = \begin{cases} 3K(1 - \frac{3}{2}\pi^2 K f^2), & |f| < f_0, \\ 0, & |f| > f_0, \end{cases}$$

where K, a normalizing constant, is related to f_0 according to

$$K \equiv \frac{2}{3\pi^2 f_0^2}.$$

This window has certain optimal properties when used with two-dimensional processes. Specifically, it furnishes a minimum-bias spectral estimate for a fixed f_0. Problem 9-15 deals with spectral smoothing. The smoothed spectrum is seen to be free of meaningless amplitude fluctuations; the inherent shape of the spectrum, however, has been preserved for further analysis.

An alternative to Eq. (9.5-30) is to start with the *sample autocorrelation* function

$$\mathcal{R}_T(\tau) = \begin{cases} \dfrac{1}{2T - |\tau|} \displaystyle\int_{-T+|\tau|/2}^{T-|\tau|/2} x\left(t - \frac{|\tau|}{2}\right) x\left(t + \frac{|\tau|}{2}\right) dt, & |\tau| < 2T, \\ 0, & |\tau| > 2T. \end{cases} \tag{9.5-34}$$

The peculiar limits on the integral are not so peculiar after a moment's reflection: Recall that the sample function is observed only over $[-T, T]$. We now form the product

$$\mathcal{R}_T(\tau)b(\tau)$$

and compute the inverse Fourier transform,

$$\mathcal{W}_B(f) = \int_{-\infty}^{\infty} \mathcal{R}_T(\tau)b(\tau)e^{-j2\pi f\tau}\, d\tau. \tag{9.5-35}$$

Equation (9.5-35) is equivalent to Eq. (9.5-30) if

$$\mathcal{R}_T(\tau) = \mathcal{F}^{-1}[\mathcal{W}_T(f)], \tag{9.5-36}$$

$$b(\tau) = \mathcal{F}^{-1}[B(f)]. \tag{9.5-37}$$

A third method avoids convolution† and takes advantage of the existence of the fast Fourier transform algorithms (FFT) discussed in Chapter 4. First, we compute $\mathcal{W}_T(f)$ as in Eq. (9.5-27). Second, we use the FFT to compute $\mathcal{R}_T(\tau)$ and form the product $\mathcal{R}_T(\tau)b(\tau)$. Finally, we use the FFT to compute $\mathcal{W}_B(f)$ from Eq. (9.5-35).

Since the available record is observed only over $[-T, T]$, the sample correlation function does not contain all possible lags τ. For this reason, $b(\tau)$ is usually of finite support (i.e., it is zero for τ outside some interval). Hence we write

$$b(\tau) = 0, \qquad |\tau| > \tau_b.$$

Because $\mathcal{R}_T(\tau)$ is even, $b(\tau)$ is even. Also, to avoid scale changes in the estimated spectrum, $b(\tau)$ is normalized so that $b(0) = 1$. Specification of either the lag window *or* spectral window uniquely defines the other because of the Fourier transform relation that exists between them. The choice of $b(\tau)$ [or $B(f)$] is determined by the criterion of fidelity and/or stability specified by the user.

The method of spectral estimation discussed here emphasized the smoothing of sample spectra using *windows*. More modern methods try to find mathematical models that generate waveforms that have the same statistics as the observed data. The reader may have encountered such terms as *moving-average* (MA), *autoregressive* (AR), or autoregressive-moving average (ARMA). These are the names of several such models. Classical techniques of spectral estimation are discussed in [9-4] and [9-5]. Model-based techniques are discussed in [9-6] and in research journals, for example, [9-7].

† It is not always desirable to avoid convolution. For example, the discrete Hamming window [Eq. (5.5-4)] involves little "number crunching" on a computer.

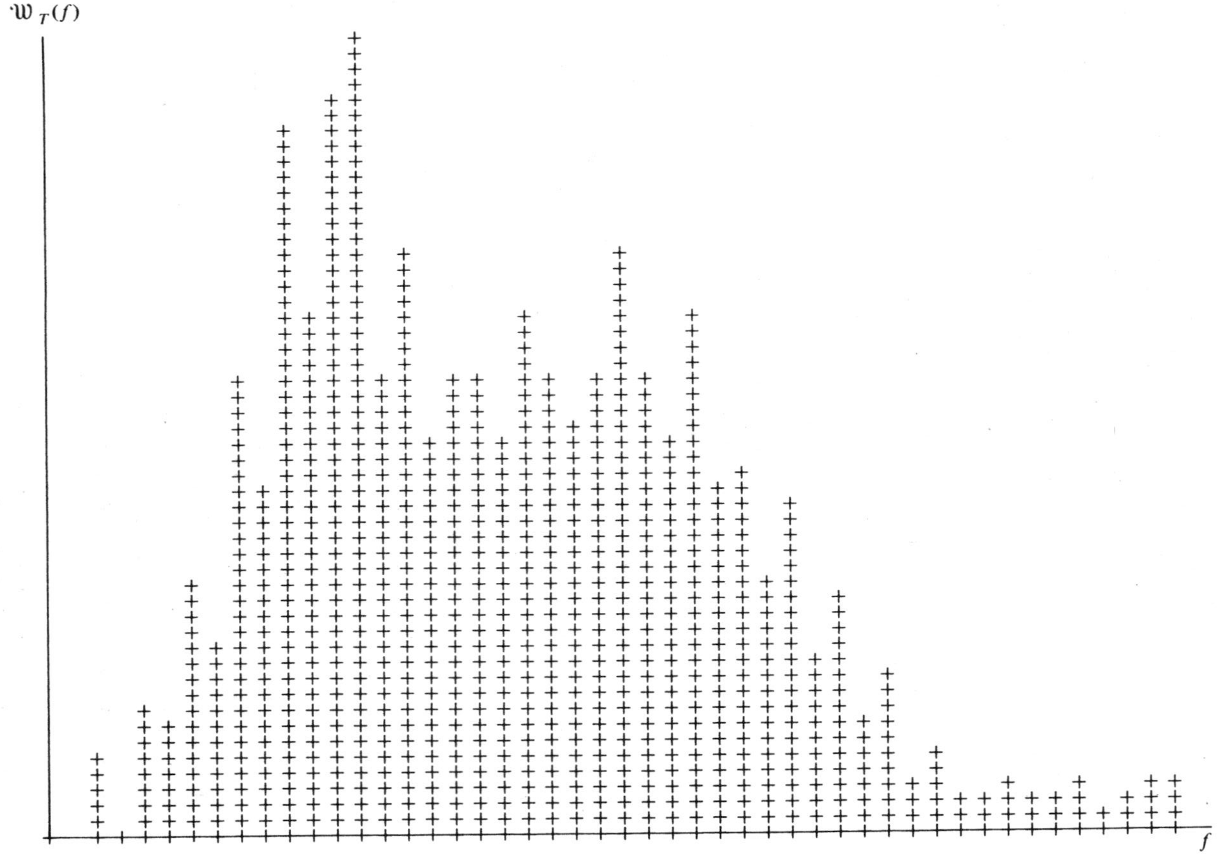

Figure 9.5-3. (a) Unsmoothed power spectral estimate.

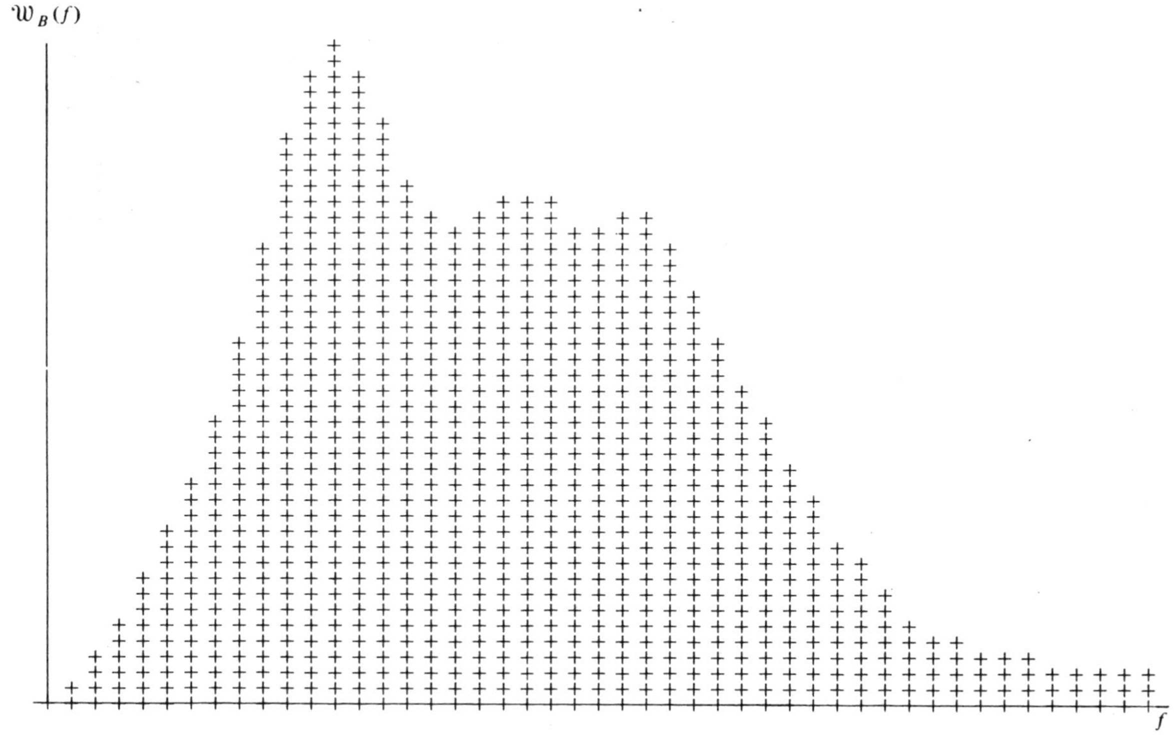

Figure 9.5-3 (continued). (b) Smoothed power spectral estimate.

9.6 INPUT–OUTPUT RELATIONS FOR RANDOM PROCESSES IN LINEAR SYSTEMS

The *black-box* representation of a linear, time-invariant system is shown in Fig. 9.6-1. The response of the system to the excitation $x(t)$ is given by

$$y(t) = \int_{-\infty}^{\infty} h(u)x(t-u)\,du. \tag{9.6-1}$$

We denote the correlation function associated with $y(t)$ as $R_y(\tau)$. Similarly, the correlation function associated with $x(t)$ will be written $R_x(\tau)$. For $Ry(\tau)$, we compute

$$\begin{aligned} R_y(\tau) = \overline{y(t)y(t+\tau)} &= E\left[\int_{-\infty}^{\infty} h(u)x(t-u)\,du \int_{-\infty}^{\infty} h(v)x(t+\tau-v)\,dv\right] \\ &= \int_{-\infty}^{\infty}\int_{-\infty}^{\infty} h(u)h(v)\overline{x(t-u)x(t+\tau-v)}\,du\,dv \\ &= \int_{-\infty}^{\infty}\int_{-\infty}^{\infty} h(u)h(v)R_x(\tau+u-v)\,du\,dv. \end{aligned} \tag{9.6-2}$$

If we let $z = v - u$, Eq. (9.6-2) can be written as

$$R_y(\tau) = \int_{-\infty}^{\infty} dz\, R_x(\tau - z) \int_{-\infty}^{\infty} du\, h(u)h(z+u). \tag{9.6-3}$$

Equation (9.6-3) is a basic result and can be concisely written as

$$R_y(\tau) = \int_{-\infty}^{\infty} R_x(\tau - z)g(z)\,dz, \tag{9.6-4}$$

where

$$g(z) \equiv \int_{-\infty}^{\infty} h(z+u)h(u)\,du. \tag{9.6-5}$$

The function $g(z)$ is known as the filter autocorrelation function. The type of integral in Eq. (9.6-5) is sometimes called a correlation integral. Equation (9.6-4) says that the output correlation function of the process $y(t)$ is the convolution of the input-process correlation function and the filter autocorrelation. $g(z)$ is uniquely determined by $h(z)$, but the converse is not true. The output correlation function is seen to be a weighted average of the input correlation function. Equations (9.6-3) to (9.6-5) are the basic input–output relations for the second moments in the time (actually *time lag*) domain. Frequently, it is simpler and more instructive to relate the input and output *power spectra*.

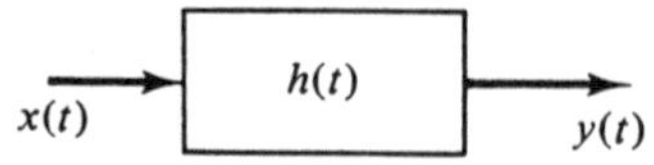

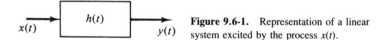

Figure 9.6-1. Representation of a linear system excited by the process $x(t)$.

The Fourier transform of Eq. (9.6-4) furnishes

$$W_y(f) = W_x(f)G(f), \tag{9.6-6}$$

where $W_y(f) = \mathcal{F}[R_y(\tau)]$, $W_x(f) = \mathcal{F}[R_x(\tau)]$, and $G(f) = \mathcal{F}[g(\tau)]$. $G(f)$ is computed from

$$\begin{aligned} G(f) &= \int_{-\infty}^{\infty} d\tau\, e^{-j2\pi f\tau} \int_{-\infty}^{\infty} du\, h(u)h(u+\tau) \\ &= \int_{-\infty}^{\infty} du\, h(u) \int_{-\infty}^{\infty} d\tau\, h(u+\tau)e^{-j2\pi f\tau} \\ &= \int_{-\infty}^{\infty} h(u)e^{j2\pi fu}\, du \int_{-\infty}^{\infty} h(\alpha)e^{-j2\pi f\alpha}\, d\alpha \\ &= |H(f)|^2. \end{aligned} \tag{9.6-7}$$

Equations (9.6-6) and (9.6-7) are among the most important results in linear systems theory. They indicate that the output spectral density is simply the input spectral density multiplied by the squared magnitude of the transfer function of the system. Equation (9.6-6) is of course completely equivalent to Eq. (9.6-4). However, the fact that $g(z)$ is not uniquely related to $h(z)$ is more easily discernible from Eq. (9.6-7). Two transfer functions with the same magnitudes but different phases give rise to the same $G(f)$ and hence the same $g(z)$.

Sometimes one is interested in the correlation of two different processes. Such a quantity is called the *cross correlation* and, for real processes, is given by

$$R_{xy}(\tau) = \overline{x(t)y(t+\tau)}. \tag{9.6-8}$$

The cross-correlation function does not generally satisfy the conditions listed in Eq. (9.3-12). For the linear system with input response $h(t)$, the cross correlation for a delay τ between input, $x(t)$, and output, $y(t+\tau)$, is

$$\begin{aligned} R_{xy}(\tau) &= \int_{-\infty}^{\infty} \overline{x(t)x(u)}h(t+\tau-u)\, du \\ &= \int_{-\infty}^{\infty} R_x(\alpha)h(\tau+\alpha)\, d\alpha, \qquad (9.6\text{-}9a) \\ &= \int_{-\infty}^{\infty} R_x(\alpha)h(\tau-\alpha)\, d\alpha. \qquad (9.6\text{-}9b) \end{aligned}$$

In Eq. (9.6-9a) we let $\alpha = t - u$; in Eq. (9.6-9b) we let $\alpha = u - t$. The cross correlation between two outputs $y(t)$ and $z(t)$ stimulated by the same input (Fig. 9.6-2) is simularly computed to be

$$R_{yz}(\tau) = \int_{-\infty}^{\infty} R_x(\tau - z)g_{12}(z)\, dz, \tag{9.6-10}$$

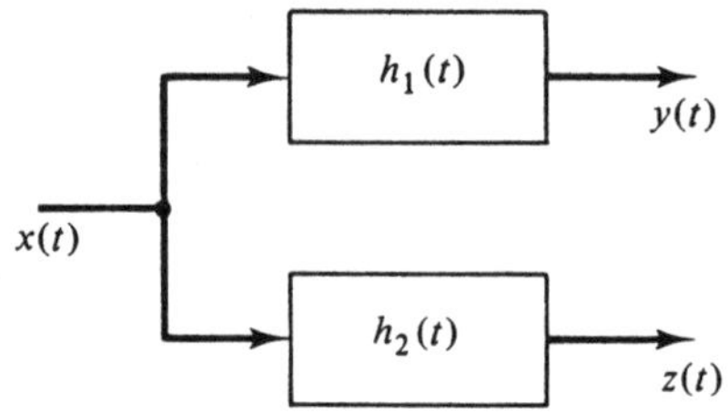

Figure 9.6-2. Two outputs produced by the same input.

where

$$g_{12}(z) = \int_{-\infty}^{\infty} h_1(u)h_2(u + z)\, du. \tag{9.6-11}$$

The quantity $g_{12}(z)$ is known as the filter cross-correlation function, and its Fourier transform, $G_{12}(f)$, relates the *cross-spectral density* $W_{yz}(f)$ to $W_x(f)$ according to

$$W_{yz}(f) = G_{12}(f)W_x(f) \tag{9.6-12}$$

From Eq. (9.6-11), $G_{12}(f) = H_1^*(f)H_2(f)$. In the special but important case of the cross-spectral density between the input and output of a system with transfer function $H(f)$, we obtain from Eq. (9.6-9b),

$$W_{xy}(f) = H(f)W_x(f). \tag{9.6-13}$$

Table 9.6-1 summarizes the most important results.

TABLE 9.6-1 BASIC INPUT-OUTPUT RELATIONS FOR RANDOM PROCESSES IN LINEAR SYSTEMS

Relations	Time domain	Frequency domain
Input–output signals	$y(t) = \int_{-\infty}^{\infty} x(\tau)h(t-\tau)\, d\tau$	$Y(f) = H(f)X(f)$
	$= \int_{-\infty}^{\infty} x(t-\tau)h(\tau)\, d\tau$	
Input–output second-order moments and spectra	$R_y(\tau) = \int_{-\infty}^{\infty} R_x(u)g(\tau - u)\, du$	$W_y(f) = G(f)W_x(f)$
	$g(\tau) = \int_{-\infty}^{\infty} h(u)h(u+\tau)\, du$	$G(f) = \lvert H(f)\rvert^2$
Cross-correlation function and cross spectra	$R_{xy}(\tau) = \int_{-\infty}^{\infty} R_x(u)h(\tau - u)\, du$	$W_{xy}(f) = H(f)W_x(f)$

9.7 THE GAUSSIAN RANDOM PROCESS

Consider a random process $x(t)$. Suppose we choose any k instants $t_1, \ldots, t_k$. The corresponding k random variables $x(t_1), \ldots, x(t_k)$ constitute a random vector $\mathbf{X}$ given by

$$\mathbf{X} = [x(t_1), \ldots, x(t_k)]^T, \tag{9.7-1}$$

where T here denotes transpose. Let $\mathbf{x}$ be a vector of k numbers in the range of $\mathbf{X}$ so that the event $\{x(t_1) \le x_1, x(t_2) \le x_2, \ldots, x(t_k) \le x_k\}$ is written $\{\mathbf{X} \le \mathbf{x}\}$. Then $x(t)$ is a Gaussian process if $\mathbf{X}$ has a jointly (multivariate) Gaussian pdf for every finite set of $\{t_i\}$ and every k.

The multivariate Gaussian pdf is given by

$$p(\mathbf{x}) = \frac{1}{(2\pi)^{k/2}|\det \mathbf{K}|^{1/2}} \exp\left[-\frac{1}{2}(\mathbf{x}-\boldsymbol{\mu})^T\mathbf{K}^{-1}(\mathbf{x}-\boldsymbol{\mu})\right], \tag{9.7-2}$$

where $\mathbf{K}$ is the covariance matrix and $\boldsymbol{\mu}$ is the vector of means. Thus

$$\boldsymbol{\mu} = E[\mathbf{X}] = [\mu_1, \ldots, \mu_k]^T$$

$$\mathbf{K} = \begin{bmatrix} r_{11} & r_{12} & \cdots & r_{1k} \\ r_{12} & r_{22} & \cdots & \\ \cdot & & & \\ \cdot & & & \\ \cdot & & & \\ r_{1k} & & & \end{bmatrix}, \tag{9.7-3}$$

where $r_{ij} \equiv r(t_i, t_j) = E\{[x(t_i) - \mu_i][x(t_j) - \mu_j]\}$.

Observation: Suppose the k variates are uncorrelated; that is,

$$\overline{[x(t_i) - \mu_i][x(t_j) - \mu_j]} = 0, \qquad \text{for } i \neq j.$$

Then

$$\mathbf{K} = \begin{bmatrix} \sigma_1^2 & & & & 0 \\ & \sigma_2^2 & & & \\ & & \cdot & & \\ & & & \cdot & \\ & & & & \cdot \\ 0 & & & & \sigma_k^2 \end{bmatrix} \equiv \text{diag}(\sigma_1^2, \ldots, \sigma_k^2), \tag{9.7-4}$$

where $\sigma_i^2 = \overline{[x(t_i) - \mu_i]^2}$ and all off-diagonal entries are zero. In this case, the joint pdf $p(\mathbf{x})$ can be written as

$$p(\mathbf{x}) = \frac{1}{(2\pi)^{k/2}} \frac{1}{[\sigma_1^2 \cdots \sigma_k^2]^{1/2}} \exp\left[-\frac{1}{2}\sum_{i=1}^{k} \frac{x_i^2}{\sigma_i^2}\right]$$
$$= \frac{1}{(2\pi\sigma_1^2)^{1/2}} \exp\left[-\frac{1}{2}\frac{x_1^2}{\sigma_1^2}\right] \cdots \frac{1}{(2\pi\sigma_k^2)^{1/2}} \exp\left[-\frac{1}{2}\frac{x_k^2}{\sigma_k^2}\right] \tag{9.7-5}$$
$$= p(x_1)\,p(x_2)\cdots p(x_k).$$

Thus, in the Gaussian case, when random variables are uncorrelated they are also independent. This is in general not true for other probability laws.
Some well-known properties of the Gaussian process are

1. $p(x)$ depends only on $\boldsymbol{\mu}$, $\mathbf{K}$.
2. If $x(t_i)$, $i = 1, \ldots, k$, are jointly Gaussian, then each $x(t_i)$ is individually Gaussian.
3. If $\mathbf{K}$ is diag $(\boldsymbol{\sigma}_1^2, \ldots, \boldsymbol{\sigma}_k^2)$, then the $x(t_i)$, $i = 1, \ldots, k$, are independent.
4. Linear transformations on Gaussian r.v.'s yield Gaussian r.v.'s.
5. A wide-sense stationary Gaussian process is always strict-sense stationary.

Property 4 is a very important result. It says that, if a Gaussian process is acted upon by a linear system, the output will be Gaussian. We demonstrate this in the following example.

Example

Show that if $\mathbf{X}$ is Gaussian then for nonsingular $\mathbf{A}$ the random vector $\mathbf{Y} = \mathbf{AX} + \mathbf{b}$ is Gaussian.

Let $\mathbf{X}$ be any k r.v.'s from the process $x(t)$. Let $\mathbf{Y}$ be any k r.v.'s from the process $y(t)$. Then $\mathbf{X} = [X_1, \ldots, X_k]^T$, $\mathbf{Y} = [Y_1, \ldots, Y_k]^T$. Let $F_{\mathbf{Y}}(\mathbf{y}) \equiv \mathbf{P}(\mathbf{Y} \le \mathbf{y})$. Then

$$F_{\mathbf{Y}}(\mathbf{y}) = P[\mathbf{X} \le \mathbf{A}^{-1}(\mathbf{y} - \mathbf{b})]$$
$$= F_{\mathbf{X}}[\mathbf{A}^{-1}(\mathbf{y} - \mathbf{b})],$$

where $F_{\mathbf{X}}(\mathbf{x}) = P(\mathbf{X} \le \mathbf{x})$. $F_{\mathbf{Y}}(\mathbf{y})$ has the same form as $F_{\mathbf{X}}(\mathbf{x})$ except that the quadratic form is now

$$[\mathbf{A}^{-1}(\mathbf{y} - \mathbf{b}) - \boldsymbol{\mu}]^T\mathbf{K}^{-1}[\mathbf{A}^{-1}(\mathbf{y} - \mathbf{b}) - \boldsymbol{\mu}].$$

If we let $\boldsymbol{\alpha} \equiv \mathbf{A}^{-1}\mathbf{b} + \boldsymbol{\mu}$, the preceding has the form

$$(\mathbf{y}^T\mathbf{A}^{-1^T} - \boldsymbol{\alpha}^T)\mathbf{K}^{-1}(\mathbf{A}^{-1}\mathbf{y} - \boldsymbol{\alpha}) = [\mathbf{y}^T - \boldsymbol{\alpha}^T\mathbf{A}^T]\mathbf{A}^{-1^T}\mathbf{K}^{-1}\mathbf{A}^{-1}(\mathbf{y} - \mathbf{A}\boldsymbol{\alpha})$$
$$= (\mathbf{y} - \mathbf{A}\boldsymbol{\alpha})^T(\mathbf{A}^{-1^T}\mathbf{K}^{-1}\mathbf{A}^{-1})(\mathbf{y} - \mathbf{A}\boldsymbol{\alpha}) \tag{9.7-6}$$
$$= (\mathbf{y} - \boldsymbol{\mu}')^T\mathbf{K}'^{-1}(\mathbf{y} - \boldsymbol{\mu}').$$

Hence the new mean is $\boldsymbol{\mu} = \mathbf{b} + \mathbf{A}\boldsymbol{\mu}$, and the new covariance matrix is $\mathbf{K}' = \mathbf{AKA}^T$. Note that for a stationary Gaussian process $\boldsymbol{\mu}(t) = \boldsymbol{\mu}$ and $\mathbf{K}(t) = \mathbf{K}$; that is, these quantities are not functions of time.

9.8 THE NARROWBAND GAUSSIAN PROCESS

When a sample function of the narrowband Gaussian process (NGP) is viewed on an oscilloscope, what appears is a waveform that resembles a sine wave with slowly varying amplitude and phase. For this reason, the NGP can be conveniently described† by

$$x(t) = V(t) \cos [2\pi f_c t + \phi(t)], \tag{9.8-1}$$

where $V(t)$ is the slowly varying envelope and $\phi(t)$ is the slowly varying phase. A way to generate an NGP process is shown in Fig. 9.8-1.

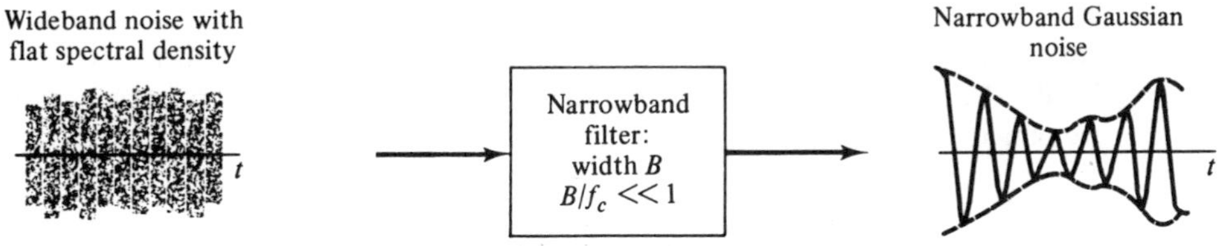

Figure 9.8-1. Generation of the NGP.

Equation (9.8-1) can be written as

$$x(t) = x_c(t) \cos 2\pi f_c t - x_s(t) \sin 2\pi f_c t, \tag{9.8-2}$$

where

$$x_c(t) = V(t) \cos \phi(t) \text{ (in-phase component)},$$
$$x_s(t) = V(t) \sin \phi(t) \text{ (quadrature component)}.$$

The zero-mean process $x(t)$ is band pass with bandwidth B and centered at center frequency f_c. The modulating functions $x_c(t)$ and $x_s(t)$ are low pass with bandwidth $B/2$. The process $x(t)$ can be viewed as the difference of two DSB waves in quadrature.

There are several ways to show that if $\overline{x^2(t)} \equiv \sigma^2$ then $\overline{x_c^2(t)} = \overline{x_s^2(t)} = \sigma^2$ and $\overline{x_c(t)x_s(t)} = 0$. One way is through the classical Rice approach described in [9-2, p. 158], in which $x(t)$ is written as a Fourier series over $[-T/2, T/2]$ and advantage is taken of the fact that the Fourier coefficients are Gaussian r.v.'s that become uncorrelated as $T \to \infty$.

Another approach is to assume the existence of fixed time T such that $1/f_c \ll T \ll 1/B$. This is not a rash assumption since in ordinary broadcast AM and FM there are, roughly, two orders of magnitude between the IF center frequency and the filter bandwidth. Under this assumption, we can compute $x_c(t)$ and $x_s(t)$ to a good approximation by treating them as Fourier coefficients and computing them likewise; that is, to compute $x_c(t)$, multiply both sides of Eq. (9.8-2) by $\cos 2\pi f_c t$ and integrate over $[t - (T/2),$

† This description is actually quite general; but if we want to associate $V(t)$ and $\phi(t)$ with a slowly varying envelope and phase, respectively, of a high-frequency carrier, then $x(t)$ should be a narrowband process.

$t + (T/2)]$; to compute $x_s(t)$, multiply both sides of Eq. (9.8-2) by $\sin 2\pi f_c t$ and integrate over $[t - (T/2), t + (t/2)]$.† Thus

$$x_c(t) \simeq \frac{2}{T}\int_{t-T/2}^{t+T/2} x(u)\cos 2\pi f_c u\, du \tag{9.8-3a}$$

and

$$x_s(t) \simeq -\frac{2}{T}\int_{t-T/2}^{t+T/2} x(u)\sin 2\pi f_c u\, du. \tag{9.8-3b}$$

Now we can make several interesting observations. First, by taking the expectation of both sides in Eqs. (9.8-3), we obtain $\overline{x_c(t)} = \overline{x_s(t)} = 0$. Second, since $x_c(t)$ and $x_s(t)$ are obtained by *linear* operations on a Gaussian process, they are themselves Gaussian. Finally, by using Eqs. (9.8-3a) and (9.8-3b) to directly compute $\overline{x_c^2(t)}$, $\overline{x_s^2(t)}$, and $\overline{x_c(t)x_s(t)}$, we obtain

$$\overline{x_c^2(t)} = \overline{x_s^2(t)} = \overline{x(t)^2} = \sigma^2 \tag{9.8-4}$$

and

$$\overline{x_c(t)x_s(t)} = 0. \tag{9.8-5}$$

(See Prob. 9-20.) We shall omit the details since the evaluation is a straightforward, albeit tedious, exercise in integration; the hardy reader is urged to check these results on his or her own. Putting all our results together, including the fact that $x_c(t)$ and $x_s(t)$ are zero-mean Gaussian processes, we find that their joint pdf is

$$f(x_{ct}, x_{st}) = \frac{1}{2\pi\sigma^2}\exp\left[-\left(\frac{x_{ct}^2 + x_{st}^2}{2\sigma^2}\right)\right], \tag{9.8-6}$$

where x_{ct} and x_{st} are numbers in the range of $x_c(t)$ and $x_s(t)$, respectively. The joint distribution function of $V(t)$ and $\phi(t)$ is obtained from

$$F(v, \phi) = P[V(t) \le v, \phi(t) \le \phi] = \int_D\int f(x_{ct}, x_{st})\, dx_{ct}\, dx_{st}, \tag{9.8-7}$$

where D is the shaded region in Fig. 9.8-2. The result is

$$\begin{aligned} F(v, \phi) &= \int_0^{\phi} d\phi' \int_0^{v} \frac{1}{2\pi\sigma^2} e^{-\xi^2/2\sigma^2}\xi\, d\xi \\ &= \begin{cases} \left(\dfrac{\phi}{2\pi}\right)(1 - e^{-v^2/2\sigma^2}), & v \ge 0, \quad 0 < \phi \le 2\pi, \\ 0, & \text{otherwise.} \end{cases} \end{aligned} \tag{9.8-8}$$

To derive Eq. (9.8-8), we used the transformation

$$\xi = (x_{ct}^2 + x_{st}^2)^{1/2} \quad \text{and} \quad \phi' = \tan^{-1}\frac{x_{st}}{x_{ct}}.$$

† In carrying out this procedure, the integrals that are discarded can be shown to be very small in comparison, in the mean-square sense, to the integral that is retained.

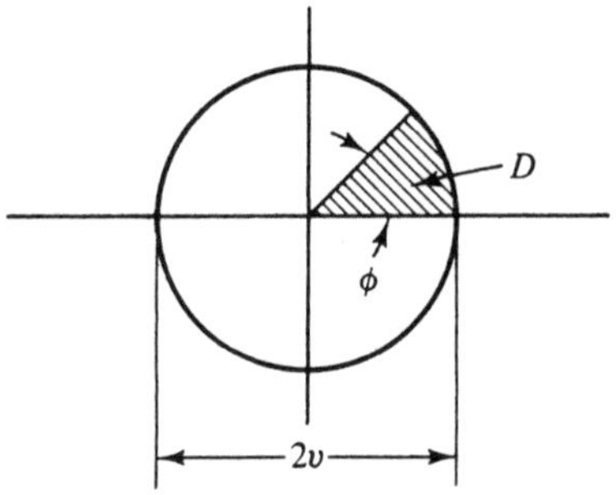

Figure 9.8-2. Region D required to compute $F(v, \phi)$.

Equation (9.8-8) clearly shows that $\phi(t)$ and $V(t)$, for fixed t, are independent r.v.'s. The joint pdf is

$$f(v, \phi) = \frac{\partial^2 F(v, \phi)}{\partial v \, \partial \phi} = f(v)g(\phi) = \left(\frac{1}{2\pi}\right)\frac{v}{\sigma^2}e^{-v^2/2\sigma^2}u(v) \cdot [u(\phi) - u(\phi - 2\pi)], \qquad (9.8\text{-}9)$$

where, for any ξ, $u(\xi)$ is the unit step starting at $\xi = 0$. We conclude that, for fixed but arbitrary t, $\phi(t)$ is an r.v. uniformly distributed in $[0, 2\pi]$ and $V(t)$ is an r.v. that obeys the Rayleigh probability law [Eq. (8.14-8), Chapter 8]. Figure 9.8-3 shows the pdf's.

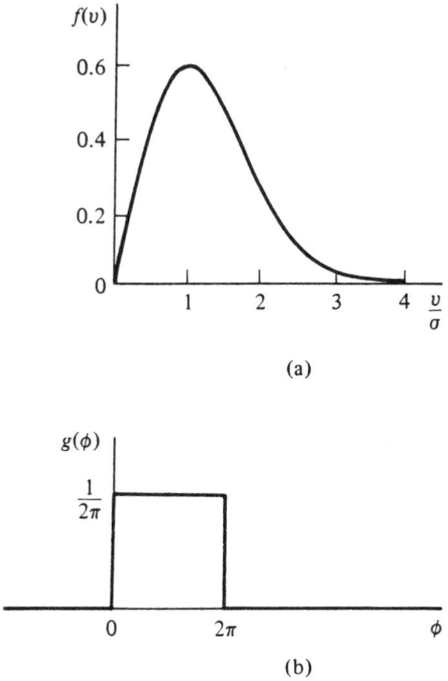

Figure 9.8-3. Probability density functions of (a) the envelope and (b) the phase of the NGP.

For a discussion and derivation of the joint densities of $V(t)$ and $V(t + \tau)$ and of $\phi(t)$ and $\phi(t + \tau)$, the reader is urged to consult [9-2, p. 161]. The solution of this more difficult problem shows that although, for fixed t, $V(t)$ and $\phi(t)$ are independent random variables, the joint pdf of $V(t)$, $V(t + \tau)$, $\phi(t)$, and $\phi(t + \tau)$ *cannot* be factored into pdf's depending only on the V's and ϕ's individually. Hence $V(t)$ and $\phi(t)$ are *not independent processes*.

9.9 GAUSSIAN WHITE NOISE

White noise is an unrealizable process, full of inherent contradictions, yet widely used in communication theory. The power spectrum of white noise is constant,† that is,

$$W(f) = N_0, \qquad -\infty < f < \infty, \tag{9.9-1}$$

and its autocorrelation function is

$$R(\tau) = N_0\delta(\tau). \tag{9.9-2}$$

For a stationary, zero-mean process, the total power is given by

$$\sigma^2 = \int_{-\infty}^{\infty} W(f)\,df.$$

Hence white noise has infinite power.‡ Furthermore, if $x(t)$ is a white-noise process, $x(t)$ and $x(t + \epsilon)$, for any fixed t and $\epsilon > 0$, are always independent r.v.'s no matter how small ϵ is. Thus any finite interval of a white-noise process contains an infinite number of independent random variables. From this observation, all types of bizarre implications follow upon which we shall not dwell. Nevertheless, despite all the inadequacies of this model, white noise is an extremely useful concept. A wideband Gaussian process that has uniform spectral density over the transmittance window of a filter can be considered a white-noise process for the sake of computations.

Many naturally occurring processes are modeled as white noise. Shot noise, semiconductor noise, and the spatial grain noise of high-resolution photographic film can all be modeled as white noise in many instances. However, in communication systems, it is thermal noise that is most often modeled as white noise. In fact thermal noise and white noise are frequently used interchangeably. The presence of thermal noise imposes fundamental limits on the performance of a communication system. For this reason we shall discuss it briefly next.

† Another commonly used definition is $W(f) = N_0/2$ for all f, where N_0 is in watts per hertz measured over positive frequencies.

‡ The infinite power resulting from the postulation of white noise is called the *ultraviolet catastrophe*.

Thermal Noise

Thermal noise is generated by thermally induced interactions between charges flowing in conducting media. The most common situation is that of electrons in random motion in a resistor. J. B. Johnson and H. Nyquist [9-8] studied thermal noise and showed, both from experimental and theoretical considerations, that the mean-squared noise voltage across a resistor of resistance R is given by

$$\overline{v^2(t)} = 4kTRB, \tag{9.9-3}$$

where T is the temperature in degrees Kelvin of the resistor, k is the Boltzmann constant (1.38×10^{-23} J/°K), and B is any arbitrary bandwidth. The spectral density is then

$$W_0(f) = \frac{\overline{v^2(t)}}{2B} = 2kTR, \tag{9.9-4}$$

which is seen to be flat.

A more careful calculation that includes quantum-mechanical effects shows that the spectral density of the thermal voltage is ([9-9], p. 551)

$$W(f) = 2\left(\frac{hf}{2} + \frac{hf}{e^{hf/kT} - 1}\right)R, \tag{9.9-5}$$

where f is frequency and h is Planck's constant, $h = 6.6257 \times 10^{-34}$ J/s. The first term is negligible at frequencies $f \ll kT/h \simeq 10^{13}$ Hz. For $f \ll 10^{13}$ Hz, $\exp[hf/kT] \simeq 1 + hf/kT$ and $W(f)$ can be approximated by

$$W(f) = 2kTR. \tag{9.9-6}$$

This is the same result furnished by Eq. (9.9-4). Thus, for frequencies in use in normal communications, but not including optical or laser frequencies, the white-noise approximation is excellent. Because thermal noise is the result of a very large number of essentially independent interactions, its statistics tend to be Gaussian. This is implied by the *central-limit theorem* of probability ([9-2], p. 81).

9.10 THE BILATERAL CLIPPER; VAN VLECK'S THEOREM

Van Vleck's theorem states that the correlation function (and therefore the power spectrum) of a Gaussian process can be determined from the second-order statistics of the zero crossings of the process [9-10]. The result has significant implications for practical power spectra computations and can save much computer time. The computing burden can be reduced by a factor of from 5 to 7 [9-11]. To develop the main theorem, we first develop some preliminary results.

Let $x(t)$ be a zero-mean Gaussian random process. If $X \equiv x(t)$ and $Y \equiv x(t + \tau)$, then the correlation coefficient, or normalized covariance, ρ, is given by [cf. Eq. (8.16-4)]

$$\rho = \frac{\overline{XY}}{(\overline{X^2}\,\overline{Y^2})^{1/2}} = \frac{R(\tau)}{R(0)}. \tag{9.10-1}$$

From Fig. 9.10-1, we see that an *odd* number of zeros in the interval $(t, t + \tau)$ is simply the event $XY < 0$. Hence the probability $P_0(\tau)$ of an odd number of zeros in the interval $(t, t + \tau)$ is simply $P(XY < 0)$. The probability of the event $XY < 0$ or, equivalently, $X/Y < 0$ for X and Y Gaussian variates was treated in Sec. 8.18. Use of Eq. (8.18-15) gives

$$\begin{aligned} P_0(\tau) &= \frac{1}{\pi}\cos^{-1}\rho \\ &= \frac{1}{\pi}\cos^{-1}\frac{R(\tau)}{R(0)}. \end{aligned} \tag{9.10-2}$$

The probability of an *even* number of zeros is

$$P_e(\tau) = 1 - P_0(\tau) = 1 - \frac{1}{\pi}\cos^{-1}\frac{R(\tau)}{R(0)}. \tag{9.10-3}$$

Equations (9.10-2) and (9.10-3) will be useful in what follows.

To remove random amplitude variations in waveforms in which information is stored in the zero crossings, a hard clipper can be used to generate square pulses, which can subsequently be filtered to produce a uniform amplitude wave. A hard clipper is a nonlinear device whose action is shown in Fig. 9.10-2. Its transfer characteristic is described by

$$y = \begin{cases} 1, & x \geq 0, \\ -1, & x < 0. \end{cases} \tag{9.10-4}$$

Let the input be a zero-mean Gaussian random process $x(t)$, and let $y(t)$ denote the output. Also, let $R_y(\tau)$ and $R_x(\tau)$ denote the correlation functions of $y(t)$ and $x(t)$, respectively. Let $Y_1 \equiv y(t)$ and $Y_2 \equiv y(t + \tau)$. Then

$$\begin{aligned} R_y(\tau) = {} & P(Y_1 = 1, Y_2 = 1) + P(Y_1 = -1, Y_2 = -1) \\ & -P(Y_1 = 1, Y_2 = -1) - P(Y_1 = -1, Y_2 = 1). \end{aligned} \tag{9.10-5}$$

Now $P(Y_1 = 1, Y_2 = 1) + P(Y_1 = -1, Y_2 = -1)$ is simply the probability $P_e(\tau)$ that $x(t)$ makes an even number of zero crossings in τ. Similarly, $P(Y_1 = -1, Y_2 = 1) +$

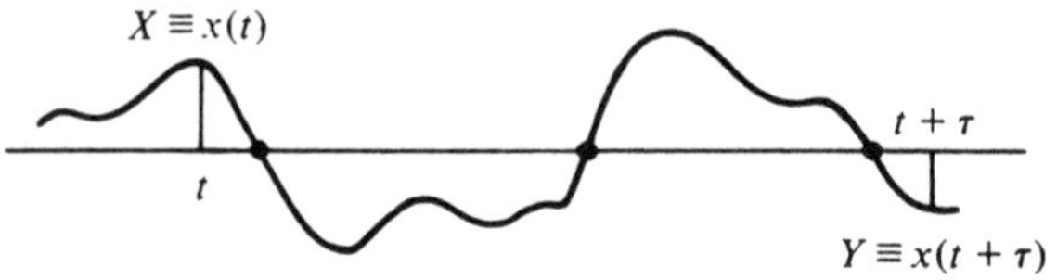

Figure 9.10-1. Zero crossings of a random process.

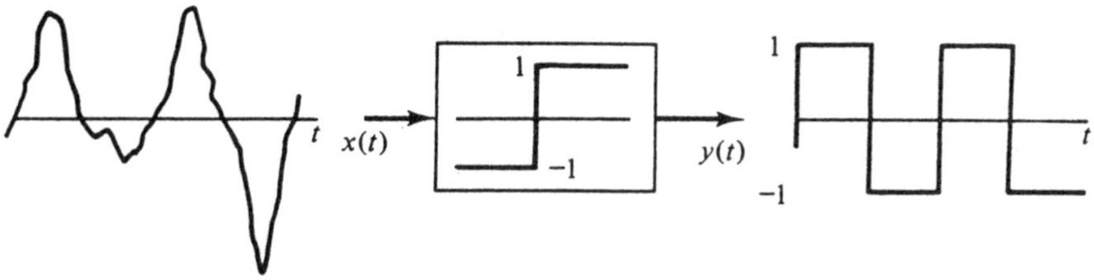

Figure 9.10-2. Bilateral hard clipper.

$P(Y_1 = 1, Y_2 = -1)$ is the probability that $x(t)$ makes an odd number of zero crossings in τ. Hence from Eqs. (9.10-2) and (9.10-3) we obtain

$$R_y(\tau) = \left[1 - \frac{1}{\pi}\cos^{-1}\frac{R_x(\tau)}{R_x(0)}\right] - \frac{1}{\pi}\cos^{-1}\frac{R_x(\tau)}{R_x(0)}$$
$$= 1 - \frac{2}{\pi}\cos^{-1}\frac{R_x(\tau)}{R_x(0)}. \tag{9.10-6}$$

With the help of a simple diagram such as in Fig. 8.18-1, Eq. (9.10-6) can be written as

$$R_y(\tau) = \frac{2}{\pi}\sin^{-1}\left[\frac{R_x(\tau)}{\sigma^2}\right]. \tag{9.10-7}$$

where $\sigma^2 \equiv R_x(0)$. Hence

$$R_x(\tau) = \sigma^2 \sin\left[\frac{\pi}{2}R_y(\tau)\right]. \tag{9.10-8}$$

Equation (9.10-8) is the key result of the theorem; it states that the correlation function $R_x(\tau)$ of a Gaussian process can be computed directly from the correlation function of the hard-clipped waveform $y(t)$.

9.11 NOISE IN AM AND DERIVED SYSTEMS

In this and subsequent sections we shall compute the signal-to-noise ratios for several different modulation and detection schemes. It will be convenient to consider the information-bearing signal, $s(t)$, as an ergodic process so that the time average of $s^2(t)$, denoted by $\langle s^2(t)\rangle$, is equal to $\overline{s^2(t)}$. The average power associated with a signal such as $s(t)\cos 2\pi f_c t$ is then

$$\lim_{T\to\infty}\frac{1}{2T}\int_{-T}^{T} s^2(t)\cos^2 2\pi f_c t\,dt = \frac{\overline{s^2(t)}}{2}.$$

Although the input noise to the receiver need be neither Gaussian nor band limited (in fact, before the RF stage it typically is very broadband, and the signal-to-noise ratio there is poorly defined), the post-IF predetection noise is normally assumed narrowband and Gaussian. Depending on whether the modulation is AM (including DSB) or SSB, the effective IF bandwidth, B_T, is $2W$ or W Hz, respectively. Since the carrier frequency is $f_c \gg B_T$ in either case, the narrowband condition holds. In what follows we shall ignore the RF stage since its primary function is the rejection of image frequencies. The two important signal-to-noise ratios are those that are computed (1) after the IF stage but before detection and (2) after detection. The former will be called the input signal-to-noise ratio and the latter the output or postdetection signal-to-noise ratio.

Synchronous Detection of DSBSC

The detection of DSBSC is shown in Fig. 9.11-1. The reader will recall from Chapter 6 that the spectrum of a DSBSC signal is essentially that of AM and consists of the two sidebands centered around the carrier frequency f_c [Fig. 6.2-2(b) with the carrier component removed]. For simplicity, we assume that the filter has the ideal band-pass characteristic shown in Fig. 9.11-1; hence the spectral density of the noise inside the passband is constant. We let N_0 denote the two-sided noise spectral density and assume that the bandwidth, B_T, of the IF stage is approximately $B_T \simeq 2W$. At point (b), the signal is a mixture of modulated wave and band-limited noise, $n(t)$. The signal $z(t)$ is given by

$$z(t) = A_c s(t) \cos 2\pi f_c t + n(t). \tag{9.11-1}$$

As already stated, because $f_c/B_T \gg 1$, $n(t)$ can be considered to be an NGP. The noise power is given by†

$$\overline{n^2(t)} = \int_{-\infty}^{\infty} |H_{\text{IF}}(f)|^2 N_0 \, df = 2B_T N_0 = 4N_0 W, \tag{9.11-2}$$

and the input signal-to-noise ratio is

$$\left(\frac{\mathcal{S}}{\mathcal{N}}\right)_i = \frac{P_s}{4N_0 W}, \tag{9.11-3}$$

where P_s, the power associated with the carrier-modulated signal, is given by

$$P_s = A_c^2 \overline{s^2(t)}/2 \qquad \text{(DSB)}. \tag{9.11-4}$$

At (c) the signal is given by

$$w(t) = \frac{A_c s(t)}{2}(1 + \cos 4\pi f_c t) + n(t) \cos 2\pi f_c t. \tag{9.11-5}$$

However, in Sec. 9.8 we found that the NGP could be written as

$$n(t) = n_c(t) \cos 2\pi f_c t - n_s(t) \sin 2\pi f_c t. \tag{9.11-6}$$

† Gain constants are arbitrarily set equal to unity because we are interested in *ratios* of signal power to noise power.

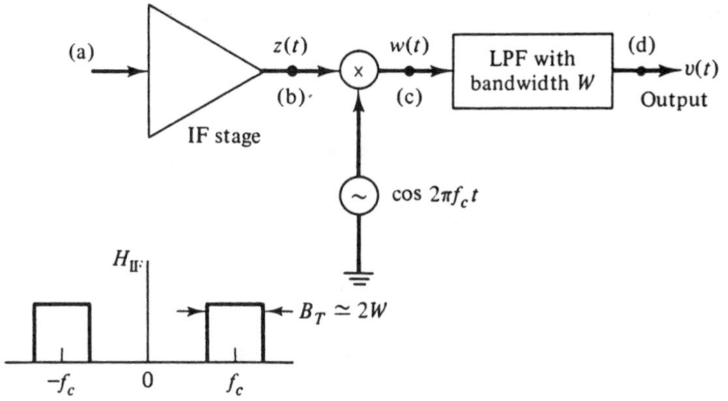

Figure 9.11-1. Synchronous detector.

Hence $w(t)$ can be written as

$$w(t) = \left[\frac{A_c s(t)}{2} + \frac{n_c(t)}{2}\right](1 + \cos 4\pi f_c t) - \frac{n_s(t)}{2}\sin 4\pi f_c t. \qquad (9.11\text{-}7)$$

After the low-pass filter, the signal at (d) is

$$v(t) = \frac{A_c s(t)}{2} + \frac{n_c(t)}{2}, \qquad (9.11\text{-}8)$$

so only the in-phase noise component figures in the signal-to-noise ratio. The output, or postdetection, signal-to-noise ratio is

$$\left(\frac{\mathcal{S}}{\mathcal{N}}\right)_0 = \frac{A_c^2\overline{s^2(t)}}{\overline{n_c^2}} = \frac{P_s}{2N_0W} = 2\left(\frac{\mathcal{S}}{\mathcal{N}}\right)_i, \qquad (9.11\text{-}9)$$

where $\overline{n_c^2} = \overline{n_s^2} = \overline{n^2}$. Hence, in the case of DSBSC modulation, the output signal-to-noise ratio is *twice* that of the input. Although DSB demodulation is simply a shifting of the sidebands to the origin of the frequency axis, the signal components add as amplitudes, while the noise adds as power. We say that the signals add *coherently*, while the noise adds *incoherently*. This is what accounts for the 3.0-dB increase in $\mathcal{S}/\mathcal{N}$ ratio.

Synchronous Detection of SSB, VSB

The synchronous detector for SSB or VSB is the same as for DSBSC and is shown in Fig. 9.11-1. If the SSB signal is generated by admitting the upper sideband of the DSB signal $x(t) = A_c s(t) \cos 2\pi f_c t$ (the upper sideband is chosen for specificity), then the modulated SSB signal is

$$x(t) = \frac{A_c}{2}[s(t)\cos 2\pi f_c t - \hat{s}(t)\sin 2\pi f_c t], \qquad (9.11\text{-}10)$$

where $\hat{s}(t)$ is the Hilbert transform of $s(t)$; that is,

$$\hat{s}(t) = \frac{1}{\pi} \int_{-\infty}^{\infty} \frac{s(\lambda)}{t - \lambda} d\lambda. \tag{9.11-11}$$

For simplicity, we again assume that the transmittance window of the IF stage is rectangular, as shown in Fig. 9.11-2. Referring now to Fig. 9.11-1, we find that the signal at (b) is

$$\begin{aligned} z(t) &= \frac{A_c}{2} [s(t) \cos \omega_c t - \hat{s}(t) \sin \omega_c t] \\ &\quad + n_c(t) \cos \omega_0 t - n_s(t) \sin \omega_0 t, \end{aligned} \tag{9.11-12}$$

where the switch to radian frequencies was made for brevity and where

$$\omega_0 = \omega_c + \pi W. \tag{9.11-13}$$

At point (c) we have

$$\begin{aligned} w(t) &= \frac{A_c}{4} s(t)(1 + \cos 2\omega_c t) - \frac{A_c}{4} \hat{s}(t) \sin 2\omega_c t \\ &\quad + \frac{n_c(t)}{2} [\cos (\omega_0 + \omega_c)t + \cos (\omega_0 - \omega_c)t] \\ &\quad - \frac{n_s(t)}{2} [\sin (\omega_0 + \omega_c)t + \sin (\omega_0 - \omega_c)t]. \end{aligned} \tag{9.11-14}$$

The low-pass filter rejects all components at frequencies greater than W Hz. Hence at (d) we obtain

$$v(t) = \frac{A_c}{4} s(t) + \frac{n_c(t)}{2} \cos (\omega_0 - \omega_c)t - \frac{n_s(t)}{2} \sin (\omega_0 - \omega_c)t. \tag{9.11-15}$$

Since $\overline{s(t)n_c(t)} = \overline{s(t)n_s(t)} = \overline{n_c(t)n_s(t)} = 0$, the mean-square value of $v(t)$ is

$$\overline{v^2(t)} = \frac{A_c^2 \overline{s^2(t)}}{16} + \frac{\overline{n_c^2(t)}}{4}, \tag{9.11-16}$$

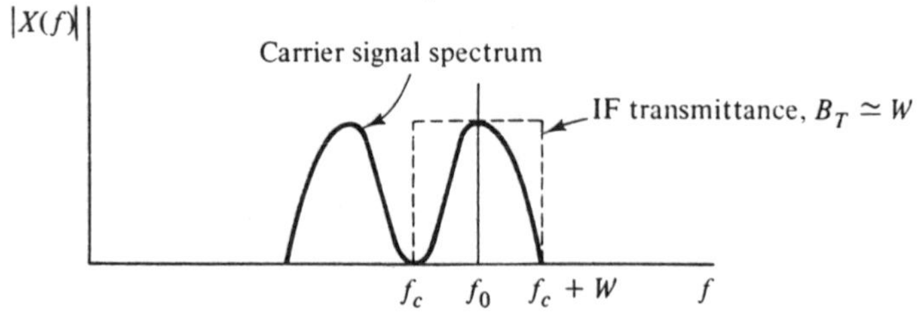

Figure 9.11-2. Transmittance of IF stage for SSB demodulation.

where we have used the fact that $\overline{n_c^2(t)} = \overline{n_s^2(t)}$. For SSB, $\overline{n_c^2(t)} = 2B_T N_0$, so that the output signal-to-noise ratio is

$$\left(\frac{\mathcal{S}}{\mathcal{N}}\right)_0 = \frac{A_c^2\overline{s^2(t)}/16}{2B_T N_0/4} \simeq \frac{P_s}{2WN_0} \qquad \text{(SSB)}, \tag{9.11-17}$$

where, in this case, $P_s = A_c^2\overline{s^2(t)}/4$ and $B_T \simeq W$.

The result for VSB is very similar to Eq. (9.11-17). That this is so can be seen by considering the equation of a VSB wave,

$$x(t) = \frac{A_c}{2}\,[s(t)\cos\omega_c t - q(t)\sin\omega_c t], \tag{9.11-18}$$

where $q(t)$ depends on the transmittance of the VSB filter [Eq. (6.9-7)]. The term proportional to $q(t)\sin\omega_c t$ is removed by the synchronous detector in much the same way as is the term proportional to $\hat{s}(t)\sin\omega_c t$ in the SSB case. If the width of the vestigial band is small compared to W, then the signal power and noise power admitted by the synchronous detector are about the same as in SSB. Hence, to a good approximation,

$$\left(\frac{\mathcal{S}}{\mathcal{N}}\right)_0 \simeq \frac{P_s}{2WN_0} \qquad \text{(VSB)}. \tag{9.11-19}$$

The input signal-to-noise ratio for SSB is computed at point (b) in Fig. 9.11-1. The mean-square value of $z(t)$ is

$$\overline{z^2(t)} = \frac{A_c^2}{4}\,[\overline{s^2(t)}] + 2WN_0, \tag{9.11-20}$$

so

$$\left(\frac{\mathcal{S}}{\mathcal{N}}\right)_i = \frac{P_s}{2WN_0} = \left(\frac{\mathcal{S}}{\mathcal{N}}\right)_0 \qquad \text{(SSB, VSB)}. \tag{9.11-21}$$

Hence, in SSB/VSB modulation, the output signal-to-noise ratio is identical to the input signal-to-noise ratio. Although less noise is admitted by the reduced bandwidths in SSB/VSB, the output signal-to-noise ratios are no greater than in DSBSC. The reason for this has, in effect, already been mentioned: The coherent addition of signal in DSBSC offsets the additional noise power admitted by the IF stage.

In deriving Eq. (9.11-20), use was made of the fact that $\overline{s(t)\hat{s}(t)} = 0$ and $\overline{s^2(t)} = \overline{\hat{s}^2(t)}$. To show the former is straightforward, we write

$$\begin{aligned}\overline{s(t)\hat{s}(t)} &= \frac{1}{\pi}\int_{-\infty}^{\infty}\frac{\overline{s(t)s(\lambda)}}{t-\lambda}\,d\lambda \\ &= \frac{1}{\pi}\int_{-\infty}^{\infty}\frac{R(t-\lambda)}{t-\lambda}\,d\lambda \\ &= \frac{1}{\pi}\int_{-\infty}^{\infty}\frac{R(u)}{u}\,du = 0 \qquad \text{(i.e., integrand is odd).}\end{aligned} \tag{9.11-22}$$

To show that $\overline{\hat{s}^2(t)} = \overline{s^2(t)}$, we need only take the expectation of the square of the integral in Eq. (9.11-11). Thus

$$\overline{\hat{s}^2(t)} = \frac{1}{\pi^2}\int_{-\infty}^{\infty}\int_{-\infty}^{\infty}\frac{\overline{s(\lambda)s(\xi)}}{(t-\lambda)(t-\xi)}\,d\xi\,d\lambda$$
$$= \int_{-\infty}^{\infty} d\alpha\, R_s(\alpha)\left[\frac{1}{\pi^2}\int_{-\infty}^{\infty}\frac{d\zeta}{\zeta(\zeta+\alpha)}\right], \tag{9.11-23}$$

where we let $\alpha \equiv \lambda - \xi$ and $\zeta = t - \lambda$. It is not difficult to show that the integral in brackets is $\delta(\alpha)$. Hence

$$\overline{\hat{s}^2(t)} = \int_{-\infty}^{\infty} R_s(\alpha)\,\delta(\alpha)\,d\alpha = R_s(0) \tag{9.11-24}$$
$$= \overline{s^2(t)}.$$

A somewhat less esoteric way of arriving at the same result is to recall that $\hat{s}(t)$ is formed from $s(t)$ by shifting all Fourier components in $s(t)$ by 90°. The mere shifting of the spectrum by a constant phase does not alter the net power in the waveform. Since the power depends on the square magnitude of the spectrum, and since only the phase is affected in constructing the Hilbert transform, Eq. (9.11-24) is established.

The synchronous detection of AM is left as an exercise (Prob. 9-24). AM, however, is usually detected with an envelope detector, which we shall consider next.

Detection of AM with an Envelope Detector

The AM envelope detector is shown in Fig. 9.11-3.

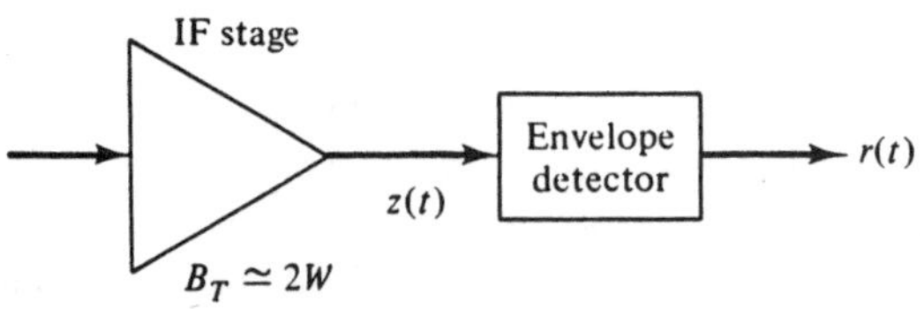

Figure 9.11-3. Envelope detector.

Signal Dominance

The simplest situation to deal with is when the input signal-to-noise ratio is high and the envelope detector is ideal. We shall assume this situation. After emerging from the IF stage, the signal is given by

$$z(t) = A_c[1 + ms(t)]\cos\omega_c t + n_c(t)\cos\omega_c t - n_s(t)\sin\omega_c t. \tag{9.11-25}$$

We can analyze the effect of the noise by considering a phasor representation for $z(t)$. This is furnished by

$$z(t) = \mathrm{Re}[Z(t)]$$
$$Z(t) = Y(t)e^{j\omega_c t} \tag{9.11-26}$$
$$Y(t) = A_c[1 + ms(t)] + n_c(t) + jn_s(t).$$

The phasor diagram is shown in Fig. 9.11-4. Only the complex amplitude $Y(t)$ is shown since $e^{j\omega_c t}$ conveys no information other than the fact that there is a superimposed rotation on $Y(t)$ of ω_c radians/s. From Fig. 9.11-4, it is easily seen that $z(t)$ can be written as

$$z(t) = r(t)\cos[\omega_c t + \phi(t)], \tag{9.11-27}$$

where

$$r(t) = (\{A_c[1 + ms(t)] + n_c(t)\}^2 + n_s^2(t))^{1/2} \tag{9.11-28}$$

and

$$\phi(t) = \tan^{-1}\frac{n_s(t)}{A_c[1 + ms(t)] + n_c(t)}. \tag{9.11-29}$$

For large signal-to-noise ratios (also called *signal dominance*), we have $\overline{A_c^2[1 + ms(t)]^2} \gg \overline{n^2(t)}$. Under this condition, we can write

$$r(t) = \{A_c[1 + ms(t)] + n_c(t)\} \cdot \left(1 + \frac{n_s^2(t)}{\{A_c[1 + ms(t)] + n_c(t)\}^2}\right)^{1/2} \tag{9.11-30}$$

$$\simeq A_c[1 + ms(t)] + n_c(t), \qquad \left(\frac{\mathcal{S}}{\mathcal{N}}\right)_i \gg 1. \tag{9.11-31}$$

The dc term in Eq. (9.11-31) does not convey information and is usually removed. Hence the detected signal is $A_c ms(t)$, the detected noise is the *in-phase* component of the Gaussian narrowband noise, and the output signal-to-noise ratio is

$$\left(\frac{\mathcal{S}}{\mathcal{N}}\right)_0 = \frac{A_c^2 m^2 \overline{s^2(t)}}{\overline{n_c^2(t)}} \simeq \frac{A_c^2 m^2 \overline{s^2(t)}}{4WN_0}, \tag{9.11-32}$$

where we assumed, as in the DSB case, that the IF bandwidth $B_T \simeq 2W$.

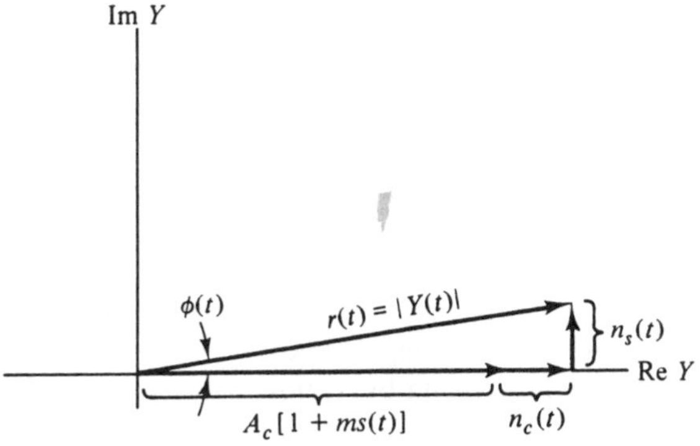

Figure 9.11-4. Phasor diagram for AM when $(\mathcal{S}/\mathcal{N})_i \gg 1$.

The input signal-to-noise ratio is computed before the envelope detector. The mean-square value of $z(t)$ is

$$\overline{z^2(t)} = \tfrac{1}{2}A_c^2[1 + m^2\overline{s^2(t)}] + 4WN_0. \tag{9.11-33}$$

The input signal-to-noise ratio is therefore

$$\begin{aligned}\left(\frac{\mathcal{S}}{\mathcal{N}}\right)_i &\simeq \frac{\tfrac{1}{2}A_c^2\,[1 + m^2\overline{s^2(t)}]}{4WN_0} \\ &= \frac{P_c + 2P_{SB}}{4WN_0},\end{aligned} \tag{9.11-34}$$

where P_c is the carrier power $= A_c^2/2$ and P_{SB} is the power in a sideband $= A_c^2m^2\overline{s^2(t)}/4$. From Eqs. (9.11-32) and (9.11-34), we can write

$$\left(\frac{\mathcal{S}}{\mathcal{N}}\right)_0 = \left(\frac{4P_{SB}}{P_c + 2P_{SB}}\right)\left(\frac{\mathcal{S}}{\mathcal{N}}\right)_i = \frac{2m^2\overline{s^2(t)}}{1 + m^2\overline{s^2(t)}}\left(\frac{\mathcal{S}}{\mathcal{N}}\right)_i. \tag{9.11-35}$$

Clearly, $(\mathcal{S}/\mathcal{N})_0$ cannot be larger than $(\mathcal{S}/\mathcal{N})_i$. The maximum sideband power for a full-load modulating tone [i.e., $ms(t) = \cos \omega_m t$] is $A_c^2/8$. In this case

$$\left(\frac{\mathcal{S}}{\mathcal{N}}\right)_0 = \frac{2}{3}\left(\frac{\mathcal{S}}{\mathcal{N}}\right)_i. \tag{9.11-36}$$

It is more common, however, that $(\mathcal{S}/\mathcal{N})_0 \ll (\mathcal{S}/\mathcal{N})_i$. Comparing these results with the results for DSB, we see that, because at least 50% of the radiated power goes into the carrier, more than twice as much AM power must be radiated to achieve the same $(\mathcal{S}/\mathcal{N})_0$ as in DSB. Hence AM is inferior in this respect to DSB.

Noise Dominance

When $(\mathcal{S}/\mathcal{N})_i \ll 1$, the envelope of the total wave is primarily determined by the envelope of the noise signal alone (Fig. 9.11-5). From the diagram, it can be seen that the envelope of the resultant is approximately given by

$$r(t) \simeq r_n(t) + A_c[1 + ms(t)] \cos \phi_n(t), \tag{9.11-37}$$

where

$$\phi_n(t) = \tan^{-1}\frac{n_s(t)}{n_c(t)}. \tag{9.11-38}$$

Interestingly, in this case we do not have a signal plus noise situation, and a conventional signal-to-noise calculation is meaningless. Since $\cos \phi_n(t)$ is a function of the noise angle $\phi_n(t)$ and is therefore a form of noise itself, the signal is corrupted by *multiplicative* noise and undergoes significant distortion. For example, when $\cos \phi_n(t) = 0$, the signal is not present at all. The conversion from masking by additive noise (which keeps the signal separate and intact) to corruption by multiplicative noise when $(\mathcal{S}/\mathcal{N})_i \ll 1$ is

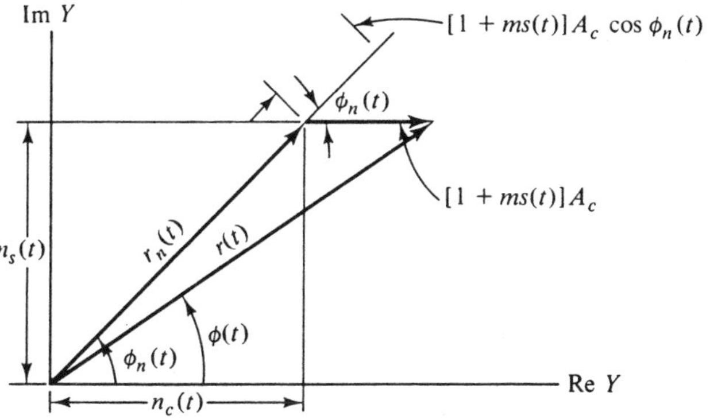

Figure 9.11-5. Phasor diagram for AM when $(\mathcal{S}/\mathcal{N})_i \ll 1$.

sometimes referred to as the *threshold effect* in AM envelope detection. It is possible to gain greater understanding of the threshold effect by considering what happens when an unmodulated carrier signal plus noise is detected by a square-law detector (Fig. 9.11-6). The noise is assumed to be NGP. The total input is written as

$$x(t) = [n_c(t) + A_c] \cos \omega_c t - n_s(t) \sin \omega_c t. \tag{9.11-39}$$

The output of the square-law detector is proportional to the squared envelope, which is

$$r^2(t) = [n_c(t) + A_c]^2 + n_s^2(t), \tag{9.11-40}$$

and the mean-square value of the squared envelope is

$$\overline{r^4(t)} = \overline{\{[n_c(t) + A_c]^2 + n_s^2(t)\}^2}. \tag{9.11-41}$$

How do we evaluate Eq. (9.11-41)? We already know from Sec. 9.8 that $\overline{n_c^2(t)} = \overline{n_s^2(t)} = \overline{n^2(t)} = \sigma^2$, that $n_c(t)$ and $n_s(t)$ are uncorrelated, and that $n_c(t)$ and $n_s(t)$ are Gaussian. But these facts imply that $n_c(t)$ and $n_s(t)$ are independent, and therefore so are $n_c^2(t)$ and $n_s^2(t)$ (Prob. 9-27). Hence $\overline{n_c^2(t)n_s^2(t)} = \overline{n_c^2(t)} \cdot \overline{n_s^2(t)} = \sigma^4$. Also, from Eq. (8.17-46) with $n = 4$, we obtain $\overline{n_c^4(t)} = \overline{n_s^4(t)} = 3\sigma^4$. Putting all these results together and because the odd moments are zero, we can write

$$\overline{r^4(t)} = 8\sigma^4 + 8\sigma^2 A_c^2 + A_c^4. \tag{9.11-42}$$

A reasonable definition of signal power for the square-law detector is the value of $\overline{r^4(t)}$ when noise is absent. This value is A_c^4. The noise power is then identified from Eq. (9.11-42) as

$$\overline{r^4(t)} - \overline{r_s^4(t)} = 8\sigma^4 + 8\sigma^2 A_c^2, \tag{9.11-43}$$

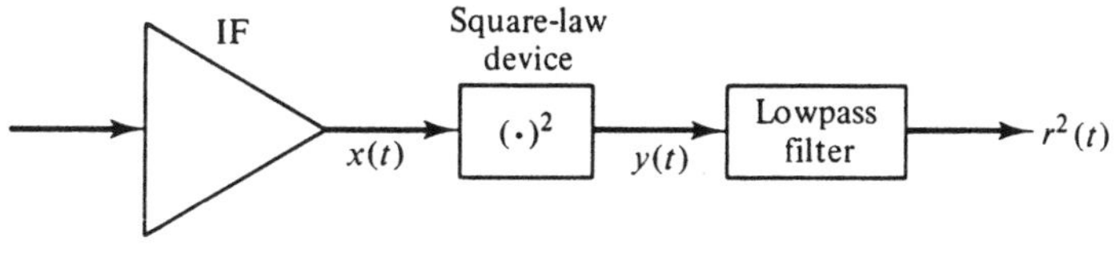

Figure 9.11-6. Block diagram of a square-law detector.

where $\overline{r_s^4(t)} = A_c^4$. The noise power thus consists of two terms: $8\sigma^4$ is the result of noise beating with noise, while $8\sigma^2 A_c^2$ is the result of noise beating with signal. Note that in envelope detection the presence of signal causes an increase in the noise. With the preceding definitions of signal and noise powers, we obtain

$$\left(\frac{\mathcal{S}}{\mathcal{N}}\right)_0 = \frac{A_c^4}{8\sigma^4 + 8A_c^2\sigma^2}.$$

However,

$$\frac{A_c^2}{2\sigma^2} = \frac{A_c^2/2}{4WN_0} = \left(\frac{\mathcal{S}}{\mathcal{N}}\right)_i,$$

from which it follows that

$$\left(\frac{\mathcal{S}}{\mathcal{N}}\right)_0 = \frac{1}{2} \cdot \frac{(\mathcal{S}/\mathcal{N})_i^2}{1 + 2(\mathcal{S}/\mathcal{N})_i}. \tag{9.11-44}$$

Equation (9.11-44) sheds a good deal of light on the AM threshold effect. For $(\mathcal{S}/\mathcal{N})_i \ll 1$, we have

$$\left(\frac{\mathcal{S}}{\mathcal{N}}\right)_0 \simeq \frac{1}{2}\left(\frac{\mathcal{S}}{\mathcal{N}}\right)_i^2, \tag{9.11-45}$$

that is, a quadratic dependence on the input signal-to-noise ratio. For $(\mathcal{S}/\mathcal{N})_i \gg 1$, we have

$$\left(\frac{\mathcal{S}}{\mathcal{N}}\right)_0 \simeq \frac{1}{4}\left(\frac{\mathcal{S}}{\mathcal{N}}\right)_i, \tag{9.11-46}$$

that is, a linear dependence. The factor of $\frac{1}{4}$ results from considering a square-law device. A factor of $\frac{1}{2}$ would result from a piecewise linear detector. The threshold effect can thus be interpreted as a transition in $(\mathcal{S}/\mathcal{N})_0$ from a linear to a quadratic dependence $(\mathcal{S}/\mathcal{N})_i$. In other words, things get suddenly much worse when the input signal-to-noise ratio falls off below the threshold. This phenomenon is discussed at greater length in [9-2, p. 265], and [9-12, p. 103]. Carlson ([9-13], p. 274) shows that a reasonable value of the threshold input signal-to-noise ratio is around 10. For other definitions of output signal-to-noise ratios including a discussion of the linear amplitude detector, see [9-14, p. 228].

The quadratic dependence of $(\mathcal{S}/\mathcal{N})_0$ on $(\mathcal{S}/\mathcal{N})_i$ when the input signal-to-noise ratio is small is known as the *small-signal suppression* of envelope detectors. Ordinary AM radios will sound very bad when $(\mathcal{S}/\mathcal{N})_i$ is anywhere near threshold; hence the large $(\mathcal{S}/\mathcal{N})_i$ situation is most common during normal listening.

Performance of AM and Derived Systems

We shall end this section by comparing and pointing out some of the salient features of the systems that have been discussed. In DSBSC, we found that synchronous detection furnished a factor of 2 (3 dB) in signal-to-noise improvement over the input signal-to-

TABLE 9.11-1 COMPARISON OF AM-TYPE SYSTEMS

System	P_s	$r_i \equiv (\mathcal{S}/\mathcal{N})_i$	$r_0 \equiv (\mathcal{S}/\mathcal{N})_0$	r_0/r_i
DSBSC (synchronous det.)	$A_c^2\overline{s^2}/2$	$P_s/4N_0W$	$P_s/2N_0W$	2
SSB, VSB (synchronous det.)	$A_c^2\overline{s^2}/4$	$P_s/2N_0W$	$P_s/2N_0W$	1
AM (envelope det.)	$A_c^2(1 + m^2\overline{s^2})/2$	$P_s/4N_0W$	$m^2\dfrac{\overline{s^2}A_c^2}{4N_0W}$	0.67 (max)

noise ratio, while no such improvement was manifest in SSB or VSB. Other things being equal, then, DSBSC would seem to be slightly superior to SSB, assuming the power in the SSB signal is not raised. However, other things are not always equal, and SSB furnishes a significant bandwidth saving over DSB, as well as greater ease of detection (see Secs. 6.7 and 6.8).

In ordinary AM there is actually a reduction in the output signal-to-noise ratio. In fact, the latter, even under conditions of a full-load modulating signal, is only two-thirds as large as the input signal-to-noise ratio. More commonly, the modulation index satisfies $m \ll 1$, and the signal-to-noise performance of AM is much worse than that of either SSB or DSBSC, especially when a peak power constraint is applied. AM also suffers from a threshold effect, which, although interesting, is not important under ordinary listening conditions. On the other hand, AM is easily detected with an envelope detector, a fact that accounts in great part for its leading role in commercial broadcasting. Table 9.11-1 summarizes some of the results of this section.

9.12 NOISE IN ANGLE-MODULATED SYSTEMS

A general treatment of the effect of noise on angle-modulation systems is an exceedingly tedious affair, especially when input signal-to-noise ratios are not too large or too small. However, at the two extremes (i.e., very large or very small signal-to-noise ratios), a number of qualitative and quantitative results can be stated. We shall begin our analysis by recalling that an angle-modulated wave can be written as

$$x(t) = A_c \cos [\omega_c t + \phi(t)] \qquad \text{(FM or PM)}, \tag{9.12-1}$$

where

$$\phi(t) = K's(t) \tag{9.12-2}$$

in the case of PM and

$$\phi(t) = 2\pi K \int^t s(\lambda)d\lambda \tag{9.12-3}$$

in the case of FM.

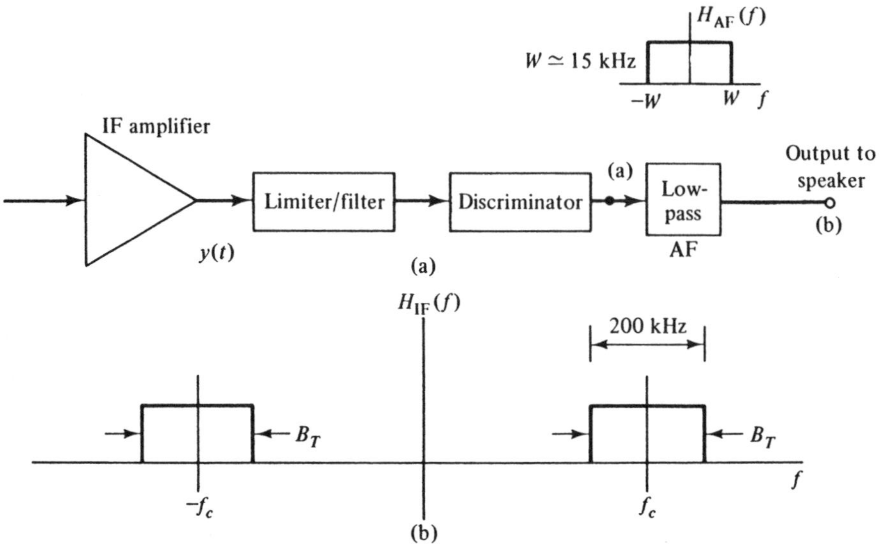

Figure 9.12-1. (a) Demodulator for FM. (b) Assumed rectangular characteristic of IF filter.

The important circuits in an FM demodulator are shown in Fig. 9.12-1. The signal $y(t)$ is angle-modulated carrier plus noise; that is,

$$y(t) = A_c \cos [\omega_c t + \phi(t)] + n(t) \tag{9.12-4}$$

The noise $n(t)$ is still an NGP since $B_T/f_c \ll 1$. The input signal-to-noise ratio is therefore

$$\left(\frac{\mathcal{S}}{\mathcal{N}}\right)_i = \frac{A_c^2/2}{2B_T N_0}, \tag{9.12-5}$$

which is independent of the signal $s(t)$. The computation of the output signal-to-noise ratio is more difficult. It is considered next.

Because $n(t)$ is narrowband, it is useful to write

$$n(t) = r_n(t) \cos [\omega_c t + \phi_n(t)], \tag{9.12-6}$$

where, as usual, $r_n(t)$ is Rayleigh distributed and $\phi_n(t)$ is uniformly distributed in $[-\pi, \pi]$. For $y(t)$ we write

$$y(t) = r(t) \cos [\omega_c t + \psi(t)], \tag{9.12-7}$$

where

$$r(t) = \{[A_c \cos \phi(t) + r_n(t) \cos \phi_n(t)]^2 + [A_c \sin \phi(t) + r_n(t) \sin \phi_n(t)]^2\}^{1/2} \tag{9.12-8}$$

and

$$\psi(t) = \tan^{-1} \frac{A_c \sin \phi(t) + r_n(t) \sin \phi_n(t)}{A_c \cos \phi(t) + r_n(t) \cos \phi_n(t)}. \tag{9.12-9}$$

As a check, we see that if there is no noise [i.e., $r_n(t) = 0$] then $\psi(t) = \phi(t)$. An ideal limiter would remove the envelope fluctuations in the input so that $r(t)$ is of no consequence. Hence, in angle modulation, signal-to-noise ratios are derived from consideration of $\psi(t)$ only. Unfortunately, the expression for $\psi(t)$ is too unwieldy for analysis. For this reason, we shall do what we did in the envelope detector analysis—consider the two separate cases of signal dominance and noise dominance individually.

Signal Dominance, $A_c^2/2 \gg \overline{r_n^2}$

A phasor diagram (Fig. 9.12-2) for this situation is obtained from

$$y(t) = \text{Re}[Y(t)e^{j\omega_c t}], \tag{9.12-10}$$

where $Y(t)$ is given by

$$Y(t) = A_c e^{j\phi(t)} + r_n(t)e^{j\phi_n(t)}.$$

From Fig. 9.12-2, it is clear that the length, L, of arc AB is

$$L = Y(t)[\psi(t) - \phi(t)]. \tag{9.12-11}$$

But $Y(t) \simeq A_c + r_n(t)\cos[\phi_n(t) - \phi(t)] \simeq A_c$, and $L \simeq r_n(t)\sin[\phi_n(t) - \phi(t)]$. Hence, from Eq. (9.12-11), we obtain

$$\psi(t) = \phi(t) + \frac{r_n(t)}{A_c} \sin [\phi_n(t) - \phi(t)]. \tag{9.12-12}$$

Since, for fixed t, $\phi_n(t)$ is an r.v. uniformly distributed in $[-\pi, \pi]$, we argue that $\phi_n(t) - \phi(t)$ is also uniformly distributed over $[-\phi(t) - \pi, -\phi(t) + \pi]$. For the

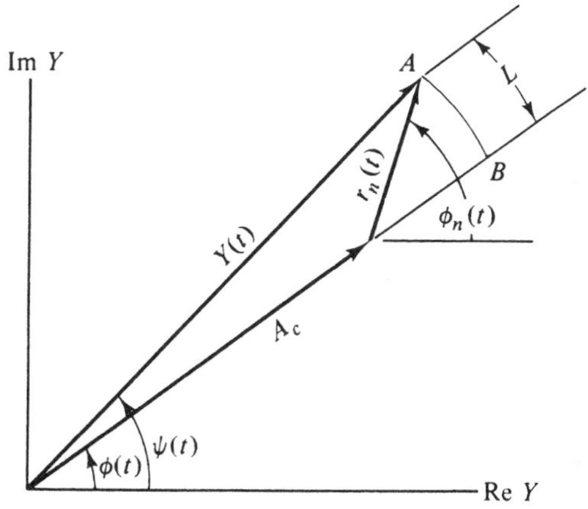

Figure 9.12-2. Phasor diagram for angle-modulated wave plus noise. The length of the vectors reflects the signal-dominance case.

purpose of computing an output signal-to-noise ratio, replacing $\phi_n(t) - \phi(t)$ by $\phi_n(t)$ does not affect the computation. Therefore, we permit ourselves to write

$$\begin{aligned}\psi(t) &= \phi(t) + \frac{r_n(t)}{A_c}\sin\phi_n(t)\\ &= \phi(t) + \frac{n_s(t)}{A_c},\end{aligned} \tag{9.12-13}$$

where $n_s(t) \equiv r_n(t)\sin\phi_n(t)$ is the quadrature component of Gaussian narrowband noise. The detected signal is

$$v(t) = \psi(t) = K's(t) + \frac{n_s(t)}{A_c} \qquad \text{(PM)}, \tag{9.12-14a}$$

and

$$v(t) = \frac{1}{2\pi}\dot{\psi}(t) = Ks(t) + \frac{1}{2\pi A_c}\dot{n}_s(t). \qquad \text{(FM)}. \tag{9.12-14b}$$

Power Spectrum of $n_s(t)$ and $\dot{n}_s(t)$.

To proceed further it is helpful to consider the power spectrum of $n_s(t)$. Starting with Eqs. (9.8-3) and identifying $n_c(t)$ and $n_s(t)$ with $x_c(t)$ and $x_s(t)$, respectively, it is a simple matter to show that the autocorrelation of $n_s(t)$ is (Prob. 9-26)

$$\begin{aligned}R_{n_s}(\tau) &= 2\int_0^\infty W_n(\xi)\cos 2\pi(\xi - f_c)\tau\, d\xi\\ &= R_{n_c}(\tau),\end{aligned} \tag{9.12-15}$$

where $W_n(\xi)$ is the power spectrum of $n(t)$ and $R_{n_c}(\tau)$ is the autocorrelation function of $n_c(t)$. If we multiply both sides of Eq. (9.12-15) by $e^{-j2\pi f\tau}\,d\tau$ and integrate, we obtain [$u(f)$ is the unit-step function]

$$\begin{aligned}W_{n_s}(f) &= W_n(f+f_c)u(f+f_c) + W_n(f_c - f)u(f_c - f)\\ &= W_{n_c}(f).\end{aligned} \tag{9.12-16}$$

Equation (9.12-16) can be used to confirm that $n_s(t)$ is a low-pass process. Thus, representing the power spectrum of $n(t)$ by

$$W_n(f) = N_0\,\text{rect}\left(\frac{f-f_c}{B_T}\right) + N_0\,\text{rect}\left(\frac{f+f_c}{B_T}\right), \tag{9.12-17}$$

we obtain, by direct substitution into Eq. (9.12-16),

$$W_{n_s}(f) = 2N_0\,\text{rect}\left(\frac{f}{B_T}\right), \tag{9.12-18}$$

so

$$\overline{n_s^2(t)} = 2N_0B_T,$$

as expected.

The power spectrum of $\dot{n}_s(t)$ can be determined from an elementary property of Fourier transforms. Recall that, if an arbitrary function $x(t)$ has Fourier transform $X(f)$, then at points where $x(t)$ is continuous,

$$\dot{x}(t) \longleftrightarrow +j2\pi f X(f).$$

Hence $\dot{x}(t)$ can be obtained by passing $x(t)$ through a network with transfer function $H(f) = +j2\pi f$. From Table 9.6-1 we obtain, with $y \equiv \dot{x}$ and $H(f) = j2\pi f$,

$$\begin{aligned} W_{\dot{x}}(f) &= |H(f)|^2 W_x(f) \\ &= (2\pi f)^2 W_x(f). \end{aligned} \tag{9.12-19}$$

Equation (9.12-19) enables us to write

$$W_{\dot{n}_s}(f) = 4\pi^2 f^2 W_{n_s}(f),$$

which, in the case of a rectangular band-pass characteristic, gives

$$W_{\dot{n}_s} = 8\pi^2 f^2 N_0 \operatorname{rect}\left(\frac{f}{B_T}\right). \tag{9.12-20}$$

Hence $\overline{\dot{n}_s^2(t)} = \frac{2}{3}\pi^2 N_0 B_T^3$.

Output Signal-to-Noise Ratio $(\mathcal{S}/\mathcal{N})_0$

With $v_1(t) \equiv n_s(t)/A_c$ and $v_2(t) \equiv \dot{n}_s(t)/2\pi A_c$, we can write Eqs. (9.12-14a) and (9.12-14b) as

$$v(t) = K's(t) + v_1(t), \qquad \text{(PM)}, \tag{9.12-21a}$$

$$v(t) = Ks(t) + v_2(t), \qquad \text{(FM)}, \tag{9.12-21b}$$

where the power spectra of $v_1(t)$ and $v_2(t)$ are, respectively,

$$W_1(f) = 2\frac{N_0}{A_c^2}\operatorname{rect}\left(\frac{f}{B_T}\right) \qquad \text{(PM)} \tag{9.12-22}$$

and

$$W_2(f) = 2\left(\frac{f}{A_c}\right)^2 N_0 \operatorname{rect}\left(\frac{f}{B_T}\right) \qquad \text{(FM)}. \tag{9.12-23}$$

These are the spectra that would be observed at point (a) in Fig. 9.12-1. The spectra are shown in Fig. 9.12-3.

Since $B_T/2$ is generally much greater than the highest frequency, W, in $s(t)$, additional filtering helps to increase $(\mathcal{S}/\mathcal{N})_0$. Thus the final output of the demodulator is

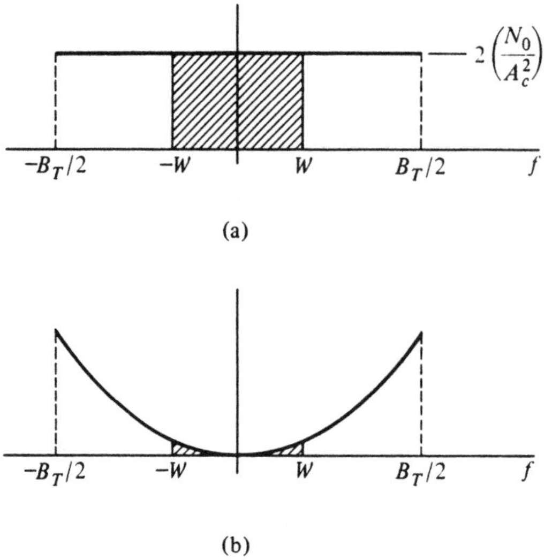

Figure 9.12-3. Detected noise spectra at output of discriminator: (a) PM noise spectrum; (b) FM noise spectrum. W is the bandwidth of the final audio low-pass amplifier.

obtained at point (b) in Fig. 9.12-1(a) after passing the signal through a low-pass filter whose bandwidth is essentially W. Hence the noise power at the speaker input is

$$\mathcal{N}_0 = \int_{-W}^{W} W_1(f)\,df = \frac{4N_0W}{A_c^2} \qquad \text{(PM)}, \tag{9.12-24}$$

$$\mathcal{N}_0 = \int_{-W}^{W} W_2(f)df = \frac{4N_0W^3}{3A_c^2} \qquad \text{(FM)}. \tag{9.12-25}$$

With $P_c \equiv A_c^2/2$ representing the power in the input carrier signal, the final signal-to-noise ratios are

$$\left(\frac{\mathcal{S}}{\mathcal{N}}\right)_0 = P_c\frac{K'^2\overline{s^2(t)}}{2N_0W} \qquad \text{(PM)}, \tag{9.12-26}$$

$$\left(\frac{\mathcal{S}}{\mathcal{N}}\right)_0 = 3P_c\left(\frac{K}{W}\right)^2\frac{\overline{s^2(t)}}{2N_0W} \qquad \text{(FM)}. \tag{9.12-27}$$

We already know that because $|s(t)| \leq 1$, K represents the maximum frequency deviation, in hertz, of the FM signal. The ratio K/W is the parameter β^* of Eq. (7.4-10), i.e., the modulation index that is associated with maximum frequency deviation and highest signal frequency, W. Hence Eq. (9.12-27) can be written as

$$\left(\frac{\mathcal{S}}{\mathcal{N}}\right)_0 = 3P_c\beta^{*2}\frac{\overline{s^2(t)}}{2N_0W}. \tag{9.12-28}$$

The transmission bandwidth, B_T, is given by

$$B_T = 2W(2 + \beta^*),$$

using a conservative measure (Sec. 7.4). For wideband FM, $\beta^* \gg 1$ and $B_T \simeq 2W\beta^*$. Hence Eq. (9.12-27) can also be written in the following revealing form:

$$\left(\frac{\mathcal{S}}{\mathcal{N}}\right)_0 \simeq \frac{3}{4}\left(\frac{B_T}{W}\right)^2 \frac{P_c\overline{s^2(t)}}{2N_0W} \qquad \text{(WBFM)}. \tag{9.12-29}$$

Equation (9.12-29) states that, provided $(\mathcal{S}/\mathcal{N})_i \gg 1$, the output signal-to-noise ratio is a quadratic function of the *bandwidth expansion ratio* B_T/W. Increasing B_T furnishes a corresponding *quadratic* increase in $(\mathcal{S}/\mathcal{N})_0$. This is why wideband FM is so superior to narrowband FM. In commercial broadcast FM, $\beta^* \simeq 5$, so the results for WBFM are basically applicable to commercial broadcasting.

Last, it is instructive to compare the performance of FM with AM schemes. In particular, we found that in the case of DSB

$$\left(\frac{\mathcal{S}}{\mathcal{N}}\right)_{0,\text{DSB}} = \frac{(A_c^2/2)\overline{s^2(t)}}{2N_0W} = P_c\frac{\overline{s^2(t)}}{2N_0W}.$$

Accordingly, from Eq. (9.12-28),

$$\left(\frac{\mathcal{S}}{\mathcal{N}}\right)_{0,\text{FM}} = 3\beta^{*2}\left(\frac{\mathcal{S}}{\mathcal{N}}\right)_{0,\text{DSB}}. \tag{9.12-30}$$

Hence WBFM can furnish a very significant increase in performance over AM systems under the constraint of equal power. The penalty is, of course, that FM requires much greater bandwidth.

Further Signal-to-Noise Improvement in FM by Preemphasis/Deemphasis (PDE) Filtering

In Sec. 7.11 we discussed the use of preemphasis (PE) and deemphasis (DE) filtering to reduce the effects of high-frequency interference and noise. We are now in a position to be somewhat more quantitative about the effects of PDE on noise in FM.

The preemphasis filtering distorts the signal somewhat but does not significantly change the signal bandwidth from W. The deemphasis filter restores the signal by high-frequency attenuation. The reader will recall that the transfer function of the deemphasis filter is proportional to

$$H_{\text{DE}}(f) = \frac{1}{1 + jf/f_1},$$

where $f_1 = (2\pi R_1C)^{-1}$ and R_1C is the time constant of the filter shown in Fig. 7.11-2. In the presence of a DE filter, Eq. (9.12-25) must be modified to

$$\begin{aligned}[\mathcal{N}_0]_{\text{DE}} &= \int_{-W}^{W} W_2(f)|H_{\text{DE}}(f)|^2\,df \\ &= \frac{N_0f_1^3}{P_c}\int_{-W/f_1}^{W/f_1}\frac{f^2}{1+f^2}\,df \\ &= \frac{2N_0f_1^3}{P_c}\left(\frac{W}{f_1} - \tan^{-1}\frac{W}{f_1}\right).\end{aligned} \tag{9.12-31}$$

Standard values for f_1 and W are 2.1 Hz and 15 kHz, respectively. Therefore, $W/f_1 \simeq 7.2$ and $\tan^{-1} W/f_1 \simeq \pi/2$. Hence, to a first approximation, we can assume that $W/f_1 \gg \tan^{-1} W/f_1$. The output noise power with deemphasis filtering is then given by

$$(\mathcal{N}_0)_{\text{DE}} = \frac{2N_0 f_1^2 W}{P_c}. \tag{9.12-32}$$

The ratio of signal-to-noise power with PDE to signal-to-noise power without PDE is

$$\frac{(\mathcal{S}/\mathcal{N})_{0,\text{PDE}}}{(\mathcal{S}/\mathcal{N})_0} = \frac{1}{3}\left(\frac{W}{f_1}\right)^2. \tag{9.12-33}$$

For the numbers given, this amounts to $\simeq$ 12 dB, an impressive gain in signal-to-noise ratio by any standard.

Noise Dominance, $A_c^2/2 \ll \overline{r_n^2}$

Returning now to Eq. (9.12-10), we find that when $A_c \ll [2\overline{r_n^2}(t)]^{1/2}$ the resulting phasor diagram is essentially dominated by the term $r_n(t)e^{j\phi_n(t)}$ (Fig. 9.12-4). Following the same line of reasoning that led up to Eq. (9.12-12), we obtain, for this case,

$$\psi(t) = \phi_n(t) + \frac{A_c}{r_n(t)} \sin[\phi(t) - \phi_n(t)]. \tag{9.12-34}$$

The dominant term is $\phi_n(t)$, the uniformly distributed phase of the narrowband noise. The second term is a highly nonlinear function of signal and noise, and since the signal does not stand alone, no meaningful signal-to-noise ratio can be computed. This case corresponds then to signal *obliteration*. The transition from signal obliteration for low $(\mathcal{S}/\mathcal{N})_i \ll 1$ to a linear dependence of $(\mathcal{S}/\mathcal{N})_0$ on P_c [Eq. (9.12-29)] when $(\mathcal{S}/\mathcal{N})_i \gg 1$ is referred to as the FM threshold effect. It is easy to gain the impression from Eq. (9.12-29) that indefinite improvement in $(\mathcal{S}/\mathcal{N})_0$ is possible by simply increasing B_T

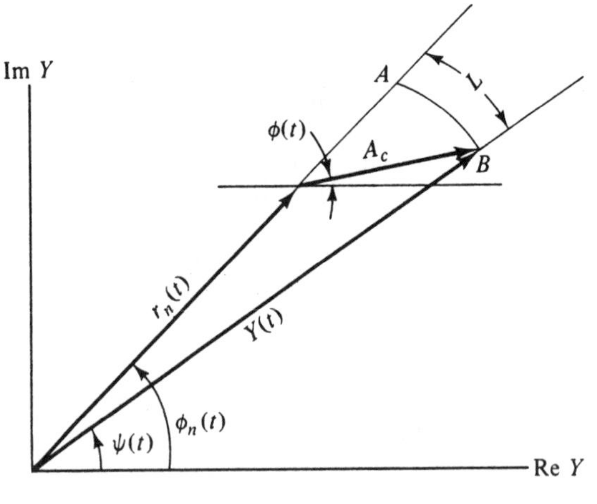

Figure 9.12-4. Phasor diagram for angle-modulated wave plus noise when noise dominates.

[and hence K since $B_T \simeq 2(K + 2W)$] without bound. However, increasing B_T also increases $\mathcal{N}_i = 2B_T N_0$, and eventually $(\mathcal{S}/\mathcal{N})_i$ will fall below the threshold, in which case signal obliteration occurs.

The precise threshold signal-to-noise value depends on the criterion that is used. For example, suppose that the threshold criterion is determined by that level of carrier amplitude that exceeds the noise amplitude 99% of the time. Then

$$P[r_n(t) \leq A_c] = 1 - P[r_n(t) > A_c] = 0.99$$

or

$$P[r_n(t) > A_c] = 0.01 = e^{-(\mathcal{S}/\mathcal{N})_i}. \tag{9.12-35}$$

Equation (9.12-35) follows from the fact that $r_n(t)$ has a Rayleigh distribution, that

$$P[r_n(t) > A_c] = 1 - (1 - e^{-A_c^2/2\sigma^2}),$$

and that

$$\frac{A_c^2}{2\sigma^2} = \frac{A_c^2/2}{2B_T N_0} = \left(\frac{\mathcal{S}}{\mathcal{N}}\right)_i. \tag{9.12-36}$$

The solution to Eq. (9.12-35) gives a 5-dB threshold level for the input carrier-to-noise power. Experimental results show that the threshold level is in the vicinity of 12–15 dB (Fig. 9.12-5).

Techniques for lowering the threshold and thereby increasing the performance of FM receivers have been proposed. These techniques, called FM threshold extensions (FMTE), use frequency feedback or phase-locked loop detectors to extend the threshold. A thorough discussion of FMTE techniques is furnished in [9-14, p. 478].

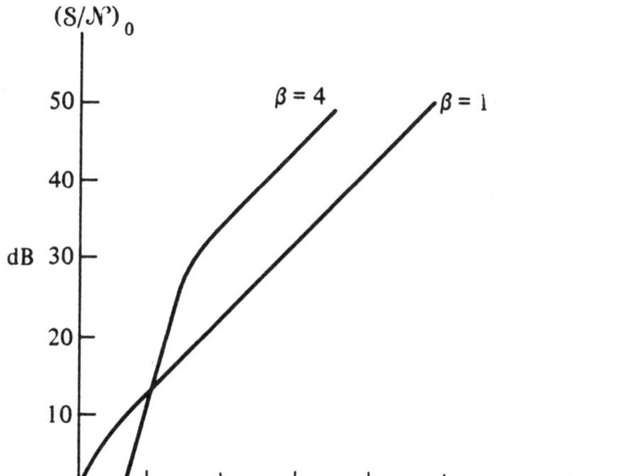

Figure 9.12-5. Output signal-to-noise ratio as a function of input carrier-to-noise ratio. *Adapted from M. G. Crosby [9-15], with permission.*

9.13 Summary

In this chapter we have studied random signals and considered the response of linear, time-invariant (LTI) systems to such signals. We introduced two extremely important functions, the correlation function and the power spectrum, and showed that they form a Fourier pair. We derived the input–output relations for correlation functions and power spectra for LTI systems.

The Gaussian process and the narrowband Gaussian process (NGP) were carefully studied. The latter figures prominently in the signal-to-noise analysis of communication systems. The signal-to-noise ratios for various types of AM and angle-modulated systems were considered. The quadratic noise characteristic at the output of FM discriminators was derived.

We ended the chapter by showing how FM preemphasis/deemphasis filtering, introduced in Chapter 7, can be used to improve the signal-to-noise ratio. Finally, the FM threshold effect was considered. FM systems are discussed in a number of books; among them are the very readable books by Schwartz [9-9] and Carlson [9-13]. Panter's book [9-14] has a very complete discussion of FM interference and other important problems.

PROBLEMS

9-1. In the experiment of throwing a fair die, define a random process $x(t)$ by

$$x(t) = kt^2,$$

where k is the outcome of a throw, $k = 1, \ldots, 6$. Compute the probability distribution function for $x(1)$ and $x(2)$. Is $x(t)$ a stationary process?

9-2. With reference to the die-throwing experiment, define a process by

$$x(t) = \cos\left(\frac{2\pi k}{6}\right) t,$$

where $k = 1, \ldots, 6$. Compute the probability distribution function for $x(1)$ and $x(2)$. Is $x(t)$ a stationary process?

9-3. Let $x(t)$ denote the "telegraph" signal of Sec. 9.4. Let α be a random variable that is independent of $x(t)$. Compute the autocorrelation function of $y(t) = \alpha x(t)$.

9-4. Let $x(t)$ be a random process given by

$$x(t) = A \cos(\omega t + \Theta),$$

where A is a constant and Θ is a uniformly distributed random variable in $(0, \pi/2)$. Compute $\overline{x(t)}$ and $R_{xx}(t, t + \tau) = \overline{x(t)x(t + \tau)}$. Is $x(t)$ a wss process?

9-5. A random process has sample functions $x(t) = X$, where X is a random variable not indexed by time. Let X be defined by

$$X = \begin{cases} \alpha, & \text{with probability } P(\alpha) = p, \\ \beta, & \text{with probability } P(\beta) = 1 - p. \end{cases}$$

(a) With $\hat{\mu}$ and μ denoting the time and statistical averages of $x(t)$, respectively, show that $\hat{\mu} \neq \mu$ and that therefore $x(t)$ is not ergodic. Are the time averages for different sample functions of $x(t)$ the same?

(b) Compute the correlation function of $x(t)$.

9-6. A warehouse contains a very large number of oscillators that fall into one of two classes: those whose mean frequency is 1.0 kHz and those whose mean frequency is 1.0 MHz. Let $x(t)$ denote the output of an oscillator. Is $x(t)$ a stationary process? (Assume oscillators have been on for a very long, i.e., infinite time.) Is $x(t)$ an ergodic process?

9-7. Let $x(t)$ be a (real) stationary random process with correlation function $R_{xx}(\tau)$. Assuming that the various derivatives exist, show that

$$R_{x\dot{x}}(\tau) \equiv \overline{x(t)\dot{x}(t + \tau)}$$

is given by

$$R_{x\dot{x}}(\tau) = \frac{dR_{xx}(\tau)}{d\tau}.$$

9-8. Show that the correlation function of the process $\dot{x}(t)$, obtained by differentiating $x(t)$ in Prob. 9-7, is given by

$$\begin{aligned} R_{\dot{x}\dot{x}}(\tau) &\equiv \overline{\dot{x}(t)\dot{x}(t + \tau)} \\ &= -\frac{d^2R_{xx}(\tau)}{d\tau^2}. \end{aligned}$$

9-9. Generalize the result in Prob. 9-8 to the process

$$x^{(n)}(t) = \frac{d^n x(t)}{dt^n}.$$

9-10. Generalize the result in Prob. 9-8 to include nonstationary processes by writing

$$R_{\dot{x}\dot{x}}(t_1, t_2) = \overline{\dot{x}(t_1)\dot{x}(t_2)}$$

and showing that

$$R_{\dot{x}\dot{x}}(t_1, t_2) = \frac{\partial^2 R_{xx}(t_1, t_2)}{\partial t_1 \partial t_2}.$$

9-11. The truncated time average of a process $x(t)$ is

$$\hat{\mu}_T \equiv \frac{1}{2T}\int_{-T}^{T} x(t)\, dt.$$

(a) Show that the variance of the r.v. $\hat{\mu}_T$ is given by

$$\sigma^2_{\hat{\mu}_T} = \frac{1}{T}\int_0^{2T}\left(1 - \frac{\tau}{2T}\right)[R(\tau) - \mu^2]d\tau,$$

where μ is the expected value of $x(t)$. If $\sigma^2_{\hat{\mu}_T} \to 0$ at $T \to \infty$, we say that $x(t)$ *is ergodic in the mean.*

(b) Let $x(t) = \cos(\omega t + \Theta)$, where Θ is uniformly distributed in $[0, \pi]$. Is $x(t)$ ergodic?

9-12. Show that proving the ergodicity of the "correlation function"

$$\hat{R}_T(\tau) = \frac{1}{2T}\int_{-T}^{T} x(t)x(t+\tau)\,dt$$

requires knowledge of the fourth-order moments of $x(t)$.

9-13. Use the Wiener–Khintchine theorem to compute the power spectra of the random telegraph process of Sec. 9.4.

9-14. Consider a random process with autocorrelation function $R(\tau) = \exp(-a|\tau|)$ with $a > 0$.
(a) Describe the asymptotic behavior of $W(f)$.
(b) Repeat part (a) when $R(\tau)$ is given by

$$R(\tau) = \frac{2a}{a^2 + (2\pi\tau)^2}.$$

***9-15.** A random process $x(t)$ is observed over $[-T, T]$. By directly computing the expected value of

$$\mathcal{W}_T(f) = \frac{1}{2T}\left|\int_{-T}^{T} x(t)e^{-j2\pi f t}\,dt\right|^2,$$

show that

$$W_T(f) \equiv \overline{\mathcal{W}_T(f)}$$
$$= 2T\int_{-\infty}^{\infty} W(\zeta)\,\text{sinc}^2\, 2T(f-\zeta)d\zeta.$$

Hence the series truncation due to the finite observation time produces an inherent "smoothing."

9-16. Explain why the sample correlation function is given by

$$\mathcal{R}_T(\tau) = \frac{1}{2T-|\tau|}\int_{-T+|\tau|/2}^{T-|\tau|/2} x\left(t-\frac{|\tau|}{2}\right)x\left(t+\frac{|\tau|}{2}\right)dt, \qquad \tau < 2T,$$

instead of

$$\mathcal{R}_T(\tau) \overset{(?)}{=} \frac{1}{2T}\int_{-T}^{T} x\left(t-\frac{|\tau|}{2}\right)x\left(t+\frac{|\tau|}{2}\right)dt, \qquad \tau < 2T,$$

when $x(t)$ is observed over $[-T, T]$.

***9-17.** Let $\phi(t)$ be a zero-mean Gaussian random process with correlation function $R_\phi(\tau)$. Compute the autocorrelation function of $x(t) = e^{j\phi(t)}$. The process $e^{j\phi(t)}$ occurs in angle modulation and coherent optics.

9-18. Let X_c and X_s be two independent zero-mean Gaussian random variables with equal variance σ^2. Show that the autocorrelation function of

$$x(t) = X_c \cos\omega_c t - X_s \sin\omega_c t$$

is given by

$$R_x(\tau) = \sigma^2 \cos\omega_c\tau.$$

***9-19.** (Sec. 9.8, the NGP) Justify that $x_c(t)$ and $x_s(t)$ are given, to a very good approximation, by Eqs. (9.8-3a) and (9.8-3b), respectively.

***9-20.** Use Eqs. (9.8-3a) and (9.8-3b) to show that $\overline{x_c^2(t)} = \overline{x_s^2(t)} = \overline{x^2(t)}$ and that $\overline{x_s(t)x_c(t)} = 0$. *Hint*: Several of the resulting integrals can be simplified by using the appropriate Fourier transform pairs.

9-21. Starting out with Eq. (9.8-2), which describes the NGP, that is,

$$x(t) = x_c(t) \cos \omega_c t - x_s(t) \sin \omega_c t,$$

show that

$$R_x(\tau) = R_c(\tau) \cos \omega_c \tau - R_{cs}(\tau) \sin \omega_c \tau,$$

where $R_c(\tau) = \overline{x_c(t)x_c(t+\tau)}$ and $R_{cs}(\tau) = \overline{x_c(t)x_s(t+\tau)}$.

9-22. In Sec. 9.10, the bilateral clipper was used. Consider a variation of such a clipper defined by

$$y(t) = \begin{cases} 1, & x(t) \le x, \\ 0, & x(t) > x. \end{cases}$$

Show that $\overline{y(t)} = F_x(x, t)$, where the latter is the probability distribution function (PDF) of $X = x(t)$.

9-23. For Prob. 9-22, show that $R_y(\tau)$ defined by $R_y(\tau) \equiv \overline{y(t+\tau)y(t)}$ is given by

$$R_y(\tau) = F_{X_1X_2}(x, x; t, t+\tau),$$

where $F_{X_IX_2}$ is the joint PDF of the r.v.'s $X_1 \equiv x(t)$ and $X_2 \equiv x(t+\tau)$.

9-24. Compute the output signal-to-noise ratio in the case of synchronous detection of an AM wave.

***9-25.** An AM wave is detected by the square-law detector shown in Fig. P9-25. Let $s(t)$ be a stationary Gaussian process with power spectrum

$$W(f) = \begin{cases} N_0, & |f| < W, \\ 0, & |f| > W. \end{cases}$$

With $W < f_c/2$ and $H(f)$ given by

$$H(f) = \begin{cases} 1, & |f| < W, \\ 0, & |f| > W, \end{cases}$$

compute the output signal-to-noise ratio if the only "noise" is the nonlinear terms generated by the square-law device. Why should $m^2 \ll 1$?

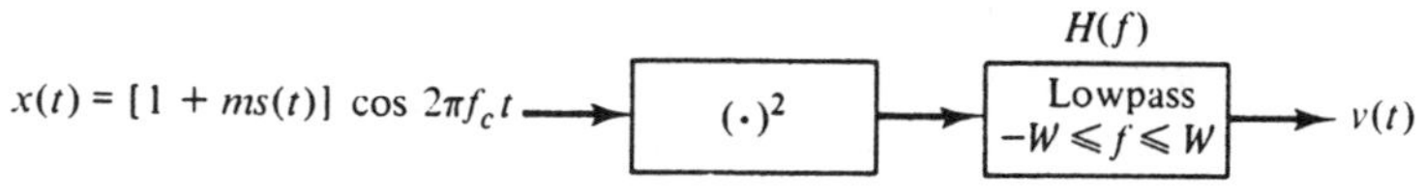

Figure P9-25

9-26. Let $n(t) = n_c(t)\cos 2\pi f_c t - n_s(t)\sin 2\pi f_c t$ represent an NGP where $n_c(t)$ is $x_c(t)$ and $n_s(t)$ is $x_s(t)$ of Eq. (9.8-2). Show that

$$\overline{n_c(t)n_c(t+\tau)} = 2\int_0^\infty W_n(\xi) \cos 2\pi(\xi - f_c)\tau \, d\xi$$
$$= \overline{n_s(t)n_s(t+\tau)},$$

where $W_n(\xi)$ is the power spectrum of $n(t)$.

9-27. Use the method of moment generating or characteristic functions discussed in Sec. 8.17 to prove that the r.v.'s $X_1 \equiv n_c^2(t)$ and $X_2 \equiv n_s^2(t)$ are independent; $n_c(t)$ and $n_s(t)$ are the in-phase and quadrature components of NGP noise. Recall that this result is required to obtain Eq. (9.11-42).

9-28. Work out the details leading to the mean-square value of the envelope square as given by Eq. (9.11-42).

9-29. Show that narrowband FM offers no signal-to-noise improvement over AM.

9-30. What signal-to-noise improvement is obtained with a PDE system characterized by a DE filter with transfer function

$$|H_{\text{DE}}(f)| = e^{-|f|/f_1}.$$

Hint: Follw the technique leading to Eqs. (9.12-31) to (9.12-33).

REFERENCES

[9-1] A. Papoulis, *Probability, Random Variables and Stochastic Processes*, McGraw-Hill, New York, 1965.

[9-2] W. B. Davenport, Jr., and W. L. Root, *Introduction to Random Signals and Noise*, McGraw-Hill, New York, 1957.

[9-3] M. Rosenblatt, "Some Comments on Narrow Band-Pass Filters," *Q. Appl. Math.*, **18,** No. 4, p. 387, 1961.

[9-4] M. G. Jenkins and D. G. Watts, *Spectral Analysis and Its Applications*, Holden-Day, San Francisco, 1968.

[9-5] R. B. Blackman and J. W. Tukey, *The Measurement of Power Spectra*, Dover, New York, 1958.

[9-6] Donald G. Childers, ed., *Modern Spectrum Analysis*, IEEE Press, New York, 1978.

[9-7] Steven M. Kay and Stanley L. Marple, Jr., "Spectrum Analysis—a Modern Perspective," *Proc. IEEE*, **69,** No. 11, pp. 1380–1419, Nov. 1981.

[9-8] H. Nyquist, "Thermal Agitation of Electric Charge in Conductors," *Phys. Rev.*, **32,** pp. 110–113, July 1928. Also J. B. Johnson, "Thermal Agitation of Electricity in Conductors," *Phys. Rev.*, **32,** pp. 97–109, July 1928.

[9-9] M. Schwartz, *Information Transmission, Modulation and Noise*, 2nd ed., McGraw-Hill, New York, 1970.

[9-10] J. H. van Vleck and D. Middleton, "The Spectrum of Clipped Noise," *Proc. IEEE*, **54,** pp. 2–19, Jan. 1966.

[9-11] P. I. Richards, "Computing Reliable Power Spectra," *IEEE Spectrum*, **4,** pp. 83–90, Jan. 1967.

[9-12] M. Schwartz, W. R. Bennett, and S. Stein, *Communication Systems and Techniques*, McGraw-Hill, New York, 1966.

[9-13] B. A. Carlson, *Communication Systems: An Introduction to Signals and Noise in Electrical Communications*, 2nd ed., McGraw-Hill, New York, 1975.

[9-14] P. F. Panter, *Modulation, Noise, and Spectral Analysis*, McGraw-Hill, New York, 1965.

[9-15] M. G. Crosby, "Frequency Modulation Noise Characteristics," *Proc. IRE*, **25**, pp. 472–514, Fig. 10, April 1937.

CHAPTER 10

Digital Modulation

10.1 INTRODUCTION

The subject of this chapter is schemes that transmit symbols of information. In Chapters 6 and 7 we analyzed methods that modulate a continuous function onto a carrier by varying its amplitude or phase. Digital modulation contains similarities and differences. A digital scheme still works with information-bearing signals that are continuous functions of time—in the physical world of voltages, fields, and waves, there are no other kind! But these signals now represent a small set of abstract symbols. Often the set is the binary symbols, which we ordinarily label 0 and 1. It is tempting to think of these as numbers, that is, points on the real line, but they are purely symbolic. To transmit one of these, a digital modulation allocates a piece of time called a *signal interval* and generates during the interval a continuous function that represents the symbol. Although the function may spill outside its interval, we always think of a function as belonging to one interval and to its symbol.

The time signal made up of those functions that represent bits is called a *baseband signal*. In a radio communication system, a second part of the modulator converts the baseband signal to an RF signal, modulating the phase, frequency, or amplitude of a carrier. In a wire or cable system, however, the baseband signal may be sent as it is or modulated onto a carrier that is not much larger in frequency than the rate of arrival of the symbols. Even if there is no carrier, we still call the equipment a ''modulator.''

The receiver end of a digital-modulation system consists of a circuit to convert the RF signal to a baseband signal and circuits to decide which symbol is represented by the baseband signal during each signal interval. A digital demodulator differs from an analog one in that it puts out a perfectly clean symbol, which is right or wrong, whereas an analog demodulator produces an output that approximately, but never exactly, matches the modulator input. These facts are important in deciding whether to use a digital transmission scheme.

Digital transmission systems are steadily replacing the older analog methods. This revolution is occurring for a variety of reasons, both economic and scientific. A digital communications network can carry a wide variety of signals by means of the same switches and transmission equipment, provided only that each signal is converted to binary symbols.† Engineering advances have made digital signals particularly easy to switch and to transmit error free. In the marketplace, most new demands for transmission service, for instance, electronic banking or airline seat reservation, are inherently nonanalog and require digital circuitry. The cost of digital components continues to drop, to the point where some circuit functions cost virtually nothing in comparison to an analog alternative. Finally, the basic nature of channels like the satellite channel, which combines low power with wide bandwidth, favors digital transmission even for signals that are originally analog.

It is easy to confuse digital *conversion* with digital *modulation*. The former, which we have discussed in Chapter 4 and will visit again in Chapter 11, changes the form of information from analog to digital; it is unnecessary if the information is already symbolic in form. The purpose of digital modulation is to transmit symbols through a medium for use at another place or another time.

The major purposes of this chapter are (1) description of some important modulation schemes and the circuits that implement them, (2) description of the effects of physical channels on these signals, and (3) analysis of the symbol error probability for these schemes. We shall begin with the last since it stems directly from earlier parts of the book. Specifically, the analysis of symbol error is based on representation of modulated signals by vectors, the subject of Sec. 2.7. It is with the geometric concepts of that section that we shall begin.

Almost all digital receivers are based on the maximum-likelihood receiver concept, with which we shall begin in Sec. 10.2. A true maximum-likelihood receiver is often easy to construct. Even when electronics or system cost compromises dictate another kind of receiver, it will be inspired by the maximum-likelihood design.

10.2 THE MAXIMUM-LIKELIHOOD RECEIVER

Our object in this section is to design a receiver that decides which digital symbol is the most likely cause of a signal that the receiver observes. This is called a *maximum-likelihood* receiver. A schematic diagram of a digital communications link is shown in Fig. 10.2-1. The nth of a sequence of symbols has been presented to the transmitter; this symbol takes on one of M values $m_1, m_2, \ldots, m_M$, and the transmitter converts it to one of the set $s_1(t), s_2(t), \ldots, s_M(t)$. Noise $\eta(t)$ is added to the transmitted time

† Binary signals imply 2 symbols; some systems use more, but all alphabets are convertible to one another.

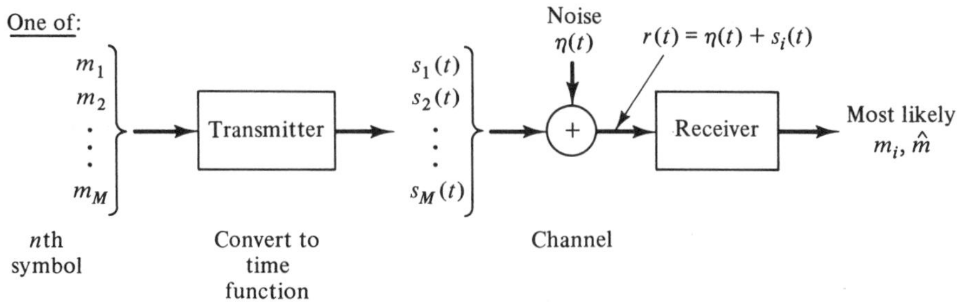

FIGURE 10.2-1. Functional diagram of a digital transmitter and receiver for sending one of M symbolic messages.

functions $s_i(t)$ to form the received signal $r(t) = s_i(t) + \eta(t)$. To decide which is the most likely symbol, the receiver must decide which is the most likely signal among the $\{s_i(t)\}$ from the sole article of information that it has, the received signal $r(t)$.

What is the most likely transmitted signal? To determine this, we must make some assumptions. The first of these is that the source of symbols is *random*, with successive symbols independent and distributed according to a distribution $P(m_1)$, $P(m_2)$, . . . , $P(m_M)$. Many real sources of data do not generate independent symbols, nor do their symbols take on a consistent distribution; but at the same time we do not wish to redesign the receiver at each change in the nature of the data. A practical compromise is to assume the data are random with some long-term distribution. To find out which transmitted signal is most likely, given the observed $r(t)$, we must calculate the conditional probabilities $P(s_1(t)|r(t))$, $P(s_2(t)|r(t))$, . . . , $P(s_M(t)|r(t))$ and find the largest. From Sec. 8.3, the ith of these probabilities is given by

$$P(s_i(t)|r(t)) = \frac{P(s_i(t) \text{ sent and } r(t) \text{ received})}{P(r(t) \text{ received})}.$$

The answer will be easier to calculate if we express the numerator as $P(r(t)|s_i(t)) \cdot P(s_i(t))$. (This is really an application of Bayes's rule from Sec. 8.4.) Then the index of the most likely symbol is the i that solves

$$\max_i \frac{P(r(t)|s_i(t))P(s_i(t))}{P(r(t))}. \qquad (10.2\text{-}1)$$

The advantage of this form is that $P(r(t)|s_i(t))$ is simply the probability that the noise $\eta(t)$ is equal to the difference $r(t) - s_i(t)$; since we assume that the statistical properties are known, $P(r(t)|s_i(t))$ is a known quantity for each i. Another interesting fact is that the same $P(r(t))$ appears in each of the elements of the maximization; it can therefore be dropped from each element without affecting which is largest. Taking into account both of these facts, we get the simpler maximization

$$\text{Find:} \qquad \max P[\eta(t) = r(t) - s_i(t)]P(s_i(t)). \qquad (10.2\text{-}2)$$

We shall denote the maximizing symbol as $\hat{m}$. Here and throughout the chapter the ^ will always mean some kind of "estimate" of a variable.

A receiver that evaluates Eq. (10.2-2) is called a *maximum a posteriori*, or MAP, receiver, which in probability theory means that the most likely symbol is chosen, considering both the probabilities of the noise and of the symbols before transmission. Unfortunately, Eq. (10.2-2) cannot be evaluated directly because it contains a fatal difficulty: There is ordinarily no way to assign probabilities in a consistent way to functions of time like $\eta(t)$. For some noise mechanisms, the stochastic process $\eta(t)$ might take on discretely many realizations, in which case we could assign distinct probabilities to each. But for most cases, including the Gaussian case, which is of most interest to us, $\eta(t)$ is a process with far too many realizations to allow probability assignment to any one outcome. The way out of this difficulty is to construct an orthogonal basis for the functions in Eq. (10.2-2), in the manner of Sec. 2.7, and then *deal with the vector space components of the functions*.

Before turning to this, we need to resolve another more practical difficulty, the fact that the symbol probabilities $\{P(m_i)\}$, which are the same as the probabilities $\{P(s_i(t))\}$, are often not reliably known. This is a nagging problem, so much so that the MAP receiver (10.2-2) is seldom used. Instead, most designers make the assumption that all symbols are equally likely. Consequently, the factor $P(s_i(t))$ is a constant during the maximization over i and may be dropped, just as was the constant $P(r(t))$. This leaves

$$\text{Find:} \quad \max_i P[\eta(t) = r(t) - s_i(t)]. \tag{10.2-3}$$

This receiver is called a *maximum-likelihood* receiver. It considers only the noise probabilities in finding $\hat{m}$. For equally likely symbols, it is also the MAP receiver. Otherwise, it performs more poorly than the MAP receiver, but the performance of the latter can only be attained by accurately tracking the symbol probabilities.

Signal Space

To analyze the probability of decision error in the MAP or maximum-likelihood receiver, we must express their noise and signal functions of time as components in a vector space. Communications engineers call this space the *signal space*. Points in this space, basis vectors, and vector distances are part of the everyday language of the communications designer. We shall limit ourselves to the case of white Gaussian noise, because this is the case of greatest practical interest and because the results take on a particularly simple form, but the kind of analysis here can be extended to many other kinds of noise (see [10-1]).

To set up our signal space, we express all functions of interest in the receiver as vectors in the manner of Sec. 2.7. The vector representation of the transmitted signal $s_i(t)$ is†

† Here, as in several other places in the book, we find it convenient to use row vectors.

$$\mathbf{s}_i = (s_{i1}, \ldots, s_{iJ}), \qquad i = 1, \ldots, M. \tag{10.2-4}$$

The jth component here is the inner product $\int_T s_i(t)\phi_j^*(t)\,dt$ of the signals $s_i(t)$ with the orthonormal basis function $\phi_j(t)$. There are J basis functions, $J \leq M$, and the integration time T is over the time duration of the M signals. The basis functions can be obtained by the Gram–Schmidt process of Sec. 2.7, acting on the M signals to produce a J-dimensional space; but in most of this chapter the signaling scheme itself will suggest an obvious orthonormal basis. In either case, the M signals are exactly given by

$$s_i(t) = \sum_{j=1}^{J} s_{ij}\phi_j(t), \qquad i = 1, \ldots, M. \tag{10.2-5}$$

We shall give examples of signal-space formation in the discussion of basic modulation schemes in Sec. 10.3.

In the same way we can express the additive white Gaussian noise $\eta(t)$ as

$$\boldsymbol{\eta} = (\eta_1, \ldots, \eta_J, \eta_{J+1}, \ldots). \tag{10.2-6}$$

The noise is *not* confined to the J signal set dimensions, and so its vector expansion needs many more components. In fact, a basis for all the Gaussian noise realizations that is "complete" in the sense of Eq. (2.6-7) requires infinitely many functions, but it will turn out shortly that we need not concern ourselves about the identity of these beyond the Jth. The received signal $r(t)$ consists of a signal plus this same noise, and so it has the expansion

$$\mathbf{r} = (r_1, \ldots, r_J, r_{J+1}, \ldots). \tag{10.2-7}$$

Again, we do not need to know the basis functions beyond the Jth.

Figure 10.2-2 expresses the receiver of Fig. 10.2-1 in vector terms for the transmission of the kth message. The vector that represents the transmitted signal has zero components after the Jth. Infinite-dimensional noise is added component-wise to produce the received signal. Each component of this satisfies $r_j = \eta_j + s_{kj}$. Because of the completeness of the basis, the vectors of Fig. 10.2-2 represent the functions in Fig. 10.2-1 to within a mean-square error of zero.

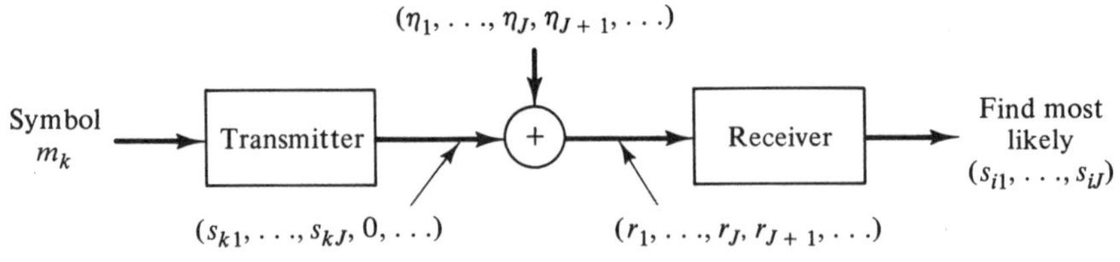

Figure 10.2-2. Vector version of Fig. 10.2-1. The Gram–Schmidt procedure guarantees that the transmitted signal $s_k(t)$ is perfectly represented.

Signal Space under White Gaussian Noise

Our aim now is the evaluation of Eq. (10.2-3), which defines the maximum likelihood receiver. A formal substitution of vectors for functions in Eq. (10.2-3) gives

$$\max_i P(\boldsymbol{\eta} = \mathbf{r} - \mathbf{s}_i),$$

the meaning of which is

$$\max_i P\left(\prod_j \{\eta_j = r_j - s_{ij}\}\right), \tag{10.2-8}$$

that is, the vectors in the event must match component-wise. It may seem precipitous to make this vector substitution, but it should be kept in mind that events like those in Eqs. (10.2-2) and (10.2-3) do not have a defined probability, at least for continuously valued noise; we may only analyze the probabilities of signals that are first converted to vector components.

Proceeding with this analysis, we note first that the components η_j are Gaussian variables because they are obtained by a linear operation (integration) from a Gaussian random process $\eta(t)$. This property of Gaussian processes is discussed in Sec. 9.7. All the rest of the properties of η_j that we need are summarized in the following lemma.

Lemma

$$E[\eta_j] = 0, \qquad \text{all } j \tag{10.2-9a}$$

$$\operatorname{cov}(\eta_j, \eta_k) = \begin{cases} 0, & j \neq k, \\ \dfrac{N_0}{2}, & j = k, \end{cases} \tag{10.2-9b}$$

where $N_0/2$ is the power spectral density of the white Gaussian noise.†

Proof. The properties of linear operations on stochastic processes give us the desired proofs. For Eq. (10.2-9a),

$$E[\eta_j] = E[\textstyle\int_T \eta(t)\phi_j^*(t)\,dt] = \textstyle\int_T E[\eta(t)]\phi_j^*(t)\,dt = 0$$

since $\eta(t)$ is a zero-mean process. Here the order of integration and expectation can be reversed because both are linear operations. For Eq. (10.2-9b), the covariance is given by

$$\operatorname{cov}(\eta_j, \eta_k) = E[\textstyle\int_T \eta(t)\phi_j^*(t)\,dt \int_T \eta(u)\phi_k(u)\,du].$$

† In digital communications, it is traditional to count the noise power spectral density as $N_0/2$, unlike the case of analog communication, where we set it to N_0. To do otherwise is not to speak the language of practicing engineers.

Reversing operations again, we get

$$\text{cov}(\eta_j, \eta_k) = \int\int E[\eta(t)\eta(u)]\phi_j^*(t)\phi_k(u)\, dt\, du.$$

Section 9.9 allows us to evaluate this as

$$\int\int \frac{N_0}{2}\delta(t-u)\phi_j^*(t)\phi_k^*(u)\, dt\, du = \int \frac{N_0}{2}\phi_j^*(t)\phi_k(t)\, dt$$
$$= \begin{cases} \dfrac{N_0}{2}, & j = k, \\ 0, & \text{otherwise.} \end{cases}$$

An immediate consequence of Eq. (10.2-9b) is that the components η_j are independent variables; this is because they are both uncorrelated and Gaussian. Thus we can rewrite Eq. (10.2-8) as

$$\max_i P\left(\prod_j \{\eta_j = r_j - s_{ij}\}\right) = \max_i \prod_j P(\eta_j = r_j - s_{ij})$$
$$= \max_i \prod_{j=1}^{J} P(\eta_j = r_j - s_{ij}) \prod_{j>J} P(\eta_j = r_j). \tag{10.2-10}$$

In the last line we use the fact that $\mathbf{s}_i$ has no components beyond the Jth. As written, this line contains a crucial fact: The second product of factors does not depend on the transmitted signal, and, as we have done twice before, we may ignore terms like it while evaluating the maximum over i. This important result, sometimes called the *theorem of irrelevance*, states that *only noise in the dimensions of the signals affects detection*. This statement is true also for analog receivers and is invoked every time we "tune out" other signals while tuning in a desired transmission. We see in Eq. (10.2-10) the digital version of this fact. Dropping the factor in Eq. (10.2-10), we get the still simpler definition of the maximum-likelihood receiver:

$$\max_i \prod_{j=1}^{J} P(\eta_j = r_j - s_{ij}). \tag{10.2-11}$$

From the lemma, Eq. (10.2-11) is simply the Gaussian expression

$$\max_i (\pi N_0)^{-J/2} \exp\left[-\sum_{j=1}^{J} (r_j - s_{ij})^2\right].$$

To simplify this further, we can take natural logs; since log is a monotone increasing function, this will not affect the outcome of the maximization. The result is

$$\max_i \left[-(J/2)\ln(\pi N_0) - \sum_{j=1}^{J} (r_j - s_{ij})^2\right],$$

which is the same as performing the minimization

$$\min_i \sum_{j=1}^{J} (r_j - s_{ij})^2. \tag{10.2-12}$$

The minimization of Eq. (10.2-12) is startling in its simplicity. It says that the maximum-likelihood receiver simply finds the nearest signal $\mathbf{s}_i$ to $\mathbf{r}$ in the sense of *ordinary Euclidean distance*.

Before continuing, let us summarize our discoveries so far. By expressing signals as component vectors, we have been able to measure the probabilities of functions that occur in useful receivers. This leads to a concrete design for the maximum-likelihood receiver. For the important case of white Gaussian noise, this receiver simply measures Euclidean distances among the vectors. In more advanced texts, such as [10-1], the signal-space method is extended to the case of colored Gaussian noise and to non-Gaussian noises like impulse noise. These results are considerably more complicated than Eq. (10.2-12).

Since the white-noise receiver in signal space mimics geometry and distance in the physical world, through intuition we can gain much from the signal-space model. Communications engineers habitually think of receivers in signal space for this reason. Fortunately, the intuitions gained there carry over to other types of noise as well. Figure 10.2-3 illustrates some of the important consequences of Eq. (10.2-12) as it applies to the reception of a set of three signals, $\mathbf{s}_1$, $\mathbf{s}_2$, and $\mathbf{s}_3$, that lie in a two-dimensional space. A plot like this, which shows the points in signal space that correspond to signals, is called a *signal constellation*. The signals here have the special characteristic that they are of equal energy E; thus the points in the constellation all lie on a circle (more generally, a sphere) of radius $\sqrt{E}$.

What is a geometrical construction that performs the minimization in Eq. (10.2-12)? To find the locus of points r that are closer to $\mathbf{s}_1$ than to $\mathbf{s}_2$, we draw the perpendicular bisector of the line joining $\mathbf{s}_1$ and $\mathbf{s}_2$. Repeating this for the pair $\mathbf{s}_1$ and $\mathbf{s}_3$ and for the remaining pair $\mathbf{s}_2$ and $\mathbf{s}_3$ delineates the three *decision regions* $\mathcal{S}_1$, $\mathcal{S}_2$, and $\mathcal{S}_3$ shown in the figure. Here $\mathcal{S}_i$ is the set of those received vectors $\mathbf{r}$ that lie closest to $\mathbf{s}_i$. Clear

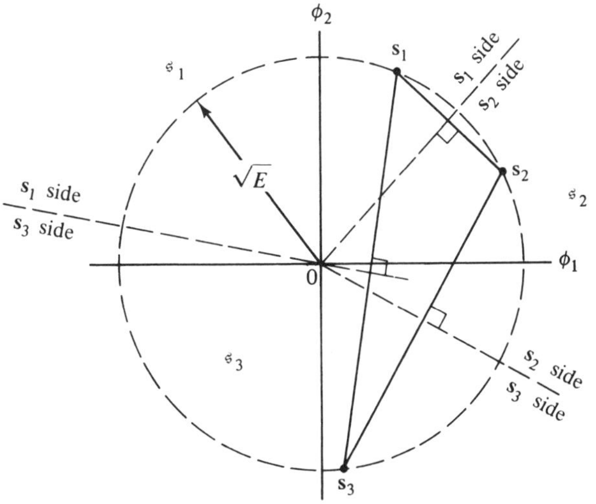

Figure 10.2-3. Example of two-dimensional signal space with three signal points, showing decision region construction. The signals are equal energy.

from the figure is another property of equal-energy signals: their decision region boundaries always terminate at the origin. A receiver that obeys decision regions like those in the figure is called a *minimum-distance* receiver. For white Gaussian noise, it is also maximum likelihood, but for other types it may not be. It is nonetheless often used because of its simplicity.

Figure 10.2-4 shows several basic types of signal constellations. All the signals in these constellations have the same energy, E. The points $(\sqrt{E}, 0, 0, \ldots)$ and $(-\sqrt{E}, 0, 0, \ldots)$ shown in Fig. 10.2-4(a) are called antipodal signals, because the points are equal and opposite to each other. All the energy of these signals appears in the first signal-space dimension; the two antipodal signals are simply $\sqrt{E}\phi_1(t)$ and its negative, $-\sqrt{E}\phi_1(t)$. Figure 10.2-4(b) depicts the two orthogonal signals $(\sqrt{E}, 0, 0, \ldots)$ and $(0, \sqrt{E}, 0, \ldots)$, with dimensions beyond the second suppressed. These signals have inner product zero. The idea of orthogonal signals can be extended to many dimensions by placing one signal vector along each dimension. Figure 10.2-4(c)

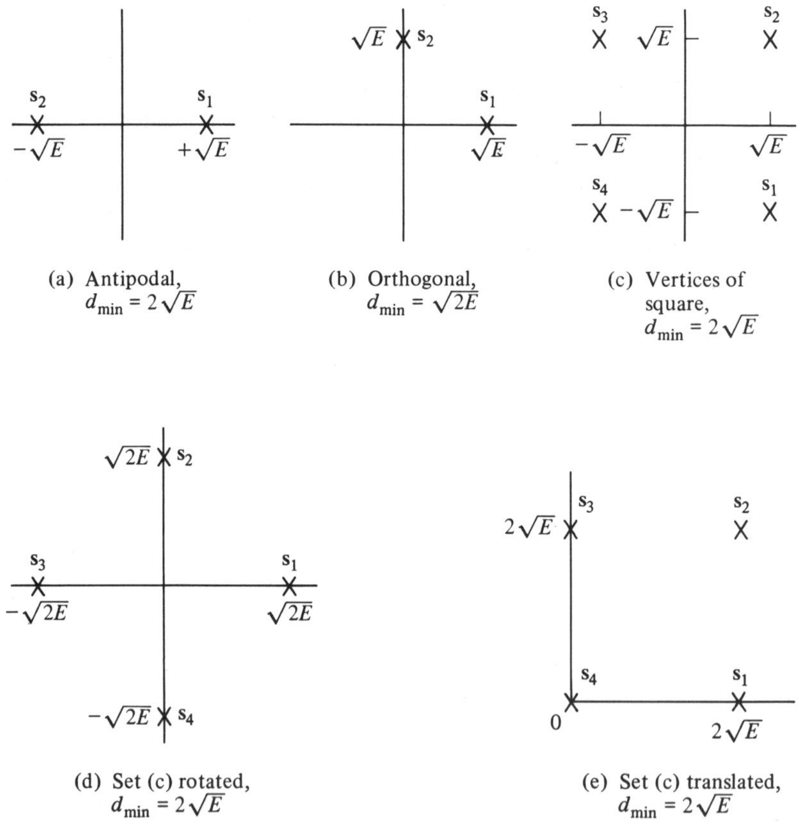

Figure 10.2-4. Some simple signal constellations in two dimensions. In Sec. 10.4, (a) is BPSK, (b) is incoherent FSK, and (c) is QPSK.

shows two sets of antipodal signals orthogonal to each other, and hence combines the ideas in Figs. 10.2-4(a) and 10.2-4(b). The signals are located at $(+\sqrt{E}, +\sqrt{E})$, $(+\sqrt{E}, -\sqrt{E})$, $(-\sqrt{E}, +\sqrt{E})$, and $(-\sqrt{E}, -\sqrt{E})$. Figure 10.2-4(d) is a rotation of Fig. 10.2-4(c). As will be shown in Sec. 10.3, rotations do not change the probability of receiver error for a constellation. Nor do they change the energies of the signals, since their distances from the origin are unaffected.

Figure 10.2-4(e) illustrates a *translation* of Fig. 10.2-4(c). The error probability of Fig. 10.2-4(e) will again prove to be the same as that of Figs. 10.2-4(c) and (d), but its energy will be larger. The average energy of the set of equally probable signals $\mathbf{s}_i$, $i = 1, \ldots, M$, is

$$\bar{E} = \frac{1}{M} \sum_{i=1}^{M} E_i = \frac{1}{M} \sum_{i=1}^{M} \|\mathbf{s}_i\|^2. \tag{10.2-13}$$

Here we have introduced the notation $\|x\|$ to represent the *norm* of $\mathbf{x}$ in the signal space;

$$\|\mathbf{x}\| = \left(\sum_{j=1}^{J} x_j^2 \right)^{1/2}. \tag{10.2-14}$$

In our space the square norm can be interpreted as energy. If each signal is translated by the vector $\boldsymbol{\alpha}$, the average energy of the new set is

$$E(\boldsymbol{\alpha}) = \frac{1}{M} \sum_{i=1}^{M} \|\mathbf{s}_i - \boldsymbol{\alpha}\|^2. \tag{10.2-15}$$

Some calculus shows that the translation vector that minimizes this is

$$\boldsymbol{\alpha}_0 = \frac{1}{M} \sum_{i=1}^{M} \mathbf{s}_i, \tag{10.2-16}$$

that is, the centroid or mean value of the M points weighted equally.

The signal set $\{\mathbf{s}_i - \boldsymbol{\alpha}_0\}$ is a *minimum-energy* version of the signals $\{\mathbf{s}_i\}$. Such signals represent the most efficient allocation of energy for a given error probability and are thus the most desirable signal set, unless some other design consideration intervenes. The constellations in Figs. 10.2-4(c) and (d) are both minimum energy, since their centroid is at the origin, but Fig. 10.2-4(e) is not. In fact, its average energy is two times that of the other two!

Matched-Filter Receiver

Before leaving the subject of maximum-likelihood receivers for white Gaussian noise, we need to investigate another startling fact about these receivers: they can be implemented by a properly designed bank of linear filters. It is a striking fact of signal theory that these receivers can be built out of such familiar elements and that the receivers need not be nonlinear. Furthermore, the filters are very simple ones for commonly used modulations.

To begin our discussion, we return to the minimization over Euclidean distance in Eq. (10.2-12). From Eq. (2.7-10), we can write the distance in terms of the original time functions as

$$\min_i \|\mathbf{r} - \mathbf{s}_i\|^2 = \min_i \sum_{j=1}^{J} (r_j - s_{ij})^2 = \min_i \int_T [r(t) - s_i(t)]^2 \, dt. \qquad (10.2\text{-}17)$$

The integral can be expanded as

$$\int_T r(t)^2 \, dt + \int_T s_i(t)^2 \, dt - 2\int_T r(t)s_i(t) \, dt,$$

the first term of which is constant throughout the minimization. So the maximum-likelihood receiver of Eq. (10.2-17) becomes

$$\min_i \left\{ \int_T s_i(t)^2 \, dt - 2\int_T r(t)s_i(t) \, dt \right\}. \qquad (10.2\text{-}18)$$

The first term here is the *energy* in the *i*th signal; it depends on i, but not on the received signal $r(t)$. The term could be precomputed for all the M signals, and we shall call it ξ_i. If all the transmitted signals have the same energy, as often happens, then the first term in Eq. (10.2-18) is constant with i and can be dropped from the minimization. Expression (10.2-18) is then equivalent to just

$$\max_i \int_T r(t)s_i(t) \, dt. \qquad (10.2\text{-}19)$$

This integral, as well as the second one in Eq. (10.2-18), is a correlation integral, like those in Sec. 9.6. It evaluates the degree of similarity between the received signal and the possible transmitted signals. A receiver built up of these operations is called a

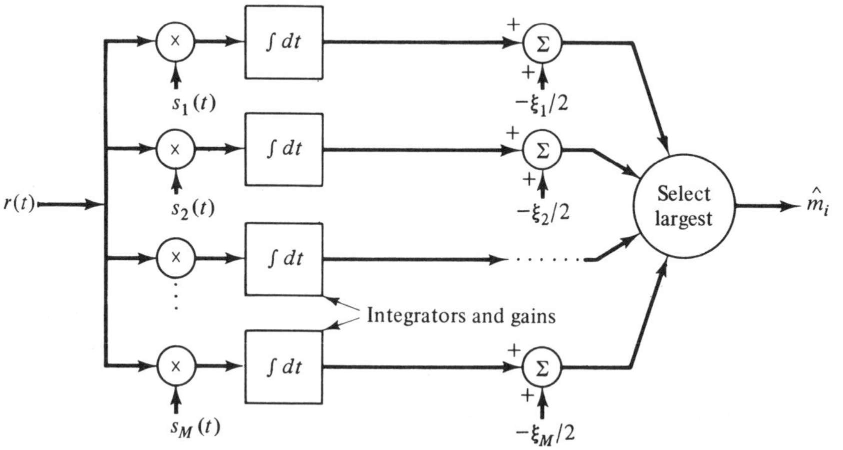

Figure 10.2-5. The correlator receiver, a maximum-likelihood receiver.

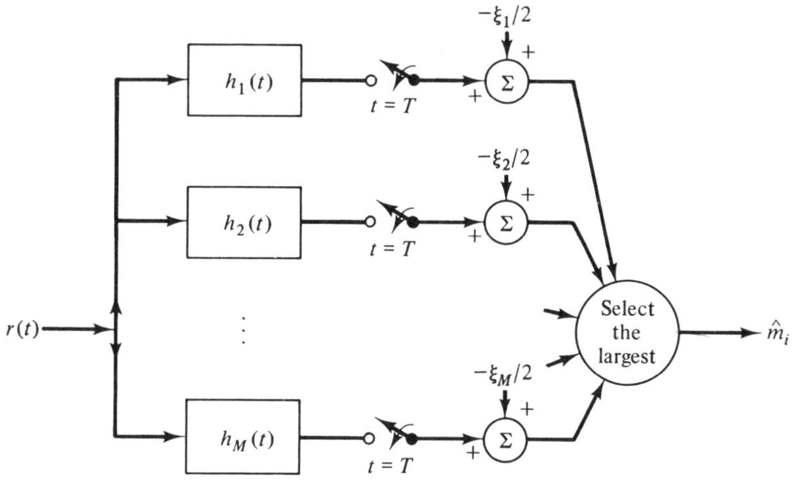

Figure 10.2-6. Bank of matched-filters receiver. The effect of this circuit is the same as that in Fig. 10.2-5.

correlator receiver, a diagram of which appears in Fig. 10.2-5. The offsetting constants ξ_i may be left out in the case of equal-energy signals. The object of this receiver is to find the transmitted signal having the highest correlation with $r(t)$, and by means of the preceding paragraphs we have shown that such a receiver is maximum likelihood.

With a little more manipulation, the multiply-and-integrate circuitry of Fig. 10.2-5 can be replaced by a considerably simpler alternative, a linear filtering operation. Let the integration time be the interval $[0, NT]$. By the properties of convolution, the integral $\int_0^{NT} r(t)\, s_i(t)\, dt$ has the same value as the convolution $r*h$ evaluated at NT, if $h(t) = s_i(NT - t)$. This $h(t)$, which we can view as a filter impulse response, is the function $s_i(t)$ time-reversed and shifted by NT. Having set up this filter, we can find the correlation between $r(t)$ and $s_i(t)$ by simply filtering $r(t)$ and sampling the output at NT. The filter $h(t)$ is called a *matched filter*. A bank of M filters and samplers make up the *matched-filter receiver* in Fig. 10.2-6. Like the correlator receiver, this receiver is maximum likelihood for noise that is Gaussian and white. Matched filters, even complicated ones, are often closely approximated in receivers by using special devices like surface acoustic wave devices or by a software implementation based on the digital techniques of Chapter 5.

10.3 ERROR PROBABILITY IN MAXIMUM-LIKELIHOOD RECEIVERS

Our aim now is to calculate the error probability of the receivers in Sec. 10.2. In the practical world, a receiver may be suboptimal by design for reasons of cost, or it may be optimal for the signal set $\{s_i(t)\}$, but the signals as received are modified by filtering or amplitude limiting. As well, the noise may not be the white Gaussian noise of Sec. 10.2. Nonetheless, the error rate of the maximum-likelihood receiver in white Gaussian noise serves as the communications engineer's benchmark, against which all practical designs are compared.

Two-Signal Error Probability

We shall begin with the case where the receiver must distinguish only two signals and then extend that result to the case of many signals. Consider the two signals that appear in Fig. 10.3-1, which are separated by a distance d. The probability distribution of white noise is such that it is uniformly and independently distributed along any set of orthogonal axes in signal space. For the purposes of our error calculation, it will be convenient to place one of these axes, the first, along the line that connects $\mathbf{s}_i$ and $\mathbf{s}_k$. If $\mathbf{s}_i$ is sent, a maximum-likelihood receiver decides $\mathbf{s}_k$ if the point $\mathbf{r}$ lies on the far side of the perpendicular bisector of this line. This event occurs if and only if the first component of noise exceeds $d/2$, using the sign convention in the figure.

The first noise component, like the others, is an independent zero-mean Gaussian variate, whose variance from Eq. (10.2-9b) is $N_0/2$, the power spectral density of the noise. Thus the probability, $p_e(k; i)$, of receiving $\mathbf{s}_k$, given that $\mathbf{s}_i$ is transmitted, is

$$p_e(k; i) = \int_{d/2}^{\infty} \frac{1}{\sqrt{\pi N_0}} \exp\left(\frac{-\eta_1^2}{N_0}\right) d\eta_1, \qquad (10.3\text{-}1)$$

which we shall call the two-signal error probability.

Expression (10.3-1) is an ordinary Gaussian integral like those in Sec. 8.10. By habit, communications engineers express it in terms of the *Q function*, defined by, assuming $x > 0$,

$$Q(x) = \frac{1}{\sqrt{2\pi}} \int_{x}^{\infty} \exp\left(\frac{-u^2}{2}\right) du. \qquad (10.3\text{-}2)$$

$Q(\cdot)$ can be obtained from the widely tabulated error function by using

$$Q(x) = \frac{1}{2} - \text{erf}(x); \qquad (10.3\text{-}3)$$

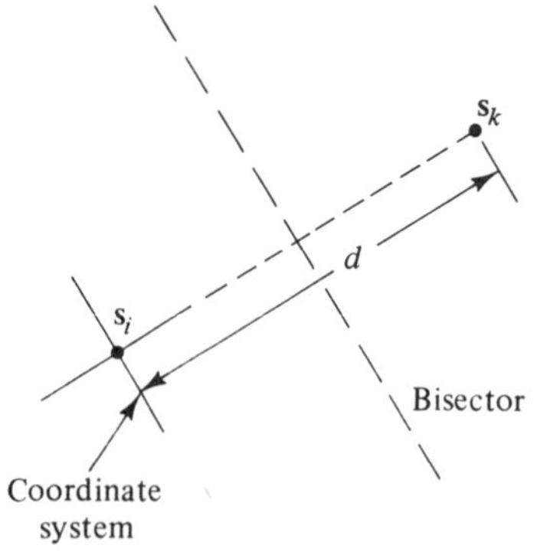

Figure 10.3-1. Setup for calculating two-signal error probability.

erf(x) is defined in Eq. (8.10-8). A simple and most useful bound on $Q(x)$ is

$$Q(x) \leq \tfrac{1}{2} \exp\left(\frac{-x^2}{2}\right). \tag{10.3-4}$$

For $x < 0$ we use the property of $Q(x)$ that $Q(-x) = 1 - Q(x)$. The utility of this bound comes from the fact that it retains the correct *exponential* behavior of the true Q function integral.

If we write Eq. (10.3-1) in terms of the Q function, the result of the variable change is

$$p_e(k; i) = Q\left(\frac{d}{\sqrt{2N_0}}\right), \tag{10.3-5}$$

and the bound (10.3-4) gives the simpler expression

$$p_e(k; i) \leq \tfrac{1}{2} \exp\left(\frac{-d^2}{4N_0}\right). \tag{10.3-6}$$

Next we would like to evaluate d for some of the constellations in Fig. 10.2-4, many of which will appear repeatedly in the rest of the chapter. The effect of this will be to convert the argument of Q in Eq. (10.3-5) from d to *average symbol energy*. To avoid confusion, we shall give this important quantity the special notation E_s. Beginning with Fig. 10.2-4(a), we see that the average energy of these two antipodal signals is E (which we relabel E_s) and d is $2\sqrt{E_s}$. Consequently,

$$p_e(k; i) = Q\left(\sqrt{\frac{2E_s}{N_0}}\right) \qquad \text{(antipodal signals).} \tag{10.3-7}$$

For Fig. 10.2-4(b), E_s is still E, but d for two equal-energy orthogonal signals is now $\sqrt{2E_s}$. Thus

$$p_e(k; i) = Q\left(\sqrt{\frac{E_s}{N_0}}\right) \qquad \text{(orthogonal signals).} \tag{10.3-8}$$

Equation (10.3-8) also applies to any pair out of a many-dimensional orthogonal set. Both constellations 10.2-4(a) and (b) consist of only two signals, $\mathbf{s}_1$ and $\mathbf{s}_2$, and from symmetry $p_e(1; 2) = p_e(2; 1)$; so the overall probability of error in these cases is

$$p_e = p(\mathbf{s}_1)p_e(2; 1) + p(\mathbf{s}_2)p_e(1; 2) = p_e(1; 2), \tag{10.3-9}$$

(i.e., our two-signal probability of error). From Eqs. (10.3-7) to (10.3-9) we can draw an important conclusion: *For the same error probability, orthogonal signals require twice the energy of antipodal signals.*

The probabilities (10.3-7) and (10.3-8) constantly recur in the analysis of digital modulations, and they are plotted in Fig. 10.3-2. Appearing there also is the Q function bound of the form (10.3-4) to probability (10.3-7), which has the particularly simple form $p_e \leq \frac{1}{2}\exp(-E_s/N_0)$. This bound curve converges to the true antipodal error probability as p_e drops toward zero.

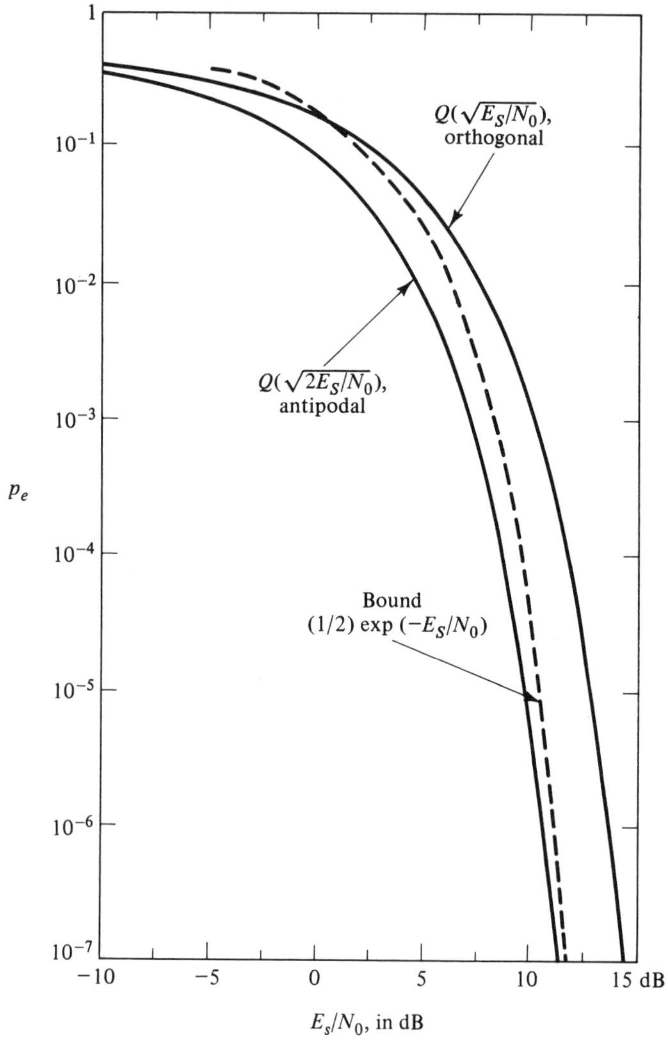

Figure 10.3-2. Basic orthogonal and antipodal binary modulation error probabilities.

Multisignal Probability of Error

The next signal constellation, Fig. 10.2-4(c), consists of four signals at the vertices of a square. The average energy of these signals is $E_s = 2E$. We can calculate $p_e(k; i)$ between any two signals in the usual way and then combine these probabilities into a four-signal error probability. This is done as follows. The four-signal probability is defined to be the probability of error given that there are four signals and $\mathbf{s}_i$ is sent. This is written in shorthand as

$$p_e(i) = P\{\text{error} \mid 4 \text{ signals}, i \text{ sent}\} = P\left\{\bigcup_{k\neq i} [\mathbf{r} \text{ in } \mathcal{S}_k \mid 4 \text{ signals}, i \text{ sent}]\right\}$$

$$= \sum_{k\neq i} P\{\mathbf{r} \text{ in } \mathcal{S}_k \mid 4 \text{ signals}, i \text{ sent}\},$$

where the last line holds because the events of receiving the messages k are disjoint. Now let $\mathcal{S}_k$ (2) be the decision region for message k if *only two messages* k and i can be transmitted. From the discussion of Fig. 10.2-3, it can be seen that region $\mathcal{S}_k(2)$ must be at least as large as $\mathcal{S}_k$. Thus our last line must be over bounded by

$$\leq \sum_{k\neq i} P\{\mathbf{r} \text{ in } \mathcal{S}_k(2) \mid 2 \text{ signals } i \text{ and } k, i \text{ sent}\}$$

$$= \sum_{k\neq i} p_e(k; i).$$

In summary,

$$p_e(i) \leq \sum_{k\neq i} Q\left(\frac{\|\mathbf{s}_k - \mathbf{s}_i\|}{\sqrt{2N_0}}\right), \tag{10.3-10}$$

where $\|\mathbf{s}_i - \mathbf{s}_k\|$ is the distance from $\mathbf{s}_i$ to $\mathbf{s}_k$.

For a maximum-likelihood receiver, we assume that all signals are equally likely, and for Fig. 10.2-4(c) this means they all have probability $\frac{1}{4}$. In addition, Fig. 10.2-4(c) is completely symmetrical, with

$$p_e(i) \leq 2Q\left(\frac{2\sqrt{E}}{\sqrt{2N_0}}\right) + Q\left(\frac{2\sqrt{2E}}{\sqrt{2N_0}}\right), \qquad \text{all } i,$$

the first term representing two signals distant $2\sqrt{E}$ and the second the signal distance $2\sqrt{2E}$. The overall error probability in terms of E_s (which equals $2E$) is thus

$$p_e \leq 2Q\left(\sqrt{\frac{E_s}{N_0}}\right) + Q\left(\sqrt{\frac{2E_s}{N_0}}\right) \qquad \text{[error bound for Fig. 10.2-4(c)]} \tag{10.3-11}$$

An important fact about Eq. (10.3-11) is that the second term tends to zero much faster than the first as the ratio E_s/N_0 grows. The bound (10.3-4) can be used to substantiate this: The second term is approximately $\frac{1}{2}\exp(-E_s/N_0)$, versus $\exp(-E_s/2N_0)$ for the first term. Figure 10.3-2 shows that a useful range of E_s/N_0 begins at 8 dB or so, at which the second term is a factor 10^{-2} smaller than the first. The calculation illustrates a basic fact about multipoint signal constellations: *the nearest neighbors to a transmitted signal dominate the error probability.*

The rotated constellation in Fig. 10.2-4(d) has the same average energy and the same structure of distances as in Fig. 10.2-4(c), and so Eq. (10.3-11) gives its error probability bound. This is not the case with the translated constellation of Fig. 10.2-4(e). Although the intersignal distances are the same in terms of E, it is no longer true that the average energy E_s is equal to $2E$. In fact, $E_s = 4E$, and when all the substitutions are made into Eq. (10.3-10), the error probability of Fig. 10.2-4(e) becomes

$$p_e \leq 2Q\left(\sqrt{\frac{E_s}{2N_0}}\right) + Q\left(\sqrt{\frac{E_s}{N_0}}\right) \qquad \text{[error bound for Fig. 10.2-4(e)]} \qquad (10.3\text{-}12)$$

This is much worse than in Eq. (10.3-11); twice as much energy on the average is needed to produce the same p_e. This shows the importance of designing *minimum-energy* signal sets.

Each of the four-point constellations in Fig. 10.2-4 has a symmetry in the sense that each signal point sees an arrangement of the same number of neighbors at the same distances as do other points in the constellation. This and the equal-probability signal assumption allowed us to equate $p_e(i)$ and p_e. For signal sets that do not have this symmetry of neighbors, for instance, that in Fig. 10.2-3, we must use the full expression

$$p_e = \frac{1}{M}\sum_{i=1}^{M} p_e(i) \leq \frac{1}{M}\sum_{i=1}^{M}\sum_{k\neq i} Q\left(\frac{\|\mathbf{s}_k - \mathbf{s}_i\|}{\sqrt{2N_0}}\right), \qquad (10.3\text{-}13)$$

where we have used the fact that each signal has probability $1/M$. This is a sum of $M(M-1)$ exponentials, each of which rapidly decays as E_s/N_0 grows, and it is clear that the term containing the least distance $\|\mathbf{s}_k - \mathbf{s}_i\|$ dominates the entire sum as E_s/N_0 grows. This signal space distance is called the *minimum distance* $d_{\min}$ of the signal set. In Fig. 10.2-3 it is the distance $\|\mathbf{s}_1 - \mathbf{s}_2\|$. A tight overbound to Eq. (10.3-13) is

$$p_e \leq \frac{2}{M} KQ\left(\frac{d_{\min}}{\sqrt{2N_0}}\right), \qquad (10.3\text{-}14)$$

where K is the number of distinct pairs of signals in the set that lie separated by the minimum distance. For the four-point sets in Figs. 10.2-4(c) and (d), bound (10.3-14) is just the first term of Eq. (10.3-11).

10.4 BASIC PHASE-SHIFT AND QUADRATURE-AMPLITUDE MODULATIONS

The great majority of digital modulations that find practical use employ two-dimensional signal constellations in which the first dimension is the cosine carrier signal of frequency ω_0 and the second dimension is the sine carrier signal of the same frequency. These modulation systems depend on an accurate phase reference, that is, a replica of one dimension's basis function, in order to set up the basis functions electrically and keep them orthogonal. A scheme like this that needs a reference is called a *coherent* modulation. Any two-dimensional scheme that can be viewed as sending independent signals in the orthogonal directions set up by $\sin \omega_0 t$ and $\cos \omega_0 t$ is called a *quadrature* modulation.

During a given signal interval T of a quadrature modulation, we may transmit any one of M signals. Each interval makes available M new signals. As electrical signals, these are sinusoids of frequency ω_0 having certain amplitudes and phases; seen in signal space, they are M points in a two-dimensional space. It is important to keep in mind both parts of this dual model of quadrature signals, since both models are used interchangeably by engineers. Ordinarily, the term quadrature implies that a new two-dimensional space is created by each signal interval and that detection and like activities in one interval do not interfere with those in others. This independence does not hold for other, advanced modulations, nor does it apply, strictly speaking, when a quadrature modulation is filtered; but we shall normally assume it is true in what follows.

Other ways exist to set up a two- or higher-dimensional signal space within one signal interval. The most common of these is the use of two or more signals with different carrier frequencies widely enough spaced so that they behave orthogonally. Knowledge of phase is not required here to achieve orthogonality. A scheme that can work without phase knowledge is called an *incoherent* modulation.

Our plan now is to describe the signals present in these basic modulations and calculate their error probability in terms of the energy invested in each data bit. Later sections will develop modulator and demodulator circuits and sophistications of these schemes.

Baseband and RF Signals for Quadrature Modulations

The expression for a general RF-modulated signal $s(t)$ that carries M-ary symbols is

$$s(t) = A(t) \cos(\omega_0 t + \psi(t)),$$

where $A(t)$ is the signal amplitude and $\psi(t)$ is its phase. We can rewrite this in the form

$$s(t) = \sqrt{\frac{2E_s}{T}}\,[I(t) \cos \omega_0 t - Q(t) \sin \omega_0 t] \tag{10.4-1}$$

with

$$I(t) = \frac{A(t)}{\sqrt{2E_s/T}} \cos \psi(t),$$
$$Q(t) = \frac{A(t)}{\sqrt{2E_s/T}} \sin \psi(t). \tag{10.4-2}$$

The truth of Eq. (10.4-1) follows from the trigonometric identity $\cos(A + B) = \cos A \cos B - \sin A \sin B$. In addition, we have normalized $I(t)$ and $Q(t)$ in a special way. Suppose that the average symbol energy E_s in Eq. (10.4-2) is set in such a way that

$$E\left\{\left(\frac{1}{T}\right) \int_{(n-1)T}^{nT} [I^2(t) + Q^2(t)]\, dt\right\} = 1, \tag{10.4-3}$$

with the expectation E over all the signals that can appear in the nth signal interval, an interval we assume is typical of all the others. Then the RF signal in (10.4-1) will invest an average per-interval energy of E_s in each M-ary symbol. This fact is easily verified by evaluating $\int s^2(t)\, dt$, with $s(t)$ equal to Eq. (10.4-1) and the range of integration the same as in Eq. (10.4-3).

The signals $I(t)$ and $Q(t)$ in Eq. (10.4-2) are called the *baseband components* of the total RF signal in (10.4-1). $I(t)$ is commonly referred to as the *in-phase* component, or signal, because it multiplies the original carrier $\cos \omega_0 t$, and $Q(t)$ is called the *quadrature* component. That there are two of these signals is a consequence of our use of a two-dimensional signal space with orthonormal basis functions

$$\phi_1(t) = \sqrt{\frac{2}{T}} \cos \omega_0 t,$$
$$\phi_2(t) = \sqrt{\frac{2}{T}} \sin \omega_0 t. \qquad (10.4\text{-}4)$$

Note, however, that $I(t)$ and $Q(t)$ are not components of signal-space vectors (even though they are sometimes called components!), but time functions. We will develop the vector components later, in the discussion of each modulation.

The process of passing from an RF signal to the baseband components in Eq. (10.4-2) is called *converting to baseband.* Not just a mathematical identity, this process forms the basis of most demodulator circuits. As we will see in Sec. 10.5, receivers commonly work by extracting the baseband signals from an RF signal and performing detection on these. Transmitters work by a reverse process, converting the baseband signals to an RF signal. Occasionally, transmission systems employ only baseband signals, without a conversion to RF.

The quantity

$$\sqrt{I^2(t) + Q^2(t)} \qquad (10.4\text{-}5)$$

is called the *envelope* of the RF signal. It is a normalized version of the instantaneous rms power in the signal. A constant-envelope modulation is one for which this expression is constant with time. This kind of modulation has advantages in a channel with nonlinear amplifiers.

A circuit that instantaneously converts a signal with a varying envelope to one with a constant envelope is called a *hard limiter*. Its effect on the I and Q components is to perform a conversion to $I_h(t)$ and $Q_h(t)$ given by

$$I_h(t) = \frac{I(t)}{\sqrt{I^2(t) + Q^2(t)}},$$
$$Q_h(t) = \frac{Q(t)}{\sqrt{I^2(t) + Q^2(t)}}. \qquad (10.4\text{-}6)$$

As in the case of FM in Chapter 7, digital signals that are purely phase modulated are not damaged by hard limiting.

Before turning to practical modulation schemes, we need formulas for the signal-space distance between two signals that are in baseband or RF form. The distance will directly give an error probability from the formulas of Sec. 10.3. Suppose $s_1(t)$ and $s_2(t)$ are the two RF signals

$$s_1(t) = A_1(t)\cos(\omega_0 t + \psi_1(t)),$$

$$s_2(t) = A_2(t)\cos(\omega_0 t + \psi_2(t)).$$

Calculating the distance between these, we get

$$\begin{aligned}\|\mathbf{s}_1 - \mathbf{s}_2\|^2 &= \int [s_1(t) - s_2(t)]^2\,dt \\ &= \tfrac{1}{2}\int A_1^2(t)\,[1 + \cos(2\omega_0 + 2\psi_1(t))]\,dt \\ &\quad + \tfrac{1}{2}\int A_2^2(t)\,[1 + \cos(2\omega_0 + \psi_2(t))]\,dt \\ &\quad - \int A_1(t)A_2(t)\,[\cos(2\omega_0 + \psi_1(t) + \psi_2(t)) + \cos(\psi_1(t) - \psi_2(t))]\,dt.\end{aligned} \tag{10.4-7}$$

Here the integral is over the interval of time that the signals operate, and we have used some standard trigonometric identities. Equation (10.4-7) can be simplified by a calculus fact that states

$$\int f(t)\cos(\omega_0 t + \theta(t))\,dt \longrightarrow 0, \qquad \text{as } \omega_0 \longrightarrow \infty$$

for any functions $f(t)$ and $\theta(t)$ that are slowly varying with respect to the changes in the signal $\cos \omega_0 t$. If we apply this to the ω_0 and the $2\omega_0$ terms in Eq. (10.4-7), the result is

$$\begin{aligned}&\|\mathbf{s}_1 - \mathbf{s}_2\|^2 \\ &= \int [\tfrac{1}{2}A_1^2(t) + \tfrac{1}{2}A_2^2(t) - A_1(t)A_2(t)\cos(\psi_1(t) - \psi_2(t))]\,dt, \text{ as } \omega_0 \longrightarrow \infty.\end{aligned} \tag{10.4-8}$$

Although we shall use Eq. (10.4-8) to compute distance from here on, it is well to remember that it applies precisely only in the limit of large carrier frequency. Otherwise, (10.4-7) provides the exact distance. Such a perturbation of distance actually occurs in systems whose carriers are not much larger than the symbol rate.

It remains to express the distance in terms of I and Q components. Starting from Eqs. (10.4-1) and (10.4-2) and carrying out the calculation as in Eq. (10.4-7), we find that

$$\begin{aligned}&\|\mathbf{s}_1 - \mathbf{s}_2\|^2 \\ &= \frac{E_s}{T}\int [I_1(t) - I_2(t)]^2 + [Q_1(t) - Q_2(t)]^2\,dt, \qquad \text{as } \omega_0 \longrightarrow \infty.\end{aligned} \tag{10.4-9}$$

Note the Pythagorean form of Eq. (10.4-9): Because the I and Q signals act along orthogonal directions, their square differences add to form the total distance. By substituting the definitions in Eq. (10.4-2) into Eq. (10.4-9), Eq. (10.4-8) can be derived.

Pure Phase-Shift-Keyed Schemes

Roughly speaking, a phase-shift-keyed (PSK) modulation is one in which only the phase is modulated by the data symbols. In some systems that are called phase-shift keyed, the phase of a *sinusoidal pulse* is modulated, and in these the envelope of the RF signal varies as well. We shall take these matters up in Sec. 10.6. The modulations that will be discussed now will have no envelope variation. We shall call these *pure* PSK systems.

The most important PSK modulations are binary and quaternary PSK, which carry two- and four-level symbols, respectively, in each signal interval. These modulations are commonly called by their acronyms, BPSK and QPSK. The RF signal of BPSK in the nth interval $(n-1)T \le t < nT$ is

$$\begin{aligned} \text{Send 0:} \quad s(t) &= \sqrt{\frac{2E_s}{T}}\cos(\omega_0 t), \\ \text{Send 1:} \quad &= \sqrt{\frac{2E_s}{T}}\cos(\omega_0 t + \pi) = -\sqrt{\frac{2E_s}{T}}\cos(\omega_0 t). \end{aligned} \tag{10.4-10}$$

Here the amplitude of the modulated cosine is chosen to give energy E_s to each binary symbol. In terms of baseband signals, BPSK has an I-component only, which is

$$\begin{aligned} \text{Send 0:} \quad I(t) &= \cos(0) = +1, \qquad (n-1)T \le t < nT, \\ \text{Send 1:} \quad &= \cos(\pi) = -1, \qquad (n-1)T \le t < nT. \end{aligned} \tag{10.4-11}$$

To express BPSK in signal space, we need to evaluate the vector components given by the inner product (s, ϕ_i), where the $\phi_i(t)$ are the basis functions of Eq. (10.4-4). This inner product is the integral $\int s(t)\phi_i(t)\,dt$ over the nth signal interval; its value in the first dimension is

$$\begin{aligned} \text{Send 0:} \quad s_1 &= +\sqrt{E_s}, \\ \text{Send 1:} \quad &= -\sqrt{E_s}, \end{aligned} \tag{10.4-12}$$

as $\omega_0 \longrightarrow \infty$. In the second quadrature dimension, the component turns out to be zero, from which we conclude the BPSK is a one-dimensional modulation. The signal constellation of BPSK in each signal interval is precisely that of Fig. 10.2-4(a), the antipodal modulation signal set. In addition, we can see from Eq. (10.4-11) that, except for a scaling by $\sqrt{E_s}$, the baseband signal can be interpreted directly as the first signal-space component.

Pure QPSK modulation carries a fourfold, or quaternary, symbol in one of four phases. These can be any set spaced 90° apart, but by convention the set $\psi = \{45°, 135°, 225°, 315°\}$ is assumed† ; this corresponds to the RF signal $\sqrt{2E_s/T}\cos(\omega_0 t + \psi)$, which is

$$\begin{aligned} \text{Send 00:} \quad s(t) &= \sqrt{\frac{2E_s}{T}}\left[+\frac{1}{\sqrt{2}}\cos(\omega_0 t) - \frac{1}{\sqrt{2}}\sin(\omega_0 t)\right], \\ \text{Send 10:} \quad &= \sqrt{\frac{2E_s}{T}}\left[-\frac{1}{\sqrt{2}}\cos(\omega_0 t) - \frac{1}{\sqrt{2}}\sin(\omega_0 t)\right], \end{aligned} \tag{10.4-13}$$

† The convention probably stems from the fact that QPSK can be realized as a superposition of two BPSKs.

$$\text{Send 11:} \quad = \sqrt{\frac{2E_s}{T}}\left[-\frac{1}{\sqrt{2}}\cos(\omega_0 t) + \frac{1}{\sqrt{2}}\sin(\omega_0 t)\right],$$

$$\text{Send 01:} \quad = \sqrt{\frac{2E_s}{T}}\left[+\frac{1}{\sqrt{2}}\cos(\omega_0 t) + \frac{1}{\sqrt{2}}\sin(\omega_0 t)\right]$$

in the nth interval $(n - 1)T \leq t < nT$. With the RF signal written in this form, the baseband signals are clear:

	$I(t)$	$Q(t)$	Phase
Send 00:	$+1/\sqrt{2}$	$+1/\sqrt{2}$	45°
10:	$-1/\sqrt{2}$	$+1/\sqrt{2}$	135°
11:	$-1/\sqrt{2}$	$-1/\sqrt{2}$	225°
01:	$+1/\sqrt{2}$	$-1/\sqrt{2}$	315°

(10.4-14)

As usual, E_s is the energy in one four-level symbol.

By the same process that led to Eq. (10.4-12), we can find the signal-space components of pure QPSK. These turn out to be identical to Eq. (10.4-14), with the first dimension component equal to the value of $I(t)$ and the second equal to the value of $Q(t)$, similar to the case of BPSK. The signal-space constellation that corresponds to Eq. (10.4-14) is Fig. 10.2-4(c); Fig. 10.2-4(d) also represents QPSK, an unconventional one, that has phases 0°, 90°, 180°, and 270°. In Eq. (10.4-14) we have chosen the double-binary notation for the quaternary transmitted symbol because QPSK can be viewed as sending two binary symbols, one in each of its signal dimensions. Because of the orthogonality of the two dimensions, these symbols can be thought of as independently transmitted by the independent signals $\cos(\omega_0 t)$ and $\sin(\omega_0 t)$. This is easiest to see from the constellation in Fig. 10.2-4(d).

We have already given the error probabilities of BPSK and QPSK in terms of the constellations in Figs. 10.2-4(a) and (c) as (10.3-7) and (10.3-11). The form of these is not the most convenient one because of the presence of the signal energy E_s. In the practical world, the measure of data is the number of *bits* it contains, since we receive a per-bit payment for each bit we accept for transmission and we incur energy cost when each bit is sent. The amount of energy invested in each data bit is called the *bit energy* E_b. This is related to the symbol energy of an M-ary modulation by $E_b = E_s/\log_2 M$, since (see Sec. 11.2) there are $\log_2 M$ binary symbols in an M-ary one.

With this substitution in Eqs. (10.3-7) and (10.3-11), our error probabilities become

$$p_e = Q\left(\sqrt{\frac{2E_b}{N_0}}\right) \qquad \text{(BPSK)},$$
$$p_e \leq 2Q\left(\sqrt{\frac{2E_b}{N_0}}\right) + Q\left(\sqrt{\frac{4E_b}{N_0}}\right) \qquad \text{(QPSK)}. \tag{10.4-15}$$

As was discussed in Sec. 10.3, these two are essentially the same, since both have the same exponential rate of decay, $\exp(-E_b/N_0)$, in terms of E_b. At an error rate of 10^{-5}, for instance, BPSK requires $E_b/N_o = 9.6$ dB (see Fig. 10.3-2), while QPSK needs slightly more, 9.9 dB. The bound $\frac{1}{2}\exp(-E_b/N_0)$ in Fig. 10.3-2 predicts about 10.3 dB for both modulations. At lower error probabilities, all three of these figures converge to the same one.

QPSK is the most commonly used PSK scheme. As will become clear in Sec. 10.5, it consumes the same bandwidth in sending 2 bits that BPSK consumes sending 1. At the same time, its error probability is virtually as good as BPSK's for the same data bit energy E_b. Constant-envelope phase modulations like QPSK find application wherever a radio or other transmission fails to preserve amplitude relationships in the signals. Satellite transmission, which ordinarily employs hard-limiting RF amplifiers, is a classic example.

When the number of phases is larger than four, M-ary PSK has a signal constellation consisting of M points spaced uniformly around a circle of radius $\sqrt{E_s}$. The only one of these to find any use is *octal* PSK ($M = 8$), which is shown in Fig. 10.4-1(a). It can be shown that the nearest-neighbor signal point in these constellations lies at distance

$$d = \frac{\sqrt{E_s}\sin(2\pi/M)}{\sin(\pi/2 - \pi/M)}, \qquad M > 4.$$

If this is substituted in Eq. (10.3-14), we get a tight overbound to the error probability. A good approximation to Eq. (10.3-14) for $M \geq 8$ is

$$p_e \approx 2Q\left(\frac{2\pi}{M}\sqrt{\frac{E_b \log_2 M}{2N_0}}\right). \tag{10.4-16}$$

Distance calculations show that E_b must be about 3.6 dB larger with octal PSK compared to BPSK and QPSK to achieve the same bit error rate. In return for this extra energy, octal PSK carries 50% more bits per signal interval than does QPSK.

Quadrature-Amplitude Modulation Schemes

A quadrature-amplitude modulation (QAM) scheme utilizes the same quadrature basis functions as do PSK schemes, but both phase and amplitude are modulated by the data. As with PSK, we must distinguish between pure QAM, which will be discussed now, and pulse QAM schemes, in which the phase and amplitude of a sinusoidal pulse are modulated. Pulse QAM will be deferred to Sec. 10.6.

The most straightforward kinds of QAM are those having RF signal

$$\sqrt{\frac{2E_s}{T}}\,[I(t)\cos\omega_0 t - Q(t)\sin\omega_0 t]$$

with

$$\begin{aligned} I(t) &= a_i\frac{\epsilon_0}{\sqrt{2}}, \qquad a_i = \pm 1, \pm 3, \pm 5, \ldots, \\ Q(t) &= b_i\frac{\epsilon_0}{\sqrt{2}}, \qquad b_i = \pm 1, \pm 3, \pm 5, \ldots. \end{aligned} \tag{10.4-17}$$

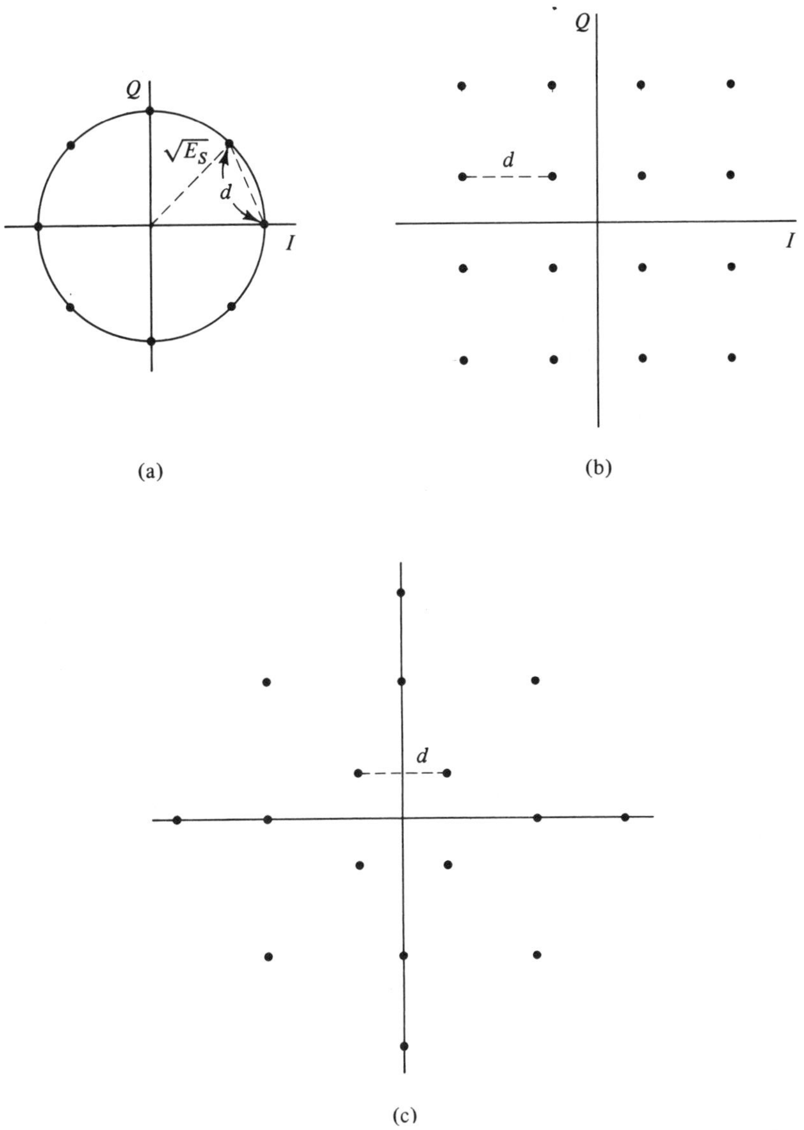

Figure 10.4-1. Two-dimensional signal constellations with more than four points. (a) 8-PSK. (b) 16-QAM with the same minimum distance. (c) 16-point phase error resistant constellation, U.S. Patent 3,887,768. (Courtesy V. Eyuboglu, Codex Corp., Mansfield, Mass.)

Here the constant ϵ_0 is chosen so that the energy normalization (10.4-3) holds. The symbols a_i and b_i carry the data; for an M-ary scheme, a_i and b_i must each take on $\sqrt{M}$ values. The result of this is a rectangular grid of points in signal space, a 16-ary example of which appears in Fig. 10.4-1(b). Note that QPSK can be viewed as a 4-ary QAM.

As with pure PSK, the QAM signal-space vector components are in general identical to the baseband signal values in Eq. (10.4-17). The term QAM, when applied to a specific modulation rather than a class, usually means the rectangular array of points like that in Fig. 10.4-1. The number of signal points often appears as a prefix; for instance, Fig. 10.4-1(b) is denoted 16-QAM. As in any modulation, the minimum distance between points in a QAM constellation has a dominating effect on the scheme's error probability. If this distance is not to shrink, the size, and hence the average symbol energy E_s, of the constellation must grow rapidly with the number of points. This is illustrated by Figs. 10.4-1(a) and 1(b).

QAM schemes are useful when the transmission channel preserves signal amplitudes and when enough signal energy is available. A standard example of this is the ordinary telephone channel. By definition, this is a channel having a bandwidth running from 300 to 3300 Hz and enough energy to give a ratio of E_b/N_0 in the range of 30 to 35 dB. 4-QAM (i.e., QPSK) and 16-QAM are commonly used to transmit data over telephone channels; schemes as large as 64-QAM are used. A carrier of about 1800 Hz is employed, and pulse shaping in the style of Sec. 10.6 is an important part of the design. Other important channels over which QAM schemes are used include coaxial cable and earth-bound microwave channels, since these can be designed to have sufficient energy and to preserve amplitudes.

Many modulations that are generically QAM do not have the perfectly rectangular pattern of Fig. 10.4-1(b). An example is a 16-point constellation patented by Codex Corporation, illustrated in Fig. 10.4-1(c). Its points have been shifted somewhat to provide extra protection from phase irregularities in the telephone channel, while keeping at the same time good distance properties and a low average energy. These signals are somewhat more complex to generate, since their baseband components do not satisfy the simple relationships in Eq. (10.4-17).

Frequency-Shift-Keyed Schemes

Frequency-shift-keyed (FSK) modulation is one in which the frequency of a carrier is modulated by the data. Since frequency is the derivative of phase, we can also take the view that the data modulate the slope of the phase in a phase-modulation system. FSK systems can be coherent or incoherent; that is, they can operate with or without a phase reference. A general expression for FSK signals is

$$s(t) = \sqrt{\frac{2E_s}{T}} \cos\left(\omega_0 t + a_n h[t - (n-1)T] \frac{\pi}{T} + \psi_0(n)\right), \qquad (n-1)T < t < nT, \tag{10.4-18}$$

where a_n is the nth data symbol, represented by $\pm 1, \pm 3, \ldots$, and ψ_0 is an optional phase offset at nT, the ends of the intervals. The constant h, called the *modulation index*, is the normalized slope, or "frequency," of the data-driven part of the signal phase. A plot of the phase in Eq. (10.4-18) (except for the carrier $\omega_0 t$) would show a piecewise-linear evolution, with phase discontinuities at the ends of the intervals in some schemes.

FSK modulations can be wideband ($h \gg 1$) or narrowband ($0 < h < 1$). Wideband modulations are incoherent, meaning that they work without a phase reference. Here the offsets ψ_0 in Eq. (10.4-18) can be ignored. These schemes are best viewed as digitally driven FM with the two frequencies in the binary case being $\omega_0 \pm h\pi/T$ rad/s. These widely spaced frequencies are virtually orthogonal, and the signal-space basis functions are $\sqrt{2/T}\cos(\omega_0 t + h\pi t/T)$ and $\sqrt{2/T}\cos(\omega_0 t - h\pi t/T)$. The applicable signal constellation is that of Fig. 10.2-4(b), which has error probability $Q(\sqrt{E_b/N_0})$ in this binary case. This means that wideband FSK requires about 3 dB more energy than QPSK or BPSK for the same error rate.

Wideband FSK is the modulation of choice in low-cost systems in which both energy and bandwidth (relative to the symbol rate) are plentiful. Examples are 300 symbol/s digital telephone modems like those used to access computers and radioteletype transmission in the shortwave band. In both cases, the rate of transmission falls far below that obtainable with more expensive equipment or with a modulation that needs a phase reference.

Narrowband FSK schemes can be coherent or incoherent. The signal space analysis of coherent schemes is beyond our scope, but as a rule they have considerably better error probability than their incoherent cousins. The exact probability depends on h, with some schemes equaling the $Q(\sqrt{2E_b/N_0})$ attained by QPSK and some performing even better than this. The price paid is the more complex detection and the acquisition of a phase reference. Although narrowband FSK can be incoherent, with some loss of error probability, it is still important to maintain phase continuity across symbol interval boundaries. This can be done by setting the phase offset term $\psi_0(n)$ equal to the sum of all phase changes prior to the nth interval. Without continuity, the signal bandwidth will be much increased.

Coherent, narrowband FSK finds use in certain advanced microwave and satellite systems in which its good energy efficiency and relatively narrow bandwidth are important. These modulations are also important in digital mobile radio where the radio spectrum is often very crowded.

10.5 QUADRATURE MODULATOR AND DEMODULATOR CIRCUITS

In this section we shall describe the circuitry that is used to transmit and receive modulations of the type in Sec. 10.4. Despite the simplicity of these circuits, they form a *maximum-likelihood* receiver for the "pure" PSK and QAM schemes of Sec. 10.4 and for a number of other useful schemes as well, including those that follow in Sec. 10.6. Often the circuits in this section are employed even when they are not a maximum-likelihood demodulator for the scheme at hand because of their simplicity and low cost.

We shall then turn to the spectra of the signals generated by the circuits. Usually, the bandwidth of a digital modulation is too wide for efficient use, so the RF signal must be bandpass filtered. This filtering has some unexpected effects.

When receiver and transmitter circuits appear together in the same box, the combination is called a *modem*. This term is a contraction of *mo*dulator and *dem*odulator.

Modulator/Demodulator Circuits: The Linear Receiver

The fundamental quadrature-modulation circuit is shown in Fig. 10.5-1. Its purpose is to carry out the baseband relation given in Eq. (10.4-1). When $I(t) = \cos \psi(t)$ and $Q(t) = \sin \psi(t)$, its effect is to create $A \cos(\omega_0 t + \psi(t))$, which is to say that the circuit modulates the carrier phase $\omega_0 t$ by the phase $\psi(t)$. Some typical baseband signals are shown in the figure for pure QPSK modulation. These signals are random square waves, taking values ± 1 for T seconds, with independent signals in the I and Q channels of the figure. These square-wave baseband signals are sometimes called NRZ pulses (for *nonreturn to zero*) to distinguish them from pulses used in other types of PSK.

A modulator for BPSK consists of just the top half of Fig. 10.5-1, the in-phase channel; this will generate a carrier shifted by 0° or 180°. QAM signals are generated through application of the proper I and Q signals, like those in Eq. (10.4-17). The heart of the quadrature modulator is the $\sin \omega_0 t / \cos \omega_0 t$ multiplication, which is carried out by a balanced modulator circuit like those in Sec. 6.3. An RF power amplifier and often a band-pass filter follow, although these are left off the figure.

Figure 10.5-2 shows a general quadrature demodulator circuit. This begins with a balanced modulator like the one in Fig. 10.5-1, the effect of which is to extract the baseband signals from the RF signal. We can see this from the trigonometric identities

$$A[I(t) \cos \omega_0 t - Q(t) \sin \omega_0 t] 2 \cos \omega_0 t = AI(t) + AI(t) \cos 2\omega_0 t - AQ(t) \sin 2\omega_0 t \qquad (10.5\text{-}1a)$$

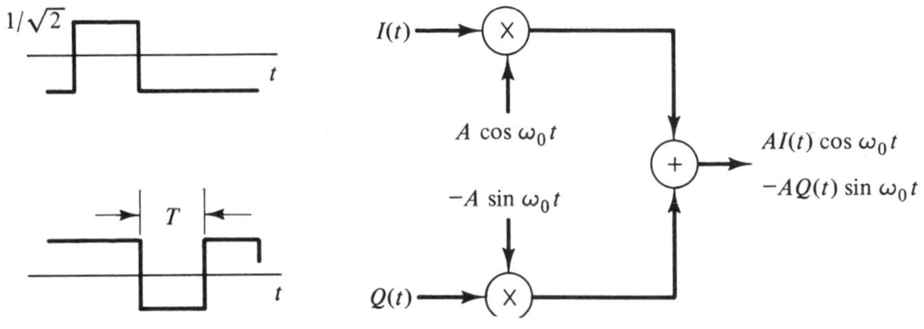

Figure 10.5-1. Basic quadrature modulation circuit with typical inputs for QPSK.

and

$$-A[I(t)\cos\omega_0 t - Q(t)\sin\omega_0 t]2\sin\omega_0 t = AQ(t) - AQ(t)\cos 2\omega_0 t - AI(t)\sin 2\omega_0 t. \tag{10.5-1b}$$

The first terms in each of these are the low-pass baseband signals I and Q, while all the remaining terms are RF signals centered at twice the carrier. A simple low-pass filter (labeled LPF) removes these, leaving only I and Q. These recovered baseband signals are seldom identical to the original I and Q because of channel noise and the effects of filtering and limiting, so we shall use the notations $\hat{I}$ and $\hat{Q}$, for the "estimated" signals, to refer to these at the receiver.

The next elements in the demodulator are the filters $H(f)$, one each in the in-phase and quadrature channels. After these are samplers that observe the filter output at the end of each signal interval. Finally, their samples are compared to thresholds, and from this it is decided what transmitted symbols are present in each channel. In designing this circuit, we are free to choose the filter $H(f)$ and the set of thresholds so as to minimize the probability of detected symbol error. The technical name for this demodulator is the *linear receiver*, after the fact that except for a frequency conversion to baseband the receiver consists only of a linear filter, whose output it observes.

The two remaining blocks in the figure are the carrier and bit timing synchronizers. The first of these must regenerate the exact unmodulated carrier $\cos\omega_0 t$, without which the balanced modulator circuit cannot obtain $\hat{I}$ and $\hat{Q}$. The second sets the optimal moment to sample the output of the $H(f)$ filters. These two synchronizers comprise a large part of the design of a demodulator. We shall return to them in more detail in Sec. 10.7.

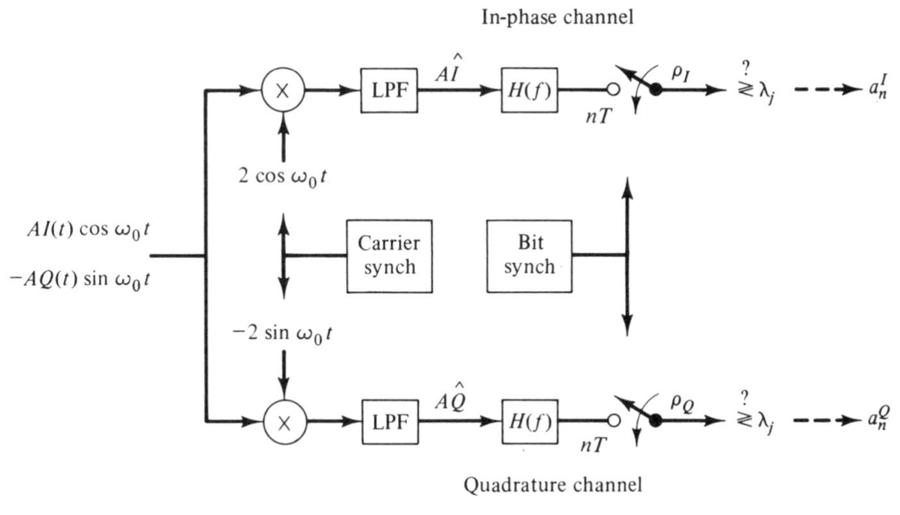

Figure 10.5-2. General quadrature demodulator. A detector of this form is called a linear receiver.

For an application of the linear receiver, we turn first to the important case of pure QPSK. Our plan is to investigate first the matched filter receiver, a maximum-likelihood design, and then show that its form is that of the linear receiver just described. In QPSK, the square-wave baseband signals $I(t)$ and $Q(t)$ are independent of each other, and they take on values $\pm 1\sqrt{2}$ during a signal interval, independently of the other intervals. We shall assume that I and Q are corrupted by white Gaussian noise (it is not hard to show that this is the case if the RF signal has the same noise) and proceed to design a matched-filter receiver like Fig. 10.2-6 for the *baseband* signals.

Focusing on just the I channel, we see that in theory two matched filters are needed, since $I(t)$ takes on the constant values $\pm 1/\sqrt{2}$ during an interval. But we can obtain what the output of one filter would have been by inverting the sign of the other's output, so only one is really required. Let this filter be matched to $+1/\sqrt{2}$ in the first interval. Then its impulse response should be $I(T - t)$, which is

$$h(t) = \begin{cases} \dfrac{1}{\sqrt{2}}, & 0 \le t < T, \\ 0, & \text{otherwise.} \end{cases}$$

This filter has no response outside the first interval and so is unaffected by signals there. In fact, it behaves as an *integrator*, integrating $\hat{I}(t)$ over just the present interval. The matched filter for the next interval can be the same integrator if its contents are dumped before the next interval begins.

To finish adapting Fig. 10.2-6 to QPSK, we remove the offset summing junctions, since QPSK consists of equal-energy signals. The "select the largest" node can be replaced by a threshold test of whether the single matched-filter output is above or below zero. This is because, with one filter matched to $+1/\sqrt{2}$, the output is $+\frac{1}{2}$ for a match to signal $1/\sqrt{2}$ and $-\frac{1}{2}$ for an "antimatch" to signal $-1/\sqrt{2}$. A complete diagram of this QPSK receiver appears in Fig. 10.5-3. At the end of each interval, the integrators are sampled and their contents are dumped just after to prepare for the next interval. This kind of receiver is called an *integrate-and-dump*, or I-*and*-D, receiver. Its integrator can in principle be implemented by a single charging capacitor with a diode switch to provide the dumping; but, despite this simplicity, the I-*and*-D *receiver is maximum likelihood for pure QPSK*. Being so simple, it is also commonly used with other schemes for which it is not optimal.

We can understand the operation of the I-and-D receiver just from common sense. The integrator averages the noisy $\hat{I}$ and $\hat{Q}$ waveforms during each interval, and the threshold finds whether the average was above or below zero. The signal-space analysis proves that one cannot do better than this for square pulses and Gaussian noise. As always, the BPSK receiver is just the top half of Fig. 10.5-3.

Receivers for QAM

A matched-filter receiver for square-pulse "pure" QAM with a rectangular constellation like that in Fig. 10.4-1 is almost the same as the receiver in Fig. 10.5-3. This is because the I and Q channel signals are multiples [see Eq. (10.4-17)] of the same

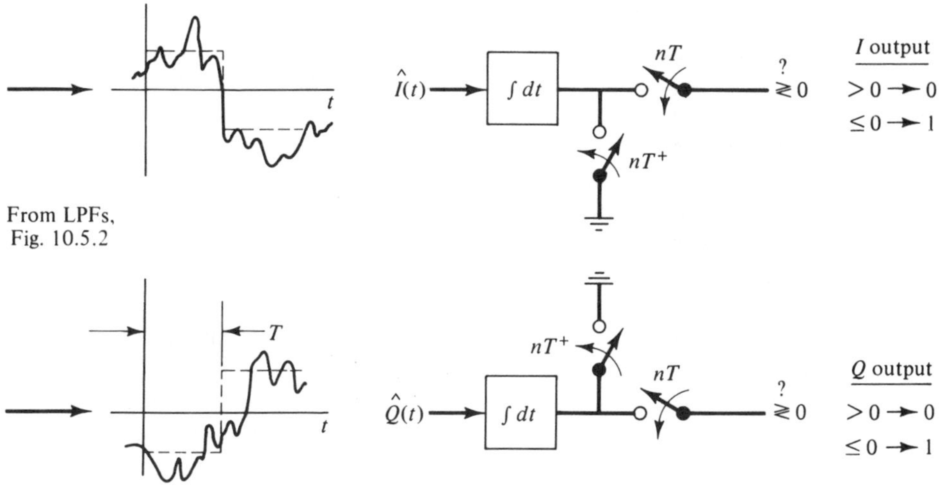

Figure 10.5-3. Integrate-and-dump receiver, showing typical noise $\hat{I}$ and $\hat{Q}$ inputs for pure QPSK. The balanced modulator and synchronizers of Fig. 10.5-2 are left off for clarity.

basic square pulse as in QPSK. Furthermore, the rectangular array implies that the symbols in the *I* and *Q* channels can be viewed independently. Thus all matched filters in a channel can be represented by one filter, the integrator, whose output is appropriately scaled. Rather than scale, we can compare the output of the standard filter to a series of thresholds as depicted in Fig. 10.5-4(a). Figure 10.5-4(b) shows the decision regions that this creates in signal space; these are clearly the regions for a maximum-likelihood receiver.

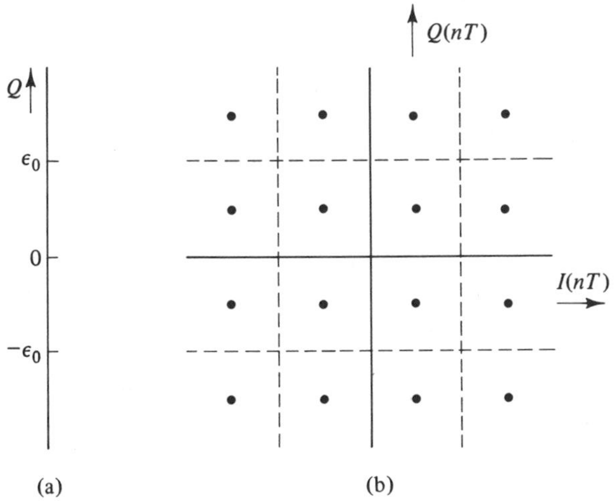

Figure 10.5-4. (a) *Q*-channel thresholds for QAM receiver, assuming the signals in Eq. (10.4-17) and the standard *Q*-channel signal height $1/\sqrt{2}$. (b) Decision regions in signal space implied by *I* and *Q* threshold decisions that act independently.

When the QAM scheme is nonrectangular [Fig. 10.4-1(c)] or when octal PSK [Fig. 10.4-1(a)] is transmitted, the integrator is still a matched filter, since the baseband pulses are still square pulses. But simple observation of separate I and Q thresholds will not lead to a maximum-likelihood receiver. Instead, a more complex processor, working from both the $\hat{I}$ and $\hat{Q}$ samples for an interval, must determine in which decision region these received coordinates lie.

I-and-D receivers are suboptimal for FSK detection. A true FSK matched filter receiver must provide filters matched to all the signals of the kind [Eq. (10.4-18)] that can appear in a single interval. For binary coherent continuous-phase FSK, a Gram–Schmidt orthogonalization of these produces in general a four-dimensional signal space. Two distinct matched filters are needed in each of the I and Q channels. The detailed design of these receivers is beyond our scope and we refer to more advanced texts ([10-3] and [10-4]). For wideband, incoherent FSK, only one matched filter for each frequency is needed; in practice, these are just high-Q resonant circuits tuned to each frequency.

Modulation Spectra

It is important to calculate the spectra of digital modulations because transmission bands are strictly regulated by government authorities. Radiation outside a fixed bandwidth must fall below a certain fraction of the total signal power to avoid interference with transmissions in neighboring channels. The modulated spectra of the "pure" schemes in this section rarely meet these restrictions, and so the schemes must be band-pass filtered.

Figure 10.5-5 shows the measured power spectra of several binary modulations, including continuous-phase FSK, pure BPSK, and BPSK that has been filtered. We use the term spectrum here for what is really the average of a number of relatively short term Fourier power spectra taken by the measuring instrument. This is similar but not identical to the notion of power spectral density introduced in Sec. 9.5. When speaking of modulation spectra, we shall always mean the first of the two, the average of the Fourier spectra.

The spectra in the figure consist of a strong central portion, called the *main lobe*, and a number of lesser peaks, called *side lobes*. As a rule, the main lobe carries most of the content of the modulation, and as long as it is received intact, the error probability of the scheme will be not much degraded. The chief significance of the side lobes is that they interfere with transmissions in neighboring channels. An important aim of signal design is to reduce these. It is clear from the figure, for instance, that BPSK obstructs neighboring channels far more than does FSK, despite the fact that all modulations in the figure carry the same 16 kbits/s and all have the same optimal-receiver error probability.

It is quite easy to calculate the spectra of modulations in this section and later in Sec. 10.6. The key requirement is that the signals in successive signal intervals be phase shifts of each other. To begin, let us review the spectra given rise to by transmission of a single symbol. The symbol causes a single square pulse of duration T seconds in

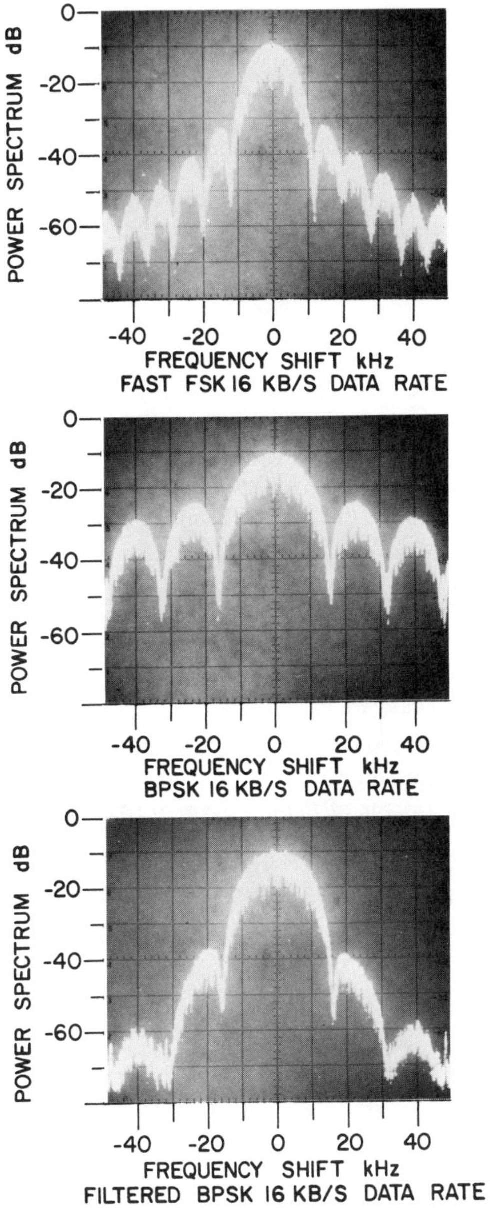

Figure 10.5-5. Measured spectra of some 16-kbits/s binary modulations. Coherent continuous-phase FSK, Eq. (10.4-18), with $h = \frac{1}{2}$ (top); BPSK (middle); BPSK with four-pole filtering, corner frequency near 16 kHz (bottom). Note the reduced spectral side lobes of FSK, compared to unfiltered BPSK. (Courtesy J. L. Pearce, Communications Research Centre, Government of Canada, Ottawa.)

the I and/or Q channels; the response to the symbol outside its interval is zero. The Fourier transform of $I(t)$ [or of $Q(t)$] is

$$\tilde{I}(f) \equiv \mathcal{F}\{I(t)\} = e^{-j\pi fT} KT \operatorname{sinc}(fT) \tag{10.5-2}$$

with the factor K the height of the square pulse. Assuming this pulse is RF-modulated, we get for the RF spectrum either one of

$$\begin{aligned} \mathcal{F}\{I(t)\cos\omega_0 t\} &= \frac{1}{2}\tilde{I}(f-f_0) + \frac{1}{2}\tilde{I}(f+f_0), \qquad f_0 = \frac{\omega_0}{2\pi} \\ \mathcal{F}\{Q(t)\sin\omega_0 t\} &= -j\frac{1}{2}\tilde{Q}(f-f_0) + j\frac{1}{2}\tilde{Q}(f+f_0). \end{aligned} \tag{10.5-3}$$

Here the tilde denotes Fourier transform of the function with the same symbol.

Next we show that the power spectrum of the entire modulated signal is just a scaled version of the one-symbol spectrum. To do this, we make a list $s_1(t)$, $s_2(t)$, . . . of all the one-symbol signals, including as separate entries the I- and Q-channel signals for each interval. We shall state the desired result in a more general form so that it can be used in later sections.

Lemma. Suppose a transmitted signal $s(t)$ is the superposition of pulses

$$s(t) = \sum_{i=1}^{N} a_i s_i(t), \qquad E[a_i] = 0$$

where $\{a_i\}$ are zero mean, independent, and identically distributed data symbol values, and the magnitudes of all the Fourier transforms $S_1(f)$, $S_2(f)$, . . . , $S_N(f)$ are the same. Then the average square magnitude of the transform $S(f)$ of $s(t)$ is

$$E[|S(f)|^2] = NE[|a_1|^2]|S_1(f)|^2. \tag{10.5-4}$$

Proof. The transform of $s(t)$ is simply

$$S(f) = \sum_{i=1}^{N} a_i S_i(f),$$

which has square magnitude

$$\begin{aligned} |S(f)|^2 &= \left[\sum_{i=1}^{N} a_i S_i(f)\right]\left[\sum_{k=1}^{N} a_k S_k(f)\right]^* \\ &= \sum_{i=1}^{N}\sum_{k=1}^{N} a_i a_k^* S_i(f) S_k^*(f). \end{aligned}$$

Our measure of spectrum is the expectation of $|S(f)|^2$. Equation (10.5-4) follows from the observation that

$$E[a_i a_k^* S_i(f) S_k^*(f)] = \begin{cases} |S_i(f)|^2 E[|a_i|^2], & i = k, \\ 0, & i \neq k, \end{cases}$$

where the latter case follows from the independence and zero-mean properties of the symbol levels.

Now we apply Eq. (10.5-4) to pure QPSK. The modulation consists of I and Q pulse trains made up of time-shifted pulses; furthermore, the I and Q spectra in Eq. (10.5-3) differ only by a phase shift j; finally, all the symbols in QPSK, whether I or Q, are assumed independent. We conclude that the lemma applies, and with the energy normalization in Eq. (10.4-3) we get, for the average power spectrum,

$$E[|S(f)|^2] = \frac{NE_s}{2}[T\,\mathrm{sinc}^2(T(f-f_0)) + T\,\mathrm{sinc}^2(T(f+f_0))], \qquad (10.5\text{-}5)$$

where N is the number of transmitted intervals. The middle spectrum in Fig. 10.5-5 is this formula, plotted in decibels. It has zeros, called *spectral nulls*, at $|f - f_0|$ equal to each integer multiple of $1/T$. For pure BPSK, only $I(t)$ is active, but the shape of the power spectrum, expression (10.5-5), is the same. We must keep in mind, however, that QPSK carries two data bits each T seconds, so a 16-kbits/s QPSK system plotted on the same scale as Fig. 10.5-5 would have a spectrum compacted by a factor of 2. This spectral advantage of QPSK over BPSK is overwhelming, and QPSK has displaced BPSK in most applications.

For a modulation in which the responses to successive data bits are not orthogonal, the spectral calculation is quite a difficult procedure. A modern, relatively simple approach appears in [10-4]. Fortunately, the method we have shown here adapts to many modulations in common use, like those in Sec. 10.6.

As mentioned before, spectra like those defined by Eq. (10.5-5) are ordinarily much too wide for commercial use, and they must be filtered, sometimes quite severely. This can have surprising effects. Pure BPSK and QPSK, for instance, which are constant-envelope modulations, develop large RF envelope variations when filtered. This is demonstrated in Fig. 10.5-6; all the bandwidths shown are in common use. The effect shown extends more or less to all constant-envelope modulations and all filters. If the filtered waveform passes through a nonlinearity, like a saturating class C RF amplifier or a hard limiter, the envelope variations are distorted and the design of a good detector becomes more difficult. In applications like satellite transmission, a saturating amplifier is almost a necessity because of its inherently higher efficiency. These matters are treated further in [10-3].

As a general rule, the effects of band-pass-filtering a pure PSK modulation become noticeable when the product BT falls below 2, where B is the double-sideband RF bandwidth of the filter. BT in effect measures the bandwidth consumption of the scheme in cycles per transmitted symbol. Filter effects become severe when BT falls below 1.4; 4 to 8 dB more signal energy is needed to overcome the disturbances caused by the filter, and detection with a linear receiver soon becomes impossible. Filtering effects are made worse by amplitude nonlinearities in the signal path. As an example of filtering effects, consider a 1 Mb/s pure QPSK system with a symbol time $T = 1/500000$ s. When this system occupies a total RF bandwidth of about 1 MHz, it will show little degradation from the ideal, and its error rate will be near the value in (10.4-15). Its bandwidth can be reduced to about 700 kHz with significant, but acceptable losses.

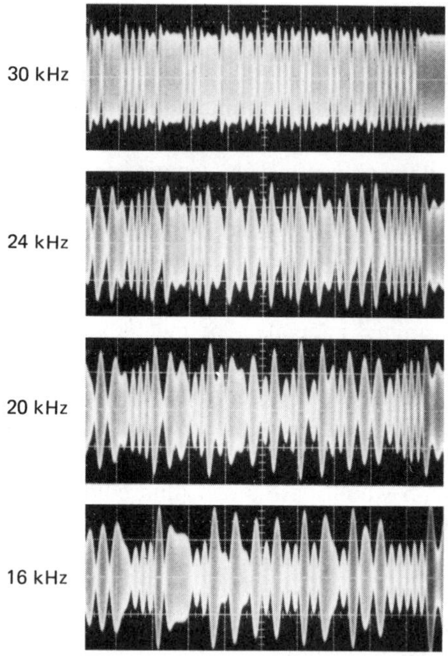

Figure 10.5-6. Variations in the envelope of the 16-kbits/s pure BPSK of Fig. 10.5-5, caused by a Chebychev filter with 3-dB double-sideband RF bandwidths shown. (Courtesy J. L. Pearce, Communications Research Centre, Government of Canada, Ottawa.)

This is a bandwidth consumption of 0.7 cycles/data bit. The RF signal envelope will resemble the 20 kHz picture in Fig. 10.5-6.

10.6 PULSE-SHAPED MODULATIONS

In Sec. 10.5 we saw one method to reduce the bandwidth of a pure quadrature-modulation scheme, band-pass filtering of the RF signal. A second way is to abandon the square baseband pulses used in the pure schemes in favor of pulses with smoother transitions. According to Eq. (10.5-3), the RF spectrum of an orthogonal-pulse modulation is just the spectrum of I and Q translated to the carrier frequency. By choosing I and Q pulses with narrow spectra, we reduce the RF spectrum as well. In actual fact, a pulse-shaped modulation can be realized as a pure modulation with RF filtering, and vice versa, although in a practical system one approach is usually easier.

We shall concentrate on the most important class of smoothed pulses, a class based on the Nyquist sampling pulses that were introduced in Chapter 4. These can produce a double-sideband RF spectrum limited to as little as $1/T$ hertz, that is, a spectrum equal to the modulation's symbol rate. A striking and eminently useful fact is that the maximum-likelihood receiver for these modulations is *still the linear receiver*. With other pulse shapings, a much more complex receiver is needed for optimal performance. This one, which is similar to the Viterbi algorithm of Sec. 11.7, is beyond our scope.

Nyquist-Pulse Modulations

We say a pulse satisfies the *Nyquist pulse criterion* if it passes through zero at times nT for some T, with $n = \pm 1, \pm 2, \ldots$, but not at time zero. An example of such a pulse is $\text{sinc}(t/T)$, which equals 1 at $t = 0$ and 0 at all other multiples of T. We have already seen the function sinc(·) in Sec. 4.2, where it played the role of interpolating function in the recovery of a waveform from ideal samples.

The role of sinc(t) and other Nyquist pulses in modulation is somewhat different from what it was in sampling. Rather than recovering an arbitrary waveform from samples, we create a band-limited waveform from samples that are restricted to M discrete values. Only certain band-limited waveforms can so be created, and the receiver tries to determine which one it was. Consider transmission of the sequence of M-ary real numbers $a_0, a_1, a_2, \ldots$, which represent a digital data sequence; in a binary scheme, for instance, the data symbols 0 and 1 are represented by the real numbers +1 and −1. Let the sequence be carried by superposing Nyquist pulses in the manner

$$v(t) = \sum_{n=0}^{\infty} a_n \sqrt{\frac{1}{T}} \text{sinc}\left(\frac{t - nT}{T}\right), \tag{10.6-1}$$

which is illustrated in Fig. 10.6-1. Note the difference between this binary modulated waveform and the sample-recovery waveform in Fig. 4-2.1.

It is clear that the data carried by $v(t)$ can be recovered exactly in the absence of noise by sampling $v(t)$ at times nT. Historically, the motivation to use Nyquist pulses was just this property. A receiver that works by sampling a baseband waveform and comparing to a threshold (zero, in the binary case), without the $H(f)$ filter in Fig. 10.5-2, is called a *sampling receiver*. Its performance in noise is much inferior to the full linear receiver.

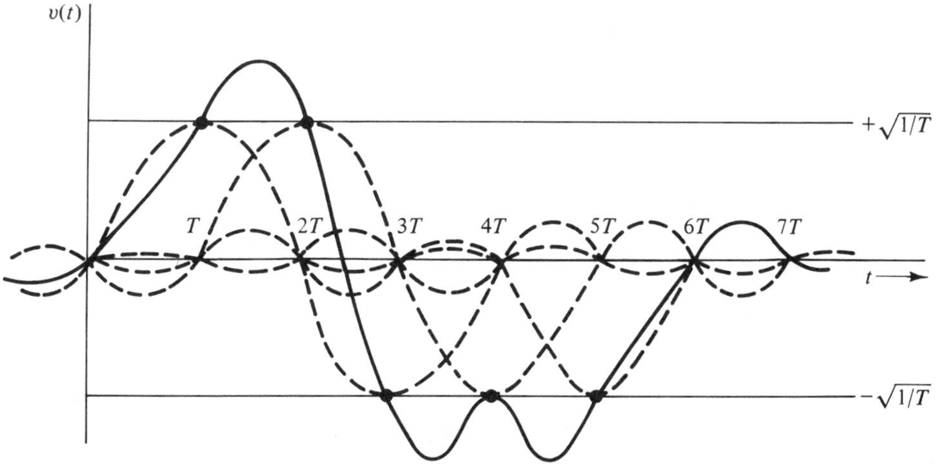

Figure 10.6-1. Sinc pulse baseband waveform generated by the symbols +1, +1, −1, −1, −1. The solid curve is the sum of the dashed pulses.

The set of time-shifted sinc(t) functions in Eq. (10.6-1), $\{\sqrt{1/T}\,\mathrm{sinc}((t - nT)/T)\}$ can be shown to be an orthonormal set over the real line. Consequently, the energy in $v(t)$ is

$$\int_{-\infty}^{\infty} v^2(t)\,dt = \sum_{n=0}^{\infty} a_n^2. \tag{10.6-2}$$

If we transmit the signal $s(t) = \sqrt{E_s}v(t)$ and require that $E[a_n^2] = 1$, then E_s is the per-symbol energy of the signal. This kind of signal is called a digital *pulse amplitude modulated*, or PAM, signal. It is simply a superposition of pulses that satisfy the Nyquist pulse criterion, without modulation to a carrier. PAM signaling finds some use in wire and coaxial cable systems.

If two PAM signals like $v(t)$ are applied to the I and Q inputs of Fig. 10.5-1, we create a *Nyquist pulse QPSK* modulation. Omitting the Q channel creates Nyquist BPSK. We can once again show orthonormality, this time for the joint set of functions $\{\sqrt{2/T}\,\mathrm{sinc}((t - nT)/T)\cos\omega_0 t\} \cup \{-\sqrt{2/T}\,\mathrm{sinc}((t - nT)/T)\sin\omega_0 t\}$ in the limit of large ω_0. The output of the modulator will be [Eq. (10.4-1)]

$$s(t) = \sqrt{\frac{2E_s}{T}}\,[I(t)\cos\omega_0 t - Q(t)\sin\omega_0 t]$$

with

$$\begin{aligned} I(t) &= \sum_{n=0}^{\infty} a_n^I\,\mathrm{sinc}\frac{t - nT}{T}, \\ Q(t) &= \sum_{n=0}^{\infty} a_n^Q\,\mathrm{sinc}\frac{t - nT}{T}. \end{aligned} \tag{10.6-3}$$

This quantity E_s is the symbol energy if $E[(a_n^I)^2 + (a_n^Q)^2] = 1$; for Nyquist QPSK, this means that all the real-number symbols are $\pm 1/\sqrt{2}$, as they were in Sec. 10.5.

Since the pulses here are time shifts of each other, Eqs. (10.5-3) to (10.5-4) give the power spectrum for sinc-pulse QPSK. It is proportional to

$$\frac{E_sT}{2}\,[\mathrm{rect}(T(f - f_0)) + \mathrm{rect}(T(f + f_0))], \qquad f_0 = \frac{\omega_0}{2\pi}. \tag{10.6-4}$$

This is a spectrum with double-sideband RF bandwidth $1/T$ Hz, with energy E_s. Sinc-pulse and pure QPSK are time-and-frequency duals of each other.

The signal-space expression for sinc-pulse QPSK turns out to be identical to that of pure QPSK, the four-point square constellation of Fig. 10.2-4(c), once for each signal interval. To verify this, we evaluate the inner product (s, ϕ_i) for a signal of the form of Eq. (10.4-1) or Eq. (10.6-3), where $\phi_i(t)$ is one of the orthonormal basis functions for the modulation. $\phi_i(t)$ will be orthogonal to all the pulses that make up $s(t)$ except for one, the $\mathrm{sinc}((t - nT)/T)\sin(\omega_0 t)$ or $\mathrm{sinc}((t - nT)/T)\cos(\omega_0 t)$ that is identical to it. The inner product with that will be $a_n^I\sqrt{E_s}$ or $a_n^Q\sqrt{E_s}$. The example of sinc-pulse QPSK illustrates that a modulation with complicated pulses may have quite a simple expression in signal space. Our signal-space result here has an important consequence: We have shown that sinc-pulse QPSK has error probability identical to pure QPSK, that is, Eq. (10.4-15).

Other Nyquist Pulses.

The use of sinc(t) to carry data leads to certain difficulties. An immediate problem is the fact that the pulse has infinite time duration and is noncausal. An acceptable solution to this is to truncate sinc(t) to a total width of perhaps $10T$ and then delay it half this width, to produce a causal pulse. A more serious problem is the size of the side lobes of sinc(t) in the time domain; the heights of these decay only as fast as $T/(t - nT)$, an inverse-linear rate. The practical result is that very accurate symbol timing recovery is required at the receiver. We would prefer to have pulses with smaller side lobes.

A commonly used alternative to the sinc(t) pulse is the *spectral raised cosine* pulse. This is defined to be the impulse response of $H(f)$, where

$$H^2(f) = \begin{cases} 1, & 0 < |f| < \dfrac{1-\alpha}{2T}, \quad 0 < \alpha < 1, \\ \cos^2\left[\dfrac{\pi T}{2\alpha}\left(f - \dfrac{1-\alpha}{2T}\right)\right], & \dfrac{1-\alpha}{2T} < |f| < \dfrac{1+\alpha}{2T}, \\ 0, & \text{elsewhere.} \end{cases} \tag{10.6-5}$$

We have already seen a raised cosine pulse in Eq. (4.4-8), in the context of recovering a function from its samples. A proof is given in Sec. 4.4 that $H(f)$ must satisfy Eq. (4.4-6) in order for the pulse $h(t)$ to satisfy the Nyquist pulse criterion; a study of Eq. (10.6-5) will show that $H(f)$ here indeed satisfies Eq. (4.4-6). An additional proof, which we omit, shows that all time shifts $h(t - nT)$ are orthogonal, just as were the shifts of the sinc(t) pulse. Consequently, a raised cosine QPSK has the usual four-point square constellation and error rate [Eq. (10.4-15)].

With these properties in hand, we can once again use Eqs. (10.5-3) and (10.5-4) to show that the spectral raised cosine QPSK has RF power spectrum proportional to $H^2(f - f_0) + H^2(f + f_0)$; that is, its spectrum is flat out to $|f - f_0| = (1 - \alpha)/2T$, after which it rolls off to zero at $(1 + \alpha)/2T$. The parameter α is called the *excess bandwidth factor*, because the quantity $1 + \alpha$ is the factor by which the bandwidth of raised cosine pulses exceeds that of sinc(t) pulses. It can be shown that the temporal side lobes of $h(t)$ steadily decrease as α grows, giving us a bandwidth/side lobe trade-off. Figure 4.4-5 plots the spectrum and pulse shape of the $\alpha = 1$ pulse.

In the design of a raised cosine QPSK modulator, the filter H is approximated with a 6- to 12-pole filter; filters with excess bandwidth factors larger than about 0.3 are relatively easy to realize in this way. In a 1 Mbit/s QPSK system like the one at the end of Sec. 10.5, $\alpha = 0.3$ would yield a double-sideband RF bandwidth of about $2(1 + \alpha)(1/2T) = 650$ kHz, instead of the 700 kHz found in Sec. 10.5. These are comparable bandwidths. Pure QPSK needs an RF filter to achieve them, while raised cosine QPSK requires a somewhat more complex baseband pulse-shaping filter. Neither modulation has constant envelope. We shall see shortly, however, that raised cosine QPSK has a better probability of error.

Another type of Nyquist pulse that finds use is the time-domain (or *temporal*) raised cosine pulse

$$h(t) = \begin{cases} 1 + \cos \dfrac{\pi t}{\beta T}, & 0 < |t| < \beta T, \qquad \text{for some } \beta > 0 \\ 0, & \text{otherwise.} \end{cases} \tag{10.6-6}$$

With $\beta = 1$, this pulse is identical to Fig. 4.4-5(a), except that the frequency axis is replaced by time and the pulse ends at T. Raising a cosine cycle gives a rounded pulse that merges with the time axis in a very smooth way, a feature that reduces the pulse's spectral side lobes. A disadvantage of temporal raised cosine pulses is that successive pulses $h(t - nT)$ are not orthogonal. The practical effect of this is that the linear receiver is not always maximum likelihood.

A pulse that trivially meets the Nyquist pulse criterion (and is orthogonal) is the rectangular pulse that we used in pure QPSK schemes.

Receiver Eye Diagrams

Much insight into detector performance can be obtained from plotting a superposition of all the I channel (or Q channel) waveforms. This kind of plot is called an *eye diagram*, because it will have the appearance of an open eye when the waveforms are such that we can expect good linear receiver performance. There is an easy way to imagine the construction of an eye diagram, which is also the method universally employed in field measurements. The total baseband waveform is connected to an oscilloscope whose time base is triggered by the receiver bit timing once each T seconds, while the transmitter is fed pseudorandom data that will drive the baseband waveform through all possible patterns. The result for a temporal raised cosine pulse with $\beta = 1$ is the top trace in Fig. 10.6-2.

Every waveform that can appear during a signal interval is superposed in this figure. Among the traces is the complete pulse $h(t)$ from Eq. (10.6-6) and the same pulse upside down, which was created by the antipodal data stream. The straight lines across the top and bottom were created by the streams . . . $+1, +1, +1,$ . . . and . . . $-1, -1, -1,$ . . . . A good exercise is to match all the trajectories in the figure with the proper ± 1 data streams. Note that Fig. 10.6-2 is not a superposition of separate *pulses*, like Fig. 4.4-2, but of total waveforms, $\Sigma\, a_n h(t - nT)$.

An important characteristic of Fig. 10.6-2 is the values of the trajectories at the receiver sampling instant nT: In the top eye diagram they are all exactly $+1$ or -1. These are the data symbols in the waveforms that correspond to the nth interval, a fact that is guaranteed by the Nyquist pulse criterion, whatever the symbols are in neighboring intervals.

When these waveforms are filtered by a 1-pole filter, the bottom eye diagram in Fig. 10.6-2 results. It would be very tedious to compute all the filtered trajectories

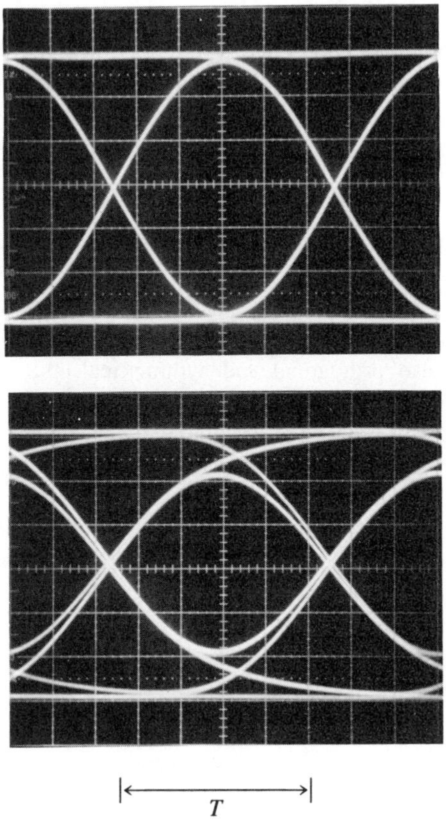

Figure 10.6-2. Eye diagrams at baseband for 9.6-kbits/s binary scheme with Nyquist pulses. Unfiltered eye (top); filtered with one-pole filter, corner frequency at 5 kHz (bottom). (Courtesy G. J. Saulnier, General Electric Corporate Research and Development Center, Schenectady, N.Y.

here, but a simple laboratory measurement easily shows that the filter's effect is the distortions in the diagram. Most of the trajectories no longer pass through $+1$ or -1 at the sampling instant, but all are still on the correct side of the zero line. Thus the sampling receiver will always make the correct decision if it compares its sample to threshold 0, despite the filtering. So will an integrate-and-dump receiver that averages the trajectories over interval n. Correct decisions will always be made as long as the eye-shaped region in the middle of the diagram is open.

Eye diagrams can diagnose many effects besides filtering. Channel noise widens the filtered trajectories into fuzzy regions because the noise makes formerly identical trajectories slightly different from each other. If the receiver's recovery of the interval timing is faulty, some of the trajectories will slip to the left or right. Excessive time-domain side lobes, like those in the sinc(t) pulse, will cause the open region to narrow. Whatever the source of these reductions in the eye, a reduction in eye height makes the receiver more sensitive to noise, whereas a reduction in width makes it more sensitive to timing errors. If the eye is reliably open about the point $(0, nT)$, the receiver will have a low bit error rate.

The filtering effects in Fig. 10.6-2 are called *intersymbol interference* (ISI). We encountered this term in the theory of sampling in Sec. 4.4 and we shall see it again in Chapter 12 in an optical communication context. We have learned here that parts of a pulse outside its own interval are relatively harmless if the pulse obeys the Nyquist pulse criterion. Filtered pulses tend not to obey the criterion. They interfere with each other, as the eye diagram shows, and reduce the resistance of the transmission to further disturbances, like noise.

Linear Receivers for Nyquist Pulse Modulations

We saw in Sec. 10.5 that the linear receiver is a maximum-likelihood receiver for pure QPSK, with the receiver filter $H(f)$ being an integrator. We shall now find that, when Nyquist criterion pulses are also orthogonal and symmetrical, the linear receiver is maximum likelihood for this kind of QPSK, too.

To demonstrate this, we shall proceed to design the matched-filter receiver for such a pulsed transmission. In a PAM system, the signal will be like that in Eq. (10.6-1), and in a PSK system, we shall attempt to match baseband signals like that in Eq. (10.6-3). The pulses $h(t - nT)$ can be the sinc$(t - nt)$ pulses in these, the spectral raised cosine pulses, or any other pulses that are symmetrical and orthogonal. Before starting, we must modify our convention of centering these pulses at time zero, because we cannot design a receiver with noncausal pulses or with pulses of infinite duration. We truncate the basic pulse to the time interval $(-T_0, T_0)$ and shift it right by T_0 to obtain the new pulse $h_T(\cdot)$; the $n = 0$ pulse is

$$h_T(t) = \begin{cases} h(t - T_0), & 0 < t < 2T_0, \\ 0, & \text{otherwise.} \end{cases} \tag{10.6-7}$$

We can make h_T as close to h in its behavior as we like by increasing T_0. We assume that h_T has all the spectral and orthogonal properties of h.

The matched filter in Fig. 10.2-6 for $h_T(t)$ is a filter with impulse response $h_T(2T_0 - t)$ whose output is to be sampled at time $2T_0$. Because $h_T(t)$ is symmetrical about T_0, $h_T(2T_0 - t)$ is in fact identical to $h_T(t)$, which means that the receiver matched filter is *identical to the pulse generation filter*. Furthermore, the response of the $h_T(2T_0 - t)$ matched filter to all the other pulses $h_T(t - nT)$, $n \neq 0$, is zero at time $2T_0$ because of the orthogonality property of the pulses.

Since all these arguments remain true for a time shift of nT, we have shown that only a single matched filter, sampled at times $nT + 2T_0$, is needed to detect all the impulses in a baseband channel. The Nyquist-pulse QPSK transmitter/receiver system that results from our discussion appears in Fig. 10.6-3. For simplicity the synchronization blocks are removed. A BPSK system deletes the Q channel, while the balanced modulators and LPFs are removed in a baseband QAM system. The matched filter sampling takes place at a delay at $2T_0$, compared to Fig. 10.5-2, but this has no consequence other than to delay the data stream. Note that the filters are sampled, but not dumped, at the sample times.

Figure 10.6-3 is a practical and widely used design, since the same filters are used throughout, the signal is relatively narrowband, and the receiver performance is

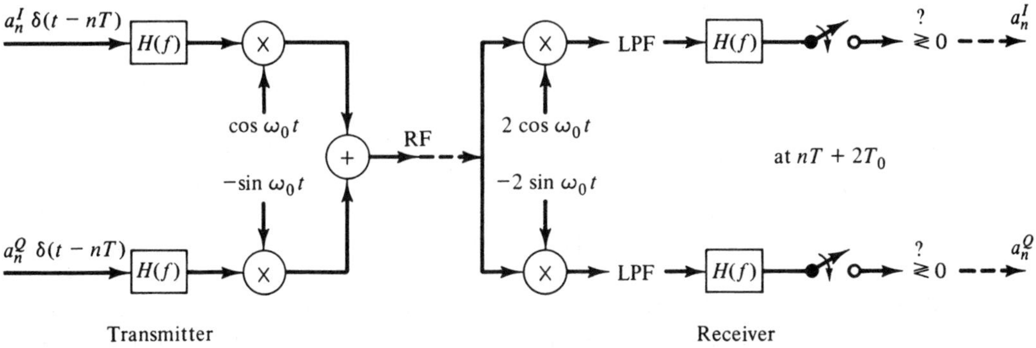

Figure 10.6-3. Transmitter/receiver for Nyquist-pulse QPSK. The receiver is maximum likelihood for sinc and spectral raised cosine pulses.

the maximum-likelihood error probability $Q(\sqrt{2E_b/N_0})$. The tails of spectral raised cosine pulses die down rapidly, so T_0 can be small, and $h_T(\cdot)$ closely resembles the true *RC* pulse. A simple integrate-and-dump receive filter sometimes replaces $H(f)$ if its performance is good enough.

10.7 SYNCHRONIZATION

It is easy to overlook that a digital demodulator cannot function without synchronization. We have already found a need for two types of synchronization. *Carrier synchronization* is needed to reproduce the exact transmitted carrier cos $\omega_0 t$ at the receiver, without which antipodal relationships and the orthogonality between the I and Q channels cannot be maintained. *Interval timing* is needed to specify the sampling instant at the receiver. A third type of synchronization, *word timing*, is needed to organize the data stream into meaningful words. In a complicated transmission network like a digital telephone trunk system, many levels of word synchronization are needed.

The analysis and construction of synchronization subsystems are probably more complex than all the rest of transmission system design put together. We can only summarize some of the main points in this book. Our discussion will jump off from the treatment of the phase-lock loop in Sec. 6.11. For a discussion beyond our level here, but still not too advanced, we recommend [10-3, Chapter 5] or [10-5, Chapters 5 and 6].

Some Basic Ideas

Although synchronization is a complex subject, there are certain rules and guidelines around which we can organize our thoughts. For the most part, we shall focus on carrier synchronization because it uses the tools already developed in the chapter, but the same ideas govern all types of synchronization.

A spectral line somehow related to the carrier must be generated. Structures like the phase-lock loop of Fig. 6.11-1 are circuits that act to slave a local oscillator, or VCO, to a phase of an oscillating input. This input, called the *reference input*, can be any repetitive waveform, although in the carrier synchronization problem it will be a sinusoid. Sometimes the reference signal is explicitly present, as a pilot carrier, or *tone*, or as a clock signal of some kind. Pilot tones are used in frequency-division multiplexed telephone trunk lines. Often the reference signal does not appear as a separate spectral impulse, or line, and in this case one must be created.

The RF modulations in this chapter are good examples of signals that lack a distinct carrier line. We may speak of QPSK as having a carrier, but a better term for ω_0 is *center frequency*, since the modulation has a continuous power spectral density without any spectral lines. A standard way to create a reference signal out of QPSK is to pass it through a nonlinearity. This produces a new signal that is distorted but contains a sine-wave component that is usable as a reference. A phase-lock structure then locks to this. We shall return to these circuits later.

Synchronization consumes energy. If a separate pilot tone or clock is provided to act as the reference signal, this pilot energy is solely used for synchronization and cannot be applied to increase the energy E_s devoted to the data symbols. When there is no separate pilot, the signal is said to be *self-synchronizing*. Here it is not so obvious that energy is being drained away, but it nonetheless is. Self-synchronization is inevitably less perfect than an externally supplied accurate carrier reference, with the result that the detector circuits have a higher probability of error. In BPSK with a carrier phase error of θ, for instance, it can be shown by signal-space analysis that the error probability increases from $Q(\sqrt{2E_b/N_0})$ to $Q(\cos\theta\sqrt{2E_b/N_0})$. To reduce the error probability back to its perfect-carrier value, a larger E_b is required. QPSK has a higher sensitivity than does BPSK.

Self-synchronization is not practical as a rule unless the symbol SNR E_s/N_0 exceeds about 6 dB, since otherwise too much noise disturbance is present after the reference signal is created. This rule presents no difficulty for PSK modulations, which need 10 to 15 dB in order to have a low enough error rate, even with an accurate carrier. Newer, more advanced modulations need less energy, and for these self-synchronization is more troublesome, and modulators that are combined with error-correcting codes (see Chapter 11) may have very low E_s/N_0, as low as 2 to 4 dB.

The amount of energy needed for synchronization is inversely proportional to signal stability. The chief types of received signal phase disturbance are background noise and carrier phase variations. The latter can come from a variety of sources, starting with irregularities in the transmitter oscillator itself. Even with a perfect oscillator, any change in the transmission path, such as vehicle motion or even warming of the medium, will change the received phase.

We can write the reference signal with these disturbances as

$$x(t) = A\cos(\omega_0 t + \theta(t)) + \eta(t), \tag{10.7-1}$$

where $\theta(t)$ represents phase variation that the synchronizer must track and $\eta(t)$ is the total noise that appears along with the reference signal. The dynamics of a phase-lock loop that observes Eq. (10.7-1) without the noise term are given by Eq. (6.11-10) or (6.11-13); the reactions of the loop to the noise are summarized in advanced texts. Without going into the mathematical details, we can say that, if the signal is subject to wide phase disturbances, the loop filter H (see Fig. 6.11-2) must be wideband and the loop gain high in order for the loop to track $\theta(t)$; this in turn means that the loop will be more disturbed by the noise term, an effect that can be reduced only by increasing the reference amplitude A relative to the noise. In short, noise and/or phase fluctuations mean the transmitted signal must be stronger for the same phase tracking error.

In the extreme, a very stable received signal requires virtually no energy in the reference signal. This is because the loop filter may be very narrowband, which means the loop will be almost unaffected by noise. The loop VCO phase can change only sluggishly, but this is acceptable because the signal is stable. At the opposite extreme, a transmission path with severe phase disturbances will make necessary a very strong reference signal, probably a separate pilot carrier. In the design of a practical system, the choice of loop filter and gain and the type of signal plan depend critically on the anticipated transmission path.

Acquisition versus tracking. The normal mode of a synchronizer is to track the received carrier phase as closely as possible, but at the beginning of the transmission, when the synchronizer has only a rough idea of the carrier frequency, a different mode is needed. This is the acquisition mode. A wider loop bandwidth is needed, and strategies like sweeping the VCO frequency may be employed, which would be inappropriate for the tracking mode.

Synchronizer Circuits

With these principles in mind, we shall now consider some common synchronizer circuits. The idea of their design is to create a reference tone and then to lock to it with a phase-lock structure.

The simplest synchronizer circuits work by *raising the modulated signal to a power*. The application of this principle to pure BPSK can be seen by squaring the signal in Eq. (10.4-10) to obtain

$$\begin{aligned} s^2(t) &= (2E_s/T)\cos^2(\omega_0 t + \psi), \qquad \psi = 0 \text{ or } \pi \\ &= E_s/T + (E_s/T)\cos(2\omega_0 t + 2\psi) \\ &= E_s/T + (E_s/T)\cos(2\omega_0 t). \end{aligned}$$

If the dc term is blocked by a band-pass filter, what remains is a sinusoid at precisely twice the frequency of the desired carrier. A phase-lock loop can lock to this, and we can obtain $\cos(\omega_0 t)$ through a frequency-divide by 2. This kind of synchronizer is called a *doubling loop*.

Figure 10.7-1 shows a *quadrupling loop*, which uses the same principle to synchronize QPSK. The modulated signal as received is

$$s(t) = A\cos(\omega_0 t + \psi(t) + \theta(t)) + \eta(t),$$

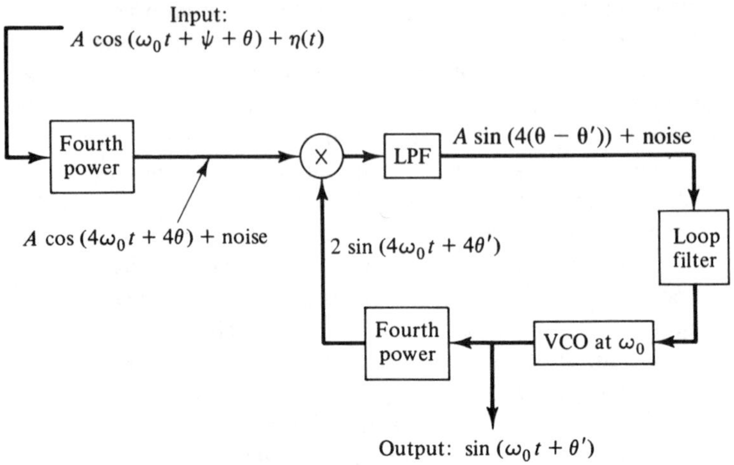

Figure 10.7-1. Frequency quadrupler circuit for QPSK carrier synchronization.

where $\psi(t)$ is the data-bearing phase, $\theta(t)$ is the time-varying carrier phase, to which we must lock, and $\eta(t)$ is additive noise. The circuit passes $s(t)$ through a fourth power nonlinearity, which also contains a band-pass filter to remove extraneous terms. The output of this is $A\cos(4\omega_0 t + 4\theta(t))$, plus a new noise term; the data phase has been canceled by the quadrupling of the phase. The rest of the circuit generates and filters a phase error signal that controls a VCO oscillating at $\sin(\omega_0 t + \theta'(t))$. The local phase $\theta'(t)$ should be close to the received reference phase $\theta(t)$. The VCO output will serve as the signal carrier in the receiver. It too must be raised to a fourth power in order to be compared to the fourth-power reference signal.

In a pure QPSK modulation, the quadrupling loop perfectly removes the data-bearing phase, leaving only the varying carrier phase. This will also occur at least roughly with Nyquist pulse QPSK or with pure QPSK that has been RF filtered. In these cases the data phase will often not quite cancel, and the phase error signal at the loop filter input will vary even though the phase error is constant. These loop disturbances are called *data noise*. The challenge in designing a synchronizer is to provide enough loop inertia to ignore data noise and "true" noise, while still providing adequate response to changes in the carrier phase.

A difficulty with the circuit in Fig. 10.7-1 is that the VCO can lock not only to the reference phase, but to $\theta(t) + k\pi/2$, $k = 0, 1, \ldots$, as well; similarly, a doubler circuit can lock to $\theta(t) + k\pi$. This *phase ambiguity* phenomenon is inherent in the process of raising to a power, which destroys the identity of k. A favored remedy to this problem is to encode the data in the *change* of carrier phase rather than the absolute phase, a process called *differential coding*. We shall return to this in Sec. 10.8.

The principle of raising to a power followed by a phase-lock loop is widely used, both in PSK synchronization and in synchronization of more sophisticated modulations. Some circuits of the latter type appear in references [10-3] to [10-5].

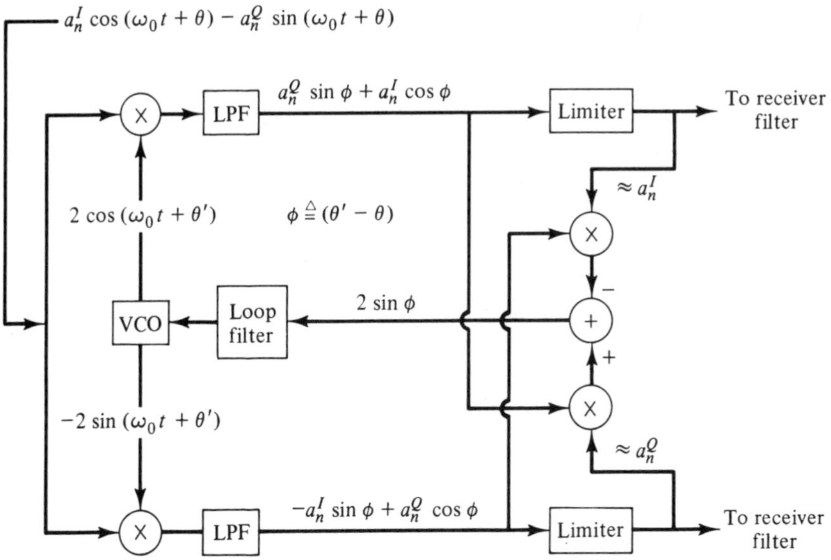

Figure 10.7-2. Simplified QPSK Costas loop. Data and phase reference are obtained from the same circuit. Operation is easiest to follow by assuming circuit is near lock and a^I and a^Q are ± 1. $\phi = (\theta' - \theta)$ is the phase error.

Another principle used in synchronizer design is the simultaneous estimation of the phase and the data in the same circuit. This is illustrated in the Costas loop of Fig. 10.7-2. This loop is an extension to QPSK of the DSB detector in Fig. 6.7-3. As a receiver, the circuit lacks only the receive filters, the threshold detectors, and bit timing recovery. The Costas loop has the same phase ambiguity problem as the quadrupling loop.

Another method of synchronizer design, present to some degree in the Costas loop, is the method of decision feedback. Here the detected data are used to re-create the transmitted signal, which is in turn compared to the received signal to produce a phase error signal.

Bit timing synchronizers are not different in principle from carrier phase synchronizers. The bit timing reference signal often consists of sign transitions in the I or Q baseband signals. These will be missing when there is no transition in the data, but those transitions that appear do so at multiples of T, offset by a constant τ_0 that has a fixed relation to the optimal data sampling instant. The inertia in the synchronizer's phase-lock loop keeps the bit timing VCO going until another transition arrives. By choosing the loop constants properly, the VCO stays locked to T rather than some other submultiple of the transitions' timing. Recovery of both carrier and bit timing is often aided by a preamble to the data signal that contains an alternating bit pattern of known phase. The preamble is sent at set intervals or whenever a data network finds that it has lost synchronization.

10.8 SOME ADVANCED TOPICS

This section expands on certain topics mentioned in the rest of the chapter. The common theme here is that our earlier, simple modulations are made more sophisticated in order to increase their noise resistance or bandwidth efficiency or to remedy an engineering difficulty that has appeared during their implementation.

Differential PSK

We saw in Sec. 10.7 that carrier synchronizers often cannot distinguish whether they are locked to the carrier phase or to the phase plus a multiple of π/M, where M is the size of the data alphabet. This makes it difficult to encode data into the absolute transmitted phase levels. In a *differential* scheme, the data are encoded into the change from interval to interval.

A differential BPSK scheme, usually abbreviated as DPSK, works as follows. If the present and previous binary data symbols are the same, phase 0 in Eq. (10.4-10) is transmitted; that is, sending $\sqrt{2E_s/T}\cos(\omega_0 t)$ means "no change" in the data symbols. If the present data symbol is opposite to the previous one, phase π is transmitted. Special differential detector circuits have been designed that detect DPSK without a separate carrier synchronizer. These work by continually multiplying together the RF signal in the present and previous intervals.

DPSK is especially useful in systems that require rapid acquisition of carrier synchronization. An example of this is a network that sends random short bursts of data. The spectrum of DPSK is identical to that of BPSK, and its error probability converges to BPSK's as the channel SNR grows. Extensions of DPSK exist to data alphabets greater than two, but these entail some sacrifice in error probability. A disadvantage of DPSK, therefore, is that it consumes wide bandwidth if it is to be power efficient.

Offset PSK Schemes

In an offset QPSK modulation the quadrature baseband signal $Q(t)$ is delayed by $T/2$ relative to $I(t)$. An example of this for pure QPSK is given in Fig. 10.8-1. Offset QPSK schemes are used because they contain no phase transmissions other than $\pm\pi/2$. Ordinary QPSK (and BPSK) both contain π transitions (i.e., sign reversals), which cause a gross discontinuity in the transmitted sinusoid. (These are visible in the top 30-kHz BPSK plot in Fig. 10.5-6). It has been found that these sign reversals are particularly badly affected by filters and nonlinearities that are present in practical systems, so offset schemes have an advantage here. An offset QPSK receiver is similar to an ordinary one except that the quadrature sampling time is delayed by $T/2$.

Pulse-shaped offset QPSK has the same format as Fig. 10.8-1, but with smooth pulses replacing the rectangular shapes there. A pulse scheme of particular interest is minimum-shift keying (MSK). This modulation is really a binary coherent continuous-phase FSK, like those described at the end of Sec. 10.4, with modulation index $h = \frac{1}{2}$ in Eq. (10.4-18). But the signals can be viewed as a type of offset QPSK, which

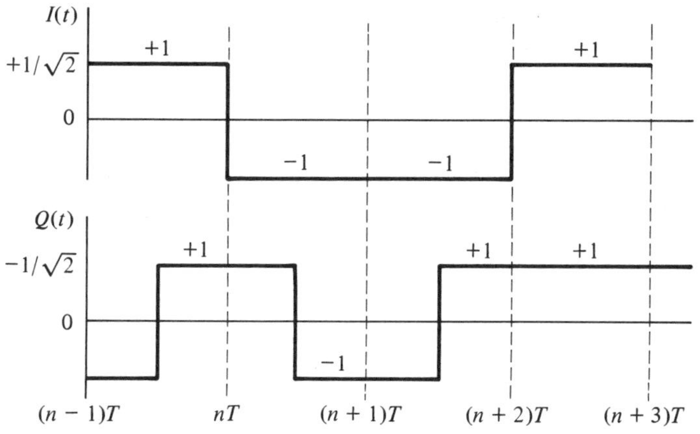

Figure 10.8-1. Baseband waveforms for pure offset QPSK. The transmitted bits, beginning with the nth, are $a^I = +1, -1, -1, +1$ and $a^Q = +1, -1, +1, +1$.

provides a simple transmitter and receiver implementation. To do this, we view a binary MSK scheme with signal interval T as a quaternary offset QPSK scheme with interval $2T$. We divide the binary data stream $a_1, a_2, \ldots$ into parallel I and Q streams by

$$\text{When } n \text{ is even, let } a^I(t) = a_n, \qquad (n-1)T < t \le (n+1)T,$$
$$\text{When } n \text{ is odd, let } a^Q(t) = a_n, \qquad (n-1)T < t \le (n+1)T.$$

The MSK signal is then

$$s(t) = \sqrt{2E_s/T}\left[a^I(t)\cos\left(\frac{\pi t}{2T}\right)\cos\omega_0 t - a^Q(t-T)\sin\left(\frac{\pi t}{2T}\right)\sin\omega_0 t\right], \qquad (10.8\text{-}1)$$

which is offset QPSK with a particular kind of sinusoidal pulse shaping.

Despite the shaping in (10.8-1), some trigonometry shows that $s(t)$ is in fact the constant-envelope FSK waveform of MSK. The linear receiver with offset Q channel provides a maximum-likelihood receiver for MSK, whose error probability is about $Q(\sqrt{2E_b/N_0})$, the same as pure BPSK and QPSK and spectral raised cosine QPSK. But the fact that MSK is really a continuous-phase FSK means that its spectral side lobes are greatly reduced from those of pure PSK, so it interferes less with adjacent channels. Unlike raised cosine QPSK, MSK has the attractive feature of a constant envelope. Special detector and synchronizer circuits have been devised for MSK.

Offset schemes, including MSK, are discussed more fully in [10-3] and in an excellent tutorial paper ([10-6]).

Partial-Response Modulations

Thus far we have studied modulations in which a data symbol has a significant effect on only a single signal interval. Although a raised cosine pulse, for instance, spreads into adjacent intervals, a matched filter receiver does not "see" it when detecting these intervals. Pulses in offset QPSK overlap in time, but they do so in orthogonal channels. Schemes in which successive data pulses overlap in a significant way are called *partial-response* schemes, after the fact that the response to a data symbol only partially lies in its own interval.

It is convenient to divide partial-response schemes into two types, AM and PM. In the AM schemes, successive pulses superpose in amplitude to build up a final baseband waveform. A simple example is the duobinary scheme, in which the response to a symbol is the double-width pulse

$$g(t) = \begin{cases} 1, & 0 < t \leq 2T, \\ 0, & \text{otherwise.} \end{cases} \tag{10.8-2}$$

Other schemes use wider pulses or pulses that are negative for some of the separate T-intervals. With Eq. (10.8-2), it is clear that the baseband signal at a given time is the sum of two data (± 1) symbols.

A spectral analysis of pulses like these shows that they can significantly reduce the spectrum of the transmitted signal. Alternatively, pulses can be designed that have a spectral null at dc, a feature that is necessary in cable systems that carry dc power and signaling on the same conductor. An optimal receiver for partial-response AM is considerably more complex than our simple linear receiver, but with modern processor chips this is often acceptable if the symbol rate is low enough. The receiver is similar to the Viterbi receiver of Sec. 11.7; its error rate is often the antipodal rate $Q(\sqrt{2E_b/N_0})$, despite the fact that the partial response bandwidth is less. Partial response AM schemes were pioneered by A. Lender in the early 1960s.

In a phase-modulation partial-response scheme, the signal phase rather than the amplitude is a superposition of pulses; that is,

$$\psi(t) = \sum_k a_k g(t - kT) \tag{10.8-3}$$

in the signal $\sqrt{2E_s/T}\cos(\omega_0 t + \psi(t))$. In these modulations, too, bandwidth can be reduced by proper choice of $g(t)$; side lobes in particular can be dramatically reduced. Unlike the case with AM schemes, pulses exist leading to error probabilities much better than even the $Q(\sqrt{2E_b/N_0})$ rate. At the same time, the PM schemes are all constant envelope. Overall, the main lobe bandwidths of PM schemes tend to be somewhat larger than their AM cousins.

The optimal receiver for a modulation like that in Eq. (10.8-3) is again the Viterbi receiver. Partial-response phase modulation is treated in [10-4]. AM schemes are discussed in [10-7].

10.9 SUMMARY

This chapter has been about the methods of analysis and the circuitry that communications engineers use when the information to be transmitted is symbolic in form. As with analog modulation, the chief quantities of interest with a digitally modulated waveform are its energy and bandwidth consumption. To calculate the first, we developed the signal space method, an analysis based on the orthogonal function expansions of Chapter 2. For Gaussian channel noise, the method led to particularly simple results that mimic the Euclidean geometry of the everyday world. To calculate bandwidth, we were able to borrow simple Fourier transform methods. Out of the many modulation schemes that have been proposed, we have discussed circuits of the simpler ones. These were based on simple changes in phase, amplitude, or frequency. Some of these use smooth pulses to conserve bandwidth; in these, it is important that successive pulses be orthogonal to each other. In Chapter 11 on information theory, we shall explore the meaning of symbolic information and the ultimate ability of transmission channels to carry it.

PROBLEMS

10-1. Prove that the maximum-likelihood decision region boundaries for equal-energy signals always terminate (or at least pass through) the origin of the signal space.

10-2. Show that the choice Eq. (10.2-16) of $\boldsymbol{\alpha}_0$ minimizes the average energy Eq. (10.2-15) of a signal constellation.

10-3. Suppose it is desired to communicate with four signals $\pm\cos(\omega_0 t)$ and $\pm\cos(\omega_0 t + ht)$. This is a form of FSK. Use the Gram–Schmidt process to find orthonormal signal-space basis functions for these four signals. Simplify your answer by assuming ω_0 is very large.

10-4. Suppose the signals in Prob. 10-3 transmit a quaternary symbol over $[0, T]$.

(a) Design a four-filter matched-filter receiver like Fig. 10.2-6 specifying the filter impulse responses.

(b) Redesign your receiver, using two filters only, plus threshold detectors.

10-5. Suppose there are two phase-modulated signals $\cos(\omega_0 t + \psi_1(t))$ and $\cos(\omega_0 t + \psi_2(t))$. Give an expression for the signal-space distance between these. Assume $\omega_0 \to \infty$.

10-6. The true error probability p_e for pure QPSK is $2Q(\sqrt{E_s/N_0}) - Q(\sqrt{2E_s/N_0})$, rather than the overbound in Eq. (10.3-11). Prove that this is so, using the independence of the I and Q channels in Gaussian noise.

10-7. Verify that Eq. (10.4-1) has energy E_s when the condition in Eq. (10.4-3) holds. (You may wish to let $\omega_0 \to \infty$ during this calculation.)

10-8. Verify the in-phase and quadrature distance expression [Eq. (10.4-9)].

10-9. Derive the error approximation [Eq. (10.4-16)] for M-ary PSK. Evaluate the energy loss that occurs in 8-, 16-, and 32-PSK systems, relative to QPSK. These losses are larger than those in QAM constellations with the same number of points. Why?

10-10. Design the baseband signal levels for pure 8-PSK modulation. (First choose a reasonable set of 8 phases.) How would you design the circuitry that follows the samplers in the 8-PSK linear receiver?

10-11. **(a)** Calculate the average energy E_s of the 16-QAM scheme in Fig. 10.4-1(b) in terms of the spacing parameter ϵ_0.

(b) 16-QAM has an error probability estimate of the form $KQ(\sqrt{d^2E_b/N_0})$. Evaluate K and d for an "interior" point in the constellation. What is the energy loss compared to QPSK?

10-12. Repeat Prob. 10-11 for 32-QAM (rectangular array). Compare energy losses of 32-QAM and 16-QAM relative to QPSK. What do you expect will be the losses of 64- and 128-QAM? (Variants of all these are used in telephone-line modems.)

10-13. Compare wideband FSK modulation with two, four, and eight frequencies.

(a) Roughly how wide a channel do they require? (This must be expressed in cycles/data bit; let the 2-FSK system consume B cycles.)

(b) What is their average signal energy E_s?

(c) What is their error probability in terms of E_b/N_0?

10-14. **(a)** Suppose a binary FSK modulation uses the two frequencies $\omega_0 \pm h\pi/T$, with $h = 3$. How close to orthogonal are the two signal-space basis functions?

(b) Repeat for $h = 3.2$.

10-15. Pure QPSK with Gaussian additive noise having power spectral density $N_0/2$ is to be detected by the sampling receiver of Sec. 10.6.

(a) Find the error probability of this receiver as a function of E_b/N_0. Compare to the maximum-likelihood detector error probability.

(b) The square pulse need actually be turned on only during a short interval surrounding the sampling time. A 10% duty cycle pulse would, for instance, reduce the signaling energy by 90%, yet such a scheme is seldom used. Why?

10-16. Suppose the square pulse in pure QPSK is replaced by the exponentially decaying pulse

$$v(t) = \begin{cases} exp\left(-\frac{t}{T}\right), & 0 < t \le T, \\ 0, & \text{otherwise.} \end{cases}$$

The linear receiver structure is a maximum-likelihood receiver for this pulse.

(a) What is the required matched-filter response $h(t)$?

(b) Find the error probability and compare to QPSK.

10-17. In Prob. 10-16, find the error probability if an integrate-and-dump receive filter is used instead of a matched filter.

10-18. Calculate the RF modulation power spectrum of the scheme in Prob. 10-16. Compare to the spectrum of square-pulse PSK with the same T.

10-19. Prove that the pulses $\sqrt{1/T}\,\text{sinc}[(t - nT)/T]$, n an integer, are orthonormal.

10-20. Prove that the RF pulses $\sqrt{2/T}\,\text{sinc}[(t - nT)/T]\cos\omega_0 t$, n an integer, are orthonormal.

10-21. Prove that the *RC* pulses generated by Eq. (10.6-5) are orthogonal. *Hint*: Use symmetries of $h(t)$.

10-22. Consider the eight 3-bit sequences made up of three symbols from ± 1. Let each group of three be preceded by the all $+1$ sequence. Match the eight sequences to the traces shown in the top eye diagram of Fig. 10.6-2.

10-23. Show that the MSK signal in Eq. (10.8-1) has a constant envelope.

10-24. Show that, in a BPSK detector with phase error θ, the error probability is $Q(\cos\theta\sqrt{2E_b/N_0})$.

10-25. Suppose that a BPSK carrier recovery PLL is such that it adapts to a sudden reference phase jump of 45° at time zero as

$$\theta'(t) = (45°)e^{-0.2t/T} \qquad t > 0,$$

where $\theta'(t)$ is the receiver phase and the signal phase has suddenly jumped to 45°. It is desired that the error rate be 10^{-6}.

(a) What E_b/N_0 is required if there is perfect phase tracking?

(b) Using the result in Prob. 10-24, how many intervals will go by until the error rate returns to within 10% of 10^{-6} if the receiver phase has the error given? (Assume the initial phase error in an interval applies to the whole interval.)

10-26. A frequency quadrupler is to be used to recover carrier phase. The loop filter is omitted [its $H(f) = 1$] and the total loop gain is K. Find the response of the local phase $\theta(t)$ to a small jump of θ_0 in the transmitted signal phase.

REFERENCES

[10-1] H. L. VAN TREES, *Detection, Estimation, and Modulation Theory*, vol. I, Wiley, New York, 1968.

[10-2] J. M. WOZENCRAFT and I. M. JACOBS, *Principles of Communications Engineering*, Wiley, New York, 1965.

[10-3] V. K. BHARGAVA and others, *Digital Communications by Satellite*, Wiley, New York, 1981.

[10-4] J. B. ANDERSON, T. AULIN, and C.-E. SUNDBERG, *Digital Phase Modulation*, Plenum, New York, 1986.

[10-5] R. E. ZIEMER and R. L. PETERSON, *Digital Communications and Spread Spectrum Systems*, Macmillan, New York, 1985.

[10-6] S. PASUPATHY, "Minimum Shift Keying: A Spectrally Efficient Modulation," *IEEE Commun. Society Mag.*, pp. 14–22, July 1979.

[10-7] S. PASUPATHY, "Correlative Coding—A Bandwidth Efficient Signaling Scheme," *IEEE Commun. Society Mag.*, pp. 4–11, July 1977.

CHAPTER 11

Information Theory

11.1 INTRODUCTION

What is information? How can it be measured? How much can be transmitted or stored in a given medium? These are the questions of information theory. As an example, consider the human voice. When stored in a high-quality medium like the compact disc, its reproduction in stereo requires 1.4 Mbits/s; good quality telephone speech, converted to bits by the best methods of Sec. 4.6, requires about 16 kbits/s; speech that is understandable but not very lifelike can be represented by 2 kbits/s or less. If instead of speaking we transmit English text by teletype at a speechlike speed, only 100 bits/s would be needed! Our example of voice illustrates several facets of information theory. The subject seeks answers to what really is the information present in speech and to whether these "bits" are a correct measure of information. It characterizes the relation between quality and information. It also calculates the capacity of a medium like the compact disc to carry bits, starting from a mathematical model of the disc.

The theory of information is full of paradoxes and delicate distinctions, and we must be careful in our development of the subject. Consider another example of information, a radio weather forecast. This time we are not concerned with whether the message is voice or text, but only with the "facts" in the message. Weathermen in general are not highly reliable, and this should somehow be reflected in a reduced information content in the "facts." The radio station may misplace or exaggerate some of the facts, reducing their real information content still further. Finally, we must consider how much of the information is already known. If it is July in the United States and the forecast states that it will not snow tomorrow, we have learned very little new knowledge. If we know that it is sunny and warm and the radio simply states that it is sunny and warm, we have learned nothing at all. Information theorists have dealt with these notions by developing a probabilistic theory about them.

Information theory stems from a single watershed paper by Shannon [11-1]. Shannon pondered these and other examples of information and information transfer and devised a totally new way to view them. The theory of information as presented in this and his later papers divides into three parts. The theory of *source coding* studies information measurement and conversion; *channel coding* calculates how much information a medium, or *channel*, can transfer. The field of *rate-distortion* theory assumes that not enough information is available about a given data source to ensure perfect reproduction of it; the theory then studies how far the distortion can be reduced at this level of information. In the next sections we shall explore these subfields one by one.

In earlier chapters we have seen examples of information sources, conversion into digital form, and information channels. The purpose of this chapter is to find the limits imposed by nature on these conversions and transmission. In Sec. 4.6, for instance, the signal-to-noise ratio of PCM was derived in terms of the number of binary symbols employed per signal sample. We shall now be able to find out the ultimate SNR that can be achieved by any process of converting samples to bits. In some cases it will turn out that linear PCM is near that limit, so it makes little sense to try to improve on it. In other cases, for example with highly correlated signal samples, PCM lies far from the ultimate performance, and we know that we should look at sophistications of PCM like differential PCM.

Another place where it would be convenient to know some ultimate limits is the digital-modulation schemes of Chapter 10. How good are PSK and FSK compared to the best schemes that can possibly exist? Are there schemes that use much less power for the same bandwidth and error probability? Is PSK wasteful of bandwidth?

Information theory has always been a provocative discipline. Its definitions and its ways of thinking will seem mathematical at first, but they provide a great many insights, not only into communications systems but into everyday life as well.

11.2 INFORMATION SOURCES

We shall begin this section by defining an information source. Our aim will be to define information in a way useful in communications theory, and not in the social or political realm, but the definition will be based on intuitions gained from everyday life. Next we shall define a measure of information, aiming again at a yardstick that will provide insight into the subjects of this book. Our discussion will end with the subject of *coding*, the conversion of information to a symbolic form, and with Shannon's first theorem, which tells us the minimum number of these symbols that are needed.

Information Sources and Entropy

In information *theory*, an information source is a *probability distribution*. This perhaps confusing notion needs some explanation. It means that a source is an object that produces an event, the outcome of which is selected at random according to a distribution. All the probability distributions given in Chapter 8 can be thought of as representing sources, and all sources, at least to an information theorist, can be thought of as the probability distributions in that chapter. In everyday life we are used to such information sources as other people, books, and newspapers, objects that at first glance might have little resemblance to a probability distribution. At second glance we might indeed categorize as a random variable an inebriated friend or a newspaper's "unidentified source within the government." Information theorists mean something more than this, however.

Let us think about the simplest of sources, one that produces one of two outputs. An everyday example is a stoplight at the intersection of two roads, for which the two outputs are red and green. To make the example more fruitful, ignore the yellow output and assume that the roads curve as they enter the intersection so that we can only see the light at the last moment. What are the characteristics of the information given by the color of the light? If the light is almost always green on our road, we approach the blind intersection with relatively little anxiety; in a sense the information presented by the light is small. If the light is green with probability $\frac{1}{2}$, we approach the intersection with a lot of uncertainty and with the feeling that we would gain very much information by knowing the light's color.

The example of the stoplight shows that information is related to *uncertainty*, to the degree of surprise that we feel in learning the color of the light. In thinking about information, this equating of information and uncertainty can be a tricky concept. Often the source outputs are plainly in front of us and offer no surprise at all. In this case it is necessary to imagine the uncertainty about the source were the outcomes not available. Our specifying the blind curves in the stoplight example was an artifice to help with this imagining. The light is the same information source with or without the curves.

Stoplights illustrate several other qualities of information sources. It often happens that we have no prior knowledge of the light's probabilities, and in this case a prudent driver sets the probabilities to $\frac{1}{2}$. After repeated use, the driver may learn that the light is normally green along his or her direction. Through the learning process, the driver redefines the probability of the green outcome to a higher probability and thus reduces the information presented by the source. This learning from past source outcomes is another tricky concept of information theory. One approach to it is the concept of conditional information, to which we shall return at the end of the section.

Another quality of stoplights illustrates the additivity of information. Suppose another stoplight in a blind intersection lies down the road from the first one. Now we have twice the uncertainty about the situation unless the lights are somehow wired together. If the lights are independent and similar in their red/green statistics, knowledge of their joint outcome provides twice the information as does a single outcome. It thus seems that information from several independent sources, or from successive outputs of the same source, must be additive.

The technical name for our stoplight information source is the *binary symmetric source* (BSS). Precisely defined, a BSS produces a sequence of independent symbols that assumes the values z_0 with probability p and z_1 with probability $1 - p$. There are many other examples of the BSS. For instance, a tossed coin is a BSS, whether or not it is fair, as long as its outcomes are independent; so also is a Bernoulli trial. The most important example for our purposes is the binary data source. In this source the independent outputs are 1 or 0 and $p = \frac{1}{2}$. We have used this source throughout the book as a kind of standardized ideal. It was used to characterize the outputs of analog-to-digital converters like PCM. In Chapter 10, it was the idealized input to a digital modulator.

The foregoing examples make it clear that a mathematical measure of information should satisfy certain axioms, which we observe in everyday life:

1. The information in independent outcomes should add.
2. Information should be proportional to the uncertainty of an outcome.
3. Information should be related to the number of symbols needed to specify an outcome.

In 1928, R. V. L. Hartley suggested the *logarithmic* measure

$$\iota(E) = \log \frac{1}{P(E)} \tag{11.2-1}$$

for the information in an outcome E with probability $P(E)$. This definition satisfies our axioms nicely. Two independent outcomes $\{EE\}$ have probability $P^2(E)$, and by Eq. (11.2-1), $\iota(EE) = 2\,\iota(E)$. The less certain an outcome is, the more information Eq. (11.2-1) imputes to it. We shall see that axiom 3 is satisfied in a sensible way if the base of the logarithm is properly defined.

Hartley's measure defines the information in a particular type of outcome of a source, but not in a source as an entity unto itself. A way to do that is to define the *expected information* in the output of a source, a quantity called the *entropy* in information theory.

Definition. The entropy of an independent-output source is given by

$$H(Z) = \sum_j P(z_j) \log \frac{1}{P(z_j)}, \tag{11.2-2}$$

where Z is a random variable that takes on values z_j with probability $P(z_j)$. The set $\{z_0, \ldots, z_j, \ldots\}$ are the possible outcomes of the source. This set is called the *alphabet* of the source, and if the set is finite, the source is a finite-alphabet source. Note that z_j is not necessarily the integer j, nor is it usually a number at all. It may be a letter of the English alphabet, a color, a weather condition, or any other outcome. The sum in Eq. (11.2-2) is over the entire alphabet, and it can be seen that Eq. (11.2-2) is simply the expectation of Hartley's information.

We have not yet specified the base of the logarithm in Eqs. (11.2-1) and (11.2-2). In reality, any base may be used, and we shall choose the one that is most convenient. Consider the binary-alphabet BSS. Its entropy is given by

$$H(Z) = P(z_0) \log \frac{1}{P(z_0)} + P(z_1) \log \frac{1}{P(z_1)} \tag{11.2-3}$$
$$= -p \log p - (1 - p) \log (1 - p).$$

When $p = \frac{1}{2}$, if we take base 2 logs, $H(Z)$ becomes unity. That is, the average information produced by the equiprobable BSS is 1 unit per source output. Focusing on a particular BSS, we consider the independent and equiprobable binary data source for which outcome z_0 is a 0 and z_1 is 1. Now we have the intriguing result that the source puts out symbolic bits at a rate identical to the entropy per symbol as measured by Eq. (11.2-3). This is a consequence of taking base 2 logs and $p = \frac{1}{2}$.

By long tradition, engineers call both the source's symbols and the units of their information measure *bits*. It is said both that the binary data source puts out bits and that its information rate is 1 bit/symbol. These are two completely different meanings of the word bit! The first meaning refers to a symbol, like A or B or, in this case, 1 or 0; the second is the accepted unit of information measure, defined by the statement that 1 bit of information is the information present in one output of the equiprobable binary data source. As well, we can now see that axiom 2 is satisfied in a reasonable way: In the case of the equiprobable BSS, the number of symbols in the source's output sequence is numerically identical to the sequence's information, either the specific information of Eq. (11.2-1) or the average information of Eq. (11.2-2).

Properties of Entropy and Examples

The relationship expressed in Eq. (11.2-3) so commonly appears that it is given a special name, the *binary entropy function*, and is defined by

$$h_B(p) \equiv -p \log p - (1 - p) \log (1 - p). \tag{11.2-4}$$

Figure 11.2-1 is a plot of $h_B(p)$.

The function $h_B(p)$ has value 0 at $p = 0$, rises to a peak of 1 at $p = \frac{1}{2}$, and falls back to 0 at $p = 1$. For the stoplight BSS, this means that the outcome of a light that is almost always green has expected information near zero; a light with equally likely colors has entropy near to 1 bit/output, and a light that is almost always red again presents almost no information.

We can generalize the properties of $h_B(p)$ and obtain useful properties of $H(Z)$ for arbitrary sources.

Property 11.2-1. Suppose a source has an alphabet of size J so that its outputs take on one of the outcomes $\{z_0, z_1, \ldots, z_{J-1}\}$. If the J outcomes are equally likely, the entropy of this source is $\log_2 J$ bits/output.

Proof. To prove the property, we simply evaluate $H(Z)$ and find that

$$H(Z) = \sum_{j=0}^{J-1} P(z_j) \log \frac{1}{P(z_j)} = -J\left(\frac{1}{J}\right)[\log (1/J)]$$
$$= \log J.$$

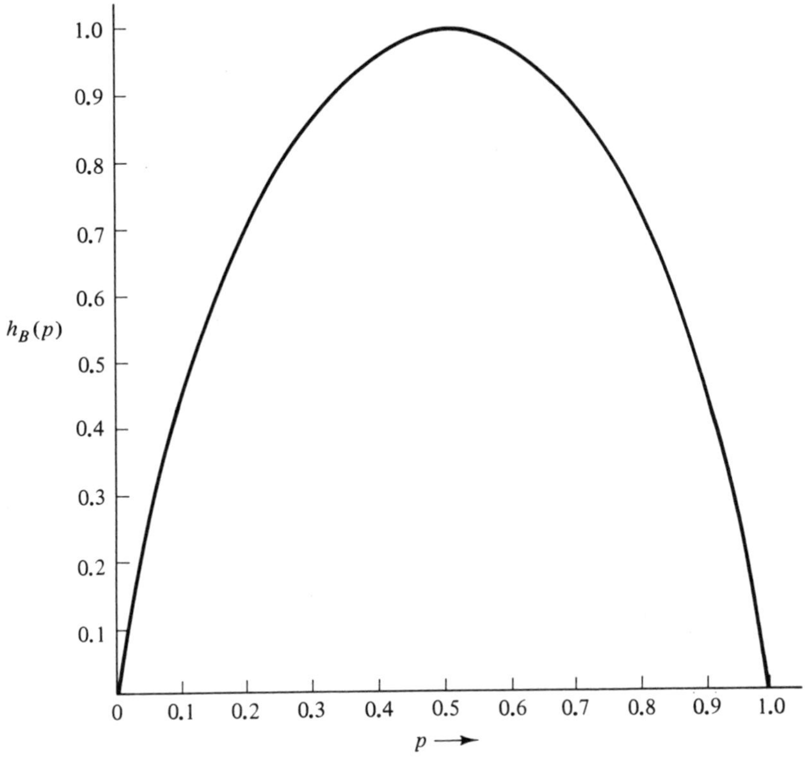

Figure 11.2-1. Binary entropy function, $h_B(p)$.

Example: A Fair Dice

A fair dice, as discussed in Sec. 8.1, has six equiprobable outcomes. The information in a dice-source output is $\log_2 6$ by Eq. (11.2-1); the entropy, or average information, of the source is this same value from Eq. (11.2-2), about 2.58 bits/output. An interpretation of this is that about 2.58 binary symbols are needed to represent the outcome of a fair dice. (The meaning of the phrase "0.58 symbol" will have to await Sec. 11.3.)

Example: Quaternary Data Transmission

As mentioned in Chapter 10, it is generally more efficient to transmit digital data as four-level symbols rather than as binary symbols. If we make the standard assumption that the quaternary symbols are independent and equiprobable, then the entropy of these is $\log_2 4$, or 2 bits/output. This property reflects our everyday notion that each quaternary symbol carries two binary symbols.

If a J-outcome source does not have uniform probabilities, then its entropy is bounded from above by $\log J$.

Property 11.2-2. If Z is a source with alphabet size J,

$$0 \le H(Z) \le \log J. \tag{11.2-5}$$

Proof. Each of the source outcomes can have a different probability $P(z_j)$. From the definition of $H(Z)$, we can write the expression

$$\log_2 J - H(Z) = \log_2 J - \sum_{j=0}^{J-1} P(z_j) \log_2 \frac{1}{P(z_j)} \cdot$$

This can be written as

$$\sum_{j=0}^{J-1} P(z_j) \log_2 J - \sum_{j=0}^{J-1} P(z_j) \log_2 \frac{1}{P(z_j)} = \log_2 e \sum_{j=1}^{J-1} P(z_j) \ln JP(z_j).$$

The well-known inequality $-\ln x \geq 1 - x$ can be applied to underbound this by

$$\log_2 J - H(Z) \geq \sum_{j=0}^{J-1} \geq \log_2 e \sum_{j=0}^{J-1} P(z_j)\left(1 - \frac{1}{JP(z_j)}\right)$$

$$= \log_2 e \left[\sum_{j=0}^{J-1} P(z_j) - \frac{1}{J} \sum_{j=0}^{J-1} \frac{P(z_j)}{P(z_j)}\right] = \log_2 e \cdot 0.$$

Thus it must be that $H(Z) \leq \log_2 J$. That $H(Z)$ is ≥ 0 follows from the form of its definition.

Concluding, we see that no J-alphabet source has entropy larger than $\log_2 J$, and this may occur only when the source outcomes are equiprobable. Returning for a moment to the dice source, we see that an unfair dice, one whose outcomes are not equiprobable, must have entropy less than 2.58 bits/throw. As the dice becomes strongly unfair and one outcome gains overwhelming probability, it is easy to calculate from Eq. (11.2-2) that the dice entropy drops to zero. That is, its outcome is no surprise!

Our final example is more complex, and it brings up the important distinction between entropy and discrimination.

Example: Distribution of Course Grades

In a large population of students that act independently of each other, the scores on a percentage-graded communications systems test will distribute according to a Gaussian, or bell, curve (or so it is alleged.) A common method of assigning the letter grades ABCDF to the percentage scores is to formulate equal-width bins, as illustrated in Fig. 11.2-2. This is done so that each grade letter covers the same fraction of a standard deviation of the test population, typically about one deviation. The motivation for this is to discriminate students into clear categories. The figure has been idealized for simplicity; in many schools the bins are shifted up to $\sigma/2$ to the left so that the distribution produces more high grades.

The figure shows the percentages of scores lying in centered equal-standard-deviation bins. One can verify that the entropy of this "grade" source is 2.04 bits/grade. By comparison, a size-5 alphabet equiprobable source has entropy $\log_2 5$, which is 2.32 bits/output. Thus equal-standard-deviation grading is suboptimal from the point of view of the information in its grades. To optimize information, we would have to assign grades in a uniform probability distribution. The resulting grade ranges are shown at the top of Fig. 11.2-2; little discrimination exists in the D-C-B range. This example shows that we cannot simultaneously optimize the clarity and the information in the categories.

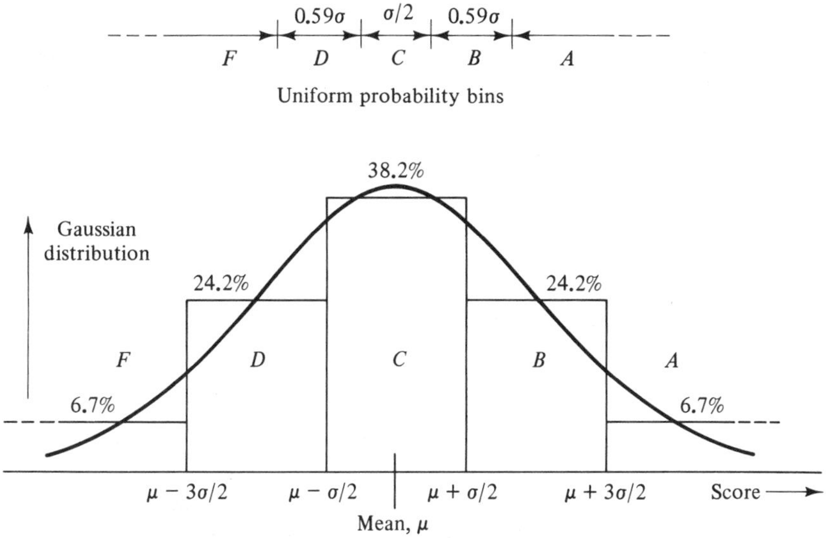

Figure 11.2-2. Idealized distribution of student test scores. Across the top is shown the grade bins that give a uniform distribution of grades.

Conditional Information

In our discussion thus far, we have assumed that successive outputs of an information source are independent. Our intuition tells us that, if instead the present output depends on previous ones, then there should be less new information in the last output. In line with axiom 2, we are less surprised by the last outcome, given knowledge of earlier ones. Most natural information sources like text or samples of speech are this kind of *conditional* source, and we need to extend our discussion of mathematical information measures to include these.

Let a source output Z depend on another random variable W. This variable could be another source, or it could be a previous output of the same source. The conditional probability distribution of Z is $P(z|w)$, where w is the outcome of the second variable. We can directly extend the definition given in Eq. (11.2-1) to get the conditional information in the outcome z,

$$\iota(z|w) = \log \frac{1}{P(z|w)} \cdot \tag{11.2-6}$$

The conditional entropy of Z, given that the outcome of W is w, is the expectation of $\iota(z|w)$:

$$H(Z|w) = \sum_j P(z_j|w) \log \frac{1}{P(z_j|w)} \cdot \tag{11.2-7}$$

To make clear the meaning of Eq. (11.2-7), we return to the example of the blind stoplight. Imagine now that there are two lights and that the color of the second is closely related to that of the first. We can see neither light until just before we reach it, but on approaching the second light, we know the color of the first. The conditional probabilities that govern the colors are given by the matrix

$$\mathbf{P}(z|w) \equiv \begin{bmatrix} P(\text{red}|\text{red}) & P(\text{grn}|\text{red}) \\ P(\text{red}|\text{grn}) & P(\text{grn}|\text{grn}) \end{bmatrix} = \begin{bmatrix} 0.9 & 0.1 \\ 0.1 & 0.9 \end{bmatrix}. \tag{11.2-8}$$

The marginal distributions are assumed to be $P(W = \text{red}) = P(W = \text{grn}) = \frac{1}{2}$, so the marginals for the second light are $P(Z = \text{red}) = P(Z = \text{grn}) = \frac{1}{2}$ as well. The conditional entropy of the second light is

$$H(Z|w) = -0.9 \log_2 0.9 - 0.1 \log_2 0.1 = 0.47 \text{ bits}$$

The same calculation holds for both outcomes w of the first light. By comparison, the entropy of the second light with no knowledge of w is 1 bit/output, more than twice the uncertainty represented by the conditional entropy. The example illustrates our intuitive feeling that knowledge of related events lessens the surprise in the outcome of an event and therefore reduces the information contained in it.

Definition (11.2-7) requires knowledge of the outcome w of the variable W. If w is unknown, it is still possible to measure the entropy of Z by means of the *equivocation*

$$H(Z|W) = \sum_i H(Z|w_i)P(w_i). \tag{11.2-9}$$

Problem 11-4 shows that this has the equivalent form

$$H(Z|W) = \sum_j \sum_i P(z_j, w_i) \log \frac{1}{P(z_j|w_i)}. \tag{11.2-10}$$

$H(Z|W)$ can be interpreted as the average uncertainty in Z after W is specified. $H(Z|W)$ for the stoplight example is the same as $H(Z|w)$, 0.47 bit, because of the symmetry of Eq. (11.2-8).

A second important quantity involving two information sources is the *mutual information*

$$I(W; Z) \equiv H(W) - H(W|Z). \tag{11.2-11a}$$

The mutual information is symmetrical with regard to W and Z, so an alternate definition is

$$I(W; Z) \equiv H(Z) - H(Z|W), \tag{11.2-11b}$$

as is shown easily in Prob. 11-6. Another definition is

$$I(W; Z) \equiv \sum_{i,j} P(w_i, z_j) \log \frac{P(w_i, z_j)}{P(w_i)P(z_j)} \tag{11.2-11c}$$

from which the symmetry in W and Z is obvious. The mutual information describes the information that one variable gives about another. We see immediately, for instance, the interesting fact that when two variables are correlated they each give the same amount of information about each other. In the case of the stoplights, the first light gives $1.0 - 0.47 = 0.53$ bit of information about the second; as well, the second gives 0.53 bit about the first, should we ever need that information! Another way of interpreting Eq. (11.2-11a) is to say that the average information about W given by Z is the uncertainty about W, which is $H(W)$, less the uncertainty that remains after Z is known.

The most important example of two correlated information sources is the input and output of a data-transmission channel. Here mutual information plays the major role, which we shall explore in Secs. 11.5 and 11.6.

Example: Conditional Entropies in Written Text

As we have remarked, natural information sources like language have successive outputs that are markedly correlated. In Sec. 11-3 we list the probabilities of letters in English text and find that their entropy is 4.03 bits/letter. In a very readable discussion of correlated sources, Abramson [11-5] gives entropies for several correlated-letter models of English text. If English is modeled as a first-order Markov source,† in which each output depends on the previous output but is independent of earlier ones, the calculated entropy drops to 3.32 bits/symbol; if each output is allowed to depend on two previous outputs, the entropy is 3.1 bits/symbol. The method of calculating these entropies is beyond the scope of this book. Shannon [11-6] once estimated that if all dependencies in English were taken into account this entropy would drop to the range of 0.6 to 1.3 bits/symbol. Other studies have placed this entropy as high as 2 [11-13]. Designing systems that exploit even simple dependencies in text is a challenging research problem.

11.3 SOURCE CODING

The word *coding* means conversion of an information source from one form to another. Sometimes this conversion is of a simple type, for instance, the conversion of some numbers from decimal to binary form. In *source coding* the aim is more sophisticated: to reduce the redundancy in the information. The consequence of this will be shorter messages so that more messages can be sent in a given time or stored in a given space. For efficient source coding, it is necessary to work with sequences of information symbols rather than isolated symbols. Shannon's source coding theorem will give us the limit imposed by nature on how much a symbol stream may be shortened by source coding.

In this section we shall set up the mathematical structure that information theorists have devised for source coding. First, let us return to the stoplight intersection in Sec. 11.2 for a simple example of this kind of coding. Suppose we wish to keep a record of the light's color at 9 A.M. each day, and suppose further that the light is green at any given time with probability 0.9 and red with probability 0.1. The simple way to record the daily color is to write down the binary outcome, either G or R, each day;

† Markov sources and random variables are discussed in [11-5] and [11-12], respectively.

TABLE 11.3-1 SOURCE CODE FOR LENGTH-3 SEQUENCES OF STOP LIGHT OUTCOMES

$\mathbf{z}^{(3)}$	$P(\mathbf{z}^{(3)})$	Code word
GGG	0.729	1
GGR	0.081	0000
GRG	0.081	0010
RGG	0.081	0011
GRR	0.009	0100
RGR	0.009	0101
RRG	0.009	0110
RRR	0.001	0111

this will consume one binary symbol per day. Another way is illustrated in Table 11.3-1. Here the information from three days is gathered into one length-3 sequence called $\mathbf{z}^{(3)}$, and each of the eight sequences is represented by a code word in the right column. The idea of this code is to represent the most common three-day outcome with a very short word, that is, 1. The average length of the code is 2.626 binary symbols per three-day outcome. This is 0.875 binary symbol per day, a reduction of 12% over simply writing down the daily outcomes.

By the end of the section, we shall have proved Shannon's theorem, which will show that reductions of as much as 53% are possible. To get the 12% reduction, we had to think of outcomes as occurring in length-3 sequences, and to get this 53% would require a much longer *sequencing*.

Sequences of Source Outputs

The idea of sequences is crucial to source coding, and indeed to all the kinds of coding in this chapter. Consider a sequence, or *block*, of K outputs $\mathbf{z}^{(K)}$ from a source Z from Sec. 11.2, which produces independent outcomes.† We could view the block as the output of a larger source $\mathbf{Z}^{(K)}$ whose single outputs are blocks of outputs of Z. Such a sequence is sometimes called the *Kth extension* of Z, and its defining probability distribution is given by

$$P(\mathbf{z}^{(K)}) = P[z(1)]P[z(2)] \cdots P[z(K)]. \tag{11.3-1}$$

Here $z(i)$ denotes the ith element of the block $\mathbf{z}^{(K)}$, and in multiplying out the single-output probabilities, we have used the independence of successive outputs of Z. If the Z alphabet is of size J, the alphabet of $\mathbf{Z}^{(K)}$ is of size J^K.‡

† We shall use the notation $\mathbf{z}^{(K)}$ to represent a sequence of length K and $\mathbf{Z}^{(K)}$ to represent a length-K random variable. Information theory texts often drop the parentheses in this notation. By z^K we will mean z raised to the Kth power.

‡ For example, suppose the original alphabet is binary with symbols 0, 1 ($J = 2$). Then, for $K = 2$, the size of the new alphabet is 2^2 and its elements are (0, 0), (0, 1), (1, 0), (1, 1).

The entropy of the extension $\mathbf{Z}^{(K)}$ is easily calculated. By definition, it is the expectation

$$\begin{aligned} H(\mathbf{Z}^{(K)}) &= E\left\{ \log \frac{1}{P(\mathbf{z}^{(K)})} \right\} \\ &= E\left\{ -\log \prod_{i=1}^{K} P[z(i)] \right\} \\ &= E\left\{ -\sum_{i=1}^{K} \log P[z(i)] \right\}. \end{aligned}$$

To interpret the last line, we use the sum-of-expectations principle of Eq. (8.15-9) and the fact that all the $z(i)$ are identically distributed; this gives

$$H(\mathbf{Z}^{(K)}) = KE\{-\log P(z(1))\} = KH(Z). \tag{11.3-2}$$

We have shown that the entropy of a block of K independent source outputs is K times the entropy of one. This is not too surprising in view of axiom 1 of Sec. 11.2.

To perform coding on these source sequences, we require a code. A code is a mapping from one set of sequences to another, done to accomplish some goal, which in our present context is the compression of source sequences into shorter runs of symbols. In technical terms, we can describe the mapping as

$$\mathscr{C}: \mathscr{Z} \longrightarrow \mathscr{W}, \tag{11.3-3}$$

Here $\mathscr{Z}$ is a set of source sequences and $\mathscr{W}$ is a new set of symbol sequences that will henceforth represent the source sequences; $\mathscr{C}$ denotes the code mapping. Often all the sequences in $\mathscr{Z}$ are the same length, and all those in $\mathscr{W}$ are of some other fixed length. In this case, $\mathscr{C}$ is called a *block code*. In other cases, either $\mathscr{Z}$ or $\mathscr{W}$ or both may contain sequences of varying lengths; these codes are called *variable-length codes*.

In this section we shall consider the case where all sequences in $\mathscr{Z}$ are uniquely represented by a sequence in $\mathscr{W}$, so a decoder exists that can reconstruct a source sequence exactly from its representative in the set $\mathscr{W}$. This kind of coding is called *noiseless* source coding. In many cases it is impossible to do source coding without noise, and we shall return to these cases in Sec. 11.4.

A simple example of noiseless coding is the alphabet converter. This encoder converts sequences of symbols from an alphabet of size A to sequences from a size B alphabet. The input to the encoder consists of the A^K sequences of length K, and its outputs are length-N sequences. The length of N is the smallest integer satisfying $B^N \geq A^K$; in fact, a little thought will show that this integer lies within the bounds

$$\frac{K \log A}{\log B} \leq N \leq 1 + \frac{K \log A}{\log B}. \tag{11.3-4}$$

We can define the ratio N/K to be the *rate* of this encoder, having the dimensions of B-ary symbols/A-ary symbols. Dividing both sides of (11.3-4) by K shows that the rate is bounded by

$$\frac{\log A}{\log B} \leq \frac{N}{K} \leq \frac{1}{K} + \frac{\log A}{\log B}, \tag{11.3-5}$$

from which it is clear that the rate can be held arbitrarily close to the ratio log A/log B as the block size K grows.

The next examples show two common noiseless coding schemes that are considerably more subtle. We are interested in what information theory has to say about these. As with most theories, we shall find that ours gives valuable insight, but that it must ignore some practical details.

Example: Morse Code

The familiar Morse code is a block-to-variable-length noiseless source code. It maps blocks of one symbol chosen from a 27-letter alphabet source (26 English letters plus the space) to binary letter sequences with lengths ranging from 1 to 4. The code, together with its word probabilities, is shown in Table 11.3-2; we have replaced the familiar dot and dash symbols with 1 and 0. Actually, the Morse code alphabet is at least ternary, since a time space is needed in this code to separate the code words from each other; in fact, Morse operators employ a variety of different spaces to mean ''end of English letter,'' ''end of English word,'' and so on, so their code is really quite complex. For simplicity, we shall represent all these spaces with the same character, # in the table.

What is the performance of this approximation to the Morse code? With the assumption that the code is ternary with symbols 0, 1, and #, the words in Table 11.3-2, each of which represents an English letter, have an average length of 3.06 ternary symbols/word. If we change them into binary words with an alphabet converter, their average length will be 4.85 bits/word. But this is only 0.15-bit improvement over simply representing each of the 27 English letters with a 5-bit binary word! It lies 20% above the entropy of the English letter source, which from the numbers in the figure is 4.03 bits/letter.

In designing his code, Morse sought to represent all the common letters, like E, T,

TABLE 11.3-2 MORSE CODE AND THE APPROXIMATE ENGLISH LETTER PROBABILITIES

Source letter	Morse code word	Approximate probability	Source letter	Morse code word	Approximate probability
A	10#	0.064	O	000#	0.063
B	0111#	0.013	P	1001#	0.015
C	0101#	0.022	Q	0010#	0.001
D	011#	0.032	R	101#	0.048
E	1#	0.103	S	111#	0.051
F	1101#	0.021	T	0#	0.080
G	001#	0.015	U	110#	0.023
H	1111#	0.047	V	1110#	0.008
I	11#	0.058	W	100#	0.018
J	1000#	0.001	X	0110#	0.001
K	010#	0.005	Y	0100#	0.016
L	1011#	0.032	Z	0011#	0.001
M	00#	0.020	Space	#	0.185
N	01#	0.057			

and A, with the shortest code words, which is the standard motive behind any block-to-variable-length code. Viewed in the right way, his code does gain great efficiency by doing this. In the early days of radio, Morse code words were sent by on/off keying an RF carrier; a 1 was a short burst, a 0 was a slightly longer burst, and the # symbol was a barely perceptible delay that consumed no RF energy. It is more reasonable under these circumstances to think of the words as binary and ignore the need for the # in both the code words and the English text words. We then find the average word's length is only 2.53 bits/word, about half of the simple 5-bit representation.

Example: Telephone Area Codes

In the telephone dialing system used in most of the world, cities or other geographical units are given numerical area or city code words, and countries are given a similar set of code words. In dialing an international call, the pattern is

$$\text{country code} + \text{area code} + \text{local number}.$$

A list of the country and area code words may be found in most telephone directories. A glance at this list shows that a variable-length code is used to represent the countries; and in most of the world outside North America, a variable-length code represents the areas within countries as well. We can use a leading 0 as a comma to indicate the start of a number. To access Lund, a medium-sized city in Sweden for instance, one encodes the desired telephone connection into the sequence of three code words 0 + 46 + 46 + local; here the first 46 means Sweden and the second means Lund (the + signs are added only for clarity and are ignored by the equipment). London, England, is reached by 0 + 44 + 1 + local, where 44 represents England and 1 London. The short word 1 is used because London is very often called. The town of Mercedes, Uruguay, a smaller city in a country with relatively few telephones, has the longer word 0 + 598 + 532 + local. Words like these need special properties to avoid confusing, for instance, a country word with other country words or with city or local words. We shall turn to this problem next.

Decodability

If a sequence of block code words is received, the symbol string may easily be parsed into individual code words; if the block length of the code is N, then every Nth symbol marks the end of a word. Having parsed the sequence, we can proceed to decode the individual words. With variable-length codes, parsing is more difficult. In the foregoing example, Lund in Sweden was accessed by 4646, by which we intended "country 46, area 46." But how is the telephone decoder to know that this does not mean "country 4, area 646"? The answer is that no country is allowed to have code word 4, nor does any country's code word start with 46, except for Sweden. With these two conditions, 4646 can only mean Lund.

Codes like this one are called *instantaneous* or instantaneously decodable codes, a definition of which is that any code word in the code may be immediately recognized and decoded as soon as its last symbol is received. Codes are instantaneous if and only if they satisfy the *prefix condition*, which states that no code word is a prefix of any other code word. A properly constructed block code, with no two words the same, obviously satisfies the prefix condition and is thus instantaneous.

TABLE 11.3-3 EXAMPLES OF BINARY LETTER CODES

$\mathscr{C}_1$ Block instantaneous		$\mathscr{C}_2$ Variable-length instantaneous	$\mathscr{C}_3$ Variable-length comma
0000	1000	1	1
0001	1001	01	01
0010	1010	0011	001
0011	1011	0010	0001
0100		0001	00001
0101		0000	000001
0110			
0111			

Some sample binary codes are shown in Table 11.3-3. Code $\mathscr{C}_1$ is clearly instantaneous because it is a block code, and $\mathscr{C}_2$ is because its words satisfy the prefix condition. Code $\mathscr{C}_3$ is a special kind of instantaneous code called a *comma* code, because it has a special symbol (or group of symbols), here the 1, that indicates the end of a word. In the Morse code example, we used the special symbol # as a comma; the Morse code is unusable as a code unless some kind of comma is employed, even if the comma is just a slight pause. Commas are sometimes placed at the beginning or even in the middle of words.

As long as there is some way to parse every sequence of code words, whether by the prefix condition or some other property, a code is said to be *uniquely decodable* or *uniquely decipherable*. Block codes are automatically so. Nonprefix codes are made decodable by restricting their words in such a way that there are no ambiguous sequences of words. The Morse code is not uniquely decodable without a comma; 100111 can be decoded for example as either AB, WS, or PI. To fix this ambiguity, we could modify the words for either W or S, and for either P or I, but the Morse code has many other ambiguities than these.

We consider next the question of how many words of a certain length a code can contain. We shall explore these limits in a graphical way with the aid of a commonly used construction called a *code tree*. Some binary trees are shown in Fig. 11.3-1. The code-word symbols appear on branches that represent them, and any complete path through a tree ending in a square block represents a code word. A full binary block code that contains 2^K words is represented by a complete tree, while other codes have some tree paths missing. A variable-length code has tree paths of different lengths. The branching character of the tree graph guarantees that no path ending in a block is a prefix of any other path. We can thus make a definitive statement about code trees and instantaneous codes: A code tree with square blocks only at the ends of its paths always represents an instantaneous code, and, conversely, any instantaneous code has such a code tree.

We are now prepared to prove a theorem called the *Kraft inequality*.

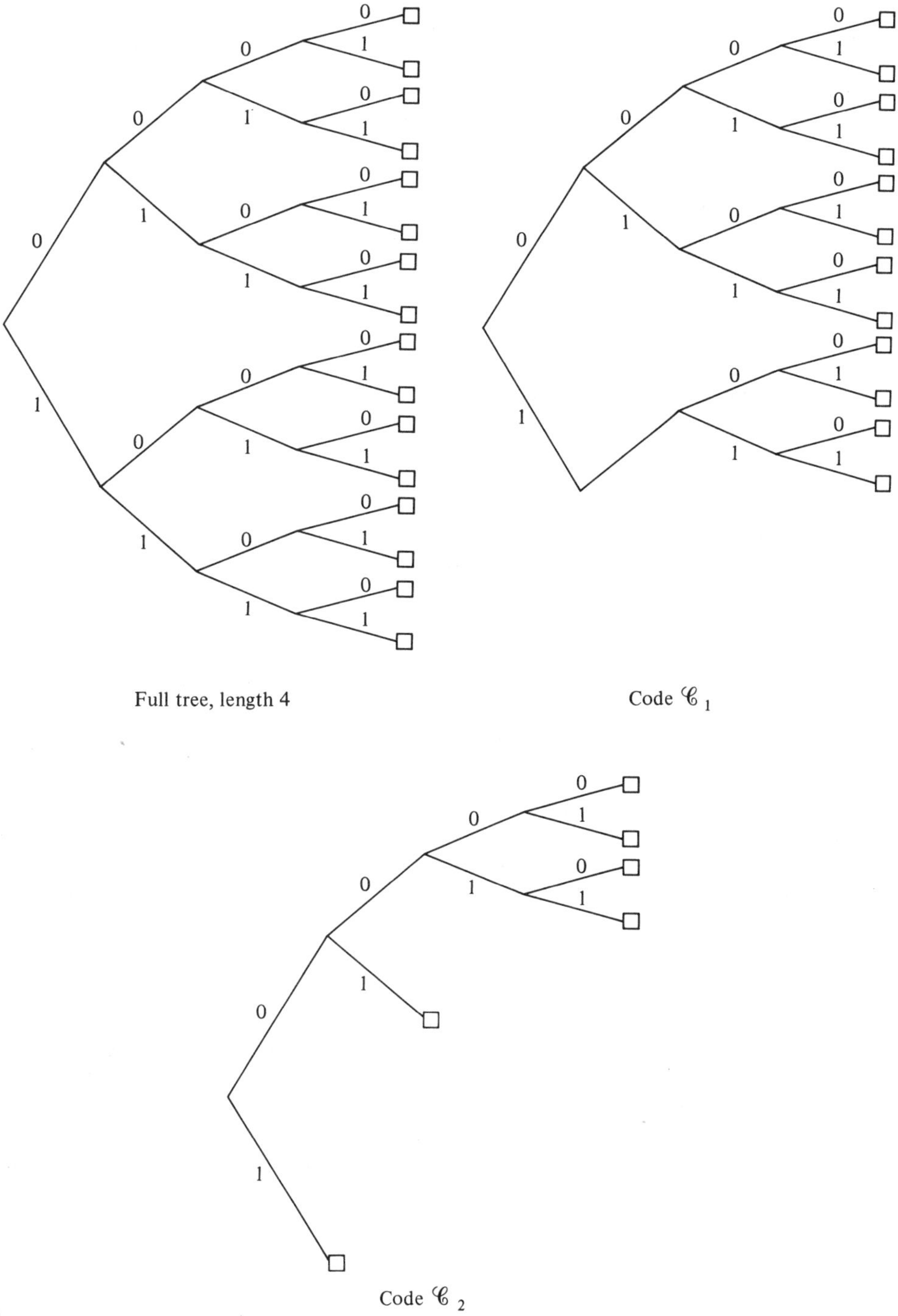

Figure 11.3-1. Code trees. A code word ends in a square block. Bits shown on branches are code-word symbols.

Theorem 11.3-1. Consider a code of m words with lengths $\ell_1, \ell_2, \ldots, \ell_m$; the symbols in the words come from an r-ary alphabet. Then a prefix code with words of these lengths exists if and only if

$$\sum_{j=1}^{m} r^{-\ell_j} \leq 1. \tag{11.3-6}$$

Inequality (11.3-6) gives an upper bound on the number of words in a code in terms of their lengths. For code $\mathscr{C}_1$ in Table 11.3-3,

$$\sum_{j=1}^{12} 2^{-\ell_j} = \tfrac{12}{16} < 1.$$

The Kraft sum [i.e., the quantity on the left of Eq. (11.3-6)] exactly equals 1 for a length-4 block code with the full 16 words or for any other full block code. For code $\mathscr{C}_2$, the sum is

$$4(\tfrac{1}{16}) + \tfrac{1}{4} + \tfrac{1}{2} = 1,$$

which from the theorem means that no words of any length can be added to this code without destroying its instantaneous character.

Proof of Theorem 11.3-1

We shall now prove the binary version of Theorem 11.3-1 with the aid of code trees; the version with $r > 2$ is not much harder. First, we observe that all prefix code trees are trimmed versions of a full block code tree with 2^k paths, where k is the length of the longest word; that is, some of their branches have been removed from the right and the block moved left to a new terminal position. The proof now follows from looking at what happens to the Kraft sum as this trimming is done. If both end branches are removed from a path, as in Fig. 11.3-2(a), then terms that total $2 \cdot 2^{-k}$ are replaced by one term of size $2^{-(k-1)}$, which means that the overall sum is unchanged. If one of the two paths is removed only, as in Fig. 11.3-2(b), then $2 \cdot 2^{-k}$ is replaced by the term 2^{-k} corresponding to the one remaining code path. This shows that any sequence of branch removals can only lessen the Kraft sum from its maximal value of 1. Thus every tree, and therefore every instantaneous code, satisfies Eq. (11.3-6).

It remains to show the "only if" assertion, that if a set of lengths $\ell_1, \ldots, \ell_m$ satisfies (11.3-6) there exists a valid prefix code with these word lengths. First, we shall reorganize the length set into the set $\{\sigma_1, \ldots, \sigma_k\}$, where σ_i is the number of words of length i. Now we shall try to construct a code with the right lengths. For the number of paths of length k, σ_k, Eq. (11.3-6) gives us that $\sigma_k 2^{-k} \leq 1$, from which we have that

$$\sigma_k \leq 2^k.$$

This states the obvious, that we cannot have more than 2^k binary code words of length k. Starting from the bottom of a code tree, we assign its length-k paths to be code

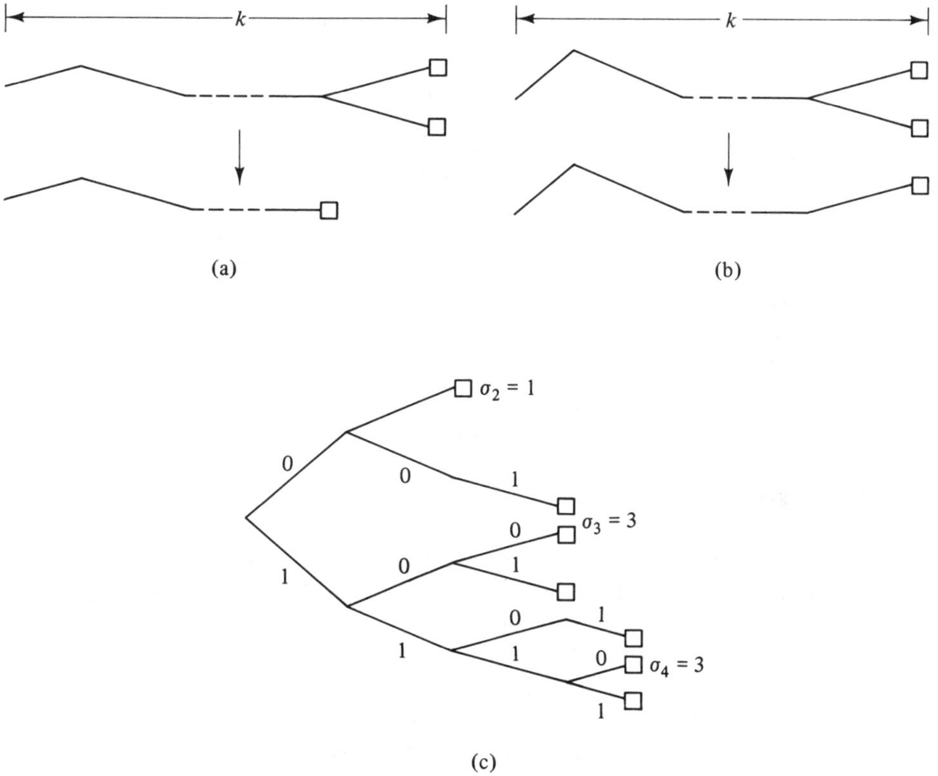

Figure 11.3-2. Code tree constructions used in the proof of Theorem 11.3-1. k is the length of the longest word in the code. (a) Replacement of two length-k words by one of length $k - 1$. (b) Replacement of two length-k words by one of length k. (c) Assignment of words and σ_i in proof of "only if" assertion.

words until σ_k are assigned. This has been done in Fig. 11.3-2(c) for a code with one word of length 2, three of length 3, and three of length 4.

Next we assign the length-$(k - 1)$ words. From Eq. (11.3-6) comes the inequality

$$\sigma_{k-1}2^{-k+1} + \sigma_k 2^{-k} \leq 1,$$

from which we have

$$2\sigma_{k-1} \leq 2^k - \sigma_k. \tag{11.3-7}$$

We can interpret Eq. (11.3-7) in terms of potential paths available at the kth level of the tree, either as code words (which are marked with a block) or as paths that are *prevented* from being words by a block earlier in the path. The left side is the number of words prevented at level k by the σ_{k-1} code words of length $k - 1$; the right side is the total potential paths available less the number of length-k code words σ_k. Inequality (11.3-7) assures us that enough paths remain in the tree to accept the needed σ_{k-1} words.

The rest of the proof consists of repeating the previous argument at each level. At level $k - 2$, Eq. (11.3-6) implies that

$$4\sigma_{k-2} \leq 2^k - \sigma_k - 2\sigma_{k-1}.$$

The left side is the number of potential words at level k that are prevented by code words at level $k - 2$ tree paths that remain.

It is actually possible to prove a stronger result, called McMillan's inequality, which states that a *uniquely decodable* code exists if and only if its word lengths satisfy the Kraft sum. This is a surprising result, because it shows that codes with the additional instantaneous condition are not any more or less available than other decodable codes. Proofs of McMillan's inequality may be found in information theory texts such as [11-3] and [11-4]. As an example of the McMillan result, we can compute the Kraft sum for the Morse code, which we saw was far from uniquely decodable. The sum is 3.75, not very near to 1, so that the Morse code fails Eq. (11.3-6), as it should.

Noiseless Source Coding Theorem

With the Kraft inequality, we can easily prove one of the important theorems of information theory, the noiseless coding theorem. Also known as Shannon's first theorem, it tells us that a source can be encoded into binary code words whose long-term average length is close to $H(Z)$ binary symbols per source output. If we wish to encode near this rate, a *long* code is ordinarily required whose input is a long string of source outputs; to see this clearly, a more sophisticated method of proof is needed than the one we shall use. As well, our proof will produce binary-alphabet source code words. For a nonbinary source code, one can simply use the alphabet converter code to change the words in the proof; their efficiency will be unaffected.

Theorem 11.3-2: The Noiseless Coding Theorem. Let a source Z have outputs from a finite alphabet. An instantaneous code exists with word lengths $\ell_1, \ldots, \ell_m$, which encodes source symbols in groups of K, whose average length per source output $E[\ell_i]/K$ satisfies

$$H(Z) \leq \frac{E[\ell_i]}{K} \leq H(Z) + \frac{1}{K}. \qquad (11.3\text{-}8)$$

Proof. We shall represent each block of source output $\mathbf{z}^{(K)}$ by a number of bits equal to the first integer larger than $\log_2 1/P(\mathbf{z}^{(K)})$. By this means, we set the lengths $\ell_1, \ldots, \ell_m$ of a binary variable-length code, with each length subject to

$$\log_2[1/P(\mathbf{z}^{(K)}(i))] \leq \ell_i \leq \log_2[1/P(\mathbf{z}^{(K)}(i))] + 1. \qquad (11.3\text{-}9)$$

Next we multiply Eq. (11.3-9) throughout by -1 and exponentiate each side to obtain the inequality

$$P(\mathbf{z}^{(K)}(i)) \geq 2^{-\ell_i} \geq \frac{P(\mathbf{z}^{(K)}(i))}{2}.$$

But if we sum the left inequality over all the code words, we simply obtain the Kraft inequality,

$$1 \geq \sum_{i=1}^{m} 2^{-\ell_i},$$

which implies that an instantaneous code exists with the desired lengths. On the other hand, if we take the expectation of Eq. (11.3-9) and make use of Eq. (11.3-2), we obtain

$$KH(Z) \leq E[\ell_i] \leq KH(Z) + 1.$$

Dividing this through by K gives Eq. (11.3-8), the desired result.

Information theorists have extended Theorem 11.3-2 in a great many ways—to sources with memory, to block codes, to sources with an infinite alphabet. These results appear in the texts listed at the end of the chapter. The result is always the same: About $H(Z)$ bits are needed on the average to encode a source output.

Getting a secure grasp of the concepts in this and the previous section can be difficult. For those in need of further examples and illustrations, we recommend the classic introductory text by Abramson [11-5].

11.4 RATE-DISTORTION THEORY

In many situations it is not possible to encode an information source without error. For the sources just discussed in Secs. 11.2 and 11.3, a storage medium or transmission channel may not be able to accept new source outputs at a rate as large as $H(Z)$. A more fundamental problem is the fact that everyday sources like speech and pictures are analog, so their data samples are continuous-amplitude real numbers. To specify even one point on the real line requires infinitely many bits! In fact, no human or machine observer measures an analog source with infinite accuracy. All measurers have finite resolution, and it often happens, as well, that the end use of the data is only worth a rough measurement. For analog sources, a mathematical theory of source coding should be one step more subtle than in the previous sections. It should accept that distortion of the source is inevitable and then ask how much the distortion can be reduced if a certain number of bits are used to encode each source output. The study of this problem has come to be called *rate-distortion* theory.

It is easy to confuse this rate-distortion viewpoint with other concepts in information theory. Some sources cannot be so encoded because the distortion of even one symbol is fatal. Examples are balances in bank accounts or personal identification numbers. Here we might even add some redundant check symbols (see Sec. 11.7) that will warn us of errors in the information symbols. Second, rate-distortion theory does not in the first instance seek to remove *redundancy* or *correlation* from data. In the design of a source encoder, removal of these rather than actual *information* must take first priority, but this is the concern of ordinary source coding theory, as we saw in examples like

English text in Sec. 11.3. The aim of rate-distortion coding is *to remove information* that is imperceptible or not worth the cost.

Distortion Measures and the Rate-Distortion Function

To set up the rate-distortion coding problem, we first need to define a distortion measure. Just as in Sec. 11.3, the code is a mapping $\mathcal{C}:\mathcal{Z} \rightarrow \mathcal{W}$, where $\mathcal{Z}$ is the set of source sequences and $\mathcal{W}$ is a set of rate-distortion code words; we are now considering only block codes, in which both the source and code sequences are of the same length K. A distortion measure is a nonnegative function $\rho(\mathbf{z}^{(K)}, \mathbf{w}^{(K)})$ that assigns a cost or distortion to encoding the sequence $\mathbf{z}^{(K)}$ with code word $\mathbf{w}^{(K)}$. We shall concentrate on *per-letter* distortion measures for which the total distortion is simply the sum of individual nonnegative letter, or symbol, distortions:

$$\rho(\mathbf{z}^{(K)}, \mathbf{w}^{(K)}) = \sum_{i=1}^{K} \rho(z_i, w_i) \geq 0. \tag{11.4-1}$$

A simple example of a per-letter distortion measure is the *probability-of-error* (PE) measure for binary sources encoded into binary words,

$$\rho(z_i, w_i) = \begin{cases} 0, & z_i = w_i, \\ 1, & z_i \neq w_i. \end{cases} \tag{11.4-2}$$

Note that Eq. (11.4-2) expresses the measure in terms of single letters and that the source and code-word alphabets are identical. The measure assesses no penalty at all if the ith source and code-word letters match and a unit penalty if they do not. The name "probability of error" comes from the fact that $(1/K)\ \Sigma_{i=1}^{K}\ \rho(z_i, w_i)$ is the relative frequency of mismatching letters in the encoding.

The most important distortion measure for use with analog sources is the squared-error measure

$$\rho(z_i, w_i) = (z_i - w_i)^2. \tag{11.4-3}$$

The sum of this distortion over a code word is the total error energy between the source sequence and its encoding. Another measure in common use is the absolute error measure

$$\rho(z_i, w_i) = |z_i - w_i|. \tag{11.4-4}$$

When the source and code-word alphabets are both $\{0, 1\}$, Eq. (11.4-4) reduces to Eq. (11.4-2).

The arguments z_i, w_i of the distortion measure ρ are in principle just abstract symbols, either source or code-word symbols, between which there is an assigned distortion. But it is convenient to adopt symbols that have some natural meaning as well. With analog sources, z_i is a real-number sample from an alphabet that is a part of the real line. The code symbols w_i are given the special name *reproducer values*, since at decoding they will reproduce, one hopes with good accuracy, the source samples. PCM conversion is a simple example of rate-distortion coding. As was developed in Sec.

4.6, all source samples falling within a line segment (bin) are represented by the same reproducer value w_i, with distortion $\rho(z_i, w_i)$. We can represent w_i by its PCM binary code, or simply consider w_i itself to be the rate-distortion code symbol and at the same time the value chosen to approximate z_i.

We shall see later in this section that coding schemes exist that perform better than PCM. It is important to define carefully the meaning of "better." We shall take it to mean that a scheme achieves lower average distortion for the same amount of information in the sequence $\{w_i\}$. From a theoretical point of view, this rate of information output is the entropy $H(W)$ per code-word letter (and hence per source letter). More practically, the rate is the average number of binary symbols needed, per source symbol, to represent the w_i. These two concepts are united by Shannon's noiseless coding theorem, Theorem 11.3-2, which states that with a proper noiseless source code, the two rates are the same.

To quantify these notions properly, we need to define the *rate-distortion function* $R(D)$. This function gives the least rate R in bits/source symbol needed by any encoder to achieve an average distortion of D. We could equally well have posed the question from the point of view of the previous paragraph, which asked what is the least distortion D that can be achieved at rate R. Both views are equivalent, and by tradition information theorists take the first view. We note that the nature of the function $R(D)$ stems from two sources, the data source and the measure of distortion; it is not related to a particular rate-distortion code or to a choice of reproducer values, since it gives the performance of the best of all these.

After the rate-distortion encoder words are stored or transmitted, they are decoded to produce an approximation to the source, and $R(D)$ gives the least rate at which information must appear for the approximation to be accurate within a tolerance D. Earlier we introduced the idea that rate-distortion coding restricts information: $R(D)$ is the rate at which a source produces information if we need only perceive the source to within the D tolerance.

The idea of a rate-distortion function was first spelled out by Shannon [11-7] some 10 years after his paper on noiseless source coding. Shannon also proved a rate-distortion coding theorem; that is, given $R(D)$, he showed that codes exist that achieve all rates and distortions larger than this R and D, and no codes exist achieving other combinations. We shall take this theorem up next and then sketch some $R(D)$ functions relevant to binary data sources and to analog-to-digital conversion.

Before studying optimal rate distortion codes, one detail remains. We have stated that the information needed to exactly specify an analog z is infinite, but we can nonetheless define the *differential entropy* of Z,

$$h(Z) \equiv E[-\log_2 p(Z)] = -\int_{-\infty}^{\infty} p(z)\log_2 p(z)\,dz, \tag{11.4-5}$$

where $p(Z)$ is the probability density† of the source output Z. Most properties of the ordinary, or *absolute*, entropy $H(Z)$ of a finite-alphabet source carry over unchanged to

† Note that $p(Z)$ is regarded here as a function of the random variable Z and is therefore itself an rv.

the differential entropy of an analog source. In particular, the lowercase $h(\cdot)$ is what is required in our coding theorem. The definitions of conditional entropy, equivocation, and mutual information, given in Eqs. (11.2-7), (11.2-9), and (11.2-11c), continue as before, but with $\int dz$ replacing the discrete summations. One small difficulty remains: The differential entropies $h(Z)$ and $h(Z|w)$ are no longer absolute measures of information, but are rather differences from the origin of a coordinate system. The origin of $h(\cdot) = 0$ in this system is the differential entropy of the uniform distribution

$$p(z) = \begin{cases} 1, & 0 \leq z \leq 1 \quad \text{(or any other unit interval),} \\ 0, & \text{otherwise,} \end{cases} \tag{11.4-6}$$

as can easily be verified from Eq. (11.4-5). Some entropies are now negative; an interpretation of this is that the source's $p(\cdot)$ is more tightly compacted than Eq. (11.4-6).

Random Codes and the Rate-Distortion Coding Theorem

The original proof of the rate-distortion coding theorem, and the simplest one to date, makes use of an original and striking idea of Shannon called *random coding*. His idea, first used in 1948 to prove his channel coding theorem (see Sec. 11.6), was to select code-word symbols at random, each letter independently according to the same probability distribution on the letters of the code alphabet. Shannon suggested choosing all the letters of all the words in this way and then showing that at least one of these code words had distortion $\rho(\mathbf{z}^{(K)}, \mathbf{w}^{(K)})$ close to D, with probability tending to 1 as the word length K grew.

Although Shannon's idea was quite rapidly accepted and gives the right answer to an astonishing number of questions, it can be frustrating to grasp at first. Isn't there some possibility that no code word will lie within D of the source sequence? What if all the randomly chosen words turn out to be identical? The answer to the first question is that we do not require an acceptable word all the time, but only almost all the time. If by chance a good code word does not exist for the present source sequence, then a bad word will be chosen; it will have a large but limited distortion, which will occur with very small probability, so the encoding distortion over many blocks is still close to D. The answer to the second question is a step more subtle. The assertion in a random-coding type of proof is that the expected encoding distortion is near to D, where the expectation is over not only the source outputs, but also the choice of the code. If the selected code is bad, we ignore it and select another. The proof states that the average selected code has average performance near D, so there must be at least one code as good as the average one.

Perhaps the most disconcerting fact about random coding is that it never actually exhibits a code. It proves only the *existence* of a code. It shows that a good code must exist in the same sense that we know that a person of at least average height must exist in a large population with a known height distribution. In fact, no one has ever seen a code with performance very close to the $R(D)$ function; the search for one has resisted determined efforts over many years.

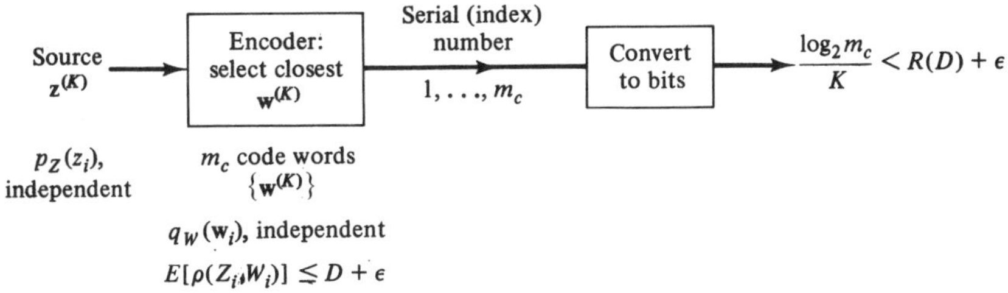

Figure 11.4-1. Schematic diagram of the rate-distortion coding process.

We shall now apply the random-coding idea to prove that codes exist with performance near the $R(D)$ function. The situation will be as diagrammed in Fig. 11.4-1. The input to the encoder is a sequence of K source outputs from a memoryless source. The source, as always, is defined by the distribution $p_Z(z)$, which may be a continuous density if the source is analog. Rate-distortion code words are of length K chosen at random by the density $\Pi_{i=1}^{K} q_W(w_i)$; that is, each code-word letter is independent and identically distributed according to $q_W(w)$, a distribution that will be specified during the proof. The number of such words selected will be m_c. The rate of our code in bits/source symbol will be

$$R \approx \frac{\log_2 m_c}{K}.$$

This is true because the code words can be serially numbered 1, 2, 3, . . . and the identity of the encoded word transmitted by converting the integer to binary form and sending the bits once for each block of K. The encoder works by finding the closest code word to $\mathbf{w}^{(K)}$ in the sense of the distortion measure. The coding theorem will show that if $E[\rho(Z_i, W_i)]$ is close to D then R is close to $R(D)$, if K is large enough.

Theorem 11.4-1: Shannon's Rate-Distortion Coding Theorem. Let $R(D)$ be the rate-distortion function for a possibly analog independent letter source defined by the density $p_Z(z)$ and a bounded distortion measure $\rho(z, w)$. Given any small $\epsilon > 0$ and $D \geq 0$, a block length K can be found such that there exists a code with

$$\frac{1}{K} E[\rho(\mathbf{Z}^{(K)}, \mathbf{W}^{(K)})] < D + \epsilon \qquad (11.4\text{-}7a)$$

and rate

$$\frac{1}{K} \log_2 m_c < R(D) + \epsilon. \qquad (11.4\text{-}7b)$$

A detailed proof of Theorem 11.4-1 for the analog case is beyond the level of this text, and the reader is referred to the classic text by Berger [11-8] or to [11-3] or [11-4] for a full proof. We shall only outline the proof now.

The first part of the proof is to define the $R(D)$ function, and the second is to show that a randomly chosen long code can approach the $R(D)$ performance in the sense of Eq. (11.4-7). We begin the first task by confining attention to only those random codes that have average distortion D or better. To obtain this average, we require the joint density $p_{Z,W}(z, w)$, or equivalently, the conditional density of w, $q_{W/Z}(w/z)$. [Note that the density $q_W(w)$ used to select the random code is not enough!] Let the set Q_D contain all the conditional densities $q(w/z)$ that satisfy the distortion criterion; that is,

$$Q_D = \left\{ q\,(w|z): \int\!\!\int p(z)q(w|z)\rho(z, w)\, dw\, dz \leq D \right\}. \tag{11.4-8}$$

Here and throughout the integral sign should be replaced by summation if the source is discrete.

The next step in defining $R(D)$ is to consider the average mutual information $I(W; Z)$ between the source and code-word symbols, a quantity that we defined in Eqs. (11.2-11). This is given by

$$I(W; Z) = \int\!\!\int p(z)q(w|z) \log \frac{q(w|z)}{q(w)}\, dw\, dz, \tag{11.4-9}$$

which like Eq. (11.4-8) requires the conditional density $q(w|z)$, not just the marginal. We can now define $R(D)$. It is

$$R(D) = \min_{q \text{ in } Q_D} I(W; Z). \tag{11.4-10}$$

That is, $R(D)$ is the least obtainable mutual information between $p(Z)$ and $q(W)$ when $q(W)$ must satisfy Eq. (11.4-8). There are a number of interpretations of Eq. (11.4-10) that try to explain why such an expression should be the limit to the performance of rate-distortion codes. A mathematical explanation, not very satisfying otherwise, is that only Eq. (11.4-10) allows the existence of codes as good as those implied by Eq. (11.4-7) during the coding theorem proof. Another explanation stems from the interpretation of mutual information as the information given about Z when W, the code symbol variable, is known. We wish to *minimize* this knowledge, that is, carry away as little information as possible, given that the encoding has average distortion D.

The second part of the theorem proof is to show that a random code, using the $q(w|z)$ that achieves the min in Eq. (11.4-10), performs near $R(D)$. A formal proof does this by defining the sets

$$S_D(\mathbf{z}) = \{\mathbf{w}: \rho(\mathbf{z}, \mathbf{w}) \leq D + \epsilon\}. \tag{11.4-11a}$$

$$S_R(\mathbf{z}) = \left\{ \mathbf{w}: \frac{1}{K} \log \frac{q(\mathbf{w} \mid \mathbf{z})}{q(\mathbf{x})} \leq R(D) + \epsilon \right\}. \tag{11.4-11b}$$

The first set is the code words that are close enough to a source word $\mathbf{z}$, while the second set collects code words with sufficiently small mutual information. The proof concludes by showing that, over the joint probability space of Z and W,

$$P(S_D \cap S_R) \longrightarrow 1, \qquad \text{as } K \longrightarrow \infty.$$

In the remote case that no code word is in the intersection of the sets in Eqs. (11.4–11a-b), then a default code word with large but bounded distortion is the encoder output; the contribution of this event to the average distortion vanishes as K grows.

Technically, Theorem 11.4-1 shows only that there exist codes at least as good as the $R(D)$ function. That there exist none that are better requires what is called a *converse* coding theorem. We shall again refer to the more advanced texts for this.

Binary Source with PE Distortion

This source/distortion measure combination, defined in Eq. (11.4-2), consists of a binary symmetric source (or BSS) encoded into reproducer words of the same length and made of the same letters; if source and reproducer letters disagree, a distortion of 1 is assessed, and otherwise 0. The rate R of the rate-distortion code must be less than or equal to 1 bit/source symbol, since the entropy and the rate of symbol production as well are unity. If the rate-distortion code has rate $\frac{1}{2}$, for example, the 2^K different source K blocks will be mapped to one of $2^{(1/2)K}$ code words. These reproducer words can in turn be represented by binary sequences of length $K/2$, so the original binary source sequences have been cut in half by the coding.

Our BSS/PE example has somewhat limited practical application, but as the simplest example of rate-distortion coding, it is a good place to begin. In the study of channel encoding, however, it provides the answer to an interesting question. Suppose we wish to transmit K binary symbols through a channel that can carry error free only RK symbols, with $R < 1$. We shall consider channel coding carefully in the next two sections; suffice it to say here that errors must necessarily occur in this channel since it is overloaded. Our question now is, what is the *least number* that must occur? We can answer this by realizing that the coding process in the channel, however it works, can be viewed as rate-distortion coding for the BSS/PE example: K binaries enter, only RK leave, the shorter output sequences can be thought of as representing a length-K reproducer sequence, and an error is counted each time a reproducer symbol disagrees with a source symbol. The function $R(D)$ gives the least average error rate D that any coding scheme can achieve whose rate is R.

Because of the simplicity of this example, it is not difficult to derive the $R(D)$ function. We first evaluate the expected distortion in Eq. (11.4-8). For the BSS, $p(0) = p(1) = \frac{1}{2}$, and from symmetry it must be that

$$\begin{aligned} &\text{Reproducer letter distribution } q(x): && q(0) = q(1) = \tfrac{1}{2}, \\ &\text{Conditional distributions } q(x|z): && q(0|1) = q(1|0) = \gamma, \\ & && q(0|0) = q(1|1) = 1 - \gamma. \end{aligned} \tag{11.4-12}$$

Here γ is a free parameter with $0 \le \gamma \le 1$ that remains to be determined. The set Q_D in Eq. (11.4-8) is defined by the condition that the expected distortion of the code must not exceed D; with the distributions given in Eq. (11.4-12) this distortion is

$$[\rho(0, 0) + \rho(1, 1)]\left(\frac{1-\gamma}{2}\right) + [\rho(0, 1) + \rho(1, 0)]\frac{\gamma}{2} \leq D,$$

which becomes

$$\gamma \leq D$$

when the PE distortion values are inserted for $\rho(\cdot, \cdot)$. Thus any probability function in Eq. (11.4-12) that satisfies this condition on γ is a member of the set Q_D. It remains to evaluate $I(W; Z)$ in Eq. (11.4-9). Substitution of the distributions yields

$$\begin{aligned} I(W; Z) &= \gamma \log_2 2\gamma + (1-\gamma)\log_2 2(1-\gamma) \\ &= 1 - h_B(\gamma). \end{aligned}$$

The function $R(D)$ at D is the minimum of $I(W; Z)$ over the allowable parameters γ, and this minimum occurs at the largest γ, which is $\gamma = D$. We have thus shown that the rate-distortion function for the BSS/PE example is

$$R(D) = 1 - h_B(D). \tag{11.4-13}$$

The calculation of $R(D)$ functions for sources without a lot of symmetry requires a full minimization over a constrained set by means of Lagrange multipliers. This is beyond our scope here, and the student is referred to [11-8] or [11-4]. An example of a more complicated source is the *unsymmetrical* binary source, with $P(1) = p$ and $P(0) = 1 - p$, combined with the PE distortion; this has

$$R(D) = h_B(p) - h_B(D).$$

$R(D)$ for the BSS is plotted in Fig. 11.4-2. It can be seen that at rate $\frac{1}{2}$, a 2:1 compression of the binary source, no coding scheme can achieve distortion below about 0.11. This means that at least 11% of the binary reproducer word letters must be erroneous in any code, on the average. At rate $\frac{1}{5}$, a 5:1 compression, no code can do better than 24% letters in error. To achieve no error at all, a code must apparently have rate 1. At rates close to 0, codes have distortion close to 0.5. It is easy to see this is true by considering that rate 0 means the code presents no information at all about the source; the decoder can do no better than guess the source outputs, a process that yields a 50% error rate. We can extend the idea of guessing to obtain a "guessing limit" to the BSS/PE $R(D)$: Imagine a crude rate R encoder that simply encodes verbatim a fraction R of the source outputs, and instructs the decoder to guess at the remaining fraction $1 - R$. The decoder will achieve 0 distortion on the first R and distortion $\frac{1}{2}$ on the last $1 - R$, for an average distortion $D = (1 - R)/2$. The guessing limit is shown for comparison in Fig. 11.4-2.

Research has shown that as the performance of a coding scheme nears the $R(D)$ function its demand for processor resources grows very rapidly. The number of computation steps needed to encode a source symbol diverges to infinity, as do the processor memory requirement and the delay between the arrival of a source output and the end of the rate-distortion code word that includes it. The limit of practical systems is shown

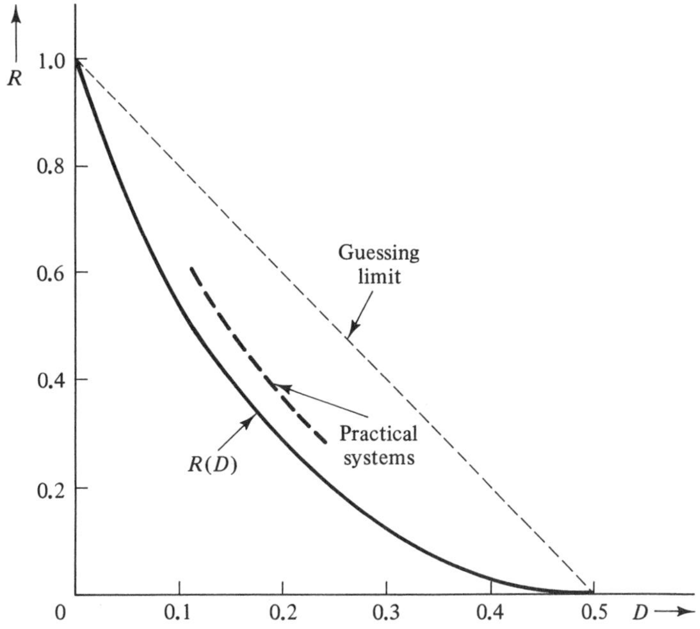

Figure 11.4-2. Rate-distortion function of a binary symmetric source with PE distortion measure.

approximately in the figure. By practical scheme is meant one equal to several commercial single-chip processors of the type available in the late 1980s.

Gaussian Source with Squared-Error Distortion

This source/distortion consists of an analog source whose outputs are a block of Gaussian random variables $\mathbf{z}$ and the square error distortion $(z_i - w_i)^2$ of Eq. (11.4-3). The block $\mathbf{w}$ is a reproducer word of the same length as the source sequence, and 2^{nR} such words are available to encode each block of n source outputs. The Gaussian source example is an intensely practical one, and it provides the standard of comparison against which are judged all encoders of the analog-to-digital conversion type. If one assumes that speech, images, and the like, are Gaussian, then $R(D)$ provides the ultimate limit to performance for the conversion methods of Sec. 4.6, including PCM, differential PCM, and any other scheme that converts analog samples to binary symbols.

The mathematics of computing this $R(D)$ are well beyond our present level, so we can give only the final result. For an independent-letter source with variance σ^2, the rate-distortion function is

$$R(D) = \tfrac{1}{2} \log_2 \frac{\sigma^2}{D}, \tag{11.4-14}$$

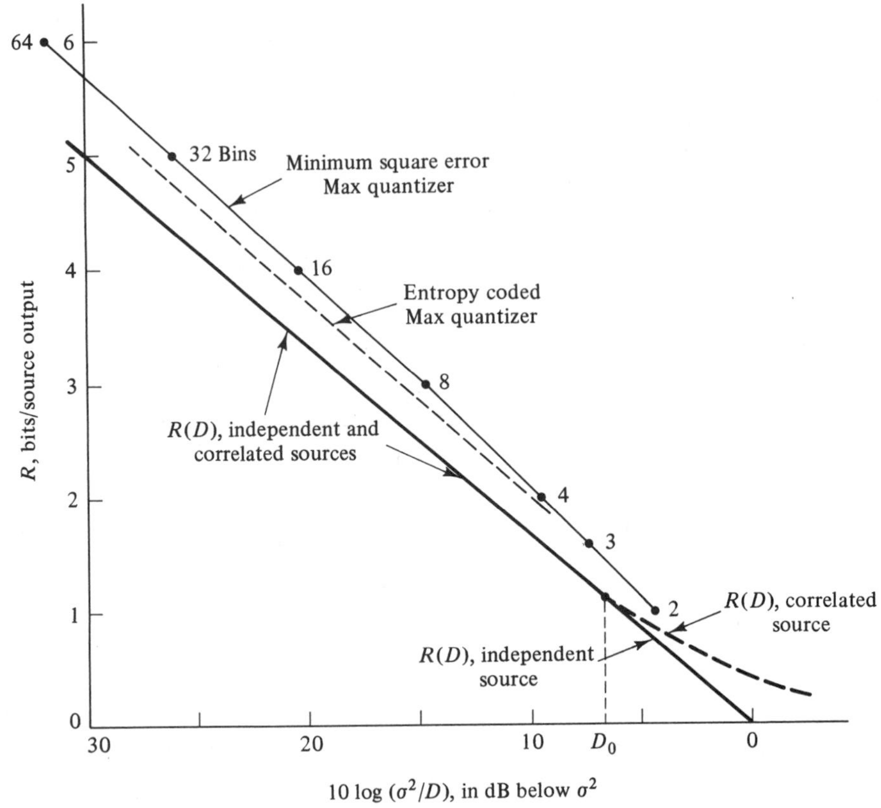

Figure 11.4-3. *R(D)* for independent and correlated Gaussian analog sources; Max quantizer performance is shown for comparison. Note that distortion in SNR terms is given by $\text{SNR}_{\text{dB}} = \text{SNI}_{\text{dB}} + 10 \log(\sigma^2/D)$.

which is plotted in Fig. 11.4-3. By choosing a decibel scale for the D axis, we get a straight line for Eq. (11.4-14). As D tends to 0, Eq. (11.4-14) shows that an infinite rate is required. At rate 0, the decoder can simply guess, as suggested before, and achieve distortion σ^2.

The best performing PCM system for our Gaussian source is the minimum-square-error Max quantizer discussed in 4.6. This is the rate in b bits/source output furnished by a PCM converter with bin spacings and reproducer values chosen to minimize the squared error. Its behavior in the *R-D* plane is also plotted in the figure. It can be seen that both the Max quantizer and the *R(D)* function obey a 6-dB rule like Eq. (4.6-4), which states that each increase of a bit in the encoder rate will decrease the distortion by 6 dB; but the Max line lies about 3.4 dB to the right, higher in distortion. We could say that use of a Max PCM scheme instead of an optimal rate-distortion encoder costs about 3.4 dB in SNR or, alternately, about 0.56 bit/source output.

It turns out that the various reproducer values of a Max quantizer are not equiprobable. We can view the quantizer as a discrete-valued data source. The entropy of this 2^b-ary finite-alphabet source that falls below the maximum (see Property 11.2-2) $\log_2 2^b = b$ bits that a 2^b-ary source can have. It follows from Sec. 11.3 that by ordinary noiseless source coding we can reduce the PCM encoder's rate of symbol output from b binary digits/source output to $H(W)$, the entropy of the quantizer reproducer values. At first glance this is a surprising result, because it seems to say that the Max quantizer is not really optimal. There is in fact no paradox, since one cannot simultaneously optimize the bins and reproducer values to minimize both square error and entropy. (We came to a similar conclusion regarding clarity and entropy in the example of Fig. 11.2-2.)

A source coder used to reduce a quantizer bit rate from b to $H(W)$ is often called an *entropy coder*. We can plot the R-D performance of the ideal entropy-coded quantizer by replacing the Max R-coordinates in Fig. 11.4-3 by $H(W)$ for each quantizer's reproducer values. This moves the quantizer performance points to the vicinity of the parallel dashed line in the figure, about 1.5 dB better than the quantizers without entropy coding.

Most real-life analog sources do not have independent outcomes, so it is of interest to find rate-distortion functions for correlated sources. One startling result of rate-distortion theory concerns Gaussian all-pole correlated sources, that is, sources that are created by passing independent Gaussian variates through a pure IIR discrete-time filter. These sources are used to model speech and images. The result is as follows.

Theorem 11.4-2. Let the input to an IIR filter be the independent sequence of Gaussian variates with variance σ^2 (and mean 0). Then the $R(D)$ function for Z, the all-pole correlated output of the filter, satisfies

$$\begin{aligned} R(D) &= \tfrac{1}{2}\log_2 \frac{\sigma^2}{D}, && D \le D_0, \\ R(D) &\ge \tfrac{1}{2}\log_2 \frac{\sigma^2}{D}, && \text{otherwise.} \end{aligned} \qquad (11.4\text{-}15)$$

In words, optimal rate-distortion coding can encode the output of the filter to an absolute criterion of accuracy D no worse than would be the case if the uncorrelated input were being encoded. The surprise in the result comes from considering that the variance of the correlated Z is ordinarily much larger than σ^2, the uncorrelated input variance! In fact, the increase in variance is the same as the SNI factor that was defined in Eq. (4.6-6). Thus the optimal distortion expressed as an SNR is 10 log(SNI) dB better for Z than it is for the IIR filter's input.

As a consequence of Theorem 11.4-2 and the choice of axes in Fig. 11.4-3, the $R(D)$ function for an all-pole correlated source is identical to the one in the figure for the uncorrelated source, at least until D exceeds D_0. The constant D_0 is a complicated function of the filter parameters and not of much concern since it usually lies in a high-distortion region that is not of much practical interest.

In Fig. 4.6-6 we illustrated the idea of differential PCM, a variant of PCM that takes advantage of correlated samples. We can imagine a Max quantizer in the Q box in this figure. Research has shown that the SNR of this differential Max quantizer is

$$\text{SNR}_{\text{dB}} = 10 \log_{10} \text{SNI} + \text{SNR}_{\text{Max,dB}}$$

when the bit rate is reasonably high, where SNR_{Max} is the quantizer performance with uncorrelated Gaussians. This demonstrates that in SNR terms the $R(D)$ function and the differential PCM performance are both offset by the same factor, 10 log SNI, from their positions with uncorrelated data. A plot of them is similar to Fig. 11.4-3, at least in the higher rate regions, and they continue to lie about 3.4 dB apart.

11.5 CHANNELS

A channel carries information. Channels such as radio links, coaxial cables, or pairs of wires carry information from one place to another. Other channels transmit information through *time*, from "now" to "then"; examples include phonograph records, computer disk files, and even punched cards. Ordinarily, these channels are not perfect, and they sometimes modify the information that they carry. Channel coding is then needed to protect the information from damage. This will be the subject of Secs. 11.6 and 11.7. In this section we shall make precise the notion of a channel and its imperfections and then define its information-carrying capacity.

It is important to make clear the differing types of information coding that we have introduced in this chapter. These are portrayed in Fig. 11.5-1. A source encoder accepts source symbols in blocks of K. Its purpose is either to remove redundant information (the noiseless encoder) or to remove nonredundant information in order to bring down the rate of information production (the rate-distortion encoder); the output of this encoding is a sequence $\mathbf{w}$, which is a length-K word of reproducer values in the rate-distortion case. There follows a possible alphabet conversion to prepare the information for channel encoding. After a PCM encoder for instance, reproducer levels are converted

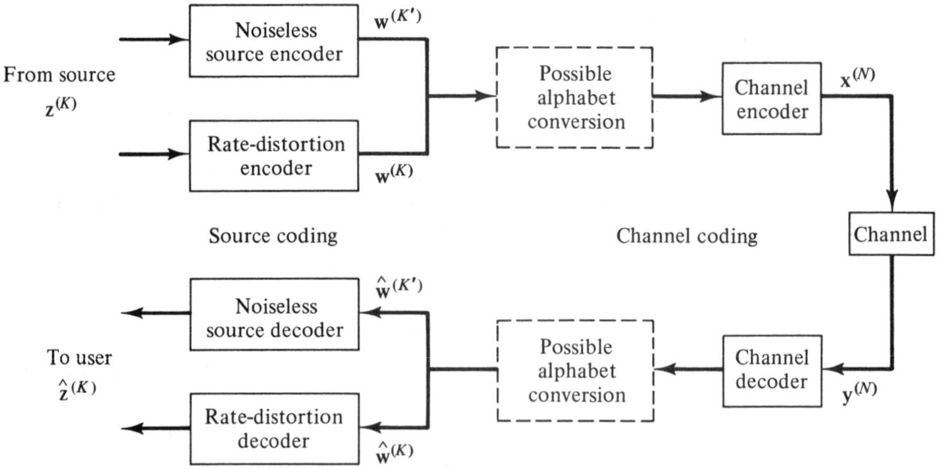

Figure 11.5-1. Schematic diagram of a complete communications link.

to binary form. Next the channel encoder maps the information into an N-block $\mathbf{x}^{(N)}$, a form whose purpose is to defeat the imperfections of the channel. The channel changes $\mathbf{x}^{(N)}$ to $\mathbf{y}^{(N)}$, a distorted version that can nonetheless be interpreted correctly by the channel decoder. Finally, a source decoder inverts the source encoder action.

Figure 11.5-1 is only a general diagram. Every practical communications link provides its own details, particularly with regard to alphabets and the conversions among them. In an application like transmission of binary source data through a binary-letter channel, there will be no conversions at all! The important point is to distinguish the kinds of coding and their purposes.

Definition of a Channel

In Sec. 11.2, we defined an information source in terms of the probability distribution of its outcomes. Information theorists continue this habit in defining a channel. A channel is defined to have an input X and an output Y, both of which are *random variables* with outcomes from input and output alphabets. The action in the channel itself is defined by a *conditional probability function* $p_{Y|X}(y|x)$ between these variables. This gives the probability that channel output y will occur, given that the input is x, during a single use of the channel.

There are several ways to portray the probabilities that make up a channel. The most common is a channel transition diagram, a general version of which appears in Fig. 11.5-2(a). Here the lines show the possible ways that an input can be converted by the channel, and the probabilities next to each line show the probability of this transition. We have assumed that there are I inputs and J outputs in the channel. Very often the input and output alphabets are identical so that transitions between identical letters represent a correct symbol transmission and all other transitions are channel errors. One can also define a channel by its transition matrix

$$[p(y_j|x_i)] \equiv \begin{bmatrix} p(y_1|x_1) & p(y_2|x_1) & \cdots & p(y_J|x_1) \\ p(y_1|x_2) & p(y_2|x_2) & \cdots & p(y_J|x_2) \\ \cdot & \cdot & \cdots & \cdot \\ \cdot & \cdot & \cdots & \cdot \\ \cdot & \cdot & \cdots & \cdot \\ p(y_1|x_I) & p(y_2|x_I) & \cdots & p(y_J|x_I) \end{bmatrix}. \tag{11.5-1}$$

Each row in Eq. (11.5-1) sums to 1, a statement of the fact that, for every input, some output must appear. If in addition to this definition, the input distribution $p_X(x)$ is known, we can calculate the output distribution from

$$p_Y(y) = \sum_{i=1}^{I} p(y|x_i)p_X(x_i). \tag{11.5-2}$$

From this, $H(Y)$ and the like may be found.

A *memoryless* channel is one for which successive uses are independent. For this

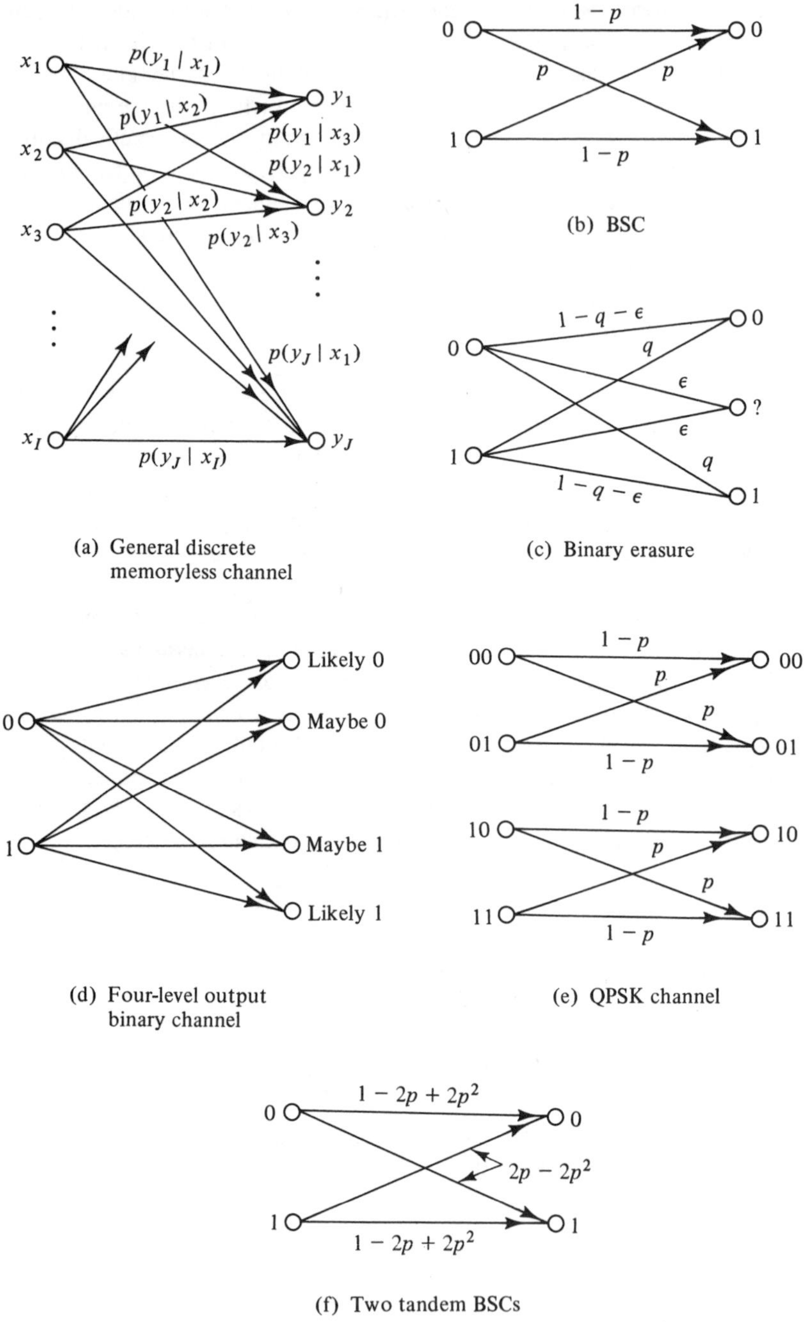

Figure 11.5-2. Examples of discrete memoryless channels.

kind of channel, the probability that a whole N-block $\mathbf{y}^{(N)}$ is received, given that the block $\mathbf{x}^{(N)}$ is sent, is the product form

$$p_{Y|X}(\mathbf{y}^{(N)}|\mathbf{x}^{(N)}) = \prod_{n=1}^{N} p_{Y|X}(y_n|x_n). \qquad (11.5\text{-}3)$$

If both the input and output alphabets are finite in size, a channel is said to be *discrete*. A channel that is both discrete and memoryless is called a *discrete memoryless channel* (DMC). DMCs are the simplest channel models and they provide useful models or partial models for a great many communication systems.

A great structure of information theory has been built on this DMC model. As communications engineers, we might reasonably ask how accurate it is. Often a commercial channel is specified by its probability of error, with nothing said about its internal workings, and in this case we lose little by viewing the channel as a DMC. But it may happen that channel errors occur in bursts, a normal occurrence in mobile radio and telephone channels. Now Eq. (11.5-3) does not hold; a Markov-chain-type channel model is more accurate, but subsequent calculations of information theoretic nature that we may wish to do will now be harder. Ordinary channel filtering, discussed in Sec. 10.5, introduces another form of memory into the channel, but in this case the subsequent calculations are of extreme difficulty.

Another liberalizing of the DMC is needed when the channel does not present discrete (e.g., binary) symbols as its output, but rather presents an analog voltage. This could occur because the channel is carrying an analog signal or because analog noise is being added to a discrete signal, as in Sec. 10.5 or 10.6. This kind of relaxation of the "discrete" constraint in the DMC is relatively easy to do.

In sum, we can see that the probabilistic channel model used in information theory can successfully model most practical situations, although occasionally the work needed to extract useful conclusions from the model will not be worth the effort.

Examples of the DMC

The *binary symmetric channel* (BSC) assumes that both input and output alphabets are binary. This simplest of all channels is portrayed in Fig. 11.5-2(b): Each symbol is carried correctly with probability $1 - p$ and erroneously with p, the *crossover* probability. The BSC is an appropriate model for binary transmission when we cannot gain access to the internal mechanism of the channel. If we know, for instance, that the channel really consists of BPSK from Chapter 10 with additive Gaussian noise, then the BSC is apt to be strongly inaccurate.

A better model in that case might be the *binary erasure channel* of Fig. 11.5-2(c). Its input is still binary, but the output now consists of 0, 1, and a new symbol, ?, which denotes erasure. The erasure channel can model a variety of situations in which the channel is allowed to announce that a symbol is highly unreliable. In the BPSK demodulator of Fig. 10.5-2, the ? symbol could be put out when the matched filter sample is near the threshold between 0 and 1. During a deep fade in a radio channel, the ? could indicate that the data should be ignored.

The four-level quantized binary channel of Fig. 11.5-2(d) is a further extension of the erasure channel idea. The channel modeled here still carries binary data, but it produces the outputs "likely 1," "maybe 1," "maybe 0," "likely 0." The added information in these will be welcome in Secs. 11.6 and 11.7, where it will help distinguish which code word a sequence of symbols might be.

The channel in Fig. 11.5-2(e) could be used to model QPSK transmission. It is two BSCs in parallel, reflecting the fact that QPSK is two independent BPSKs in phase quadrature. A common practical situation, the tandeming of two channels, is illustrated in Fig. 11.5-2(f). Some contemplation will show that its crossover probability is that of two BSCs of Fig. 11.5-2(b) in series.

Continuous-Output Channel Models

Channel transition diagrams are a clumsy way to depict a channel whose outputs are analog variables. The quadrature modulations of Chapter 10 assume a channel with discrete inputs and continuous-amplitude outputs, which should be modeled as in Fig. 11.5-3. Figure 11.5-3(a) models the *antipodal modulation channel*, the case when the channel input $\mathbf{x}^{(N)}$ is a binary vector and each use of the channel corresponds to a new orthogonal direction in signal space. The inputs $\pm\sqrt{E}$ in the model are the signal-space components in Fig. 10.2-4(a) or (c), although we may sometimes give them notations like 0 and 1. The channel action is to add an independent Gaussian variate to $\pm\sqrt{E}$, creating one time-discrete analog output for each channel use. Figure 11.5-3(a) accurately models the BPSK channel, either pure or Nyquist, as well as QPSK, offset QPSK, and MSK. For the QPSK and MSK cases, the modulation employs two such channels independently.

Transmission schemes like 8-QAM and 16-QAM place more than 1 bit in each signal-space dimension and are consequently not antipodal schemes. For these, Fig. 11.5-3(b) is a more appropriate channel model. This channel is called the *additive*

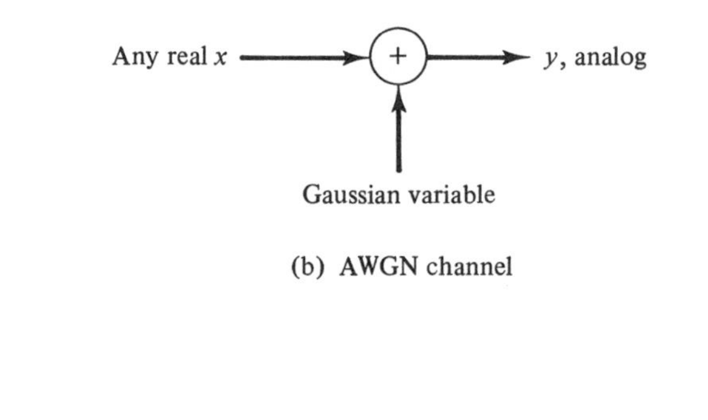

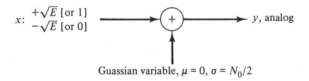

(a) Antipodal modulation channel

(b) AWGN channel

Figure 11.5-3. Additive Gaussian noise channels; the input of (a) is restricted to $\pm\sqrt{E}$.

white Gaussian noise (AWGN) model. It simply assumes a real-number input at each channel use, to which the channel adds an uncorrelated Gaussian variate. The AWGN channel is a good model for any transmission scheme in which the signal-space components of the modulated signal behave in an independent manner once they enter the channel.

Channel Capacity

Perhaps the most striking result of information theory is its derivation of the ultimate transmission capacity of a channel. For a century prior to the 1950s, communications engineers thought that the output of a noisy channel could only be imperfect data. It was known that data could be made more reliable by increasing transmitter power and, with the advent of FM, by increasing bandwidth, but perfect reliability at a fixed power and bandwidth was thought to be impossible in principle. Shannon reversed this belief in 1948 [11-1] when he showed (for the DMC) that virtually perfect transmission was possible in a given channel at any rate less than a fixed C bits/channel use and was impossible otherwise, provided one had available a large enough processor and could tolerate enough delay. He called the constant C the *capacity* of the channel. His work was quickly extended to every imaginable channel. The result was a revolution in communication theory.

We shall now define capacity and compute it for some of the channels just described. The definition of capacity, like the definition of the rate-distortion function, arises during the proof of a coding theorem, that is, a theorem proving the existence of channel codes that can carry information at rates up to C. We shall defer the coding theorem to Sec. 11.6.

The formal definition of capacity that arises is

$$C \equiv \max_{p_X(\cdot)} \sum_y \sum_x p_X(x) p_{Y|X}(y|x) \log_2 \frac{p_{Y|X}(y|x)}{\sum_{x'} p_X(x')\, p_{Y|X}(y|x')}. \tag{11.5-4}$$

Here $p_X(x)$ is the distribution of the channel input variable. If Y is a continuous variable, the same definition applies but with an integral replacing the y sum. We see that *capacity results from a maximization over the possible input distributions and that it is a function only of the channel transition probabilities*. To get the best performance out of a channel, apparently we must present it with just the right distribution of inputs, regardless of the distribution of the data; this will be a job for the channel code. The idea of mutual information provides another way to define capacity. Recall from the discussion following the definitions (11.2-11) that $I(Y; X)$ measures the information about X given by Y. Now X and Y are random variables correlated by the action of the channel. The largest information that can be carried by the channel is

$$C = \max_{p_X(\cdot)} I(X; Y), \tag{11.5-5}$$

an amount that will be carried only when X has just the right maximizing distribution. For other distributions, the channel carries simply $I(X; Y)$, without the max. It is straightforward to show (see Prob. 11-18) that Eqs. (11.5-5) and (11.5-4) are mathematically identical.

Example: Capacity of the BSC

For the BSC, it is clear from the symmetry of the channel and of Eq. (11.5-4) that the maximizing X distribution will be $p(0) = p(1) = \frac{1}{2}$. The channel transition probabilities are $p_{Y|X}(1|0) = p(0|1) = p$ and $p(1|1) = p(0|0) = 1 - p$. Substitution of these values into Eq. (11.5-4) gives

$$C(p) = 1 + p \log_2 p + (1 - p) \log_2 (1 - p),$$

which is

$$C(p) = 1 - h_B(p), \qquad p \leq \tfrac{1}{2}. \tag{11.5-6}$$

This is yet another role for the binary entropy function. C as a function of p is plotted in Fig. 11.5-4. An error-free BSC has capacity close to 1 bit/channel use, as we would expect. C falls to 0 at a 50% error rate; the channel outputs now are completely unreliable.

In the practical world, a rather poor BSC is one with a 5% error rate. From Eq. (11.5-6), we can calculate the $C(0.05) = 0.714$, which means that with a proper code this BSC can still transmit error free 71% of the capacity of the perfect $p = 0$ BSC, at least in principle. In reality, no practical channel code has been found that approaches this rate, even if a decoded error rate of 10^{-4} to 10^{-3} is acceptable. This despite 30 years of concentrated research!

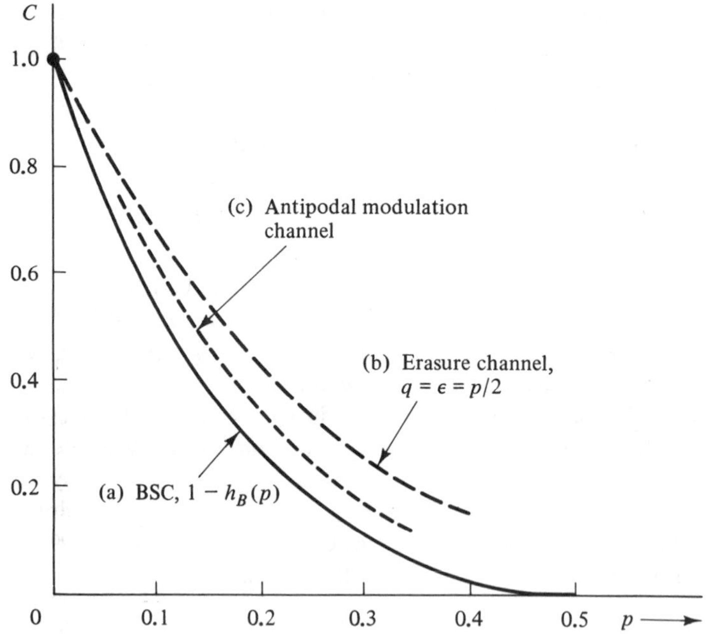

Figure 11.5-4. Capacities of several binary-input channels. (a) BSC. (b) An erasure channel. (c) The antipodal modulation (additive Gaussian noise) channel. For (c), E_b/N_0 is that which gives p for the BSC.

Example: Erasure Channel Capacity

Symmetry and substitution give the capacity of the erasure channel as well. This time the transition probabilities are $p(1|1) = p(0|0) = 1 - q - \epsilon$, $p(?|1) = p(?|0) = \epsilon$, and $p(0|1) = p(1|0) = q$, as given in Fig. 11.5-2(c). The capacity is

$$C(\epsilon, q)$$
$$= (1 - \epsilon) - (1 - \epsilon)\log(1 - \epsilon) + (1 - \epsilon - q)\log(1 - \epsilon - q) + q\log q. \qquad (11.5\text{-}7)$$

If we set $\epsilon = q = p/2$, we can get an indication of the effect on capacity of the added erasure knowledge compared to the simple BSC. With this set of parameters, incorrect transmission takes place with probability p in both channels, but the erasure channel can warn the decoding scheme with a ? symbol during half of these instances. Figure 11.5-4 shows the result, a marked increase in channel capacity compared to the BSC.

Example: Antipodal Modulation Channel Capacity

Here again the p_X function must be the symmetrical one. The conditional probability $p_{Y|X}$ that defines the channel is now continuous. Straightforward substitution into definition (11.5-4) gives

$$C = \int \sum_{x=\sqrt{E_b},\,-\sqrt{E_b}} \tfrac{1}{2} p_{Y|X}(y|x) \log_2 \left[\frac{p_{Y|X}(y|x)}{\sum_x \frac{1}{2} p_{Y|X}(y|x)} \right] dy.$$

The noise has variance $N_0/2$, and from Sec. 10.3

$$p_{Y|Z}(y|\sqrt{E_b}) = \frac{1}{\sqrt{\pi N_0}} \exp\left[\frac{(y - \sqrt{E_b})^2}{N_0}\right],$$

$$p_{Y|X}(y|-\sqrt{E_b}) = \frac{1}{\sqrt{\pi N_0}} \exp\left(\frac{(y + \sqrt{E_b})^2}{N_0}\right].$$

Some manipulation gives

$$C = -\tfrac{1}{2}\log_2(\pi e N_0) + 1 - \tfrac{1}{2}\int_{-\infty}^{\infty} [p_{Y|Z}(y|\sqrt{E_b})$$
$$+ p_{Y|Z}(y|-\sqrt{E_b})] \log_2[p_{Y|X}(y|\sqrt{E_b}) + p_{Y|X}(y|-\sqrt{E_b})]\, dy. \qquad (11.5\text{-}8)$$

We have used the bit energy E_b from Chapter 10, since in the antipodal channel one binary symbol is sent for each channel use.

We can now compare the capacity of the antipodal modulation channel with that of the cruder BSC model. To start, we assume that two transmission systems both make use of a Gaussian channel with signal-to-noise ratio E_b/N_0. One system is the BPSK or QPSK optimal modulator/demodulator of Sec. 10.5. Its bit error probability is $p = Q(\sqrt{2E_b/N_0})$; its input and output are strictly binary, and an accurate model for it is to consider the entire system as a BSC with crossover p and capacity of Eq. (11.5-6). The second system we cannot specify exactly except to say that it is a coded system whose rate is near the capacity in Eq. (11.5-8). The existence of this code is proved in the coding theorem that justifies Eq. (11.5-8). Figure 11.5-4 shows this capacity plotted against p rather than the E_b/N_0 that appears in Eq. (11.5-8), with the conversion between the two done by solving $p = Q(\sqrt{2E_b/N_0})$. By plotting in this way, we can see that the antipodal modulation channel has a somewhat higher capacity than the BSC in information bits per channel use, for the same E_b/N_0 in the channel. Note that E_b here is the energy per binary channel symbol, not per information bit.

Figure 11.5-4 also shows that the capacity of an erasure channel of the type shown with three outputs has even higher capacity than the antipodal modulation channel. This shows that erasures are a valuable kind of channel knowledge, more valuable in some cases than viewing the channel output as being continuous. Special codes have been designed to deal with erasures, which are much more efficient than ordinary codes that correct unknown errors.

In this discussion we must take care to distinguish the codes of information theory from symbol-by-symbol modulations. The codes carry data symbols in very long blocks in an essentially error free manner as long as the data rate in information bits per channel use falls below capacities like those in Fig. 11.5-4. Modulators, on the other hand, carry data symbols singly. The symbols are subject to a fixed error rate that depends on E_b/N_0.

Example: Capacity of the Additive White Gaussian Noise Channel

The capacities of both the BSC and the antipodal modulation channel must fall below 1 bit/channel use, since the input alphabets in both cases are only binary. The AWGN channel places restrictions on neither input nor output; its sole condition is that the ratio of input signal energy per channel use to N_0 be E_s/N_0. Consequently, its capacity can be much higher than unity.

The calculation of AWGN capacity requires some advanced information theory (see [11-9] or [11-4]) and quite a different method of coding theorem proof from the discrete memoryless channel. The result is

$$C = \tfrac{1}{2}\log_2 1 + \frac{2E_s}{N_0} \text{ bits/channel use.} \tag{11.5-9}$$

Figure 11.5-5 plots this capacity, together with the antipodal modulation capacity repeated from Fig. 11.5-4, but this time against the E_s/N_0 argument, with $E_b = E_s$.

We learned in Chapter 10 that a practical value for E_s/N_0 in a BPSK or QPSK modulation system is 10 dB or larger. These modulations force us to assume the antipodal modulation channel, which from the figure has a considerably lower capacity than the AWGN channel with the same physical signal and noise energies. The additional assumption of a BSC, which corresponds to ignoring the matched-filter analog outputs in the demodulators, leads to another, relatively smaller loss.

Bandwidth, Dimensionality, and Capacity

All the foregoing channel models make no mention of signal bandwidth. Information is simply considered to be transmitted in words **x**, in which each letter is an independent channel use, or in the language of signal space, an orthogonal dimension. To introduce the idea of bandwidth into information theory, it is necessary to relate dimensionality to bandwidth.

We can make a rough attempt at this by extending the ideas of Sec. 10.6. Consider an interval of bandwidth $(-W, W)$ and a time interval of duration T. How many signal-

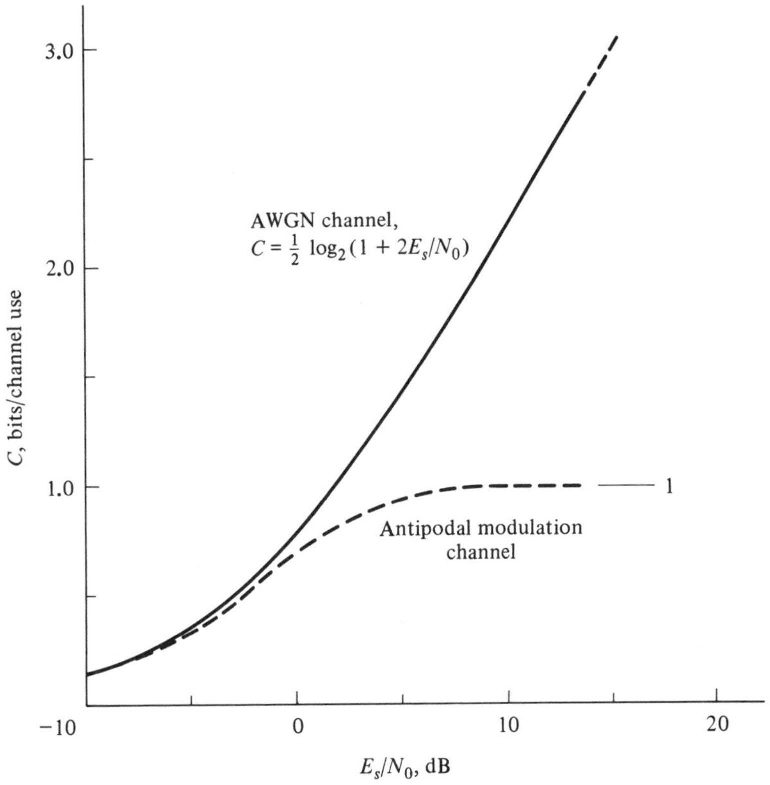

Figure 11.5-5. Capacity versus E_s/N_0 for AWGN and antipodal modulation channels. E_s is the average signal energy per dimension or, equivalently, per channel use.

space dimensions can be packed into this time and bandwidth? Let the data be carried by the function

$$\text{sinc}(t) \equiv \frac{\sin(\pi t \tau_0)}{\pi t/\tau_0}, \qquad -\infty < t < \infty.$$

That is, each τ_0 seconds a new signal a_i sinc($\cdot$) is initiated, with the variable a_i carrying information. The total transmitted signal is

$$s(t) = \sum_i a_i \,\text{sinc}(t - i\tau_0).$$

The sinc($\cdot$) functions in this sum are orthogonal, as in Sec. 10.6, and by sampling $s(t)$ at time $i\tau_0$, we can recover a_i. This transmission system is simply the Nyquist system of Sec. 10.6 with the minimum possible bandwidth. The bandwidth of a superposition of sinc(t) pulses like $s(t)$ is $(-1/2\tau_0,\ 1/2\tau_0)$ hertz. Furthermore, after a time T, T/τ_0

dimensions will have been carried by $s(t)$. By substituting $W = 1/2\tau_0$, we conclude that altogether $2WT$ *signal-space dimensions* are carried by the signal.

Our logic is not quite accurate for several reasons. The main reason is that sinc(t) is not a time-limited function, so the $s(t)$ consisting of T/τ_0 pulses will not be strictly limited to T seconds. If we attempt to limit $s(t)$, its bandwidth will be infinite. In fact, we cannot simultaneously time and band limit any function. Nonetheless, a rigorous result that close to $2WT$ dimensions will be available has been proved by allowing a vanishing fraction of bandwidth to lie outside $(-W, W)$ and by allowing the time duration to grow large. A proof of this appears in [11-9, Chapter 5].

With this result in hand, we can modify the capacity C in Eq. (11.5-9) to obtain the capacity as a function of time and bandwidth for a Gaussian channel of energy E_s per dimension. First, we equate the number of channel uses to the number of dimensions. Since $2W$ dimensions per second are available in the long term, substitution in Eq. (11.5-9) then gives

$$C = W \log_2 1 + \frac{2E_s}{N_0} \text{ bits/second.} \qquad (11.5\text{-}10)$$

This formula† allows us to compare directly the transmission rate of a physical modulation to the ultimate capacity implied by the system's energy and bandwidth.

Alternate measures of the capacity of a transmission system, not related to information theory, are of course often used. One of these is based on the interference with each other of a stream of pulses. It is used to calculate a capacity for an optical communication system in Sec. 12.7. We also took this attitude in Sec. 10.6, in saying that Nyquist-criterion pulses cannot be transmitted faster than $2W$ per second.

11.6 The Channel Coding Theorem

At this point in our discussion of information transmission we have developed the idea of capacity, the ultimate ability of a channel to carry symbolic information. To actually carry information at a rate near capacity, or at least to get a performance improvement in that direction, we shall need a *channel code*. We saw in the previous section that the code will have to accomplish several functions. Generally, the channel carries information symbols at a slower rate than the number of channel uses. The BSC, for instance, has a binary symbol channel output, but it cannot carry one error-free information symbol for each channel use. The information symbols must somehow be represented by code words so that they may be carried at a slower rate. Another function of the code will be to match the information symbol probabilities to a possibly new distribution that maximizes channel information transfer. A final function of the channel code is to

† The reader may wonder about the relation between Eq. (11.5-10) and Eq. (1.3-1). The latter can be derived quite simply from Eq. (11.5-10); see [11-9, p. 321].

realize the promise of Shannon's channel coding theorem, that information may be carried essentially error free at any rate below capacity.

Our main goal in this section will be to prove a version of the channel coding theorem. This will be done by proving the existence of a channel code with the required error performance. Although our theorem will apply only to the BSC, it contains the basic ideas of a more advanced proof that would apply to the general DMC. A full proof of the coding theorem appears in the advanced texts listed at the end of the chapter. Coding theorems can be proved as well for the antipodal modulation channel and for the AWGN channel, although the methods employed differ somewhat; these too appear in the advanced texts.

We can give a general definition of channel coding that applies to all kinds of channels: *Codes create a set of channel symbol patterns*, or "words," each representing a sequence of information symbols. The encoder generates the pattern corresponding to an information sequence input. The decoder observes the channel outputs, and if it does not see one of the set of patterns, it knows an error has occurred. In this case it finds the most likely pattern, given its observation, and releases this as its output.

Without an agreed upon set of patterns, it is clear that channel error correction is impossible, since otherwise we could not recognize the errors. In some of the more recent developments in channel coding, these patterns are imposed by restricting signal phases or amplitudes to certain values at certain times. For the simple BSC, both the data and the channel alphabets are finite and binary. A little thought will show that the length of a BSC block code word, N, must be longer than K, the length of the corresponding block of information symbols. Otherwise, all 2^N channel patterns would be used by representing one of the 2^K data patterns, and it would be impossible to detect errors in the received patterns. For this reason, code words are longer than K, and the extra channel symbols are called *parity check* symbols. It should be stressed that channel coding does not automatically imply the use of parity check symbols; they are needed only in a situation like the BSC, in which the data and channel input alphabets are the same.

In the practical world, the goal of error-free transmission at a rate near capacity is unreachable The engineering use of channel codes is not to eliminate errors in channels, but to *reduce* them. Practical schemes to do this are the subject of Sec. 11.7.

Encoding and Decoding for the BSC

Like any code, a channel code for the BSC is a mapping like that of Eq. (11.3-3).

$$\mathscr{C}: \mathscr{X} \rightarrow \mathscr{Y}.$$

$\mathscr{X}$ is a set of binary data sequences of length K and $\mathscr{Y}$ is a set of binary code words of length N, with $N > K$. Each of the 2^K data words is mapped to a unique code word. 2^{N-K} channel sequences are not code words. The *rate* of the code is K/N data bits carried per channel use.

Understanding the workings of a code is greatly aided by imagining the code

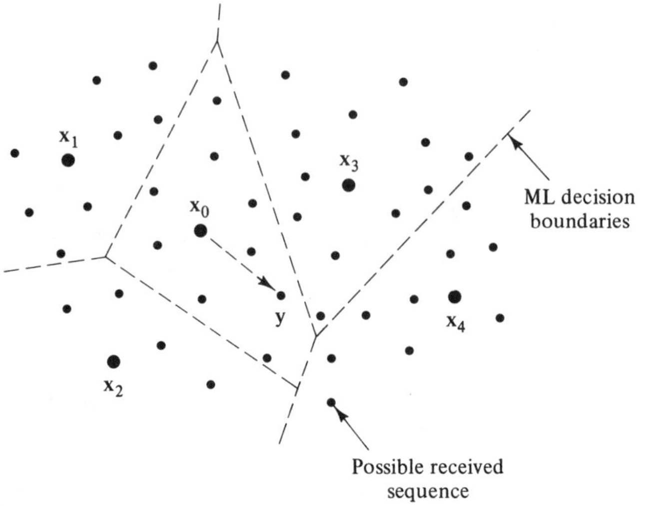

Figure 11.6-1. Arrangement of code words **x** in a geometric space. The action of the channel is to move the code word to another nearby point.

words as points scattered around a geometric space amid a sea of channel sequences. Such a space is shown in Fig. 11.6-1. The measure of distance in this space is the *Hamming distance* between two sequences **x** and **y**, defined by

$$d_H(\mathbf{x}, \mathbf{y}) = \sum_{j=1}^{N} (x_j + y_j), \tag{11.6-1}$$

where + denotes the exclusive-OR operation. In words, the distance is the number of symbols that disagree between the two words. When a code word **x** is transmitted, the action of the channel is to move **x** to a point **y** that is likely to be close by. We can divide the space into subsets of points that are closest to each of the code words. The decoder will determine in which subset a channel output sequence lies and put out the code word corresponding to the subset.

We want to show now that a decoder that works this way, decoding the closest code word in terms of Hamming distance to a received sequence **y**, is a maximum-likelihood decoder. From Sec. 10.2, a maximum-likelihood decoder is one that maximizes the conditional probability of receiving **y**, $P(\mathbf{y}|\mathbf{x}_i)$, over the choice of the transmitted code word $\mathbf{x}_i$. Suppose **y** and $\mathbf{x}_i$ are separated by Hamming distance h_i. In a BSC, the probability that a **y** is received with these h_i bit errors is

$$P(\mathbf{y}|\mathbf{x}_i) = (1-p)^{N-h_i} p^{h_i}, \qquad p < \tfrac{1}{2},$$

if p is the crossover probability. We can maximize this, then, over i, but it is more convenient to perform the equivalent operation of maximizing its log,

$$\max_i \{(N - h_i) \log (1-p) + h_i \log p\}.$$

The term $N \log (1 - p)$ is constant during the maximization, so it can be dropped from the expression. The operation that remains is

$$\max_i \{h_i \log p - h_i \log (1 - p)\}$$

or

$$\max_i \left\{h_i \log \frac{p}{1 - p}\right\}. \tag{11.6-2}$$

By assumption, $p/(1 - p)$ is less than 1, so expression (11.6-2) is maximized when h_i is least. Thus the ML decoder will put out the closest **x** to **y**.

We will return to this minimum-Hamming-distance decoder in the discussion of practical decoders in Sec. 11.7. For our proof of the channel coding theorem, a simpler, suboptimal decoder will do. This decoder forms a "ball" around the received sequence **y** whose radius is $N(P + \epsilon)$, as shown in Fig. 11.6-2. Here Np is the approximate number of channel errors that occur in a length-N code word sent through the BSC. By the central limit theorem of probability, the actual number of channel errors becomes increasingly concentrated near Np as N grows, and with high probability the code word **x** lies in a thin shell at Np. To be more sure that the ball does include **x**, a small positive ϵ has been added to p. The ball decoder will work by decoding **y** as the **x** that lies in the ball. If no **x** or if more than one **x** lie in the ball, the decoder will declare itself to have failed, and we count the event as a decoding error.

Before turning to the coding theorem, we need to reintroduce one other concept, the random code. These codes were used in Sec. 11.4 in the proof of the rate-distortion coding theorem. Historically, Shannon's use of them for the channel coding theorem came first and was a watershed for the rest of information theory. For a BSC, a random code word consists of independently chosen binary symbols. The symbols 0 and 1 are chosen with probability 1/2, which is the capacity-achieving channel distribution in definition given by Eq. (11.5-4). 2^K $(= 2^{RN})$ such words are chosen, all independently, to make up the code.

As with the rate-distortion random codes, some of these codes are poor, and a small number even have many code words that are all the same! But the typical random

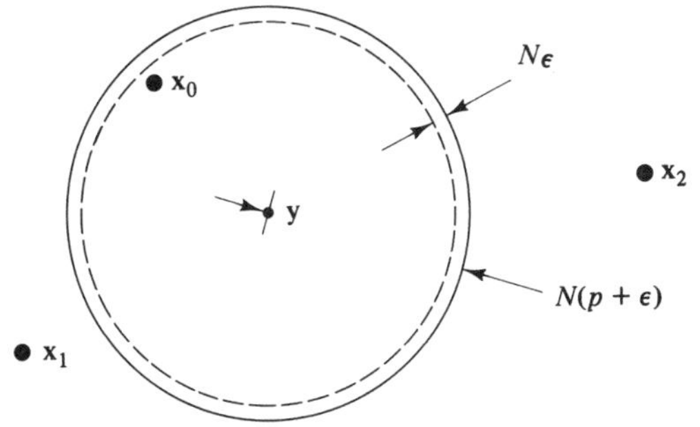

Figure 11.6-2. Decoding region of the ball decoder, from the point of view of the received sequence **y**. As N grows, the sent code word $\mathbf{x}_0$ is very likely to lie near the shell. $\mathbf{x}_1$ and $\mathbf{x}_2$ are neighboring code words.

code has good performance, so good that it achieves channel capacity. We shall be able to show that the expected error performance, averaged over the collection of all the codes, tends to zero as N grows for any rate $R < C$.

In the geometric space of Fig. 11.6-1, the words of a random code would appear to be scattered at random through the space. With such irregularity, the ball decoder in Fig. 11.6-2 could well find more than one code word in the ball surrounding $\mathbf{y}$, but this event in fact becomes very unlikely as N grows. Phenomena like these for the BSC occur in all types of channels.

Proof of the BSC Channel Theorem

First, a formal statement of the theorem.

Theorem 11.6-1. Suppose equiprobable binary data are to be sent through a BSC. At any $R < C$, there exists a channel code with rate R data bits/channel use and block length N whose probability of decoding error tends to zero as N grows.

To prove the theorem, we first select a random code of rate R. One of these words $\mathbf{x}_0$ is transmitted, and the ball decoder is used to decode the received sequence $\mathbf{y}$. There are two ways that a decoding error can occur. If more than $N(p + \epsilon)$ channel errors occur, then $\mathbf{x}_0$ does not lie in the ball centered at $\mathbf{y}$; since the decoder does not look outside the ball, it must make an error. The second way, disjoint from the first, occurs when $N(p + \epsilon)$ or fewer errors occur, but more than one code word lies in the ball. This latter event would not happen in a sensible code, but it could happen in a random code.

Using the laws of probability, we can express the probability of error as

$$p_e = P(E_0) + P(E_0^c)P\left[\bigcup_{i \neq 0} (V_i)\right], \tag{11.6-3}$$

where E_0 denotes the event that more than $N(p + \epsilon)$ channel errors occur, E_0^c is its complement, and V_i is the event that the random code word $\mathbf{x}_i$ lies in the ball around $\mathbf{y}$. The union operation here constructs the event that one or more code words other than $\mathbf{x}_0$ lie in the ball. The probability measure $P(\cdot)$ assigns probability to events that are *generated jointly* by the code selection and the BSC error mechanism. This leads us to quite a subtle point. p_e is actually the probability of decoding error when a code is selected at random and a word is then sent through the BSC. It is an average over both these processes.

Now we need to calculate the probabilities in Eq. (11.6-3), or at least find sufficiently tight bounds to them. The event E_0^c is independent of all the V_i since the BSC mechanism is independent of the code selection. Therefore,

$$p_e = P(E_0) + P(E_0^c)P\left(\bigcup_{i \neq 0} V_i\right).$$

Furthermore, it is always true that $P(E_0^c) \leq 1$, so

$$p_e \le P(E_0) + P\left(\bigcup_{i \ne 0} V_i\right).$$

The second term can be overbounded once again using the union bound of probability to obtain

$$p_e \le P(E_0) + \sum_{i \ne 0} P(V_i). \tag{11.6-4}$$

The rest of our proof consists of evaluating the probabilities in Eq. (11.6-4). Despite the preceding overbounds, and some additional ones to come, our calculations will be precise enough to establish the theorem.

An Estimate to $P(E_0)$. This is simply the probability that more than $N(p + \epsilon)$ errors occurred. It does not depend on which word was sent. The weak law of large numbers (see [11-12]) states that the sum of independent experiments, in this case the BSC errors, exceeds the mean of the sum, Np, with vanishing probability as N grows. This means that for any $\epsilon \ge 0$

$$P(E_0) \rightarrow 0, \qquad \text{as } N \rightarrow \infty.$$

An Overbound to $\Sigma P(V_i)$. We consider first just one of the V_i, that is, the event that code word $\mathbf{x}_i$, selected at random, lies within the ball around $\mathbf{y}$. For this to happen, the Hamming distance between $\mathbf{x}_i$ and $\mathbf{y}$ must be no more than $N(p + \epsilon)$. Restated, it must have happened during the selection of $\mathbf{x}_i$ that no more than $N(p + \epsilon)$ binary symbols differed from those in $\mathbf{y}$. From the binomial law of Sec. 8.5, the probability that k symbols differ is $\binom{N}{k} p^k (1 - p)^{N-k}$. The probability that no more than $N(p + \epsilon)$ differ is

$$P(V_i) = \sum_{k=0}^{N(p+\epsilon)} \binom{N}{k} p^k (1 - p)^{N-k} \le \sum_{k=0}^{N(p+\epsilon)} \binom{N}{k} \left(\frac{1}{2}\right)^N \tag{11.6-5}$$

Our next step is to resort to yet another overbound on Eq. (11.6-5). This time we use an inequality from combinatorial analysis, which states that†

$$\sum_{k=0}^{\alpha N} \binom{N}{k} \le 2^{Nh_B(\alpha)}, \qquad \text{any fixed } \alpha < \frac{1}{2}. \tag{11.6-6}$$

This means that the left side of Eq. (11.6-5) is overbounded by $2^{Nh_B(p+\epsilon)-N}$.

To overbound the sum of all the $P(V_i)$, we make use of the fact that all the code words were selected in an identical fashion. Thus all the numbers $P(V_i)$ are identical. If there are m_c words in the code, then it must be that

$$\sum_{i \ne 0} P(V_i) = (m_c - 1)P(V_i) \le m_c 2^{Nh_B(p+\epsilon)-N}. \tag{11.6-7}$$

As long as

$$\log_2 m_c < N[1 - h_B(p + \epsilon)], \tag{11.6-8}$$

† For example, see R. Ash, *Information Theory*, Wiley, New York, 1965.

the bound in Eq. (11.6-7) will tend to zero as N grows. The rate of the random code is $(1/N) \log_2 m_c$, so Eq. (11.6-8) is a statement that p_e tends to zero as N grows whenever $R \leq 1 - h_B(p + \epsilon)$. Since we can choose any $\epsilon > 0$, even one very small, we can essentially replace $1 - h_B(p + \epsilon)$ with $1 - h_B(p)$; this is Eq. (11.5-6), the capacity of the BSC.

This completes the proof that p_e tends to zero at any rate below capacity for the average of all random codes. There must exist at least one code that performs as well as the average.

As with the rate-distortion coding theorem, a converse theorem is needed to show that $1 - h_B(p)$ really is the ultimate rate at which the BSC can carry information. Such a theorem would show that at rates above this the decoding error probability cannot be driven to zero. We shall leave the converse proof to more advanced texts.

11.7 PARITY CHECK CODES

When a transmission medium is modeled as a binary symmetric channel, the only means to improve the reliability of the transmission is through the addition of extra symbols to the data sequence. These are called parity check symbols. By analyzing the patterns of these symbols, an algorithm or circuit can correct errors in the data. Alternately, errors can be *detected*, but not corrected, and a warning issued to the user. The BSC coding theorem gives a theoretical estimate of the ratio $(N - K)/N$ of parity check symbols that are needed for error-free performance in a long block code of word length N that carries K data symbols. Despite the fact that such codes are known to exist, practical parity check codes perform less well than these. In the engineering world, the aim of a parity check code is simply to gain some improvement in the BSC error rate.

We shall introduce the basic ideas behind the parity check codes that are now in use and feature two especially important classes, Hamming codes and convolutional codes. Codes for control of errors form a large subject that is confined for the most part to books at an advanced level. Most codes are based on the algebraic theory of finite groups and fields. Many clever decoding schemes have been devised, and more sophisticated channel models than the BSC are used. The interested reader is referred to the two recent texts ([11-10] or [11-11]).

Basic Ideas of Binary Error Correction

Within each code word of length N, we will assume that the first K binary symbols are the data symbols. The last $N - K$ are parity check symbols, which will depend somehow on the data symbols. The rate of the code is K/N data bits per channel use. By observing the parity check symbols, we shall be able to correct or detect errors in the entire code word.

As we deduced in Sec. 11.6, a maximum-likelihood decoder for any BSC code is one that finds the closest code word in Hamming distance to the received word. Suppose that we wish to guarantee the correction of any pattern of t or fewer errors in a received word. This will require that no code words be closer to each other than $2t + 1$. If two words lie $2t$ apart, a received word can lie "in between" the two, distant t from each, and we cannot know unambiguously which code word was sent. If two words lie less than $2t$ apart, it will be possible for a received word with t errors to lie closer to the wrong code word.

These facts are easy to see with a simple parity check code called a *repetition* code. This code transmits only 1 data bit per word, and it has just two words, 000. . .00 and 111. . .11; that is, the data symbol is simply repeated N times. The code's rate is $1/N$ (a very poor rate). The Hamming distance between the code's two words is N, from which we can conclude by the preceding logic that the code will correct up to $(N - 1)/2$ errors. If $N = 9$, for example, in a BSC with $p = 0.01$, the probability of erroneous decoding is the probability that five or more errors occur, which is

$$\binom{9}{5} p^5(1-p)^4 + \binom{9}{6} p^6(1-p)^3 + \binom{9}{7} p^7(1-p)^2 + \binom{9}{8} p^8(1-p) + \binom{9}{9} p^9,$$

which is about 1.21×10^{-8}. This is an excellent decoder error probability, but the price paid is the very low rate. Repetition coding is used in many data communication systems, typically to send a small number of symbols like identifiers or synch bits that are of particular importance.

The opposite of the repetition code is the *single parity check* code. This code carries $N - 1$ data bits in each word and a single check symbol that is set so that the total number of 1's in the word is always even. The code's rate is $(N - 1)/N$. If an odd number of channel errors occurs, the number of 1's in the word will be odd; the word is said to have *odd parity*. Through this means, an odd number of errors is detected, but not corrected. An even number of errors is undetectable. Single parity check codes have been used for many years to detect errors in computer memories and data transmission links. A commonly used block length is 9, with the parity bit checking 8 data bits. The effectiveness of such a scheme is more than it would at first seem in a system that allows retransmissions. Consider again the BSC medium used with the repetition code; it had $p = 0.01$, which represents quite a poor computer memory or data link. The probability that an 8-bit data word passes in any way incorrectly through this BSC is $1 - (1 - p)^8$, or about 0.077. By adding a single parity check, we can be warned about the most likely error patterns and can ask for a retransmission. With this length-9 code, 9 bits are received correctly with probability $(1 - p)^9$, and a single error has probability $9p(1 - p)^8$; what remains, about 0.0035 in our example, is dominated by the probability that a double error occurs, an event that we cannot perceive and from which we cannot recover. In summary, the single parity check code improves the probability of unwitting error from 0.077 to 0.0035. The improvement is much more dramatic in a BSC with a practical p!

In these two simple examples, the closest that any two words lie to each other is N for the repetition code and 2 for the single parity code. This means that we can guarantee correction of $(N - 1)/2$ errors in the first case and detection of at most 1 error in the second. The nearest distance that any pair of words in a code lies apart is called the code's *minimum distance* $d_{\min}$. The best error correction that can be guaranteed for a code, regardless of the code word transmitted, is that all combinations of $(d_{min} - 1)/2$ or fewer errors can be corrected.

In the BSC the noise mechanism inverts the data bits when an error occurs. A convenient way of keeping track of this is by introducing another length-N binary vector, the *error sequence* **e**. If an error has occurred in the jth position, $e_j = 1$; otherwise, $e_j = 0$. The received sequence **y** can now be obtained by adding **e** to the code word **x** in a bitwise exclusive-OR fashion; that is, the BSC's effect is described by $\mathbf{y} = \mathbf{x} + \mathbf{e}$. The sequence **e** is sometimes called the error pattern for this transmitted word.

Linear Parity Check Codes

In creating more effective codes than these simple ones, we need a sensible method of relating the extra parity check bits to the data bits. The method should be easy to explain, easy to implement in hardware, and effective at correcting errors at a high code rate and a short block length. Virtually all parity check codes in use employ a method of relating the check bits to the data bits through multiplication by a matrix. The elements of the matrix, like those of the words, are binary, and the rules of addition and multiplication are the exclusive-OR and the AND operation, respectively. The codes so generated have a wealth of algebraic properties, the most important of which is that the code words form an algebraic group. The properties lead to strong error-correcting abilities and efficient decoding schemes, all of which are described in detail in more advanced texts. Because this kind of code is obtained by a linear transformation of the data bits, it is called a *linear* code. An older name is group code.

We shall now define a given linear code in terms of the matrix that creates it. Code words will be written as row vectors. Specifically, the words of a linear code are all **x** that satisfy

$$\mathbf{0}^{(N-K)} = \mathbf{xH}, \tag{11.7-1}$$

which is to say, they are the null space of **H.** Here $\mathbf{0}^{(N-K)}$ denotes the length $N - K$ all-zero row vector. **H,** an $N \times N - K$ matrix, is called the *parity check matrix* for the code, and Eq. (11.7-1) states that for each of the $N - K$ columns of **H** the dot product of **x** and that column is **0.** This is a direct extension of the idea in the single parity check code that all digits of **x** must sum to an even number, that is, 0 in exclusive-OR terms; now we have $N - K$ parity relationships instead of just one. **H** for the single parity check code is a single column of 1's.

At this point we shall make a simplification in the structure of the code word **x:** We set the first K bits of **x** equal to the data bits $a_1, a_2, \ldots, a_K$ and the last $N - K$ bits equal to the parity check bits $c_1, c_2, \ldots, c_{N-K}$, whatever **H** implies that these may be. This kind of code, in which the data bits explicitly appear in the words, is called a *systematic* code. It can be shown that no loss in error-correction ability comes of our restricting all codes to be systematic ones.

The form of **H** for a systematic code turns out to be

$$\mathbf{H} = \begin{bmatrix} \mathbf{P} \\ \text{---} \\ \mathbf{I}_{N-K} \end{bmatrix} = \begin{bmatrix} p_{11} & p_{12} & \cdots & p_{1,N-K} \\ p_{21} & p_{22} & \cdots & p_{2,N-K} \\ \cdot & \cdot & & \cdot \\ \cdot & \cdot & & \cdot \\ \cdot & \cdot & & \cdot \\ p_{K,1} & p_{K,2} & \cdots & p_{K,N-K} \\ \hline 1 & 0 & \cdots & 0 \\ 0 & 1 & \cdots & 0 \\ \cdot & & & \\ \cdot & & & \\ \cdot & & & \\ 0 & 0 & \cdots & 1 \end{bmatrix} \qquad (11.7\text{-}2)$$

Here $\mathbf{I}_{N-K}$ is the identity matrix of size $N - K$, and $\mathbf{P}$ is a matrix that we are free to choose in the design of the code. Thus Eq. (11.7-1) separates as

$$\mathbf{0} = \mathbf{xH} = (a_1, \ldots, a_k)\,\mathbf{P} + (c_1, \ldots, c_{N-K}). \qquad (11.7\text{-}3)$$

It is clear from this that, once $\mathbf{P}$ is designed and the data symbols accepted, then each element of $c_1, \ldots, c_{N-K}$ must be set in such a way that $\mathbf{a}\,\mathbf{P} + \mathbf{c} = \mathbf{0}$. For instance, it must be that

$$c_1 + \sum_{i=1}^{K} a_i p_{i,1} = 0$$

for the first element. In this way the elements in the code word $\mathbf{x}$ are set.

If a BSC channel error occurs, the product $\mathbf{x}\ \mathbf{H}$ will generally not be $\mathbf{0}$. By analyzing the nonzero product, we can deduce where the error lies. Consider first the case of a single error in the jth position; the received sequence $\mathbf{y}$ is $\mathbf{x} + \mathbf{e}$, where only the jth element of $\mathbf{e}$ is a 1. Then we can form

$$\mathbf{y}\ \mathbf{H} = (\mathbf{x} + \mathbf{e})\ \mathbf{H} = \mathbf{0} + \mathbf{e}\ \mathbf{H},$$

where $\mathbf{eH}$ must be the jth row of $\mathbf{H}$. We know $\mathbf{H}$, and the error position can be identified by comparing $\mathbf{eH}$ to its rows. The computed quantity $\mathbf{eH}$ is called the *syndrome* of $\mathbf{x}$, and decoding by this simple comparison method is called syndrome decoding. Just as a syndrome in the medical world is a set of disease symptoms, the syndrome $\mathbf{eH}$ shows what is diseased about $\mathbf{x}$.

If two channel errors occur, say in the ith and jth places, then $\mathbf{eH}$ is the sum exclusive-OR of rows i and j of $\mathbf{H}$. Can these errors be successfully identified? The answer is yes if the sum of the rows is not equal to any other row in $\mathbf{H}$; otherwise, we could not distinguish the double error from a single error in some other position. A general condition for correcting all single and double errors is as follows: All N rows of $\mathbf{H}$ and all $N(N - 1)/2$ sums of two rows must be distinct. To correct triple errors, $\mathbf{H}$ must have distinct sums of three rows, and so on. In every case, the all-zero syndrome indicates that $\mathbf{y}$ is already a code word and is presumably correct.

Aside from **H,** a code may also be defined by its *generator matrix* **G,** which relates the data sequence and its code word by

$$\mathbf{x} = (a_1, \ldots, a_k)\,\mathbf{G}. \tag{11.7-4}$$

For a systematic code, the general form of **G** is

$$\mathbf{G} = [\mathbf{I}_K \,\vdots\, \mathbf{P}] = \left[\begin{array}{cccc:cccc} 1 & 0 & \ldots & 0 & p_{11} & p_{12} & \cdots & p_{1,N-K} \\ 0 & 1 & \ldots & 0 & p_{21} & p_{22} & \cdots & p_{2,N-K} \\ \cdot & & & \cdot & \cdot & & & \\ \cdot & & & \cdot & \cdot & & & \\ \cdot & & & \cdot & \cdot & & & \\ 0 & 0 & \ldots & 1 & p_{K,1} & p_{K,2} & \cdots & p_{K,N-K} \end{array}\right], \tag{11.7-5}$$

where **P** is the same matrix as in the definition of **H.** It is clear from Eq. (11.7-5) that the first K bits of **x** are the data symbols and that the parity check bits in **x** are a superposition of those rows of **P** corresponding to 1's in the data bits. With some contemplation, one can see that the words in the code consist of all 2^K linear combinations of the K rows of **G.**

Example: Length-7 Hamming Code

The Hamming codes are a class of codes that correct all single errors. They were among the first codes discovered and they are of great practical importance. By definition, a Hamming code **H** matrix consists of *all* nonzero rows of length $N - K$. There are $2^{N-K} - 1$ such sequences, and consequently the block length N of a Hamming code is equal $2^{N-K} - 1$. Setting $N - K$ to 2 creates a length-3 code that turns out to be the length-3 repetition code. Setting $N - K$ to 3 creates a length-7 code that is the first interesting Hamming code. To make up the **H** for this code, we list as rows all the nonzero sequences of length 3, being careful to adhere to the style of Eq. (11.7-2):

$$\mathbf{H} = \begin{bmatrix} 1 & 1 & 0 \\ 0 & 1 & 1 \\ 1 & 1 & 1 \\ 1 & 0 & 1 \\ 1 & 0 & 0 \\ 0 & 1 & 0 \\ 0 & 0 & 1 \end{bmatrix}.$$

There are 16 words that satisfy $\mathbf{x}\,\mathbf{H} = \mathbf{0}^{(N-K)}$. Placing the four data bits first, we can write these as in Table 11.7-1. The all-zero data sequence, as with all linear codes, corresponds to the all-zero code word. The next four words, indicated by *, correspond to data sequences with a single 1, and these comprise the generator matrix **G.** The remainder of the words are superpositions of two or more of these four generator words.

The length-7 Hamming code will correct any single error in the received sequence **y.** A bit error in the first position, for instance, will lead to the syndrome $\mathbf{yH} = (1\ 1\ 0)$, the first row of **H,** so we correct the first bit. But errors in both the first and third positions

TABLE 11.7-1 LENGTH-4 HAMMING CODE, WITH $K = 4$ DATA BITS. THE ROWS MARKED * MAKE UP THE GENERATOR MATRIX.

(a_1, a_2, a_3, a_4)	(c_1, c_2, c_3)	. . .		
0 0 0 0	0 0 0		0 1 0 1	1 1 0
1 0 0 0	1 1 0	*	0 0 1 1	0 1 0
0 1 0 0	0 1 1	*	1 1 1 0	0 1 0
0 0 1 0	1 1 1	*	1 1 0 1	0 0 0
0 0 0 1	1 0 1	*	1 0 1 1	1 0 0
1 1 0 0	1 0 1		0 1 1 1	0 0 1
1 0 1 0	0 0 1		1 1 1 1	1 1 1
0 1 1 0	1 0 0			
1 0 0 1	0 1 1			

will lead to $\mathbf{yH} = (1\ 1\ 0) + (1\ 1\ 1) = (0\ 0\ 1)$, which is the last row of $\mathbf{H}$. The most likely error sequence to cause $(0\ 0\ 1)$ is a single bit error in the last position, not the double-error pattern. A maximum-likelihood decoder must thus decide in favor of this incorrect pattern and so make a decoding error.

The rate of the length-7 code is $R = \frac{4}{7}$ data bits per channel use. The next longer Hamming code, created by $N - K = 4$, has length 15 and rate $R = \frac{11}{15}$, and, as always, corrects single errors. By switching to a longer, more complex code, we obtain a higher data throughput for the same error correction capability. Linear codes generally show a performance increase with block length.

Some study of the words in Table 11.7-1 will show that the minimum distance between any two words is 3. A maximum-likelihood decoder will correct $(d_{\min} - 1)/2$ errors, that is, the single error that was claimed for Hamming codes.

A large number of linear codes have been discovered in the last 30 years. The most important of these are the convolutional codes, whose words are produced by a convolution of the data bits with a generator sequence, the cyclic codes, whose words have a cyclic shift property, and the BCH codes, which are a generalization of the Hamming codes and form a subclass of the cyclic codes. BCH codes, named after their discoverers Bose, Chaudhuri, and Hocquenghem, are representative of the most powerful linear codes. This means that, for a given block length, they have the best combination of high rate and high error correction. Table 11.7-2 lists the capabilities of some of these best codes.

One can always implement an encoder for a linear code by multiplying the data sequence by the generator matrix $\mathbf{G}$. The maximum-likelihood syndrome decoder can be implemented through a table look-up in which the syndrome $\mathbf{yH}$ serves as an address to a location that has information on the error location. When the number of syndromes 2^{N-K} becomes large, a cheaper, perhaps nonmaximum likelihood decoder becomes desirable. A great many of these have been devised and are described in advanced texts. Typically, these are constructed out of shift registers and simple logic arrays or are software algorithms based on the algebraic properties of the code.

TABLE 11.7-2 A SELECTION OF THE BEST ERROR-CORRECTING CODES AT BLOCK LENGTH *N*, *K* DATA BITS, AND ERROR-CORRECTING ABILITY *T* OR FEWER.

N	K	t	N	K	t	N	K	t
7	4	1	255	247	1	1023	1013	1
				239	2		1003	2
15	11	1		231	3		993	3
	7	2		215	5		973	5
	5	3		179	10		923	10
				139	15		873	15
31	26	1					828	20
	21	2	511	502	1		778	25
	16	3		493	2		728	30
				484	3		698	35
63	57	1		466	5		648	41
	51	2		421	10		608	45
	45	3		376	15		573	50
	36	5		340	20		523	55
				304	25		483	60
127	120	1		259	30			
	113	2						
	106	3						
	92	5						
	64	10						

Codes are shown only for $t = 1, 2, 3, 5, 10, 15, \ldots$, and for rates greater than about 0.45. These codes are all binary BCH codes, whose lengths are restricted to $2^N - 1$.

Convolutional Codes

Perhaps the most commonly used of the multiple-error correcting linear codes are the convolution codes. Although these are linear codes, they are often encoded and decoded without reference to their algebraic properties. In addition, they need not be used in a block fashion, but can accept as input a free-running stream of data bits. We shall now explore some of the basics of these codes from a "nonalgebraic" point of view.

For simplicity, we shall discuss only convolutional codes with rates that are $1/m$, where m is an integer; in reality, codes exist at all rates of the form $(\log_2 q)/m$, q and m integers. Second, we shall consider only systematic codes, and unlike the earlier codes, which had the data bits grouped at the beginning of a word, our systematic convolutional codes will *interleave* the data bits every mth code-word symbol. The generator matrix for this kind of convolutional code takes the form

$$\mathbf{G} = \begin{bmatrix} 1 & g_{2,1} & \cdots & g_{m,1} & 0 & g_{2,2} & \cdots & g_{m,2} & 0 & g_{2,3} & \cdots & g_{m,3} & \text{etc.} \\ & & & & 1 & g_{2,1} & \cdots & g_{m,1} & 0 & g_{2,2} & \cdots & g_{m,2} & \text{etc.} \\ & & & & & & & & 1 & g_{2,1} & \cdots & g_{m,1} & \text{etc.} \end{bmatrix} \qquad (11.7\text{-}6)$$

Elements not shown are 0's. The structure of **G** consists of a large number of rows, with each row side-slipped ot the right by m elements, but otherwise identical to the one above it. The structure of a row consists of an initial run of 0's followed by ν groups of m digits, followed by another run of 0's; the νm digits have the form $(1\ g_{2,1} \cdots g_{m,1})(0\ g_{2,2} \cdots g_{m,2}) \cdots (0\ g_{2,\nu} \cdots g_{m,\nu})$. ν is called the *constraint length* of the code, and this sequence of bits is the *generator sequence*. The number of rows in **G** may extend indefinitely. In this case the block length of the code may be thought of as growing as well, but the structure of the code is really set by the length-νm generator sequence. The code turns out m bits for each data bit in a lockstep fashion, with each data bit affecting, or "constraining," νm code word bits.

Actually, the **G**-matrix description of a convolutional code is usually avoided in favor of a diagram like Fig. 11.7-1, which defines the code in terms of a bank of m tapped shift registers. Here $m = 3$ and the rate is $\frac{1}{3}$. Initially, the shift registers contain all zeros. Data bits shift into the registers from left to right. In a systematic code, the top shift register is vestigial, and it simply passes the data bit to the commutator. The

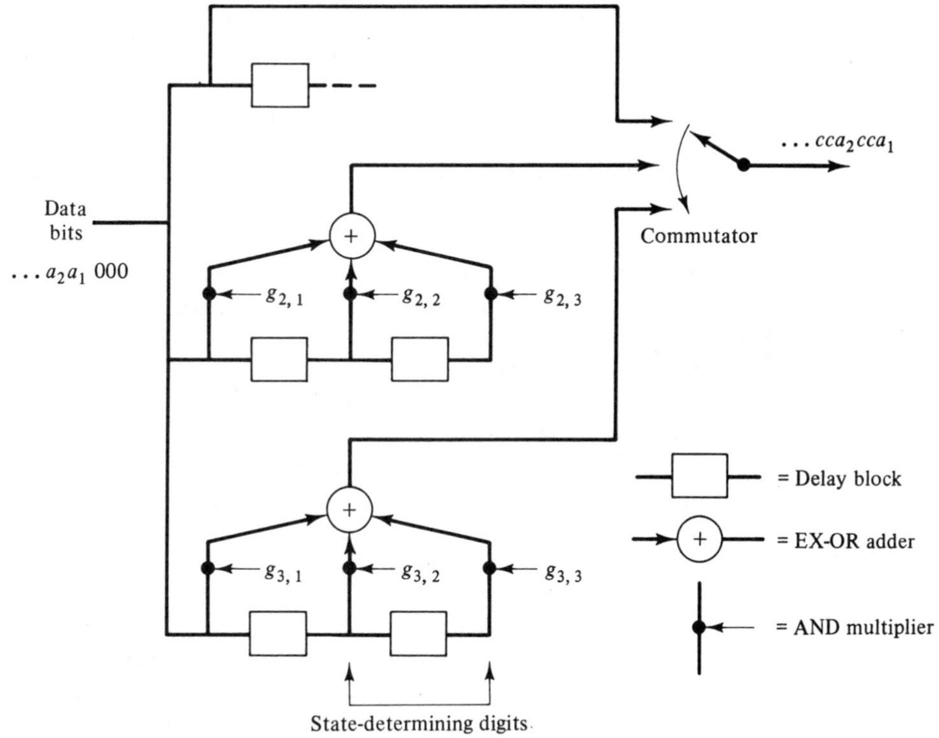

Figure 11.7-1. Shift register generation of a rate $\frac{1}{3}$ convolutional code with $\nu = 3$. Since the code is systematic, the top shift register is vestigial. 0's are present in the shift registers before the data bits arrive. c denotes a parity check; a denotes a data bit.

lower two registers are connected to arrays of exclusive-OR adders and AND multipliers that produce the two parity checks in accordance with the generator sequence. The commutator packs the outputs into groups of three in the pattern parity/parity/data. Data bits simply flow through this circuit, and it is clear that no concept of a block is needed.

An immense body of literature exists about convolutional codes, most of it beyond our scope. We shall focus briefly on one topic, the state-space description of the codes and a related decoding algorithm.

The shift register circuit is a finite-state machine, and, as such, we can describe its activity by means of state transitions. The state of this machine can be defined by two successive data bits, since all the register contents are tied together; these are the rightmost 2 bits, as indicated in Fig. 11.7-1. We can name the states in decimal form by making the natural binary conversion (00, 01, 10, 11) $\longrightarrow$ (0, 1, 2, 3) and arrive at the state transition diagram shown in Fig. 11.7-2(b). Some useful further information appears on this diagram. In parentheses, ( ), is shown the data bit that drives the machine from the existing state to the next state. This bit should be thought of as present at the left entrance to the registers in Fig. 11.7-1, with the registers shifting right during the transition to make the bit part of the next state. Also shown along each transition is the encoder output ($m = 3$ digits) that is the product of the transition. These are generated from the data bit and the old state bits.

A common way of portraying the encoder progress is by the *state trellis* diagram in Fig. 11.7-2(a). A state trellis plots the states versus time for an encoder. All paths in time are shown, even though the encoder progresses along only one. Once again, the driving data bits and the encoder outputs are indicated. It is assumed that the encoder starts in state 0. At first, the trellis is not fully drawn in, a reflection of the fact that the encoder cannot immediately reach all states from state 0. By the third (in general the νth) data bit, the trellis is filled in, and it henceforth repeats exactly.

After the νth time unit, every state in the trellis has two transitions entering it and two leaving. This fact gives rise to a computationally efficient scheme called the Viterbi algorithm for decoding which path an encoder has followed through the trellis. The algorithm is a maximum-likelihood decoder. To understand its principle of operation, imagine that a noisy channel sequence has been received and that the Hamming distances from each group of three code bits in the trellis to the corresponding channel bits have been calculated. Now the trellis branches are populated with numbers, as in the example of Fig. 11.7-3, which assumes that the channel sequence 111 101 111 000 110 . . . has been received. The most likely path through the trellis is the one with the lowest Hamming distance, the shaded one in the figure. This path, which corresponds to the code word 111 110 111 000 110 . . . , lies at distance 2 from the channel sequence.

The Viterbi algorithm works by calculating recursively the distance up to each state at a given time. These accumulated distances are shown in Fig. 11.7-3 in brackets, [ ]. Consider the calculations at time 3. At this time, two paths enter each state; for instance, the state sequences 0–2–3–3 with distance 2 and 0–0–2–3 with distance 5 both enter state 3 after 3. The critical step in the Viterbi algorithm is *to delete the path out of each pair that has the larger distance*, leaving only one path. Any minimum-

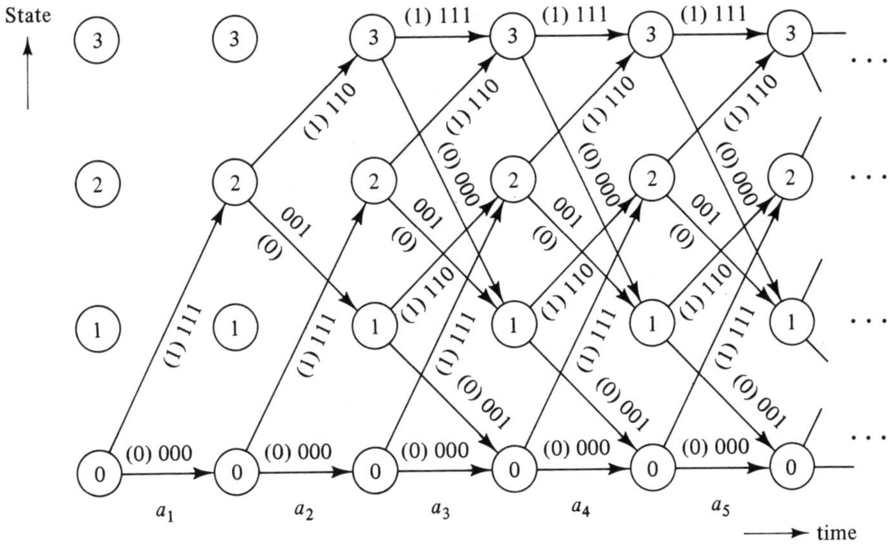

(a) State trellis

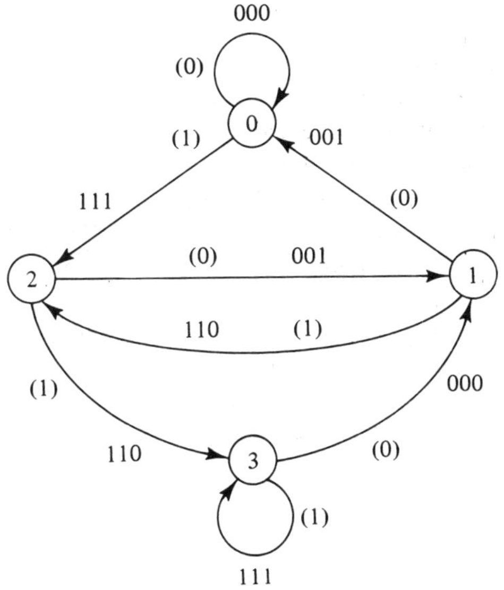

(b) State transition diagram

Figure 11.7-2. State transition diagram and trellis diagram for rate $\frac{1}{3}$ $\nu = 3$ convolutional code of Fig. 11.7-1, with generator sequence (111)(001)(001); that is, (111) is the first "column" of taps in Fig. 11.7-1, and the second and third columns are (001) and (001). A symbol shown in parentheses, ( ), is an input data bit; groups of three symbols shown on branches are encoder outputs as in Fig. 11.7-1.

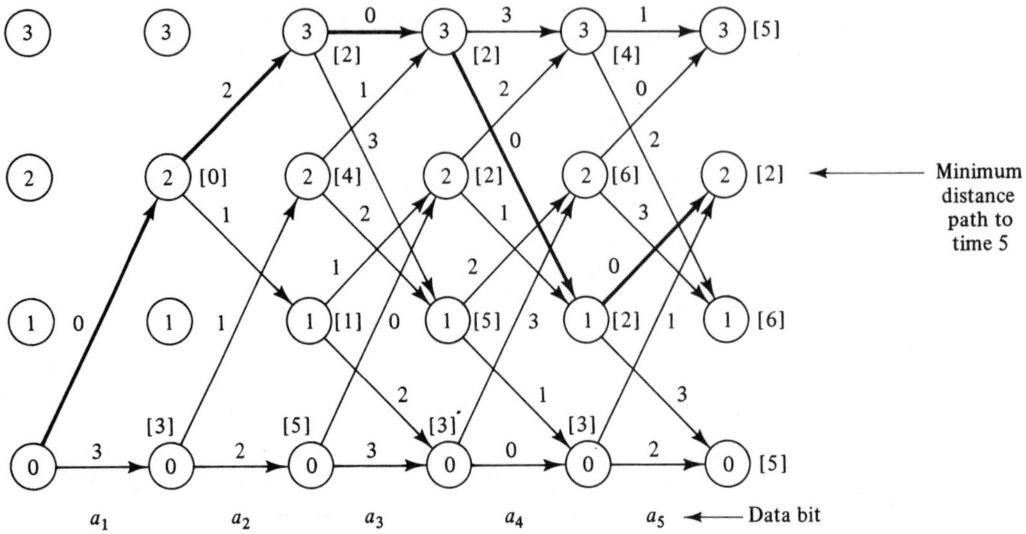

Figure 11.7-3. Examples of trellis distance increments, shown on branches, for the code of Fig. 11.7-1 when the channel sequence 111 101 111 000 110 . . . is received. [ ] indicates total Hamming distance to that point. The code in Fig. 11.7-2 is used.

distance trellis path, no matter how long, that passes through this state at this time must have the kept path as an antecedent. This step is repeated for each state position at time 3, leaving only single paths entering each state.

The Viterbi algorithm repeats this path-removal procedure at each time epoch, and after each time it always has the least-distance trellis path leading to each state. At time 5 in Fig. 11.7-3, it knows that the minimum-distance path in the whole tree is the path through states 0–2–3–3–1–2 ending in state 2. In the basic Viterbi algorithm, the process is continued until the message ends, and the best trellis path remaining is the decoded message.

The code in Fig. 11.7-2 has $d_{\min} = 5$, which means that it should correct up to 2 bit errors in a local region of the message. In the example of Fig. 11.7-3, such a double-bit error seems to have occurred. The best convolutional code of rate $\frac{1}{3}$ and constraint length 3, whether systematic or not, has $d_{\min} = 8$.

11.8 SUMMARY

This chapter has introduced the field of information theory, a subject that measures information, measures the ability of media to carry information, and studies codes that convert information into more efficient forms for storage and transmission. Information in symbolic form has been a continuing theme of this book. We first encountered it in Chapter 4 in the study of sampling and quantization of continuous waveforms. The theme reappeared in Chapter 10, where we investigated methods of transmitting symbolic, or "digital," data. Information theory is essentially a mathematical structure that seeks to explain the observed behavior of these and other symbolic data systems.

In this chapter we have seen several distinct parts of information and coding theory. Source coding was concerned with the measurement of information and with its reduction to more compact form. Rate-distortion coding considered the special case of information that was not discrete valued. Channel coding studied channels, their modeling, and their capacity to carry information, and schemes to code information so that it will pass more reliably through channels.

PROBLEMS

11-1. **(a)** Verify that the entropy of the distribution in Fig. 11.2-2 is 2.04 bits/grade.
(b) Shift the bins left by $\sigma/2$ so as to produce higher grades. What is the entropy now?

11-2. Suppose that weather falls into three types: sunny, cloudy, rain. The probabilities of these occurring in the absence of a forecast are 0.5, 0.4, and 0.1.
(a) Find the entropy of this "weather" source of information.
(b) A weatherman tries to predict this weather, but is wrong with probability $\frac{1}{2}$; if he is wrong, the other outcomes are equiprobable. What is the entropy of the "weather" source given that the weatherman's forecast is known?

11-3. **(a)** A fair coin is tossed until a tail appears. L is the length of the run of heads. Find $H(L)$.
(b) Repeat for a coin with $P(\text{head}) = 0.9$.

11-4. Prove Eq. (11.2-10).

11-5. A communications systems course has two tests. The grades on the first are distributed as in Fig. 11.2-2. On the second test, students get the same grade with probability $\frac{1}{2}$ and one grade higher with probability $\frac{1}{2}$ (students with A's get the same grade).
(a) Find the entropy in the outcomes of a student's second grade if nothing is known about his or her first grade.
(b) Find the equivocation $H(Z|W)$, where W is the first grade and Z the second.

11-6. Prove that Eqs. (11.2-11a) and (11.2-11b) are equal, and that both equal Eq. (11.2-11c).

11-7. Prove the relationship in Eq. (11.3-4).

11-8. From a telephone book that lists some of the codes, draw a code tree for the city codes of the United Kingdom. How do you know this is an instantaneous code? Compute the Kraft sum in Eq. (11.3-6).

11-9. Construct a code tree for the Morse code of Table 11.3-2, ignoring the comma symbol. Place a block at the end of each word in your tree. How do you know this code is not instantaneous? Compute the Kraft sum in Eq. (11.3-6).

11-10. Rework the proof of the Kraft inequality so that it applies to codes with symbols from an alphabet of arbitrary size r.

11-11. **(a)** Find the differential entropy of the analog source whose outputs x are distributed according to the exponential density function $p_X(x) = e^{-x}$, $x \geq 0$. Compare the entropy of the uniform distribution of Eq. (11.4-6).
(b) Repeat, if $p_X(x) = \frac{1}{2}e^{-|x|}$, all x.

11-12. For an analog source and its differential entropy, show that relations equivalent to Eqs. (11.2-10) and (11.2-11a) hold. For Eq. (11.2-11a), assume that $I(W; Z)$ is defined by Eq. (11.2-11c).

11-13. It is desired to send binary data at 10 kbits/s through a channel with the fixed ability to carry just 7.5 kbits/s in an error-free manner. According to Shannon's theorems, what is the least average number of reconstruction errors attainable by a coding scheme that implements this transmission? How many errors occur if the scheme sends 7.5 kbits/s and guesses at the rest?

11-14. In Sec. 4.6, the compact disc was discussed. According to Eq. (4.6-4), its SNR is estimated to be 88.7 dB.

(a) Assume music can be modeled as having independent Gaussian samples. Calculate from $R(D)$ for this source the SNR that can be expected at an encoding rate of 16 bits/sample.

(b) By extrapolating the curve of Fig. 11.4-3, estimate the SNR of the 16-bit optimal (Max) PCM-type quantizer.

(c) How many bits is the saving of part (a) over part (b) equivalent to? If this were implemented, how much longer would a disc play?

11-15. **(a)** Figure 11.5-2(f) shows two BSCs in series. Derive the transition probabilities shown there.

(b) Derive the general expression for L BSCs connected in tandem. Your result is a model for a communications system with repeaters.

11-16. A binary channel has transition matrix

$$[p(y|x)] = \begin{bmatrix} 0.9 & 0.1 \\ 0.4 & 0.6 \end{bmatrix}.$$

(a) Using Eq. (11.5-2), find $p_Y(y)$ if the inputs are equiprobable.

(b) Find the mutual information $I(X; Y)$ when the inputs are equiprobable. What is the meaning of your result?

11-17. Find the capacity of the channel in Prob. 11-16.

11-18. Show that definitions in Eqs. (11.5-5) and (11.5-4) are identical.

11-19. Derive Eq. (11.5-7), the capacity of the binary erasure channel. (Use symmetry!).

11-20. **(a)** Carry out the manipulations that derive Eq. (11.5-8) from the capacity definition in Eq. (11.5-4).

(b) Consider one of the quadrature channels in 16 QAM of the type defined in Eq. (10.4-17). The channel carries a four-level data symbol. Give an expression for the capacity of this channel in a form similar to Eq. (11.5-8).

11-21. A QPSK-modulated transmission system uses Nyquist pulses with 30% excess bandwidth. With Gaussian noise and a linear receiver, it operates at 10^{-5} error rate. The rate of transmission is 1 Mbit/s.

(a) Using Eq. (11.5-9), find the capacity in bits/channel use for the AWGN channel used here.

(b) Using Eq. (11.5-10), find the capacity in bits/s for the channel.

11-22. Prove inequality (11.6-6). *Hint*: Use the Stirling estimate for factorials.

11-23. A reasonably good BSC is one with error probability $p = 10^{-4}$.

(a) Find the capacity of this channel.

(b) The single parity check code of length 9, illustrated in Sec. 11.7, is used over this channel. Find the error-detection probability, the undetected-error probability, and the probability of correct reception.

(c) The rate of the code in part (b) is $\frac{8}{9}$. For what crossover probability p is this the capacity of a BSC?

11-24. Consider the case of two BSCs from Prob. 11-23 in tandem. (See also Prob. 11-15.)

(a) Find the capacity of the tandem channel.

(b) Find the probabilities in Prob. 11-23(b), but now for the tandem channel.

11-25. Three identical BSCs in parallel all carry the same symbols. At the output of the three, a majority-logic decision is made to decide which symbol was carried by the three.

(a) If $p = 10^{-4}$ in each BSC, what is the error probability for the three acting together in this way?

(b) What is their capacity?

(c) If the three carry information independently (each taking a third of the symbols), what is the capacity of this configuration?

11-26. Suppose a BSC is implemented by BPSK in which $E_b/N_0 = 10$ dB.

(a) Find the crossover p.

(b) Find the conversion factor between the signal-space distance $||s_1 - s_2||^2$, as used in Eq. (10.2-17), and Hamming distance. (Note that Hamming distance is of the "square" type.)

11-27. A BSC has $p = 0.01$.

(a) Find its capacity.

(b) Find the error probability of the length-7 rate $\frac{4}{7}$ Hamming code when used with this BSC.

(c) Compare to the error probability of the length-3 repetition code, whose rate is approximately the same. Which code is better?

11-28. Consider the rate $\frac{1}{2}$ convolutional code with generator sequence 11 01 00 01.

(a) Draw the encoder circuit in the style of Fig. 11.7-1.

(b) Give the generator matrix **G** in the form of Eq. (11.7-6) for the case where the code is used in block form with $N = 10$ bit code words.

(c) From this, find the parity check matrix **H** of an equivalent code. (Note that any rearranging of the columns of **G** will give an equivalent code.)

(d) What is $d_{\min}$ for this code?

REFERENCES

[11-1] C. E. Shannon, "A Mathematical Theory of Communication," *Bell System Technical Journal*, **27**, pp. 379–423 and 623–656, July 1948. This paper is reprinted in the very readable book *Mathematical Theory of Communication*, by C. E. Shannon and W. Weaver, University of Illinois Press, Urbana, Ill., 1963.

[11-2] R. V. L. Hartley, "Transmission of Information," *Bell System Technical Journal*, **7**, pp. 535–563, 1928.

[11-3] R. G. Gallager, *Information Theory and Reliable Communication*, Wiley, New York, 1968.

[11-4] A. J. Viterbi and J. K. Omura, *Principles of Digital Communication and Coding*, McGraw-Hill, New York, 1979.

[11-5] N. Abramson, *Information Theory and Coding*, McGraw-Hill, New York, 1963.

[11-6] C. E. Shannon, "Prediction and Entropy of Printed English," *Bell System Technical Journal*, **30,** pp. 50–64, Jan. 1951.

[11-7] C. E. Shannon, "Coding Theorems for a Discrete Source with a Fidelity Criterion," *IRE Natl. Conv. Record*, Part 4, pp. 142–163, 1959.

[11-8] T. Berger, *Rate Distortion Theory*, Prentice-Hall, Englewood Cliffs, N.J., 1971.

[11-9] J. M. Wozencraft and I. M. Jacobs, *Principles of Communication Engineering*, Wiley, New York, 1965.

[11-10] S. Lin and D. J. Costello, Jr., *Error Control Coding: Fundamentals and Applications*, Prentice-Hall, Englewood Cliffs, N.J., 1983.

[11-11] A. M. Michelson and A. H. Levesque, *Error-Control Techniques for Digital Communication*, Wiley, New York, 1985.

[11-12] H. Stark and J. W. Woods, *Probability, Random Processes, and Estimation Theory for Engineers*, Prentice-Hall, Englewood Cliffs, N.J., 1986.

[11-13] W. R. Bennett, Jr., *Scientific and Engineering Problem Solving with the Computer*, Prentice-Hall, Englewood Cliffs, N.J. 1976.

CHAPTER 12

Fiber-Optical Communications

12.1 INTRODUCTION

A spectacular recent development in communications technology is the tremendous growth of fiber-optic communications (FOC). While numerous FOC links already exist in Europe and North America, the number of systems planned for installations in the next few years will increase the length of FOC lines by many thousands of kilometers. For example, in the coming decade the American Telephone and Telegraph company (AT&T), in conjunction with European interests, plans to install four transatlantic cables linking the United States and Canada with England, France, and Spain. AT&T's domestic fiber-optic lines have expanded from 2100 km in 1985 to 17,700 in 1987.

Major FOC links exist in the United States, such as those connecting San Francisco to Los Angeles and Boston to Richmond, Virginia. Fiber-optical communication links are in use abroad, especially in the United Kingdom and France. As a case in point, the Ministry of Post, Telecommunications and Broadcasting of the city of Biarritz, France, inaugurated a city-wide FOC network on May 21, 1984, enabling that city to claim, with some justification, that it is the world's first ''fiber-optical'' city. The growth of FOC systems is rapid. If anything, there have been claims by specialists that FOC is expanding *too fast*: that the available bandwidths will shortly exceed demand, and that FOC will displace communication satellites for point-to-point communications. There have even been predictions that FOC will put many satellite companies out of business.[†]

To what properties of FOC systems can one attribute this extraordinary activity and interest, especially at the expense of other, competing point-to-point systems? A

[†] See the article by F. Guterl and G. Zorpette in the *IEEE Spectrum*, August 1985, pp. 30–37.

TABLE 12.1-1 COMPARATIVE CAPABILITIES: FIBER OPTICS VERSUS SATELLITES

Characteristic	Fiber-optic systems	Satellites	Comments
Bandwidth	Limited only by electronics at terminals; theoretical bandwidth of fiber is 1 terahertz	Most transponders have bandwidths of 36, 54, or 72 MHz	565 Mb/s currently available on fiber-optic lines; 1.7 Gb/s recently announced by AT&T for 1987; satellite bandwidth depends on frequency reuse and number of spot beams
Immunity to interference	Immune to electromagnetic interference	Transmission subject to interference from various sources, including microwave	
Durability of links	Storms can knock down overhead lines	Storms can disable individual antennas but leave network intact	
Security	Difficult to tap without detection	Signals must be encoded for security	
Multipoint capabilities	Primarily a point-to-point medium	Point-to-multipoint communications easily implemented	Large areas of coverage makes satellites only cost-effective means of reaching sparsely populated regions; multipoint-to-single-point communications also useful for data collection
Flexibility	Difficult to reconfigure to meet changing demand	Easy to reconfigure if hardware has been appropriately designed	
Connectivity to customer site	Local loops required	With antenna installed on customer premises, as with 14/12-GHz band, local loops not required	

Source: F. Guterl and G. Zorpette, ''Fiber Optics: Poised to Displace Satellites,'' *IEEE Spectrum*, pp. 30–37, August 1985. Used with permission of the IEEE.

TABLE 12.1-2 SOME TECHNICAL DETAILS OF THE FOC SYSTEMS IN THE UNITED KINGDOM

	8 Mbit/s short-haul	8 Mbit/s long-haul	34 Mbit/s	140 Mbit/s
Network applications	Junction	Trunk	Trunk	Trunk
Typical maximum system length	25 km	280 km	280 km	280 km
Repeater spacing	Up to 12 km	Up to 12 km	Up to 14 km	8 km
Power feed station spacing	Not applicable	Up to 60 km	Up to 40 km	Up to 32 km
Operational wavelength	←850 or 900 nm→		850 or 1300 nm	850 nm
Opto-electronic components				
Transmit	LED or laser	LED or laser	LED or laser	Laser
Receive	APD	APD	APD or PIN	APD
Cable				
Number of fibers	8	8	8	8
Fiber type	←graded index→			
Fiber dimensions	←15/125 μm→			
Optical attenuation	Up to 8 dB/km	2–8 dB/km	2–8 dB/km	3 dB/km
Power feed conductors	No	Yes	Yes	Yes

Source: J. E. Midwinter, "Optical Fiber Communications Systems Development in the U.K.," *IEEE Communications Magazine*, January 1982. With permission.

brief list of fiber-optic virtues are that they are light and durable; they are not affected by weather (satellite signals often are); they are remarkably immune to electromagnetic interference and other sources of noise; messages transmitted on FOC links are secure since they cannot be easily intercepted or tapped; they do not require large structures built in remote locations, as terrestrial microwave systems do; they can be made extremely low-loss and nondispersive (e.g., 4-gigabit/second transmission rate over 117 km without a repeater); unlike geosynchronous satellites,[†] they do not induce an annoying delay in signal propagation; and, finally, their intrinsic bandwidth is enormous, being limited in practice only by the electronics at the terminals. Against these virtues are drawbacks: once in place, FOC systems cannot easily be redirected; they are less cost-effective in sparsely populated areas; and compared to satellites, they are much less useful in point-to-multipoint communications. Table 12.1-1 lists some of the capabilities of FOC systems versus communication satellites. Table 12.1-2 lists some of the technical details of FOC links in the United Kingdom.[‡] In Table 12.1-1 the abbreviations LED, APD, and PIN stand for, respectively, light-emitting diode, avalanche photodiode, and intrinsic *p-n* photodiode. The term "graded-index" refers to a particular type of fiber discussed in Sec. 12.2.

The main components of FOC systems are, typically, a modulator and light source (e.g., LED or laser), the fiber, couplers and switches, repeaters, a receiver (essentially a photodetector and amplifier), and the receiver decoder logic (Fig. 12.1-1). Dramatic

[†] These orbit 36,000 km above the earth and, in effect, hover over a point on the earth.

[‡] Any FOC system mentioned in this book at the time of writing will probably not represent the state of the art at the time of printing.

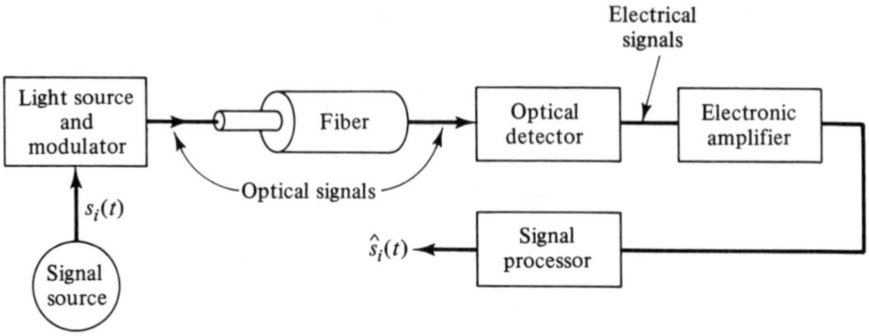

Figure 12.1-1. Optical communication system. Not shown is a repeater that would be inserted in the fiber link if the fiber cable were too long. The repeater would regenerate a strong, sharp optical pulse from a weaker, smeared optical output.

advances in component technology have greatly increased the commercial and technological appeal of FOC. For example, with regard to the fiber itself, much higher bit rates and longer distances between repeaters are now possible because of several new developments: *single-mode* fibers virtually eliminate a chief source of pulse broadening; low-loss fibers can be manufactured on a *production basis*; and *material dispersion* (another source of pulse broadening due to the dispersive nature of the medium) is reduced by shifting the wavelength band of minimal chromatic dispersion to overlap with the band of minimum attenuation. The development of electrooptic switches using lithium niobate crystals to switch bit streams from one fiber channel to another will increase the bit rate still more. These and other developments promise to make FOC systems even more competitive in the future than they are at present.

Entire books have been written on FOC systems ([12-1] to [12-6]). Clearly, in one chapter we can but barely scratch the surface of this subject. Rather than use what little space we have on a study of the technology and fabrication of FOC components,† we shall instead review the underlying physics by which these devices work and concentrate on FOC from a systems point of view.

12.2 OPTICAL FIBERS

An optical fiber is a dielectric waveguide that is normally used to transmit light energy from one point to another. The typical fiber is cylindrical in shape and consists of three parts: an inner *core*, a core-surrounding *cladding*, and an outside buffer *coating* that protects, isolates, and strengthens the fiber. The wave-guiding properties of the fiber are primarily determined by the core and cladding. Henceforth we shall ignore the buffer coating in our discussion since it plays little role in the underlying theory.

The light-guiding properties of a fiber can be explained in terms of a fundamental

† Of which many will be obsolete, no doubt, in the near future.

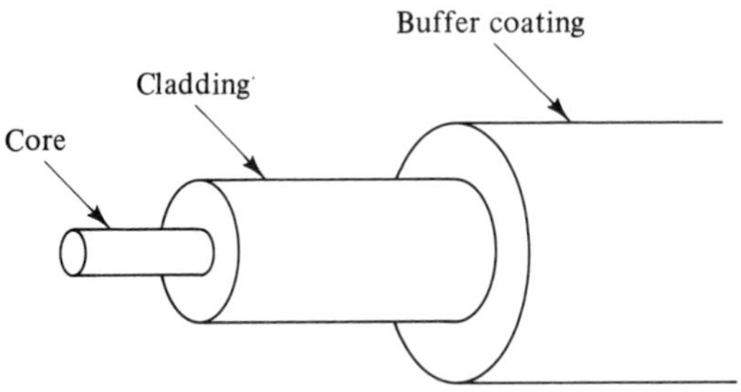

Figure 12.2-1. Structure of a single optical fiber.

optical parameter n known as the *index of refraction* of a medium. For our purposes, it suffices to define n by the relation

$$n = \frac{c}{v}, \tag{12.2-1}$$

where c is the speed of light in free space and v is the speed of light in a medium. Thus n is a characteristic parameter of a medium and accounts for many of that medium's optical properties. Some typical values of n are 1.0 for air and 1.5 for glass. The refractive index† usually varies with wavelength. For example, the refractive index of fused bulk quartz varies from 1.54 to 1.40 in the wavelength range from 0.2 micrometers (μm) to 4.0 μm ([12-2], p. 44). Figure 12.2-1 shows the structure of a fiber.

The index of refraction n_1 of the core is greater than the index of refraction n_2 of the cladding. While, in theory, a cladding is not necessary for wave propagation in the core, in practice the cladding serves several purposes: it reduces losses from the core, adds strength to the fiber, and reduces time dispersion in the transmitted pulse. The light-guiding properties of fiber are based on the principle of total internal reflection: when a ray passing through a medium M_1 of higher refractive index n_1 is incident upon a medium M_2 of lower refractive index n_2, it will experience total reflection at the boundary between M_1 and M_2 if the angle of incidence θ of the ray exceeds the critical angle θ_c. When $\theta < \theta_c$, the ray continues into medium M_2 and undergoes *refraction* (i.e., bending) according to Snells law: $n_1 \sin\theta = n_2 \sin\tilde{\theta}$, where $\tilde{\theta}$ is the refracted angle in medium M_2. The various situations are shown in Fig. 12.2-2.

Figure 12.2-3 shows the structure of a simple type of fiber called a *step-index* fiber. The choice of name, *step index*, should be obvious from the refractive-index profile across the fiber. From the preceding discussion if $\theta > \theta_c$, the ray will be confined inside the core and it will bounce back and forth (i.e., propagate) between the M_1-M_2 interface. The type of ray shown is called a *meridional ray* because it passes through the axis of the fiber. The paths of nonmeridional rays are more complex and we shall ignore them in this survey chapter. Assume that a ray is launched into the fiber from an outside medium M_0 (e.g., free space). It is natural to ask, what is the maximum value of the launch angle ϕ, say ϕ_m, that will still give rise to a ray confined within

† *Index of refraction* and *refractive index* are used interchangeably.

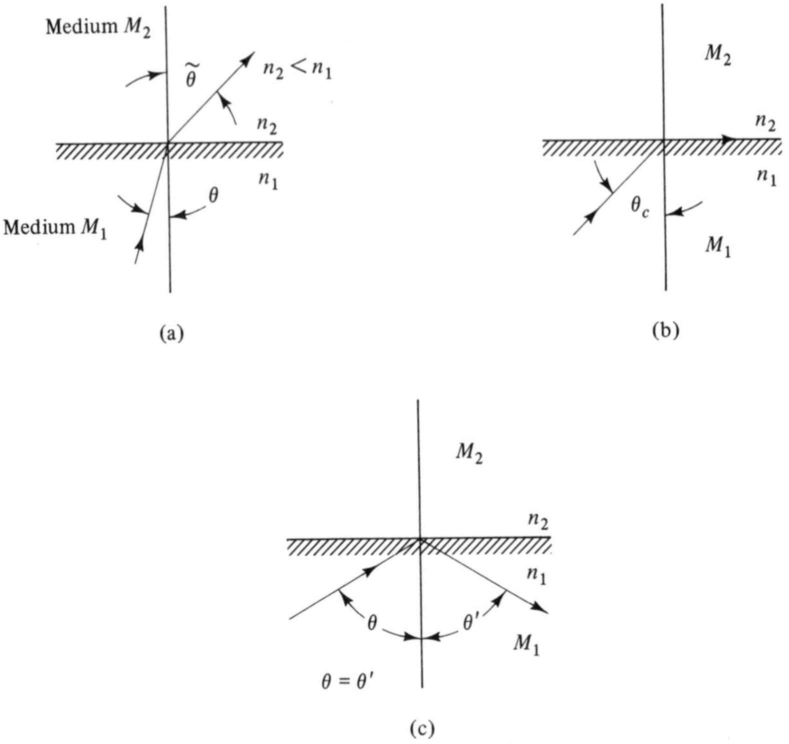

Figure 12.2-2. Refraction and reflection at the boundary between two dielectric media. In (a), $\theta < \theta_c$ and the incident ray is refracted in medium M_2 according to the law $n_1 \sin \theta = n_2 \sin \tilde{\theta}$; in (b) $\theta = \theta_c$ and the ray propagates along the boundary between M_1 and M_2; in (c) $\theta > \theta_c$ and the ray experiences total internal reflection. Case (c) applies in optical fiber propagation.

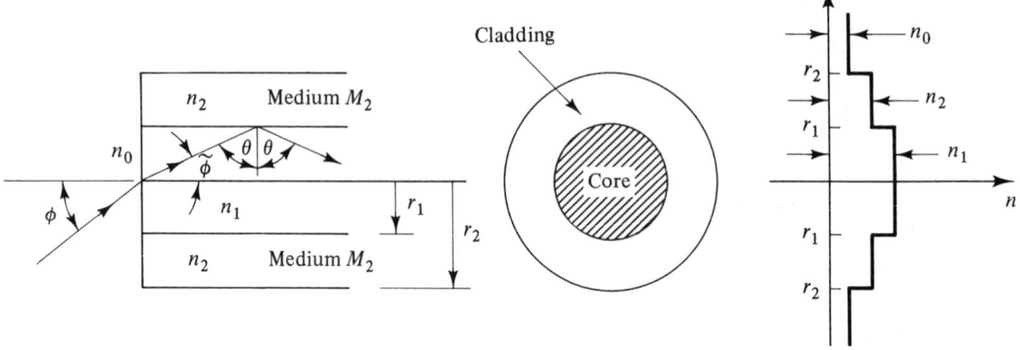

Figure 12.2-3. Ray tracing in a step-index fiber. Actually, the cladding thickness is generally much greater than the core thickness.

the core? To answer this important question, we note that as ϕ increases, θ decreases; thus for ϕ greater than ϕ_m, θ will be less than θ_c and lossless propagation will cease. Thus assume that the reflection angle θ assumes its limiting value θ_c. Then

$$n_1 \sin \theta_c = n_2, \qquad \sin^2 \theta_c = \frac{n_2^2}{n_1^2} \tag{12.2-2}$$

or

$$\begin{aligned} \cos \theta_c &= [1 - \sin^2 \theta_c]^{1/2} \\ &= \frac{(n_1^2 - n_2^2)^{1/2}}{n_1}. \end{aligned} \tag{12.2-3}$$

But $\cos \theta_c = \sin \tilde{\phi}_m$ and

$$\begin{aligned} n_0 \sin \phi_m &= n_1 \sin \tilde{\phi}_m \\ &= (n_1^2 - n_2^2)^{1/2}. \end{aligned}$$

If M_0 is free space, then $n_0 = 1$ and

$$\sin \phi_m = (n_1^2 - n_2^2)^{1/2}. \tag{12.2-4}$$

Thus all rays satisfying $\phi < \phi_m$ will couple into the fiber core; those for which $\phi > \phi_m$ will escape. The quantity $\sin \phi_m$ is often called the *numerical aperture* (NA) of the fiber. It is a measure of the light-collecting ability of the fiber. In fact, the maximum fraction of source power that a small diffuse Lambertian radiator (defined later) can transmit into the fiber cannot exceed $\sin^2 \phi_m$. To demonstrate this fact, let the source power per unit solid angle at angle ϕ, measured relative to the normal of the emitting surface, be denoted by $W(\phi)$. A Lambertian (Fig. 12.2-4) source satisfies $W(\phi) = W_0$

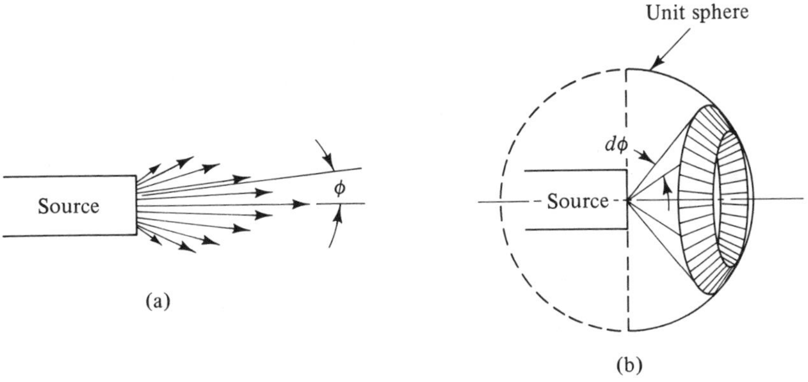

Figure 12.2-4. (a) A small diffuse Lambertian source $W(\phi) = W_0 \cos \phi$. (b) Computing the total power radiated by a Lambertian source through a spherical surface encapsulating the source.

$\cos\phi$ for $|\phi| < \pi/2$ and zero otherwise. Then the total source power output P_0 emitted by the source is

$$P_0 = \int_\Omega W(\phi)\, d\Omega, \tag{12.2-5}$$

where $d\Omega$ is the solid angle subtended by rays between the angles ϕ and $\phi + d\phi$. From Fig. 12.2-4(b), it can be shown that for the unit sphere

$$d\Omega = dA = 2\pi \sin\phi\, d\phi, \tag{12.2-6}$$

where dA is the area of the infinitesmal ring. Hence,

$$\begin{aligned} P_0 &= 2\pi \int_0^{\pi/2} W_0 \cos\phi \sin\phi\, d\phi \\ &= \pi W_0. \end{aligned} \tag{12.2-7}$$

However the amount of power, P_f, entering the fiber is limited by the angle ϕ_m. Hence

$$\begin{aligned} P_f &= 2\pi \int_0^{\phi_m} W_0 \cos\phi \sin\phi\, d\phi \\ &= \pi W_0 \sin^2\phi_m \end{aligned}$$

and

$$\sin^2\phi_m = (\text{NA})^2 = \frac{P_f}{P_0}. \tag{12.2-8}$$

A typical value of the NA is 0.3. Hence for a free-space interface between source and fiber, only 9% of the total radiation output by the source will enter the fiber.

To maximize the amount of power that a fiber can collect, we obviously would like to make $\sin^2\phi_m$ as large as possible. Returning to Eq. (12.2-4) and defining $n_1 - n_2 \triangleq \Delta n$ and $(n_1 + n_2)/2 \triangleq \bar{n}$, we can write

$$\sin^2\phi_m = 2\bar{n}\,\Delta n,$$

which suggests that the difference in refractive indexes between core and cladding be made as large as possible (e.g., by using a glass with very high n_1 as a core and no cladding at all; that is, the "cladding" is then free space with $n_2 = n_0 = 1$). However, for various reasons this is not done. Two compelling reasons have to do with (1) *radiation losses* at the core surface and (2) *multipath dispersion*. The former refers to the fact that, when electromagnetic radiation is reflected at a surface, nonpropagating *evanescent* waves penetrate the surface. Without a thick cladding to contain these evanescent waves, nonuniformities in the core and at the boundary are more likely to convert evanescent waves into propagating waves, a process that results in attentuation of the confined fiber signal. *Multipath or modal*† *dispersion* produces pulse spreading and is aggravated when there is no cladding.

† *Multipath*, *intermodal*, and *modal* dispersion are often used interchangeably. This we shall do here as well.

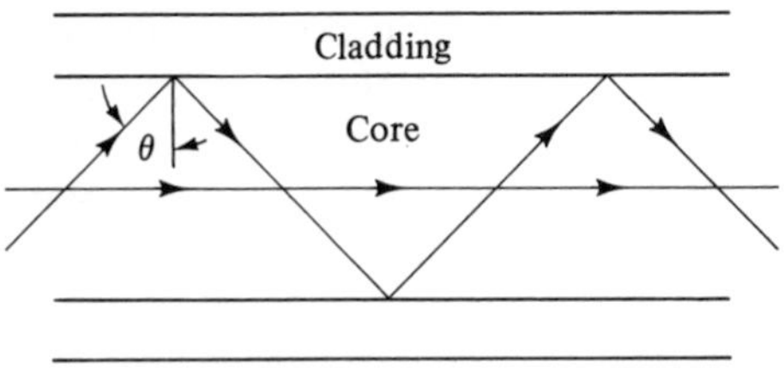

Figure 12.2-5. Oblique and axial rays in a fiber. The optical path length for the oblique ray is longer than it is for the axial ray.

Single-Mode and Multimode Fibers

Figure 12.2-5 shows two types of rays in a fiber: an axial ray and an oblique ray with internal reflection angle θ. If L is the length of the fiber, then—from simple geometry—the transit times for photons following the axial and oblique rays are

$$\tau_a = \frac{L}{c/n_1} \qquad \text{(axial ray)}, \tag{12.2-9}$$

$$\tau_0 = \frac{L}{(c/n_1)\sin\theta} \qquad \text{(oblique ray)}. \tag{12.2-10}$$

The fact that the transit times for the two rays are different leads to a phenomenon called *modal dispersion*. Modal dispersion causes pulse broadening and reduces the information capacity of the waveguide channel. Strictly speaking, the term *mode* refers to a particular harmonic solution to Maxwell's equations when these are applied to the geometry of the fiber. *Bound* or *trapped* modes are those that are guided by the fiber. We shall avoid discussing propagation in fibers from a mathematically rigorous viewpoint (i.e., by Maxwell's equations). Instead, we shall use a geometrical model of modes, which will suffice to explain most mode-related phenomena. A geometrical model of a mode is shown in Fig. 12.2-6.

The two oblique rays together represent a mode. High-ordered modes are those for which θ is small; low-ordered modes are those for which θ is large. The highest-order modes are associated with the most oblique rays, and the lowest-order modes with the least oblique ones. Clearly, the axial mode is the lowest-ordered mode. The axial mode is quite important and we shall say more about it shortly.

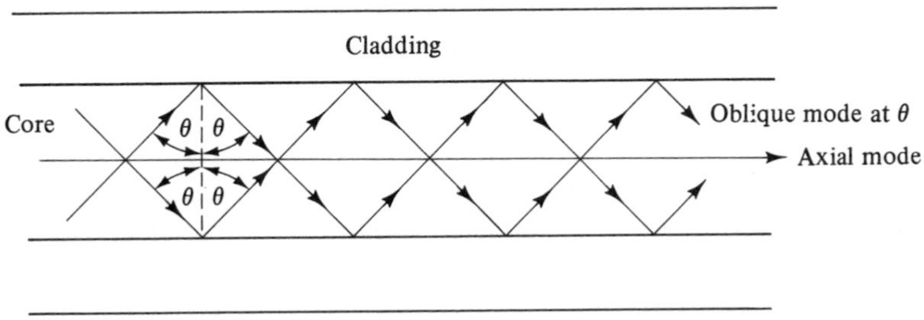

Figure 12.2-6. Two modes in a fiber. The transit time of the oblique mode is greater than that of the axial mode. Each internal reflection angle θ represents a different mode.

When a pulse of light is launched into a fiber, its transmission through the fiber must satisfy Maxwell's equations, and hence each pulse gives rise, generally, to many trapped modes. However, if these modes experience different delays, as shown in Figs. 12.2-5 or 12.2-6, they will not recombine at the fiber end in the same fashion in which they were launched. The result is a broadened pulse much as when an electrical pulse goes through a low-pass filter.

From the simple ray theory reviewed here, it would seem that a fiber could support an uncountably infinite number of modes, one for each θ in the range $\theta_c \leq \theta < \pi/2$. In fact, this is not so. That any given optical waveguide can only support a finite number of modes is one of the most important results of optical communication theory. It is this fact that accounts for the importance of *single-mode* fibers in optical communication systems. Later we consider a planar dielectric waveguide consisting of a slab of dielectric with refractive index n_1 and thickness d sandwiched between two semi-infinite regions of dielectric material of refractive index n_2 (Fig. 12.2-7). This single optical waveguide will enable us to compute the maximum number of modes explicitly. Had we considered a circular waveguide, the principles would have been the same but the resulting equations would have been more involved.

Number of Modes in a Planar Waveguide

To demonstrate that a planar waveguide can support only a finite number of modes, one must borrow two results from electromagnetic theory. The first is that when Maxwell's equations are solved for the two-dielectric configuration, as shown in Fig. 12.2-2, we find that, when total internal reflection occurs, the reflected wave undergoes a phase change (increase) each time it is reflected from the boundary. The phase changes are given by

$$\tan\left(\frac{\Phi_N}{2}\right) = \frac{\sqrt{r^2 \sin^2\theta - 1}}{r\cos\theta}, \tag{12.2-11a}$$

$$\tan\left(\frac{\Phi_p}{2}\right) = \frac{r\sqrt{r^2 \sin^2\theta - 1}}{\cos\theta}, \tag{12.2-11b}$$

where $r \equiv n_1/n_2$ and Φ_N and Φ_p are the phase changes in the wave components normal and parallel to the reflecting surface. The second borrowed result is that the electric field in the fiber is made up of components of the form

$$E(\mathbf{r}, t) = A(\mathbf{r})e^{j(\omega t - \mathbf{k}\cdot\mathbf{r})},$$

where $A(\mathbf{r})$ is the amplitude of the wave, ω is the temporal radian frequency, t is time, $\mathbf{k}$ is the wave vector, and $\mathbf{r}$ is the position vector. In this discussion, neither the amplitude variation $A(\mathbf{r})$ nor the time-harmonic phasor $\exp[j\omega t]$ is of interest, so we focus only on the term

$$E(\mathbf{r}) \triangleq e^{-j\mathbf{k}\cdot\mathbf{r}}. \tag{12.2-12}$$

In general, the $\mathbf{k}$ vector can be written as, using the unit Cartesian vectors $\hat{\imath}_x$, $\hat{\imath}_y$, $\hat{\imath}_z$,

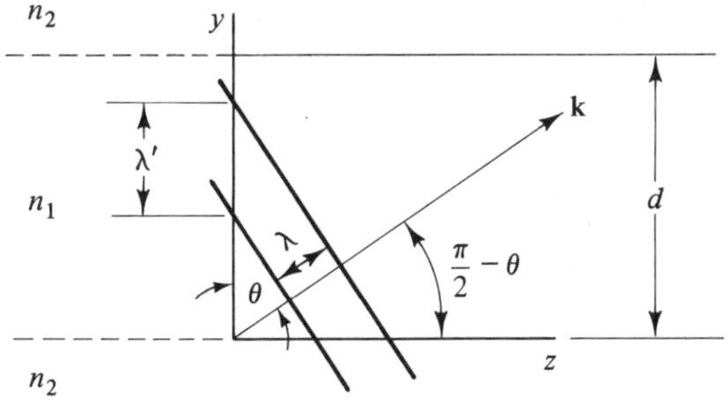

Figure 12.2-7. A ray with direction angles $\alpha = 0$, $\beta = \theta$, $\gamma = \pi/2 - \theta$ in a planar dielectric waveguide.

$$\mathbf{k} = k_x\hat{\imath}_x + k_y\hat{\imath}_y + k_z\hat{\imath}_z, \tag{12.2-13}$$

where $k_x = k \cos \alpha$, $k_y = k \cos \beta$, and $k_z = k \cos \gamma$, and $k \triangleq |\mathbf{k}| = 2\pi/\lambda$; λ is the wavelength in the medium and α, β, and λ are the *direction cosines* of the vector $\mathbf{k}$ with the x, y, and z axes. For a wave traveling as shown in Fig. 12.2-7 with the $\mathbf{k}$ vector confined to the y-z plane, the angles α, β, and γ are 0, θ, $\pi/2 - \theta$, respectively. Hence

$$\mathbf{k} = \frac{2\pi}{\lambda}\cos\theta\,\hat{\imath}_y + \frac{2\pi}{\lambda}\sin\theta\,\hat{\imath}_z.$$

Also, $\mathbf{r} = x\hat{\imath}_x + y\hat{\imath}_y + z\hat{\imath}_z$ and

$$\mathbf{k}\cdot\mathbf{r} = \frac{2\pi}{\lambda}y\cos\theta + \frac{2\pi}{\lambda}z\sin\theta.$$

Thus Eq. (12.2-12) can be rewritten as

$$E(r) = \exp(-j2\pi/\lambda[y\cos\theta + z\sin\theta]). \tag{12.2-14}$$

Equation (12.2-14) suggests that the zigzag notion of a ray can be split up into a motion parallel and perpendicular to the axis of the waveguide (i.e., the z axis). In the y direction, the wave is confined between $y = 0$ and $y = d$ and, for a steady-state mode, must set up a steady-state pattern called a *standing wave*. This requires that the pattern as viewed in the y direction remain invariant with respect to the number of bounces of the wave between $y = 0$ and $y = d$. Thus, after a complete pass, the total phase change impressed upon the wave must be a multiple of 2π. In the z direction there are no boundary-induced constraints, with the result that in that direction we observe a traveling wave. From Fig. 12.2-7 we see that the effective wavelength along the y direction is $\lambda' = \lambda/\cos\theta$. At frequency $\nu \triangleq \omega/2\pi$ hertz, the wavelength in the medium allowing light propagation at speed v is

$$\lambda = \frac{v}{\nu} = \frac{v}{c}\frac{c}{\nu} = \frac{1}{n_1}\lambda_0$$

and

$$\lambda' = \frac{\lambda_0}{n_1\cos\theta}.$$

The optical-path-induced phase delay ψ associated with a complete vertical-travel cycle (i.e., a two-reflection pass starting at y and returning to y after traversing a distance $2d$) is

$$\begin{aligned}\psi = k_y 2d &= \frac{4\pi d \cos\theta}{\lambda} \\ &= \frac{4\pi d n_1 \cos\theta}{\lambda_0} \\ &= \frac{4\pi d}{\lambda'}.\end{aligned} \tag{12.2-15}$$

However, the *total* phase delay, ψ_T, induced in traversing the distance $2d$ must include the phase advance Φ_N experienced by the normal component of the wave at each reflection.† Hence, with M a positive integer, we obtain for the steady state

$$\psi_T = 2\pi M = \frac{4\pi d n_1 \cos\theta}{\lambda_0} - 2\Phi_N. \tag{12.2-16}$$

Solving for $\cos\theta$ in Eq. (12.2-16), we obtain

$$\cos\theta = \frac{(\pi M + \Phi_N)\lambda_0}{2\pi d n_1}. \tag{12.2-17}$$

Thus we see from Eq. (12.2-17) that, since M is an integer, only certain discrete values of $\theta = \theta_M$ are allowed. The integer M is called the *mode number*, and θ_M is the characteristic angle associated with the mode. In particular, there is a maximum value of M (i.e., a highest mode number that depends on d). To see this, we solve Eq. (12.2-17) for M to obtain

$$M = \frac{2dn_1 \cos\theta_M}{\lambda_0} - \frac{\Phi_N}{\pi}.$$

But for total internal reflection we must have

$$\sin\theta_M \geq \sin\theta_c = \frac{n_2}{n_1},$$

whence it follows that

$$\cos\theta_M \leq \cos\theta_c = \left[1 - \left(\frac{n_2}{n_1}\right)^2\right]^{1/2}.$$

Hence

$$\begin{aligned}M_{\text{MAX}} &\leq \frac{2dn_1 \cos\theta_c}{\lambda_0} - \frac{\Phi_N}{\pi} \\ &\leq \frac{2dn_1[1 - (n_2/n_1)^2]^{1/2}}{\lambda_0} - \frac{\Phi_N}{\pi}.\end{aligned} \tag{12.2-18}$$

† The total phase advance is therefore $2\Phi_N$.

In the literature on optical fiber communication, wide use is made of the *normalized thickness* parameter, V, defined by

$$V \triangleq \frac{2\pi d(n_1^2 - n_2^2)^{1/2}}{\lambda_0}. \tag{12.2-19}$$

Hence, for the planar waveguide,

$$M_{\text{MAX}} \leq \frac{V}{\pi} - \frac{\Phi_N}{\pi}. \tag{12.2-20}$$

Finally, we note from Eq. (12.2-11a) that, when $\theta = \pi/2$, $\Phi_N = \pi$, which is the largest value that Φ_N can take. Equations (12.2-19) and (12.2-20) enable us to compute the maximum possible number of modes in the waveguide.

Example 12.2-1

Let $n_1 = 1.42$ and $n_2 = 1.40$. Take $\lambda_0 = 0.5\ \mu\text{m}$ and $d = 70\ \mu\text{m}$. Then $V = 209$ and, estimating Φ_N by π, we get

$$M_{\text{MAX}} \leq \frac{209}{\pi} - 1 = 66.$$

Thus about 66 modes can be accommodated by planar waveguide with the given parameters.

We point out that it is always possible to make d and hence V small enough so that only a single mode can be supported by a fiber. When this is the case, the fiber is said to be a *single-mode fiber*. For the case of the step-index circular waveguide, this occurs for $V < 2.405$. Fibers for which $M > 1$ are called *multimode fibers*. The number of modes M_{MAX} in a multimode step-index fiber can be well approximated by the relation $M_{\text{MAX}} \simeq V^2/2$ ([12-6], p. 18).

Despite the many practical advantages of multimode over single-mode fibers with respect to handling and energy-carrying capacity, the latter do not suffer from the phenomenon of *modal dispersion*. We discuss this important phenomenon next.

Modal Dispersion

To make the point as simply as possible, we return to the simple planar waveguide and ray model used in Fig. 12.2-6. Let L denote the length of the fiber. The total effective path for the optical ray is $L' = L/\sin\theta$ and the total delay is

$$\tau = \frac{L'}{v} = \frac{L'}{c/n_1} = \frac{n_1 L}{c \sin\theta}. \tag{12.2-21}$$

The mode angle θ can vary as $\theta_c < \theta \leq \pi/2$; for modes close to θ_c the delay is, approximately,

$$\tau_{\theta_c} = \frac{n_1 L}{c \sin\theta_c} = \frac{n_1^2 L}{c n_2}, \tag{12.2-22}$$

while for modes close to $\pi/2$ the delay is

$$\tau_{\pi/2} = \frac{n_1 L}{c \sin \pi/2} = \frac{n_1 L}{c}. \tag{12.2-23}$$

The absolute difference in delays is

$$\begin{aligned} |\Delta\tau| &= |\tau_{\pi/2} - \tau_{\theta_c}| = \left|\frac{n_1 L}{c}\left(1 - \frac{n_1}{n_2}\right)\right| \\ &= \frac{n_1 L(n_1 - n_2)}{n_2 c}. \end{aligned} \tag{12.2-24}$$

Thus the various modes do not recombine in the same phase with which they were launched. This causes a spreading of the signal pulse. This undesired effect, called variously *multipath* dispersion, *intermodal* dispersion, or just *modal* dispersion, is characteristic of multimode fibers and can occur even when the source bandwidth is zero (i.e., a monochromatic source).

Example 12.2-2

Consider a 1-km fiber with $n_1 = 1.42$ and $n_2 = 1.40$ as in Example 12.2-1. Then

$$|\Delta\tau| = \frac{1.42 \times 10^3(1.42 - 1.40)}{1.40 \times 3 \times 10^8} = 68 \text{ ns}.$$

Thus the maximum modal dispersion for this fiber is 68 ns/km. However, the reader is cautioned that this computation is only an estimate since we ignored, among other things, the phase shifts introduced by reflections at the boundaries. A detailed solution based on Maxwell's equations is furnished in [12-6, Chapter 1].

Some typical dimensions for single-mode and multimode fibers are shown in Fig. 12.2-8.

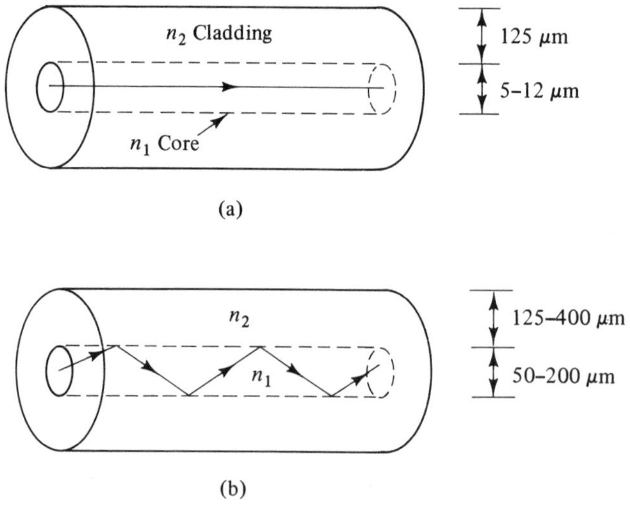

Figure 12.2-8. Typical dimensions of fibers. (a) Single mode. (b) Multimode.

In summary, multimode fibers suffer from modal dispersion; single-mode fibers do not. However, multimode fibers are easier to manufacture, splice, interconnect, and couple to light sources. In particular, multimode fibers can be used with incoherent sources such as light-emitting diodes (LEDs). Single-mode fibers require the highly directional output (coherence) of a laser source for efficient energy coupling into the fiber core. This should be regarded as an advantage for multimode fibers, because LEDs are simpler, cheaper, and more rugged than lasers ([12-3], Chapter 4). The lower information capacity of multimode fibers can, to a considerable extent, be increased by carefully designing the profile of the refractive index in the core. We discuss this point next.

Graded Index Fibers

So far the discussion has assumed a waveguide with constant core index n_1 and cladding index n_2. Such waveguides are called *step-index* guides. To ameliorate the effect of modal dispersion in multimode step index fibers, graded index (GRIN) fibers are frequently used. The structure of a GRIN fiber is shown in Fig. 12.2-9. The index of refraction of a GRIN fiber is typically described by

$$n(r) = \begin{cases} n_1\left[1 - 2\Delta\left(\frac{r}{a}\right)^{\alpha}\right]^{1/2} & r < a, \\ n_1(1 - 2\Delta)^{1/2}, & r \geq a, \end{cases} \tag{12.2-25}$$

where a is the radius of the core, α is a parameter called the profile parameter, and Δ is defined as

$$\Delta \equiv \frac{n_1 - n_2}{n_1}$$

(i.e., the fractional difference of n_1 and n_2). To understand how the GRIN fiber works, refer to Fig. 12.2-10, which shows the path of two rays, an axial ray and a meridional

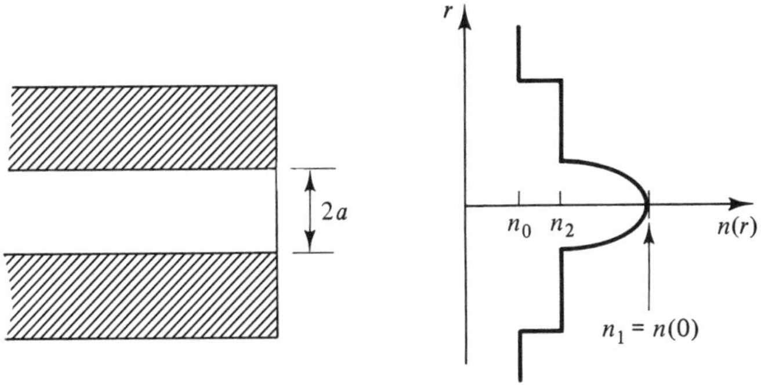

Figure 12.2-9. Structure of the graded-index fiber.

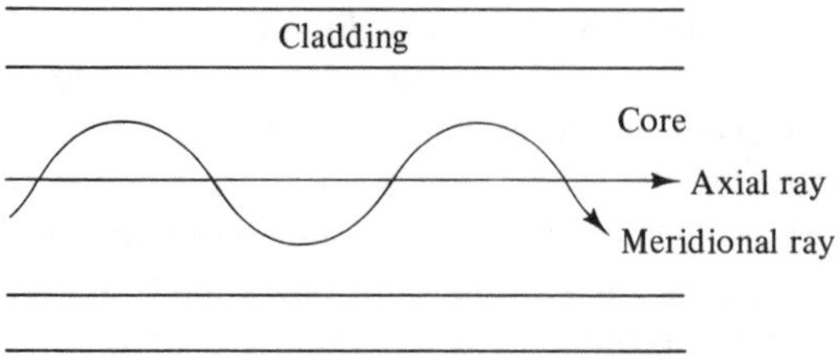

Figure 12.2-10. Path of two rays in a GRIN fiber.

ray. While the axial ray travels the shortest path, it also travels the most slowly since $n(0) > n(r)$ for $r \neq 0$. The oblique meridional ray travels farther, but much of its path lies in a medium of lower refractive index and thus its average speed is much greater. If the index profile is properly chosen, it is possible for all the rays in the core to undergo nearly the same delay and thus arrive at the receiver at approximately the same time.

Note that the quasi-parabolic index profile confines the ray to the core in a manner analogous to how quasi-parabolic thickness variations of a lens tend to focus rays of light toward the lens axis. While the GRIN fiber cannot completely remove pulse broadening due to modal dispersion, it can greatly reduce it. The following formula ([12-1], p. 354) gives the differential delay between the fastest and the slowest mode when the GRIN fiber is excited by a monochromatic source:

$$\Delta\tau_G = \frac{Ln_1\Delta^2}{8c}. \tag{12.2-26}$$

In deriving Eq. (12.2-26), it is assumed that $\alpha = 2(1 - 1.2\Delta)$. This value of α furnishes the minimum value of intermodal dispersion [12-7]. For the step-index fiber, the maximum dispersion time $\Delta\tau_{SI}$ is given by Eq. (12.2-24) [i.e., $\Delta\tau_{SI} = n_1L(n_1 - n_2)/n_2c$]. Hence

$$\Delta\tau_{SI} = \frac{n_1^2L\Delta}{cn_2} \simeq \frac{L}{c}n_1\Delta \tag{12.2-27}$$

and

$$\frac{\Delta\tau_G}{\Delta\tau_{SI}} = \frac{\Delta}{8}. \tag{12.2-28}$$

Example 12.2-3

As in Example 12.2-2, consider a 1-km fiber with $n_1 = 1.42$ and $n_2 = 1.40$. Assume that a GRIN fiber with $n(r)$ as in Eq. (12.2-25) has $\alpha = 2(1 - 1.2\Delta)$. Then

$$\frac{\Delta\tau_G}{\Delta\tau_{SI}} = 1.76 \times 10^{-3}$$

and the GRIN time dispersion is 568 times less than that of the step-index guide.

In practice, small deviations from the optimum GRIN profile can greatly increase the effects of intermodal dispersion ([12-1], p. 355). Also, the optimum value of α for one wavelength will typically not be optimum at another wavelength. Finally, the effective numerical aperture of GRIN fibers can be shown to be less than those of step-index fibers ([12-3], Sec. 2-5) so that the ultimate choice of which fiber type to choose involves numerous trade-offs.

A study of multimode propagation in graded-index fibers is found in [12-7].

Material Dispersion

A second important reason why pulses broaden as they travel through a fiber is the dependence of the refractive index of the fiber material on the wavelength of the source ([12-3], p. 62). The actual pulse-broadening mechanism is the following: the envelope of a pulse injected into the fiber travels essentially at a velocity called the *group velocity*. As we shall see, the group velocity depends on the refractive index n, which in a dispersive medium depends on wavelength λ [i.e., $n = n(\lambda)$]. A Fourier resolution of the pulse produces spectral components over a range of wavelengths (frequencies). Each of these components travels at a different speed. The net result is a broadening of the pulse and a reduction of the information-handling capacity of the fiber.

To understand the mechanics of material dispersion, consider the sum of two sine waves a small amount $\Delta\omega$ above and $\Delta\omega$ below a carrier frequency ω_0; that is,

$$f(t) = \sin(\omega_0 + \Delta\omega)t + \sin(\omega_0 - \Delta\omega)t. \tag{12.2-29}$$

When $f(t)$ is launched as a wave along the z direction in an infinitely extended dielectric medium with unconstrained propagation constant β ($\beta = 2\pi n/\lambda$), we obtain the wave

$$\begin{aligned} f(t, z) &= \sin[(\omega + \Delta\omega)t - (\beta + \Delta\beta)z] + \sin[(\omega - \Delta\omega)t - (\beta - \Delta\beta)z] \\ &= 2\cos(\Delta\omega t - \Delta\beta z)\sin(\omega t - \beta z). \end{aligned} \tag{12.2-30}$$

The *phase velocity* v_p is the velocity at which one must travel to see a constant phase in the carrier $\sin(\omega t - \beta z)$. Clearly,

$$v_p \equiv \frac{dz}{dt} - \frac{\omega}{\beta}. \tag{12.2-31}$$

The envelope or signal term $2\cos(\Delta\omega t - \Delta\beta z)$ modulates the carrier; it travels at the *signal* or *group velocity* v_g given by

$$v_g = \lim_{\Delta\beta\to 0} \frac{\Delta\omega}{\Delta\beta} = \frac{d\omega}{d\beta}. \tag{12.2-32}$$

Thus, from Eqs. (12.2-31) and (12.2-32), we obtain

$$v_g = \frac{v_p}{1 - (\omega/v_p)(dv_p/d\omega)}, \tag{12.2-33}$$

which relates the group velocity to the phase velocity in a dispersive medium.

Consider now propagation in a dispersive medium. Our aim is to compute the group delay per unit distance, τ_g, and the differential group delay per unit distance, $\Delta\tau_g$, for wave components $\Delta\lambda/2$ above and below the mean wavelength λ. Since the refractive index now depends on λ (for convenience we refer back to the free-space wavelength λ), we can write

$$v_p = \frac{c}{n(\lambda)} = v_p(\lambda)$$

and

$$\beta = \frac{\omega}{v_p} = \frac{2\pi}{\lambda_m} = \frac{2\pi n(\lambda)}{\lambda}, \tag{12.2-34}$$

where λ_m is the wavelength in the medium. The group delay for signal components in the infinitesmal band $d\omega$ about ω or $d\lambda$ about λ is, after traveling a distance L,

$$\begin{aligned} T_g &= \frac{L}{v_g} = L\frac{d\beta}{d\omega} \\ &= -\frac{\lambda^2 L}{2\pi c}\frac{d\beta}{d\lambda} = -\frac{2\pi c L}{\omega^2}\frac{d\beta}{d\lambda}. \end{aligned}$$

The group delay per unit distance is

$$\tau_g = \frac{T_g}{L} = -\frac{\lambda^2}{2\pi c}\frac{d\beta}{d\lambda}. \tag{12.2-35}$$

Next consider the computation of $\Delta\tau_g$; this is the difference in group delay for components separated by $\Delta\lambda$ in wavelength:

$$\begin{aligned} \Delta\tau_g &= \frac{d\tau_g}{d\lambda}\Delta\lambda \\ &= -\frac{\Delta\lambda}{2\pi c}\left[2\lambda\frac{d\beta}{d\lambda} + \lambda^2\frac{d^2\beta}{d\lambda^2}\right]. \end{aligned} \tag{12.2-36}$$

The second line of Eq. (12.2-36) was obtained with the help of Eq. (12.2-35). Since $\beta = 2\pi n(\lambda)/\lambda$, we can rewrite Eq. (12.2-36) in terms of $dn/d\lambda$ as

$$\Delta\tau_g = -\frac{\lambda\,\Delta\lambda}{c}\frac{d^2n}{d\lambda^2}. \tag{12.2-37}$$

If the source bandwidth, expressed in wavelengths, is σ_λ, then the effective differential group delay is σ_g and

$$\sigma_g = (\Delta\tau_g)_{\max} = -\frac{\lambda\sigma_\lambda}{c}\frac{d^2n}{d\lambda^2}. \tag{12.2-38}$$

Equation (12.2-38) is a central result of the theory and extremely important to the design engineer. It shows that the material dispersion depends on the source bandwidth

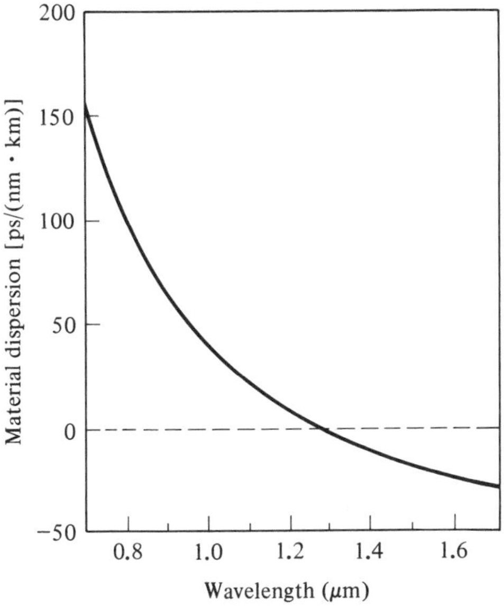

Figure 12.2-11. Material dispersion for silica. (Reproduced from J. W. Fleming, *Electron. Lett.*, **14,** p. 326, May 1978, by permission.)

σ_λ and the *second derivative* of the refractive index. It also shows that if $n(\lambda)$ is a constant (i.e., it does not depend on λ) then there is no pulse spreading due to material dispersion. Indeed, we can relax this constraint and allow $n(\lambda)$ to vary as a linear function of λ and *still not obtain pulse spreading*. Equation (12.2-38) can be interpreted as the amount of spreading that a very narrow pulse emitted by a σ_λ-bandwidth source would undergo per unit distance in a medium with a prescribed $n(\lambda)$. Figure 12.2-11 shows the material dispersion as a function of wavelength for silica; it is a graph of σ_g, normalized with respect to σ_λ, versus wavelength.

There are other pulse-broadening mechanisms. *Waveguide* or *intramode* dispersion is a consequence of the *normalized thickness* V [Eq. (12.2-19)] dependence on λ even when the medium is nondispersive. It generates a broadening within each mode if the source is not monochromatic. Normally, it is much smaller than material dispersion. A fairly thorough discussion of this type of dispersion is given in [12-2, Sec. 5.5]. *Profile dispersion* occurs in GRIN fibers and refers to the wavelength dependence of the parameter α in Eq. (12.2-25). Keiser in [12-3, p. 71] gives a discussion of this phenomenon. In most instances, however, it is modal and material dispersion that are the primary broadening mechanisms ([12-6], Chapter 1). We discuss their combined effect next.

Pulse Broadening Due to Intermodal and Material Dispersion

In general, to compute the total pulse broadening due to the major dispersive mechanisms, we must make assumptions regarding a variety of factors, including the pulse shape, the coherence of the source, the type of fiber, and so on. As a first approximation,

TABLE 12.2-1 COMBINED EFFECTS OF INTERMODAL AND MATERIAL DISPERSION

Type of source	λ (μm)	$\Delta\lambda/\lambda$	σ_m (ns/km)	σ_T (SI)[a] (ns/km)	σ_T (GRIN)[b] (ns/km)
LED	0.9	0.033	2.1	15	2.2
Laser	0.9	0.003	0.2	15	0.5
LED	1.3	0.04	0.1	15	0.5
Laser	1.3	0.004	0.01	15	0.5
LED	1.55	0.04	1.2	15	1.3
Laser	1.55	0.004	0.1	15	0.5

Source: John Gowar, *Optical Communication Systems*, Prentice-Hall, Englewood Cliffs, N.J., 1984. By permission.

[a] σ_i for a step-index fiber is assumed to be 15 ns/km.

[b] σ_i for a graded-index fiber is assumed to be 0.5 ns/km.

however, we can ignore the coupling between the broadening phenomena and assume that the different delay mechanisms operate independently. Based on this assumption, we can proceed by treating the broadening introduced by each separate dispersion mechanism as *independent* random phenomena. Thus let $\bar{X}$ denote the average propagation time of a pulse and let ΔX_m and ΔX_i denote the variations in propagation times due to material dispersion and intermodal dispersion, respectively. Then, with $\Delta X_T \equiv \Delta X_m + \Delta X_i$ representing the total variation in propagation time about the average, we obtain

$$\begin{aligned}\sigma_T^2 \equiv \overline{\Delta X_T^2} &= \overline{[\Delta X_m + \Delta X_i]^2} \\ &= \overline{\Delta X_m^2} + \overline{\Delta X_i^2} \qquad \text{(by independence).}\end{aligned} \tag{12.2-39}$$

Hence, with $\sigma_m \equiv [\overline{\Delta X_m^2}]^{1/2}$ and $\sigma_i \equiv [\overline{\Delta X_i^2}]^{1/2}$ denoting the rms pulse width for material and intermodal dispersion, respectively, we obtain

$$\sigma_T = \sqrt{\sigma_i^2 + \sigma_m^2} \tag{12.2-40}$$

and, more generally, if $\Delta X_j j = 1, \ldots, N$ are the propagation-time variations introduced by N independent mechanisms, then $\sigma_T = \sqrt{\sigma_1^2 + \cdots + \sigma_N^2}$. What justification do we have for treating ΔX_i and ΔX_m as independent r.v.'s? In the case of material dispersion, we use the fact that the source is a random emitter of radiation, whose spectral density represents the *average* power emitted at wavelength λ (or frequency ν). The actual emitted power is a random process whose statistical properties are best described using the *coherence function*.[†] Since the differential delay will depend on the instantaneous power distribution as a function of λ, we treat ΔX_m as a r.v. In the case of intermodal dispersion, we treat the distribution of power among the various modes as a random phenomenon independent of the source coherence, thus allowing ΔX_i to be viewed as a r.v. independent of ΔX_m.

Table 12.2-1 gives some values of σ_T under the combined effects of intermodal and material dispersion in step-index (SI) and graded-index (GRIN) fibers at different

[†] Discussed in Sec. 12.5.

wavelengths. We see that for the step-index fiber the primary mechanism for pulse broadening is intermodal dispersion; this holds at all frequencies for both LED and laser sources. For GRIN fibers used with LEDs, material dispersion dominates except at 1.3 μm; at that wavelength, the material dispersion of silica is so low that *intramode* dispersion may become the most significant pulse-distortion mechanism ([12-3], p. 66) in single-mode fibers.

Attenuation in Fibers

Attenuation losses in fibers as low as 0.2 dB/km and even less under laboratory conditions have been obtained. Nevertheless, the combination of attenuation and pulse broadening necessitates the periodic insertion of repeaters in long trunk lines. Attenuation in fibers results primarily from (1) absorption and (2) scattering. Absorption losses occur when the optical energy in the fiber interacts either with the intrinsic fiber materials or impurities such as iron, chromium, cobalt, copper, and OH^- ions. These interactions involve electronic transitions in the intrinsic material or the impurities. Eventually, there is, within the material, a conversion of optical energy to thermal energy.

Scattering losses result from microscopic variations about the average material density and local microscopic variations in the composition of the fiber. These variations cause minute variations in the index of refraction n and lead to Rayleigh scattering. In this type of scattering, optical energy radiates away from the fiber. Other scattering losses result when the fiber is strained as in bending. Losses in fibers are discussed in many places; see [12-6, p. 26] and [12-5, Chapters 2, 3].

12.3 SOURCES FOR OPTICAL FIBER COMMUNICATION

In this section we shall briefly discuss, in semiqualitative fashion, the basic principles of operation of optical sources. There is an extraordinarily large volume of literature on this subject,† and the rapid rate of progress in the technology almost assures that any particular configuration discussed here will be outmoded in the near future. It is assumed here and in Sec. 12.4 on optical detectors that the reader is familiar with the basic principles of semiconductor physics.

Principle of Light Emission

When an atom or electron makes a transition from an energy level E_2 to an energy level E_1, the radiation at frequency

$$\nu = \frac{E_2 - E_1}{h} \tag{12.3-1}$$

† For example, see Chapter 16 and its references in [12-5]. A good discussion is also furnished in [12-4, Chapter 5].

is either absorbed or emitted. In Eq. (12.3-1), h is Planck's constant (6.625×10^{-34} J-s). In optical communications the wavelengths of greatest interest are in the range of 0.8 to 0.9 μm and 1.0 to 1.6 μm. In these ranges, there exist either good sources or low-loss optical fibers. Moreover, at 1.3 μm material dispersion is absent in silica, and the potential for large information capacity exists at that wavelength using this substance.

To be specific, assume that $E_2 > E_1$. Then an atom of energy level E_2, upon decaying spontaneously to energy level E_1, emits a photon of frequency $(E_2 - E_1)/h$. This process is called *spontaneous emission* and forms the basis for radiation from thermal sources.

It is also possible for an atom at level E_1 to absorb a photon and move to a higher energy level E_2. This process is called *resonant absorption*. Finally, it is possible for an incident photon to induce an atom at E_2 to decay to E_1 and emit a photon. This new photon has the same phase and frequency of the incident photon and is said to be *coherent* with the incident photon. This process is called *stimulated emission*. If this process is repeated many times, it leads to *light amplification*. However, light amplification cannot be sustained in thermal equilibrium because the number of atoms at E_2 would be rapidly depleted. At thermal equilibrium, the number of atoms, N_2, in level E_2, is related to the number of atoms, N_1, in level E_1 by the Boltzmann distribution

$$N_2 = N_1 \exp\left[\frac{-\Delta E}{kT}\right], \tag{12.3-2}$$

where k is the Boltzmann constant (1.38×10^{-23} J/K), T is the absolute temperature in degrees Kelvins, and $\Delta E \equiv E_2 - E_1$. From Eq. (12.3-2), we see that $N_2 > N_1$.

A requirement for laser radiation is *population inversion* (i.e., to make $N_2 > N_1$). One method of achieving a population inversion is to apply a large amount of external energy and obtain $N_2 > N_1$ through resonant absorption (i.e., excite the atoms at E_1 to level E_2).

To generate and maintain a large photon flux by stimulated emission, optical feedback is used. This feedback is realized by a resonant cavity with mirrors at each end; one of the mirrors (the output aperture) is partially transmitting. Photons bouncing back and forth between the mirrors generate new photons by stimulated emission. The steady state is reached when the internal losses plus the transmitted energy equals the energy supplied by the external source.

In optical communications the two principal source types are semiconductor light-emitting diodes (LED) and semiconductor lasers. In a semiconductor, the higher energy state, E_2, is associated with the conduction band, while the lower energy state, E_1, is associated with the valence band. Between them is a largely unoccupied or "forbidden" energy region called the *bandgap* or *energy gap* E_g. When an electron in E_2 decays to E_1, it produces radiation at an approximate wavelength

$$\lambda = \frac{hc}{E_g} \sim \frac{1.2398}{E_g}, \tag{12.3-3}$$

where λ is in micrometers (μm) and E_g is in electron volts (eV). In the valence band, the electron recombines with a *hole*, the whole process being called *radiative electron transfer*.

Radiative electron transfer is much more probable when the minimum energy of the conduction band occurs at the same electronic wave number as the maximum of the valence band (direct transition). Radiative electron transfer is highly improbable when this condition is not met, such as in materials like silicon (Si) or germanium (Ge). The direct transition condition is met in gallium arsenide (GaAs) and the ternary (three elements) mixed crystals such as $Ga_{1-x}Al_xAs$ ($0 \leq x \leq 1$). Here x is called the mole fraction, and for, say, $x = 0.1$ the group III elements consist of 90% gallium and 10% aluminium that combines with 100% of the group V element arsenic. By varying x and/or changing elements (e.g., replacing aluminium by indium) different emission frequencies can be obtained ([12-4], p. 54].

Light-Emitting Diodes

The structure of a light-emitting diode is shown in Fig. 12.3-1. This type of LED is called a double heterostructure (or *double heterojunction*), *double* because there are two junctions and *hetero* because the materials on either side of the junctions are different. The principle of operation is as follows. Under forward bias, the majority carriers (holes) in the *p*-type $Ga_{1-x}Al_xAs$ move toward the junction region from the left. The junction

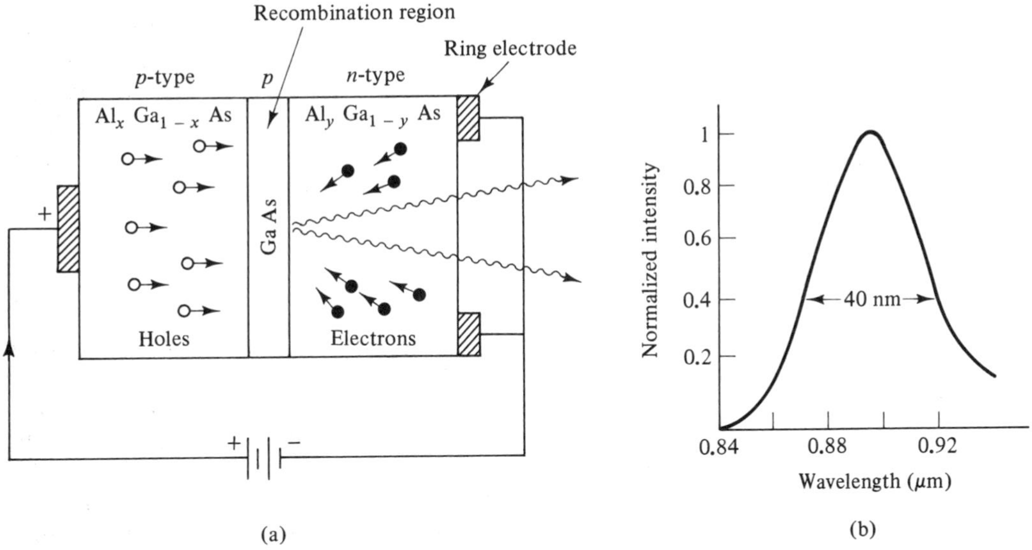

Figure 12.3-1. (a) Structure of a double-heterostructure LED. (b) A typical spectrum of a GaAs LED emission. The center frequency is 0.89 μm and the width is around 40 nm. (From Y. Suematsu and K.-I. Iga, *Introduction to Optical Fiber Communications*, Wiley, New York, 1982. By permission.)

TABLE 12.3-1 STRUCTURE OF A GaAlAs LED

Function	Material	Thickness (μm)
Metal contact layer		
Used for better metal contact	*p*-type GaAs	~1
Light guiding and carrier confinement	*p*-type $Ga_{1-x}As_xP$	~1
Recombination region	*n*-type $Ga_{1-y}As_yP$	~0.3
Light guiding and carrier confinement	*n*-type $Ga_{1-x}As_xP$	~1
Substrate	*n*-type GaAs	
Metal contact layer		

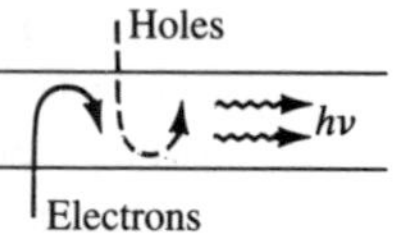

From G. Keiser, *Optical Fiber Communications*, McGraw-Hill, New York, 1983. By permission).

region is *p*-type GaAs. The majority carriers (electrons) in *n*-type $Al_yG_{1-y}As$ move toward the junction region from the right. Radiative recombination takes place in the junction because electrons cannot diffuse into the *p*-type $Ga_{1-x}Al_xAs$ due to the barrier potential at the $Ga_{1-x}Al_xAs$/GaAs junction. Hence luminescence occurs only in the GaAs layer due to *carrier confinement. Optical confinement* is achieved by adjusting the indexes of refraction of the layers so that the total internal reflection is realized in the recombination region.† The wavelength of radiation is that associated with the bandgap energy of GaAs. The emitted photons are not reabsorbed because the bandgap of GaAs is small compared to the bandgap of the GaAlAs layers. Table 12.3-1 lists the adjoining layers, their functions, and typical thicknesses of a GaAlAs double heterostructure light emitter. In the structure, $x > y$ to allow for carrier and optical confinement. Figure 12.3-2 shows a schematic of a double heterostructure LED coupled to a fiber.

Semiconductor Laser Sources

For optical communication systems requiring very large bandwidths (say > 50 MHz) and/or those using single-mode fibers, semiconductor injection lasers are preferred to LEDs. A major difference between LED emission and that of semiconductor lasers is that with the latter the emission spectrum is very narrow and the radiance much more directional. Typically, the spectral width of a LED might be 30 to 40 nm, while for a laser above threshold (to be explained shortly) the width might be less than 3 nm. The reason for this is that laser emission involves stimulated emission in a resonant cavity, which is an extremely frequency selective process. Earlier we stated that lasing required

† An optical waveguide is formed by keeping the indexes of refraction on either side of the active (recombination) region a few percent (5 to 10) less than that of the active region.

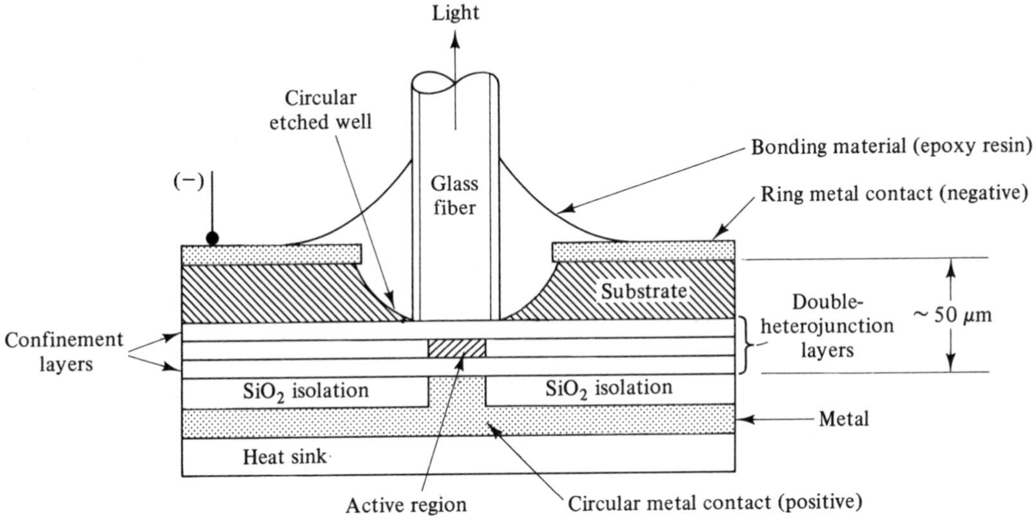

Figure 12.3-2. A high radiance surface emitting LED. The active region is 50 μm in diameter and up to 2.5 μm thick. The emission pattern is Lambertian (spatially incoherent). (Adapted from Chapter 16 of reference [12-5], by permission.)

optical feedback and a population inversion. In a semiconductor laser, optical feedback is achieved with parallel mirrors at each end of a tiny optical cavity within the material, and population inversion is achieved by injecting a current into the material, which fills the lower-energy states of the conduction band with carriers. The structure of the semiconductor laser has many similarities to that of the semiconductor LED. Like the LED, an often-used configuration is that of a layered double heterostructure, as in Fig. 12.3-2. A major difference is the presence of the parallel mirrors in the laser, which are realized by making two parallel cleaves (one for each mirror) along natural cleavage planes of the semiconductor crystal.

When a small external "pump" (i.e., bias current) is applied to a semiconductor laser, lasing does not begin immediately. Lasing begins only when the pump energy generates enough new photons to account for all the losses within the material. Thus laser emission does not occur until the bias current reaches a threshold value. The emitted radiation below the threshold current is broadband and Lambertian (incoherent); above the threshold value it is narrowband and directional (coherent).

As photons bounce back and forth between the cleaved mirrors, new photons are generated by stimulated emission. As already stated, these new photons are coherent with the incident photons (same frequency and phase). Other photons are absorbed or radiated from the cavity. When the light beam makes a complete pass, the number of gained photons must at least equal the number of lost photons. This is a necessary condition for lasing.

The required fractional gain in energy can be easily computed. Let the energy reflectivities of the two mirrors be $r_1 = r_2 \equiv r$ and assume a cavity of length L. If α is the linear absorption coefficient, then the fraction of energy lost per round trip is

$$r^2 e^{-2\alpha L}.$$

If g denotes the linear gain coefficient, then the fractional energy gain per round trip is

$$e^{2gL}.$$

Hence, for the onset of oscillations we require that

$$r^2 e^{2g_{th}L} \cdot e^{-2\alpha L} = 1$$

or

$$g_{th} = \alpha - \frac{1}{2L} \ln r^2. \tag{12.3-4}$$

Typical values of α, L, and r for a GaAs semiconductor laser are 10 cm^{-1}, 300 μm, and 0.31, respectively, yielding a threshold gain of 49 cm^{-1} ([12-4], p. 59). Figure 12.3-3 shows the structure of a double-heterostructure semiconductor laser, while Fig. 12.3-4 compares the radiance output as a function of current input of a typical LED with that of a semiconductor laser. Scanning electron microscope images of the junction region of semiconductor lasers are given in [12-4, p. 66].

Modulation of optical sources is done either *directly* or by *external* means. Direct modulation means that the injection current to the LED or semiconductor laser is varied in proportion to the applied signal. If these variations are small and confined to the linear region of the optical-output, current-input response curve, a linear modulation of light intensity is possible. When fiber-optical systems are used in the digital mode, the source output consists of on–off pulses and linearity is less of an issue. In external

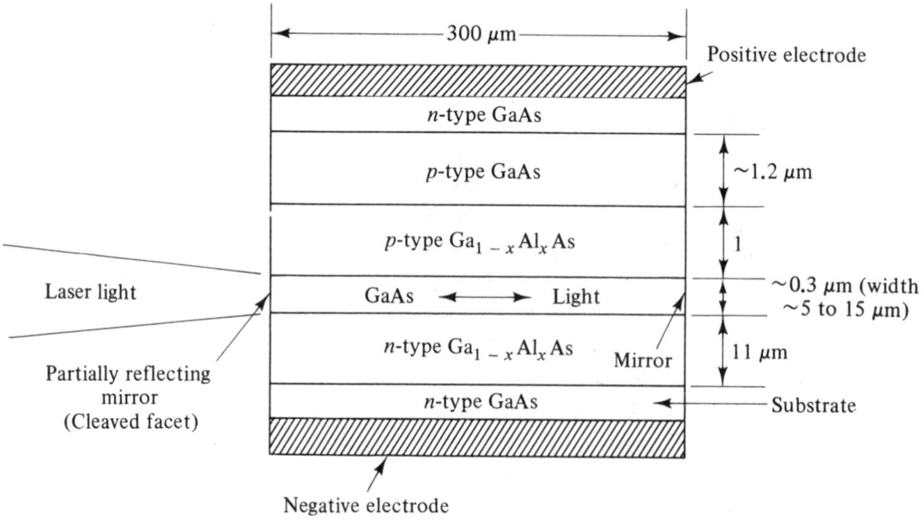

Figure 12.3-3. Structure of a double-heterostructure semiconductor laser using a stripe geometry.

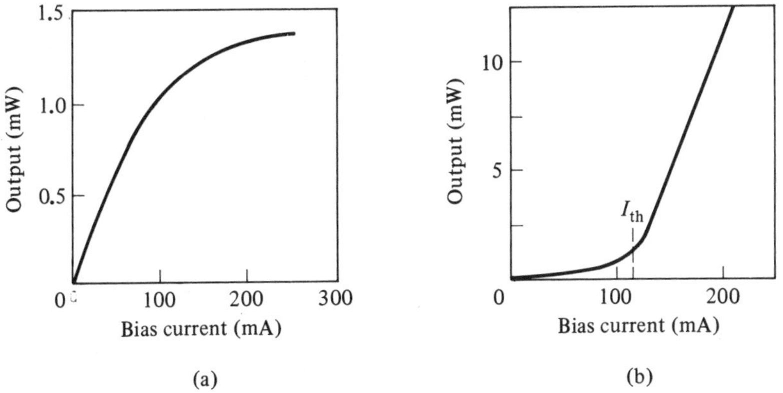

Figure 12.3-4. Radiance versus applied current for (a) a semiconductor LED and (b) a semiconductor double-heterostructure laser. (From Y. Suematsu and K.-I. Iga, *Introduction to Optical Fiber Communications*, Wiley, New York, 1982. By permission.)

modulation, the output power of the source is generally not varied by the injection current. Instead, an external modulator modulates the light beam emitted from the source. The external modulator is controlled by the signal source. External modulation of light can be done in a variety of ways. We give an example of *electrooptical amplitude modulation*. When an electrical signal $s(t)$ (e.g., the modulating signal) is applied to an electrooptical crystal, the polarization[†] of a beam of light passing through the crystal is rotated in accordance with the amplitude of the applied electrical signal. This phenomenon is sometimes called electrically induced birefringence. A polarizer plate at the output end of the crystal only passes light that is polarized in a given direction. Thus the variations in $s(t)$ cause a corresponding variation in the polarization rotation of the light passing through the crystal. In turn, the intensity of the light emerging from the polarizer plate varies with $s(t)$. Maximum light is transmitted when the polarization induced by $s(t)$ is parallel to the fixed polarization direction of the polarizer plate; minimum light is transmitted when $s(t)$ induces a polarization orthogonal to that of the plate. If the signal level is not too large, the output light intensity can be made to linearly follow the input modulation amplitude.

The higher speed and greater flexibility allowed by external modulators are balanced by the simplicity and low power requirements of direct modulation.

12.4 OPTICAL DETECTORS

Optical detectors in fiber-optical communication systems should have the following characteristics: high responsivity[‡] at the operating wavelength; high bandwidth; minimum

[†] Polarization refers to the direction of vibration of the electric field of an electromagnetic wave (the light beam). A good discussion of this subject is furnished in Amnon Yariv, *Optical Electronics*, 2nd ed., Holt, Rinehart and Winston, New York, 1976.

[‡] To be defined shortly.

additional noise currents added by the detector; low susceptibility of performance characteristics to changes in ambient conditions; and small size and low power. The two main types of detectors of use in optical fiber communication systems are semiconductor *p-i-n photodiodes*† and semiconductor *avalanche photodiodes*. In certain wavelength ranges, these devices possess many of the desirable characteristics just listed. Among the most important wavelengths are those in the range from 0.8 to 0.9 μm, where AlGaAs lasers and LEDs have their emission lines, and the range from 1.1 to 1.6 μm, where optical fibers exhibit low losses and very low material dispersion. Silicon detectors are quite responsive to wavelengths in the range from 0.4 to 0.9 μm but are not suitable above 1.0 μm. For operation near 1.3 μm (recall the extremely low material dispersion of silica fibers near this wavelength), germanium and *ternary* or *quaternary*‡ semiconductors such InGaAsP and GaAsSb are suitable.

Two important figures of merit in use in evaluating photodiodes are (1) the quantum efficiency η and (2) the responsivity R. The quantum efficiency η is defined as

$$\begin{aligned}\eta &\equiv \frac{\text{number of carrier pairs generated}}{\text{number of incident photons}}\\ &= \frac{I_p/e}{P_o/h\nu}\\ &\leq 1,\end{aligned} \tag{12.4-1}$$

where I_p is the primary current generated by the absorption of incident photons, P_o is the incident optical power, e is the electronic charge (1.602×10^{-19} coulombs), and $h\nu$ is the energy of an incident photon ($= 1.24/\lambda$ eV, where λ is expressed in micrometers). Typical values of η range from 0.3 to 0.95.

The responsivity, R, is defined as

$$\begin{aligned}R &\equiv \frac{\text{primary photocurrent}}{\text{incident optical power}} = \frac{I_p}{P_o}\\ &= \frac{\eta e}{h\nu}.\end{aligned} \tag{12.4-2}$$

For an *ideal* photodiode, $\eta = 1$ and $R = \lambda/1.24$ amperes per watt (A/W) when λ is in micrometers. Typical values of R are 0.65 μA/μW for silicon at 0.8 μm and 0.45 μA/μW for germanium at 1.3 μm ([12-3], p. 149).

p-i-n *Photodiodes*

The principle of the *p-i-n* photodiode (PIN-PD) operation can be understood by referring to Fig. 12.4-1. Incident photons of energy $h\nu \geq E_g$ (E_g is the bandgap of the semiconductor material) generate hole–electron photocarriers by breaking predominantly covalent

† *p*-type, *i*ntrinsic, *n*-type, written either as *p-i-n* or *PIN* in the literature.

‡ Crystal components consisting of three (ternary) or four (quaternary) elements.

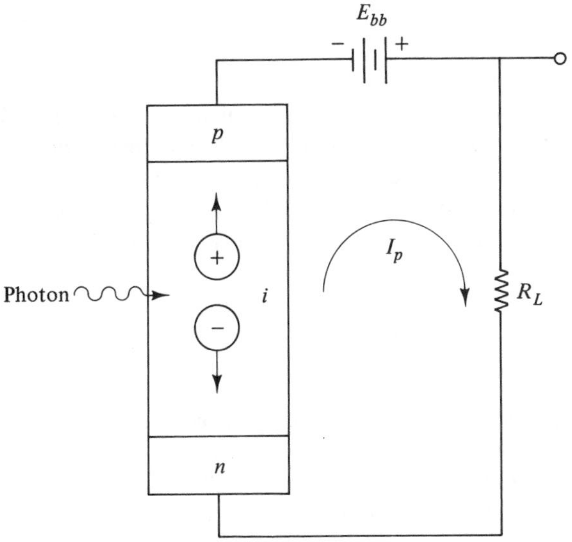

Figure 12.4-1. Circuit diagram showing reverse-biased *p-i-n* photodiode.

bonds. The generation of photocarriers occurs primarily in the intrinsic layer, which also coincides with the *depletion layer*. The depletion region is so called because, under normal operating conditions of reverse bias, any free carriers there are removed by the action of the attracting electric field generated by the applied bias voltage E_{bb}. The photocarriers generated by the incident optical field drift across the junctions and constitute the received signal (i.e., the photocurrent I_p).

Avalanche Photodiode

The main difference between the PIN-PD and avalanche photodiode (APD) is that the latter incorporates an internal gain mechanism that greatly increases the number of free carriers above that of the original photocarriers generated by the incident illumination. The gain mechanism is realized by generating an extremely high internal electric field that imparts to the photocarrier enough energy so that when it collides with a bound electron in the valence band it can convert it to a free electron. This phenomenon is known as *impact ionization*. The free electron in turn can, by impact ionization, generate other free electrons, and so on, thereby generating a large current gain (the *avalanche* effect). A measure of the average current gain is the amplification G defined† as

$$G \equiv \frac{I_M}{I_P},$$

where I_M is the average value of the secondary (i.e., multiplied) current and I_p is initial photocurrent. G can be estimated from

† The internal gain g is a random process. Because g is not constant, it leads to fluctuations in the secondary current, which is called *excess gain noise*. We think of G as an average gain.

$$G = \frac{1}{1 - (V_{bb}/V_B)^k} \qquad (k \text{ between 3 and 6}),$$

where V_{bb} is the reversed biased voltage across the junction of the APD, V_B is the breakdown voltage ($V_{bb} < V_B$), and k is a fitting factor that depends on the material and the structure. Typical values of G are in the range from 30 to 100, although gains as high as 10,000 have been observed ([12-5], p. 605). A thorough discussion of photodetectors in optical fiber communication systems is furnished in ([12-5], Chapter 18).

12.5 COHERENCE

The information capacity of the fiber-optical channel depends on the coherence of the source illumination. Continuously operating gas lasers (e.g., the helium–neon laser) are among the most coherent of sources, whereas extended broadband thermal sources are among the least coherent. Among the semiconductor sources discussed in Sec. 12.3, the LED is spatially and temporally much less coherent than the semiconductor laser. As we shall see in Sec. 12.6, the broadening of a pulse propagating in a fiber that exhibits *material dispersion* depends inversely on the coherence of the source (i.e., the more coherent the source, the less broadening is observed). This would seem to speak well for using coherent sources to maximize the information capacity of the fiber. On the other hand, when highly coherent sources are used, adjacent pulses can interact to produce *coherence-induced intersymbol interference* whose intensity may be *greater* than the sum of the intensities of the pulses taken separately and thus may mask the true level of the signal. Since intersymbol interference is a form of noise, this phenomenon tends to reduce the information capacity of the fiber. Thus we see that there are at least two competing mechanisms at work, one tending to increase the information capacity of a fiber, the other to decrease it.

To calculate the effect of source coherence, we must first briefly discuss the notion of coherence. For computing pulse broadening in a fiber, we need only to understand the notion of *temporal coherence*; however, in optical communication systems using free-space propagation, *spatial coherence* is also very important. Hence we shall briefly discuss both.

Analytic Signals

In Chapter 2, we introduced the *complex analytic signal* and showed that, for a narrowband signal, the complex analytic signal could be factored into a complex amplitude, representing the variations in amplitude and phase, and an informationless carrier term $\exp[j\omega_0 t]$, representing a constant frequency sinusoid. Thus a source signal of the form $x(t) = A(t)\cos[\omega_0 t + \phi(t)]$ can be represented by the complex analytic signal

$$u(t) = a(t)e^{j\omega_0 t}, \tag{12.5-1}$$

where $a(t) \equiv A(t) \exp[j\phi(t)]$ is the complex envelope or, as is more common in optics, the *complex amplitude*. Clearly, $x(t) = \text{Re } u(t)$, so no information is lost (or gained) by using the complex analytical signal.†

Primarily for mathematical and expository reasons, it is advantageous to describe optical waves by complex analytic signals. This is widely done in the literature and will be done here as well. Another point is that optical signals are inherently four dimensional; that is, they depend on position P (e.g., in Cartesian coordinates x, y, and z) and time t. Hence, to describe an optical signal, we usually write $u(P, t)$ meaning the optical signal at position P and time t.

An extensive discussion of the theory of coherence is furnished in [12-9].

Temporal Coherence

Consider $u(P, t)$ and let it represent an optical disturbance with time-varying complex amplitude $a(P, t)$. We consider the narrowband case where the source has bandwidth $\Delta\nu \ll \nu_0$, where ν_0 is the carrier frequency in hertz. Thus we expect the amplitude and phase of $a(P, t)$ to change at the rate $1/\Delta\nu$. Consider the complex amplitude at two separate instants $a(P, t)$ and $a(P, t + \tau)$. For values of $\tau \ll 1/\Delta\nu$, we expect $a(P, t)$ and $a(P, t + \tau)$ to be highly correlated. For values of $\tau \gg 1/\Delta\nu$, we expect $a(P, t)$ and $a(P, t + \tau)$ to be uncorrelated. We shall refer to $\tau_c \equiv 1/\Delta\nu$ as the *coherence time* of $u(P, t)$†:

$$\tau_c \equiv \frac{1}{\Delta\nu} \tag{12.5-2}$$

For example, a laser with bandwidth 100×10^6 hertz has $\tau_c = 10^{-8}$ second.

The correlation of complex amplitudes and coherence is virtually synonymous. The correlation of complex amplitudes can be measured by a device called a *Michelson interferometer*, and the Fourier transform of the correlation yields the power spectrum of the source. This procedure for determining the emission spectrum of the source is called *Fourier spectroscopy*.

Spatial Coherence

Consider the field $u(\cdot, t)$ at two points P_1, P_2. Assume that the delay of a wave in reaching P_1 is about the same as that for P_2. Or, more precisely, if τ_1 is the delay in reaching P_1 and τ_2 is the delay in reaching P_2, that $\Delta\tau \equiv \tau_1 - \tau_2 \ll \tau_c$. If P_1 and P_2 approach each other, we would expect $u(P_1, t)$ and $u(P_2, t)$ to become more strongly correlated. As they move farther apart, they become less correlated. If P_1 and P_2 are

† Complex analytic signals are sometimes defined in terms of the conjugation of $a(t)$ and $\exp[j\omega t]$; there is no particular advantage to this. For a discussion on the use of analytic signals in optics, see [12-8].

‡ A more precise definition of τ_c exists in [12-10]. Using this precise definition, we would obtain $\tau_c = 1/\Delta\nu$ only for a source with uniform power spectrum over $\Delta\nu$ and $\tau_c = 0.664/\nu$ for a source with a Gaussian-shaped spectrum. For our purposes, $\tau_c = 1/\Delta\nu$ is adequate.

in a plane, this leads to the concept of the *correlation area* of $u(P, t)$ or, more generally, to the concept of *spatial* or *transverse* coherence.

Point sources generate spatially coherent fields; extended sources generally do not. Laser light generally has much higher spatial coherence than extended thermal sources. The spatial coherence of semiconductor laser sources is useful for directing a large amount of energy into a narrow beam that can be coupled efficiently to a single-mode fiber. The mathematical theory of coherence, both spatial and temporal, is based on the *mutual coherence function*.

Mutual Coherence Function

The *mutual coherence function* $\Gamma_{12}(\tau)$ is defined as

$$\Gamma_{12}(\tau) \equiv E\{u[P_1, (t+\tau)]u^*[P_2, t]\}, \tag{12.5-3}$$

where $E(\cdot)$ is the expectation operator. The reader will recognize Eq. (12.5-3) as an extension of the ordinary temporal covariance function of Eq. (9.3-9) except that the mutual coherence function depends on *position* as well as time. Thus it is the cross-correlation of the field at P_1 and time $t + \tau$ with the field at P_2 and time t. If the optical path from the source to P_1 and P_2 is almost the same and we let $\tau = 0$, then $\Gamma_{12}(0) = E\{u(P_1, t)u^*(P_2, t)\}$ and is a measure of the spatial coherence of the field. If, on the other hand, we let $P_1 = P_2 = P$, then $\Gamma_{11}(\tau) = E\{u^*(P, t)u(P, t + \tau)\}$ is a measure of the temporal coherence properties of the field (or source). The function $\Gamma_{11}(\tau)$ is called the *self-coherence function*.

In many instances, the self-coherence function is independent of spatial position P and depends only on time delay τ. We shall assume that this holds for light sources of optical communication systems.† Then we can drop the symbol P and write

$$\Gamma(\tau) \equiv E[u^*(t)u(t+\tau)] \tag{12.5-4}$$

or better yet, after normalization

$$\gamma(\tau) \equiv \frac{\Gamma(\tau)}{\sigma_u^2}, \tag{12.5-5}$$

where $\gamma(\tau)$ is called the *complex degree of coherence*,

$$\sigma_u^2 \equiv E[|u(t)|^2] = E[|u(t+\tau)|^2],$$

and a zero-mean, stationary field is assumed.

From Eq. (12.5-5) we see that $|\gamma(\tau)| \leq 1$. Most sources of use in communication systems are narrowband. Hence, in Eq. (12.5-1), $a(t)$ represents the relatively slow variations in the amplitude and phase fluctuations. From Eqs. (12.5-1), (12.5-4), and (12.5-5) we obtain

† This is, essentially, the property of *cross-spectral purity*, which states that in many instances $\Gamma_{12}(\tau)$ can be factored into a product of two functions, one depending only on P_1 and P_2 and the other depending only on τ ([12-11], Chapter 5).

$$\gamma(\tau) \equiv \rho(\tau)e^{j\omega\tau}, \tag{12.5-6}$$

where

$$\rho(\tau) \equiv E\left[\frac{a*(t)a(t+\tau)}{\sigma_a^2}\right] \tag{12.5-7}$$

and $\sigma_a^2 = E[|a(t)|^2] = E[|a(t+\tau)|^2] = \sigma_u^2$. The quantity $\rho(\tau)$ is identical with the complex degree of coherence except for the factor $\exp[j\omega\tau]$ and represents the normalized correlation of the envelope. It is useful in describing the *source coherence*. Also, σ_a^2 (or σ_u^2) is sometimes called the *intensity* of the field and denoted by the symbol I. Both $\gamma(\tau)$ and $\rho(\tau)$ of Eqs. (12.5-5) and (12.5-7), respectively, are special cases of the *correlation coefficient* introduced in Eq. (8.16-4).

With the help of the material in this section, we can now investigate the factors that affect broadening in a single-mode fiber.

12.6 PULSE TRANSMISSION THROUGH A SINGLE-MODE FIBER

As we discussed in Sec. 12.2, in the fiber-optical channel, pulses broaden primarily because of two phenomena: (1) material dispersion and (2) modal dispersion. Material dispersion refers to the fact that the index of refraction n depends on the wavelength λ, and therefore the speed and, consequently, the delay of the various Fourier components in the fiber are slightly different. Modal dispersion refers to the fact that the group velocities for the different modes differ, and therefore the different modes of the pulse experience different delays. In a single-mode fiber the primary mechanism for pulse broadening is material dispersion. We shall consider the effect of this phenomenon from a systemic point of view.

Transfer Functions

A single-mode time-harmonic wave at frequency ω with transverse variation $B(x, y; \omega)$ propagating down a fiber is described, at a point z, by

$$u_z(x, y, t) = B(x, y; \omega)e^{j(\omega t - Kz)}, \tag{12.6-1}$$

where K is the complex propagation constant given by

$$K = \beta - j\alpha. \tag{12.6-2}$$

In Eq. (12.6-2), β is a real propagation constant or phase constant and α is the attenuation constant. At $z = 0$,

$$u_0(x, y, t) = B(x, y; \omega)e^{j\omega t} \tag{12.6-3}$$

and at $z = L$ (the fiber is assumed to be L meters long),

$$u_L(x, y, t) = B(x, y; \omega)e^{j(\omega t - KL)}. \tag{12.6-4}$$

We are not usually interested in the x, y variation, so we suppress the x, y arguments. Then, at $z = 0$,

$$u_0(t) = B(\omega)e^{j\omega t},$$

and at $z = L$,

$$u_L(t) = B(\omega)e^{j\omega t} \cdot e^{-jKL}.$$

We define the transfer function $H(\omega)$ by $u_L(t)/u_0(t)$ when the input is given by Eq. (12.6-3). Thus

$$H(\omega) = e^{-jKL}. \tag{12.6-5}$$

and

$$u_L(t) = B(\omega)H(\omega)e^{j\omega t}. \tag{12.6-6}$$

From Eq. (12.6-6), it should be clear why $H(\omega)$ is called the transfer function. Now consider a signal $f(t)$ injected into the fiber. We resolve $f(t)$ into harmonic components by using the Fourier transform; that is,

$$f(t) = \int_{-\infty}^{\infty} F(\omega)e^{j\omega t}\frac{d\omega}{2\pi}. \tag{12.6-7}$$

Each harmonic component $F(\omega)e^{j\omega t}$ is acted upon by $H(\omega)$, so the received pulse $q(t)$ at the exit of the fiber is given by

$$q(t) = \int_{-\infty}^{\infty} F(\omega)H(\omega)e^{j\omega t}\frac{d\omega}{2\pi}. \tag{12.6-8}$$

In optical communications, the signal is often a baseband modulation of a high-frequency carrier $e^{j\omega_0 t}$. Then

$$f(t) = a(t)e^{j\omega_0 t}. \tag{12.6-9}$$

The Fourier transform of $f(t)$ is

$$\begin{aligned} F(\omega) &= \int_{-\infty}^{\infty} a(t)e^{-j(\omega-\omega_0)t}\,dt \\ &\equiv A(\omega - \omega_0), \end{aligned}$$

where $A(\omega) = \mathcal{F}[a(t)]$.[†] Thus the received signal is

$$\begin{aligned} q(t) &= \int_{-\infty}^{\infty} A(\omega - \omega_0)H(\omega)e^{j\omega t}\frac{d\omega}{2\pi} \\ &= \left[\int_{-\infty}^{\infty} A(\omega)H(\omega + \omega_0)e^{j\omega t}\frac{d\omega}{2\pi}\right]e^{j\omega_0 t} \qquad (12.6\text{-}10a) \\ &= b(t)e^{j\omega_0 t}. \qquad (12.6\text{-}10b) \end{aligned}$$

[†] Do not confuse $A(\omega)$ in this section with $A(t)$ used in Sec. 12.5 to denote the time-varying amplitude of a carrier signal $x(t)$.

The quantity $b(t)$ is the envelope of the received signal $q(t)$. Hence $H(\omega + \omega_0)$ can be interpreted as the *transfer function for the envelopes of the signals*.

First, we note that if K is constant over supp[†] $A(\omega - \omega_0)$ then $H(\omega)$ will be constant and there will be *no* pulse broadening, although for $\alpha \neq 0$ there will be a uniform attenuation given by $\exp[-\alpha L]$. However, there will always be some material dispersion because the index of refraction n and therefore K or β do depend on frequency [i.e., $K = K(\omega)$ or $\beta = \beta(\omega)$]. Why $\beta(\omega)$ depends on ω if n depends on ω will be shortly discussed. We shall consider only low-loss fibers, so we set $\alpha = 0$ and see how $\beta(\omega)$ affects the transmission capacity of the channel. We write

$$H(\omega) = e^{-j\beta(\omega)L},$$
$$H(\omega + \omega_0) = e^{-j\beta(\omega+\omega_0)L},$$

and

$$\beta(\omega) = \beta(\omega_0) + \beta'(\omega_0)(\omega - \omega_0) + \frac{1}{2!}\beta''(\omega_0)(\omega - \omega_0)^2 + \cdots. \tag{12.6-11}$$

Equation (12.6-11) is a Taylor series expansion of $\beta(\omega)$ about $\omega = \omega_0$. From Eq. (12.6-11), we obtain

$$\beta(\omega + \omega_0) = \beta(\omega_0) + \omega\beta'(\omega_0) + \frac{\omega^2}{2!}\beta''(\omega_0) + \cdots + \frac{\omega^m\beta^{(m)}(\omega_0)}{m!} + \cdots. \tag{12.6-12}$$

Most of the time (but not always) the significant dispersion can be accounted for by approximating $\beta(\omega + \omega_0)$ with the first three ($m = 2$) terms. Thus, with $H_e(\omega) \equiv H(\omega + \omega_0)$ denoting the envelope transfer function, we obtain

$$H_e(\omega) = \exp\left[-j[\beta_0 + \beta_0'\omega + \frac{1}{2}\beta_0''\omega^2]L\right], \tag{12.6-13}$$

where $\beta_0 \equiv \beta(\omega_0)$, $\beta_0' \equiv \beta'(\omega_0)$, and so on.

Propagation Constants

The propagation constant of guided modes can be shown to lie in the range ([12-5], p. 39)

$$n_2 k < \beta < n_1 k,$$

where n_2 is the index of refraction of the cladding, n_1 is the index of refraction of the core, and k is the propagation constant of plane waves in a vacuum (i.e., $k = 2\pi/\lambda$). Optical fibers used for communication purposes usually satisfy $(n_1 - n_2)/n_1 \ll 1$. In such cases, it has been shown ([12-12]) that

$$\beta \simeq \frac{2\pi n_2}{\lambda}.$$

[†] The support is the set of ω's such that $A(\omega - \omega_0) \neq 0$.

Hence, if n_2 depends on ω, then β will depend on ω, and so material dispersion induces a frequency dependence in the propagation constant of guided modes.

Effective Envelope Transfer Function

When we use Eq. (12.6-13) in Eq. (12.6-10a), we obtain

$$q(t) = e^{j(\omega_0 t - \beta_0 L)} \int_{-\infty}^{\infty} A(\omega) \exp\left[-j\left(\beta_0'\omega + \frac{1}{2}\beta_0''\omega^2\right)L\right] e^{j\omega t} \frac{d\omega}{2\pi}. \tag{12.6-14}$$

The factor outside the integral represents the carrier delay and is of no consequence. Recalling that the group velocity is $v_g \equiv [d\beta_0/d\omega]^{-1}$, we can rewrite the integral as

$$\int_{-\infty}^{\infty} A(\omega)\, e^{-j\beta_0'' L\omega^2/2}\, e^{j\omega(t-L/v_g)} \frac{d\omega}{2\pi}. \tag{12.6-15}$$

Thus the only effect of the $\beta_0'\omega$ term in Eq. (12.6-13) is to introduce an envelope delay of $\tau_g \equiv L/v_g$. Hence the broadening of the envelope results only from the effective envelope transfer function:

$$\tilde{H}(\omega) \equiv e^{-j\beta_0'' L\omega^2/2}. \tag{12.6-16}$$

The impulse response of the envelope is directly obtained from Eq. (12.6-14). Recalling that $q(t) = b(t)e^{j\omega_0 t}$, we obtain from Eqs. (12.6-14) and (12.6-15)

$$b(t) = e^{-j\beta_0 L} \int_{-\infty}^{\infty} A(\omega) e^{-j\beta_0'' L\omega^2/2}\, e^{j\omega(t-L/v_g)} \frac{d\omega}{2\pi}.$$

Since we are only concerned with the effect of the fiber on the broadening of the pulse, we ignore the delay L/v_g and replace $t - L/v_g$ inside the integral† by t; also $e^{-j\beta_0 L}$ is a constant factor of no consequence. Hence the undelayed but broadened envelope is computed from

$$\begin{aligned} b(t) &= \int_{-\infty}^{\infty} A(\omega) e^{-j\beta_0'' L\omega^2/2}\, e^{j\omega t} \frac{d\omega}{2\pi} \\ &= \int_{-\infty}^{\infty} a(\xi)\tilde{h}(t-\xi)\, d\xi, \end{aligned} \tag{12.6-17a}$$

where $\tilde{h}(t) \equiv \mathcal{F}^{-1}[\tilde{H}(\omega)]$ and is given by

$$\tilde{h}(t) = K_1 e^{jt^2/2\beta_0'' L}. \tag{12.6-17b}$$

The constant K_1 is unimportant in the analysis, and we set it to unity as we do with other constants of no consequence.

† Replacing $t - L/v_g$ by t is, in effect, setting the delay to zero.

Received Pulse Intensity

The average signal current generated at the detector is proportional† to the received optical intensity defined as

$$I_o(t) = E[|b(t)|^2], \tag{12.6-18}$$

where the expectation operator $E[\cdot]$ is required because the source signal is a random process. From Eqs. (12.6-17) and (12.6-18), we get

$$I_o(t) = \int_{-\infty}^{\infty}\int_{-\infty}^{\infty} E[a(t')a^*(t'')]h(t - t')h(t - t'')\,dt'\,dt''. \tag{12.6-19}$$

It is important to recall that $a(t)$ is the envelope of a *modulated carrier*. Let $a_s(t)$ be the envelope of the *unmodulated carrier*, and let $z(t)$ be the modulating signal. For convenience, we assume $z(t)$ is real (i.e., amplitude modulation); then the intensity of the modulating signal is $m(t) \equiv z^2(t)$. Thus

$$a(t) = z(t)a_s(t)$$

and

$$E[a(t')a^*(t'')] = z(t')z(t'')E[a_s(t')a_s^*(t'')].$$

With

$$\sigma_s^2 \triangleq E[|a_s(t')|^2] = E[|a_s(t'')|^2]$$

defined as the source intensity, we obtain, from Eq. (12.5-7),

$$E[a(t')a^*(t'')] = \sigma_s^2\rho_s(t' - t'')z(t')z(t'')$$

and, with σ_s^2 set equal to unity for convenience, we can rewrite Eq. (12.6-19) as

$$I_o(t) = \int_{-\infty}^{\infty}\int_{-\infty}^{\infty} [m(t')]^{1/2}[m(t'')]^{1/2}\rho_s(t' - t'')\tilde{h}(t - t')\tilde{h}^*(t - t'')\,dt'\,dt'' \tag{12.6-20a}$$

$$= \int_{-\infty}^{\infty}\int_{-\infty}^{\infty} z(t')z(t'')\rho_s(t' - t'')\tilde{h}(t - t')\tilde{h}^*(t - t'')\,dt'\,dt''. \tag{12.6-20b}$$

Equation (12.6-20) is a fundamental result; it shows that the received pulse intensity variation depends on the source modulation $z(t)$, the source coherence $\rho_s(\tau)$, and the envelope-related impulse response of the fiber $\tilde{h}(t)$. For $\tilde{h}(t)$, we take Eq. (12.6-17b), that is, $\tilde{h}(t) = e^{jt^2/2\beta_0''L}$. To evaluate Eq. (12.6-20), we must assume a particular modulating intensity $m(t) = z^2(t)$ and a particular source coherence function $\rho_s(\tau)$. It is common to choose a Gaussian profile for both. Thus

$$\rho_s(\tau) = e^{-(\tau^2/2\tau_c^2)} \tag{12.6-21}$$

† This assumes that the detector area is very small (a point detector) or that the intensity is uniform across the surface of the detector. Otherwise, the signal current is proportional to the integral of $I_o(t)$ across the detector surface.

and

$$z(t) = e^{-(t^2/4\sigma_m^2)} = [m(t)]^{1/2}. \tag{12.6-22}$$

As already stated, the source coherence time is $\tau_c = [\Delta\nu]^{-1}$ (i.e., the reciprocal of the bandwidth). If we now insert Eqs. (12.6-21) and (12.6-22) into Eq. (12.6-20), we obtain the result we seek:

$$I_o(t) = K_2 e^{-(t^2/2\sigma_R^2)}, \tag{12.6-23}$$

and where σ_R denotes the width of the received pulse and is given by

$$\sigma_R^2 = \sigma_m^2 + \frac{(\beta_0'' L)^2}{4\sigma_m^2} + \frac{(\beta_0'' L)^2}{\tau_c^2} \tag{12.6-24}$$

and K_2 is a constant. This result shows that one of the components of the pulse-broadening phenomenon is due to the bandwidth/coherence properties of the source. Several interesting points can now be deduced from Eq. (12.6-24): (1) The spread of the pulse always increases when the finite bandwidth of the source is not ignored;† (2) incoherent sources (τ_c small) broaden the signals more than coherent sources (τ_c large); and (3) broadening depends strongly on the product of the length of the fiber and the material dispersion coefficient β_0''. In theory, if a frequency can be found at which $\beta_0'' = 0$, then no pulse broadening would occur even for noncoherent sources.

It is somewhat disconcerting to observe that the pulse broadening goes to infinity if $\tau_c \to 0$ (i.e., an absolutely incoherent source). However, such sources do not exist in nature and if they did they could not radiate net energy according to the theory of wave propagation ([12-11], p. 206).

While Eq. (12.6-24) describes the pulse broadening in single-mode or material-dispersion-dominated fibers it does not yield the capacity of the single-mode fiber channel. To do this, we must consider a sequence of pulses. This calculation is done in the next section.

12.7 CAPACITY OF THE FIBER-OPTICAL CHANNEL

We consider a digital fiber-optical system in which material dispersion is the phenomenon that constrains the fiber capacity. Figure 12.7-1 shows such a system: a sequence of *ones* and *zeros* are encoded by a pulse modulator that imparts on–off pulses to a carrier signal. The carrier signal is generated by a LED or laser dioder. After the optical signal is detected by a photodiode, the induced electric signal is amplified, shaped, and compared with a threshold. A decision is then made whether a *one* or a *zero* was sent.

There are several ways that the capacity of a single-mode fiber can be defined.

† To ignore the bandwidth of the source, set $a_s(t) = 1$ for all t (i.e., a monochromatic source). Such a source has zero bandwidth and $\tau_c = \infty$. A gain-stabilized continuous gas laser is among those sources coming closest to this idealization.

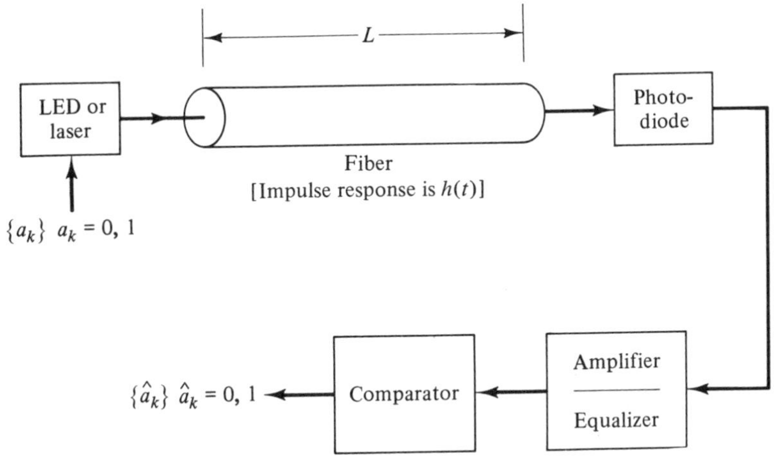

Figure 12.7-1. Fiber-optical digital channel.

One way to compute the capacity is in information-theoric terms in the manner of Sec. 11.5, but such a calculation is beyond the scope of this discussion. Another way is to send pulses closer and closer together until the error probability (i.e., decoding a zero as a one, or vice versa) becomes unacceptable. However, this calculation is quite lengthy. Still another, easier way is to push the pulses closer and closer together until the intersymbol interference becomes unacceptable. This is the approach we take here. It is essentially the same approach as that used in Sec. 10.6 to discuss Nyquist signaling in which intersymbol interference is the main concern. Also, we shall disregard thermal noise generated in the receiver. While such noise is important in limiting the overall capacity of the channel, it does not affect directly the capacity of the fiber.

Consider sending the sequence 101. A *one* is manifest by the presence of a pulse, a *zero* by the absence of a pulse (Fig. 12.7-2). Assume that at the receiver the received waveform is sampled at the instants $t = 0, T/2, T$, and so on. If the sample value exceeds a threshold, then we say a one is received; otherwise we say a zero is received. Now we see that because of pulse spreading, the intersymbol interference (ISI) that occurs at $t = T/2$ could cause a transmitted *zero* to be decoded as a one. While it is true that coherent sources produce less pulse broadening than incoherent sources, the coherent intersymbol interference may in fact be larger than that of the incoherent pulses. Thus it is not clear from these considerations alone what degree of source coherence maximizes the fiber capacity.

Another important practical consideration is that we must allow for some jitter in the sampling times. Thus, while the coherency-induced ISI might be zero exactly at $t = T/2$, it might be quite large at, say $t = T/2 \pm \epsilon$, where ϵ is a small amount of random jitter about the theoretical sampling instant. Hence, in computing the capacity of the fiber based on ISI, we should search for the largest value of ISI in the *vicinity* of the sampling instant. Jitter in the sampling times is of much less concern in the incoherent case because the ISI does not vary rapidly about the sampling instant.

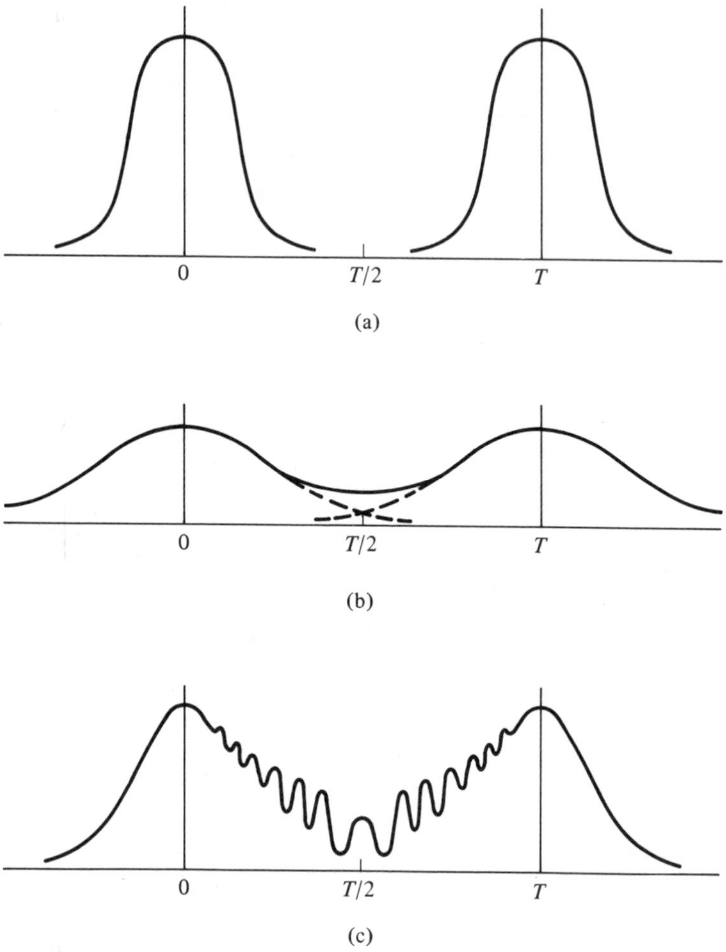

Figure 12.7-2. (a) Transmitted pulses at $z = 0$. (b) Possible configuration of received pulses when source is incoherent. (c) Possible configuration of received pulses when source is coherent. A 101 sequence is being transmitted.

Based on this line of reasoning, we shall compute the maximum ISI at $T/2$ that occurs when two *one* pulses, separated in time by T, are transmitted. The reciprocal of the smallest value of T, consistent with an ISI criterion, is proportional to and indeed will be taken as the transmission capacity.

Transmission Capacity Based on Two-Pulse Intersymbol Interference [12-13]

Consider the transmission of two pulses, separated in time by T so that

$$m(t) = z^2(t) + z^2(t - T). \tag{12.7-1}$$

If the overlap between the two transmitted pulses is insignificant, then

$$m^{1/2}(t) \simeq z(t) + z(t - T). \tag{12.7-2}$$

To compute the received intensity, I_R, we need only insert Eq. (12.7-2) into Eq. (12.6-20a). A straightforward calculation furnishes

$$I_R(t) = I_o(t) + I_o(t - T) + I_I(t), \tag{12.7-3}$$

where $I_o(t)$ is given by Eq. (12.6-23) and $I_I(t)$ is the additional interference term resulting from the coherency of the source and is given by

$$I_I(t) = 2\,\mathrm{Re} \int_{-\infty}^{\infty} \int_{-\infty}^{\infty} z(t')z(t'' - T)\rho_s(t' - t'')\tilde{h}(t - t')\tilde{h}^*(t - t'')\, dt'\, dt''. \tag{12.7-4}$$

In Eq. (12.7-4), Re as always stands for "real part." When Eqs. (12.6-17b) and (12.6-21) are used in Eq. (12.7-4) with $z(t) = e^{-t^2/4\sigma_m^2}$ we obtain [12-13].

$$I_I(t) = 2C_I I_o\left(t - \frac{T}{2}\right) \cos\left[W_I\left(t - \frac{T}{2}\right)\right], \tag{12.7-5}$$

where

$$C_I \triangleq \exp\left[-\xi \frac{T^2}{8\sigma_R^2}\right] \tag{12.7-6}$$

$$\xi = 1 + \frac{(\beta_0'' L)^2}{\sigma_m^2 \tau_c^2} \tag{12.7-7}$$

and

$$W_I = \frac{T\beta_0'' L}{4\sigma_m^2 \sigma_R^2}. \tag{12.7-8}$$

σ_R^2 is given by Eq. (12.6-24). We note several points from this result: (1) When the source is highly incoherent ($\tau_c \to 0$), $C_I \to 0$ and the interference fringes vanish. In that case the ISI is the additive power of the neighboring pulses. (2) The coherence-induced ISI is an oscillating function that has a Gaussian envelope centered midway between the two pulses. The maximum value occurs at $t = T/2$. (3) The frequency W_I of the oscillating function increases with increasing separation between T. We note that, when $T = 0$, $I_R(t)$ is simply $4I_o(t)$ (i.e., when the amplitude is doubled, the intensity increases by a factor of 4).

Let us now compute the transmission capacity of the system using an ISI threshold criterion. We shall argue that the ISI becomes unacceptable if the ISI signal at $t = T/2$ exceeds the level δ. Thus, calling $T_{\min}$ the smallest spacing between the pulses prior to ISI violating the overlap criterion, we obtain the equation

$$I_R\left(\frac{T_{\min}}{2}\right) = I_o\left(\frac{T_{\min}}{2}\right) + I_o\left(-\frac{T_{\min}}{2}\right) + I_I\left(\frac{T_{\min}}{2}\right) = \delta. \tag{12.7-9}$$

From Eqs. (12.6-23) and (12.7-5), we obtain

$$2e^{-(T_{\min}/8\sigma_R^2)}\left[1 + \exp\left[-(\xi - 1)\frac{T_{\min}^2}{8\sigma_R^2}\right]\right] = \delta$$

and

$$T_{\min}^2 = 8\sigma_R^2\left\{\ell n\frac{2}{\delta} + \ell n\left[1 + \exp\left[-(\xi - 1)\frac{T_{\min}^2}{8\sigma_R^2}\right]\right]\right\}. \qquad (12.7\text{-}10)$$

This equation cannot be solved explicitly for $T_{\min}$. However, solutions can be obtained in (1) the coherent limit ($\tau_c \to \infty$) and (2) the incoherent limit ($\tau_c \to 0$). In the former we obtain

$$T_{\min}^2 = 8\sigma_R^2\,\ell n\frac{4}{\delta} \qquad \text{(coherent limit)}, \qquad (12.7\text{-}11)$$

while in the latter we obtain

$$T_{\min}^2 = 8\sigma_R^2\,\ell n\frac{2}{\delta} \qquad \text{(incoherent limit)}. \qquad (12.7\text{-}12)$$

The transmission capacity, N, is given by $T_{\min}^{-1}$. The maximum transmission capacity $N = N_{\max}$ is reached when we choose σ_m to satisfy

$$\frac{dN}{d\sigma_m} = 0$$

or

$$\frac{dN}{d\sigma_R^2}\frac{d\sigma_R^2}{d\sigma_m} = 0. \qquad (12.7\text{-}13)$$

Since $dN/d\sigma_R^2 \neq 0$ for the N's resulting from Eqs. (12.7-11) and (12.7-12), the optimum pulse width results from solving $d\sigma_R^2/d\sigma_m = 0$. Fron Eq. (12.6-24), this occurs when $\sigma_m^2 = \beta_0''L/2$, and $N_{\max}$ results when this value of σ_m^2 is used in σ_R^2. For the coherent limit, the smallest σ_R^2 is $\sigma_R^2 = \beta_0''L$. Then

$$N_{\max} = \left[8\beta_0''L\,\ell n\frac{4}{\delta}\right]^{-1/2} \qquad \text{(coherent limit)}. \qquad (12.7\text{-}14)$$

For the incoherent limit, assume, to get a reasonable answer, that $\tau_c \ll \beta_0''L/\sigma_m$. Then $\sigma_R^2 = (\beta_0''L)^2/\tau_c^2$ and

$$N_{\max} = \left[2\Delta\nu\beta_0''L\left(2\ell n\frac{2}{\delta}\right)^{1/2}\right]^{-1} \qquad \text{(incoherent limit)}, \qquad (12.7\text{-}15)$$

where $\Delta\nu$ is the source bandwidth in Hertz. Figure 12.7-3 shows $N_{\max}$ computed from Eq. (12.7-10) (for $\sigma_m^2 = \beta_0''L$) for various values of $\Delta\lambda$ and L. The results shown in Fig. 12.7-3 are for a $\beta_0'' = 3 \times 10^{-26}$ s^2/m, $\lambda = 1$ μm, and $\delta = 0.20$. Here δ is the ratio of the ISI at a point midway between the two pulses to the peak pulse amplitude.

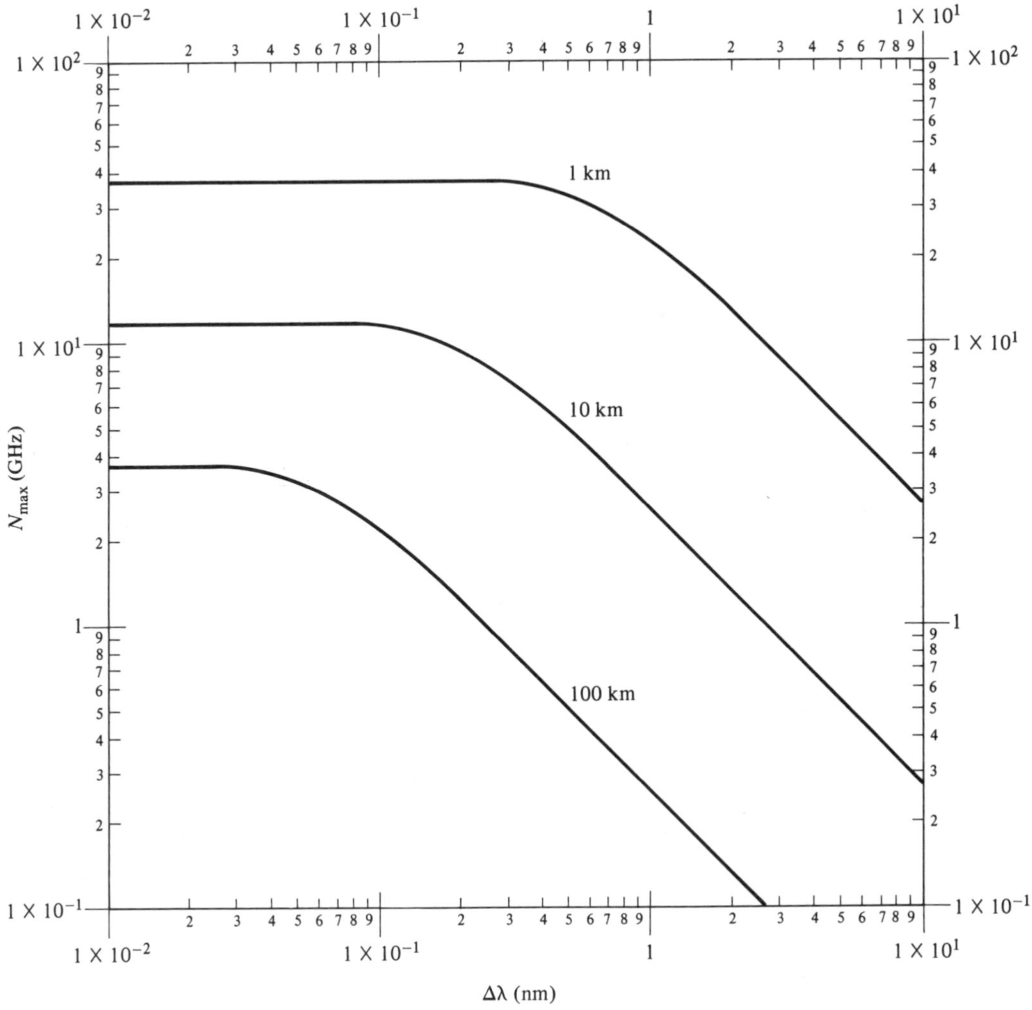

Figure 12.7-3. Maximum permissible pulse rate versus source line width $\Delta\lambda$ for the shown fiber lengths L. $\delta = 0.20$, $\beta = 3 \times 10^{-26}$ s^2/m, and $\lambda = 1$ μm.

12.8 OPTIMUM RECEPTION OF OPTICAL SIGNALS

Consider a rather typical scenario by which optical radiation gets converted to an electrical signal at a *p-n* junction: (1) optical illumination falls upon the depletion region of a *p-n* junction operated in a reverse-bias mode; (2) absorbed photons increase the energy of bound electrons from the valence to conduction bands. The free electron–hole pairs, known as *photocarriers*, flow into an external circuit and constitute the *photocurrent*; and (3) the photocurrent generates a voltage across a load or bias resistor, which constitutes the raw signal from which information is ultimately extracted.

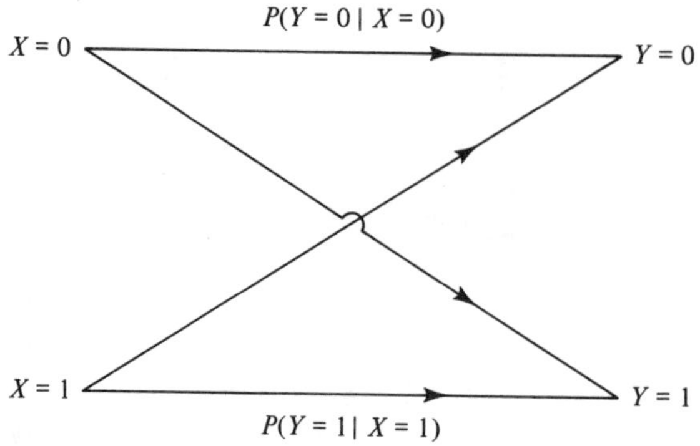

Figure 12.8-1. Binary symmetric channel.

When the signals are weak and/or corrupted by noise, the question arises as to how best to process the detected electrical signal so as to minimize some criterion. Often this criterion is the maximum a posteriori probability (MAP) rule. We have already developed the MAP receiver in Sec. 10.2. However, the case of principal interest there was Gaussian noise. In optics, Gaussian noise is of less interest than Poisson noise. Thus we consider the MAP receiver for this case.

We shall assume that the signal photoelectron (PE) current can be modeled as a Poisson process. This is a good approximation when the source illumination is monochromatic or near-monochromatic laser light ([12-11], Chapter 2). The fluctuations in the signal resulting from the random generation of photoelectrons is known as *quantum* or *shot noise*. Thus the signal that contains the information is intrinsically noisy and represents a fundamental limit of nature. The main other source of noise in a well-designed photodetector is dark current;[†] it results from the thermal generation of photoelectrons or other carriers in the detector when no light is incident on the diode. Since it is unrelated to the signal light, it is a form of noise.

The MAP Rule

Optical fiber channels can be modeled as the binary symmetric channel shown in Fig. 11.5-2(b) and redrawn here for convenience as Fig. 12.8-1. We shall calculate the MAP receiver for this kind of a channel.

Assume that a zero ($X = 0$) or a one ($X = 1$) is sent with probabilities $P[X = 0]$ and $P[X = 1]$, respectively. An output $Y = y_i$ is observed. How should we decide whether a zero or one was sent?

[†] There are other sources of noise: background radiation noise is negligible in fiber optics; *excess gain noise* in avalanche photodiodes results from fluctuations in the internal gain parameter g that represents the ratio of output to primary current. *Leakage current* refers to carriers leaking across the surface of the diode and depends on geometry, bias voltage, and other factors.

Solution

The MAP strategy is to choose that value of $X = x^*$ such that $P(X = x^*|Y = y_i)$ is a maximum. For the sake of specificity, let $y_i = 0$. Then the MAP rule is

$$\text{if} \quad \frac{P[X = 0|Y = 0]}{P[X = 1|Y = 0]} \begin{cases} \geq 1, & \text{then assume } X = 0, \\ < 1, & \text{then assume } X = 1. \end{cases} \tag{12.8-1}$$

If we use Bayes's formula,

$$P[X = x_i|Y = y_j] = \frac{P[Y = y_j|X = x_i]P[X = x_i]}{P(Y = y_j)}, \tag{12.8-2}$$

in Eq. (12.8-1), we obtain the MAP rule in the following form:

$$\frac{P[Y = 0|X = 0]}{P[Y = 0|X = 1]} \begin{cases} \geq K, & \text{assume } X = 0, \\ < K, & \text{assume } X = 1, \end{cases} \tag{12.8-3}$$

where $K \equiv P[X = 1]/P[X = 0]$. The ratio on the left is called the *likelihood ratio* (*LR*) or likelihood function. The value of K is called the threshold. The test $LR \gtrless K$ is called a *likelihood ratio test*. We can now proceed to determine the optimum receiver strategy for a restricted class of signals.

Optimum Demodulation for Piecewise Constant Modulation [12-14]

We observe at the receiver the Poisson rate parameter $\lambda(t)$ over the interval $(-T, T)$. We assume that this rate is the photoelectron rate. At the end of the interval we decide which one of two messages m_1 or m_2 was sent.† At the transmitter, the modulating signal is a *sequence of piecewise constant* Poisson rate parameters $\lambda^{(k)}(t) = \{\lambda_i^{(k)}\}$, $i = 1, \ldots, N$, if message m_k, $k = 1, 2$, is said to be transmitted. By piecewise linear modulation, we mean that

$$\lambda(t) = \lambda_i, \qquad t_{i-1} \leq t < t_i, \qquad i = 1, \ldots, N, \tag{12.8-4}$$

where $-T = t_0 \leq t_1 \leq t_2 \leq \cdots \leq t_N = T$. Figure 12.8-2 shows two possible modulations for two messages m_1 and m_2.

Since Poisson photoelectrons are assumed, we must assume single-mode laser illumination for the results to be valid. As is discussed in several places, not all illuminations generate Poisson photoelectrons ([12-11], p. 470; [12-15], p. 60).

Assume that dark current is the only source of noise and that its rate parameter is λ_o.‡ If the modulation is $\lambda(t)$, the overall photoelectron rate is $\tilde{\lambda}(t) \equiv \lambda(t) + \lambda_o$

† For example, m_1 and m_2 could denote a pair of binary signals.

‡ This restriction is somewhat artificial. Various independent Poisson noise sources can be combined into a single noise source with overall rate parameter λ_o.

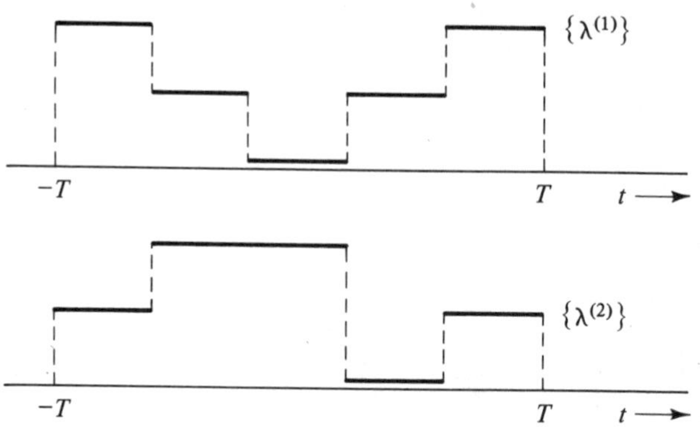

Figure 12.8-2. Two possible rate parameter sequences $\{\lambda^{(1)}\}$ and $\{\lambda^{(2)}\}$ for messages m_1 and m_2, respectively.

since the signal and dark current noises are independent Poisson processes. If $\lambda(t)$ is piecewise linear, so will be $\tilde{\lambda}(t)$, assuming the values of $\lambda_o + \lambda_1$ in (t_0, t_1), $\lambda_o + \lambda_2$ in (t_1, t_2), and so on. The probability $P[n_i]$ of n_i PEs in $\Delta t_i \equiv t_i - t_{i-1} = \Delta t$ (independent of i) is

$$P[n_i] = \frac{[(\lambda_o + \lambda_i)\,\Delta t]^{n_i}}{n_i!} e^{-(\lambda_o+\lambda_i)\Delta t}. \tag{12.8-5}$$

Since the number of emitted PEs in any Δt is independent of the number of emitted PEs in any other Δt (the independent increment assumption of the Poisson process), the joint probability function $P\{n_1 \text{ in } (t_0, t_1); n_2 \text{ in } (t_1, t_2); \ldots n_N \text{ in } (t_{N-1}, t_N)\} \equiv P[\mathbf{n}]$ is given by

$$P[\mathbf{n}] = \prod_{i=1}^{N} \frac{[(\lambda_i + \lambda_o)\,\Delta t]^{n_i}}{n_i!} e^{-(\lambda_i+\lambda_o)\Delta t}. \tag{12.8-6}$$

In the *binary communication system* the observed PE count comes from one of two a priori known signals. Thus, if m_1 is the messages, $\boldsymbol{\lambda}^{(1)} \equiv \{\lambda_i^{(1)}\}$ is the modulation, and if m_2 is sent, $\boldsymbol{\lambda}^{(2)} \equiv \{\lambda_i^{(2)}\}$ is the modulation. The likelihood ratio is

$$LR = \frac{P(\mathbf{n}|\boldsymbol{\lambda}^{(1)})}{P(\mathbf{n}|\boldsymbol{\lambda}^{(2)})} \tag{12.8-7}$$

and its logarithm (the log-likelihood ratio) is

$$\Lambda \equiv \ell\text{n}\,\frac{P(\mathbf{n}\,|\boldsymbol{\lambda}^{(1)})}{P(\mathbf{n}|\boldsymbol{\lambda}^{(2)})},$$

where, for $k = 1, 2$,

$$P(\mathbf{n}|\boldsymbol{\lambda}^{(k)}) = \prod_{i=1}^{N} \frac{[(\lambda_i^{(k)} + \lambda_o)\,\Delta t]^{n_i}}{n_i!} e^{-(\lambda_i^{(k)}+\lambda_o)\Delta t}. \tag{12.8-8}$$

Thus

$$LR = \prod_{i=1}^{N} \left[\frac{\lambda_i^{(1)} + \lambda_o}{\lambda_i^{(2)} + \lambda_o}\right]^{n_i} e^{-\Delta t(\lambda_i^{(1)}-\lambda_i^{(2)})}. \tag{12.8-9}$$

However the log-likelihood ratio is just as effective in decision making† and is easier to use; so we obtain

$$\Lambda = \sum_{i=1}^{N} n_i\alpha_i - \Delta t \sum_{i=1}^{N} \beta_i, \tag{12.8-10}$$

where

$$\alpha_i \equiv \ell n \frac{\lambda_i^{(1)} + \lambda_o}{\lambda_i^{(2)} + \lambda_o} \tag{12.8-11}$$

and

$$\beta_i \equiv \lambda_i^{(1)} - \lambda_i^{(2)}. \tag{12.8-12}$$

The decision rule is that if $\Lambda > 0$ then the observed vector $\mathbf{n}$ implies m_1; otherwise, $\mathbf{n}$ implies m_2. The second term in Eq. (12.8-10) is independent of the signal; it can be computed beforehand. Hence with

$$\phi \equiv \sum_{i=1}^{N} n_i\alpha_i \tag{12.8-13}$$

and

$$K \equiv \Delta t \sum_{i=1}^{N} \beta_i \tag{12.8-14}$$

the decision rule becomes

$$\text{If} \quad \phi \begin{cases} > K, & \text{then } \mathbf{n} \rightarrow m_1, \\ < K, & \text{then } \mathbf{n} \rightarrow m_2. \end{cases} \tag{12.8-15}$$

The object ϕ, sometimes called a *test* or *detection statistic*, is a random variable whose mean and variance depend on whether $\boldsymbol{\lambda}^{(1)}$ or $\boldsymbol{\lambda}^{(2)}$ was the modulation. Define

$$\begin{aligned} \mu_k &\equiv E[\phi|m_k] \qquad (k = 1, 2) \\ &= \sum_{i=1}^{N} \alpha_i E[n_i|m_k]. \end{aligned} \tag{12.8-16}$$

Since n_i is a Poisson r.v., it follows that $E[n_i|m_k] = \Delta t(\lambda_i^{(k)} + \lambda_o)$. Hence

$$\mu_k = \sum_{i=1}^{N} \alpha_i(\lambda_i^{(k)} + \lambda_o)\,\Delta t \qquad (k = 1, 2). \tag{12.8-17}$$

The mean square of ϕ is computed as follows:

† Because the log function is a monotonic function of its argument, it is often used in problems involving comparing two levels. It was used in Chapter 10 in determining the optimum receiver strategy.

$$\overline{\phi^2} \equiv E[\phi^2] = \sum_i \sum_j \alpha_i \alpha_j \overline{n_i\, n_j}$$

$$= \sum_i \alpha_i^2 \overline{n_i^2} + \sum_{i \neq j}\sum \alpha_i \alpha_j \overline{n}_i\, \overline{n}_j \tag{12.8-18}$$

$$= \sum_i \alpha_i^2 \overline{n_i^2} + \left[\sum_i \alpha_i \overline{n}_i\right]^2 - \sum_i \alpha_i^2\, [\overline{n}_i]^2.$$

Finally, the variance of ϕ is given by

$$\text{var}(\phi) = \overline{\phi^2} - (\overline{\phi})^2 = \sum_{i=1}^{N} \alpha_i^2\, [\overline{n_i^2} - (\overline{n}_i)^2]. \tag{12.8-19}$$

For *any* Poisson n_i with parameter $(\lambda_o + \lambda_i)\, \Delta t$,

$$(\overline{n}_i)^2 = [\Delta t(\lambda_i + \lambda_o)]^2 \tag{12.8-20a}$$

and

$$\overline{n_i^2} = [\Delta t(\lambda_i + \lambda_o)]^2 + \Delta t(\lambda_i + \lambda_o). \tag{12.8-20b}$$

Hence, using Eqs. (12.8-20a) and (12.8-20b) in (12.8-19), we obtain

$$\text{var}(\phi) = \sum_{i=1}^{N} \alpha_i^2\, \Delta t(\lambda_i + \lambda_o) \tag{12.8-21}$$

and, conditioning upon m_1 or m_2,

$$\sigma_k^2 \equiv \text{var}(\phi | m_k) = \sum_{i=1}^{N} \alpha_i^2\, \Delta t[\lambda_i^{(k)} + \lambda_o], \qquad k = 1, 2. \tag{12.8-22}$$

If $\sigma_1^2 = \sigma_2^2 \equiv \sigma_e^2$, then a reasonable SNR is given by

$$\text{SNR} = \frac{(\mu_1 - \mu_2)^2}{\sigma_e^2} \tag{12.8-23}$$

$$= \frac{\Delta t \left[\sum_{i=1}^{N} \alpha_i(\lambda_i^{(1)} - \lambda_i^{(2)})\right]^2}{\sum_{i=1}^{N} [\lambda_i^{(1)} + \lambda_o]\alpha_i^2}. \tag{12.8-24}$$

If $\sigma_1^2 \neq \sigma_2^2$, then a good choice of σ_e^2 is $\sigma_e^2 = \sigma_1^2 P[m_1] + \sigma_2^2 P[m_2]$, where $P[m_i]$ is the a priori probability of m_i. Equation (12.8-24) can be used to compute (1) the maximum achievable SNR, and (2) the optimum signaling under weak signal–high noise conditions, that is, when $\lambda_i^{(k)}/\lambda_o \ll 1$, $i = 1, \ldots, N$, $k = 1, 2$.

Maximum Achievable SNR When $\lambda_i^{(k)}/\lambda_0 \ll 1$, All *i*, *k*

We use the fact that $\ell n(1 + x) \simeq x$ for $x \ll 1$ [e.g., $\ell n(1 + 0.2) \simeq 0.18$, $\ell n(1 + 0.09) \simeq 0.086$]. Then, from Eq. (12.8-11),

$$\alpha_i = \ell n\left(1 + \frac{\lambda_i^{(1)}}{\lambda_o}\right) - \ell n\left[1 + \frac{\lambda_i^{(2)}}{\lambda_o}\right]$$
$$\simeq \frac{\lambda_i^{(1)} - \lambda_i^{(2)}}{\lambda_o}. \tag{12.8-25}$$

When this result is used in Eq. (12.8-24), we obtain

$$\text{SNR} = \frac{(\mu_1 - \mu_2)^2}{\sigma_e^2}$$
$$\simeq \frac{\Delta t}{\lambda_o}\sum_i [\lambda_i^{(1)} - \lambda_i^{(2)}]^2. \tag{12.8-26}$$

To maximize the SNR, we observe that, for any i where $\lambda_i^{(1)} + \lambda_i^{(2)} > 0$ (there must be at least one such i if the source is emitting an average energy $E > 0$ per message), one of the two rate parameters must be larger than the other, say $\lambda_i^{(1)} > \lambda_i^{(2)}$. Thus

$$[\lambda_i^{(1)} - \lambda_i^{(2)}]^2 = |\lambda_i^{(1)}|^2\left[1 - \frac{\lambda_i^{(2)}}{\lambda_i^{(1)}}\right]^2 \leq |\lambda_i^{(1)}|^2 \tag{12.8-27}$$

with equality if and only if $\lambda_i^{(2)} = 0$. Since this argument can be repeated for each i, $i = 1, \ldots, N$, *one of the modulating waveforms should be set equal to zero*. Therefore $\lambda_i^{(2)} = 0$ all i, and the two modulating signals $\lambda^{(1)}(t)$ and $\lambda^{(2)}(t)$ are *orthogonal*.

The nonzero waveform should be a narrow pulse, no wider than Δt. To see this, observe that, for $\lambda_i^{(1)} \equiv \lambda_i \geq 0$, all i,

$$\text{SNR} = \frac{\Delta t}{\lambda_o}\sum_{i=1}^{N} \lambda_i^2 \leq \frac{\Delta t}{\lambda_o}\left[\sum_{i=1}^{N} \lambda_i\right]^2$$
$$= \frac{\Delta t}{\lambda_o}\left[\sum_{i=1}^{N} \lambda_i^2 + \sum_{i=j}^{N}\sum^{N} \lambda_i\lambda_j\right] \tag{12.8-28}$$

with equality if and only if all $\lambda_i = 0$ except one. Hence $\lambda^{(1)}(t) = \lambda$ for some $t_{i-1} \leq t < t_i$ and zero everywhere else, and $\lambda^{(2)}(t) = 0$. Thus, in the weak signal–high noise case,

$$\text{SNR} = \frac{\Delta t}{\lambda_o}\sum_{i=1}^{N} [\lambda_i^{(1)} - \lambda_i^{(2)}]^2$$
$$= \frac{\lambda^2 \Delta t}{\lambda_o} = \Delta t\left(\frac{\lambda}{\lambda_o}\right)\lambda \tag{12.8-29}$$
$$= (\text{SNR})_{\max}.$$

Since $\lambda/\lambda_o \ll 1$ in the case considered, we might ask if it is possible to obtain a large SNR. The answer is yes provided that the quantum efficiency† of the detector is high and $\lambda\, \Delta t \gg \lambda_o/\lambda$.

† For a definition of quantum efficiency, see Eq. (12.4-1). In this analysis we assumed, implicitly, that the quantum efficiency, η, was 1. If the λ rate parameters refer to photons, then the photoelectron rate parameters at the receiver are $\eta\lambda$, where η is the quantum efficiency of the detector (Prob. 12.27). Of course, if $\eta = 1$, the rate parameters for photons and photoelectrons are equal.

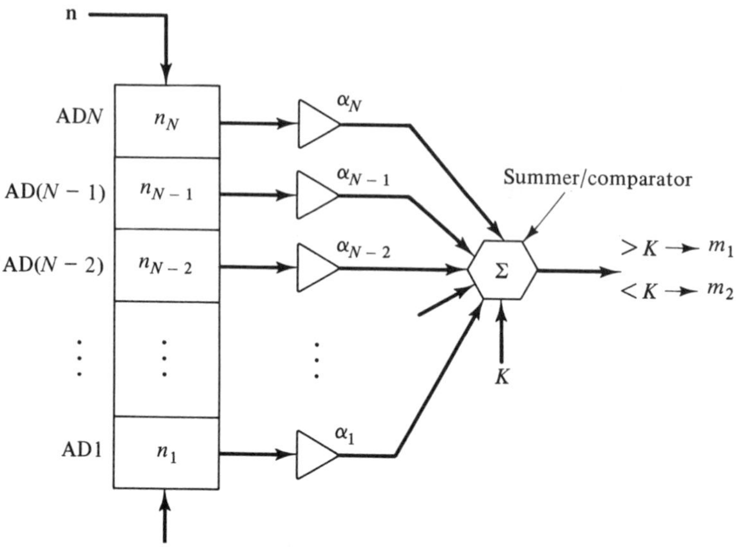

Figure 12.8-3. Implementation of the optimum receiver.

Design of the Optimum Receiver

There are at least two ways of implementing the optimum receiver ([12-14]). We discuss here only a single, convenient architecture. The detection statistic in the general case is

$$\begin{aligned}\phi &= \sum_{i=1}^{N} n_i \,\ell\text{n}\, \frac{\lambda_i^{(1)} + \lambda_o}{\lambda_i^{(2)} + \lambda_o} \\ &= \sum_{i=1}^{N} n_i \,\ell\text{n}(\lambda_i^{(1)} + \lambda_o) - \sum_{i=1}^{N} n_i \,\ell\text{n}(\lambda_i^{(2)} + \lambda_o) \\ &\equiv \phi^{(1)} - \phi^{(2)}.\end{aligned} \tag{12.8-30}$$

Thus, if $\phi^{(1)} - \phi^{(2)} > K$, choose m_1; if $< K$, choose m_2.

A possible implementation of the optimum receiver for the binary case is shown in Fig. 12.8-3. The photoelectron count n_1 obtained during the first counting interval is stored at address AD1, that obtained during the second at AD2, and so forth. At the end of the interval $(-T, T)$, a timing pulse opens the buffer gates and a signal proportional to $\Sigma_i \alpha_i n_i$ is transmitted to a comparator. If the signal exceeds a threshold K, then the message is assumed to be m_1; otherwise, m_2 is the assumed message.

12.9 EQUALIZERS AND REPEATERS IN FIBER-OPTICAL DIGITAL COMMUNICATION SYSTEMS

Fiber-optical digital communication (FODC) systems use repeaters in much the same way as other long-line digital communication systems use repeaters. Repeaters are necessary to prevent pulses from attenuating and interfering too much over long distances; they regenerate weak signals into strong ones. Table 12.1-2 shows some typical receiver spacings on early FODC systems in the United Kingdom.

A repeater converts the optical signal to an electrical one, "equalizes" and amplifies the pulses, makes a decision whether a zero bit or one bit was sent, and then uses this information to regenerate a strong, noise-free pulse of light that is transmitted to the next repeater. When making the decision regarding the binary state of the pulse, it is important to have as little ISI interference at the sampling instant. The equalizer ameliorates the interference from nearby pulses by filtering the received signal in a way to make the *overall* impulse response of the line have the desired transmission properties for digital communication. Equalizers are extensively used in digital communication systems to overcome intersymbol interference of the kind discussed here and in Chapter 10.

The electrical portion of a repeater is shown in Fig. 12.9-1. The input–output relations for the equalizer are

$$y(t) = \int_{\infty}^{\infty} x(t')h_e(t - t')\, dt', \tag{12.9-1}$$

$$Y(\omega) = X(\omega)H_e(\omega). \tag{12.9-2}$$

To illustrate the use of the equalizer, assume that the electrical pulses that are generated by the received light intensity are described by

$$x(t) = x(0)\, e^{-t^2/2\sigma_R^2} \tag{12.9-3}$$

Suppose we wish to generate a new signal $y(t)$ from $x(t)$ that has a smaller width, say σ_e^2, where $\sigma_e^2 < \sigma_R^2$. Then

$$y(t) = y(0)\, e^{-t^2/2\sigma_e^2} \tag{12.9-4}$$

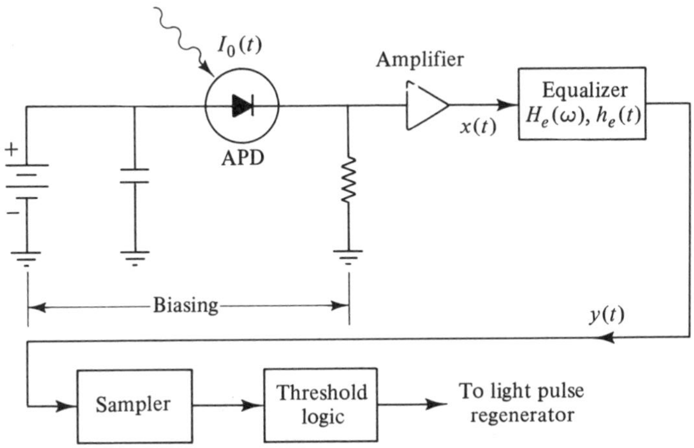

Figure 12.9-1. Block diagram of electrical portion of a repeater.

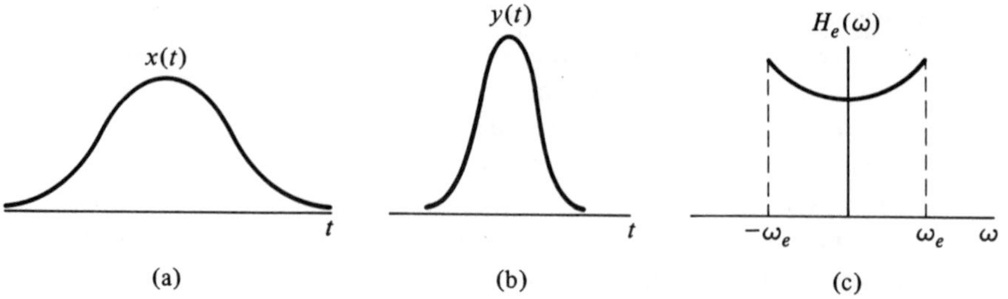

Figure 12.9-2. (a) Input to equalizer. (b) Output from equalizer. (c) Transfer function of equalizer.

and

$$H_e(\omega) = \frac{y(0)}{x(0)} e^{\omega^2(\sigma_R^2 - \sigma_e^2)/2}. \qquad (12.9\text{-}5)$$

Clearly, Eq. (12.9-5) cannot be valid for all ω. In practice, the equalizer operates only over a narrow band of frequencies.

In the presence of strong noise, the equalizer should not be used without some prior noise smoothing filtering; some equalizers, such as the one pictured in Fig. 12.9-2, enhance high frequencies and therefore will tend to amplify the high-frequency components of broadband noise to the detriment of the signal.

The ideal overall impulse response of the channel should generate the following situation: at the sampling instant $t = mT$, the pulse being sampled should have its peak value, while all its neighbors should be zero (Fig. 12.9-3). While in practice it is impossible to set all the ISI equal to zero, a properly designed equalizer can greatly reduce the influence of adjacent pulses while the test pulse is being sampled.

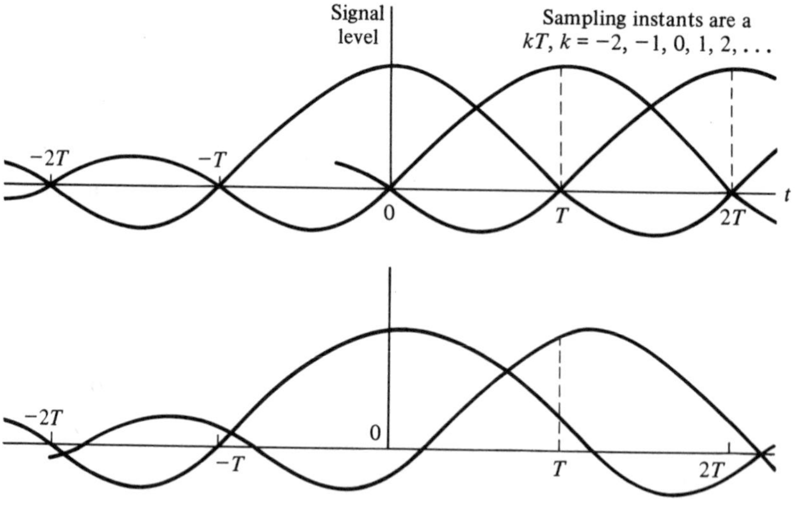

Figure 12.9-3. (a) Ideal (zero ISI) pulse locations. (b) Significant ISI.

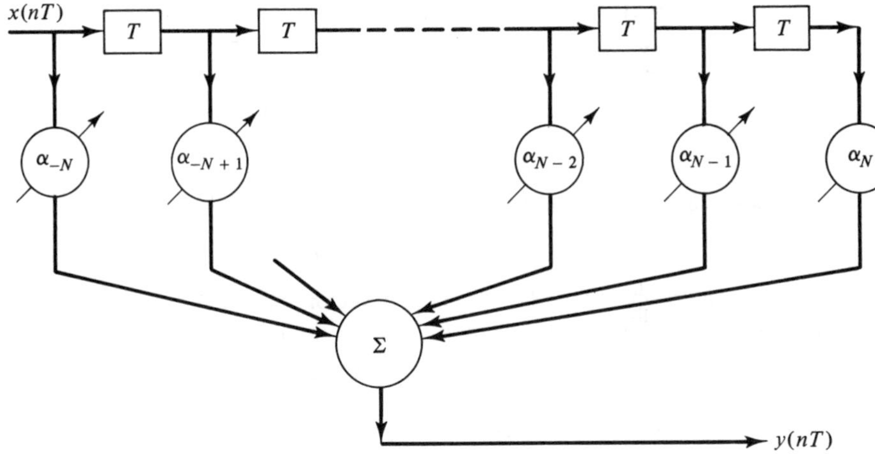

Figure 12.9-4. Transversal equalizer with $2N + 1$ taps.

A common equalizer is the zero-forcing equalizer, which can be realized with a transversal filter. An example of such a filter using $M = 2N + 1$ taps is shown in Fig. 12.9-4. From the diagram we see that at time t the output is given by

$$y(t) = \sum_{k=0}^{2N} a_{-N+k}x(t - kT) \tag{12.9-6}$$

$$= \sum_{j=-N}^{N} a_j x(t - NT - jT). \tag{12.9-7}$$

Let $y_n \equiv y[(n + N)T]$, that is, the output sampled at $t = (n + N)T$. Then at the sampling instants

$$y_n = \sum_{j=-N}^{N} a_j x_{n-j}, \tag{12.9-8}$$

where $x_{n-j} \equiv x(nT - jT)$.

The ideal situation is achieved when at the sampling instant, say $t = nT = 0$, we obtain

$$y_n = \begin{cases} 1, & n = 0, \\ 0, & n \neq 0. \end{cases} \tag{12.9-9}$$

Obviously, this cannot be realized for all n since we have only $2N + 1$ tap gains to adjust. However, we can achieve

$$y_n = \begin{cases} 1, & n = 0, \\ 0, & n = \pm 1, \ldots, \pm N. \end{cases} \tag{12.9-10}$$

Using Eq. (12.9-10) in Eq. (12.9-8) enables us to write the system

$$
\begin{aligned}
y_{-N} &= a_{-N}x_0 + \cdots + a_N x_{-2N} = 0, \\
y_0 &= a_{-N}x_N + \cdots + a_N x_{-N} = 1, \\
y_N &= a_{-N}x_{2N} + \cdots + a_N x_0 = 0.
\end{aligned}
\tag{12.9-11}
$$

Equation (12.9-11) can be written in matrix form as

$$\mathbf{y} = \mathbf{X}\,\mathbf{a}, \tag{12.9-12}$$

where $\mathbf{y}$ and $\mathbf{a}$ are $2N + 1 \times 1$ column vectors and $\mathbf{X}$ is the $(2N + 1) \times (2N + 1)$ matrix of pulse values. The solution to Eq. (12.9-11) is

$$\mathbf{a} = \mathbf{X}^{-1}\mathbf{y} \tag{12.9-13}$$

and when $\mathbf{y}$ assumes its ideal value (i.e., all zeros except for a 1 at $n = 0$), then

$$\mathbf{a} = \mathbf{X}^{-1}\boldsymbol{\phi}, \tag{12.9-14}$$

where $\boldsymbol{\phi} = (0, 0, \ldots, 0, 1, 0, \ldots, 0)^T$.

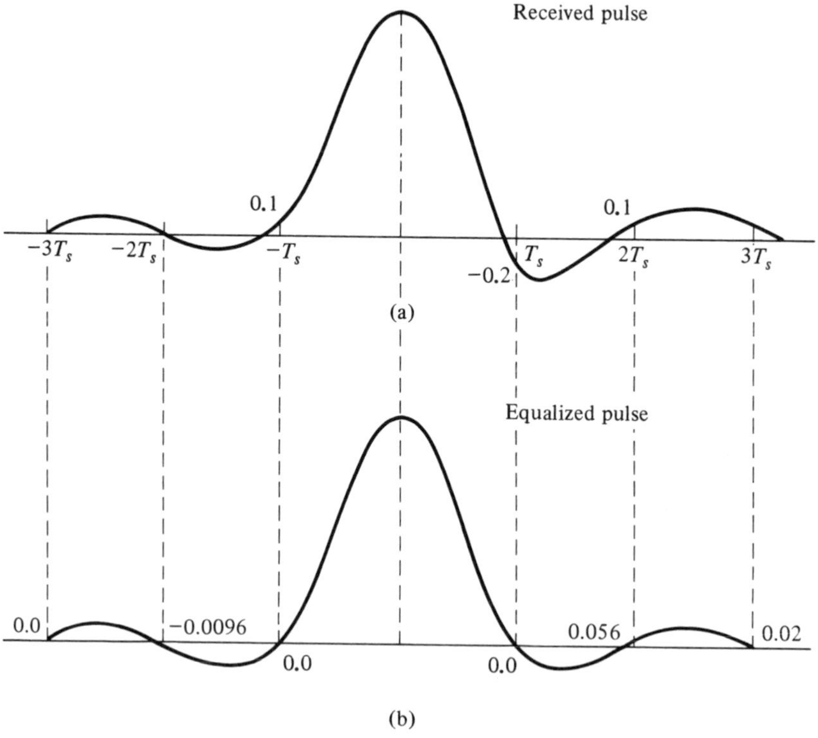

Figure 12.9-5. Reduction of ISI using three-tap equalizer. (a) Original pulse. (b) Equalized pulse. (Adapted from K. S. Shanmugam, *Digital and Analog Communication Systems*, Wiley, New York, 1979. By permission.)

Example [12-16, p. 232]

Assume x_{-2}, x_{-1}, x_0, x_1, and x_2 are respectively given by 0.0., 0.1, 1.0, −0.2, and 0.1. Then

$$\mathbf{a}_{\text{opt}} = \begin{bmatrix} 1.0 & 0.1 & 0.0 \\ -0.2 & 1.0 & 0.1 \\ 0.1 & -0.2 & 1.0 \end{bmatrix}^{-1} \begin{bmatrix} 0 \\ 1 \\ 0 \end{bmatrix}$$

$$= [-0.096, 0.96, 0.20]^T.$$

With this value of $\mathbf{a}_{\text{opt}}$, we obtain one zero on either side of the sample instant. However, although the ISI has been reduced at other sampling instants, it has not been set to zero everywhere (see Fig. 12.9-5).

In practice, an elegant technique is used to automatically adjust the gains of the taps by sending test pulses. The procedure is based on the observation that $\mathbf{X}$ is nearly diagonal (i.e., the ISI is relatively small); therefore, $\mathbf{X}^{-1}$ is nearly diagonal. An algorithm to realize this "preset" equalization is given in ([12-16], p. 234). Adaptive equalization is discussed in [12-17].

12.10 PHOTODETECTOR SIGNAL-TO-NOISE RATIO

Before concluding this chapter, it is useful to consider some of the noise considerations peculiar to optical systems that we did not encounter in electrical communication systems, such as were discussed in Chapter 9.

The normal photodetector biasing arrangement is shown in Fig. 12.10-1(a). In Fig. 12.10-1(b) is shown the equivalent circuit: C_d represents the junction capacitance of the diode, R_L is the load resistance, and R_A and C_A denote the input resistance and capacitance, respectively, of the amplifier. We shall ignore the noise associated with the amplifier and consider only the signal-to-noise ratio S/N at the amplifier input.

As stated in Sec. 12.8, the most important noise sources related exclusively to the photodetector are the quantum noise associated with the generation of photocarriers, the dark current noise, and the avalanche gain or excess-gain noise. Another source of noise, surface leakage noise, can be controlled by careful design and fabrication ([12-3], p. 155). Both quantum and dark-current noise are often modeled as shot noise processes (i.e., they obey Poisson statistics). Assuming a Poisson process type of current $i(t)$ with expected value I, it is well known (see, e.g., [12-18, Chapter 7]) that the mean-square (m.s.) value of the current fluctuations about I is given by

$$\overline{i_{ms}^2} \equiv \overline{[i(t) - I]^2} = 2eIB, \tag{12.10-1}$$

where e is the electronic charge† and B is the effective bandwidth of the system. In an avalanche photodetector, the internal gain g is a randomly fluctuating parameter. The m.s. value of g can be written as

$$\overline{g^2} = G^2F(G), \tag{12.10-2}$$

† For its value, see Sec. 12.4.

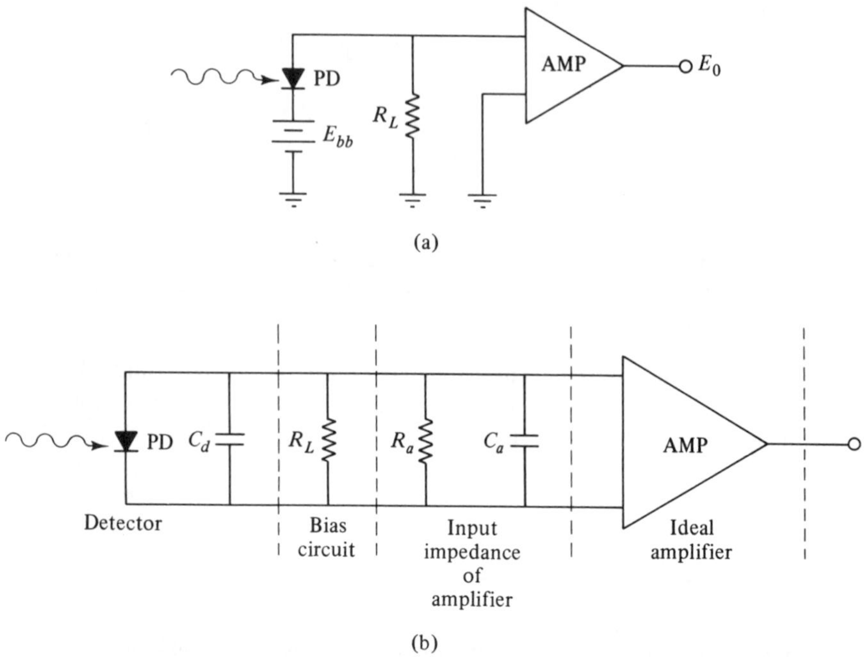

Figure 12.10-1. (a) Normal biasing of photodiode. (b) Equivalent circuit.

where $G \equiv \bar{g}$ and $F(G) \equiv \overline{g^2}/G^2$ is the *noise factor*. The noise factor is a measure of the degradation due to the avalanche multiplier compared to an ideal noiseless amplifier of gain G. The *primary photocurrent* $i_p(t)$ is proportional to the total number of incident photons per second. Thus

$$i_p(t) = \frac{\eta e p(t)}{h\nu}, \tag{12.10-3}$$

where $h\nu$ is the energy per photon,† $p(t)$ is the incident optical power, and η is the quantum efficiency. The signal portion of $i_p(t)$ is a constant or "slowly varying" dc component I_p. The shot noise associated with the primary photocurrent is

$$\overline{(i_p(t) - I_p)^2} = 2eI_pB \tag{12.10-4}$$

and that associated with the dark current $i_d(t)$ is

$$\overline{(i_d(t) - I_d)^2} = 2eI_dB, \tag{12.10-5}$$

where I_d is the dc value of the dark current. After avalance multiplication, we obtain

$$\overline{[g(i_p(t) - I_p)]^2} = 2eI_pG^2F(B)B \tag{12.10-6}$$

and

† As always, ν is the frequency in hertz and h is Plank's constant, $h = 6.626 \times 10^{-34}$ J-s.

$$\overline{[g(i_d(t) - I_d)]^2} = 2eI_dG^2F(B)B. \qquad (12.10.7)$$

Since the two effects [i.e., the generation of $i_p(t)$ and $i_d(t)$] are independent, they add to produce a total m.s. shot current of

$$\overline{i_{ST}^2} = 2e(I_p + I_d)G^2F(G)B. \qquad (12.10\text{-}8)$$

In addition, there is an m.s. component of thermal noise current given by

$$\overline{i_{Th}^2} = \frac{4k_BT_A B}{R_L}, \qquad (12.10\text{-}9)$$

where k_B is Boltzmann's constant ($k_B = 1.38 \times 10^{-23}$ J/K) and T_A is the absolute temperature in degrees Kelvin. Thus the total m.s. noise current is

$$\overline{i_{Th}^2} + \overline{i_{ST}^2} = 2e(I_p + I_d)G^2F(G)B + \frac{4k_BT_AB}{R_L}.$$

The average value of the signal portion of the photocurrent after avalanche multiplication is $\bar{g}I_p = GI_p$. Thus the signal-to-noise ratio at the amplifier input $(\mathcal{S}/\mathcal{N})_i$ is given by

$$(\mathcal{S}/\mathcal{N})_i = \frac{G^2I_p^2}{2e(I_p + I_d)G^2F(G)B + 4k_BT_A/R_L}. \qquad (12.10\text{-}10)$$

If I_p contains a time-varying signal $i_s(t)$ superimposed on a dc pedestal, one should replace I_p^2 in the numerator by $<i_s^2(t)>$, where $<\ >$ denotes a long-term time average. The dc pedestal should also be included in the I_p when computing the shot noise.

12.11 SUMMARY

One of the most significant events in communications technology in the last quarter of the twentieth century is the development of fiber-optical communications on a grand scale. In this chapter we have attempted to introduce the reader to the basic operating principles of such systems. We discussed both components (fibers, sources, and detectors) and systems. To discuss the phenomenon of broadening in single-mode fibers, we had to introduce the basic notions of the theory of coherence. We made some elementary estimates of the information-handling capacity of single-mode fibers based on intersymbol interference and discussed optimum detection strategies when optical signals with Poisson statistics were being transmitted. We discussed the fundamentals of repeaters and equalizers and closed the chapter with some basic signal-to-noise calculations.

Clearly, our brief discussion has only scratched the surface. An excellent reference book (705 pages) on this subject is [12-5]. Any of the first six references, however, will contain much additional information to that presented here. With the material in this chapter well in hand, the reader should have no difficulty in understanding more detailed or advanced literature on this subject.

PROBLEMS

12-1. Consider the interface between two media M_1 and M_2 as shown in Fig. P12-1. An incident electric field E_i impinges on the interface. Let $E_{i,n}$, $E_{i,p}$, $E_{t,n}$, $E_{t,p}$, $E_{r,n}$, and $E_{r,p}$ denote the normal and parallel components of the incident, transmitted, and reflected fields, respectively. By *normal* and *parallel* we mean the components of the electric field vector that are normal and parallel to the plane of incidence. From Maxwell's equations, we get

$$\frac{E_{r,n}}{E_{i,n}} = \frac{n_1 \cos\theta_i - n_2 \cos\theta_t}{n_1 \cos\theta_i + n_2 \cos\theta_t},$$

$$\frac{E_{r,p}}{E_{i,p}} = \frac{n_1 \cos\theta_t - n_2 \cos\theta_i}{n_1 \cos\theta_t + n_2 \cos\theta_i},$$

and

$$\frac{E_{t,p}}{E_{i,p}} = \frac{2n_1 \cos\theta_i}{n_1 \cos\theta_t + n_2 \cos\theta_i}.$$

Show that if $\theta_i > \theta_c$ where

$$\theta_c \equiv \sin^{-1}\left(\frac{n_2}{n_1}\right)$$

then

$$\left|\frac{E_{r,n}}{E_{i,n}}\right| = 1 \quad \text{and} \quad \left|\frac{E_{r,p}}{E_{i,p}}\right| = 1.$$

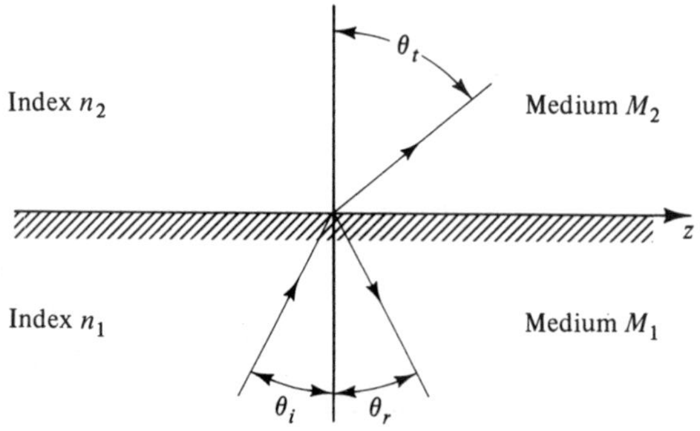

Figure P12-1

12-2. In Prob. 12.1, show that

$$\frac{E_{r,n}}{E_{i,n}} = e^{2j\psi}, \qquad \frac{E_{r,p}}{E_{i,p}} = e^{2j\delta},$$

where

$$\tan\psi \equiv \frac{(r^2\sin^2\theta_i - 1)^{1/2}}{r\cos\theta_i}$$
$$\tan\delta \equiv \frac{r(r^2\sin^2\theta_i - 1)^{1/2}}{\cos\theta_i}$$

and $r \equiv n_1/n_2$.

12-3. In Prob. 12.1, show that the transmitted wave decays as e^{-u}, where

$$u \equiv \frac{2\pi n_2}{\lambda_o}[r^2\sin^2\theta_i - 1]y.$$

12-4. In Probs. 12-1 and 12-2, you found that when $\sin\theta_i > n_2/n_1$ then

$$\cos\theta_t = \pm j\,[r^2\sin^2\theta_i - 1]^{1/2}.$$

Explain why only the solution with the minus sign is physically meaningful. *Hint*: Use the results of Prob. 12-3.

12-5. In Prob. 12-1, take $n_1 = 1.52$ (glass), $n_2 = 1$ (air), and $\theta_i = 60°$.
(a) Is $\theta_i > \theta_c$?
(b) If the answer to part (a) is yes, what reduction in the amplitude of the electric field in medium 2 is observed in a distance into medium 2 equal to one wavelength?

12-6.† Consider the dielectric rod of radius a and index of refraction n_1 shown in Fig. P12-6. Let the rod be surrounded by an infinite dielectric medium of refractive index n_2. Use Maxwell's equation or the wave equation in cylindrical coordinates to compute the electric fields inside the fiber.

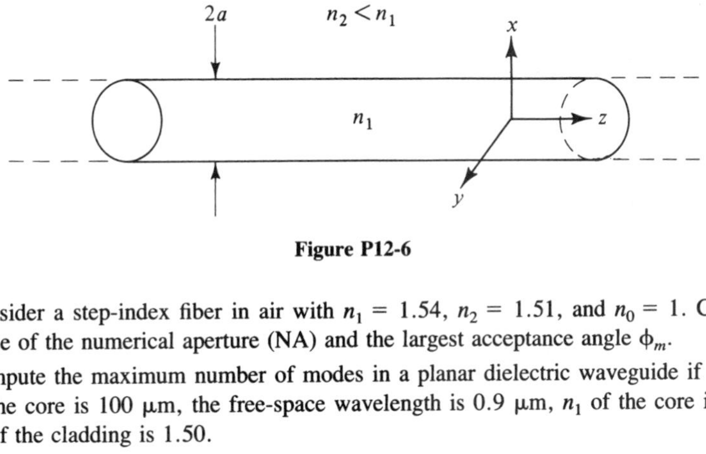

Figure P12-6

12-7. Consider a step-index fiber in air with $n_1 = 1.54$, $n_2 = 1.51$, and $n_0 = 1$. Compute the value of the numerical aperture (NA) and the largest acceptance angle ϕ_m.

12-8. Compute the maximum number of modes in a planar dielectric waveguide if the width d of the core is 100 μm, the free-space wavelength is 0.9 μm, n_1 of the core is 1.52, and n_2 of the cladding is 1.50.

12-9. **(a)** Explain the difference between intermodal dispersion, material dispersion, and waveguide dispersion.
(b) At a single frequency operation, which is the key factor in reducing transmission capacity?
(c) Estimate the mode-dispersion time $|\Delta\tau|$ for a 10-km fiber for which $n_1 = 1.52$ and $n_2 = 1.50$. What approximation is being made in using Eq. (12.2-24)?

† This problem may require some research and/or more effort.

12-10. From the solution of Maxwell's equation in a fiber, it can be shown that the propagation constant, β, for any mode must satisfy

$$kn_2 \leq \beta \leq kn_1,$$

where $k = 2\pi/\lambda_0$ and λ_0 is the free-space wavelength. It is customary to express the intramode or waveguide dispersion in terms of the parameter Vb, where V is the normalized thickness given in Eq. (12.2.-19) and

$$b \equiv \frac{(\beta^2/k^2) - n_2^2}{n_1^2 - n_2^2}.$$

Show that if the source width, in wavelength, is $\Delta\lambda$, there will result a differential delay for the Fourier components within a mode given by

$$|\Delta\tau|_{WG} = \frac{nL\,\Delta\lambda}{c\lambda} V \frac{d^2(Vb)}{dV^2},$$

where the subscript WG refers to waveguide dispersion. Assume a weakly guiding case, where $n_1 - n_2 \ll n_1$ or n_2 and $n_1 \simeq n_2 \equiv n$.

12-11. Optical radiation is absorbed in semiconductor materials according to the exponential law

$$P(x) = P_0(1 - e^{-\alpha x}),$$

where $P(x)$ is the absorption at a depth x and α is the absorption coefficient (which depends strongly on wavelength and therefore makes a particular semiconductor suitable only over a limited wavelength range).

(a) Keeping in mind that the recombination of photogenerated hole–electron pairs is very fast at the surface, why would both α too large and α too small be undesirable?

(b) If the surface (power) reflectivity is ρ, what is the effective absorbed power?

12-12. Derive an expression for the primary photocurrent I_p in terms of the effective absorbed power if the width of the depletion (absorption) region is w. Use this to compute the quantum efficiency of η.

12-13. Consider an idealized binary communication system in which all sources of external noise are ignored. A zero or a one is sent. The receiver is sensitive enough to detect a single photoelectron. A zero is sent by sending no optical power. A one is detected whenever there is at least one photoelectron. Since there is no noise, a zero can never be detected as a one. When a one is sent, the mean number of photoelectrons is μ and the probability of k photoelectrons is the Poisson law; that is,

$$P(k|\mu) = \frac{e^{-\mu}\mu^k}{k!}.$$

(a) Show that the probability of error, $P(\epsilon)$, assuming an equal number of *zeros* and *ones*, is

$$P(\epsilon) = \frac{1}{2}e^{-\mu}.$$

(b) Show that, for $p(\epsilon) < 10^{-9}$, μ must be greater than 20.

(c) Show that the minimum average received power P_R must satisfy

$$P_R > \frac{20h\nu}{T\eta},$$

where $h\nu$ is the photon energy, T is the bit duration, and η is the quantum efficiency. This minimum represents an absolute limit on quantum detectability.

12-14. Compute the complex degree of coherence for a source radiating a complex analytic field $u(t) = a(t)\, e^{j\omega_0 t}$, where ω_0 is the carrier frequency and $a(t) = Ae^{j\Theta}$. Let A and Θ be independent random variables not dependent on time t with probability densities

$$f_A(\alpha) = \begin{cases} \alpha e^{-\alpha^2/2} & \alpha \geqslant 0 \\ 0 & \alpha < 0 \end{cases}$$

$$f_\Theta(\theta) = \begin{cases} \dfrac{1}{2\pi}, & 0 \leq \theta < \pi, \\ 0, & \text{elsewhere} \end{cases}$$

What is the magnitude of the complex degree of coherence? Assume that $u(t)$ is a wide-sense stationary random field; that is, $\overline{u(t)} = \overline{u(t + \tau)}$ for all t, τ and $\sigma^2_{u(t)} = \sigma^2_{u(t+\tau)}$ for all t, τ.

The next six problems are linked together.

12-15. Assume a source with bandwidth $\Delta\nu$ generating radiation with mean frequency $\bar{\nu}$. Show that if the source produces incident power $p(t)$ on a point-size detector D then, at time t, the average number of photoelectrons per second, $\lambda(t)$, generated at D is given by

$$\lambda(t) = \frac{\eta p(t)}{h\bar{\nu}},$$

where η is the quantum efficiency of the detector.

12-16. In Prob. 12-15, assume that $p(t)$ is constant and independent of time t [$p(t) = p_0$]. Assume that the probability $P[K; t, t + \tau]$ of generating K photoelectrons in the interval $(t, t + \tau)$ is the Poisson law. Write an explicit expression for $P[K; t, t + \tau]$ in terms of the parameters of Prob. 12-15. Show that $P[K; t, t + \tau]$ in this case depends only on τ and hence may be written as $P[K; \tau]$.

12-17. In general, the incident power $p(t)$ on the photodetector varies randomly with time. Thus at any instant of time we can model $p(t)$ as a random variable with probability density function (pdf) $f_p(\alpha)$. It therefore follows that any function of $p(t)$ will also be a random variable.

(a) Compute the pdf of

$$Z \equiv \int_t^{t+\tau} \lambda(\xi)\, d\xi,$$

assuming $\tau \ll [\Delta\nu]^{-1}$ and that $\lambda(t)$ is as in Prob. 12-15.

(b) Write an expression for the unconditional (i.e., averaged over Z) probability mass function $P[K; t, t + \tau]$ of observing K photoelectrons in time $(t, \;\; \tau)$.

12-18. In the case of a coherent, gain-stabilized light source such as a laser, the incident power $p(t)$ is a constant, say p_0. Use the result of Prob. 12-17, part (b), to compute $P[K; t, t + \tau]$. This result should be the same as in Prob. 12-16. Compute the average number of photoelectrons $\mu_K \equiv E[K]$ and the variance σ_K^2 in terms of p_0, η, and $h\bar{\nu}$.

12-19. Assume that the incident illumination falling on a point detector is polarized thermal light for which it is known that the pdf, $f_p(\alpha)$, of the power obeys

$$f_p(\alpha) = \frac{1}{\mu_p} e^{-\alpha/\mu_p},$$

where μ_p is the average power incident on the point detector. Assume a point detector and that $\tau \ll [\Delta\nu]^{-1}$.

(a) Compute $f_Z(z)$ (Z as in Prob. 12-17).

(b) Compute $P[K; t, t + \tau]$.

12-20. Compare the fluctuations in the photoelectrons K produced in $\tau \ll [\Delta\nu]^{-1}$ when **(a)** the illumination is from a coherent, gain-stabilized source, and **(b)** when the illumination is from a polarized thermal source. Which source produces more fluctuations (i.e., noise)? To what property of the illumination can this be attributed?

12-21. Derive Eqs. (12.6-23) and (12.6-24).

12-22. Derive Eqs. (12.7-5) to (12.7-8).

12-23. In computing the capacity of a fiber, why must a sequence of pulses be considered? Why, for example is it not enough to simply calculate the broadening of a single pulse, as in Eq. (12.6-24), and then take the inverse of the pulse width as the capacity of the fiber channel?

12-24. It is suggested that the fiber-optical channel capacity be estimated as follows: Use Eq. (12.6-17a) to compute the output pulse envelope $b(t)$ in terms of the input pulse envelope $a(t)$ and the envelope point-spread function $\tilde{h}(t)$; that is,

$$b(t) = \int_{-\infty}^{\infty} a(\xi)\tilde{h}(t - \xi)\, d\xi.$$

Compute the intensity $r^2(t) = |b(t)|^2$. Repeat the calculation for a pulse delayed by T and obtain $r^2(t - T) = |b(t - T)|^2$. Next reduce T until the sum of the two intensities increases to δ at $T/2$, the maximum permissible ISI; that is,

$$r^2(t) + r^2(t - T)\,\big|_{t=T_{\min}/2} = \delta.$$

Finally, take N as $1/T_{\min}$ as the capacity of the channel. What is fundamentally inconsistent about this approach?

12-25. Derive Eq. (12.7-10) and apply the coherent and incoherent limits to obtain Eqs. (12.7-14) and (12.7-15). Why can't we let $\tau_c = 0$ in the incoherent limit?

12-26. Describe the MAP strategy for the ternary communication channel shown in Fig. P12-26. The following probabilities are given:

$$P(Y = 2|X = 1) = P(Y = 3|X = 1) = \frac{\alpha}{2},$$

$$P(Y = 1|X = 2) = P(Y = 3|X = 2) = \frac{\beta}{2},$$

$$P(Y = 1|X = 3) = P(Y = 2|X = 3) = \frac{\gamma}{2},$$

as well as $P[Y = 1|X = 1] = 1 - \alpha$, $P[Y = 2|X = 2] = 1 - \beta$, and $P[Y = 3|X = 3] = 1 - \gamma$. Also, $P[X = 3] = \frac{1}{2}$, $P[X = 2] = \frac{1}{3}$, and $P[X = 1] = \frac{1}{6}$.

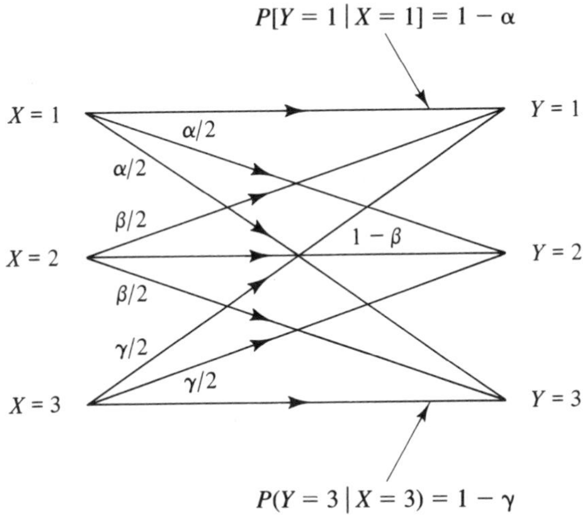

Figure P12-26

12-27. Obtain the maximum SNR when the quantum efficiency is not unity and when the modulation is piecewise constant, as in Sec. 12.8. Assume that the rate parameters describing the modulation refer to photon rates, not photoelectron rates.

12-28. Design a three-tap zero-forcing equalizer for the received signal

$$x(t) = e^{-(1/2)t^2}.$$

The sampling interval is $T = 2$.

REFERENCES

[12-1] J. Wilson and J. F. B. Hawkes, *Optoelectronics: An Introduction*, Prentice-Hall, Englewood Cliffs, N.J., 1983.

[12-2] J. Gowar, *Optical Communication Systems*, Prentice-Hall, Englewood Cliffs, N.J., 1984.

[12-3] G. Keiser, *Optical Fiber Communications*, McGraw-Hill, New York, 1983.

[12-4] Y. Suematsu and Ken-Ichi Iga, *Introduction to Optical Fiber Communications*, Wiley, New York, 1982.

[12-5] S. E. Miller and A. G. Chynoweth (eds.), *Optical Fiber Telecommunications*, Academic, Orlando, Fla., 1979.

[12-6] Michael K. Barnoski (ed.), *Fundamentals of Optical Fiber Communications*, Academic, Orlando, Fla., 1976

[12-7] D. GLOBE and E. A. J. MARCATILI, "Multimate Theory of Graded-Core Fibers," *Bell Syst. Tech. J.*, **52,** pp. 1563–78, 1973.

[12-8] M. BORN and E. WOLF, *Principles of Optics*, 2nd rev. ed., Macmillan, New York, 1964, Chapter 10.

[12-9] M. J. BERAN and G. B. PARRENT, *Theory of Partial Coherence*, Prentice-Hall, Englewood Cliffs, N.J., 1964.

[12-10] L. MANDEL "Fluctuations of Photon Beams: The Distribution of the Photo-Electrons," *Proc. Phys. Soc.* (London) **74,** p. 223, 1959.

[12-11] J. W. GOODMAN, *Statistical Optics*, Wiley, New York, 1984.

[12-12] A. W. SNYDER, "Asymptotic Expressions for Eigenfunctions and Eigenvalues of a Dielectric or Optical Waveguide," *Trans. IEEE Microwave Theory Tech.*, **MTT-17,** pp. 1130–1138, 1969.

[12-13] B. E. A. SALEH and M. I. IRSHID, "Coherence and Intersymbol Interference in Digital Fiber Optic Communication Systems," *IEEE J. Quantum Electronics*, **QE-18,** pp. 944–951, June 1982.

[12-14] B. REIFFEN and H. SHERMAN, "An Optimum Demodulator for Poisson Processes: Photon Source Detectors," *IEEE Proc.*, **51,** pp. 1316–1320, 1963.

[12-15] H. STARK and J. W. WOODS, *Probability, Random Processes, and Estimation Theory for Engineers*, Prentice-Hall, Englewood Cliffs, N.J., 1986.

[12-16] K. S. SHANMUGAM, *Digital and Analog Communication Systems*, Wiley, New York, 1979.

[12-17] S. U. H. QURESHI, "Adaptive Equalization," *Proc. IEEE*, **73,** pp. 1349–1387, Sept. 1985.

[12-18] W. B. DAVENPORT, JR., and W. R. ROOT, *An Introduction to the Theory of Random Signals and Noise*, McGraw-Hill, New York, 1958.

Index

A

B

C

D

J

K

L

M

Q

W

Z